T0406380

Science Networks. Historical Studies

Founding Editors
Erwin Hiebert, Basel, Switzerland
Hans Wußing, Leipzig, Germany

Volume 64

This series is devoted to historical studies in the fields of mathematics, physics, and astronomy, including their applications. It is mainly composed of monographs, although it may occasionally include critical editions of primary sources and biographies of important actors. It aims to attract a multiple audience including scientists, historians, philosophers, and graduate students throughout the academic world. Submitted projects are examined by an international and multidisciplinary committee of experts. The publication language is preferentially English, and the diffusion is international.

Erika Luciano

The Jewish Mathematical Diaspora from Fascist Italy

Looking for a Space of Intellectual Survival

 Birkhäuser

Erika Luciano
Department of Philosophy and Education Sciences
University of Turin
Turin, Italy

ISSN 1421-6329 ISSN 2296-6080 (electronic)
Science Networks. Historical Studies
ISBN 978-3-031-64895-3 ISBN 978-3-031-64896-0 (eBook)
https://doi.org/10.1007/978-3-031-64896-0

Mathematics Subject Classification: 01A60; 01A70; 01A74; 14-03

This book is published under the imprint Birkhäuser, www.birkhauser-science.com by the registered
company Springer Nature Switzerland AG
The registered company address is: Gewerbestrasse 11, 6330 Cham, Switzerland

If disposing of this product, please recycle the paper.

Agli amori della mia vita: Luca e i nani Ivan, Amos, David
Nulla è possibile senza di voi.

Preface

When Italian emigration is mentioned, people generally think of the Great Migration wave that involved 14 million citizens between 1876 and 1915. These included legions of peasants, small landowners, and labourers, mostly from the regions of Southern Italy, who left the country en masse in search of fortune in the United States, South America, and Australia, as well as in other European countries (Belgium, France, Germany, etc.). Sometimes there were young men alone, at other times there were whole families, often very numerous ones. They were the so-called *cafoni* or *pezzenti*: people who sold everything they owned to buy a one-way ticket; individuals 'not white, but not completely black', often labelled as *mafiosi*, for whom integration was a challenge and who often preferred to ghetto-ise themselves in Italian neighbourhoods.[1]

However, there was also a completely different type of Italian emigration: emigration determined by the anti-Jewish legislation of 1938. Quantitatively and qualitatively, the two are not comparable. The flow of Jewish emigration was minimal, compared to the previous exodus, and affected about 6000 individuals, which constituted 12.7% per cent of the Italian Jewish population in 1938. The number refers to Italian Jewish citizens and does not include the 9000 foreign Jews residing in Italy at the time of promulgation of the *Measures for the Defense of the Race*, who from after 1919 lost Italian citizenship. These emigrants were, predominantly, members of the upper middle class of Central and Northern Italy: artisans, industrialists, owners of commercial enterprises, State officials, skilled workers, and the wealthy and culturally well-equipped class of self-employed individuals (physicians, lawyers, notaries, engineers, architects, etc.). The migrant community was transversal in relation to social, political, and religious orientation and included

---

[1] Immigration from Northern Italy, for example from the Veneto, should not be forgotten. These immigrants were neither peasants nor beggars.

individuals with a bland, if not non-existent, sense of the roots of their identity, belonging to all leanings: anti-fascists and fervent fascists, nationalists and communists, old liberals and socialists, as well as many who had never been involved in politics. Many professed themselves as completely secular, while others had had a religious education but had become non-practicing over time. Very few were orthodox practitioners. A total of 68.2% had contracted mixed marriages and had educated their children either as Catholics or in no religion at all.

Jewish emigration was a form of mass expatriation determined by three factors: a cause of a political nature (persecution); the impossibility of providing for one's self and family, since they could no longer exercise their own professions; the inability to tolerate the loss of civil and political rights along with reduction to a caste of pariahs (social and professional downgrading, marginalisation, the injury to their own dignity as men and citizens, etc.); the urgency of ensuring school education and/or a future career for their children. It was a 'semi-forced' exodus, since the racial laws did not introduce the expulsion from the country of Jews of Italian nationality, officially encouraged by the regime until 1941 (date of the last project to build a Jewish enclave in Italian East Africa) but, in reality, hampered by inefficiency and rampant corruption in the Ministry of Foreign Affairs, embassies, and consular offices.

The emigration took place in three waves: the first and most consistent soon after introduction of the racial laws (in the final months of 1938-September 1939), one of smaller size in the years 1940–1941, with increasingly rushed and perilous departures after the entry into war of Italy (June 10, 1940) and a last wave between October 1943 and the first months of 1944, when, as a result of the Nazi-Fascist occupation, some 4000 Jews fled from mass roundups and deportations to their only remaining land of refuge: Switzerland. The last wave had the typical characteristics of illegal immigration: entry in violation of Swiss immigration laws, recourse to *passeurs* (smugglers) and false identity documents, internment in camps with the status of illegal immigrants and, after an initial period of stay, with that of refugees. In 1942 the migratory flow remained almost stationary apart from the return of some displaced people from abroad to Italy, some family reunifications, and the lucky few who still managed to leave the country.

This exodus had some similarities with the Great Italian migration, which can be summarised as follows. It hinged on networks of intra- and inter-family solidarity, which could be formed and act with remarkable effectiveness by virtue of the small numerical size of the Italian Jewish population. The function and role of these chains were decisive not only on departure, but also on arrival in the host countries. In South America as in the States or in Australia, Little Italy-s were created of Italian Jews, ready to help each other:

These Italian Jews were entirely different. They formed a special emigration. When they arrived, they did not ask for help, and even gave it instead. They knew how to sort themselves out, helping each other. They did not mix easily with the Jews of New York or even with the Italo-Americans. They stayed preferably together, with each other. [. . .] More or less, they remained isolated. For the Jews they were Italians. For Italo-Americans, they were Jews. For the Americans, they were a subject of wonder and sympathy.[2]

As with the historical emigration, the Jewish exodus was planned as temporary: those who left were hoping to stay abroad only for a limited period until the repeal of the racial laws (a future event that was simply taken for granted). Like the peasants of the Great Migration, the Italian Jews were also greeted by prejudices, different but equally rooted: the stereotypes of the Jewish Freemason, lobbyist, plutocrat, etc. were affirmed tremendously quickly in Uruguay as in the United States, in Argentina as in Australia.

A very large group of emigrants was made up of intellectuals. The latter were among the professional categories most affected by the anti-Jewish legislation which prevented them from almost any form of activity, both in the public and private sectors. University professors were prevented not only from teaching but also from publishing, from attending institutes and libraries, and from participating in the activities of academies and scientific societies. Journalists were banned from any collaboration, even as freelancers. The theatre, the cinema, and the radio purged all professionals of Jewish origin. Jewish intellectual emigration involved in particular the university, which in 1938 counted 99 university professors of Jewish origin out of 1250, and at least two hundreds of assistants and lecturers (*assistenti, liberi docenti*), often awarded teaching assignments.[3] 116 scholars emigrated, including 75 full professors, subdivided by disciplines and by university of provenance as in the tables.[4] An entire generation of scholars was removed.

---

[2]Prezzolini (1950), pp. 249–250: *Questi ebrei italiani erano interamente differenti. Formarono un'emigrazione speciale. Arrivati non si misero a chiedere soccorsi, e persino invece ne dettero. Sapevano mettersi a posto da soli, aiutandosi gli uni con gli altri. Non si mescolaron facilmente con gli Ebrei di New York e nemmeno con gli Italo-Americani. Stavan di preferenza fra di loro. [. . .] Più o meno, rimasero isolati. Per gli Ebrei erano italiani. Per gli Italo-Americani erano ebrei. Per gli Americani un soggetto di meraviglia e di simpatia.*

[3]Ventura (1997).

[4]Both the tables derive from the database of the Italian academic refugees given in Appendix of this volume. The list is based on unpublished and published sources (e.g. Ventura 1997; Capristo and Fabre 2018; Guarnieri 2019a and Guarnieri 2019b). All the scholars listed in Appendix are named in at least one of these sources. University of provenance is intended to designate the institution where the scholar was serving at the time the Racial Laws were introduced. The large number of sources used to create the list ensures the completeness and reliability of the information. Occasional omissions or errors cannot, however, be completely excluded.

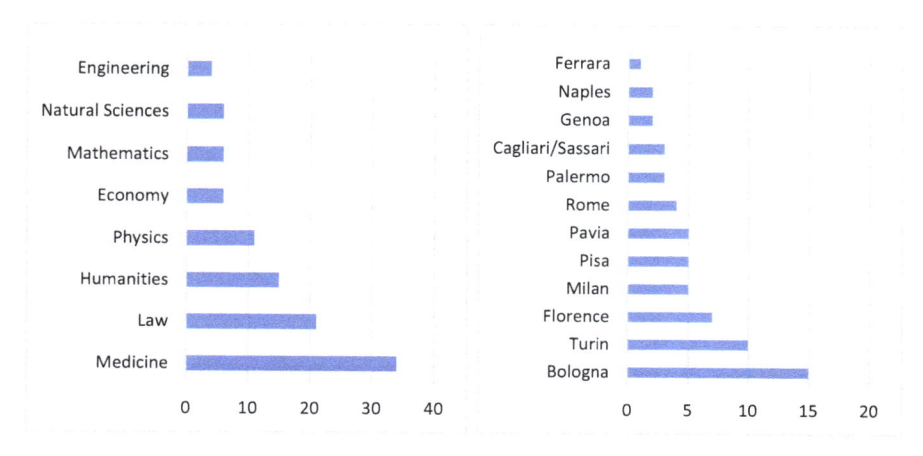

Figure 1 Distribution of émigrés scholars according to faculties and to university of provenance

The fact that the percentage of university professors among emigrants was so high is understandable. First, very few scholars could survive without receiving a salary, drawing only on savings and personal assets. The pension granted to the purged professors was in fact very low, except for those who had many years of service seniority. Moreover, for people who had made culture their life mission, the banishment from all scientific arenas and exclusion of their children from any form of education and instruction were intolerable.

Jewish academic emigration from racist Italy has two basic elements of analogy with Jewish emigration tout-court. In the first place, it was also facilitated by networks of solidarity which, unlike links based on family ties and affinities of geographical origin, were grounded on the web of international relations that individual scholars had woven in the years of the Belle Époque of scientific internationalism through congresses, study stays and research trips abroad, the exchange of publications, etc. Cecil Roth recalled with subtle humour, in his *Reminescenze sugli ebrei italiani durante le loro traversie*:

> As October 1938 approached, several refugees began to arrive. I remember one of them who told me that, before leaving Rome, he had gone to visit Rabbi David Prato to ask him for a presentation for me. Prato, in taking his notebook of addresses, said to him, "Please wait: I will look in the list of the Elders of Zion". After the racial laws were established, Jewish refugees from Italy continued to come to England in considerable numbers. In various cases, I was the only Englishman they knew, or with whom they had had some previous contact, or from whom they could get a presentation; and a very large part of them came to visit me and asked me for advice. I was told that this saying circulated: "The first day in Bow Street (to register with the police); the second to the Roths." As time went by, we made it a habit to set up regular meetings for them every Sunday evening; we tried to introduce them to English Jews – especially those who could help them – and to broaden the circle of their connections, sometimes with great success.[5]

---

[5]Roth C. (1965), pp. 204–205: *Con l'avvicinarsi dell'ottobre 1938, cominciarono ad arrivare parecchi altri rifugiati. Uno di essi ricordo che mi riferì che, prima di lasciar Roma, era andato a*

In the case of academics, alongside the solidarity chains there were also various international organisations to help the victims of political and racial persecution, such as the Society for the Protection of Science (SPSL) and Learning and the Emergency Committee in Aid of Displaced German Scholars (ECADFS), and philanthropic foundations such as the Rockefeller Foundation and the Guggenheim Foundation which had supported the flight of scholars from Central and Eastern Europe since 1933. Since the end of 1938, not without having to overcome considerable and stubborn internal conflicts, the agencies and the foundations included Italian scholars in their rescue agendas. By contrast, the fact that the vast majority of Italian scholars were made up of 'Yom Kippur Jews'—i.e., those Jews who enter the synagogue just one time per year, on the occasion of Yom Kippur—meant that the contribution of Italian Jewish institutions assisting migrants was minimal. The scant sympathy that Zionism had gathered among the ranks of Italian university professors—some of whom, indeed, openly rejected it—made the British mandate of Palestine a migratory route of secondary importance.[6]

Academic emigration was also a 'family affair', at least for certain disciplines such as mathematics, physics, and physiology. In this case, the family did not constitute blood ties but consisted of the research School. This signified belonging to a collective of scholars engaged in developing a common research project under the guidance of a leader identified as a common mentor or Master (*Maestro*), and sharing political, social ideals, and whole traits of their own personal and work lives. If, as was the case with the above-mentioned disciplines, the research Schools were ethnically homogeneous, teachers and pupils, Masters and protégés emigrated together.

As the emigration of a specific professional sector (that of university workers), academic exodus from racist Italy embodied some peculiar characteristics. The choice between leaving or staying was conditioned not only by individual financial possibilities, but also by the age of the potential refugee, by gender, and by the breadth and effectiveness of the social and relational capital which he/she could raise. Adaptability and flexibility of the migration project were central, as were

---

far visita al rabbino David Prato per chiedergli una presentazione per me. Prato, nel prendere il suo taccuino di indirizzi, gli disse: "Aspetti che cerco nell'elenco dei Savi Anziani di Sion". Approvate le leggi razziali, i rifugiati ebrei d'Italia seguitarono a venire in Inghilterra in numero considerevole. In vari casi, io ero l'unico inglese che o conoscevano, o con cui avevano avuto un qualche precedente contatto, o da cui potevano ottenere una presentazione; e una grandissima parte di essi venne a farmi visita e a chiedermi consiglio. Mi fu raccontato che era invalso questo modo di dire: "Il primo giorno a Bow Street (per registrarsi presso la polizia); il secondo dai Roth". Con l'andare del tempo prendemmo l'abitudine di metter su per loro delle riunioni regolari ogni domenica sera; cercavamo di presentarli a ebrei inglesi – specialmente quelli che li potevano aiutare – e di allargare la cerchia delle loro conoscenze; talvolta, con grande successo.

[6] Fano A. (1955). Between July 1938 and June 1940, 504 Italian *olim* arrived in Eretz Israel, of whom 84 left again after a short time, after 1945 or even after 1948. These numbers were confirmed by Della Pergola and Tagliacozzo (1978). Literally, *olim* means 'immigrants on *aliyah* to Israel'. *Aliyah* (literally 'ascent') is a Hebrew term which evokes the 'return to roots' that every believer must accomplish, referring to the dispersion of the Jewish people.

prestige, skills, and competences. It was difficult, if not impossible, for a scholar at the beginning or at the end of a career to find a position abroad, unless it were a matter of exceptional talents; men had better chances than their female colleagues, even if occupying equal positions before discrimination; since in 1938 women mainly occupied minor positions in academia (only Anna Foà had won a chair), they had very little chance of taking the path of expatriation. Scholars who for various reasons had remained secluded from international research trends or who had not created strong and continuous relationships with foreign colleagues had almost no prospect of leaving. In particular, those who had scant or distant connections with the English-speaking intellectual world saw their opportunities reduced to a minimum.

The choice of destination depended not only on immigration quota systems and success in obtaining visas and affidavits (from this point of view, faculty members were a privileged category with regard to US immigration legislation, for example) but also on the possibility of finding a university, research institution, laboratory, archive, library, or school willing to hire them. In other words: a general strategy for living and surviving was required. Few university professors demonstrated a willingness to 'change job' or had the skills to embark on new professional adventures. Even transfers from academia to other sectors of knowledge workers were not frequent.

The previous relations with fascism of the aspiring expat also influenced the outcome of attempts to leave Italy. While a freelancer or entrepreneur forced to emigrate because of the racial laws was not called upon to document whether he had been fascist, anti-fascist, or a-fascist, a scholar seeking an opportunity abroad was more frequently required to do so, both before departure and after settling in the host country. Often this was not a mere self-declaration about one's political past or present beliefs, but a choice of field. Suffice it to mention the case of physicists and engineers such as E. Fermi, E. Segré, E. Fubini, U. Fano, etc. who had to decide whether or not to collaborate in Allied research programmes. The inner conflict between loyalty to the homeland and gratitude to the country of adoption, although it may seem over-sentimental to our eyes, was often lacerating.[7]

The emigration of persecuted Jewish scholars in 1938 does not fully coincide with what is usually called the 'brain drain'. In some ways it was similar, because just like the brain drain that so much worries Italian governments today, it forced two types of scholars to emigrate: people of exceptional talent such as Nobel laureates like Enrico Fermi, emigrated for love because his wife Laura Capon was a Jew, and young people who had just entered the world of research (postgraduate students, lecturers, assistants, etc.) such as the future Nobel laureates Salvatore Edoardo Luria, Franco Modigliani, and Rita Levi-Montalcini. The loss of their human capital in some sectors slowed down, and even halted, Italy's cultural, technological, and

---

[7]For some, there was much more involved: the aim to prevent a world war from being won by the 'forces of evil'. The mere thought of what the world would be like after an Axis victory necessarily prompted which side to be on. This also would result in the liberation of one's own country.

economic progress, with serious short- and long-term implications. Unlike the classic brain drain, however, Jewish academic emigration involved dozens of university professors and schoolteachers with stable positions who in the normal circumstances of the rule of law would never have thought of leaving.

Jewish academic emigration from racist Italy posed various specific obstacles:

- linguistic difficulties, all the more important for people who worked with knowledge and words, and who wanted to express their thoughts with a certain precision and lexical finesse. In a country like Italy where the learning of English had not been promoted at all by the fascist regime, the language handicap was painfully felt;
- competition with incomers from other totalitarian regimes, and in particular with those fleeing the Third Reich, who in the five years preceding the Italian racial laws had occupied many positions in English-speaking academia;
- the (understandable) perplexities of foreign institutions in welcoming into their staff men who had been notoriously fascist until 1938 and sometimes remained so;
- the embarrassment of the newcomers in interacting with Italians who had left the country before 1938 for political reasons (the so-called *fuoriusciti*);
- the difficulty in entering scientific and institutional contexts characterised by forms, methods, and dynamics of organisation of research and teaching significantly different from the Italian ones (Segré called it 'the university minuet that one danced at Berkeley, as one did in all universities'.[8]) Globally, it must be frankly said, Italians struggled to adopt new ways of thinking and making culture. They were very able to export their know-how, but rarely re-targeted their profiles to detach from the traditions of study in which they had been trained and in which they had worked until expatriation.

Contextualisation of Research Almost completely ignored by historians for 50 years, Jewish emigration from fascist Italy is a phenomenon still largely to be studied, and all the more so regarding Jewish intellectual emigration. With the exception of a few notes contained in De Felice's *Storia degli ebrei italiani sotto il fascismo*,[9] no one dealt with the subject until 1988 when Toscano was the first to establish the quantitative dimension of the phenomenon and some of its global traits.[10] In the nineties, some aspects were explored further: the relationship between Italian Jewish refugees and foreign anti-fascism;[11] their contributions to anti-fascist movements such as the Mazzini Society or *Italia Libre*, in specific geographical areas such as the United States, Latin America, and Australia;[12] the exodus to

---

[8] Segré (1993), p. 133.

[9] De Felice (1961), pp. 284–290, 417–433.

[10] Toscano (1988).

[11] Fanesi (1994); Sarfatti (2007a).

[12] Parshall (2022), pp. 191–231, 297–335; Zevi (1984); Montagnana (1987).

Switzerland during the Nazi-Fascist occupation;[13] the personal and professional stories of some illustrious displaced scholars.[14] The latter, after all, had been the perspective of a well-known volume by Laura Capon Fermi *Illustrious Immigrants: The Intellectual Migration from Europe, 1930-41*, an engaging and moving book that reads like an adventure novel, and of various autobiographical writings in which the filter of memories and affections hindered, however, any objective evaluation of events and their actors.[15]

Overall, the area most investigated in the literature was that of the United States, with works dedicated to the emigration of the Masters, to the presence of women in the migration waves and, recently, a book on the diaspora of Italian psychologists.[16] Meanwhile, historians dealt episodically with the large number of scholars and scientists who emigrated to Latin America, and, to date, there are practically only two contributions: one centred on the encounter of Italian culture with Latin American culture in the years 1938–1945, and one devoted to the 'undesirables', that is, the intellectuals who arrived in Argentina to escape racial persecution and whose inclusion in university life was opposed by the Argentine authorities (who were, in some cases, clearly sympathetic to fascism).[17]

Already in 2010, Capristo acutely noted two facts: studies were needed on archival sources, focusing on specific academic and professional categories.[18] The eightieth anniversary of the promulgation of the racial laws, re-launching the studies on anti-Jewish persecution in Italy, stimulated new investigations precisely in these two directions. The University of Florence promoted the research prosopography project *Intellectuals fleeing from fascist Italy. Migrants, exiles and refugees for political and racial reasons* and a related website, edited by Guarnieri.[19]

Having ascertained that nobody had dealt with the Jewish mathematical diaspora from fascist Italy and inspired by Siegmund-Schultze's book *Mathematicians fleeing from Nazi Germany* (2009), the rational concept of this book was born. Literature on Jewish mathematical emigration from fascist Italy was basically non-existent. The works related to intellectual emigration mentioned the mathematicians *en passant*, citing only three names (Fubini, Terracini, and Levi), and without exploring

---

[13] Broggini (1993, 1999).

[14] Pontecorboli (2013).

[15] Capon Fermi (1968); Colonnetti (1973); Mortara (1985); Bravo and Jalla (1988); Vita-Finzi (1989); Terracini L. (1989); Rosenthal Fuà (2004); Pincherle (2011); Thomson (2014), etc. The work [Terracini L. (1989)] is a critical study of university immigration in Argentina, enriched by personal and friendship memories, from conversations and interviews, and from articles that appeared in local newspapers such as *Los Tucumanos de Italia* (La Gaceta, Tucumán 15.11.1970) and *Los Argentinos de Italia* (La Nación, Buenos Aires, 1.8.1971).

[16] Camurri (2009); Gissi (2008) (English translation Gissi 2010); Gissi 2015 (English translation Gissi 2016); Guarnieri 2016.

[17] Treves R. (1985); Korn (1988).

[18] Capristo 2010 (English translation Capristo 2014), p. 181, 196.

[19] Guarnieri ed. https://intellettualinfuga.fupress.com/; Guarnieri 2019a.

individual and collective stance in face of expatriation at all. The Italian historians of mathematics and science who first addressed racial persecution and its consequences had focused on other issues: the quantitative dimension of persecution, the implicit and explicit motivations of fascist racial politics, the similarities and differences between anti-Semitic persecutions in Italy and Europe, persecution and gender, etc. Unlike physicists, expat mathematicians had neither given interviews nor produced memoirs.[20] Compared with the large and very detailed autobiographical writings by Emilio Segré, Sergio de Benedetti, and Edoardo Amaldi,[21] only a few pages of the *Ricordi di un matematico* by Alessandro Terracini were available.[22] Written many years after the facts, they contained obvious inaccuracies, confusions of dates and people, omissions and misleading information. In obituaries as well as in biographies, the migrant experiences were dismissed in a handful of lines. Some useful information could be obtained from some papers published by their children and grandchildren: the biography of Guido Fubini's son, Eugenio (written by his nephew David), the *Memories of an atomic physicist for my children and grandchildren* (written in old age by Gino Fano's son, Ugo) and the booklet of family memories published by Beppo Levi's daughter, Laura. This is the reasoning behind our choice to work on archival sources mainly, and with the correspondence above all.[23]

Turin, Italy   Erika Luciano

---

[20] See for example Israel and Nastasi (1998); Israel (2010).

[21] Segré E. (1993); De Benedetti S. (1965); Battimelli and De Maria (1997) (eds.); Battimelli and Paoloni (1998) (eds.).

[22] Terracini A. (1968).

[23] Fubini D.G. and Brown (2015); Fano U. (2000); Levi L. (2000).

Acknowledgements

This book would not exist without the immense help I received from so many institutions and individuals during the five years of its writing. I have a large number of people to thank heartily.

My gratitude goes first and foremost to the relatives and descendants of the protagonists of this book, who opened the doors of their private archives to me and allowed me to enter their homes and the lives of G. Ascoli, B. Colombo, G. Colonnetti, B. Levi, T. Levi-Civita, G. Fubini and A. Terracini: Cristina Ascoli, Davide Ascoli, Irene Ascoli, Maria Bolgiani, Guido Bolgiani, Paola, and Marco Salbol; Emma Kursner and Marina Colombo Fubini; Margherita Colonnetti; Emilia Levi; Pier Vittorio and Tullio Ceccherini; Laurie Jacobs and David G. Fubini; Benedetto Terracini.

I am also intensely grateful to the Directors and the staff of all the Archives and Libraries who provided me with digital copies of the archive documents, during the lockdown period due to the Covid pandemic and that granted publication rights for various illustrations: Albert Einstein Archives, Hebrew University of Jerusalem; Archive of the Jewish traditions and customs Benvenuto and Alessandro Terracini; Archive of The Society for the Protection of Science and Learning; Archives of the Emergency Committee in Aid of Displaced Foreign Scholars, The New York Public Library; Archives of the National Academy of Lincei; Archive of the National Institute for the Application of the Calculus (courtesy of Dr. A. Celli); Archives of Words, Images and Publishing Communication, University of Milan; Center for Contemporary Jewish Documentation in Milan; Central State Archive; State Archives of Turin; Enrico Persico Archive, Department of Physics, University of Rome La Sapienza; Historical Archive of the Italian Mathematical Union; Historical Archive of the Polytechnic University of Turin; Historical Archive of the University of Genoa; Historical Archive of the University of Turin; Institute for Advanced Study, Princeton, Shelby White and Leon Levy Archives Center, School of Mathematics Records; Oswald Veblen Papers, Library of Congress Manuscript Division, Washington, D.C.; Special Mathematics Library, Department of Mathematics, University of Turin; Swiss Federal Archives; The Caltech Archives.

My special thanks go to Profs. Reinhard Siegmund-Schultze, Christopher Hollings, and Salvatore Coen, who read the first manuscript, and who provided many valuable comments, suggestions, and advice. Their annotated versions are now, and will always be, part of my personal archive!

A huge thank you to Victoria Clifford, for the steadfast patience she always maintained through the process of linguistic revision. Her help was simply invaluable. My deepest thanks also go to Emma Sallent del Colombo, who reviewed the transcriptions of the letters in Spanish, and to Elena Scalambro, for her rereading of the final manuscript and help in its editing.

A heartfelt thank you to the colleagues who invited me to present some chapters of this volume at conferences and workshops that they organised: M. Ash (Prague 2016), Academy of Sciences of Turin (2018), Archivio Terracini (Torino 2018), C. Cattarulla (Buenos Aires 2018), M. Adamson and S. Turchetti (London 2018), Mathesis Torino (2018), M.T. Borgato and C. Phili (Athens 2019), G. Santi (Bolzano 2020), Mathesis Pavia (2020), C. Fontanari (Trento 2021), W. Wójcik (Krakow 2021), Archivio Terracini (Torino 2021), L. Giacardi, R. Tazzioli and Laurent Mazliak (Trento 2021), G. Bini (Rome 2021), S. Despeaux, D. Dumbaugh and J. Lorenat (Oberwolfach 2022); R. Siegmund-Schultze (Aalborg 2023), A. Demuro, T. Préveraud, G. Jouve and R. Tazzioli (Lille 2023), Seminario FINO (Torino 2024). I owe much to the discussions and exchange of thoughts that took place on these occasions, and I profited from them in order to clarify various aspects of my historical reconstructions. Thanks to all the colleagues who, over the years, have discussed the topics addressed in the book with me and, in particular, thanks to: D. Aubin, J. Barrow-Green, D. Ciesielska, C. Ciliberto, A. Conte, H. Durnová, C. Eckes, P. Freguglia, L. Giacardi, H. Gispert, A. Guerraggio, S. Kennedy, M. Marchisio Conte, M. Mattaliano, A. Millán Gasca, R. Nossum, L. Pepe, A. Piazza, C.S. Roero, B. Santirocco, P. Valabrega, and L. Vieira Souza da Silva.

To the Directors of the *Science Networks series. Historical Studies*, Profs. O. Darrigol and P. Ullrich, to the anonymous reviewers, to Drs. S.A. Goob, C. Tominich, F. Trotter, F. Ferrari, and D.I. Jagadisan, I express my sincerest gratitude for having believed in this book project and helped me in every possible way to achieve it.

When Prof. Giorgio Israel passed away, this research project was still only a sketch, devoid of any consistency as only dreams are. Constantly, while I was writing the book, I wondered what he would say about it. Now that the book is finished, I can only hope that he would have liked it. At all events, it has been written in loving memory of him.

Five years of work are also five years of life. Thanks to my family (my husband Luca and our sons Ivan, Amos, David, my father Spirito with Alessandra, my in-laws Laura and Giorgio, Sabrina, Marco, and Chiara, and my nephews Samuele and Niccolò). Thanks to the '*dwarfs*' support team': Zia Dada (Silvana Bonda), Elisa

Salonio, and Sara Vietti. Without their invaluable help, without these people who always back and support me, nothing would be possible. Finally, the friends, who surrounded me with affection and made me smile even in the most tiring moments: Chiara and Roberto Dutto, Daniela and Luca Selvini, Gabriella and Lorenzo Tropini-Bassino, Carmela and Walter Donna.

About the Book

The Structure of the Book This book deals with Jewish mathematical emigration from Italy after 1938, and its protagonists are six mathematicians who left the country to rebuild a life abroad (Guido Fubini, Gino Fano, Beniamino Segre, Alessandro Terracini, Beppo Levi, and Bonaparte Colombo), five who tried to emigrate without success (Guido Ascoli, Arturo Maroni, Bruno Tedeschi, Nella Friberti, and Giulio Bemporad), and four who chose to stay in Italy and dedicated themselves to helping colleagues leave and return (Tullio Levi-Civita, Guido Castelnuovo, Federigo Enriques, and Vito Volterra). It is structured in two parts, one dealing with the global phenomenon and the other looking at individual destinies. The first part provides a substantial introduction to the topic and is followed by the discussion of six case studies, dedicated to the six immigrant scholars previously mentioned. While allowing space for historical exposition, each paragraph is followed by the relevant, annotated archive documents.

Italian Jews, distinct from Sephardi and Ashkenazi, have a particular history and characteristics, including their own customs and rites (the so-called Italian rite, with its Roman subvariant) and a variety of Judeo-Italian languages such as Bagitto in Livorno and Roman Judaic (*giudaico romanesco*) in Rome. Without taking into account the delicate path traced between the Risorgimento and WWI, from emancipation to assimilation through integration, it is impossible to understand their stances during the fascist rule, and their reactions to religious and racial anti-Semitism. Chapter 1 provides the description of this general historical background, describing the Jewish mathematical micro-society that was created in Italy starting from 1848: a world of scholars, selective but not aristocratic, limited in its interests but fervent; patriotic but also with a lucid, international, cultural conscience; liberal in the best sense, even if not strictly democratic.

The mathematical diaspora from fascist Italy is the result of the *Measures for the Defense of the Race* issued by the regime in the autumn of 1938, since it derives closely from the institutional, epistemic, and social upheavals that they triggered. Italy—it must be said—has come to terms with its history at a relatively late date, and it was only in 1988 that the Lincei Academy lifted the veil of silence on this chapter. Since then, however, the literature concerning the legal frame of the anti-Jewish persecution, the quantitative dimension, the assumptions underlying the regime's racial policy and its implications for Italian culture and science has been enormously enriched. If it is true that it would be enough to refer to the works of Israel and Nastasi on science and race in fascist Italy, it is equally certain that some basic information on the anti-Jewish persecution carried out by the fascist rule can be useful to the reader. Chapter 2 is dedicated to this topic. Racial laws are here explained in some detail, and the relationships between racial and academic anti-Semitism are discussed at some length, taking into special consideration institutions such as the Italian Mathematical Union and fascist agents such as Francesco Severi and Enrico Bompiani. Due to the fact that for the six mathematicians who managed to emigrate, there were at least 47 others who decided or were forced to stay, we opted to dedicate space to those stories, too (Sects. 2.10, 2.11, Chaps. 4 and 5). Particular emphasis was given to the counter-discrimination procedure, which already at the time seemed incomprehensible abroad and which slowed down Jewish intellectual emigration. The counter-discrimination applications submitted by mathematicians published in Sect. 2.11 provide an interesting picture of how these scholars paraphrased *ex post* their existential, professional, and political trajectories both before and during the fascist twenty-year period and help us to understand the difficulties that they would have met in defining their status as expats for a political cause but not as political exiles in the strictest sense.

Chapters 3, 6, and 7 constitute the core of the book and set out to answer a question that can be formulated in these terms: what features are specific to mathematicians and to mathematics in the phenomenon of Jewish intellectual emigration from fascist Italy? The specific traits of the Italian mathematical diaspora, which differentiate it from other contemporary flights like that from Nazi Germany, can be identified in the following eight points.

It was a diaspora of quantitatively significant dimension, since it involved six professors of mathematical disciplines out of a total of 86 faculty members in service in 1938. The six refugee mathematicians all had a very strong bond with a single University that of Turin, where they had been trained and/or in which they had worked for a number of years, varying from 7 in the case of Beniamino Segre to the 41 of Gino Fano. They were all specialists in a single field, geometry, and they all belonged to a single research School, the so-called Italian School of Algebraic Geometry, founded by Luigi Cremona and brought to international excellence by Corrado Segre, Guido Castelnuovo, Federigo Enriques, and Francesco Severi. The Italian Geometric School, in both its Turin and Roman 'branches', was an intellectual community made up of 60% Jews, including not only university professors, but also middle and secondary school teachers, many of them women.

Jewish mathematical emigration from fascist Italy received little support by international rescue organisations for a variety of reasons explained in the book (Sects. 3.2, 3.3, 3.17, and 3.18). None of the applications submitted to the SPSL and ECADFS led to the awarding of a grant. On the other hand, the role of solidarity networks was pivotal, when planning expatriation, finalising the transfer abroad, and fitting into new academic environments (Sects. 3.9, 3.10, and 3.11). All the mathematicians fleeing from Italy did so with their families; being linked by family or friendship ties, and/or by the fact of having been fellow students and colleagues, they planned the exodus together. Their emigration was therefore, in a certain way, a family affair (Sects. 3.7 and 3.8). Mutual advice on destinations, tips on how to collect departure documents, tackling the maze of impossible bureaucracy, information on how to prepare an application and how to access a job market that often appeared opaque were all crucial factors in ensuring the success of the expat operation. The Masters, the great names of Italian mathematics who had represented the field abroad since the end of the nineteenth century, mobilised their networks of international connections. For various reasons, Levi-Civita, Castelnuovo, Enriques, and Volterra did not consider leaving but helped their colleagues to do so. Levi-Civita was, in particular, the 'patriarch' of mathematicians who were intent on emigrating. The networks of solidarity crossed and reached beyond disciplinary boundaries and generational barriers: those who first succeeded went on to help the others, sensitising foreign mathematicians towards the 'Italian question'. Oswald Veblen and Max Ascoli in the United States, Julio Rey Pastor, Cortés Pla, and Alfred Rosenblatt in Latin America all facilitated the exodus and resettlement of Italian mathematicians overseas.

Billy Wilder is credited with the somewhat brutal sentence: 'the optimists died in the gas chambers; the pessimists have pools in Beverly Hills.' This does not apply to the Italian case. The mathematical diaspora from fascist Italy was not an escape, because in 1938 nobody could foresee that the dispossession of rights would have preceded the deprivation of lives; nor was it a political emigration in the classical sense, because none of the would-be refugee mathematicians had carried out anti-fascist political activity. Theirs was a voluntary shift, a painful step that involved difficult decisions. For this reason, in this book it has been decided to leave some room for discussion of the motivations that led to the choice of whether to stay or leave in search of a space for intellectual survival (Sect. 3.6).

While the United States was the final destination for most of the intellectuals who fled from Central Europe and for other groups of Italian scholarly expats, for Italian mathematicians it was not so. Only Guido Fubini managed to land at the Institute for Advanced Study, thanks to the joint intervention of Levi-Civita and Veblen. The failure of the attempts to find an opening in the US was due to various factors, analysed in Sects. 3.13, 3.19, and 3.20; these factors included competition with mathematicians who had fled from Central and Eastern Europe, a mediocre knowledge of English, and the common style and approach in algebraic geometry typical of the Italian School, which was not universally appreciated in English-speaking countries. The South American route was decidedly more successful: greater linguistic and cultural affinities, the presence of two Italian mathematical missions in Brazil, the long-lasting imprint left by the stays as visiting professors of Levi-Civita and Enriques in Argentina, Brazil, and Peru, and the need to recruit teaching staff for two newly-created Mathematics Departments (Rosario and Tucumán) laid the foundations for a targeted strategy of recruitment of Italians (Sect. 3.15).

Italy's entry into the war and consequent rupture of diplomatic relations with many countries were experienced dramatically by Italian mathematicians who, despite immigration, had continued to feel Italian and had not cut cultural, linguistic, ideal, and affective ties with their homeland. For men who had remained alien to Italian political life, who had not woven any connection with the political expats, nor approached anti-fascist organisations abroad, the transition of status from emigrants to enemy aliens was decisively traumatic (Sect. 3.22). The choice between loyalty to the homeland or to the host country was difficult, especially for the elderly (Fano, Fubini, and Levi). The absence of any trace of denunciation or critique of the fascist rule, in any document, is quite surprising. The mathematicians expressed only words of sorrow, daring amazement and anguish for those who had remained in Italy and for the consequences that the war would have on the country.

In the period of the Nazi occupation and the Salò Republic, out of the forty mathematicians who remained in Italy, five were deported to Auschwitz; two fled to Switzerland and the others lived in hiding. In Switzerland, in the context of a last migration wave made up not only of Jews, but also of anti-fascists, deserters, disbanded soldiers, draft evaders, etc., a teaching experience unique in its genus took place: the creation by the Turin applied mathematician and engineer G. Colonnetti of a University in exile, an internment camp reserved for Italian university students confined in Switzerland. Sections 3.23 and 3.24 are dedicated to this story.

The inclusion of Italians in the new scientific frameworks followed a peculiar dynamic. Under the umbrella of belonging to a unique mathematical School, they imported from Italy into their new settings a way of doing, teaching, and communicating mathematics that was typical of this group (Sect. 6.1). Where, as in Argentina, the conditions made it possible to recreate the Italian models of organisation of mathematical life, their work was widely appreciated and left a remarkable imprint (Sects. 6.3, 6.5, and 6.6). In contexts such as the United States and the United Kingdom, interactions between Italians and local colleagues were much more problematic (Sect. 6.2). In general terms, and with the necessary distinctions, a key component of their commitment was represented by dissemination and popularisation activities, with various cycles of conferences, seminars, broadcast transmissions, and editorial initiatives dedicated to promoting the best Italian mathematical traditions, in the field of geometry specifically (Sect. 6.4).

During their time abroad, émigré mathematicians did not establish relations with local Jewish communities, nor did they acquire or recover a feeling of religious or cultural identity. On the other hand, until the armistice, they maintained contacts with relatives, friends, and colleagues in Italy, even if they were fascists, and they resumed these relationships as soon as possible after the liberation. Upon the repeal of the racial laws (January 1944), all but Levi decided to return (Sects. 7.1 and 7.2). The speed with which they reconnected with their ex-fascist colleagues (Severi, Bompiani, Picone, etc.) and the desire to contribute to the cultural reconstruction of a country that had persecuted them are surprising and have no analogues concerning either émigrés from Central and Eastern Europe or other communities of Italian intellectuals expatriated in 1938.

The backbone of the book is provided by **archival documents,** which have given many new facts and interesting biographical and institutional details on the Italian Jewish mathematicians who chose to play the emigration card. It is a somewhat unique corpus because, until now, nobody had collected such material for any category of scholars displaced from Italy. The documents gathered in the public and private archives of Oxford, New York, Pasadena, Princeton, Turin, Bologna, Rome, etc. amounted to just under 2000, between letters and documents. Making a selection of these was difficult, especially as far as the second part of the book is concerned. Firstly, it was necessary to avoid a difference in treatment between the individuals. Secondly, objectivity and detachment in exposition, fundamental requirements of any historical narrative, were put to the test when, through the reading of the documents, we accessed to the private universe of these people. There are families who opened their homes and archives, who asked me to tell their stories of humiliation and redemption, nostalgia and pride, sense of uprooting, and will for recovery. Others exercised a sort of right to oblivion, sometimes out of prudence, others in order to respect the 'paternal silence surrounding mortal injuries that was part of an educational strategy intended to contain the right to resentment'.[24] In the light of these considerations and starting from the conviction that the duty to document historical theses must never encroach on voyeurism, I excluded a priori all those documents whose contents were eminently private. In the documents selected for publication, I omitted every reference to patrimonial issues, health, and private affairs. The cuts were indicated with [. . .]. Addresses have also been systematically omitted. The documents collected, crossed with other types of sources and integrated into a unitary narrative, have allowed us—we hope—to provide an objective but not unempathetic description of the Italian Jewish mathematical micro-society when faced with persecution and emigration.

The English **translation** of the documents presented various difficulties, both because the dictionary and formulas of fascist rhetoric cannot be properly rendered in another language, and because the protagonists of this book, as members of a School, used a family lexicon whose nuances escape the rigid geometry of scientific language and sometimes pass unnoticed even to the Italian reader. How was it possible to restitute the university 'baron' prose of Severi (typical of the powerful and famous Italian academics), or the ambiguously mellifluous style of Bompiani, or the boorish anti-Semitic rhetoric of Demorazza's anonymous reports? We have done our best, but we are under no illusions that we have completely succeeded in our intent. We hope that the decision to provide the Italian text with parallel translation into English will allow a direct approach to the documents for readers who have some knowledge of Italian language and, why not, act as a stimulus to others to approach this language. The critical apparatus of notes is very concise. To avoid repetition, basic biographical information on the persons mentioned in correspondence is provided in the footnotes.

---

[24]Terracini Benedetto (2020), p. 100: *Forse il silenzio paterno sulle ferite mortali faceva parte di una strategia educativa intesa a contenere il diritto al risentimento.*

Some clarifications on lexical solutions are necessary. The terms *ebreo* and *israelita*, synonyms in Italian, were translated with *Jew*; the term was used in the sense established by art. II, 8 of racial laws: 'a Jew is one who was born from both parents of Jewish race, even if he belongs to a religion other than the Jewish one; one who was born from a mother of the Jewish race and from a father unknown; one who, despite being born from parents of Italian nationality, of which only one is Jew, belongs to the Jewish religion, or is enrolled in an Israelite community, or has made, in any other way, manifestations of Judaism'.[25] The term *giudeo*, translated into *Juda*, which appears in various documents, is vulgar and derogatory. For semantic and historical reasons, the adjective *razziale* has been translated into *racial* instead of *racist*, corresponding to the Italian *razzista*. In Italian, *racial* means 'which concerns race, which is based on race';[26] *racist*, on the other hand, means 'proponent of racism'.[27] In official and private documents, moreover, the adjective *racist* almost never occurred. To avoid ambiguity, the two Italian synonyms for Jewish communities (*comunità* and *università*) have been rendered with a single word: *community*. In the text, particular attention has been paid in handling the many terms that indicate the condition of those who leave the country where they were born to move abroad. The term *emigration* (*emigrazione*) indicates this phenomenon in a neutral way; *exile* (*esilio*) denotes a voluntary move for political reasons (persecution); *flight* (*fuga*) indicates a mass and rapid emigration of individuals in a situation of serious risk to life. The term *dispatrio* has not been translated. It defines a condition of the soul and not only of the body: the sense of displacement and loss of roots. A similar polysemy denotes the subjects: *expats*, *émigré*, and *emigrants* (*emigrati*); *exils* and *exiled* (*esuli*), and refugees (*rifugiati*). In the documents, obviously, the terms used by the authors were faithfully translated, even when the lexical choices were improper and even if the semantic shift was neither accidental nor secondary.

Research Methodology This is a research study within the social history of mathematics, encompassing a broad perspective which entangles the institutional and political history and the history of international relations. The approach is mainly descriptive in the first part and prosopographic in the second. Palladio's digital humanities software seemed appropriate for visualising complex networks of historical data. The typical interpretative tools and categories of the social history of

---

[25] Regio Decreto Legge 17.11.1938, n. 1728 *Provvedimenti per la difesa della razza italiana*, Capo II, Art. 8 *Degli appartenenti alla razza ebraica: Agli effetti di legge: a) è di razza ebraica colui che è nato da genitori entrambi di razza ebraica, anche se appartenga a religione diversa da quella ebraica; b) è considerato di razza ebraica colui che è nato da genitori di cui uno di razza ebraica e l'altro di nazionalità straniera; c) è considerato di razza ebraica colui che è nato da madre di razza ebraica qualora sia ignoto il padre; d) è considerato di razza ebraica colui che, pur essendo nato da genitori di nazionalità italiana, di cui uno solo di razza ebraica, appartenga alla religione ebraica, o sia, comunque, iscritto ad una comunità israelitica, ovvero abbia fatto, in qualsiasi altro modo, manifestazioni di ebraismo.*

[26] *Il grande dizionario Garzanti della lingua italiana*, ed. 1987, p. 1568: *che concerne la razza; che si basa sulla razza.*

[27] *Il grande dizionario Garzanti della lingua italiana*, ed. 1987, p. 1569: *fautore del razzismo.*

science (namely, the notions of research School, tradition, network, style, circulation of knowledge, vectors of scientific sociability, etc.) have been used without further clarification, according to the authors cited in the footnotes. In particular, according to Rowe (2002) a mathematical School has been defined as a group led by one or more mathematicians, localised within a single institutional setting, and comprising a significant supply of advanced-level students, whereas a mathematical tradition implies that one can find a common research orientation among different actors who do not share a common institutional site, but are linked by traceable influences on each other.

Motives of the Research   This research emerged from historical interest and was developed with proper historical purposes, such as the evaluation of the influence that the emigration of Jewish mathematicians had on the scientific and cultural fabric of Italy and of the host nations, and the understanding of the transformations that mathematics has undergone in Italy since the post-war period. Neither of these facts can be detached from the critical knowledge of fascism and racial politics. However, the topic is not of interest for experts interested solely in Italian history of mathematics. In fact, it is not improper to hypothesise a socio-educational implication of this material. The growth of migration flows and the persistent economic difficulties have raised concerns at all levels regarding the return of racist and xenophobic attitudes or position among the Italian population. In the last few years, the monitoring centre of the *Commissione speciale per il contrasto ai fenomeni dell'intolleranza, del razzismo, dell'antisemitismo e dell'istigazione all'odio e alla violenza* vehemently denounced the spread and multiplication of acts of violence and intolerance against those who are regarded as 'different' for reasons of ethnicity, race, religion, gender, sexual orientation, physical and/or mental disability, and fragility. A substantial percentage of these acts took place in schools. Under the assumption that racist preconceptions essentially spring from ignorance, school and out-of-school activities were offered, aimed at exploring and stimulating historical and critical reflection on the concepts of race and discrimination. Accepting the national guidelines *Per una didattica della Shoah a scuola,* elaborated by the Italian delegation of the International Holocaust Remembrance Alliance in 2017, the experts in didactics of STEM disciplines have started to build courses and laboratories on the history of racism and anti-Semitism. Teaching and public engagement activities dedicated to Italian mathematicians who were victims of racial persecution form part of this framework. Since 2017, many schools adhering to the National Project for Scientific Degrees have worked in class on this topic, rediscovering the history and the stories of mathematicians who fled racist Italy.

Contents

Abbreviations

AI	Reale Accademia d'Italia; Royal Academy of Italy
AMS	American Mathematical Society
ANL	Accademia Nazionale (già Reale) dei Lincei, Roma; National (before Royal) Academy of Lincei, Rome
BSM	Biblioteca Speciale di Matematica, Dipartimento di Matematica, Università di Torino; Special Mathematics Library, Department of Mathematics, University of Turin
CNR	Consiglio Nazionale delle Ricerche; National Research Council
DAGR	Divisione Affari Generali e Riservati; Political and General Affairs Office
DELASEM	Delegazione per l'Assistenza degli Emigranti Ebrei; Delegation for the Assistance of Jewish Emigrants
Demorazza	Direzione generale per la demografia e la razza; General Office of Demography and Race
DGPS	Direzione Generale della Pubblica Sicurezza; General Office of Public Security, Police Division
ECADFS	Emergency Committee in Aid of Displaced Foreign Scholars
EGELI	Ente di Gestione e Liquidazione Immobiliare; Agency for Real Estate Management and Liquidation
FESE	Fond Européen de Secours aux Étudiants; European Student Service Fund
IAS	Institute for Advanced Studies
ICM	International Congress of Mathematicians
INAC	Istituto Nazionale per le Applicazioni del Calcolo; National Institute for the Applications of the Calculus
INDAM	Istituto Nazionale Di Alta Matematica; National Institute of Advanced Mathematics
IUC	Campo Universitario Italiano, Losanna; Italian University Camp in Lausanne

MEN	Ministero dell'Educazione Nazionale; Ministry of National Education
MFA	Ministero degli Affari Esteri; Ministry of Foreign Affairs
MI	Ministero dell'Interno; Ministry of Interior
Minculpop	Ministero della Cultura Popolare; Ministry of Popular Culture
MIT	Massachusetts Institute of Technology
MPI	Ministero della Pubblica Istruzione; Ministry of Public Instruction
PNF	Partito Nazionale Fascista; National Fascist Party
Pol. Pol.	Polizia Politica; Secret Political Police
RSI	Repubblica Sociale Italiana; Fascist Italian Social Republic
SPSL	Society for the Protection of Science and Learning
UCII	Unione delle Comunità Israelitiche Italiane; Union of Italian Jewish Communities
U.K.	United Kingdom
UMI	Unione Matematica Italiana; Italian Mathematical Union
U.S.	United States of America
u.s.	*ultimo scorso*
UTo	Università di Torino; University of Turin
[. . .]	omitted text
[ ]	addition of editor
b.	busta; folder
fol.	folium
fols.	folios
n.	number
nos.	numbers
n.d.	no date
n.e.	no publishing house
n.p.	no place
n.n.	not numbered
fasc.	fascicolo; file
p.c.	postal card
r	recto
v	verso

Photographs Index and Credits

Part I
The Migration Phenomenon

The liner Conte di Savoia, 1940 ca

Chapter 1
From the Ghetto to the City, and Thence to the Country

1.1 Emancipation, Integration, Assimilation

The Jewish presence in Italy is very ancient, dating back to the Maccabean period (II BC). The *Giudei der Cupolone*, as they were nicknamed in Roman vernacular, settled in Rome at that time and included (among others) the Alatris, maternal ancestors of the architect and mathematician Mario G. Salvadori.[1] In Piedmont, the main settlements began to be established in the fifteenth century and consisted of Jews escaping from Provence, attracted by the prospect of moderately good living conditions. Through the sixteenth, seventeenth, and eighteenth centuries, many important communities (then called *università*, universities) formed throughout the Peninsula. Their presence was concentrated in some regions (Piedmont, Tuscany, Veneto, Lazio), and more sporadically in the south.

In pre-unification Italy, social and living conditions of the Jewish minority differed greatly from place to place.[2] In the Kingdom of Sardinia, for example, Jews were segregated in specific areas (ghettos), and suffered numerous deprivations, including the prohibitions of attending schools, belonging to art and trade guilds, and owning land.[3] In the Papal States, the plight was even worse: poverty was widespread, and Jews were forced to perform humiliating rites such as annual supplication to the Pope on Carnival Saturday. However, the conditions in the Grand Duchy of Tuscany and the Kingdom of Lombardy-Venetia were better. In the first, although there were Jewish neighbourhoods (that of Livorno, above all) where community life revolved around the temple, the school, the ritual bath (*Mikveh*) and the butcher's shop, there was no obligation to reside in the ghetto. Jews were also allowed access to studies and the practice of certain liberal professions.

---

[1] Salvadori (1997), p. 390. See also Celli et al. (2013) (eds.).
[2] D'Azeglio (1848), pp. 33–42.
[3] Segre R. (1986–1990).

E. Luciano, *The Jewish Mathematical Diaspora from Fascist Italy*, Science Networks. Historical Studies 64, https://doi.org/10.1007/978-3-031-64896-0_1

From a cultural point of view, despite the precariousness of segregated life, the rate of illiteracy among Jews was much lower than in of the rest of the population. Since ancient times, and even in the smallest communities, a wide and ramified network of Jewish schools had flourished for the purpose of putting individuals in a position to read the Torah independently.[4] In Jewish schools, the teaching had a strong identity (with several hours per week devoted to Hebrew, history, and duties, i.e., the catechism) but the rudiments of Italian language, mathematics, history, geography, and science were also imparted. The traditional schools, duly reorganised, thus became a place of both religious and secular training and constituted one of the pivots on which the communities were based, to the point that some scholars would postulate a 'continuity between traditional rabbinic education and modern humanistic and scientific training ... in those contexts where there were the greatest freedom and well-being'.[5] The Jewish educational network, significantly large before emancipation (1848), included kindergartens and primary schools, which enjoyed a good reputation during the Risorgimento period for the methods, structures, and educational models there adopted, and drew the interest of politicians, intellectuals, and educators such as L. Vigna, V. Aliberti, F. Aporti, and P. Baricco.[6] Professional and technical schools, colleges, boarding schools, and institutes of arts and crafts also sprang up, thanks to private charitable funds. Children of both sexes were admitted, free of charge if they came from poor families. Their doors were often opened also to Catholic students. Schooling was not limited to boys; indeed, Jewish girls were in a privileged position compared to their classmates and in the city of Turin alone young women have been able to count on five female Jewish boarding schools since the early nineteenth century.[7]

At the beginning of the nineteenth century, motivated by the conviction that 'an age that censures life sentences cannot approve of ghettos',[8] Camillo Benso (the Count of Cavour) and many intellectuals, including brothers M. and R. D'Azeglio, V. Alfieri, and V. Gioberti, pleaded for the extension of rights of freedom and equality to oppressed minorities in the Kingdom of Sardinia. Eventually, King Carlo Alberto made a parliamentary decision (*Statuto Albertino*) in March 1848 permitting the full extension of civil and political rights to non-Catholics, i.e. the Waldensians (February 17, 1848) and Jews (March 29 and June 19, 1848).[9] The gratitude of the Jews for this so-called second emancipation was prompt and explicit. It became a leitmotif of nineteenth-century rabbinical rhetoric and translated into

---

[4]Artom E.S. (1913); Colombo (1925); Giribaldi Sardi (1993); Guetta Sadun (1997); Morpurgo (2016).

[5]Momigliano (1987), pp. 134–135: *continuità tra la tradizionale educazione rabbinica e la moderna formazione umanistica e scientifica ... in quei luoghi in cui maggiori erano libertà e benessere.*

[6]Vigna and Aliberti (1848), pp. 64–74, 154–155; Sacchi (1845); Baricco (1865).

[7]Terracini Benvenuto (1932); Maida (2001), pp. 72–74, 305–306; Luciano (2013–14), pp. 326–327.

[8]Gioberti (1848), p. 44: *un secolo che biasima gli ergastoli non può approvare i ghetti.*

[9]Arian Levi and Disegni (1998); Della Peruta (1983); Luzzato Voghera (2000).

solemn celebrations in honour of the King, several acts of devotion to the new Italy of Risorgimento, and many charitable donations to the poor, the children, and the philanthropic institutions of every faith.

As a result of this emancipation, Jews began a new life: they could practise any profession or commercial activity and participate actively in political life, which they did with great determination and success.[10] Several wealthy and distinct families, such as those of some mathematicians whom we will mention later (including Corrado and Beniamino Segre, Gino Fano, Vito Volterra, Beppo, and Eugenio Elia Levi) left their hometown communities and moved to major towns where they promptly fitted into the climate of freedom sanctioned by emancipation. Thanks to the high cultural levels achieved in the pre-Risorgimento period, they began to establish themselves in a series of contexts ranging from the liberal professions to industry, from manufacturing to academia.[11] The Segre family started, for example, a series of activities in Saluzzo that revolved around the spinning of silk, the Almagiàs affirmed themselves in finance and the Fano family in agriculture.[12] Right from the start, young Jews also enlisted in the army, and engaged in the Risorgimento struggles and wars of independence, on all battle fronts.

The unification of Italy, in 1861, extended emancipation to the entire Italian Jewish population (except for residents in the Papal States), for a total of 39,182 individuals, divided into 300 settlements. The phenomenon of urbanisation (i.e. the decision to move from small villages to towns), increasingly massive since the second half of the nineteenth century, emptied the vast majority of small communities in favour of large concentrations in the urban centres of Rome, Milan, Livorno, Turin, Florence, Trieste, Bologna, Venice, etc.[13] This can be easily ascertained by comparing two Italian demographical maps of Italian Judaism before and after emancipation (Fig. 1.1a and b).

Starting from the assumption that 'education is synonymous with emancipation, and emancipation is synonymous with education',[14] in the eyes of the Savoy authorities, access to study was a key tool in the process of integration, i.e. in the delicate shift from the ghetto to the town, and thence to the nation. The Albertine Statute had granted Jews the possibility of enrolling in state institutions. This was a concession of maximum importance since exclusion from education had represented one of the most sinister forms of discrimination, and the first post-emancipation generations immediately benefitted. Among the first Jews who undertook studies in mathematics, Simeone Levi (1843–1913) stands out. Having received his earliest

---

[10] Foà (1978); *Italia Judaica* (1993); *Risorgimento e minoranze religiose*, (1998); Formiggini (1998); Beer and Foà (2013).

[11] Bernardini (1996).

[12] Segre B. (1963), p. 9; Goodstein (2007), pp. 11–18; Janovitz and Mercanti (2008), pp. 43–46; Fano U. (2000), pp. 176–178.

[13] Servi (1871); Della Pergola (1993).

[14] Berti (1849–1850), p. 724: *educare è sinonimo di emancipare, ed emancipare è sinonimo di educare.*

a　　　　　　　　　　　　　　　　　b

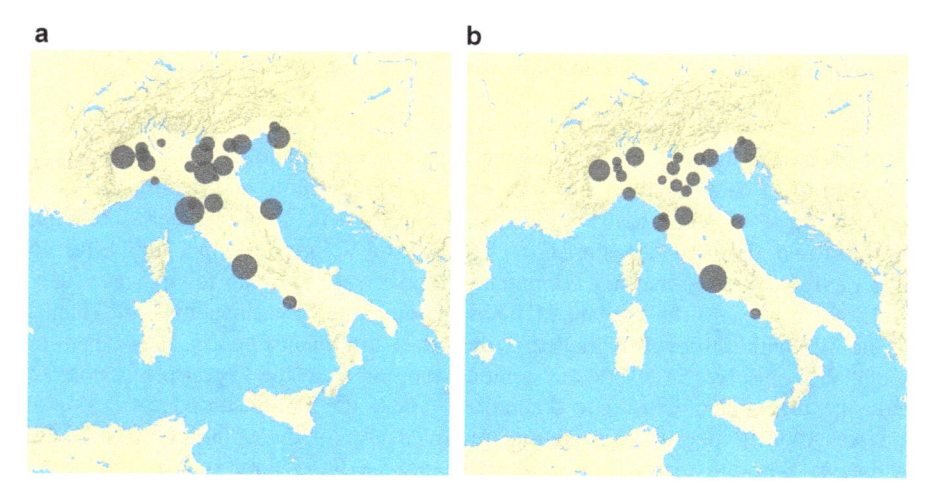

Fig. 1.1 (**a**) (left) and (**b**) right Demography of Italian Judaism before and after emancipation

education at home and at the Jewish primary school in Moncalieri, Levi enrolled in state education immediately after emancipation: firstly, at the Gioberti lyceum in Turin, then in the degree course of mathematics at the University of Turin, before moving to Pisa, where he graduated in 1864 under the direction of E. Betti.[15] In 1869 Levi tried to obtain the degree of aggregate doctor at the University of Turin. The discussion of the thesis, entitled *Dell'equilibrio di un corpo elastico*, took place amid a tense atmosphere in which G. B. Erba, president of the jury, 'prevented the candidate from fully developing the theme'.[16] Although embittered by the experience, Levi continued his research until 1876, publishing an article on trigonal coordinates and a textbook of arithmetic and algebra.[17] From this date onwards, he devoted himself to studies of Egyptology and linguistics. In 1886 the Lincei Academy awarded him the Royal Prize for linguistics for the monumental *Vocabolario Geroglifico Copto Ebraico*. Levi was the first of many Jewish mathematicians pushed to attend state schools from childhood, by parents who entertained an ambition of seeing them quickly and completely integrated.

Social affirmation passed (also) through the achievement of higher education degrees, which is why, from the 1870s onwards, the number of Jews enrolling in Italian universities constantly grew, especially in certain locations such as Turin, Pisa, and Padua. As far as the faculties of sciences are concerned, most of these individuals who dedicated themselves to scientific studies completed only the first two years of the university programmes to obtain the license, or diploma (*licenza*), and subsequently went to the School of Applications (*Scuola di Applicazione*),

---

[15] Levi S. (1864–65) [2001], pp. 293–294.

[16] Arian Levi and Viterbo (1999), p. 66: *impedisce al candidato di svolgere completamente il tema.*

[17] Levi S. (1876); Levi S. (1871).

Fig. 1.2 Turin, 1900 ca., courtesy of Livia Giacardi

aspiring to careers as architects or engineers, surveyors, and accountants. Many stopped their studies after gaining the license. What oriented their choices was a complex mix of family and economic factors, as well as cultural models. The testimonies of many mathematicians born in the 1860s confirm the impression that the so-called *humanitas scientifica* was a category of thought that was almost foreign to Jewish families, which generally considered scientific training in terms of the employment prospects that it presented. Segre, Fano, and many others had to face serious generational clashes with their fathers, who pushed them in the direction of studies in engineering, finance, administration, industry, and trade, while they themselves were inclined towards pure research undertaken 'out of love for knowledge'.[18]

This situation gradually changed in the following years: the trend of enrolment in scientific degree courses continued to be remarkable (a hundred freshmen in Turin alone between 1880 and 1900), but the number of Jews who graduated in pure mathematics while aspiring to a career in education or academia also grew. Among these were Giulio Ascoli, Salvatore Pincherle, Giulio Vivanti, Vito Volterra, Corrado Segre, and Guido Castelnuovo, followed by those born in the seventies: Gino Fano, Federigo Enriques, Tullio Levi-Civita, Adolfo Viterbi, Beppo Levi, Aldo Finzi, Guido Fubini, and many others.

Thanks to having grown up in a context that offered genuine equality in educational opportunities, there were also many young Israelite women who devoted themselves to scientific studies, so much so that V. Ravà reported in 1902 that, out of 20 mathematics graduates between 1877 and 1900, four were of Jewish origin: Ida Maestro in Padua, Emilia Tagliacozzo in Pisa, Ida Terracini and Costantina Levi in Turin (the first and second student to obtain doctorates in Mathematics at this University in 1892 and 1893, respectively).[19] In the period of 1892–1938, Jewish female students constituted 5% of graduates from the Faculty of sciences of Turin[20] (Fig. 1.2).

---

[18]Luciano (2013a), pp. 335–336.

[19]Ravà (1902), p. 641.

[20]Luciano (2013–14), p. 328.

In Pisa, the percentage was even higher.[21] The majority, after having obtained the degree and diploma of the Teachers Training School (*Scuola di Magistero*), opted for careers in schools as teachers, headmistresses, and educators. In such a circumstance, gender played a part. In fact, such professions were considered particularly suited to women, being conceived as natural extensions of parental care.[22]

Integration, desired and sought, generally resulted in the loss of some identity traits: no longer attending Jewish schools, learning Jewish language, and going to pray or study in community places, Italian Jews lost the sense and habit of religious practices. Integration became assimilation. While continuing, at least nominally, to profess the Jewish religion, to subscribe to communities of birth or residence, and to financially support their institutions, the mathematicians we mentioned (the men slightly more than the women) personified the adage according to which 'a Jew is an Italian who does not go to church on Sunday' and were almost foreign to Jewish traditions, which they considered with a kind of benevolent irony. Many passages of Corrado Segre's correspondence with his fiancé, later to become his wife, Olga Michelli, testify to this:

> Then, a thought, among others, came to me the same evening that I left you, a tormenting thought . . . This: that the ideals on which I feed myself and that I strongly support, my views so different from those of the majority, alien from social and religious prejudices, in short, my complete system of values, could end up unsympathetic to you, who has been clearly educated by parents with all sorts of respect for ancient customs.[23]
> Speaking of Jews, I forgot to write to you that there is even a *Kosher* hotel-restaurant. Your mom will think that I should go there.[24]
> You will write to me the simple truth about how you pass *Yom Kippur*. I am not affected yet. Maybe tonight I will go to the theatre with [Eugenio] Bertini.[25]

The number of observant Jews in the intellectual environment is infinitesimal, particularly among the men of science. As a matter of fact, even those who partially preserved their identity roots would find it increasingly difficult to describe and

---

[21]Galoppini (2011).

[22]Dolza (1987), pp. 27–34.

[23]UTo-ACS: C. Segre to O. Michelli Segre, Turin 18.2.1893: *Poi, fra gli altri, un pensiero che m'era venuto la sera stessa che ti lasciai, un pensiero tormentoso . . . Questo: che le idealità di cui io mi pasco e che tanto mi piacciono, le mie vedute diverse da quelle della maggioranza, aliene da pregiudizi sociali, religiosi, ecc. tutto quanto insomma il mio modo di sentire, dovesse alla lunga riuscire poco simpatico a te che evidentemente sei stata allevata da genitori che hanno invece ogni sorta di riguardi per gli usi antichi.* See also Segre Fuà (1952).

[24]UTo-ACS: C. Segre to O. Michelli Segre, St. Moritz 31.7.1905: *A proposito di ebrei, mi scordai di scriverti che qui vi è persino un albergo-ristorante cascèr. La mamma tua penserà che io dovrei andare lì.*

[25]UTo-ACS: C. Segre to O. Michelli Segre, Rome 28.9.1906: *Mi scriverai tutta la verità sul modo come passi il Chipur. Io per ora non me ne risento. Forse stasera, a Chipur finito, andrò con Bertini a teatro.* Other family letters contain similar comments. For example, inviting colleague and friend Guido Castelnuovo to his wedding, Segre did not communicate the date of the ceremony in the synagogue but only that of the civil rite 'because there is no doubt about the greater importance of it' (14.3.1893).

define, rationally and historically, their concept of being Jewish. As Emanuele Artom, a keen interpreter of Italian Judaism, wrote:

> Judaism is not a religion, because many Jews consider themselves as such without believing in God, or believing in God in a manner different from what is prescribed by Jewish theology, providing that this exists; it is not a race, because ethnologists say the contrary; it is not a homeland, because we feel bound to the land of birth; it is a fourth thing, unique in humanity: we are bound by a tradition, as one may be by a solidarity of faith, blood or places. Precisely because it is unique in the world, there is no common name that serves to indicate entities of this genre.[26]

Solicited to consider themselves as Italian citizens, regardless of the community they belonged to, Jewish mathematicians manifested a sort of indifference also to the events of international Judaism. A statement in favour of Alfred Dreyfus was signed by the anthropologist Cesare Lombroso, the economist Achille Loria, the physiologist Giulio Fano, but by only two mathematicians, both non-Jews: Luigi Cremona and Eugenio Beltrami.[27] Again, in 1916 Sylvain Lévi contacted Volterra, through Jacques Hadamard, to ask him if it was feasible to create in Italy an analogue of the French Information and Action Committee Among Jews in Neutral Countries. Volterra did not provide any feedback, nor showed any knowledge of the activities of the *Alliance Israelite Universelle* and the powerful Jewish American community.[28]

Zionism also had very little hold on Italian mathematicians, with some sporadic exceptions such as Alessandro Padoa, statisticians Riccardo and Roberto Bachi, and the astronomer Giulio Bemporad. Indeed, intervening in the debate on the condition of Jews in Eastern Europe and on Zionism sparked by the emergence of national issues following the First World War, Beppo Levi undersigned a harsh article on *Israel*.[29] Here he argued that the Jewish national question could not be resolved with the foundation of a State in Palestine because this would have presented the diaspora with the dilemma of choosing between two homelands, and therefore between two identities. Moreover, he continued, a Jewish State would have nullified the fundamental characteristic of the Jewish people, namely that of being a nation even without being rooted in a territory. To preserve and respect the uniqueness of the Jewish people, according to Levi, Jewish nationalism had to express itself through a

---

[26] Emanuele Artom (1940–44), in Schwarz (2008), p. 13: *l'ebraismo non è una religione, perché molti ebrei si considerano tali senza credere in Dio o credendovi in modo diverso dalla teologia ebraica, dato che questa ci sia; non è una razza, perché gli etnologi affermano il contrario; non è una patria, perché noi ci sentiamo legati alla terra di nascita; è una quarta cosa, unica tra gli uomini; siamo avvinti da una tradizione, come lo si può essere da una solidarietà di fede, di sangue o di luoghi; appunto perché è unico al mondo non ha un nome comune, che serve per indicare le entità dello stesso genere.*

[27] *I letterati italiani a Emilio Zola*, Il Vessillo Israelitico, XLVI, 1898, pp. 38–40.

[28] S. Lévi to V. Volterra, Paris 12.4.1916 in Mazliak and Tazzioli (2009), pp. 165–169.

[29] Levi B. (1918), p. 2. Levi would return to the theme in his correspondence with the brother Giulio Augusto in 1930, coinciding with incidents in Palestine between Arabs and settlers and the publication of I. Sciaky's article *Sionismo in crisi* in the journal *Civiltà moderna* (II, 2, 15.8.1930). See Momigliano Levi (2016), p. 277.

cultural movement that would allow Jews to rediscover their identity as a nation within the diaspora, without renouncing the rights they enjoyed in their respective homelands.

Italian Judaism tried (without success) to stem the assimilatory process. *Il Vessillo Israelitico*, one of the main periodicals in Jewish Italy, suggested for example:

> Let us bring education among us to the degree of civilization and progress of the times but, as far as we can, and up to a certain age, this education be given in our schools, if we do not want the new generation, defrauded of all religious education, to grow neither Israelite, nor Catholic, but instead atheist.[30]

The same magazine recorded with maniacal precision the student performances by young Jews, honours degree graduates or winners of prizes or scholarships, and the career successes achieved by Segre, Castelnuovo, Volterra, sifting through the official *Bollettini* of the Ministry of Public Instruction every month.[31] Three of these amply exemplify the overall strategy:

> Venice. Castelnuovo, a beautiful name: his father, Professor Enrico, a poet, one of the most fruitful and empathetic novelists in Italy; his son Dr. Guido, a mathematician, of high intelligence, very cultivated, who promises to go much further in exact sciences. Guido Castelnuovo, just graduated from the University of Padua, and who has already published learned essays of Geometry, obtained from the Ministry one of the post-doc fellowships in higher mathematics. Guido Castelnuovo honours the beautiful name he bears.[32]

> Saluzzo. Prof. Corrado Segre, who teaches at the University of Turin, is without exaggeration a genius in mathematics. Although young, he has already had a wonderful career. He is the son of the well-known industrialist of this city …, the lamented Abram Segre, who renovated the S. Martino silk factory, now managed by the company Mana & Demartini. Professor Segre, though young, made extremely rapid progress and is already reputed to be, without exaggerating, an eminent figure in science.[33]

---

[30] *Conferenze di Maestri, Scuola Superiore Femminile Ebraico Italiana*, Il Vessillo Israelitico, III, 1855, pp. 365–366: *Portiamo l'istruzione fra noi al grado della civiltà e del progresso de' tempi ma, per quanto si può, e fino a una certa età, questa istruzione sia data nelle nostre scuole, se non si vuole che la nuova generazione, defraudata d'ogni istruzione religiosa, venga a crescere né israelita, né cattolica, ma atea.*

[31] See e.g. *Notizie Diverse—Italia*, XLVII, 1899, pp. 109–110, 155–156. Reports often close with comments such as: 'Out of 15 awardees, 4 Israelites!' (*Su 15 premiati, 4 Israeliti!*) (Il Vessillo Israelitico, XLIV, 1896, p. 208).

[32] *Notizie Diverse—Italia*, Il Vessillo Israelitico, XXXIV, 1886, p. 377–378: *Venezia. Castelnuovo, un bel nome: il padre Professor Enrico, un poeta, uno dei più fecondi e dei più simpatici novellieri d'Italia; il figlio dottor Guido, un matematico, ingegno vivacissimo, coltissimo, che promette di andare molto innanzi nella scienza esatta. Guido Castelnuovo, testé laureato nell'Università patavina, e che ha già pubblicato dotte memorie di Geometria, ottenne dal Ministero uno dei posti di perfezionamento nelle matematiche superiori. Guido Castelnuovo fa onore al bel nome che porta.*

[33] *Notizie Diverse—Italia*, Il Vessillo Israelitico, XLII, 1894, pp. 36–37: *Saluzzo. Il prof. Corrado Segre, che insegna alla R. Università di Torino, è senza esagerazione un genio per la matematica. Quantunque giovane ancora egli ha già percorso una splendida carriera. Egli è figlio del noto industriale di questa città . . . , del compianto Segre Abram, che ridusse alla moderna il setificio di*

Turin. The Turin Academy of Sciences welcomed with pleasure and great appreciation the tribute of the first volume of the *Bollettino di bibliografia e storia delle matematiche* by prof. Gino Loria, a brilliant and very useful work for lovers of this science. A noteworthy paper by prof. Tullio Levi-Civita was also received in the academic *Atti*. As in Turin, also in Rome and Milan, at the Lincei academy and the *Accademia Scientifico Letteraria*, the papers of many illustrious Israelites were approved and praised.[34]

The disappearance of small and medium-sized communities, combined with the decline of Jewish schools—reduced to being schools *of* and *for* the poor, attended by students who came from the most observant or needy families—meant that the assimilatory phenomenon was complete by the end of the 1910s. In the census of 1911, many Jews specify that they were of Israelite origin but did not belong to any religion. On the other hand, the custom of registering with the community and the practice of making donations and bequests to Jewish welfare institutions (orphanages and hospitals, mutual aid societies, etc.) was generally maintained. Marriage, moreover, remained prevailingly endogamous, with few mixed unions, even at the cost of having to resort to matchmakers for the arrangement of engagements and marriages, as in the case of Segre:

> Dearest, I thank you for your letter [...] and friendly advice. But, while I deplore, like you, the method used by the Chosen People (a method to which, however, can be added reasonable precautions to mitigate its dangers, as I do), I must recognise that it cannot be done otherwise, because of the low population of the People themselves. Here in Turin, for example, I know, directly or indirectly, the young ladies to whom I could aspire but, among them, no-one is right for me! Making the social rounds here, in a city of 330,000 inhabitants, would never have solved my problem! For some years, going to the baths or in the mountains, I looked around ...: nothing for me! Believe me: without pitiful souls acting as intermediaries, it is very unlikely that our co-religionists will mate![35]

S. Martino ora esercito dalla Ditta Mana e Demartini. Il prof. Segre, quantunque giovane, fece dei progressi rapidissimi ed è già reputato, senza adulazione, un'illustrazione della scienza.

[34] *Notizie Diverse—Italia*, Il Vessillo Israelitico, XLVII, 1899, pp. 109–110: Torino. *L'Accademia delle scienze di Torino nello scorso mese accolse con piacere e molto lodò il primo volume del* Bollettino di bibliografia e storia delle matematiche *del prof. Gino Loria, opera geniale ed utilissima ai cultori di questa scienza. Accolse poi anche negli atti accademici una pregevole memoria del prof. Tullio Levi-Civita. Come a Torino, così pure all'Accademia dei Lincei a Roma, all'Accademia scientifico letteraria di Milano si approvarono e si lodarono gli scritti di molti illustri israeliti.*

[35] *ANL-Castelnuovo*: C. Segre to G. Castelnuovo, Turin 23.9.1892: *Carissimo, Ti ringrazio della tua lettera [...] e degli amichevoli consigli. Ma, mentre deploro anch'io come te il metodo che si usa dal popolo eletto (al quale però si possono aggiungere tali precauzioni da attenuarne i pericoli: e così faccio io), debbo riconoscere che non si può fare altrimenti, in causa della poca densità del popolo medesimo. Qui a Torino per es° io so, direttamente od indirettamente, quali sono le signorine a cui potrei aspirare: fra esse nessuna fa per me! Il frequentar la società qui, in una città di 330 000 abit[i] non m'avrebbe mai servito a risolvere il mio problema! Da qualche anno, andando ai bagni od in montagna, mi guardo attorno ...: niente che faccia per me! Credi che senza le anime pietose che facciano da intermediari, è ben poco probabile ai nostri correligionari di accoppiarsi!*

1.2   The Three Pillars of Life: Family, Country, and Profession

The scholars we deal with are mathematicians, but their profession is not the only thing they share. There are at least two other elements at the base of the system of values of Italian Jews belonging to the post-Risorgimento generations, who lost their identity roots: homeland and family. Brought up on bread and Italian-ness, in families that were increasingly emancipated, in which religion blended with the cult of the State, they followed a strange secular cult, evocatively described by Emanuele Artom:

> Both father [Emilio Artom] and mother [Amalia Segre] were endowed with a strong sense of Italian-ness and adhered to monarchical-liberal ideals. My father's devotion to the House of Savoy was boundless. Mom followed the same currents sentimentally and connected them with religion. I will never forget that she taught us that whoever dies fighting for the Fatherland goes to heaven, according to the teaching of the second book of Maccabees.[36]

In light of this common background, the First World War appeared, in the eyes of many young Jews, to be the epilogue of the Risorgimento struggles that had led to emancipation, a sort of fourth war of independence.[37] Beppo Levi, in particular, was among those who had the profound conviction that the intervention had to take place, not so much for an international political calculus but for a moral principle, that is, the defence of European civilisation from the violence of Germany: 'we live in days of waiting for the decisions of our government: hoping that the decision will be to attack the barbarians while they attack others'.[38] Since 1915 onwards Eugenio Elia Levi, Alessandro Terracini, Emilio Artom, Adolfo Viterbi, Ermanno Senigaglia, Benedetto Calò, and Paolo Michel fought at the front. Viterbi, Senigaglia, Calò, Michel, Eugenio Elia Levi, and his brother Decio Valerio lost their lives in war.[39] Artom and the brother of Alessandro Terracini, the glottologist Benvenuto, were seriously wounded and decorated with the silver cross for military valour. Volterra and Fano, who were well over the age for being called up, stopped their research

---

[36] Emanuele Artom (1940–44), in Schwarz (2008), p. 13: *Tanto il babbo* [Emilio Artom] *quanto la mamma* [Amalia Segre] *erano dotati di un forte senso di italianità, e aderivano alle idealità monarchico-liberali. La devozione di mio padre verso Casa Savoia era illimitata. La mamma seguiva sentimentalmente le stesse correnti, e le collegava con la religione. Non dimenticherò mai che ella ci insegnava che, chi muore combattendo per la Patria, va in paradiso, secondo l'insegnamento del secondo libro dei Maccabei.*

[37] Ferrara degli Uberti (2011), pp. 223–236.

[38] B. Levi to G. A. Levi [1915] in Momigliano Levi (2016), p. 264: *viviamo in giorni d'attesa sulle decisioni del nostro governo: augurando che la decisione sia di picchiare sui barbari mentre picchiano gli altri.* Beppo Levi's letters to his brother Giulio Augusto, a man of letters and a philosopher, are preserved in the Momigliano Levi's private family archive, Fondo Giulio Augusto Levi.

[39] See e.g. Cattaneo (1917–1918); Sittignani (1917–1918); *Necrologio Ermanno Senigaglia, caduto in guerra*, 1916. On the case of E. E. Levi, by far the most celebrated, see Celli and Mattaliano (2015).

activity to devote themselves exclusively to the war effort. Volterra devoted himself to applied research and participated in action in the field; Fano dedicated himself to propaganda within the Mobilisation Committee. Terracini, in addition to fighting on the Karst, developed mathematical research activities in photometry, ballistics, and geodesy during his period of service in the Fifth Artillery Division under the command of General Roberto Segre, cousin of mathematician Corrado.

Participation in the war further attenuated—if it were ever necessary—the perception of a Jewish otherness. As Artom recollected: 'in a field full of remnants of the battle I found together a page of the Missal and a *Tefillà* containing the *Shema*. I did not collect them because they were bloody, but I will always remember them when I hear the Hitlerian speak'.[40] On the other hand, war reinforced patriotic-nationalistic feelings, even among those who had been politically neutral. Segre, for example, at first anti-interventionist and criticized by colleagues who mistook his pacifism for pro-Germanism,[41] from 1915 systematically closed his letters with exclamations such as: 'Long live Italy! Hurrah for the war!'[42] The cosmopolitanism of Castelnuovo and Enriques, similarly, wavered. Both abhorred war as a means of resolution to political-economic conflicts and Enriques even resigned from the board of directors of *Scientia* because he refused the idea of publishing in the journal contributions of political character, whether for or against Italian neutrality.[43] However, both Castelnuovo and Enriques believed that, as mathematicians, they could and must give their contribution to the benefit of the war, which was topmost in all their thoughts, and took an active part in propaganda, managing the distribution of interventionist pamphlets by Pietro Fedele and Attilio Tamaro:

> Have you seen the speech to Italy from the English intellectual world? Would it not be appropriate for the Italian-English Committee to initiate an answer? [...] Would it not be the case for that part of the Italian intellectuals who think more freely, to emphasise the high ideals of our war, regarding the civilisation that will prevail in the world, ideals certainly not inferior to those, to which government propaganda has almost exclusively restricted itself?[44]

---

[40] Emilio Artom (1940–41), in Treves (1954), p. 58: *ricordo un campo pieno di resti della battaglia ove trovai insieme un foglio di un libro da messa e la pagina della Tefilà contenente lo Scemà. Non li raccolsi perché lordi di sangue, ma li ricorderò sempre quando sentirò parlare gli hitleriani.* Tefillà means prayer. It is common in Jewish prayer books to find the Shema (literally 'Listen') which is undoubtedly the fundamental Jewish prayer.

[41] Luciano and Roero (2016), pp. 140–145.

[42] ANL-*Volterra*: C. Segre to V. Volterra, Turin 27.5.1915 and Turin 13.1.1917: *viva l'Italia!*; *W [viva] la guerra!*

[43] Linguerri (2005), pp. 63–70.

[44] ANL-*Volterra*: G. Castelnuovo to V. Volterra, Oriolo Romano 31.7.1915: *Hai visto l'indirizzo all'Italia del mondo intellettuale inglese? Non sarebbe il caso che il Comitato Italo-Inglese si facesse iniziatore di una risposta? [...] Non sarebbe il caso che quella parte degli intellettuali italiani che pensa più liberamente, mettesse in rilievo le alte idealità della nostra guerra, riguardo alla civiltà che dovrà prevalere nel mondo, idealità certo non inferiori a quelle, a cui la propaganda governativa si è quasi esclusivamente ristretta?* Castelnuovo referred to the following works: P. Fedele, *Perché siamo entrati in guerra*, Roma: Tipografia Nazionale di G. Bertero e

Salvatore Pincherle, despite having had his daughter Edvige widowed by war and two orphaned grandchildren, wrote enthusiastically to Volterra: 'in the exultation of liberation, I lived the brightest hours of my life: for 60 years I had been waiting for the moment of redemption of my Trieste! And now it is united to the great motherland'.[45]

Family was the third 'pillar' of these men. Since Jews were a small minority of the Italian population, and given that mixed marriages were comparatively few, the mathematicians were united by an inextricable network of kinships. Fathers and sons, mothers and daughters, in-laws and sons-in-law, uncles and cousins, met day after day, year after year, not only in university classrooms, scientific laboratories, or research centres but also in many other contexts: evenings in their living rooms, holidays in the mountains and at the sea, hiking and skiing, moments of private life in which the 'old' watched the 'young' grow personally and professionally, saw them marry and become parents in turn, establishing themselves in academia or a profession (Figs. 1.3 and 1.4). There are dozens of testimonies of this social reality, unique in its kind, which would be swept away by persecution:

> S[t] Moritz is always full of Jews, in all hotels. Indeed (Dr Levi told me) it was said that these 2 main resorts of the Engadine, namely *Saint Maurice* and *Pontresina*, should be called instead: *Saint Moise* and *Palestine*.[46]

> The Kosher restaurant-hotel is the *Edelweiss* owner Bergmann. Following my letter, Volterra decided to come to St Moritz. He immediately telegraphed me to reserve a single room for him. [. . .] I suggested he go to the hotel Calonder, where the Fano are staying (they are very satisfied). I have conferred repeatedly with the hotelier; let's hope it will go better than for the Foà family! [. . .] Here is another Lincean member: Senator [Oreste] Tommasini of Rome. So, there will be 5 of us here in St Moritz. We could have a meeting![47]

C., 1915 and A. Tamaro, *L'Adriatico—Golfo d'Italia. L'italianità di Trieste*, Milano: Fratelli Treves, 1915. See also ANL-*Volterra*: G. Castelnuovo to V. Volterra, Rome 9.9.1915.

[45] ANL-*Volterra*: S. Pincherle to V. Volterra, Bologna 5.11.1918: *nell'esultanza della liberazione, ho vissuto le ore più fulgide della mia vita: da 60 anni aspettavo il momento della redenzione della mia Trieste! Ed ora essa è unita alla grande madre.* See also S. Pincherle to V. Volterra, Bologna 20.11.1918: 'I return from a short trip to Trieste, where the sight of our flag waving over the town hall and in the former lieutenancy gave me an indescribable emotion. It will take a lot of wisdom to give the great Change a smooth and frictionless progression! But, in the city, the joy is absolutely immense'. (*Ritorno da una breve gita a Trieste, dove la vista della nostra bandiera sventolante sul municipio e nell'ex-luogotenenza mi ha procurato una emozione indescrivibile. Occorrerà molta sapienza per dare al grande Cambiamento un andamento regolato e privo di attriti! Ma, in città, il giubilo è assolutamente immenso.*)

[46] UTo-ACS: C. Segre to O. Michelli Segre, St. Moritz 29.7.1905: *St Moritz è sempre pieno di ebrei, in tutti gli alberghi. Anzi (mi diceva quel Dr Levi) è stato detto che questi 2 paesi principali dell'Engadina, cioè* Saint Maurice *e* Pontresina, *dovrebbero chiamarsi invece:* Saint Moise *e* Palestina.

[47] UTo-ACS: C. Segre to O. Michelli Segre, St. Moritz 7.8.1905: *Il ristorante-albergo cascèr si chiama* Edelweiss, *proprietario Bergmann. In seguito alla mia lettera Volterra s'è deciso a venire a St Moritz. Mi telegrafò subito di fissargli una camera ad un letto. [. . .] Lo faccio andare all'hôtel Calonder, dove sono i Fano (i quali sono soddisfattissimi). Ho conferito ripetutamente col*

Fig. 1.3 From left to right: Emma and Gina Castelnuovo (daughters of Guido Castelnuovo), Libera and Cornelia Trevisani (wife and sister-in-law of Levi-Civita), Enrico Fermi and Enrico Persico, Ceccherini Silberstein Family Archive

> We have been up here for about twenty days, and we are very comfortable, for the place is beautiful and the company is nice and varied. A stone's throw from our house there are the Fubini, the Fermi and the Terracini families; at a short distance, Gino Fano; on the other side of the Dora, in the main hamlet of Courmayeur, a quarter of an hour from here, there are Enriques and Persico.[48]

1.3   The Italian School of Algebraic Geometry

From 1883 to 1938, several dozen Jews graduated in Mathematics in various Italian universities; a dozen of these became full professors (*professori ordinari*) of various mathematical disciplines, while many others entered the academic staff as lecturers (*assistenti*, *liberi docenti*) and tenured or associate professors (*professori incaricati*).

---

proprietario; e speriamo che la cosa vada meglio che per i Foà! [. . .] *Vi è qui un altro linceo: il senatore Tommasini di Roma. Così verremo ad essere a St Moritz 5. Potremo fare una seduta!*

[48] ANL-*Volterra*: G. Castelnuovo to V. Volterra, Courmayeur 2.8.1931: *Noi siamo quassù da una ventina di giorni e ci troviamo molto bene, per il posto che è bellissimo e per la compagnia simpatica e varia. A due passi da casa nostra stanno i Fubini, i Fermi e i Terracini; a poca distanza Gino Fano; dall'altra parte della Dora, nella frazione principale di Courmayeur, a un quarto d'ora di qua, vi sono Enriques e Persico.*

Fig. 1.4 Tullio Levi-Civita (first at left) and Guido Fubini (center), in Sulden (Bolzano), Ceccherini Silberstein Family Archive

A further thirty became teachers or headmasters in state middle and secondary schools. Most of these scholars belong to a unique research School.

The nature of a School is very difficult to explain. The Italian term is in fact, untranslatable, as is its meaning.[49] A School is neither an institution nor a Faculty, although it is usually attached to a centre, a campus, i.e. a physical place in which a group of scholars work side by side for long periods of time consecutively. A School can be defined as a particular way of organising scientific activity. It is a community

---

[49] On the notion of research School, see Geison (1981); Geison and Holmes (1993); Rowe (2002).

of individuals engaged in developing a collective research project under the guidance of one or more 'directors' who are recognised as Masters (*Maestri*) by the entire team.[50]

The Italian School of Algebraic Geometry was an academic-based scholarly group consisting of 35 university professors,[51] and a dozen teachers in middle and secondary schools.[52] It mostly developed across three locations (Turin, Bologna, and Rome), and over four generations: the first and second (which chronologically fall outside our narrative) identified its leaders in L. Cremona, G. Battaglini, and E. D'Ovidio, the third in C. Segre and E. Bertini, while the fourth basically grew under the mastership of F. Enriques and F. Severi (Fig. 1.5).

As is evident from its name, the School's research programme focussed entirely on algebraic geometry in two main components: classical algebraic geometry and projective-differential geometry, both declined according to a typically Italian approach and characterised by three essential elements: (1) a large heritage of geometric culture, especially in hyperspatial projective geometry; (2) a particular style of treatment and exposition, pervaded by frequent appeal to intuition and synthetic arguments; (3) from the 1920s onwards, a weak preparation in abstract algebra and algebraic topology, cultivated only by a few members (G. Scorza, A. Bassi, P. Buzano, and F. Conforto).[53] From the ranks of Italian geometers, number theory too counted a small handful of specialists, mainly trained by L. Bianchi in Pisa. Among them A. M. Bedarida and G. Ricci can be mentioned.

---

[50] Castelnuovo G. (1929); Castelnuovo G. (1930); Fano G. (1895); Fano G. (1925); Fano G. (1934); D'Ovidio (1903); Veronese (1903).

[51] On the Italian School of Algebraic Geometry see, inter alia, Brigaglia and Ciliberto (1998); Brigaglia (2001); Conte and Giacardi (2016); Luciano and Roero (2016). A tentative list of mathematicians belonging to the Italian School of Algebraic Geometry was proposed by Brigaglia (2001, pp. 203–204): Giuseppe Battaglini (1826–1894), Luigi Cremona (1830–1903), Enrico D'Ovidio (1842–1933), Eugenio Bertini (1846–1933), Riccardo de Paolis (1854–1892), Giuseppe Veronese (1854–1917), Mario Pieri (1860–1913), Corrado Segre (1863–1924), Guido Castelnuovo (1865–1952), Federigo Enriques (1871–1946), Beppo Levi (1875–1961), Michele de Franchis (1875–1946), Carlo Rosati (1876–1929), Luigi Brusotti (1877–1959); Francesco Severi (1879–1961), Guido Fubini (1879–1943), Giovanni Giambelli (1879–1935), Ruggiero Torelli (1884–1915), Annibale Comessatti (1886–1945), Alessandro Terracini (1889–1968), Oscar Chisini (1889–1967), Eugenio Togliatti (1890–1977), Giacomo Albanese (1890–1947), Beniamino Segre (1903–1977), Luigi Campedelli (1903–1978), Fabio Conforto (1900–1954). Other names should be added: Federico Amodeo (1859–1946), Gino Loria (1862–1954), Luigi Berzolari (1863–1949), Gino Fano (1871–1952), Gaetano Scorza (1876–1939), Arturo Maroni (1878–1966), Enrico Bompiani (1889–1975), Achille Bassi (1907–1973), and Piero Buzano (1911–1993).

[52] Pilo Predella (1863–1939), Francesco Palatini (1865–1940), Alessandro Padoa (1868–1937), Roberto Bonola (1874–1911), Ugo Amaldi (1875–1957), Alberto Tanturri (1877–1924), Matteo Bottasso (1878–1918), Luisa Viriglio (1879–1955), Alice Osimo (1886–195?), Emilio Artom (1888–1952), Vittorina Segre (1891–1944), Elsa Bachi (1894–1972), Annetta Segre (1897–1944), Umberto Bini, Maria Mascalchi (1902–1976), Adriana Enriques (1902–1994), Emma Castelnuovo (1913–2014), etc.

[53] On the Italian style in algebraic geometry see Brigaglia and Ciliberto (2004); Brigaglia et al. (2004); Schappacher (2015).

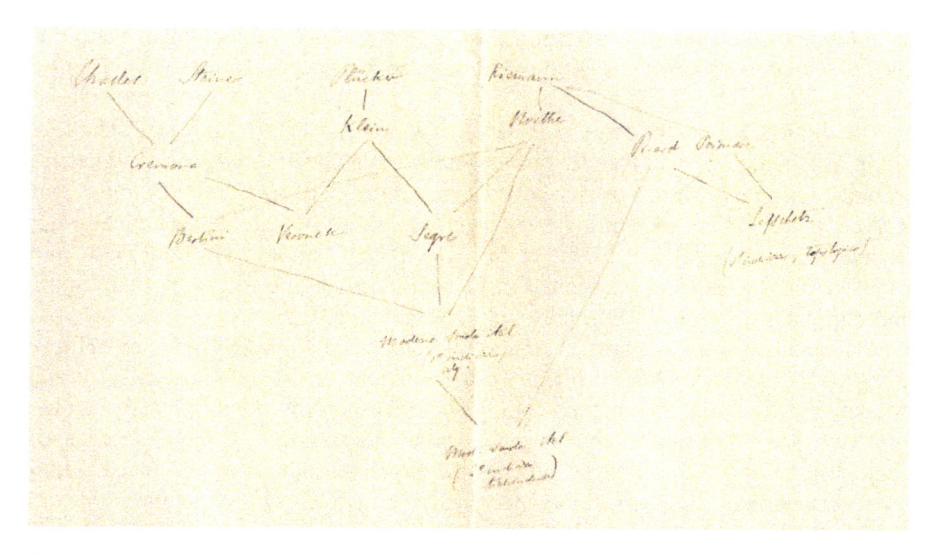

Fig. 1.5 Outline of the plenary lecture *La geometria algebrica e la scuola italiana*, ICM Bologna 1928, ANL-*Castelnuovo*, in Gario 1999

The Italian School of Geometry reached its apogee—a *führende Stellung* as it was labelled by F. Meyer and H. Mohrmann—[54] in the first years of the twentieth century when the theory of algebraic surfaces was developed through a joint research venture between Segre, Castelnuovo, Enriques, and Severi. The leadership position achieved by Italian geometers at national and international levels is undisputable: from 1897 to 1932, the Italian delegations at the International Congresses of Mathematicians always included at least a couple of members of this School; Segre, Castelnuovo, and Severi were plenary lecturers in Heidelberg (1904), Rome (1908), Zurich (1920), and Bologna (1928); international publishing projects such as the *Encyklopädie der mathematischen Wissenschaften* greatly benefitted from their expert collaboration.

The School's major contributions were made before the First World War. Afterwards, Italian algebraic geometry entered a period of profound crisis, in which the limits of its peculiar style plainly emerged, together with a demand for new methods and tools to carry out a complete refoundation of the theory. Despite Severi's address of some of these issues, especially regarding transcendent and topological methods, his vision remained fundamentally anchored to tradition. His lack of competence in the fields of abstract algebra and algebraic topology, combined with an autarkic vision of research organisation and a swaggering self-centredness, caused the decline of the School. From the early 1930s to the Second World War, Severi dragged his students (B. Segre, E. Martinelli, F. Conforto, G. Albanese,

---

[54] Meyer and Mohrmann (1923), p. VI.

A. Comessatti, and others) into an endless loop of attempts—all doomed to failure—to establish the theorem of characteristic linear series.[55]

The School, which was ethnically homogeneous (a large percentage of its members being Jews), would eventually be beaten by racial persecution.

Since its beginnings, the Italian Geometric School had combined the research programme in geometry with a plan of studies and activities related to the teaching of mathematics.[56] Segre, Castelnuovo, Enriques, Severi, and many of their protégés shared a genuine interest in methodological questions, teaching, teacher training and in school publishing. Castelnuovo was the Italian delegate of the International Commission on Mathematical Instruction; Enriques the editor-in-chief of one of the most important Italian journal for mathematics education (*Periodico di Matematiche*) and he had created and edited the most renowned Italian book series for teacher training of the time (*Questioni riguardanti le matematiche elementari*). Castelnuovo, Enriques, Luigi Berzolari and Severi had been Presidents of *Mathesis*, the oldest Italian association of secondary school teachers of mathematics and physics. Enriques and Severi were authors of books of geometry and arithmetic widely distributed throughout the peninsula which, due to being adopted as textbooks almost universally, had made the fortune of the Zanichelli and Vallecchi publishing houses. The sensitivity of the Masters to the instances and debates on mathematical education explains why the School counted so many teachers, including Emilio Artom, Guido Ascoli, Alberto Levi (brother-in-law of Beppo Levi), Alessandro Padoa, Elsa Bachi, Alice Osimo, Vittorina Segre and Annetta Segre (nieces of Corrado Segre)[57] (Fig. 1.6).

These were 'brokers': teachers not attached to a university institution, who were also researchers in foundations and didactics of mathematics, and who produced dozens of articles of pedagogical content and textbooks for various types of schools and educational programmes, in which one can easily grasp the legacy of Segre, Castelnuovo, Enriques and Severi, and of the suggestions they dispensed to their students: educating pupils to mathematical discovery; preference for the experimental and intuitive approach in teaching; addressing the various mathematical concepts in a historical perspective, benefitting from research on the foundations and elementary mathematics from an advanced standpoint. The fact that secondary school teachers (recruited by national selection procedures) travelled around Italy before being posted permanently in one institution, means that the pedagogical tenets of the School, its vision of mathematics education and its output (textbooks, teaching materials, etc.) were widely distributed and adopted by Italian teachers.

A School, however, is much more than a form of organisation of mathematical life. It is, as Castelnuovo wrote, a family, a circle of individuals who collectively construct and develop a research programme, but who also share socio-cultural paradigms, ideological convictions, and political ideas:

---

[55] Nastasi and Rogora (2020).
[56] See Giacardi (2010); Giacardi (2015).
[57] Luciano (2017a), pp. 194–196.

Fig. 1.6 Scientific lyceum G. Ferraris in Turin: Emilio Artom (lower left) and Guido Ascoli (top right), Archive of the scientific lyceum G. Ferraris, Turin

The School goes beyond the value of the man and the importance of a given discipline, to affect all scientific organisation. All of you know what difficulties we meet in our Latin countries, which are prevalently individualist, in constituting a scientific School, a reunion or I would almost say a family of people collaborating in developing and pursuing a well-defined project of research. [...] Now, to create a School worthy of the *Maestro* is not sufficient, nor is it sufficient that he knows how to trace out such a vast plan of research as to go beyond his own working capacity. It is also necessary that he succeed in communicating his passion and his faith to his disciples and in demanding and directing their collaboration.[58]

The Italian geometers were linked to each other by training (they were each other's pupils), because they had been fellow students and then colleagues, because they were friends or relatives. Very often, for all four reasons. Enriques, for example, was a student, a colleague, and a brother-in-law of Castelnuovo. Beniamino Segre

---

[58]Castelnuovo G. (1930), p. 615: *La scuola va oltre il valore dell'uomo e l'importanza di una determinata disciplina, per investire tutta l'organizzazione scientifica. Voi tutti sapete quali difficoltà si incontrino nei nostri paesi latini, prevalentemente individualisti, a costituire una scuola scientifica, cioè una riunione o direi quasi una famiglia di persone collaboranti nello sviluppare e proseguire un piano organico di ricerche. [...] Ora per dar vita ad una scuola non basta il valore del Maestro, né basta che egli sappia tracciare un piano di ricerche così vasto da superare la propria forza di lavoro. Occorre altresì che egli riesca a comunicare la sua passione e la sua fede ai discepoli e sappia esigerne e dirigerne la collaborazione.*

was a pupil of Fano, Fubini and C. Segre (of whom he was also a cousin) and a colleague of B. Levi, who had all in turn been students of C. Segre at an earlier moment. The Italian geometers shared an important part of their lives and careers; meeting, separating, and meeting up again in different university contexts.[59] Their wives became friends, their children became classmates and friends, and often colleagues themselves at a later date.

Sociality, for the members of this School, played an essential role. In the parlours of Enriques, Castelnuovo, Levi-Civita and Volterra, mathematicians and intellectuals gathered.[60] *Via d'Azeglio*, *via Boncompagni*, *via Sardegna*, *Ariccia* are addresses that inspired warm memories in any Italian or foreign mathematician who lived between 1880 and 1938. Great mathematicians and physicists of the time—J. Hadamard, G.D. Birkhoff, A. Einstein, N. Bohr, M. Planck, etc.—popped in there for dinner or tea and, while the parents spent a few hours in informal chatter, their children also became acquainted. Adriana and Alma Enriques, Emma and Gina Castelnuovo thus become friends with the young nuclear physicists, the so-called boys of Via Panisperna. In a page that deserves to be quoted in full, Laura Capon Fermi vividly recalled this situation:

With the young people from the Institute of Physics, I used to meet in Professor Castelnuovo's apartment. By established custom, they received guests every Saturday evening after dinner. They lived on the top floor of a newly built house in via Boncompagni 16, higher than the surrounding houses. From the terrace of the Castelnuovos' you could see the roofs of the city sloping away, unequal, like a wave of water rippled by the wind, against the dark background of the Alban Hills. [...] Professor Guido Castelnuovo, an affable man, with a silvery beard and affectionate and scrutineering eyes, stood up to meet us, and introduced us to others. The greatest mathematicians of the time gathered in that living room, many of whom, like the owner of the house, have now disappeared, while leaving a lasting memory: Volterra, Levi-Civita and Enriques, the brother of Mrs. Castelnuovo. They were a group of men bound by affective bonds which, arising from a common scientific interest, were cemented by the personal talents that they also shared: they had deep and inquiring minds, equitable and sure judgment; they were of high moral soundness and of a reserved goodness. They gathered at the Castelnuovos' with their wives and children, one Saturday after another, to spend a few hours in informal chatter among friends. They talked of the latest happenings in the Faculty: of births and deaths, marriages and flirtations, faculty policies, new discoveries, and theories, and of the rising stars in the field of physics. In the presence of these great men, I felt deeply awed, and it was almost a relief to be told by Mrs. Castelnuovo that I could go into the dining-room, where the young people were gathered. The young Castelnuovos', the Enriques cousins and some of their friends were sitting around the dinner table covered with a green cloth. They pointed me to a place, between Prof. Persico and a boy with rosy cherubic youth, Edoardo Amaldi, the same boy who in the following year was to move from engineering to physics. [...] Next autumn, the group underwent some changes. Persico left Rome, as he had obtained the chair of Theoretical Physics in Florence; Fermi and Rasetti came to sit around the Castelnuovos' dining-room

---

[59] The mobility of university professors was remarkable at the time and their first assignments were usually in peripheral locations or in a newly established Faculty.

[60] See e.g. Enriques G. (1983); Barone and Ciardi (2021) (eds.); Gerbi (2013).

table. Now and then a member of the group brought along a new friend. In this way, on a Saturday evening, Emilio Segré made his appearance among us.[61]

However, the School should not be mistaken for an idyllic reality. As happens in any family, there was no shortage of rivalries, tensions, competition, and internal clashes. In the case of Italian team, these squabbles broke out mainly after the death of Segre (1924), when Severi took the lead and 'defiled' the ideal of the School. Moved by 'a tenacity that often verges on obstinacy',[62] Severi was an outstanding scholar and an inspiring teacher, but his jealous and difficult character prevented him from accepting the fact that others could have ideas better than his own. Intending to take upon himself the leadership of the School, from the mid-1920s Severi launched on a collision course with Enriques and, to a lesser extent, with Castelnuovo. At the root of this was, initially, a matter of hierarchy of ideas. Emblematic testimony of this comes in a long and brutal letter which Severi wrote to B. Segre in 1932 when the latter was assigned the keynote lecture for the opening ceremony of the academic year 1931–32 at the University of Bologna. 'Words must be weighed!!!'[63] Severi recommended on this occasion: intellectual property must be attributed with great prudence in periods when research was often carried out in collaboration, or through

---

[61] Laura Capon Fermi (1954), pp. 53–54: *Coi giovani dell'Istituto Fisico mi ritrovavo spesso in casa dei Castelnuovo, che, per vecchia abitudine, ricevevano ogni sabato sera dopo cena. Abitavano all'ultimo piano di un casamento di recente costruzione in via Boncompagni 16, più alto delle case circostanti. Dalla terrazza dei Castelnuovo si vedevano i tetti della città allontanarsi, ineguali, come una distesa d'acqua increspata dal vento, sullo sfondo scuro dei Colli Albani. [. . .] Si alzò per venirci incontro il professor Guido Castelnuovo, uomo affabile, con la barba argentea e gli occhi affettuosi e scrutatori a un tempo; e ci presentò agli altri. Si radunavano in quel salotto i più grandi matematici del tempo, dei quali molti, come il padron di casa, sono ora scomparsi, pur lasciando un ricordo duraturo: Volterra, Levi-Civita ed Enriques, il fratello della signora Castelnuovo. Erano un gruppo di uomini legati da vincoli affettivi, che, sorti dal comune interesse scientifico, venivano cementati dalle doti personali che pure condividevano: avevano la mente profonda e indagatrice, il giudizio equanime e sicuro; erano di principi morali elevati e di una bontà riservata. Venivano in casa Castelnuovo un sabato dopo l'altro a parlare del più e del meno, di faccende della Facoltà, di nascite e di morti, di fidanzamenti e di matrimoni, delle ultime scoperte e teorie, degli astri che sorgevano sull'orizzonte della fisica. Alla presenza di questi grandi uomini mi sentii profondamente intimidita, e provai sollievo quando la signora Castelnuovo mi disse di andar pure nella stanza accanto, dove si eran riuniti i giovani. I ragazzi Castelnuovo, i cugini Enriques e alcuni loro amici erano seduti attorno alla tavola da pranzo coperta da un tappeto verde. Mi indicarono una sedia, fra il prof. Persico e un ragazzo con la rosea faccia paffuta, quell'Edoardo Amaldi che l'anno dopo doveva passare da ingegneria a fisica. [. . .] L'autunno seguente avvennero cambiamenti nel gruppo. Persico, che aveva ottenuto la cattedra di Fisica Teorica all'Università di Firenze, partì da Roma; Fermi e Rasetti vennero a sedersi alla tavola da pranzo dei Castelnuovo. Ogni tanto qualcuno portava un nuovo amico. E così un sabato sera fece la sua comparsa Emilio Segré.*

[62] Severi (1953), p. 69: *una tenacia che a volte sconfina nell'ostinazione.*

[63] CA, *BSP*: F. Severi to B. Segre, Rome 2.1.1932: *Le parole vanno pesate!!!* Segre's correspondence is kept in the California Institute of Technology Archives and Special Collections, Beniamino Segre Papers (CA, *BSP*). Documents belonged to Segre's son, Sergio.

suggestions from the seniors to the early-career scholars. Caution was especially important when it was a matter of describing the mathematical merits of a School.[64]

At the root, however, laid not only a claim of individual prestige. The Italian School of Geometry was a very powerful academic pressure group at the local and national levels. In Turin as in Bologna or Rome, geometers dominated the mathematics Departments from the late nineteenth century to the Second World War. In Turin from 1883 to 1938, D'Ovidio, Segre, Fano and Fubini held uninterruptedly two out of the three courses of higher mathematics (Advanced Analysis and Geometry) and three out of the four courses of basic mathematics (Algebra, Analytic Geometry, Descriptive and Projective Geometry), in addition to that of Didactics of Mathematics at the *Scuola di Magistero*, from its creation to suppression (1875–1920). Algebraic geometers occupied all the board roles for the Special Mathematics Library from its foundation (1883) up to the racial laws, and of the Mathematical Seminar of the University and Polytechnic of Turin, from its establishment (1928) to 1938. In Rome, the distribution of power was more balanced, thanks to the presence of Volterra (from 1900) and Levi-Civita (from 1919), but Castelnuovo, Enriques and Severi were very influential academic policymakers. Moreover, Italian geometers sat on the boards of all the leading mathematical journals (*Annali di Matematica pura ed applicata, Rendiconti del Circolo Matematico di Palermo, ...*), in the Higher Council for National Education, and they were always selected as members of the examining boards in national university competitions for geometry chairs. The exertion of power further affected the relations among the members of the School, even among the most solitary individuals. One example will suffice. In August 1936, after Pincherle passed away, it was necessary to appoint his successor on the editorial board of the *Annali*. Severi wanted at all costs that the position be awarded to his favourite pupil, Segre. Fubini, meanwhile, endorsed Levi:

> a man of vast culture, of original and profound thought; B. Segre is a very good and worthy person, but he is often more an imitator than an original mind. Therefore, also given the seniority, it seems to me that it would be to give an undeserved slap to B. Levi. He is often a mediocre expositor, but his ideas are far from being confused. I fear (and this be completely confidential) that S[everi] wants everywhere that people be absolutely devoted to him: it is a method that does not belong to him alone.[65]

Both tried to win over to their side the other co-director of the *Annali*, Levi-Civita. In the end, Severi won, and Levi-Civita offered the position to Segre.[66] Fubini swallowed the pill in bitter silence.

---

[64] CA, *BSP*: G. Castelnuovo to B. Segre, Rome 4.6.1938.

[65] G. Fubini to T. Levi-Civita, Turin 2.8.1936, in Nastasi and Tazzioli (2003), p. 139: *uomo di vasta cultura, di pensiero originale e profondo; il B. Segre è una bravissima e degnissima persona, ma è sovente più un imitatore che una mente originale. Perciò, data anche l'anzianità, mi pare sarebbe dare uno schiaffo immeritato al B. Levi. Questi non sempre espone bene, ma ha idee tutt'altro che confuse. Temo (e questo sia detto in via assolutamente confidenziale) che il S[everi] voglia dappertutto persone a lui assolutamente devote: è un metodo che non appartiene a lui solo.*

[66] CA, *BSP*: T. Levi-Civita to B. Segre, Selva in Val Gardena (Bolzano) 15.8.1936.

1.4   The Belle Époque of Scientific Internationalism

From 1858, the date of the trans-European journey of E. Betti, F. Brioschi, and F. Casorati, which marks the beginning of the so-called mathematical Risorgimento, Italian scholars wove an international web of relations. This network would play an essential role for individuals embarking on the process of emigration as it largely determined the finding of destinations and would condition the positive or negative outcome of resettlement. For simplicity's sake, three key moments can be identified: Belle Époque, 1920s, and 1930s.

As far as the first period is concerned, there were two prevailing destinations: France and Germany. For the School's members, the latter was the horizon most popular with Segre, Loria, Fano, etc. spending some very inspiring research sojourns there. The correspondence between the Italian geometers and their German colleagues (F. Klein, L. Kronecker, O. Schlömilch, D. Hilbert, A. Hurwitz, etc.), over 200 letters exchanged between 1879 and 1923, fully document this special link.[67] Klein, identified as the great 'Master at a distance',[68] exerted on Italians a profound influence, both from the point of view of research projects (the *Erlangen Program* in 1872, hyperspatial projective geometry, etc.) and from that of pedagogical thinking (the *Merano conference* in 1905, the psychological foundations of geometry, elementary mathematics from a higher standpoint). However, there was more: many aspects of the mathematical life were imported from Germany (from Göttingen, especially): the oral culture and the model of the Seminars, the organisation of mathematical libraries with reading rooms, where students and teachers could freely access journals, reprints, and the collections of lithographs and models, the tradition of summer courses for the training and refresher courses of mathematics teachers, etc.[69] The collaboration of Italian geometers on the *Encyklopädie der mathematischen Wissenschaften* and the partnerships between the Zanichelli and Teubner publishing houses for translations of Italian works into German, and vice versa, are all indicative of a network of exchanges in both directions that developed throughout the Belle Époque but was abruptly interrupted by the outbreak of the First World War.

France, the Latin sister most geographically and culturally close to Italy, was the other horizon under observation.[70] Volterra, Castelnuovo, and Enriques were regular visitors to Paris, from the late nineteenth century onwards. Volterra, particularly, counted É. Borel, É. Picard, P. Painlevé, and J. Hadamard among his closest friends. Here again dialogue did not revolve about research only. From the École Polytechnique, for example, Volterra took inspiration when in 1906 he was charged with founding the Polytechnic of Turin; similarly, in the programme that Enriques

---

[67] Luciano and Roero (2012); Luciano and Roero (2016).

[68] C. Segre to F. Klein, Turin 24.2.1921, in Luciano and Roero (2012), p. 218: *Ella è stata il mio Maestro, pur essendo noi a tanta distanza!*

[69] On the Göttingen mathematical tradition see Rowe (1986); Rowe (2004); Rowe (2018).

[70] Brechenmacher et al. (2016).

built for the journal *Scientia,* there is much of the *Revue du mois* created by Borel in 1906. Finally, it was the model of the Paris institutes that Castelnuovo and Volterra dreamed of creating in Rome:

> As for the intellectual environment, we can only envy it, thinking with regret of the difference that passes between our Italian faculties, in each of which there are few professors capable of understanding each other, and the richness of the Faculty of Paris where men like Poincaré, Picard, Darboux, Appell are gathered ..., without naming the youngest generation. I do not believe that Italy would be able to put together a group of men as complete as that in Paris, but much could already be achieved if instead of scattering our scholars around twenty universities, we tried to attract the best talents to Rome. And instead ...![71]

On the other hand, links with the Anglo-Saxon world appear to be more marginal. From the end of the nineteenth century, various scholars came to Italy for a study trip: W. H. Young and his wife G. Chisholm (Turin 1898), J. Coolidge (Turin 1903–04), C. L. E. Moore (Turin 1908), C. H. Sisam (Turin 1908), G. C. Evans (Rome 1910–12), E. Kasner (Rome 1913), E. B. van Vleck (Rome 1914), and O. Veblen (Turin 1900, Rome 1914). Inversely, Volterra travelled to London, for the first time, in 1902, and Castelnuovo in Cambridge in 1912. Fano made his first American tour in 1906; Volterra in 1910, 1912, and 1919. Acquaintance with English-speaking mathematicians, however, was generally quite superficial until the First World War, and Italians sometimes adopted a quasi-paternalistic attitude towards Americans. Some passages of the letters sent by Castelnuovo to Volterra, on the occasion of his stay at Clark University in 1909, are revealing: 'this will not be a period of rest for you, but you do well not to miss an opportunity to visit some city in the United States, and to bring there the voice of Italy'.[72]

The Great War had an impact on international relations and represented the fading of those ideals of scientific internationalism in which the mathematical community had hitherto firmly believed.[73] Exchanges with Germany ceased when Klein signed the manifesto *An die Kulturvelt,* and were restored only from 1919 onwards by Segre and Enriques. On the contrary, Volterra supported the ostracism of mathematicians arriving from the former Central Powers and their exclusion from ICMs until 1928. Relations with the French community remained very strong, all the more so because, in the post-war period the scientific interests of Castelnuovo, Enriques, and Volterra turned to subjects such as probability calculus, statistics, epistemology, philosophy

---

[71] ANL-*Volterra*: G. Castelnuovo to V. Volterra, Rome 13.2.1912: *Quanto all'ambiente intellettuale, a noi non resta che invidiarlo, pensando con rammarico alla differenza che passa tra le nostre Facoltà italiane, in ciascuna delle quali si trovano pochi professori atti a comprendersi, e la ricchezza della Facoltà di Parigi dove sono riuniti uomini come Poincaré, Picard, Darboux, Appell ..., senza nominare i più giovani. Non credo che l'Italia riuscirebbe a metter insieme un nucleo di uomini così completi come i colleghi di Parigi, ma molto si potrebbe già ottenere se invece di sparpagliare i nostri scienziati in venti Università, si cercasse di attrarre i migliori a Roma. E invece ...!*

[72] ANL-*Volterra*: G. Castelnuovo to V. Volterra, Badia Prataglia 19.7.1909: *non sarà un periodo di riposo questo per te, ma fai bene a non lasciar cadere un'occasione per visitare qualche città degli Stati Uniti, e per portare colà la voce dell'Italia.*

[73] Lehto (1998); Parshall and Rice (2002).

of science, and mathematical biology, which were intensively cultivated in Paris.[74] To the old guard (H. Poincaré, É. Borel, É. Picard, etc.), new names were added (J. Pérès, M. Fréchet, P. Lévy, P. Montel, É. Cartan, etc.). The English-speaking mathematical world remained somewhat in the shadow, although Fano was invited as lecturer to Aberystwyth in 1923 while G.D. Birkhoff, V. Snyder, and S. Lefschetz spent sabbatical semesters in Italy in the 1920s. In the meantime, in post-war Europe, new research Schools were emerging. Several early career scholars (A. Rosenblatt, O. Zariski, D. Struik, etc.), many of whom were Rockefeller fellows, arrived in the Eternal City to work with the Rome mathematical group.[75] A clear sign of the changes occurring in the international geography of mathematics is attendance at the conference which was held in Bologna in 1928. The congress, the first that readmitted German-speaking mathematicians, brought to Italy 836 participants from 37 different countries. That was the occasion for Terracini, Fano, Segre, Fubini, and Levi to become acquainted with colleagues from the Americas, who would help them to emigrate and relocate abroad.

Fascism was instrumental in a strategy of cultural centralisation—making Rome the Princeton of Italy—but, as Birkhoff acutely noted in 1926, while before the war the principal European mathematical centres were Paris, Göttingen, and Rome, it now seemed 'that Paris and Göttingen were of decidedly more importance than Rome'.[76] Severi's leadership dictated a new agenda of international relations for the Italian geometric School in the late twenties and in the thirties. Attracted by his fame, foreign students from all over the world arrived in Rome: A. Weil, L. Roth, B. Leendert van der Waerden are just a few of those names. Because of his proper vision of the future of algebraic geometry (and also slightly due to his character), Severi progressively shifted the positioning of the Italian School on the global mathematical scene: strengthening scientific relations with the German-speaking area, reducing relations with English-speaking academia (the Cambridge group excluded), expanding the Italian protectorate in some 'peripheral' countries, with protégés from Romania, Bulgaria, Spain, Argentina, etc. whom he felt he could more easily tutor. After all, Severi was a person who used to repeat:

> Today Italy occupies one of the very top places in the global mathematics ranking, if not the first. This is recognised everywhere by foreign mathematicians, and as Italians and as fascists we must be legitimately proud of it. The trade balance registers a very high surplus in this field: we export many, but many more, ideas than we import.[77]

---

[74] Mazliak and Tazzioli (2021); Israel and Millán Gasca (2002); Clery (2020).

[75] Parikh (1991), p. 17.

[76] *G.D. Birkhoff's Report to the International Education Board (Rockefeller, New York/Paris) of September 1926 Concerning His Trip to Europe*, in Siegmund-Schultze (2001), p. 268.

[77] Severi (1933), p. 644: *l'Italia occupa oggi nel mondo uno dei primissimi posti, se non il primo in fatto di matematica. Questo è dovunque riconosciuto dai matematici stranieri e come italiani e come fascisti dobbiamo esserne legittimamente orgogliosi. La bilancia commerciale segna qui un fortissimo attivo: esportiamo molte, ma molte più idee di quante non ne importiamo.*

The lack of interpenetration of classical Italian algebraic geometry with the new approach, which was being affirmed in the United States, combined with a policy of cultural autarchy, in which Severi was a master, would have a profound impact at the time of emigration: the possibilities of finding a position abroad largely depended on foreign evaluation of the quality of research by the late members of the School.

After Hitler came to power in 1933, the international relations framework changed again. The elite of Italian mathematics was then gathered in Rome, where new research institutes had been created: the *Istituto Nazionale per la Storia delle Scienze* coordinated by Federigo Enriques (1927), the *Istituto Nazionale per le Applicazioni del Calcolo* directed by Mauro Picone (Naples in 1927, in Rome from 1932), the *Istituto Centrale di Statistica* led by Corrado Gini from 1927 and, the last chronologically, Severi's creation: the *Istituto Nazionale di Alta Matematica*, inaugurated in 1939. The Nazi rise to power had a massive impact on academic mobility: scholars fleeing from the Reich began to arrive in Italy along with students who hoped to conclude their university studies in Italy.[78] From 1933 onwards, Fano, Castelnuovo, Fubini, and Terracini no longer went to Germany regularly; Levi-Civita was there for the last time in 1934.

1.5   The Jews Under Fascist Rule: The Mathematical Community

The question of the relationship between Jews and fascist rule is very complex and has been the subject of specific studies, whose results can be basically summarised in three points:[79] (1) Jews were Italian citizens, divided like others into social classes, with the typical interests of the various social classes; (2) there were fascist Jews, children of 'banal'[80] families which passed from the Risorgimento ideals to interventionism and nationalism; (3) Italian Jews were fascists like the others, but more anti-fascist than the others[81] (Fig. 1.7).

Literature on the treatment of Jews in Mussolini's Italy, however, hardly mentions mathematicians, which is why it may be appropriate to outline the variety of their attitudes towards the regime. Two phases can be distinguished. In the first, which goes from the march on Rome (October 1922) to Matteotti's murder (June 1924) some mathematicians, like Levi and Artom, disappointed by the 'mutilated

---

[78] Voigt (1999); Signori (2000); Signori (2009).

[79] See e.g. De Felice (1961); Caffaz (1978).

[80] Bassani 1938 [2014], p. 453: 'I came from a family of this type: Jewish and fascist. But let's be clear: countless other Jewish families were at that time like ours, normal (and banal) like ours'. (*Uscivo da una famiglia di questo tipo: ebraica e fascista. Ma sia ben chiaro: infinite altre famiglie ebraiche erano a quell'epoca come la nostra, normali (e banali) come la nostra.*

[81] Sarfatti (2007a), p. 25: *gli ebrei italiani erano fascisti come gli altri italiani, più antifascisti degli altri italiani.*

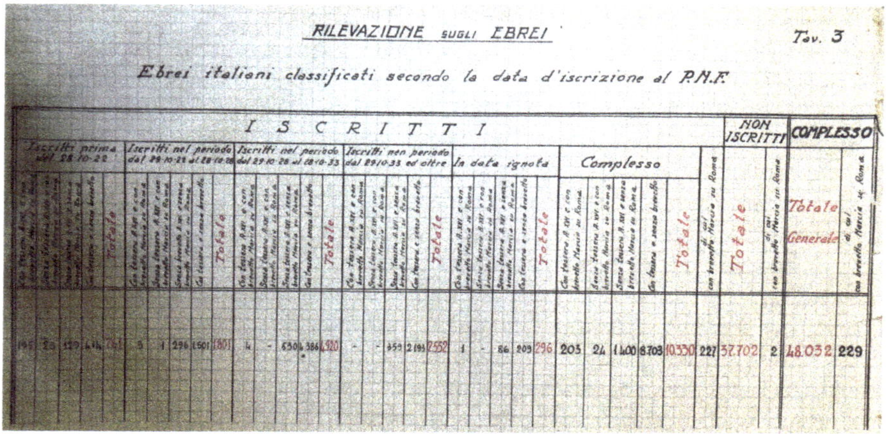

RILEVAZIONE SUGLI EBREI — Tav. 3

Ebrei italiani classificati secondo la data d'iscrizione al P.N.F.

ISCRITTI: Iscritti prima del 28-10-22					Iscritti nel periodo dal 29-10-22 al 28-10-28					Iscritti nel periodo dal 29-10-28 al 28-10-33					Iscritti nel periodo dal 29-10-33 ed oltre					In data ignota					Complesso					di cui con brevetto Marcia su Roma	NON ISCRITTI: Totale	di cui con brevetto Marcia su Roma	COMPLESSO: Totale Generale	di cui con brevetto Marcia su Roma
Con tessera A.XVI e con brevetto Marcia su Roma	Senza tessera A.XVI e con brevetto Marcia su Roma	Senza tessera A.XVI e senza brevetto Marcia su Roma	Con tessera e senza brevetto	Totale	Con tessera A.XVI e con brevetto Marcia su Roma	Senza tessera A.XVI e con brevetto Marcia su Roma	Senza tessera A.XVI e senza brevetto Marcia su Roma	Con tessera e senza brevetto	Totale	Con tessera A.XVI e con brevetto Marcia su Roma	Senza tessera A.XVI e con brevetto Marcia su Roma	Senza tessera A.XVI e senza brevetto Marcia su Roma	Con tessera e senza brevetto	Totale	Con tessera A.XVI e con brevetto Marcia su Roma	Senza tessera A.XVI e con brevetto Marcia su Roma	Senza tessera A.XVI e senza brevetto Marcia su Roma	Con tessera e senza brevetto	Totale	Con tessera A.XVI e con brevetto Marcia su Roma	Senza tessera A.XVI e con brevetto Marcia su Roma	Senza tessera A.XVI e senza brevetto Marcia su Roma	Con tessera e senza brevetto	Totale	Con tessera A.XVI e con brevetto Marcia su Roma	Senza tessera A.XVI e con brevetto Marcia su Roma	Senza tessera A.XVI e senza brevetto Marcia su Roma	Con tessera e senza brevetto	Totale					
195	23	129	414	761	3	1	296	1501	1801	4	-	530	4386	4920	-	-	359	2193	2552	1	-	86	209	296	203	24	1400	8703	10330	227	37.702	2	48.032	229

Fig. 1.7 Statistical prospect of Italian Jews enrolled in PNF, ACS-*Ebrei*

victory' and worried about the strikes and unrest of the Red Biennium, looked with sympathy at the rise of an authoritarian regime.[82] Levi admitted, in his correspondence with his brother Giulio Augusto:

> When it was born, fascism could seem and seemed to me too—and I considered it with sympathy—something like what you said a couple of years ago [a revolution], but it was a meteor of some days. It immediately went astray, and it is a form of brigandage, which does not even have the merit of the courage of the bandits, because it takes place under the protection of the State.[83]

Belle Époque gentlemen, such as Fano, considered Mussolini and his cohorts to be boors (*bifolchi*); they were disgusted by their ways, by the lack of respect for the state and democratic institutions, but they were not overly worried.

---

[82] Emilio Artom (1940–44), in Treves (1954), p. 58. The Red Biennium was a two-year period, between 1919 and 1920, of intense social conflict in Italy. The population was confronted with rising inflation and a significant increase in the price of basic goods, in a period when extensive unemployment was aggravated by mass demobilization of the Royal Italian Army at the end of the war. The revolutionary period was followed by the violent reaction of the Fascist black shirts' militia, with the support of Italian industrialists and landowners, and eventually by the march on Rome in October 1922.

[83] B. Levi to G.A. Levi, n.d. but before 1925, in Momigliano Levi (2016), p. 273: *Il fascismo poté parere e parve anche a me—e lo considerai con simpatia—qualcosa di simile a quello che tu dici un paio d'anni fa, quando esso nacque, ma fu una meteora di giorni. Ha subito tralignato ed è una forma di brigantaggio, che non ha nemmeno il merito del coraggio dei banditi, perché si svolge sotto la protezione dello Stato.* Levi would soon repent: 'Mussolini understands nothing other than the theatrically violent act [. . .]. The only thing to ask the rulers is to govern in every branch as they want, but to leave just enough freedom and continuity in the State system as to allow others to do something useful'. (*Il Mussolini non capisce altro che l'atto teatralmente violento* [. . .]. *La sola cosa da chiedere ai governanti è in ogni ramo quella di governare a loro talento, pur di lasciare quel tanto di libertà e di continuità negli ordinamenti da permettere agli altri di fare qualcosa di utile.*)

The very opposition to the Gentile Reform (1923)—the first act of Mussolini's authoritarian regime—regarded content, rather than the political nature, and attracted men with various ideologic opinions. Among the mathematicians who raised their voice of protest, there were early anti-fascists like Volterra, but also conservative liberals such as Castelnuovo and Enriques, and right-wing individuals such as Berzolari and Pincherle. Similarly, among the signatories of the *Manifesto degli intellettuali fascisti*, the so-called *Manifesto Gentile*, published on 21 April 1925, there was a Jewish mathematician (Pincherle), and among those of the *Manifesto degli intellettuali antifascisti*, the so-called *Contromanifesto Croce*, released on 1 May 1925, there were five (Castelnuovo, Levi, Levi-Civita, Volterra, and Padoa). Signatures for or against, however, were not indicative. Severi, a great fascist since 1929, signed the *Contromanifesto*.

Faced with the Fascist Laws (1925–26), which marked the beginning of the fascist, nationalist, authoritarian, centralising, and corporatist regime, a fear began to spread of a possible deprivation of individual freedoms and the risk of being denied rights conquered in the Risorgimento. The failure of an assassination attempt on Mussolini (Bologna, 1926) provided the regime with a pretext to stiffen its authority. In 1929, with the signing of the Lateran Pacts, fascism earned the allegiance of the Catholic church. In the meantime, the fascistisation of culture began, with the founding of the *Istituto fascista di cultura*, later to be known as *Istituto nazionale di cultura fascista*, in 1925, and the creation of the Royal Academy of Italy in 1926. In the mathematical world, however, a mid-way stance continued to prevail, and between the two opposite poles of the early fascists and anti-fascists, the grey zone of a-fascists widened. This comprised a wide sample of individuals, including those who, like the astronomer Azeglio Bemporad, boasted of being more fascist than Mussolini, along with many people who publicly complied but privately mocked the movement: Fubini, for example, who asked Levi-Civita if wearing a tuxedo instead of the fascist black shirt was enough to hold the Commemoration of Luigi Bianchi at the Lincei.[84]

Individual reaction towards the regime, moreover, was established and renegotiated from time to time, according to opportunities. Pincherle, for example, compromised himself with the power in exchange for crowdfunding needed for UMI and the organisation of the International Congress of Mathematicians in Bologna in 1928.[85] The most typical case, however, is that of Severi who, from 1929, reconnected with fascism. The closer he got to the Duce and the higher he climbed fascist hierarchies, the more Severi increased his power at local, national, and international levels, up to the point of becoming the man solely in command of Italian mathematics. So, from 1930 onwards Severi represented Italy at almost all formal events abroad and travelled the world as an ambassador of science and

---

[84] G. Fubini to T. Levi-Civita, [Turin] 17.4.1929, in Nastasi and Tazzioli (2003), p. 121: *Il 5 Maggio dovrò venire per la Commemorazione Bianchi: basta una giacca nera?*

[85] Giacardi and Tazzioli (2021).

Italianness, giving lectures and conferences throughout Europe, South America (1930); and Japan (1936). From the fascist authorities, Severi also obtained funding for the National Institute of Advanced Mathematics, which he conceived as the flagship of Italian science and the driving force of mathematical research, through proselytism among young talented people and reference to the most distinguished Masters.

There are two clear turning points in the narrative of the relationship between Jewish mathematicians and fascist rule: 1931 and 1933. The first is the date of the mandatory oath of loyalty to the regime by all teachers and state administrative employees. The story of the oath of loyalty is by now well known. In January 1929 Severi, a former socialist, presented the Duce with a long document, stigmatising the humiliating condition of those university professors who were erroneously considered as anti-fascists or purely and simply for a-fascists, though being 'for a long time spiritually close to fascism, much more than many others who hastily converted after the decisive triumph of the Revolution'.[86] Political opportunism, ambition, old and new grudges with his colleagues, the desire to rout Enriques' competition in the run for election as a member of the Royal Academy of Italy induced Severi to make this move. In this memoir, he stated the leading motives of the political line that he would constantly suggest: intransigence for the diehards, 'amnesty for political acts contrary to the regime by then outmoded'.[87] A few days later, Severi also suggested to G. Gentile how to definitively solve the issue of intellectuals' loyalty to the party:

a new formula of the loyalty oath would suffice for the purpose, if it were proposed and annotated by the Grand Council as the supreme political body, in which the avant-garde wings are also represented through the Party and the press [. . .]. The provision should be represented as an act of intransigence aimed at obtaining the much desired fascistization of universities; as an appeal to the loyalty of the professors, who could not fail to take the oath without incurring far more serious measures than dismissal.[88]

The epilogue is well known. Two years later (August 1931), swearing allegiance to the King and devotion to the country which had been introduced with the Unification was made compulsory with a new formula (oath of allegiance to the

---

[86] F. Severi, *Promemoria*, Rome 31.1.1929, in Guerraggio and Nastasi (2005b), p. 104, 107: *qualificati per antifascisti o puramente e semplicemente per a-fascisti* [. . .] *eppure da gran tempo spiritualmente vicini al fascismo, assai più di molti convertiti a precipizio dopo il trionfo decisivo della Rivoluzione.*

[87] F. Severi to G. Gentile, Barcelona 15.2.1929, in Guerraggio and Nastasi (2005b), p. 102: *sanatoria di atti politici ormai lontani.*

[88] F. Severi to G. Gentile, Barcelona 15.2.1929, in Guerraggio and Nastasi 2005b, p. 102: *una nuova forma di giuramento basterebbe allo scopo, s'essa fosse proposta e chiosata dal GC [Gran Consiglio] come supremo Corpo politico, in cui son rappresentate anche, attraverso al Partito e alla stampa le ali d'avanguardia; che così assumerebbero la corresponsabilità dell'atto e non potrebbero più avversarlo, apertamente o copertamente. Occorrerebbe che il provvedimento fosse rappresentato come un atto di intransigenza diretto a ottenere la tanto richiesta fascistizzazione delle Università; come un appello alla lealtà dei professori, i quali non potrebbero mancare al giuramento senza incorrere in provvedimenti ben più gravi della messa a riposo d'autorità.*

King, his descendants and the fascist regime) for university professors. Just a dozen of university professors out of 1225 refused to take it.[89] Of these, four were Jews (G. Errera, V. Volterra, G. Levi della Vida, and F. Luzzatto) and one was a mathematician: Volterra. Receiving the request to sign on 9 November, Levi-Civita drafted (but did not send) a letter to the Rector at Rome, Pietro de Francisci, inquiring if he could assume that the new oath did not explicitly preclude him from dissenting spiritually from the political ideals of the regime.[90] Many others swore the oath almost without hesitation. Indeed, Fubini, having been informed by his son Eugenio, a fellow student of Levi-Civita, that he had decided to swear the oath with inner reserve, so as not to open the academia doors to the new barbarians, wrote to him:

> I am extremely pleased that your great name, your figure, which is so gigantic in Italian mathematics, be preserved at the Italian School. And this joy of mine is shared by those who know and therefore love the illustrious Master, it is shared particularly by my whole family. My father-in-law, my wife, everyone wants me to transmit to you their feelings, a charge that I carry out with an exultant soul. I am pleased with these facts that indirectly prove how much everyone esteems in you one of the purest glories of Italian science.[91]

Was that political myopia? Disinterest? The eternal stereotype of mathematicians locked up in an ivory tower, all absorbed in their high speculations and disinterested in civil and political life? F. G. Tricomi, at an Inter-Nation conference held in Pasadena in 1951, was asked the question, 'How was it possible that in Italy the academic freedom was only slightly weakened despite more than twenty years of fascist rule, and so quickly and fully restored after the end of that period of tyranny?' He would answer:

> it has something to do with a somewhat mild character of the fascist rule in Italy, but [I feel] that the real explanation is to be found in the fact that the Italian universities were and are governed by an excellent basic law which could be changed only slightly (and formally rather than substantially) by the fascists.[92]

---

[89] Goetz (2000); Boatti (2017); Roero (2021).

[90] Goodstein (2018), p. 218. Levi-Civita thoroughly discussed the matter with a colleague from Turin, the famous pathologist Giuseppe Levi. Both finally reached the decision to swear the oath 'with grave regret' after receiving informal assurances from the Rectors of their universities that their freedom of thought would be inwardly guaranteed. See also Polverini (1991).

[91] G. Fubini to T. Levi-Civita, Turin 1.12.1931, in Nastasi and Tazzioli (2003), p. 124: *Sono oltremodo lieto che il tuo grande nome, la tua figura, che tanto è gigantesca nella matematica italiana, siano conservati alla scuola d'Italia. E questa mia gioia è condivisa da quanti conoscono e perciò amano il Maestro illustre, è condivisa particolarmente da tutta la mia famiglia. Mio suocero, mia moglie, tutti vogliono che io sia interprete presso di te dei loro sentimenti; ufficio che io compio con l'animo esultante. Io sono lieto di questi fatti che indirettamente provano quanto tutti stimino in te una delle glorie più pure della scienza italiana.*

[92] BSM-*Tricomi*: *Preparazione dell'opuscolo: A quarant'anni ...*, mss. Prot. no. 376, Turin 16.12.1935–29.4.1936, fols. 1–13. The manuscript is preserved in Tricomi's archive (see Luciano 2021).

The second moment dates to 1933, when the regime inflicts a second *vulnus* on the rule of law: the obligation to join the PNF, the so-called obligation to *prendere la tessera* (literally: take the card). In Rome or Bologna, it was no longer a game: Bompiani, Picone, Severi, Pincherle, were already members of the party, and their fascist faith had been ascertained. In Turin, by contrast, Tricomi had (up to that point) managed to strike a gentlemen's agreement among his faculty colleagues: that all would stay completely outside the party or join it all together. The Rector of the University, S. Pivano, then turned to Fano.

Together with Fubini, G. Peano, and C. Somigliana, Fano had been nominated by Severi for election as a member of the Royal Academy of Italy. After cashing in on the success of the imposition of the oath under the new formula and successfully becoming the only mathematician to be elected to the Academy at its creation (1929), Severi steered all the appointments of mathematicians running for election.[93] Nomination to the Academy of Italy was very prestigious, which is why it is not surprising that Peano, Somigliana and Fano—who all swore the loyalty oath in 1931 but none of whom was yet enrolled in the PNF—tried to position themselves tactically from a political point of view to facilitate Severi in his strategy. It was precisely in those months that Pivano asked Fano to apply pressure to his colleagues, so that the full professors at the Turin School of Mathematics would join the party. The outcome was infamous: in the space of a few months, everyone 'carried the card': Fubini and Terracini in April 1933, Somigliana and Persico in July. Tricomi was the last to 'bite the bullet' in October.[94] As Terracini would write in 1968:

> I must confess that our behaviour, and mine in particular, was not too brilliant, in the sense that we soon submitted to the pressing invitation; I am ashamed to say so. The remorse was increased by the belated observation that no damage was incurred by colleagues who refused to 'take the card', such as professors Guido Ascoli, Eligio Perucca, Mario Falco. Much more rigid had been the reaction of fascism against the very few university professors who had refused, up and down in that period (precisely, in 1931), to take the oath of allegiance.[95]

---

[93] Fubini immediately pulled out of the game, convinced that nobody would be able to avoid the election of Levi-Civita. See G. Fubini to T. Levi-Civita, Turin 23.11.1929, in Nastasi and Tazzioli (2003), p. 123: 'I believe that Volpe is convinced of the enormity; he told me to believe he will be repaired soon. He does not believe the well-known prejudice, because he was convinced that Fermi was Semitic' (*Credo che Volpe sia convinto della enormità; mi disse credere si riparerà presto. Non crede la preguidiziale ben nota, perché era convinto che il Fermi fosse semita*). Fubini was wrong. Levi-Civita and Enriques never succeeded in being elected to the *Accademia d'Italia*. On Levi-Civita's case see Capristo (2001) and Capristo (2003).

[94] Tricomi (1967), p. 35: *inghiottire il rospo*.

[95] Terracini A. (1968), p. 109: *devo confessare che il nostro, e in particolare il mio, contegno non fu troppo brillante, nel senso che presto seguimmo il pressante invito; mi vergogno a dirlo. Il rimorso fu aumentato dalla tardiva constatazione che in nessun danno incorsero i colleghi che rifiutarono di prendere la tessera, quali ad es. i professori Guido Ascoli, Eligio Perucca, Mario Falco. Ben più rigida era stata invece la reazione del fascismo contro i pochissimi professori universitari che si erano rifiutati, su per giù in quel torno di tempo (esattamente, nel 1931) di prestare il giuramento di fedeltà.*

Terracini was right because enrolment in the PNF was not mandatory for any of the Turin mathematicians, nor was it a condition for access to or advancement in the university career. In other words, since they were all faculty members, Fano, Fubini, and Somigliana (as well as Terracini and Tricomi, although they had been awarded a chair more recently) could have avoided the act.

From the end of 1933, the situation stabilised. Fascism became a dictatorial regime whose energies were aimed at a dual purpose: to strengthen its image abroad and to gain mass consensus in Italy. Card-carrying members had nothing to fear. Being Jews, or not, made no difference. Those who were publicly compliant with the regime, or those who at least did not contest it, did not encounter any kind of problem in appointments, promotions, crowdfunding, etc. Others were and/or became genuinely convinced of fascist ideology, for example the authors of the journal *La nostra bandiera*, a periodical created by the economist E. Ovazza, which was entirely dedicated to documenting the contribution of Italian fascist Jewry to the intellectual, social, and economic life of the nation.[96] Still others, and these constituted the absolute majority, were indifferent, a-fascist, continuing to live and work in universities and institutes as they had always done, possibly muttering with family and friends over the absurdity of certain rituals.

1.6   Italian Mathematicians and Anti-Semitic Persecutions in Central and Eastern Europe

The passing of the Nazi racial laws (*Law for the renewal of the Public Administration*, 7.4.1933 and *Law for the protection of hereditary characters*, 14.7.1933) ignited an unprecedented debate among Italian Jewish mathematicians regarding the opportunity to react publicly to anti-Semitic persecution. Fubini was the first to propose to his colleagues to resign *en masse* from the *Deutsche Mathematiker-Vereinigung*, a striking and particularly forceful gesture for a research team like the Italian School of Geometry, which during the Belle Époque had been clearly German-oriented:

> Given the prevailing Hitlerism, don't you think we should resign from Deutsche Math. Verein.? After what Einstein suffered, after the events in Leipzig, etc., this seems to me to be due: but I do not know if this is appropriate, and if this can harm our unfortunate colleagues. I will also write about it to my other Italian friends, but first I would like to hear your opinion, which will certainly inform my decision. And, if you approve of our resignation, I will write about it, as I told you, to Pincherle, Beppo Levi, [Giulio] Vivanti, [Gino] Loria. Prof. Terracini agrees with me; Fano is not in Turin now.[97]

---

[96] Ventura 2002.

[97] G. Fubini to T. Levi-Civita, [Turin] 14.4.1933 in Nastasi and Tazzioli (2003), pp. 125–126: *Dato l'Hitlerismo imperante, non ti pare che noi dovremmo dimetterci dalla Deutsche Math. Verein.? Dopo quanto ha subito Einstein, dopo i fatti di Lipsia ecc. a me questo pare dovuto: ma non so se è cosa opportuna, e se ciò può danneggiare i nostri disgraziati colleghi. Ne scriverò anche agli altri*

Until then, anti-Semitism in Italy had not been absent but had been mostly opportunistic, a result of power struggles between scholarly groups, personal rivalries, and protests against supposed wrongs and abuses suffered. That many members of the Italian School of Geometry were Jews, and united by family and professional ties, had led to the first accusations of lobbying against the 'Jewish camarilla' since the end of the nineteenth century. Chronologically, the first episode had occurred in relation to the competition for the chair of Projective and Descriptive Geometry at the University of Bologna (1896). The commission was set up by F. Aschieri, E. Bertini, D. Montesano, C. Segre, and G. Veronese. The candidates were F. Amodeo, E. Ciani, M. Pieri, F. Enriques, and G. Fano. Thus, among the examiners there was one Jew, and two among the candidates. The competition was won by Enriques; Amodeo, who failed, felt himself unjustly evaluated and confided with F. Gerbaldi, who replied:

> I read the report of the Bologna competition even before you wrote to me about it, and I was struck by the way in which the Commission treated you again. Everything is to be expected from the *camarilla* of Bertinis and Segres; woe to those who fall into their path! [. . .] Do not lose courage, for if there is a God not only for the Jews, one day or another justice will also come to you.[98]

This was the first of a series of slanders, the results of spite and resentment rather than grounded anti-Semitism, as displayed by P. Burgatti, R. Marcolongo, and various others.[99] Even a gentleman like Peano made a comment in bad taste on racial origins.[100] In somewhat analogous terms, when Tricomi spoke of the 'Turin Jewish group', he did so in reference to the polarised local mathematical environment he encountered upon his arrival in 1924:

> Among the colleagues I found in the new institute there were—in addition to Peano and Terracini, who arrived almost together with me—C. Somigliana, G. Fano and T. Boggio as

---

amici italiani, ma innanzi tutto desidero sentire il tuo parere, a cui senz'altro informerò la mia decisione. E, se tu approvi le nostre dimissioni, ne scriverò, come ti ho detto, a Pincherle, Beppo Levi, Vivanti, Loria. Il prof. Terracini è d'accordo con me; Fano non è ora a Torino. Regarding the resignation from the DMV of Fubini and other Italian mathematicians, see G. Fubini to T. Levi-Civita, Turin 21.2.1935 and 6.3.1935 in Nastasi and Tazzioli (2003), pp. 129, 130–131.

[98] F. Gerbaldi to F. Amodeo, Palermo 11.5.1897, in Enea (2013), p. 113: *Ho letto anche prima che tu me ne scrivessi, la relazione del concorso di Bologna, e mi ha fatto penosa impressione la maniera con cui la Commissione ti ha trattato anche questa volta. Quanto alla* camarilla *dei Bertini e Segre c'è da aspettarsi di tutto; guai a chi cade nella loro disgrazia! [. . .] Addio, non perderti di coraggio, ché se un Dio vi è non soltanto per gli Ebrei, un giorno o l'altro la giustizia verrà anche da te.*

[99] Department of Mathematics, University of Rome La Sapienza, Roberto Marcolongo Archive: P. Burgatti to R. Marcolongo, Bologna November 1909; [1910]; 2.1.1912; 1.1.1913; 1.2.1913; 19.2.1914; [1915] and 8.2.1921.

[100] G. Peano to R. Montessus de Ballore, Turin 19.11.1897, in Le Ferrand (2012), p. 10: 'A propos de l'affaire Dreyfus, et analogues, dont s'occupent tous les journaux, quelques mes amis et moi nous désirons savoir à quelle race, ariane ou sémitique, appartiennent les différents mathématiciens français. [...] Je vous remercie de la nouvelle preuve d'amitié que vous me donnerez en répondant à ma question (Je suis de race latine, la même des gaulois, des goths, des slaves etc.).'

well as G. Fubini, who had the chair at the Polytechnic but was awarded the courses of Higher Analysis at the University, and C. Burali-Forti who taught at the Military Academy. However, I must say that—as, unfortunately, often happens—these professors did not get along very well with each other, and there was on the one hand the "Jewish" group (headed by C. Segre as long as he lived) with conservative tendencies, to which Fano and Fubini adhered, and on the other hand the "vectorial" group of Peano, Boggio and Burali-Forti, in which a spirit of Frond breathed. As for the noble C. Somigliana, a direct descendant of A. Volta, who was then Dean of the Faculty, he oscillated between one and the other, but leant towards the "Jewish" group, although he was not entirely immune to a bit of anti-Semitism.[101]

Roughly in the same period, anti-Semitism and incompetence merged in the malicious and ferocious review of the arithmetic text for primary schools entitled *Abbaco*, published by Levi in Parma in 1922 (Levi B. 1922):

An edifying booklet by Shylock's race, who believes, without ever having studied Pythagoras, that everything in the Universe can be reduced to numbers. Future middlemen and swindlers will learn that the Sum is the goal of life, that Subtraction is lawful and recommendable when it is practised on others, that the Multiplication of goods (with adulteration, watering down, etc.) is the soul of commerce and that the Division of profits is a harsh necessity for the partners of a company. Recently a learned Jewish mathematician, Beppo Levi, has printed a brand-new *Abbaco*, in which we finally learn the true logical and scientific method for teaching numbering to children. Here is an essay: after 1 comes 2, after 2 comes 3, after 3 comes 4..., after 9 comes 10 etc. And after Beppo Levi, of course, comes Einstein.[102]

Nominations of Fano and Fubini to the Royal Academy of Italy revealed the first traces of academic anti-Semitism in Turin: Fano and Fubini were Freemasons, it was rumoured—with voices arriving from Somigliana and Boggio, a pupil of Peano—

---

[101] Tricomi (1967), p. 18: *Fra i colleghi che trovavo nella nuova sede c'erano—oltre al Peano e al Terracini, che arrivò quasi assieme con me—C. Somigliana, G. Fano e T. Boggio nonché G. Fubini, che era titolare al Politecnico ma all'Università aveva l'incarico di Analisi superiore e C. Burali-Forti che insegnava soltanto all'Accademia militare. Debbo però dire che—come, purtroppo, spesso succede—questi professori non andavano molto d'accordo fra loro, e si aveva da un lato il gruppo "ebraico" (capeggiato finché visse da C. Segre) con tendenze conservatrici, cui aderivano Fano e Fubini, e, dall'altro lato, il gruppo "vettorialista" di Peano, Boggio e Burali-Forti, in cui spirava invece spirito di fronda. Quanto al nobile C. Somigliana, discendente diretto di A. Volta, ch'era allora Preside della Facoltà, egli oscillava fra gli uni e gli altri, propendendo però pel gruppo "ebraico", nonostante non fosse del tutto immune da un po' di antisemitismo.*

[102] Giuliotti and Papini (1923), pp. 51–52: *Libretto edificante del figliolame di Shylock, il quale crede, senza aver mai studiato Pitagora, che tutto, nell'Universo, si riduca a numeri. I futuri sensali e civaioli v'imparano che la Somma è il fine della vita, che la Sottrazione è lecita e raccomandabile quand'è esercitata sugli altri, che la Moltiplicazione delle merci (colla sofisticazione, l'annacquatura, ecc.) è l'anima del commercio e che la Divisione degli utili è una dura necessità per i soci di un'azienda. Ultimamente un dotto matematico ebreo, Beppo Levi, ha stampato un* Abbaco *nuovo di zecca, nel quale s'impara finalmente il vero metodo logico e scientifico per insegnare la numerazione ai bambini. Eccone un saggio: dopo l'1 viene il 2, dopo il 2 viene il 3, dopo il 3 viene il 4, ..., dopo il 9 viene il 10 ecc. E dopo Beppo Levi, naturalmente, viene Einstein.*

and for this reason their candidacies had to be opposed.[103] Shortly afterwards, the first anonymous complaint against a Jewish mathematician was received by the Ministry of National Education. The target was Fubini, accused of nepotism, favouritism, and Jesuitism of the worst kind (D1.6.2).

At the beginning of 1933, however, there were few people in Italy who were really informed about anti-Jewish persecutions at an international level. Only Levi-Civita, it can be said, had a grasp of the situation, because since 1908 he had collected M. Abraham's testimonies on anti-Semitism in German universities, had followed the polemics against Jewish physics since 1921, and had been kept informed of pogroms in Central and Eastern Europe by some of his correspondents, Alfred Rosenblatt first and foremost:

> I am oppressed by the political situation in which the Jews of my country find themselves. You have certainly heard of the Steiger trial in Lvov, a new page in this martyrdom. It is good that I have a place to escape to: in the domain of ideas, in the domain of the scientists of this radiant country that is Italy.[104]

> We do not massacre the Jews in Poland, as in Romania and now in Lithuania, but we deprive them of the means to live and we force them either to convert or to leave the country. This occurs firstly in the universities which are, here as in all Eastern Europe, the cribs of anti-Semitism.[105]

Despite the interaction with Rosenblatt or Zariski, who had fled to Italy to escape persecution, Enriques and Castelnuovo, as well as Fubini and Fano, had a poorly defined perception of the violence of the anti-Semitic phenomenon in Central and Eastern Europe. Moreover, being hyper-assimilated Israelites, they did not even feel a particular affinity with Jews from Poland or Ukraine, so much so that when Levi-Civita and his wife Libera visited the Jewish quarter in Krakow, they were amazed, and Libera noted in her travel diary: 'I saw Polish Jews, with sidecurls and overcoats like in Zangwill films!'[106]

---

[103] The curious thing is that, in fact, at that moment there were only two Freemasons among the mathematicians of Turin, and they were not Jews.

[104] ANL-*Levi-Civita*: A. Rosenblatt to T. Levi-Civita, Krakow 4.12.1925: *Je suis aussi opprimé par la situation politique dans laquelle se trouvent les juifs de notre pays. Vous connaissez certainement le procès Steiger à Leopol, nouvelle page de ce martyre. Il est bien que j'aie où m'enfuir dans le domaine de cette pensée claire et profonde, dans le domaine des idées des savants de ce pays rayonnant qu'est l'Italie.*

[105] ANL-*Levi-Civita*: A. Rosenblatt to T. Levi-Civita, Krakow 16.12.1927: *On ne massacre pas les juifs en Pologne, comme en Roumanie et maintenant en Lituanie, mais on leur ôte les moyens de vivre et on les forces de cette façon ou de se convertir ou bien de quitter le pays. Cela se rapporte en premier lieu aux Universités, qui sont ici comme dans toute l'Europe orientale les centres de l'antisémitisme.*

[106] Ceccherini Silberstein Family Archive: [Travel diary of Libera Trevisani Levi-Civita], Krakow 22.9.1925: *Ho visto gli ebrei polacchi con i ricci e la palandrana come in Zangwill!* Libera Trevisani referred to the Anglo-Jewish writer Israel Zangwill. Zangwill had published very famous novels: *Children of the Ghetto* (1892), *Ghetto Tragedies* (1893 and 1899), *Dreamers of the Ghetto* (1898), *Ghetto Comedies* (1907), and *The King of Schnorrers* (1894). Several of his stories had become films.

The arrival in Rome of the first displaced German scholars gave a new dimension to the situation.[107] Distinguished mathematicians, with whom Italians had been in contact for decades, had lost their chairs and been obliged to leave their homeland and look for a position abroad. Young people had to interrupt their studies and began a *peregrinatio academica* among the universities of the countries, like Italy, which still accepted Jewish freshmen. It was the arrival of the first refugees (both teachers and students like H. Lewy, I. Opatowski, W. R. Wasow, and R. Frucht), that convinced Italian Jewish mathematicians that nothing would ever be the same again. As Castelnuovo wrote to Volterra:

> Here in Rome, little is known about what is happening in Germany; refugee professors have not yet been seen. Some vague news has reached Levi-Civita, and other information has been brought from Paris to Fubini, who I saw here days ago. What is already known is enough to predict that no one will go to study mathematics, and perhaps not even physics, in German universities. And the Göttingen Mathematics School, after an uninterrupted century of glory, ends![108]

Assistance given to foreign Jews was, however, quite inexistent: requests from the Academic Assistant Council (later SPSL) were dropped by the Union of the Italian Jewish Communities and the Comitato Italiano di Assistenza agli Emigranti Ebrei (D1.6.1 and D1.6.3).[109] Einstein's endorsement with the International Student Service in 1934 was fiercely criticised by the National Student League, which alleged Einstein of murky contacts with fascist Italy (D1.6.4).

In 1933–34, the process of fascistisation was meanwhile moving towards completion. As of 1934, with the revision of the Statutes of the academies and scientific societies, the appointment of the elected president and vice-president was subordinate to the approval of Ministry of National Education (MEN). This in fact allowed the Duce to overturn the results of consultations if the elected post-holders were undesirable. So, for example, at the renewal of the Scientific Commission of UMI in 1934, Picone, Severi, and L. Fantappié were preferred to Vivanti and Volterra, despite the ballot results. The vice-presidency went to Bompiani, who had received half the number of Fubini's votes. Those who did not join the PNF could no longer sit on examining boards for national competitions, nor be publishers. Career advancements and transfers were impossible, because before calling the winners of a competition, a preliminary monitoring of the individual's political position was conducted. G. Ascoli, for example, who wanted to transfer from Pisa to Milan, and

---

[107] On mathematical emigration from Nazi Germany see Siegmund-Schultze 2009; Ash 1996; Ash (2008); Barrow-Green et al. (2011).

[108] ANL-*Volterra*: G. Castelnuovo to V. Volterra, Rome 14.5.1933: *Di ciò che succede in Germania qui si sa poco; professori profughi a Roma non si son visti. Qualche vaga notizia ha ricevuto il Levi-Civita, qualche altra ne ha portata da Parigi il Fubini che vidi qui giorni fa. Quel che si sa basta già a far prevedere che nessuno andrà più a studiar matematica, e forse nemmeno fisica, nelle Università tedesche. E la Scuola matematica di Gottinga, dopo un secolo ininterrotto di gloria, si chiude!*

[109] The Italian Committee for Assistance to Jewish Emigrants was founded in Trieste in 1921 to assist emigrants from Eastern Europe to reach Palestine. It was active until 1940.

was among the very few mathematicians who had not joined the Party in 1933, was forced to draw up a very humbling declaration to explain the reasons that had prevented him from enrolling (D1.6.5). The few openly anti-fascists, or those intellectuals suspected of being so, were no longer allowed to represent Italy abroad in scientific meetings and began to encounter problems in the procedure for the annual renewal of passports. Until the end of 1936, however, the plight of Jewish scholars was the same as their Aryan colleagues. If they did not deal with politics, they did not face—a priori—any discrimination.

Documents

D1.6.1 W. Adams [Academic Assistance Council] to F. di L. Ravenna, London 7.9.1933

SPSL, *Italy*, fols. 426–427
Dear Sir,[110] I wish to bring to your attention the work of the Academic Assistance Council which has been formed in England for the relief of dismissed university workers. At the moment it is chiefly concerned with those academic persons who have been dismissed from positions in Germany, although it has never been the intention of the Council to confine its assistance entirely to Germans and certainly not entirely to Jews. I enclose a copy of the first appeal that this Council issued. As a result of this appeal a small sum of money was collected sufficient to enable the Council to establish an office to collect information about the university workers dismissed in Germany and to make a beginning with practical assistance. The policy of the Council is to discover offers of hospitality in universities and learned institutions and then to make a maintenance grant to the refugee to enable him to accept an offer of hospitality. The Council is not attempting to place these refugees in the ordinary teaching positions within the regular academic field. It feels that to do so would conflict with the interests of English academic workers. Along these lines the Council has already made maintenance grants to enable 28 German scholars to accept offers of hospitality within England. The Council is co-operating very fully with a similar assistance committee which has been formed in America, a copy of whose appeal leaflet I enclose. The Council is also co-operating with similar academic committees in Belgium, Holland, France, Czecho-Slovakia, Poland, Scandinavia and Switzerland. It would welcome any co-operation that you were willing to offer within Italy. At the moment the Council has no contact with an academic group or committee interested in this matter within Italy. I should be grateful therefore if you could explain what the Union of the Italian Jewish Community is doing in Italy and along what lines we

---

[110]Felice di Leone Ravenna (1869–1937), president of the Unione delle Comunità Israelitiche Italiane (UCII) from 1930 to 1937.

could assist you or you could assist us. If there is any further information that you require, I shall be very glad to provide it. Yours faithfully, Walter Adams, General Secretary[111]

D1.6.2 Anonymous Complaint Against G. Fubini, Turin 7.10.1933

ACS-*Fubini*, fol. 1r

To His Excellency the Head of Government, Duce Benito Mussolini. Your Excellency desires that honest citizens and sincere fascists come out of the Royal University, people ready to use in the service of the Fatherland, the gifts of their minds and the strength of their youth. The vast majority of students fully correspond to the will of H.E. since they are educated to acquire the love of the homeland and of the regime by worthy masters. Not in all Universities, however, it is disheartening to see, does that happen. In the University of Turin, School of Mathematics, a few Jewish teachers, Freemason-socialists headed by the all-powerful Prof. Fubini, with a skill and Jesuitism of the worst kind, resort to every measure possible to demolish that which the Regime, with titanic efforts, is building. Furthermore, said Faculty is home to tyranny of all types: favoured are the protected ones, the disciples who must one day continue the infamous work, destroyer of the homeland, whilst those who they know they cannot draw into their circle are oppressed, boycotted, and damaged in countless ways. Lawlessness, persecution, and nepotism are moved under the aegis of a good, magnificent Rector,[112] so good as to be the laughingstock of these enemies of the homeland. A severe meticulous investigation will bring to light what I have had the honour of exposing to Your Excellency. An Old Black Shirt, anonymous for obvious necessity.

A Sua Eccellenza il Capo del Governo. Cavaliere Benito Mussolini. Vostra Eccellenza desidera che dalla Regia Università, escano cittadini onesti e fascisti sinceri, pronti a mettere a disposizione della Patria, le doti della loro mente e le forze della loro giovinezza. La grande maggioranza degli studenti corrisponde in pieno al volere di V. E. poiché sono educati all'amore della patria, e del regime da degni maestri. Non in tutte le Università però, è sconfortante constatarlo, succede così. Nella R. Università di Torino, Scuola di matematica, pochi professori ebrei, social-massoni capeggiati dall'onnipossente prof. Fubini, con un'arte ed un gesuitismo della peggior specie si adoperano con ogni mezzo, per demolire quanto il Regime con titaniche imprese, sta costruendo. In detta Facoltà si verificano inoltre soprusi di ogni risma: sono favoriti i protetti, i discepoli che

---

[111] Walter Adams (1906–1975), general secretary of the Academic Assistance Council, then Society for the Protection of Science and Learning (SPSL) from 1933 to 1939.

[112] Silvio Pivano (1880–1963), Rector of the University of Turin in the years 1928-1937.

dovranno un giorno continuare la opera infame, disfattrice della patria, e sono oppressi, boicottati, danneggiati in ogni modo quelli che essi sanno di non potere attirare nella loro cerchia. Illegalità, persecuzioni e simili sono mosse sotto l'egida di un Magnifico Rettore buono, tanto buono da risultare lo zimbello di questi nemici della patria. Una severa minuziosa inchiesta metterà alla luce del sole quanto ho avuto l'onore di esporre all'Eccellenza Vostra. Vecchia Camicia Nera, anonima suo malgrado per evidente necessità.

D1.6.3 Comitato Italiano di Assistenza agli Emigranti Ebrei to Academic Assistance Council, Trieste 24.10.1933

SPSL, *Italy*, fol. 428
To the General Secretary of the Academic Assistance Council, We beg to confirm the receipt of your favour of the 13 October 1933 with the enclosed "Questionnaire" and we inform you that no displaced university or school teachers or other professional workers as such for the purples of taken up an employment and their branch of activity have applied to us up to-day. We do not see also any possibility that this applicance [sic!] could be allowed to take up any employment here in Italy as formers are forbidden to work as professionals. The German Refugees assisted by our Committee were mainly immigrants on the way to Palestine or Refugees surching [sic!] any other employment in Italy. Only few Refugee students have been assisted by our Committee. Yours faithfully, Comitato Italiano di Assistenza agli Emigranti Ebrei

D1.6.4 J. Cohen and T. Draper [National Student League] to A. Einstein, Princeton 15.10.1934

ECADFS Records, *Italy*, fols. 1r-2r-3r
Dear Professor Einstein: We noted with amazement your endorsement of the International Student Service in its appeal for funds to help refugee German students. We are not amazed because you took steps to help exiles from fascism. We are amazed at the lack of investigation implicit in your endorsement because we are certain that a careful investigation of this organization will convince you that it is a foe, not a friend, that confronts us. We do not consider ourselves the ones to instruct you that there is opposition and "opposition"—the kind that fights and the kind that serves as a camouflage for the enemy. The International Student Service has been commissioned by James G. McDonald, High Commissioner for German Refugees for the League of Nations, with the exclusive responsibility for raising money for the students. We would raise no objections if it were clear that the organization in question would carry on an honest and unremitting opposition to fascism—not only in Germany but wherever fascism raises its brutal and

bloody head. Quite the contrary can be proven of the International Student Service. In an official report entitled "Academic Relief of International Student Service for students unable to continue their studies in German universities", it is unmistakably reported that German refugees have been helped out of the Hitler frying-pan into the Mussolini fire. On page 6 of this report, it is proudly stated: "We have succeeded in finding a collaborator in Italy [Madame Catelli] with excellent relations to the University of Rome. She is able to advise students as to conditions of life and study in Italy and to assist them in carrying through the formalities necessary for enrolment in an Italian University." On page 2, it is reported that three students were actually handled through the Rome office. We can be sure that much less difficulty would have been encountered finding a collaborator in the Soviet Union, but the report does not mention them at all. We consider this question a question of principle. The number is unimportant. What is important is that an organization that will rescue students from German fascism only to put them at the mercy of Italian fascism is no friend but a dangerous enemy. Fascism everywhere, in its very essence, means war, oppression and hunger for the masses of people. Because Italy has not received the publicity in respect to violent anti-Semitism that is a special curse in Germany, shall we be blind to the indignities heaped upon such men as Arturo Toscanini, Benedetto Croce, Gaetano Salvemini and hundreds of others—to mention only intellectual atrocities? We ask you to reconsider and retract your endorsement of the International Student Service as a fit instrument given the exclusive responsibility for raising funds for our exiled German students. It must renounce its claim to such a high trust by its alliance with fascism in Italy. We must not let the left hand go ignorant of what the right hand is doing. We cannot oppose fascism in Germany if we accept fascism in Italy as our ally. Your voice is a powerful one. A word from you endorsing this organization strengthens it. A word from you renouncing it will strip it of the pretensions without which it would be powerless. Sincerely yours Joseph Cohen Executive Secretary National Student League; Theodor Draper Editor, Student Review

October 17, 1934
My dear Dr Einstein: A news release of the National Student League has just been called to my attention in which they protest your support of International Student Service because of our alleged contact with fascist Italy. I regret exceedingly that you are bothered with this matter because you were so generous in your willing-ness to have us use your testimonial. I do not think I need to tell you much about ISS because you know a great deal about its work in Europe during the last ten years. Furthermore, you know Dr. Walter Kotschnig, Kurt Berlowitz and many of your colleagues in Germany who are active in ISS work. However, I had breakfast this morning with Walter Kotschnig and he asked me to tell you that ISS does not have a committee in Italy because there can be no committee organized there on the basis of the principles of ISS. To this extent ISS does not approve of what is happening in Italy. On the other hand, inasmuch as ISS is a

non-political organization trying to deal as impartially as possible with the relief problem as far as it relates to students, it did assist three German students who were in Italy, in finding places in an Italian University. This was definitely in response to their own request. The vast majority of students aided by ISS are placed in France, England, Holland, Palestine and the United States. We do not have any office in Rome as is implied in the release of the National Student League. If you decide that the letter from the National Student League is worth answering, I hope you will send us a copy of your answer. If there are any other questions you would like us to answer on this question, we would be glad to do so. Every good wish. Sincerely yours Francis A. Henson

D1.6.5 Declaration by G. Ascoli, Milan 1.12.1934

APICE-*Ascoli*, fol. 1r-v
In submitting the joint application for enrolment in the university Section of the A.F.S. [Associazione Fascista della Scuola] I would like to declare that, in the impossibility of obtaining today (due to the suspension of enrolment) the membership card of the P.N.F., as would be my very fervent desire, I ask that this application be considered as a testimony and guarantee of my full adherence to the fascist regime and of my commitment to being in all cases a faithful and disciplined soldier. Although I have always nourished, as every good citizen must do, a keen interest in public affairs, I never took an active part in the political life of the nation, for reasons of attitude well common to men of study. However, I strongly sympathised with the tendencies directed towards the elevation of the humblest classes (an indispensable elevation functional to the formation of a solid civil and national conscience), and I have always deplored the fact that they often take on partisan and deleterious stands, and that, on the other hand, the ruling classes do not understand the fundamental importance of these problems that the war and the post-war period have tragically brought to light. I declare with profound satisfaction that the work of Fascism in this field, aimed at introjection in consciences of the true unity of Italy, a multifaceted work in which nothing escaped of what is needed for social pacification and for the moral and material health of the people, a work which materialised in institutions of formidable grandeur, has convinced me, destroying any possible prevention, that from the Revolution taking place before our eyes, a miracle that came out of the hands of a Man [Mussolini], a new civilisation will arise, in which the nagging economic and social problems of our age will be brilliantly composed in a superior unity. Entirely absorbed in my life of studies, before and after 1930, a time when, after a few years of intense work, I won a university competition, but which was also strongly occupied by the family (which I formed a few years ago), I missed the deadlines for the submission of application for enrolment in the P.N.F. Therefore, I find myself today in the impossibility of validating, with an explicit act of adhesion, my present spiritual situation, which would place me in the conditions

that the Regime requires from those who intend to collaborate effectively in the grandiose work that it develops. Allow me, however, to recall that within the limits of my modest possibilities I show my full sympathy to the institutions of the Regime by subscribing to the National Welfare Work, placing myself at the disposal of the G.U.F. [Gruppi Universitari Fascisti] of Pisa for the preparation of the *Littoriali* of culture and art, and offering (as a gift to poor students) ten copies of my lithographed *Lezioni*.[113] I hope that this frank declaration, together with my past as a public teacher, in twenty-five years repeatedly praised and rewarded, and my work as a soldier and artillery officer, on the Karst and on the Piave, decorated with the War Cross, wounded in the line of fire, as proved by the joint state of service, can be considered sufficient proof of my fervent faith as an Italian and fascist and of my decisive will to serve in Fascism, under the orders of the Duce, for the sacred future of the Fatherland, and to help produce under his guidance a higher ideal of human coexistence. 1 December 1934—XIII

Nell'inoltrare l'unita domanda di iscrizione alla Sezione universitaria dell'A.F.S. tengo a dichiarare che, nella impossibilità di ottenere oggi, per la sospensione delle iscrizioni, la tessera di appartenente al P.N.F., come sarebbe mio vivissimo desiderio, vorrei che la domanda stessa fosse considerata come testimonianza e garanzia della mia piena adesione al Regime fascista e del mio impegno di esserne in ogni caso milite fedele e disciplinato. Pur nutrendo sempre, come deve ogni buon cittadino, vivo interesse per la cosa pubblica, non ho mai, per ragioni di carattere ben comuni negli uomini di studio, preso parte attiva alla vita politica della Nazione. Ho tuttavia vivamente simpatizzato con le tendenze dirette all'elevazione delle classi più umili, elevazione indispensabile perché possa formarsi una solida coscienza civile e nazionale, e ho deplorato sempre che esse assumessero spesso forme faziose e deleterie, e che, d'altra parte, le classi dirigenti non comprendessero l'importanza fondamentale di questi problemi che alla guerra e al dopoguerra spettava portare tragicamente alla luce. Dichiaro con profonda soddisfazione che l'opera del Fascismo in questo campo, diretta a porre nelle coscienze la vera unità d'Italia, opera multiforme cui nulla è sfuggito di ciò che vale per la pacificazione sociale e per la salute morale e materiale del popolo, e che si è concretata in istituzioni di mirabile grandiosità, mi ha convinto, distruggendo ogni possibile prevenzione, che dalla Rivoluzione che sta sotto i nostri occhi attuandosi, miracolo uscito dalle mani di un Uomo, sorgerà una civiltà nuova, in cui gli assillanti problemi economici e sociali dell'età nostra saranno genialmente composti in una superiore unità. Assorbito interamente nella vita degli studi, prima e dopo il 1930, epoca in cui, dopo pochi anni di lavoro intensissimo conseguii la vittoria in un concorso universitario, occupato inoltre molto dalla famiglia, da pochi anni formata, ho lasciato trascorrere i termini per la presentazione della domanda di iscrizione al P.N. F., e mi trovo così oggi nella impossibilità di convalidare con un atto esplicito di

---

[113] G. Ascoli, F. Cecioni, O. Nicoletti, *Lezioni di Analisi Matematica*, Pisa, Litografia Gozani, 1934.

adesione quella che è la mia presente situazione spirituale, e che mi porrebbe nelle condizioni che il Regime richiede da chi intende collaborare fattivamente all'opera grandiosa che esso compie. Mi sia permesso tuttavia ricordare che nei limiti delle mie modeste possibilità ho mostrato alle istituzioni del Regime la mia piena simpatia sottoscrivendo alle Opere assistenziali, ponendomi a disposizione del G.U.F. di Pisa per la preparazione dei Littoriali *della cultura e dell'arte, e facendogli dono, per l'opera di assistenza degli studenti poveri, di dieci copie delle mie* Lezioni *litografate. Io spero che questa franca dichiarazione, insieme al mio passato di pubblico insegnante, in venticinque anni più volte lodato e ricompensato, alla mia opera di combattente, sul Carso e sul Piave, in qualità di soldato e di ufficiale di Artiglieria, decorato di Croce di guerra, ferito sulla linea del fuoco, comprovato dall'unito stato di servizio, possano al bisogno essere considerati prova sufficiente della mia fervida fede di italiano e di fascista e della mia decisa volontà di servire nel Fascismo, agli ordini del Duce, l'avvenire sacro della Patria, e di realizzare con esso un più elevato ideale di convivenza umana. 1° Dicembre 1934—XIII*

Chapter 2
The Fateful Year 1938: The Persecution of the Italian Jews

2.1 Many, Cunning and Powerful: The Words and Tropes of Fascist Propaganda

From 1936 onwards, with the colonial war being waged in Ethiopia, the first signals of biological racism began to appear on the Italian cultural scene, with regard to the danger of the *meticciato* (mestizaje). Meanwhile, fascist Italy was building closer ties with Nazi Germany. The visit of Goebbels and Hitler to Italy in September 1936 and May 1938, respectively, symbolically sealed this special relationship, sanctioned the following autumn by the signing of a treaty of friendship known as the Rome–Berlin Axis. However, the question of race regarding Jews, and in general anti-Jewish racism, remained in the background until the end of 1937.

In the first few months of 1938, a violent campaign of anti-Semitic indoctrination, orchestrated by the upper echelons of the party (the Minister of National Education G. Bottai, to name but one), began to monopolise the Italian press. Some of the most sinister intellectuals of the regime, including P. Orano, G. Preziosi, N. Pende, G. Landra, and J. Evola, specifically recruited, produced, and issued in a few weeks an avalanche of publications, including the re-edition of the *Protocols of the Elders of Zion* (in *La vita italiana*, 1938). In this mass of writings, all the traditions of anti-Semitic fraudulent thought, both ancient and recent, converged: the anti-Jewish prejudice of Catholic origin and biological racism, together with the deliria against the Jewish socialist and Masonic international demo-plutocracy.[1] Clumsily thrown together, these leitmotifs often gave rise to grotesque phenomena. Thus, for example, Orano devoted an entire chapter of his volume *Gli ebrei in Italia* (Roma, Pinciana, 1938) to demonstrating how, thanks to an extraordinary gift for the

---

[1] Israel and Nastasi (1998); Israel (2010). About the racial legislation in Italy see also, inter alia, Sarfatti (1988); Vesentini (1990); Nastasi (1998); Cavaglion and Romagnani (2002); Guerraggio and Nastasi (2005a); Sarfatti (2007b); Guerraggio and Nastasi (2007); Guerraggio and Nastasi (2018b); Vercelli (2018); Piazza (2021).

E. Luciano, *The Jewish Mathematical Diaspora from Fascist Italy*, Science Networks. Historical Studies 64, https://doi.org/10.1007/978-3-031-64896-0_2

mathematical sciences, Israelites had succeeded in monopolising this field to the detriment of Aryans. Among the names he mentioned in support of this thesis was Cremona, but the argument was so confused that Cremona's daughter, Itala, replied with indignation:

> You want to demonstrate, actually, that the contribution made to science and art by Italians of Jewish origin, in the last half century, is minimal. Very well but [...] my father was not a Jew but a Christian-Catholic.[2]

As a matter of fact, the Italian mathematical community did not introject the key instances of the *Deutsche Mathematik* or *Deutsche Physik* à la Bieberbach and Dingler.[3] It is true that some anti-Jewish stereotypes began to surface in papers by Bompiani, Ettore Bortolotti, and Severi, but in the majority of cases, it was a matter of omitting the name and contributions of a particular Jewish individual, usually considered a rival in his own field of expertise. Such omissions were generally ascribed by the injured party to academic rudeness which had little or nothing to do with the question of race. In Paris, Severi's talk *Peut-on parler d'un esprit latin même dans les mathématiques?*[4] had aroused sarcasm, more than fear. Fubini, Castelnuovo, and Volterra had commented on its many short cuts in logic and obvious stretches: the subtle logician Peano who was passed off as emblematic of the strongly synthetic Latin spirit and love for intuition! The first essays of systematic rewriting in an Aryan key of the history of recent and contemporary Italian mathematics date from 1938 onwards, and in many cases were constructed as a form of individual *damnatio memoriae*. Thus, for example, it is easy to argue that when Severi accused Jewish geometers of having filled Italian science with their ancient Eastern and Masonic traditions,[5] he was speaking of Enriques mainly. Similarly, Bortolotti did not miss any opportunity for maligning the hated Yids, while he appreciated Castelnuovo, the first who showed having read and appreciated his studies.[6]

Until February 1938, Mussolini publicly reassured the Italian Jews, some of whom occupied top positions in the State hierarchy, and foreign observers. In a renowned *Nota* on *L'Informazione diplomatica* (February 16, 1938) the argument was made that the fascist government had never thought of, nor was thinking of adopting political, economic, and moral measures against the Jews as such, except, of course, in the event that they were elements hostile to the regime, although it

---

[2] I. Cremona Cozzolino to P. Orano, end of March 1938, in Avagliano and Palmieri (2013), p. 153: *Lei vuol dimostrare, ciò che è, che il contributo dato alle scienze e all'arte dagli italiani di origine ebraica, nell'ultimo mezzo secolo, è minimo. Benissimo ma [...] mio padre non era ebreo ma cristiano-cattolico.*

[3] On mathematics and physics during Nazi Germany see, inter alia, Forman (1973); Beyerchen (1977); Segal (2003).

[4] Severi (1935), pp. 581–589.

[5] Severi (1941), p. 137–140.

[6] AS-UMI: E. Bortolotti to E. Bompiani, Bologna 13.12.1938, *gli odiati Giudei; il primo che mostri di aver letto e apprezzato quei miei studi.*

reserved the right to 'monitor the activity of the Jews recently arrived in our country and to ensure that the role of the Jews in the life of the whole nation be not disproportionate to the intrinsic individual merits and to the numerical importance of their population'.[7] This is exactly what was to become the strong theme of anti-Jewish propaganda in Italian high culture: the disproportion embodied in the high number of Jewish scholars in universities, research institutes, clinics, libraries and archives, along with its natural corollaries: their 'infiltrations' (*infiltrazioni*) into the vital ganglia of the country and the 'tampering' (*manomissione*) with Italian science, carried out by Jews in the long phase of tranquillity they had enjoyed from the Risorgimento onwards.[8] These stereotypes became recurrent in the Italian mathematical environment, particularly in contexts such as Rome, Bologna, and Turin where rumours circulated of the existence of shadowy and very powerful Jewish coteries revolving around Enriques and Castelnuovo, Levi and Segre, Fano and Fubini. It was, clearly, a form of opportunistic anti-Semitism exercised by mathematicians who had proclaimed themselves protégés of Jews, who considered them anything but inferior yet intended to appropriate themselves of their power and prestige in the academic world (chairs, positions in academies, scientific societies, editorial committees, publishing houses, etc.). So for example, in three months, from March to May 1938, the correspondence of the UMI Scientific Board (Berzolari, Bompiani, Bortolotti, and G. Sansone) accumulated dozens of comments, one more poisonous than the other, against Levi and Segre (see D2.1.1 and D2.1.2). These were accused of all sorts of offences: meddling with the elections for the renewal of officers of the UMI, transformation of the UMI into a Bolognese camarilla, plagiarism of the elderly President Berzolari, monopoly of the *Bollettino* and its destruction with a wicked editorial policy, manipulation of university calls to the detriment of Aryan candidates. To these should be added, in the course of 1938 and 1939, accusations of defeatism in the cultural field, victimisation, and boycotting of the Volta 1939 conference *Matematica contemporanea e sue applicazioni* (see §2.8).

The prevailing opportunist character of Italian anti-Semitism justifies the presence of considerations of the same tenor long after the promulgation of the racial laws. For example, venting with Somigliana about his failure to be elected at the Academy of XL, in 1942, Berzolari insinuated that it was not surprising that U. Amaldi had been preferred to him, because Amaldi, despite being scientifically inept, was the 'amanuensis and minion of some Jews, especially of Levi-Civita and Enriques, which, really, has earned him a lot of money. Well, as a prize for such activity he was appointed pontifical academician, two years ago, member of the Academy of Sciences in Turin, and recently of the Academy of XL. One of my

---

[7] *L'Informazione diplomatica* 1938, in Fabre 2007, p. 46: *vigilare sull'attività degli ebrei venuti di recente nel nostro Paese e far sì che la parte degli ebrei nella vita complessiva della Nazione, non risulti sproporzionata ai meriti intrinseci dei singoli e alla importanza numerica della loro comunità.*

[8] Landra (1939).

colleagues defined the last appointment as scandalous, but perhaps this is a bit exaggerated'.[9]

Documents

D2.1.1 G. Sansone to E. Bompiani, Florence 31.3.1938

AS-UMI, *Bompiani correspondence*, p.c.
Dear Bompiani, [. . .] I saw from a letter from Ettore to Enea that the Bolognese binomial does not back down; basically, they want to act as if there were no reactions from you, Tonelli, Scorza and myself.[10] On 24 April, we will go to Bologna and if we find that the indications for officers' elections are compromised by a prior fraudulent referendum, we will state our case in the meeting and repeat it to all honest people.[11] The so-called persecuted, when they have the opportunity, always know how to find appropriate circumventional strangulation manoeuvres. [. . .] Yours faithfully Giovanni Sansone

Caro Bompiani, [. . .] Ho visto da una lettera di Ettore B. ad Enea che il binomio bolognese non disarma; vuole agire in sostanza come se non ci fossero le lettere tue, di Tonelli, di Scorza e la mia. Il 24 aprile andremo a Bologna e se troveremo che le indicazioni per le cariche sociali sono compromesse da un preventivo referendum diremo il fatto nostro in seduta e lo ripeteremo a chi di ragione. I così detti perseguitati, quando possono, sanno sempre trovare appropriate manovre legalmente strangolatorie. [...] Affezionatissimo Giovanni Sansone

---

[9]L. Berzolari to C. Somigliana, Pavia 3.10.1942 in D'Agostino 2007, pp. 28–29: *amanuense e tirapiedi di alcuni ebrei, soprattutto Levi-Civita e Enriques, il che, veramente, gli ha fruttato molti quattrini. Ebbene, come premio di tanta attività è stato nominato accademico pontificio, due anni fa corrisp. dell'Acc. di Torino, e recentemente alla Società dei XL. Uno dei predetti colleghi ha qualificato l'ultima nomina come scandalosa, ma forse è un po' esagerato.*

[10]The Bolognese, or Judeo-Bolognese binomial, consisted of Beppo Levi and Beniamino Segre, accused by Bompiani to be 'the two real puppeteers' of the UMI. In this letter Giovanni Sansone refers to the manoeuvres which preceded the elections of the UMI Executive Board. Bompiani, Levi, Segre, Ettore Bortolotti, and Gaetano Scorza were all members of the Scientific Board. Sansone and Tonelli were running for election. Enea Bortolotti, professor of Analytical and Descriptive Geometry at the University of Florence, was Ettore's son.

[11]Finally, Luigi Berzolari was elected president and Pietro Burgatti vice-president of UMI. The latter, however, died suddenly on May 20, leaving his position vacant. The election result was submitted to the Minister of National Education, G. Bottai, who exercised his power by confirming Berzolari president, and appointing Bompiani vice-president, although the latter obtained only eight votes (Guido Fubini got 74 votes and Annibale Comessatti 61). Moreover, Bottai excluded all Jewish mathematicians from the UMI Scientific Commission: Segre, Levi, Fubini, and Levi-Civita who got the most votes.

D2.1.2 E. Bortolotti to E. Bompiani, Bologna 8.4.1938

AS-UMI, *Bompiani correspondence*, p.c.
Dear Bompiani, I had a circular letter from Berzolari, which you will also have received, and which circumvents your proposal, as regards the sitting of the 24th. Meanwhile, answers to Berzolari's letter are already arriving, together with the voting cards. As Secretary (am I this??) should I keep these letters that will be opened on the election day? [...] In any case, we must try to make sure that there is a good number of our group at the assembly, because I imagine that Judah will already have made the call and beaten the Chitet!! I am always firm in the purpose of no longer holding down the position of secretary, no matter what happens! Next year we will have Graffi here, who could be appointed as secretary.[12] He is a serious person. (I don't know if he is enfeoffed to Judaism, he must be half Jewish!!). [...] Warmest wishes by your old friend Ettore Bortolotti

Caro Bompiani, ho avuto una lettera circolare di Berzolari, che avrai avuto anche tu, e che elude la tua proposta, per quel che riguarda la seduta del 24. Intanto cominciano già ad arrivare le risposte alla lettera di Berzolari, cioè le schede di votazione. Come Segretario (??) debbo conservare io le lettere che si apriranno il giorno della votazione? [...] In ogni caso bisogna cercare di far in modo che all'assemblea ci sia un buon numero dei nostri, perché mi immagino che avranno fatto la chiamata e battuto il Chitet!! Io sono sempre fisso nel proposito di non occuparmene più dell'ufficio di segretario, qualunque cosa accada! Qui avremo l'anno venturo anche Graffi che potrebbe fare il segretario, ed è persona seria. (Non so se infeudata al giudaismo, deve essere mezzo ebreo!!). [...] Saluti affettuosi dal tuo vecchio amico Ettore Bortolotti

2.2 'In the Beginning, We Didn't Understand Where It Would Go'[13]

The anti-Semitic press campaign was closely followed by Italian mathematicians. In Turin, for example, in the Artoms' living room, faculty members and intellectuals gathered to read the crazy interventions of the regime scribblers and discussed the meaning of modern anti-Semitism, which they described as an occult international force analogous to Freemasonry or Bolshevism, trying to evaluate the difference between Italian anti-Semitism and German race prejudice:

---

[12] In fact, Bortolotti served as secretary of UMI from 1922 to 1945. Dario Graffi, professor of Rational Mechanics at the University of Bologna from 1939 to 1975, never assumed this position.

[13] Emilio Artom (1940–41), in Treves (1954), p. 62: *In principio non capivamo dove si sarebbe arrivati.*

They are trying to devalue our old irredentism, which the author [Yvon de Begnac] calls "the peaceful and Jewish irredentism of Trieste." And so, the work of Italian-ness of our *Triestini* is liquidated. [...] The article seems generally fair, moderate, honest: why did the author want to malign the Jews? To follow the current mood? But who wanted to revive it in Italy? Those who want to believe that antisemitism is imposed by Germany do not persuade me. German racism, for example, finds some defenders, but also finds many open opponents. Why do we never listen to a voice of defence? It is incomprehensible.[14]

Intellectuals who were fiercer sent newspapers and periodicals their response to attacks or intensified their activities in promoting Jewish literature, culture, and language. Artom even dedicated himself to writing a book on anti-Semitism. The writing process was quick but having concluded, he understood that 'it would be useless to publish it. Rereading some works written a century ago in our favour by non-Jews, I understood that only considerations of humanity and justice are exposed in them: today they would not find ears open to listen to them'.[15]

In Rome, Levi-Civita saw the escalation of anti-Semitic persecution through the eyes of his foreign students and their pleas for help. Rosenblatt, who had emigrated to Peru in 1936 thanks to his recommendations, kept him informed of the 'aberration of consciences' in Europe and Latin America: 'We read with emotion what is happening now in Romania, and we see that anti-Semitism is infiltrating even in Peru. The situation is, moreover, different here, given the small number of Jews, but the main newspaper, *Comercio*, practices a Nazi policy'.[16] M. Haimovici, a former pupil of Levi-Civita in the academic year 1932–33, who was continuing work in the field of the mechanics of fluids under his mentorship, told him about the situation in Romania and asked him for help to emigrate (see D2.2.1). Levi-Civita, who planned to travel to Romania in June, wasted no time and suggested that Haimovici play the Argentinian card, turning to E. Terradas. Due to the long time necessary for overseas correspondence ('a letter goes from here to Argentina—I think—in just under a month',[17]) Levi-Civita's recommendation was not successful and Haimovici would

---

[14]Emilio Artom (1938), in Treves (1954), p. 43: *si cerca di svalutare il vecchio irredentismo. E l'autore non manca di chiamarlo «il pacifico e giudaico irredentismo triestino.» E così è liquidata l'opera di italianità dei nostri Triestini. [...] L'articolo pare in fondo equo, moderato, onesto: e perché si è voluto dire una malignità contro gli Ebrei? Per seguire la corrente oggi di moda? Ma chi l'ha messa di moda in Italia? Chi vuol credere che sia imposta dalla Germania non mi persuade. Il razzismo tedesco, p. es., trova qualche difensore, ma trova anche molti apertamente contrari. Perché noi non ascoltiamo mai una voce di difesa? Non si capisce.*

[15]Emilio Artom (1938), in Treves (1954), p. 37: *non servirà a niente pubblicarlo e non lo pubblicherò. Rileggendo certe opere scritte un secolo fa a nostro favore da non Ebrei, ho capito che solo considerazioni di umanità e giustizia vengono esposte in esse: oggi non troverebbero orecchi aperti ad ascoltarli.*

[16]ANL-*Levi-Civita*: A. Rosenblatt to T. Levi-Civita, Miraflores 11.1.1938, in Nastasi and Tazzioli (2003), pp. 313–14: *Nous lisons avec émotion ce qui se passe maintenant en Roumanie et nous constatons que l'antisémitisme s'infiltre même dans le Pérou. La situation est, d'ailleurs, différente ici, vu le petit nombre des juifs, mais le journal principal Comercio pratique une politique nazi.*

[17]ANL-*Levi-Civita*: M. Haimovici to T. Levi-Civita, Jassy 4.5.1938: *una lettera va da qui fino in Argentina—credo—[in] poco meno di un mese.* He continued his academic activity until 1940, when he was forced to leave the University. He was, for a while, the Director of the Jewish high

face very difficult years of persecution and hiding.[18] In the same period, Hilda Geiringer also asked Levi-Civita and Volterra for help to find accommodation in the United States, realising that her position in Istanbul was becoming increasingly precarious (see D2.2.2).[19]

With just a few exceptions, in the beginning most Italian Jewish mathematicians did not understand which direction the situation would take. No one considered it plausible that events like those occurring in Central and Eastern Europe might take place in Italy. The widespread impression, or maybe the collective illusion, was that the anti-Semitic wave in Italy was an episode of degeneration of logic, a transitory form of collective hysteria, unpleasant, blameful but devoid of irreversible effects. The few declared anti-fascists, such as Volterra or Colonnetti or Errera, talked of their disgust for the regime, of the isolation to which they had been condemned for lack of compliance, but never spoke of anti-Semitism.[20] Volterra, in those months, was in full swing: he was expected in Scotland to attend the James Gregory tercentenary and to receive an honorary degree conferred on him.[21] In his itinerary, he counted on spending a few days in Paris, then in Cambridge, and was planning a further trip to Paris for the autumn of 1938 to attend the meeting of the *Bureau de Poids et Mesures* and to give a series of lectures at the *Institut Henri Poincaré*. In none of his letters to foreigners did Volterra mention the racial question or the spew of anti-Semitic sentiment in Italy.

Fervent fascists, whether they were so by calculation or conviction, appeared even more unaware of the transcending situation. The case of Giorgio Mortara is emblematic. An economist, full professor of Statistics at the University of Milan and the director of the *Giornale degli Economisti Italiani*, Mortara was invited by Tibal to give a lecture at the European Centre of the Carnegie Foundation in Paris. As per the law, he asked MEN for authorisation and informed the Rector of the invitation, specifying that he had written to Paris a general acceptance, subject to the possibility of going there. The opinions of both the Ministry of National Education and the Ministry of Foreign Affairs, which were perfectly aware that Mortara was a Jew, were entirely positive: 'indeed, it will be appropriate that Prof. Mortara takes advantage of the opportunity to exposit and support fascist views and arguments

---

school in Iasi. The encouragements and moral support of his professors, as well as that of his colleagues persuaded Haimovici to continue his scientific research. He returned to his position after the collapse of the pro-Nazi regime in Romania on 23 August 1944.

[18] See *Mendel Haimovici* ... 2006, pp. 233–34.

[19] Hilda Geiringer (1893–1973), also known as Hilda von Mises and Hilda Pollaczek-Geiringer, was an Austrian mathematician. Dismissed from the University of Berlin in 1933, she went to Brussels, before following von Mises to Istanbul, where she continued to research in applied mathematics, statistics, and probability theory. In 1938, Geiringer left Turkey and went to Bryn Mawr College in Pennsylvania, where she was appointed to a lecturer position.

[20] See for example, in ANL-*Volterra*, G. Errera to V. Volterra, Turin 7.4.1938.

[21] ANL-*Volterra*: V. Volterra to E.T. Whittaker, [Rome, May 1938].

in that environment.'[22] Bottai's authorisation, however, was sent late, which is why Mortara 'for not being indiscreet or late in understanding what seemed the clear meaning of silence'[23] informed his French colleagues that he must forgo the conference. When he finally received (after three months) the affirmative answer of the Minister, Mortara immediately made amends. Amid the anti-Semitic press campaign, he wrote to Tibal, apologising and asking if it was possible to return to the original plan, including his lecture in the programme. At stake, in addition to personal prestige, was his will to fulfil

> the wish expressed by his Excellency the Minister, that the views of Italy be adequately exposed: which is important because these are problems (economic autarchy) in which our policy is often exposed abroad in an imprecise form and designed to discredit Italy. Needless to say, also on this occasion – if I have the opportunity – I will speak, according to my constant habit, in order to associate the objectivity of the scholar with the Italian-ness of the fascist citizen.[24]

Eventually, Mortara went to Paris, with the regime's blessing. The lecture at Carnegie would be his last before discrimination and the last where he would be introduced as a professor at the University of Milan. At no stage of the story did Mortara's Jewish ethnicity matter: Mortara was a highly skilled scholar who could expose fascist theses on delicate issues, and who, thanks to the undisputed prestige he enjoyed, could properly defend the positions of the regime. The fact that Mortara was a Jew was of little importance, even in the eyes of Bottai who was to become one of the most ardent proponents of racial politics.

If in Italian high culture there reigned a faint perception of the imminent danger, it is not surprising that Jewish mathematicians took leave of their colleagues at the end of the semester without nourishing anxieties for the future. At the beginning of summer, Volterra, Castelnuovo, Levi-Civita, Fano, Fubini applied (as in every other year) for the renewal of their passports to go to congresses abroad, they corrected and dismissed drafts, and planned the courses for the autumn. Their letters reveal the usual concerns that derive from profession and family: Fubini asked Levi-Civita if he knew of a comprehensive book tackling the question of stability or instability because he wanted to take something on holiday to study in view of the course he would give in the autumn; Enriques was late with the delivery of his contribution to the *Encyclopaedia of Unified Science*; the Castelnuovos must delay their departure

---

[22] APICE-*Mortara*: MEN to A. Pepere, Rome 5.3.1938: *Sarà anzi opportuno che il prof. Mortara profitti dell'occasione per esporre e sostenere in tale ambiente vedute ed argomenti fascisti.*

[23] APICE-*Mortara*: G. Mortara to A. Pepere, Milan 10.3.1938: *per non parere indiscreto o tardo a comprendere quello che mi pareva il chiaro significato del silenzio.*

[24] APICE-*Mortara*: G. Mortara to A. Pepere, Milan 10.3.1938: *il desiderio espresso da S.E. il Ministro, che le vedute dell'Italia siano adeguatamente esposte: cosa importante perché si tratta di quei problemi (autarchia economica) nei quali spesso all'estero la nostra politica viene esposta in forma imprecisa e atta a screditarla. È superfluo dire che anche in questa occasione—se l'avrò— parlerò, secondo la mia costante abitudine, in modo da associare l'obiettività dello studioso alla italianità del cittadino fascista.*

for Engadin because they did not want to leave their daughter Emma alone in the city until she had completed the tests for the competition for school teachers of mathematics and physics.

Documents

D2.2.1 M. Haimovici to T. Levi-Civita, Jassy 12.3.1938

ANL-*Levi-Civita*, fol. 1r-v
Distinguished Professor, [...] After receiving your letter, I first waited a few days for Doetsch's book on Laplace's transformations (which you advised me to consult) to arrive here.[25] But then events emerged that you probably heard about and that took away from me any serenity to focus on anything.[26] Now things have calmed down a little, but only superficially. Distinguished Professor, I do not know if I can afford to beg you for anything. But there are times when we dare everything and the kindness that you always showed to me encourages me to do so now. Please excuse me if I bother you. It is about this: I am determined to do everything possible to emigrate. But I am in the dark about the possibilities. It seems that in Europe it is impossible everywhere, given the restrictions that all countries have taken to combat immigration. But perhaps in another continent it would still be possible, for example in Canada, South America, Australia or elsewhere. I would go wherever I could find a place where I could live in a modest way by my profession. I know that it is very difficult to find such a place, but I am obliged to try everything, because this situation is no longer feasible. Maybe I will be able to go to Paris at some time, at least to try to find out about the possibilities. Given your connections and authority in the scientific world, a recommendation on your part would carry considerable weight. It is for advice, and possibly for such a recommendation, that I would like to appeal to you. Please accept the expression of my deepest gratitude and my most sincere regards, M. Haimovici

Illustre Professore, [...] Dopo ricevuta la Sua lettera, ho aspettato prima alcuni giorni perché arrivasse qui il libro di Doetsch sulle trasformazioni di Laplace, che Ella mi consigliava di consultare. Ma poi sono arrivati avvenimenti di cui Ella ha probabilmente sentito e che mi hanno tolto ogni pazienza di concentrarmi sopra qualunque cosa che sia. Ora le cose si sono un po' calmate, però soltanto

---

[25] G. Doetsch, *Theorie und Anwendung der Laplace-Transformation*, Berlin, Springer, 1937.

[26] In Romania, the crisis of democracy in 1938 led to the ascent to power of authoritarian regimes who put into place anti-Semitic racial legislation. The first law that seriously affected a major part of the Jewish population was passed by the Goga government in 1938. According to this law, about 200,000 Jews lost their citizenship and a whole range of rights, included the right to employment and to property.

superficialmente. Illustre Professore, non so se mi posso permettere di pregarla qualche cosa. Ma vi sono momenti quando si osa tutto e la benevolenza che Ella mi ha sempre mostrata mi incoraggia ad osare. La prego di scusarmi se La importuno. Si tratta di questo: sono deciso a fare tutto il possibile per emigrare. Ma sono nel buio sulle possibilità. Sembra che in Europa sia dappertutto impossibile, date le restrizioni che hanno prese tutti i paesi per combattere l'immigrazione. Ma forse in un altro continente sarebbe ancora possibile, p. es. nel Canada, nell'America del Sud, nell'Australia o altrove. Io andrei dovunque mi sarebbe possibile trovare un posto dove potessi vivere in modo quantunque modesto colla mia professione. So che è molto difficile trovare un tale posto, ma sono obbligato di tentare tutto, perché così non va più. Forse potrò andare a Parigi fra qualche tempo, per cercare almeno di informarmi sulle possibilità. Date le sue relazioni e la sua autorità nel mondo scientifico, una raccomandazione da parte sua avrebbe un peso considerevole. È per un consiglio ed eventualmente per una tale raccomandazione che vorrei pregarla. La prego di gradire l'espressione della mia più viva gratitudine ed i miei distintissimi ossequi, M. Haimovici

D2.2.2 Testimonial of T. Levi-Civita for H. Geiringer, Rome 9.7.1938

ANL-*Volterra*, fol. 1r

To the Academic and Administrative Council of the University of Istanbul. It is a particular pleasure for me to express the highest esteem for the scientific work of Miss Hilda Geiringer. She demonstrates both exceptional versatility and depth, and I think that her influence on students and young researchers must be most advantageous and stimulating. Without entering into mathematical details, I would like to point out various groups of research of the highest rank: first of all (I begin with works that I used directly) papers on strictly non-deformable reticular systems, to which Mr. Amaldi and I referred to in our treatise on rational mechanics;[27] works devoted to the numerical resolution of linear equations; focused on subtle questions of probability calculus and their applications to statistical distributions, dealt with from several points of view; finally numerous articles on the theory of plasticity, recently merged into a very remarkable organic treatise, where we find the theory of Saint-Venant clearly exposed, and opportunely linked to its complements, fully developed in recent times. The fruitful activity of Miss Geiringer and her commendable results confirm that it is in the interest of every advanced scientific institution to guarantee her permanent collaboration. Distinguished regards, Levi-Civita

Au Conseil Académique et d'Administration de l'Université de Istanbul. Il m'est particulièrement agréable de témoigner la plus haute estime pour l'œuvre

---

[27] T. Levi-Civita, U. Amaldi, *Lezioni di Meccanica Razionale*, Bologna, Zanichelli, 2nd ed. 1938.

scientifique de M.me Hilda Geiringer. Elle y fait épreuve à la fois d'une versatilité et d'une profondeur hors ligne, et je pense que son influence sur les étudiants et sur les jeunes chercheurs doit être des plus heureuses et stimulantes. Sans entrer nullement en détails d'ordre mathématique, je me permets de signaler divers groupes de recherches de tout premier ordre: tout d'abord (je commence par des travaux que j'ai utilisés directement) sur les systèmes réticulaires strictement indéformables, auxquels M. Amaldi et moi avons fait des emprunts dans notre traité de mécanique rationnelle ; sur la résolution numérique des équations linéaires ; sur des questions subtiles de calcul des probabilités et sur leurs applications aux distributions statistiques sous plusieurs points de vue ; enfin des nombreux articles sur la théorie de la plasticité, fondus récemment dans un traité organique très remarquable, où l'on trouve nettement exposée, et heureusement rattachée aux compléments mûris dans ces derniers temps, la théorie de Saint-Venant. L'activité féconde de M.me Geiringer et les beaux résultats qui lui sont dus ne rendent pas douteux qu'il soit dans l'intérêt de toute haute institution scientifique de s'assurer sa collaboration permanente. Sentiments distingués, Levi-Civita

2.3   The Racial Census

On 14 July 1938, ten scientists published a piece in *Giornale d'Italia* entitled *Il fascismo e i problemi della razza*.[28] Also known by the title of *Manifesto of racist scientists* or, in abbreviated form, *Manifesto of the race*, it can be considered to be the theoretical starting point of the racist campaign in Italy. The statement affirmed for the first time the existence of a pure Italian race, defined as being of Aryan origin, and proclaimed the extraneity of the Jewish element with respect to the Italic race, and the non-assimilability of the Jews. Consequently, there were to be no mixed marriages with Jews. These tenets were quite new both for fascism and for Italian culture, which displayed racist currents itself but had not yet assimilated the theoretical elaborations of Nazism. For these reasons, the *Manifesto* aroused strong perplexities among Italian intellectuals and in the Catholic world. To propagate the racial fiction, a special journal was created, *La Difesa della Razza*, under the direction of T. Interlandi, with the first issue published on 5 August 1938.[29]

The *Manifesto* cautioned that further specific political action would be taken. Its doctrinal formulation provided, in fact, the foundation for the census of the Jewish

---

[28] From the University of Rome: Lino Businco (General Pathology), Guido Landra (Anthropology), Nicola Pende (Endocrinology), Marcello Ricci (Zoology), Franco Savorgnan (Demography), Sabato Visco (Physiology), and Edoardo Zavattari (Zoology); from Florence: Lidio Cipriani (Anthropology); from Bologna: Arturo Donaggio (Neuropsychiatry); from Milan: Leone Franzi (Paediatrics). On the *Manifesto of the race* see, inter alia, Gillette (2002).

[29] Loré (2008); Cassata (2008).

minority, which in turn paved the way for the infamous season of racial policy. In the middle of summer, surpassing colleagues in solicitude and zeal, the Minister of Public Instruction Bottai sent over 2000 forms to the employees of his dicastery and to the members of academies and scientific societies, asking them to fill in a sort of racial charter. Each person was called to declare his or her own ancestors by paternal and maternal lineages, the race of the spouse, professed religion, and registration with a Jewish community.

The racial census constituted a most serious laceration of the Italian cultural world, a fracture in some ways even deeper than that inflicted by the enforced oath of allegiance to the fascist regime. While Jews, many of whom had lost the sense of their ethnic roots, were unready and unable to meet the terms of such an inquiry, the vast majority of their Aryan fellows rushed to declare their belonging to the pure Italic race. For a scholar like B. Croce who replied indignantly: 'the only effect of the requested declaration would be to make me blush, forcing me, who goes by the surname Croce (crux), to the hateful and ridiculous act of protesting that I am not a Jew, just when these people are persecuted',[30] there were dozens of intellectuals who carefully documented that they did not have even a drop of Jewish blood. Very few refused to return the completed form, even though they would not have suffered any harm for refusing to do so. Unlike the oath of 1931, in fact, no disciplinary or administrative sanction was placed on those who did not return the completed chart.

For Italian Jews, answering the census was, at the same time, something incomprehensible and humiliating. Artom, for example, replied: 'I do not understand the meaning of the expression Jewish race, but for centuries my ancestors were Jews. Descent from biblical patriarchs is not certain'.[31] Mortara explained that he could not answer the question of whether or not he belonged to a race of which he denied the existence on the basis of scientific evidence. Edoardo Volterra, the eldest son of Vito, historically set his answer about the race of his parents and wife Nella Mortera: 'the Volterra, Almagià and Mortera families had left Palestine and settled on Italian soil in ancient times'. Fano and Mortara added a comment on the religious faith. Fubini, Terracini, Enriques, Levi-Civita, Segre, and Levi limited themselves to ticking the boxes. (Fig. 2.1)

The racial census supplied these data: on 28 October 1938 there were 39,000 Italian Jews (out of a population of 42,398,489 individuals); 6820 mixed families (4000 with Aryan wife and Jewish husband); 3500 out of 13,000 children born from mixed families professed themselves to be of Jewish religion, the others were Catholic or atheist. In total, 31,200 Jews resulted as belonging to a Jewish community, and therefore apparently professed the Jewish religion. Special caution is required as far as this last statistic is concerned. According to a law promulgated

---

[30] *l'unico effetto della richiesta dichiarazione sarebbe di farmi arrossire, costringendo me, che ho per cognome Croce, all'atto odioso e ridicolo insieme di protestare che non sono ebreo, proprio quando questa gente è perseguitata*, in Capristo (2002), p. 38.

[31] CDEC: Emanuele Artom Archive, 1934–1972, fol. 1r: *non comprendo il significato dell'espressione «razza ebraica», ma da secoli i miei antenati erano ebrei. Non certa la discendenza dai patriarchi biblici.*

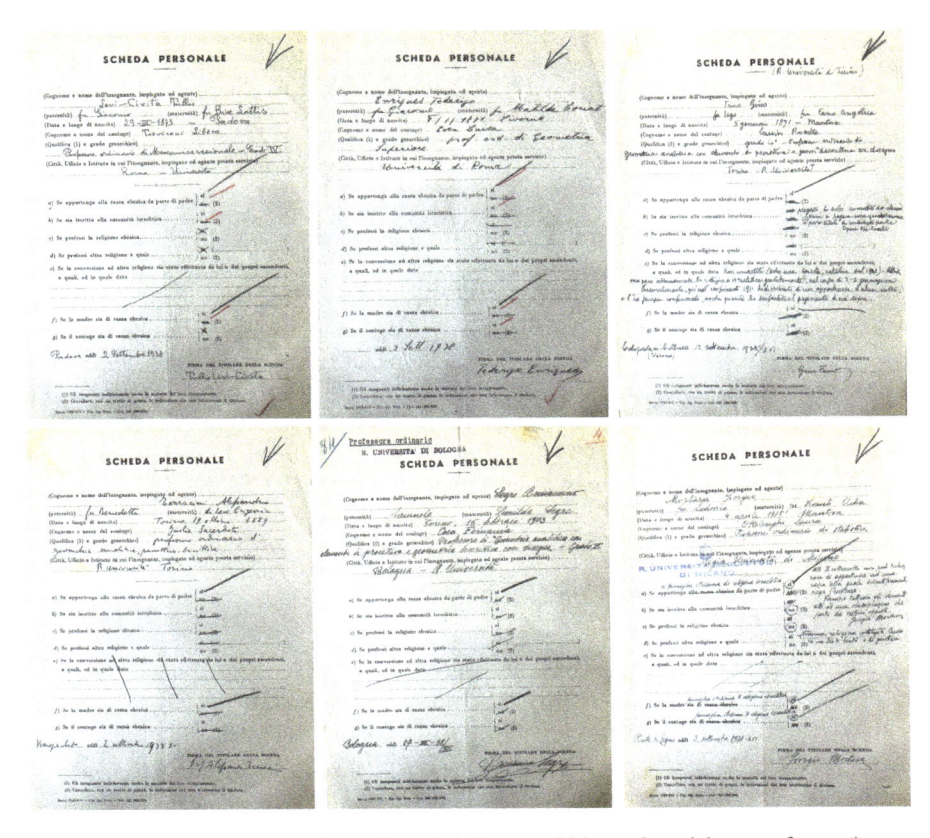

Fig. 2.1 Levi-Civita, Enriques, Fano, Terracini, Segre and Mortara's racial census forms, August 1938, ACS-*Ebrei*

in 1931, enrolment with the Jewish community was mandatory but the majority of Italian Jews, registered by parents at birth, had completely forgotten this status and did not record any religious leaning. Moreover, according to the *Demorazza* Institute, there had been many faked conversions and false baptisms. The phenomenon was apparently massive in the countryside, more limited in towns. In total, 2009 abjurations were recorded in three months (August–October 1938).[32]

At the beginning of August, Bottai subverted academia with another unprecedented act. He sent the Rectors of all Italian universities, through the prefectures, the following letter encrypted in Dante's Cypher:

---

[32] A comparison with the situation in Nazi Germany can be of some interest. In Germany, there were about a half million 'Jews' according to Nazi records in 1933 from among a population of 67 millions of individuals. Consequently, the impression is that there were (percentagewise) much fewer conversions in Italy before 1938 than in Germany before 1933.

I warn that authorisations which have been so far granted to Professors and Assistants of non-Italian Race to intervene in any capacity at congresses and similar events abroad must be intended completely revoked. Please conform in accordance with this directive and report to me by telegraph that this is complete. I also call for rigorous compliance with provisions already given, by virtue of which any person wishing to participate in congresses and other international events, even in a private capacity, can in no case intervene if he has not obtained explicit Ministerial authorisation in advance.[33]

The impact on high culture was immediate. Levi-Civita was obliged to cancel his participation in the International Congress of Applied Mechanics, scheduled for September 1938 in Cambridge Mass. Volterra informed Borel and other French colleagues that his sojourn in Paris should be postponed due to problems with the issue of his and his wife's passports.[34] The Bolognese astronomer Luigi Jacchia, a member of the Commission on variable stars of the International Astronomical Union, who intended to take advantage of the association meeting in Stockholm (August 3–10, 1938) to explore the possibilities of emigrating, asked to go to Sweden as a private citizen but was denied permission (D3.8.3).

2.4   The Anti-Jewish Legislation

Inviting Italians to frankly proclaim themselves to be racist, in the infamous speech in *Piazza Unità d'Italia* in Trieste (18.9.1938), Mussolini brought the ideological path undertaken in the previous months to its logical conclusion:

With regard to domestic policy, the current burning issue is the racial question. Also in this field we will adopt the necessary solutions. Those who believe that we have obediently imitated anyone, or worse, acted on suggestions, are poor fools toward whom we do not know if we should direct our contempt or our pity. The racial problem did not suddenly burst out of nowhere, as those who are accustomed to brusque awakenings think (since they are used to long armchair naps). It is in relation to imperial conquest; because history teaches us that empires are conquered by arms but are held by prestige. And for prestige it is necessary to have a clear, severe racial consciousness, that establishes not only the differences, but also very clear superiorities. The Jewish problem is thus only one aspect of this phenomenon. Our position has been determined by these indisputable facts. World Jewry has been, for sixteen years, despite our policy, an irreconcilable enemy of fascism. In Italy, our policy has led, in the Semitic elements, to what can today be called a true rush to board the ship. However,

---

[33] APICE-*Mortara*: MEN to A. Pepere, Milan 1.8.1938: *Avverto che autorizzazioni finora eventualmente concesse a Docenti e Assistenti di Razza non italiana intervenire a qualsiasi titolo a congressi ed analoghe manifestazioni all'estero intendonsi tutte revocate. Vogliate disporre in conformità assicurandomene telegraficamente. Richiamo altresì rigorosa osservanza disposizioni già date in virtù delle quali chiunque desideri partecipare congressi ed altre manifestazioni di carattere internazionale anche a titolo privato, non può in alcun caso intervenire se non abbia ottenuto esplicita autorizzazione Ministeriale.*

[34] ANL-*Volterra*: V. Volterra to É. Borel, [Ariccia] 18.10.1938. In the following weeks, Borel was still convinced that it was a matter of mismanaged bureaucracy. It was only in November that he learned that Volterra's trip to Paris would be postponed indefinitely (ANL-*Volterra*: V. Volterra to É. Borel, Ariccia 7.11.1938 and É Borel to V. Volterra, 29.11.1938).

Jews of Italian citizenship who have unquestionable military or civil merit towards Italy and the Regime, will find understanding and justice. As for the others, a policy of separation is what awaits them. When all is said and done, the world will perhaps be surprised more by our generosity than by our severity; that is, unless the Semites beyond our borders and within our country, and above all their powerful friends and defenders, force us to radically harshen our policy.[35]

The policy of separation that Mussolini outlined was that of the racial laws. Issued between September 1938 and February 1945, these laws were a series of provisions (three main decrees, flanked by many extraordinary measures, ordinances, and administrative circulars) by which firstly the fascist government, and then the Italian Social Republic, stripped the Italian Jews of political and civil rights conquered in the Risorgimento. The first bill sanctioned the exclusion of foreign and Italian Jews from schools, academia, politics, finances, professional world, and all sectors of public and private life. Italian citizens of the Jewish religion were no longer allowed to attend school or marry non-Jews, to be drafted into the Army, to own or administer companies, to own or administer land and real estate over a certain value, to hire non-Jewish workers, or to be hired in the public administration, to participate in political parties (PNF included), in banks, insurance companies, newspapers, artistic, research, and educational institutions. Jewish students were expelled from schools of all levels, from kindergartens to university, and all teaching and technical-administrative personnel (secretaries, janitors, librarians, laboratory technicians, etc.) was removed from service. Use of textbooks written by Jewish authors was prohibited in all State institutions (the provision was renamed 'procedure of literary cleansing'), and Jews were excluded from any kind of publishing activity.[36]

A decree regulated the status of foreign Jews who had taken up residence in Italy, or in its colonies subsequent to the Great War, including also such persons who had

---

[35] *Nei riguardi della politica interna, il problema di scottante attualità è quello razziale. Anche in questo campo noi adotteremo le soluzioni necessarie. Coloro i quali fanno credere che noi abbiamo obbedito ad imitazioni, o peggio, a suggestioni, sono dei poveri deficienti, ai quali non sappiamo se dirigere il nostro disprezzo o la nostra pietà. È perché sono abituati ai lunghi sonni poltroni. È in relazione con la conquista dell'Impero, poiché la storia ci insegna che gli imperi si conquistano con le armi ma si tengono con il prestigio, e occorre una chiara, severa coscienza razziale che stabilisca non soltanto delle differenze ma delle superiorità nettissime. Il problema ebraico è dunque un aspetto di questo fenomeno. La nostra posizione è stata determinata da questa incontestabilità dei fatti. L'ebraismo mondiale è stato, durante i sedici anni, malgrado la nostra politica, un nemico irreconciliabile del Fascismo. In Italia la nostra politica ha determinato, negli elementi semiti, quella che si può oggi chiamare, si poteva chiamare, una corsa vera e propria all'arrembaggio. Tuttavia, gli ebrei di cittadinanza italiana, i quali abbiano indiscutibilmente meriti militari e civili nei confronti dell'Italia e del Regime, troveranno comprensione e giustizia. In quanto agli altri, seguirà una politica di separazione. Alla fine, il mondo dovrà forse stupirsi, più della nostra generosità che del nostro rigore, a meno che, i nemici di altre frontiere e quelli dell'interno e soprattutto i loro improvvisati e inattesi amici, che da troppe cattedre li difendono, non ci costringano a mutare radicalmente cammino.*

[36] Fabre (1998). The name *Bonifica Libraria* derived from the title of an article which appeared in *Critica Fascista* 18, n. 5, 1939, pp. 66–67.

in the meanwhile acquired Italian citizenship: 'Foreigners of the Jewish race who at the date of publication of the present decree law are within the Kingdom, Libya and the Aegean possessions and who began their sojourn therein subsequent to 1 January 1919, must leave the territory of the Kingdom, Libya and the Aegean possessions within 6 months from the date of publication of the present decree law. Those who have failed to conform to this obligation by the aforesaid deadline shall be expelled from the Kingdom in accordance with article 150 of the codified text of the Public Security laws after the application of penalties established by law'.[37] This was and would remain the only provision about forced emigration. As in Nazi Germany, the emphasis in the Italian anti-Jewish campaign was therefore placed on public administration. It was only from the first semester of 1939 onwards that a bill added restrictions on Jews owning real estate, firms, and commercial activities.[38]

Approved by the Grand Council of Fascism on 6 October 1938, the *Dichiarazione sulla razza* asserted a wider definition of Jewish race, to also include those who had a Jewish father and a mother of 'foreign nationality', and the children of mixed marriages who professed the Jewish faith on 1 October 1938.

The text *Integrazione e coordinamento in unico testo delle norme già emanate per la difesa della razza nella Scuola italiana* (November 17, 1938) removed from service teachers, faculty members and lecturers, who were pensioned off if in possession of at least ten years of seniority in the role, and dismissed without any compensation in other cases. A special indemnity was given in the presence of dependents if they had no other income. Having dependents and being enrolled to the PNF enabled (but not automatically) access to additional sums. The pensions were calculated by the Court of Audit (*Corte dei Conti*) on the basis of art. 21 of the decree: 'the minimum pension is granted to those who have completed at least ten years of service; in other cases, an indemnity is granted equal to as many twelfths of the last salary as the number of years of service that they have completed.'[39] Pensions thus depended linearly on the number of years of service.

In order to better understand the gravity of the measure, one should take into consideration that with 40 years of service, Levi-Civita in 1938 received yearly a stipend of 36,000 lire (3000 lire monthly); he was dismissed with a yearly pension of 28,000 lire (2333 lire monthly). Arnaldo Momigliano, full professor from 1936,

---

[37]Royal Law 7.9.1938, no. 1381 *Provvedimenti nei confronti degli ebrei stranieri*, Art. 4: *Gli stranieri ebrei che, alla data di pubblicazione del presente decreto-legge, si trovino nel Regno, in Libia e nei Possedimenti dell'Egeo e che vi abbiano iniziato il loro soggiorno posteriormente al 1° gennaio 1919, debbono lasciare il territorio del Regno, della Libia e dei Possedimenti dell'Egeo, entro sei mesi dalla data di pubblicazione del presente decreto. Coloro che non avranno ottemperato a tale obbligo entro il termine suddetto saranno espulsi dal Regno a norma dell'art. 150 del testo unico delle leggi di P.S., previa l'applicazione delle pene stabilite dalla legge.*

[38]Levi F. (1991, 1998).

[39]Royal Law 17.11.1938, no. 1728 *Provvedimenti per la difesa della razza italiana*, Capo III, Art. 21 *Disposizioni transitorie e finali: é concesso il trattamento minimo di pensione se hanno compiuto almeno dieci anni di servizio; negli altri casi è concessa una indennità pari a tanti dodicesimi dell'ultimo stipendio quanti sono gli anni di servizio compiuti.* See Capristo and Fabre (2018).

received a yearly stipend of 19,000 lire (1580 lire monthly). He did not have ten years of service on his record, and consequently he received no pension at all. He received just a *liquidazione* or *buona-uscita*, i.e. a lump sum of 9500 lire.[40] After this sum, which he picked up on 21 January 1939, he received no other money. An intermediate case: Terracini, who had 30 years of service on his record but just 14 as full professor, with a wife and three children as dependents, received a yearly stipend of 26,000 lire in 1938. He was discharged with a pension of 14,428 lire. Previous service as assistant, military service, and so on was taken into consideration.[41] This was of course devastating. No early career scholar could survive in absence of family assets or other financial resources (land, estates, etc.), even more so because the Court of Auditors perpetuated a whole series of abuses to deprive the purged scholars of the sums owed to them, or to greatly reduce them. Looking for other (paid) work was mandatory.

Discriminatory measures were enforced in the following years (1939–40): new bills were drafted which severely limited the mobility and rights of Italian Jews in terms of family law and estate law. For example, Jews were prohibited from staying in hotels, although the latter measure never actually came into force due to the strong protests of hoteliers. In order to coordinate and monitor the application of these measures, two government agencies were created: the General Office of Demography and Race (DGDR) and the Agency for Real Estate Management and Liquidation of Jewish assets (EGELI). To these would be added the notorious General Inspectorate for Race created by RSI on 18 April 1944.

When Italy entered the war on Germany's side (10 June 1940), Mussolini decreed the internment of foreign Jews in concentration camps and Italian Jews were prescribed forced labour. Accused of 'cheering' for the Allies and spreading mendacious and defeatist news about the course of the war (*disfattismo*), Jews were prohibited from holding radio equipment and were required to hand over any in their possession.

A report by the Ministry of Interior dated 28 October 1941 photographed the situation of Italian Jewish population after three years of persecution. In this document Italian Jews were subdivided according to profession: 4350 artisans, 5200 working freelance (lawyers, physicians, notaries, etc.), 1000 industrialists, 4200 office workers, 1450 business owners, 700 workers, and 220 landowners.[42] According to such a scheme, 4200 office workers and 700 ordinary workers were

---

[40] The law entitled him to two months' salary, but the lump sum was partly based on his previous services as assistant, lecturer, etc.

[41] To have an idea about the value of the lira at the time, these examples can be taken into consideration: the monthly salary of a worker (*net*): 300–400 lire; a Fiat *Topolino* car: 8500 lire; bread (kg): 2 lire; sugar (kg): 7–8 lire; newspapers: 0.20 lire; a transatlantic journey on Oceania, from Italy to Buenos Aires: 13,000 lire per person. The lira was worth somewhat more than a euro today, but not much more, given the calamity of wages for workers everywhere in Europe.

[42] The scheme seems consistent. This is about 16,000 people, mainly male. Together with their families and given that only a small part of the population (maybe 6000) emigrated, this seems in accordance with the figure of about 47,000 Italian Jews in 1938.

completely without resources; 5200 working freelance had very limited resources, because they could offer their services to Jewish clients only. Industrialists had reduced incomes because they had to pay an Aryan intermediary.

The Nazi occupation took Jewish persecution in Italy to its climax: from the deprivation of rights, they moved on to persecution of lives. The first acts of brutal violence were recorded on the same day of the armistice (8.9.1943), when Field Marshal Pietro Badoglio announced Italy's unconditional surrender to the Allies. From that date onwards, it was a growing crescendo of violence: vandalism and burning of synagogues, lynching, roundups, and mass deportations to the extermination camps. On 16 October 1943, Gestapo troops (with the collaboration of officials from the puppet RSI) carried out a roundup in the ghetto of Rome. In total, 1259 people were arrested in the raid, including 689 women, 363 men, and 207 children. Of them, 1023 were deported to Auschwitz, of which 16 survived.

2.5 The Quantitative Dimension

The extent of the discriminatory procedure triggered by the racial laws has now been fully ascertained, by combining census data with documentation meticulously collected by DGDR: 174 professors were dismissed, while 676 scholars were expelled from academies and scientific societies.[43] About 7000 Jewish students with Italian citizenship were purged from schools: 4000 children from primary, about 1000 from middle and secondary schools, and 2000 from university institutions. Science Faculties (i.e., the Faculty of mathematical, physical, and natural sciences, as they were and are still called today) were strongly affected. Forty-eight out of the 174 professors dismissed were mathematicians (Fig. 2.2).

Among these, there were nine full professors (Tullio Levi-Civita and Federigo Enriques [Rome]; Guido Fubini, Gino Fano, and Alessandro Terracini [Turin]; Beniamino Segre and Beppo Levi [Bologna]; Guido Ascoli [Milan]; and Arturo Maroni [Pavia]), an extraordinary professor (Ettore Del Vecchio), an assistant professor (Bonaparte Colombo [Turin]), and two non-tenured lecturers (Alberto Mario Bedarida [Genoa] and Bruno Tedeschi [Trieste]).

For mathematics, the most affected universities were therefore Turin, Bologna, and Rome, with three, two, and two faculty members dismissed out of five, seven, and four, respectively.[44] To these names should be added those of Castelnuovo, Loria, Vivanti, C. Rimini, and E. Almansi, who were not officially affected by the racist measures as they had already retired, but who were targeted by a series of small and large abuses and humiliations, such as denial of access to archives, laboratories, and institutes, and from borrowing books and reprints from university libraries.

---

[43] Finzi (1997); Galimi and Procacci (2009); Dell'Era and Meghnagi (2023). About the persecution in specific universities (Genoa, Perugia, etc.) see Rollandi (2002); Salustri (2020); Varnier (2002).

[44] Rinaldelli (1997–98); Luciano (2018); Graffone (2018); Finzi (1994); Coen (2002).

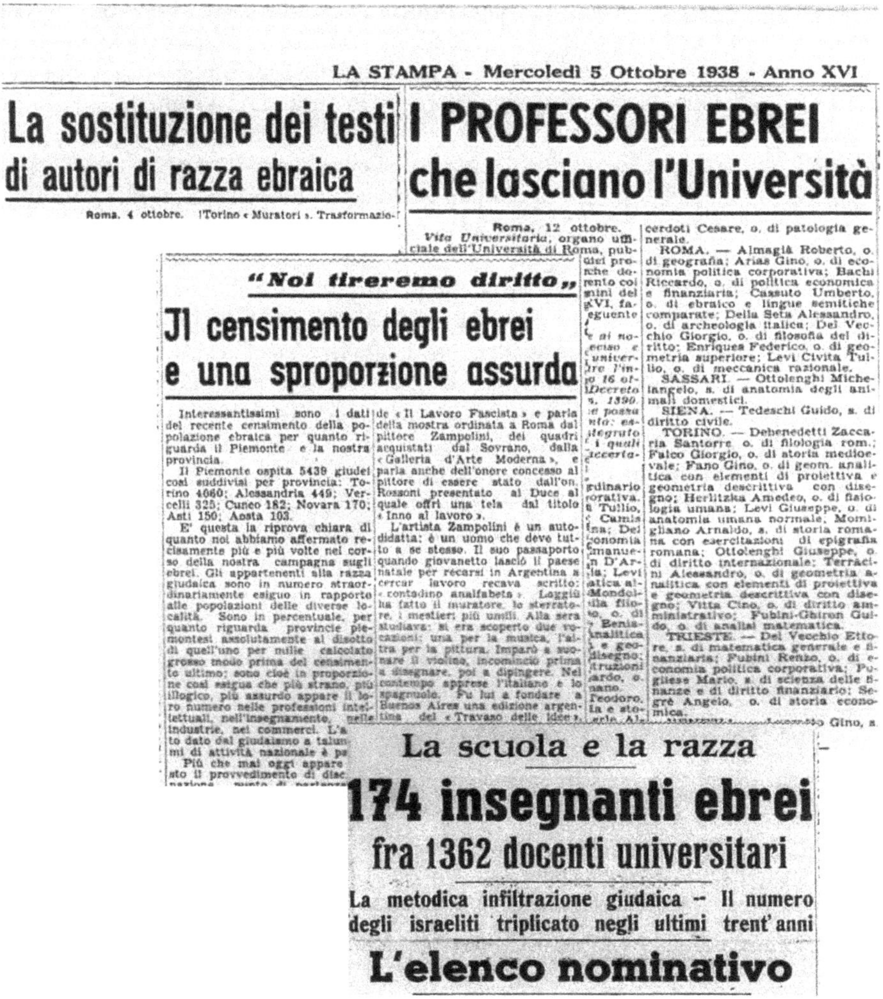

LA STAMPA - Mercoledì 5 Ottobre 1938 - Anno XVI

La sostituzione dei testi di autori di razza ebraica

Roma, 4 ottobre. (Torino « Muratori ». Trasformazio-

I PROFESSORI EBREI che lasciano l'Università

Roma, 12 ottobre.
Vita Universitaria, organo ufficiale dell'Università di Roma, pub-

cerdoti Cesare, o. di patologia generale.
ROMA. — Almagià Roberto, o. di geografia; Arias Gino, o. di economia politica corporativa; Bachi Riccardo, o. di politica economica e finanziaria; Cassuto Umberto, o. di ebraico e lingue semitiche comparate; Della Seta Alessandro, o. di archeologia italica; Del Vecchio Giorgio, o. di filosofia del diritto; Enriques Federico, o. di geometria superiore; Levi Civita Tullio, o. di meccanica razionale.
SASSARI. — Ottolenghi Michelangelo, s. di anatomia degli animali domestici.
SIENA. — Tedeschi Guido, s. di diritto civile.
TORINO. — Debenedetti Zaccaria Santorre, o. di filologia rom.; Falco Giorgio, o. di storia medioevale; Fano Gino, o. di geom. analitica con elementi di proiettiva e geometria descrittiva con disegno; Herlitzka Amedeo, o. di fisiologia umana; Levi Giuseppe, o. di anatomia umana normale; Momigliano Arnaldo, s. di storia romana con esercitazioni di epigrafia romana; Ottolenghi Giuseppe, o. di diritto internazionale; Terracini Alessandro, o. di geometria analitica con elementi di proiettiva e geometria descrittiva con disegno; Vitta Cino, o. di diritto amministrativo; Fubini-Ghiron Guido, o. di analisi matematica.
TRIESTE. — Del Vecchio Ettore, s. di matematica generale e finanziaria; Fubini Renzo, o. di economia politica corporativa; Pugliese Mario, s. di scienza delle finanze e di diritto finanziario; Segrè Angelo, o. di storia economica.

"Noi tireremo diritto"

Il censimento degli ebrei e una sproporzione assurda

Interessantissimi sono i dati del recente censimento della popolazione ebraica per quanto riguarda il Piemonte e la nostra provincia.

Il Piemonte ospita 5439 giudei così suddivisi per provincia: Torino 4060; Alessandria 449; Vercelli 325; Cuneo 182; Novara 170; Asti 150; Aosta 103.

E' questa la riprova chiara di quanto noi abbiamo affermato recisamente più e più volte nel corso della nostra campagna sugli ebrei. Gli appartenenti alla razza giudaica sono in numero straordinariamente esiguo in rapporto alle popolazioni delle diverse località. Sono in percentuale, per quanto riguarda provincie piemontesi assolutamente al disotto di quell'uno per mille calcolato grosso modo prima del censimento ultimo; sono cioè in proporzione così esigua che più strano, più illogico, più assurdo appare il loro numero nelle professioni intellettuali, nell'insegnamento, nelle industrie, nei commerci. L'a[…] to dato dal giudaismo a talun[…] mi di attività nazionale è pa[…]

Più che mai oggi appare […] sto il provvedimento di disc[…]

de « Il Lavoro Fascista » e parla della mostra ordinata a Roma dal pittore Zampolini, dei quadri acquistati dal Sovrano, dalla « Galleria d'Arte Moderna », e parla anche dell'onore concesso al pittore di essere stato dall'on. Rossoni presentato al Duce al quale offrì una tela dal titolo « Inno al lavoro ».

L'artista Zampolini è un autodidatta: è un uomo che deve tutto a se stesso. Il suo passaporto quando giovanetto lasciò il paese natale per recarsi in Argentina a cercar lavoro recava scritto: « contadino analfabeta ». Laggiù ha fatto il muratore, lo sterratore, i mestieri più umili. Alla sera studiava: si era scoperto due vocazioni: una per la musica, l'altra per la pittura. Imparò a suonare il violino, incominciò prima a disegnare, poi a dipingere. Nel contempo apprese l'italiano e lo spagnuolo. Fu lui a fondare a Buenos Aires una edizione argentina del « Travaso delle Idee ».

La scuola e la razza

174 insegnanti ebrei fra 1362 docenti universitari

La metodica infiltrazione giudaica — Il numero degli israeliti triplicato negli ultimi trent'anni

L'elenco nominativo

Fig. 2.2 Collage of news about the racial laws from La Stampa, autumn of 1938

Prizes and scholarships named after Corrado Segre, Eugenio Elia Levi, and Lazzaro Fubini (father of Guido) were renamed, although they continued to be funded by Jewish families.

In the case of physics, an entire research School (that of nuclear physics) was erased. Enrico Fermi received the Nobel Prize on December 10, 1938, and then left Stockholm to go to the United States. Fermi belonged to the 'Italic race', but he had married a Jew, Laura Capon, and although the fascist government had courted him in every way possible, he had been able to foresee the degeneration of events. His students, the so-called 'boys from via Panisperna' after the name of the street where the Institute of Nuclear Physics was located, were all young people, some of whom were already faculty members, and others still precarious researchers: Emilio Segré, Bruno Rossi, Giulio Racah, Eugenio Fubini, Ugo Fano, Leo Pincherle, Sergio de Benedetti, and Eugenio Curiel. They represented the avant-garde of physics at an international level and had in Fermi a Master of excellence who could help them to expatriate. In fact, almost all Fermi's pupils would do that very thing.

To complete the comprehensive picture of the consequences of anti-Jewish persecution in Italian science faculties, three other scholarly communities should be mentioned: naturalists, astronomers, and statisticians. The first was a gender semi-balanced community, which counted the only female full professor purged in 1938 (Anna Foà) and various graduate students and early career researchers, including Luisa Volterra and Gina Castelnuovo, daughters of Vito and Guido, respectively. Racist legislation led to the dismissal of two lecturers and three full professors.

The small group of Italian astronomers was also decimated by the racial laws. Two first-class astronomers (Guido Horn d'Arturo and Azeglio Bemporad) and two *liberi docenti* (Giulio Bemporad, Azeglio's cousin, and Luigi Jacchia) lost their positions in the observatories of Turin, Bologna, and Naples.[45]

Unlike astronomers, who generally trained in mathematics or physics and worked in scientific faculties, statisticians and demographers mostly arrived from studies of law or economics and were attached to these faculties. Three scholars who had a teaching assignment in a scientific course of studies were dismissed: father and son Riccardo and Roberto Bachi, and Giorgio Mortara.[46]

Beyond academia, there lay the world of school. Forty-seven of the 166 teachers removed from service had graduated in Mathematics or held a joint degree in Mathematics–Physics. This number does not include the five teachers of actuarial science, statistical analysis, and accounting in technical institutes with a professional or commercial track, who generally graduated from the Faculty of economics. Most purged schoolteachers had been trained in Turin, Bologna, and Rome. The impact of the racial laws was in fact particularly dramatic in these three contexts where ample transversal communities had been created, consisting of university professors engaged in the methodological debate (Segre, Pincherle, Enriques, Castelnuovo) and of teachers, who had assimilated the tenets asserted by their mentors and translated them into daily teaching practice.[47] These transversal communities, or networks, which featured a significant number of highly qualified women, were wiped out by racial politics.

---

[45] Bònoli and Mandrino (2015) (eds.); Cornell (1996).
[46] Cocchi and Favero (2009) and Giaconi (2020).
[47] Luciano (2023), pp. 89–90.

2.6 The Bureaucracy of Persecution

Although Mussolini had declared that Italy would be distinguished more by its indulgence than its rigour, the bureaucracy machine proceeded with brutal efficiency in implementation of the purge measures. Archive sources have definitively shown how inconsistent is the myth of the indulgent application, '*all'italiana*', of racial legislation. Already in mid-September the minutes of the faculty and school boards meetings recorded the expulsion of Jewish students and the purge of non-Aryan staff. Artom, for example, was considered 'justifiably absent'[48] at the first meeting at scientific lyceum G. Ferraris in Turin, convened for September 15. On October 26, in the minutes of the meetings of the teaching staff, it stated that 'Prof. Artom no longer worked in the Institute'[49] and that two Aryan colleagues, C. del Giudice Barolo and B.M. Einaudi Colla, had substituted him in teaching and in the role of school treasurer.

The replacement of books by authors belonging to the Jewish race, including some classical texts adopted throughout the whole country such as *Elementi di Geometria* by F. Enriques and U. Amaldi (Bologna, Zanichelli, 1903, 7th ed. 1938), was equally fast, to the intense satisfaction of publishing houses such as Vallecchi, which gained advantage on the 'Yid Zanichelli'. Anti-Jewish legislation did not omit the smallest of details, such as the removal of scientific instruments and geographic maps drawn up by Jews.

In the universities, the draconian activity of the offices started a couple of weeks after the introduction of the race legislation. The first measure spoke of 'dispensation' (*dispensa*) as opposed to dismissal from service. Moreover, not everyone had responded to the racial census, which is why it was necessary to ascertain the racial identity of various faculty members who had not returned the form. In Milan, for example, Bruno Finzi and Guido Finzi, professors of Rational Mechanics and Veterinary Medicine, respectively, were invited to urgently document that they were not Jews according to the definition provided by the racial laws.[50] Both replied by return of post, but it was not enough. The surname Finzi sounded typically Jewish, which is why the Ministry asked for a supplementary investigation.[51] Bruno and Guido Finzi were required to document the nationality of their fathers and wives. Both did so. Guido, in particular, declared he belonged to the Italian race, he professed himself a Catholic, enrolled in PNF since 1928, married to Maria

---

[48] Archive of the scientific lyceum G. Ferraris in Turin, [*Verbali 1938-39*]: *assente giustificato*.

[49] Archive of the scientific lyceum G. Ferraris in Turin, [*Verbali 1938-39*]: *il prof. Artom non insegna più nell'Istituto*. See also Archive of the scientific lyceum Galileo Ferraris in Turin, *Registro di carriera scolastica di Emilio Artom: stato personale* sezione *Ufficio presente nella scuola—osservazioni* and *Liceo Gioberti, Liceo D'Azeglio, Liceo Galileo Ferraris* ... 2012.

[50] APICE-*Finzi*: MEN to A. Pepere, Rome 17.10.1938; A. Pepere to B. Finzi and G. Finzi, Milan 19.10.1938; A. Pepere to MEN, Milan 22.10.1938; A. Pepere to MEN, Milan 2.11.1938.

[51] APICE-*Finzi*: MEN to A. Pepere, Rome 28.3.1939; A. Pepere to G. Finzi and B. Finzi, Milan 31.3.1939; A. Pepere to MEN, Milan 4.3.1939.

Resnati (in turn, of Catholic religion and Italian race), and father of four children, all of whom were baptised.[52]

After the Declaration on Race, the university offices speeded up their practices. Around the middle of October, the first letters of dismissal, almost identical for all people, began to be sent out (D2.6.1). In some cases, a few personal words were added by Rectors, addressing the colleagues to whom they were more attached. In Milan, for example, A. Pepere wrote personally to the discriminated colleagues, and many thanked him 'for the human words with which he wanted to make less bitter the detachment from that University that in the depths of the heart each one would always be remembered as one's own'.[53] In Turin, the Director of the Polytechnic (G. Vallauri) sent everyone the same messages of greeting, automatically drafted by their secretariats:

> Distinguished Prof. Dr. ##, On behalf of the Ministry, I inform you that, pursuant to art. 3 and 6 of the R.D.L. September 5, 1938-XVI, no. 1390, you are suspended from your office as Full Professor of ## at this Royal Polytechnic, as of 16 October. On this occasion, I am obliged to inform you that the board of Directors, in its meeting of the 14th in the current month, in acknowledging the aforesaid decision, charged me with offering you the warmest thanks, to which I add my personal gratitude, for your effective collaboration as a member of the Council itself and for your long and praiseworthy work for the benefit and decorum of this Institute. The Director G. Vallauri[54]

> Distinguished Dr. Prof. ##, I inform you that the council of this Faculty of ## in its meeting of the current ## October, having acknowledged receipt of the interruption of your office as ## Professor of ## and in remembrance of your scientific and didactic merits, has asked us to present you with its mindful greetings and to express to you the assurance that the memory of your long and precious work will always remain alive in the souls of those who have had the privilege of being your colleagues and disciples.[55]

---

[52] APICE-*Finzi*: B. Finzi to A. Pepere, Milan 21.10.1939.

[53] APICE: B. Terracini to A. Pepere, Turin 19.10.1938: *per le umane parole con cui ha voluto rendere meno amaro il distacco da quella Università che nel fondo del cuore si ricorderà sempre come propria.*

[54] ASPoliTo-*Fubini*: G. Vallauri to G. Fubini, Turin 18.10.1938: *Ch.mo Prof. Dott. Uff. ##, D'incarico del Superiore Ministero Vi comunico che, ai sensi degli art. 3 e 6 del R.D.L. 5 settembre 1938-XVI, n° 1390, siete sospeso dal Vostro ufficio di Professore Ordinario di ## presso questo R. Politecnico, a datare dal 16 ottobre corrente. Con l'occasione mi faccio dovere di parteciparVi che il Consiglio di Amministrazione, nella sua seduta del 14 corrente, nel prendere atto del provvedimento di cui sopra, mi lasciò incarico di porgerVi i più vivi ringraziamenti, ai quali aggiungo i miei personali, per la fattiva e pregiata Vostra collaborazione come membro del consiglio stesso e per la lunga e benemerita Vostra opera a vantaggio e decoro di questo Istituto. Il Direttore G. Vallauri.*

[55] ASPoliTo-*Fubini*: G. Vallauri to ##, Turin 25.10.1938: *Ch.mo Prof. Dott. Comm. ##, Vi comunico che il Consiglio di questa Facoltà di ## nella sua adunanza del ## ottobre corrente, preso atto della cessazione del Vostro Ufficio come Professore ## di ## e ricordate le Vostre benemerenze scientifiche e didattiche, mi ha lasciato incarico di porgerVi il suo memore saluto e di esprimerVi l'assicurazione che il ricordo della lunga e pregiata opera Vostra rimarrà sempre vivo nell'animo di quanti Vi furono colleghi e discepoli. See also D2.6.2, D2.6.3, and D2.6.4.*

Documents

D2.6.1 P. De Francisci to T. Levi-Civita, Rome 23.10.1938

ACS-*contro-discriminazione*, fol. 1r
Your personal census form shows that you belong to the Jewish race. Therefore, you have been suspended from service with effect from 16 October 1938-XVI in accordance with royal decree law of 5 September 1938, no. 1390. Respectfully yours, the Rector De Francisci[56]

Dalla Vostra scheda di censimento personale risulta che appartenete alla razza ebraica. Siete stato, pertanto, sospeso dal servizio a decorrere dal 16 ottobre 1938 XVI a norma del R.D.L. 5-9-1938 n° 1390. Con osservanza, il Rettore De Francisci

D2.6.2 G. Fubini to G. Vallauri, Turin 25.10.1938

ASPoliTo-*Fubini*, fol. 1r
To H.E. the Director of the R. Polytechnic of Turin. In response to the communication dated 19/10 from your High Excellency, the Director of R. Politecnico di Torino, I can only say the following:

1. I have no military merit because I was repeatedly declared unfit for military service.
2. My academic position in Italy and abroad, the work I have done in the School are facts, of which your High Excellency can be a good judge, due to your perfect knowledge of all the relative details.
3. I can only recall that Prof. Lane (of the University of Chicago) in his review of a treatise written in collaboration with a postdoc student of mine, Prof. Čech of the University of Brno, when summarising the results of a new branch of geometry (of which I was one of the founders) concluded, roughly, with the following words: "This book will be enough to remind Americans of the need to learn Italian if they want to be aware of the most important scientific advances."[57] Please accept my most respectful regards. G. Fubini Ghiron

---

[56] Pietro de Francisci (1883–1971), Rector of the University of Rome from 1935 to 1943.

[57] E. P. Lane (1886–1969), professor of mathematics at the University of Chicago, had reviewed in enthusiastic terms the book by G. Fubini and E. Čech, *Geometria proiettivo-differenziale*, Bologna, Zanichelli, vol. I 1927, vol. II 1928, in the Bulletin of the American Mathematical Society (33, 1927, pp. 113–114 and 34, 1928, pp. 382–383). The sentence that Fubini quoted off-the-cuff stated: 'This treatise fulfils a long-felt need. There is no other of its kind. Hereafter, the projective differential geometer must know his Fubini and Čech, and every student of geometry has now a new reason for acquiring at least a reading knowledge of *la lingua italiana*.' Eduard Čech (1893–1960) had studied with Fubini in Turin in 1921–22.

A S.E. il Direttore del R. Politecnico di Torino. In risposta alla comunicazione in data 19/10 di V.E. il Direttore del R. Politecnico di Torino, posso soltanto dire quanto segue:

1) Io non ho benemerenze militari perché replicatamente dichiarato inabile al servizio militare, 2) Quale sia la mia posizione accademica in Italia ed all'estero, quale sia l'opera che ho svolto nella Scuola sono fatti, di cui V.E. può essere buon giudice, perché a perfetta conoscenza di tutti i particolari relativi. 3) Posso soltanto ricordare che la recensione del prof. Lane (della Università di Chicago) di un trattato (scritto in collaborazione di un mio quasi allievo, il prof. Čech della Università di Brno), in cui si riassumevano i risultati di una nuova branca della geometria, di cui io sono stato tra i fondatori, concludeva, all'incirca, colle parole seguenti: "Questo libro basterà a ricordare agli americani la necessità di imparare l'italiano, se vogliono essere al corrente dei progressi scientifici più importanti." Gradisca V.E. i miei più distinti saluti.

D2.6.3 G. Fano to G. Vallauri, Colognola ai Colli (Verona) 29.10.1938

ASPoliTo-*Fano*, fol. 1r
Dear Sir, Director of the Royal Polytechnic of Turin, I warmly thank the council of this Faculty of Engineering for the mindful greeting that, through you, it was pleased to address to me. I have always felt, and I feel particularly attached to the Institute, for having experienced all its phases: from the initial practices for its first constitution to the most recent years of your enlightened and energetic management. The organisation and progressive improvement of the two geometry courses for the teaching licence were certainly one of the main pursuits of my (not short) career as a teacher. Kind regards, Gino Fano

Ill.mo Sig. Direttore del R. Politecnico di Torino, Ringrazio vivamente il Consiglio di cotesta Facoltà di Ingegneria del memore saluto che per mezzo Vostro si è compiaciuto rivolgermi. A cotesto Istituto mi sono sempre sentito e mi sento particolarmente legato, per averne vissute tutte le fasi, dalle pratiche iniziali per la sua prima costituzione agli anni più recenti della illuminata ed energica Vostra Direzione. L'organizzazione e il graduale perfezionamento dei due corsi di geometria del biennio sono stati certo una delle esplicazioni principali della mia non breve carriera di Insegnante. Gradite distinti ossequi, Gino Fano

D2.6.4 G. Fubini to G. Vallauri, Paris 1.11.1938

ASPoliTo-*Fubini*, fol. 1r
Dear Vallauri, I wrote to you [D2.6.2], because I believed it to be my duty. I thank you for the words you wanted to add, and even more for the affectionate phrases that you have so kindly addressed to me. *If you deem it appropriate*, please express to the Council, on my behalf, the assurance of my gratitude. Appreciate the devoted and grateful greeting of your affectionate G. Fubini

Caro Vallauri, Ti ho scritto quella lettera, perché l'ho creduto mio dovere. Ti ringrazio delle parole che hai voluto aggiungerle, e ancora più delle frasi affettuose che mi hai così cortesemente indirizzato. Se lo crederai opportuno, *ti prego di esprimere a mio nome al Consiglio Didattico i sensi della mia gratitudine. Gradisci il devoto e grato saluto del tuo aff. G. Fubini*

2.7 Silent Witnesses and Active Characters

Expulsions occurred amidst the silence of institutions, society, and high culture. The minutes of faculty meetings and sessions of academies and scientific societies, observatories, and research centres record with sterile synthesis the purge of Jewish scholars and their substitutions with Aryans. In middle and secondary schools, dismissals took place amid the indifference of colleagues who, by their attitude, ended up being not only silent witnesses but also direct accomplices:

> And in particular I did not know how to give myself peace seeing that they accepted the new state of affairs, I do not say without making opposition, because they would not have had the opportunity to do so, but without at least trying to understand the appalling phenomenon of a government that treats as criminals the best of its citizens, people that the government itself had judged worthy of esteem and honoured until a few months before. Twice I cried at that time; once when in Courmayeur I met an old colleague of twenty years previously, who as soon as she saw me expressed her indignation towards the anti-Semitic move and recalled how, among the teachers in Aosta, I had undoubtedly had the strongest of patriotic feelings, just in contrast with her who had opposite political tendencies; I got a lump in my throat and I felt tears welling up and if I spoke, I would burst into sobs. The other time I cried was when a lady came to visit me in Turin, the mother of a former student of mine, who I had helped by giving him a few hundred lire and other assistance to start his military career. When she said that no one ever had so many, and so selflessly showed, feelings of charity towards her family as me, who was a Jew, I could not restrain myself and I cried like a child.[58]

---

[58] Emilio Artom (1940–41), in Treves (1954), pp. 62–63: *E in particolare non sapevo darmi pace vedendo che essi accettavano il nuovo stato di cose, non dico senza opporsi, perché non avrebbero avuto il modo di farlo, ma senza cercare almeno di capire il fenomeno nuovo per loro di un governo che si scaglia sui migliori fra i suoi cittadini e colpisce come delinquenti individui che al governo stesso erano parsi fino a pochi mesi prima degni di stima e di onori. Due volte io piansi in quel periodo; una volta quando a Courmayeur incontrai una collega di vent'anni prima, che appena mi*

However, there were not only silent witnesses but also profiteers. Faced with persecution, appetites were unleashed. On December 10, 1938, in a meeting of the Scientific Commission that has become sadly famous, UMI expressed its strong concern about the consequences that the measures on race might have on mathematics, a sector which had been seriously affected in quantity and quality, and concluded: 'one must not give either in Italy or abroad the impression that the removal of Jewish elements has produced a decline in Italian mathematical activity. It's a matter of national pride!'[59]

Among editorial boards, there was a move to kick out the Jews, depriving editorial staff of efficient and high-skilled collaborators such as Segre and Levi. The book cleansing act, which concerned school textbooks only, was extended by tacit agreement to include articles and books.

There were those who took advantage of this for themselves and for their students, those who came to galvanise foreign colleagues. Picone, for example, turned to W. Sierpiński:

> You certainly know the anti-Jewish measures taken by our government for Universities and Academies and it is therefore urgent that scientists of Aryan race collaborate as actively as possible to show how science can equally advance even without Jewish intervention, and this will be more effective if such collaboration be international. I therefore ask you to send me your own unpublished works or papers by your followers as soon as possible for publication in the periodicals edited by the Academy of Lincei and the Royal Society of Naples. In particular, works from Arians are needed for the Academy of Lincei, in which the number of members of Jewish race has reached a very high percentage. Sure that you will certainly want to promote a great increase in Aryan-Polish scientific collaboration with the Italian Academies, and waiting for your positive response, I send you [...] my best wishes.[60]

---

vide mi espresse il suo sdegno verso il movimento antisemita e mi ricordò come fra gli insegnanti di Aosta io fossi senza dubbio il più caldo di sentimenti patriottici, proprio in contrasto con lei di tendenze assai diverse; mi venne un nodo alla gola e sentii che le lacrime mi scendevano dagli occhi, e, se avessi parlato, sarei scoppiato in singhiozzi. L'altra volta fu quando venne a trovarmi a Torino una signora di cui avevo beneficato il figlio, mio alunno, permettendogli col dono di qualche centinaio di lire e con altri aiuti di iniziare la carriera militare. Quando ella affermò che nessuno mai aveva tanto e così disinteressatamente mostrato sentimenti di carità verso la sua famiglia quanto io ebreo, non potei trattenermi e piansi come un bambino.

[59] AS-UMI: L. Berzolari to E. Bompiani, Pavia 13.12.1938: *che sia assolutamente impedito che in Italia e soprattutto all'estero, si riceva l'impressione che l'allontanamento degli ebrei abbia prodotto un declino nell'attività matematica italiana. È una questione di orgoglio nazionale!* The conclusions of the sitting of the Scientific Commission were communicated on a confidential basis, before publication, to various mathematicians. See e.g. L. Berzolari to G. Vacca, Pavia 14.12.1938 in Nastasi and Scimone (1995), pp. 11–12.

[60] M. Picone to W. Sierpiński, Rome 7.1.1939, in Guerraggio et al. (2007), pp. 57–58: *Voi conoscete certamente i provvedimenti antiebraici presi dal nostro Governo per le Università e per le Accademie ed urge, pertanto, che gli scienziati di razza ariana collaborino il più attivamente possibile per mostrare come la scienza possa egualmente progredire anche senza l'intervento giudaico, e ciò sarà tanto più efficace quanto più detta collaborazione sarà internazionale. Vi prego quindi, anche per tali ragioni, di volere al più presto possibile inviarmi Vostri lavori inediti, o di Vostri discepoli, per la pubblicazione di essi nei Rendiconti dell'Accademia dei Lincei e della Società Reale di Napoli. Specialmente occorrono lavori provenienti da ariani per l'Accademia dei*

Symbols of the 'betrayal of intellectuals' were Severi and Bompiani. In addition to preparing the agenda voted by the UMI Scientific Commission, they asked the Ministry of National Education to return all the chairs left vacant by Jewish mathematicians, reassuring the Ministry that Italian mathematics 'even after the elimination of some Jewish scholars, had preserved scientists capable of maintaining the highest tone of the discipline'.[61] Severi, a long-time friend of Levi-Civita and a pupil of Jewish Masters like Segre, Fano, Castelnuovo, and Enriques, was especially fierce against the latter. He transferred himself to the chair of Higher Geometry that had been held by Enriques and took over from him the presidency of the National Institute for the History of Sciences, closing it down in a few months. He took away from Enriques the presidency of Mathesis, and expelled from the editorial board of the *Annali di Matematica pura ed applicata* even his protégé: Beniamino Segre. The story was told with bitterness by Segre himself to Levi-Civita:

> Dear and distinguished Professor, I suppose that you too have been exempted from the duties of co-director of *Annali*. Having learned of a little double-dealing in this regard, I hasten to communicate it to you so that you can regulate yourself, but I beg you to make discreet use of this information. The initiative started from S. [Severi], who some time ago reported to the President of the A.I.E. [*Associazione Italiana Editori*] the situation in which the board of *Annali* found itself. The President then wrote to Dr. D.M. [Della Monica] requesting his opinion; and he responded by completely deferring the case to the decision of S. [Severi]. Nothing like this has ever happened before now![62]

Bompiani, a more subtle mind than Severi, succeeded in the total exclusion of Jews from UMI and the University of Rome (D2.7.1). He proposed rigid application of the legislation on texts by non-Aryan authors, refusing to publish articles of Italian and foreign Jews in the *Bollettino* of UMI, even if submitted before the introduction of the racial laws (as in the case of a paper of I. Opatowski), and also prohibited contributions by 'dubious' authors, such as Bruno Finzi, who actually was not Jewish (D2.7.2).

Lincei, nella quale i soci di razza ebraica raggiungevano una percentuale elevatissima. Sicuro che vorrete senz'altro farvi promotore di un grande incremento della collaborazione scientifica ariano-polacca con le Accademie italiane, in attesa di un Vostro cenno di assenso, vi invio [. . .] l'espressione del più cordiale saluto.

[61] *Riunione della Commissione scientifica della U.M.I.* [10.12.1938], Bollettino dell'UMI, s. 2, 1, 1939, p. 89: *anche dopo le eliminazioni di alcuni cultori di razza ebraica, ha conservato scienziati che, per numero e per qualità, bastano a mantenere elevatissimo, di fronte all'estero, il tono della scienza matematica italiana.*

[62] ANL-*Levi-Civita*: B. Segre to T. Levi-Civita, Bologna 16.10.1938: *Caro ed illustre Professore, suppongo che Lei pure sia stata esonerata dalle mansioni di Condirettore degli* Annali. *Avendo al riguardo saputo di un poco simpatico retroscena, mi affretto a comunicarglielo affinché Ella possa regolarsi, pregandola però di farne uso discreto. L'iniziativa è partita dal S. [Severi], il quale tempo addietro segnalò al Presidente dell'A.I.E. [Associazione Italiana Editori] la situazione in cui si trovava la Direzione degli* Annali. *Detto Presidente scrisse allora al dott. D.M. [Della Monica] richiedendone il parere; e questi rispose rimettendosi completamente alle decisioni del S. [Severi] In nessun altro caso è stato fatto finora nulla di simile!*

Undoubtedly, there were also acts of solidarity, which cannot and must not be forgotten. For example, G. Pompili, just after promulgation of the race provisions, reached Viareggio to work with Enriques:

> And in a warm afternoon, just before sunset, under the purplish blue of the beautiful sky of Viareggio, the Master, and the disciple hug, both with tears in their eyes. Giuseppe Pompili had, with all his sweet simplicity, performed a great and generous act; he stayed with us for a couple of days and my poor fading eyesight still glimpsed the smile of Federigo Enriques who chatted animatedly of mathematics with his precious friend, forgetting the tragedy of the moment. ... The extraordinary is to do nothing extraordinary.[63]

Much more substantial was the concrete help, even putting at risk personal safety, that mathematicians would offer to their Jewish colleagues in the period of the German occupation. Thanks to M. Puma, T. Viola, A. Frajese and F. G. Tricomi, Castelnuovo and Enriques escaped the round-up of October 1943 and would live in hiding until the liberation. E. Togliatti helped G. Loria to leave Genoa and to take refuge, at more than eighty years old, in the Pellice valley.[64]

Documents

D2.7.1 E. Bompiani to H. Geppert, Rome 15.1.1939

AS-UMI, *correspondence 1939*, fol. 1r
Dear Prof. Geppert, I was very pleased with your letter of the 14th and I respond immediately, thanking you first of all for the appointment of Severi and myself as members of the editorial board of the *Zentralblatt*.[65] [...] We are completely reorganising the U.M.I., which had been left in abandonment by the Jews who for a long time had control over it all; already from the 1st issue [of *Bollettino*], which will be released at the end of January, you will see the difference, and we already have material ready for more than one other issue. We are also giving the Union a sound financial basis that will allow us to take further initiatives. The UMI Congress, whose organisation had been badly arranged by Levi, will be held in Bologna in 1940 (at Easter);[66] but we will certainly meet first, on the occasion of the Volta

---

[63] L. Cohen Enriques (2001), pp. 84–85: *E in un tiepido pomeriggio, poco prima del tramonto, sotto l'azzurro violaceo del bel cielo di Viareggio, si abbracciarono il Maestro e il discepolo: ambedue con le lacrime agli occhi. Giuseppe Pompili aveva, con tutta la sua dolce semplicità, compiuto un atto grande e generoso; rimase con noi un paio di giorni e i miei poveri occhi semispenti rividero il sorriso di Federigo Enriques che parlava animatamente di matematiche col prezioso amico, dimenticando la tragedia del momento. ... Lo straordinario è di non fare nulla di straordinario.*

[64] Togliatti (1952).

[65] Harald Geppert (1902–1945) had been nominated as managing editor (*Generalredakteur*) of both the *Jahrbuch über die Fortschritte der Mathematik* and *Zentralblatt*. Severi and Bompiani had entered the editorial board of the *Zentralblatt* in late 1938, after the expulsion of Levi-Civita. See §3.12.

[66] The second national congress of the UMI would take place in Bologna on 4–6 April 1940.

Conference, which we are beginning to organise in these very days with Severi.[67] [...] Here we have settled beautifully, with the transfer of Severi to Higher Geometry, calling Signorini for Rational Mechanics and Tonelli for Analysis (the move will take place in October).[68] What do you think? [E. Bompiani]

Caro prof. Geppert, mi è giunta molto gradita la Sua del 14 e rispondo subito, ringraziandola anzitutto per quanto la nomina di Severi e mia a membri del Comitato di Redazione del Zentralblatt. *[...] Stiamo riorganizzando completamente l'U.M.I. (che era stata lasciata in abbandono proprio dagli ebrei che l'avevano tutta in mano); già dal 1° fasc. (che uscirà alla fine di Gennaio) vedrà la differenza (ed abbiamo già materiale pronto per più di un altro fascicolo. Inoltre stiamo dando all'Unione una solida base finanziaria che ci permetterà altre iniziative. Il Congresso, la cui organizzazione era stata male impostata dal Levi, sarà fatto a Bologna nel 1940 (a Pasqua); ma noi ci vedremo certamente prima, in occasione del Convegno Volta, di cui proprio in questi giorni iniziamo l'organizzazione con Severi. [...] Noi qui ci siamo assestati magnificamente con il passaggio di Severi alla Geom. Super., con la chiamata di Signorini per la Meccanica raz., e con la chiamata di Tonelli per l'Analisi (il trasferimento avverrà in Ottobre). Che le pare? [E. Bompiani]*

D2.7.2 E. Bompiani to L. Berzolari, Rome 8.3.1939

AS-UMI, *Bompiani correspondence*, fol. 1r
My dearest Berzolari, I have just received a letter from [Ettore] Bortolotti [...]. From the same letter I learn of an article by B. Finzi on the Institute for Applications of the Calculus.[69] If you allow me, I would like to tell you about my impressions in this regard, both on the author, and on the article. Apart from the definition, Finzi is of Jewish race; if we publish his article, we are formally right; but it is my conviction that, given the diffusion we wish to create, and that we have given to the *Bollettino* in the middle schools, we must be more cautious than the law requires. [...]

Carissimo Berzolari, ricevo ora una lettera di Bortolotti [...]. Dalla stessa lettera apprendo di un articolo di B. Finzi sull'Istituto per le Applic. del Calcolo. Se mi permetti, vorrei dirti la mia impressione in proposito, sia sull'A., sia

---

[67] See §2.8.

[68] Antonio Signorini (1888–1963) had been called from Palermo to occupy the chair of his Master Levi-Civita. During the years of 1939–1942 Tonelli was officially transferred from Pisa to Rome. However, he considered the appointment as purely nominal.

[69] B. Finzi, *L'Istituto Nazionale per le applicazioni del calcolo nel quadriennio 28 ottobre 1933-XII-27 ottobre 1937-XV, Roma, 1938*, Bollettino dell'UMI, s. 2, 1, 1939, pp. 487–489.

sull'articolo. A parte la definizione, Finzi è sicuramente di razza ebraica; se noi pubblichiamo l'art. siamo formalmente a posto; però è mia convinzione che, data la diffusione che vogliamo dare, e abbiamo dato al Boll. nelle scuole medie, dobbiamo essere anche più guardinghi di quanto la legge imponga. [...]

2.8   Case Study. Presence as Absence: The Volta 1939 Conference

Created by royal decree on 7 January 1926, following the publication of the *Manifesto of the anti-fascist intellectuals*, but officially inaugurated only on 28 October 1929, the *Reale Accademia d'Italia* was established as a central hub, designed to encourage the collaboration of pre-existing institutions towards purely national aims and to represent national culture in front of foreign countries.[70] It was an advisory body for the regime on cultural matters, a centre for the coordination of the country's intellectual forces and a propaganda authority, which was entrusted with the task of conserving the purity of the national character of research according to 'the genius and traditions of Italian noble race' and enhancing the huge wealth of treasures produced by Italian genius, 'so that it should no longer happen that Italian discoveries be awarded their first recognition and applications beyond the borders'.[71]

Chaired by well-known fascist intellectuals such as T. Tittoni, G. Marconi, G. D'Annunzio, L. Federzoni, G. Gentile, and G. Dainelli, the Academy counted 60 members, divided into four classes: Moral and Historical Sciences; Physical, Mathematical and Natural Sciences; Humanities and Arts. The appointments of the first thirty academicians dated back to March 1929; the others were made by the Duce, in agreement with the Minister of National Education, on the basis of three candidatures proposed by the Academy for each vacant seat. The appointment was for life and entailed honorary privileges and material advantages. On a par with the great dignitaries of the State, authorised to bear the title of Excellence (*Eccellenza*), the academicians enjoyed an appealing prerogative: 36,000 lire, in addition to attendance fees and mission allowances, which could be combined with salaries and pensions. For this reason, as well as for the prestige of the title, the appointments were tempting to many and caused considerable fractures in the various scholarly communities, including that of mathematicians. In fact, the three years that passed

---

[70]On the history of the *Accademia d'Italia* see, inter alia, Ferrarotto (1977); Ostenc (1994), pp. 117–118; Turi (2002, 2016).

[71]Annuario della Reale Accademia d'Italia, I, 1929, p. 35: *Come non dovrà più accadere che le scoperte dei nostri scienziati ricevano fuori dai confini il primo riconoscimento e le prime vaste applicazioni, così sarà ora possibile porre in luce e valorizzare gli immensi tesori ancora sconosciuti o sperduti, prodotti dal genio italico.*

from the creation to the inauguration of the Academy were due to the endless manoeuvres which surrounded the appointment of the first academics (16 of whom belonged to the Lincei and thus ensured a certain continuity with the traditional academic world), to the refusal of some to join the Academy and vice versa to the pressures of others to enter at all costs. It is hardly necessary to cite the battle that took place between Severi and Enriques about entering the first squad (March 1929) and those that were fought around the appointments of Fano, Fubini, Levi-Civita, Peano, Berzolari, and Marcolongo.

The undisputed protagonist of the initiatives of the Academy of Italy related to mathematics is Severi. There was no scientific or cultural aspect of mathematics addressed by this institution on which he did not intervene. The mathematical candidacies were all piloted by him, the series *Memorie* and *Atti* were dominated by his contributions and by those of his protégés. It was his idea to start a collection of books illustrating the Italian contribution to mathematics and of making the Academy's journals an icon of Italian-ness, admitting works by foreign mathematicians only in exceptional cases, and prohibiting the use of any language other than Italian. In order to 'bring the light of his knowledge and make a distinguished work of Italian-ness, with a truly fascist spirit', Severi also travelled, as an *accademico d'Italia*, all over the world, from Argentina to Japan, with mathematical missions to which the Italian press dedicated a media coverage never again replicated.

The Academy of Italy received an annual endowment of three million lire from the government; in 1930, moreover, the engineer G. Motta, executive Director of the *Società Generale Italiana Edison di Elettricità*, based in Milan, donated ten million lire to Mussolini to establish a foundation named after A. Volta, with the aim of promoting higher culture in all branches of knowledge, but with special regard to the physical, mathematical, and natural sciences. The Duce entrusted the administration and management of the newly born Foundation to the *Accademia d'Italia*.

One of the tasks with which the latter institution had been entrusted was to promote national intellectual work abroad, but this required a thorough knowledge of foreign countries. The positions diverged, however, in the end the Academy decided to entrust the Volta Foundation with this programme. The Volta Foundation thus became the task force of the Academy of Italy for international relations. It integrated the Academy in its 'highly understood work of Italian-ness' and supported it in its policy of spreading Italian culture abroad though an activity of cultural diplomacy.[72] It did so, indeed, effectively, by financing (in nine years) 61 travel fellowships abroad for young scholars and nine international conferences, entrusted in turn to each of the four classes. The first and last conferences went to the class of Sciences: the first (October 1931) was the well-known Volta congress on nuclear physics organised by Fermi (secretary general of the Academy), which brought eight Nobel Prize laureates to Rome. The last was the Volta Conference of 1939, organised by Severi and Bompiani, two scholars of the elite of Italian mathematics

---

[72]Luciano (2016).

in the fascist period. The Volta Conferences had a special scientific and political hallmark:

> By forging cordial personal relations between the representatives of our culture and those of foreign culture, they facilitate the exchange of ideas, broaden the intellectual horizon, and foster a more equitable evaluation of the individual contributions to the common culture. It is therefore a particular advantage for us Italians, who have been systematically devalued for a long time and who have everything to gain by being better known.[73]

Since they were associated with lavish visits, exhibitions and receptions, the Volta Conferences also served a propaganda function. Nuclear physics, Europe, immunology, theatre, Africa, etc. were all themes chosen specifically to highlight the strong-willed and realistic ideals that must inspire the new fascist spirit and mood, and to illustrate to the most distinguished men of study in the world the achievements of the government, the various primacies, the leadership positions historically acquired and those of more recent conquest.

Participation in the Volta Conferences was strictly by invitation. For each event, the Academy chose a president and selected about 60 participants, half Italian and half foreigners. Participants were divided into speakers and guests, who might also be joined by auditors. Participation was in a personal capacity, and there were no official delegations from the various countries until 1936. Political checks were carried out on speakers and guests by the Ministry of Foreign Affairs and the Italian diplomatic institutions; the names were then submitted to Mussolini for final approval. For the Volta Conference of 1939, the political investigations were supplemented by checks on racial identity.

Among his various assignments, Severi was an advisor for the Volta Foundation, and, after the death of O. M. Corbino, he joined the Foundation's board. It was in this period that the Academy commissioned him to organise the IX Conference. Scheduled for October 1939 (22–28), the conference was postponed *sine die* due to the outbreak of war, and finally cancelled, although the proceedings were published in the summer of 1943, shortly after the fall of fascism.[74] This event was an exercise of science anti-diplomacy from a double perspective: political and mathematical. The extensive correspondence kept in the archives of the Academy of Italy and the Italian Mathematical Union fully returns the behind the scenes to this virtual conference, a dense plot of academic opportunisms, rivalries, authentic or simulated adhesions to fascist and racist ideology, ethical and scientific ideas betrayed by conviction or calculus.

All aspects of the event, from the agenda to selection of the speakers, were managed by Severi and Bompiani, who meant to take advantage of the conference

---

[73] Marpicati (1934), p. 19: *Stringendo cordiali relazioni personali fra i rappresentanti della nostra cultura e quelli della cultura estera, facilitano lo scambio delle idee, allargano l'orizzonte intellettuale e favoriscono una più equa valutazione dell'apporto di ciascuno alla cultura comune. È quindi un vantaggio particolare per noi italiani, che siamo stati per lungo tempo sistematicamente svalutati e che abbiamo tutto da guadagnare a esser meglio conosciuti.*

[74] *Convegno di Scienze Fisiche Matematiche e Naturali, Matematica contemporanea e sue applicazioni*, Roma, Reale Accademia d'Italia, 1943.

to display their personal conception of the history and political geography of mathematics, and to exhibit the *führende Stellung* of Italian mathematics, vis-à-vis anglophone countries in particular.

The first negotiation focused on the theme: excluding physics (which had already been dealt with in 1931) the choice could fall on natural sciences or mathematics, which however appeared to the elite of the fascist rule too neutral and unfitting for a propaganda occasion. Mathematics did not seem to have great prospects of success and instead Severi, very fiercely, beat the competition by presenting himself at the meeting of the Volta Foundation on 8 May 1937 already with a title—*Matematica contemporanea e sue applicazioni*—and a rational. No other topic—he stated on this occasion—could more perfectly highlight one of the sectors in which the creative activity of Italians had been most fruitful, in which Italy held an undisputed primacy and testify 'the importance that the regime attributed to the highest scientific speculation, also in consideration of the benefits that pure science always ended up bringing to applications and therefore to social wealth and to the resolution of autarkic problems'.[75]

Cashing in on the success, and charged with the presidency of the Conference, Severi had to build the team. He did so by involving a single colleague, Bompiani, whom he admired for his cunning intelligence in moving in political and ministerial circles. Severi's self-feeling of being the man alone at the command of Italian mathematics, prevented him from trusting any applied mathematician: the applications, which had been exalted in the rational of the Conference to smooth the upper echelons of the fascist party by appealing to one of the main leitmotifs of the Latin genius, came off the radar of the Conference, never to return.

Once the question of the organising committee had also been archived, invitations had to be issued. According to the Regulation of the Volta Conferences, invitations needed to be first examined and authorised by the Ministry of Foreign Affairs and then finally issued by the embassies. In the summer of 1938, the enactment of the racial laws was now in the air. Everything suggested that the introduction of anti-Jewish measures was imminent. The presidency of the Council ordered the exclusion of Jews from delegations to international congresses on 18 June 1938. The measure had been preceded by another provision dated 18 November 1937: the obligation to request from the Presidency of the Council of Ministers the authorisation for all trips relating to scientific missions, events, and congresses abroad. At the beginning of July, the *Manifesto of the race* had been published.

The leadership of the Academy of Italy was worried. G. Vallauri, A. Bruers, and F. Pellati wondered if it was not a mistake to call a completely Aryan mathematics conference.[76] A maximum number of Jewish scholars could be set—they suggested;

---

[75] A. Conti (alias F. Severi) (1938), pp. 129–130: *l'importanza che il regime attribuisce ai problemi della più alta speculazione scientifica, in considerazione anche dei benefici che la scienza pura finisce sempre con l'arrecare alle applicazioni e quindi al benessere sociale ed alla risoluzione dei problemi autarchici.*

[76] ANL-AI: A. Bruers to F. Pellati, Rome 8.9.1938.

at least a few Jewish foreigners could be invited, and the theme of the conference could be changed. Severi, on the other hand, was quiet. Playing ahead, he had already excluded Italian Jewish mathematicians from the list of invited speakers, so when, on 19 November, two days after the promulgation of the *Measures for the Defense of the Race,* Federzoni observed that they would have to be scrupulously applied, Severi could reassure him: the conference would succeed perfectly even without the participation of the Jews.[77] Federzoni would have his appreciation of comrade Severi, who had taken on this 'noble task to show that mathematics was not a province of Israel'[78] put on record. Actually, Severi was not entirely honest with Federzoni: the racially based skimming had been carried out only for Italians, not for foreigners or for scholars of mixed race.

The exclusion of Jewish scholars 'constituted further proof of the involvement and active participation in the anti-Semitic policy launched by the regime by the leading institution of fascism in the cultural sector',[79] and was particularly delicate in a field such as mathematics in which the Jewish presence was relevant and internationally recognized. How could one talk about algebraic geometry without Castelnuovo and Enriques, about functional analysis without Volterra, about mathematical physics without Levi-Civita? Despite the precautions put in place, protests and dissociations could not be avoided. The most significant episode of scientific and human solidarity was on the part of the Dutchman J. A. Schouten, professor of mathematics at the Polytechnic University in Delft, who explicitly refused to take part in a congress that excluded Levi-Civita, Segre, D. von Dantzig, and L. Berwald (D2.8.1, D2.8.2, and D2.8.3).

The correspondence relating to the organisation of the 1939 Volta Conference, preserved in the archives of the Academy of Italy and the Italian Mathematical Union, has brought many new elements of evaluation of individual responsibilities. Letters showed that those who agreed to intervene, often with pride and satisfaction, were aware of the ostracism of Jewish colleagues and that the reactions of dissent were in the minority, not only on the part of Italians but also of foreigners and even of scholars from countries where political anti-Semitism did not exist. Of course, it is impossible to assess whether individual behaviour was dictated by an anti-Jewish prejudice or by other factors, for example the need to collaborate with Italian mathematicians in preparation for the 1940 ICM in Cambridge Mass. It is impossible to say whether the prestige of having been appointed as an invited speaker at a Volta Conference led G. Vrânceanu, or G. De Rham to overlook the fact that their old Italian Masters had been excluded. Documents prove, however, that while Severi assumed full moral responsibility for the exclusion of Jewish mathematicians, it was

---

[77] ANL-AI: *Verbale del Consiglio della Fondazione Volta*, 19.11.1938, p. 615: *dimostrerà il suo perfetto funzionamento anche senza la partecipazione dei matematici ebrei.*

[78] ANL-AI: *Verbale del Consiglio della Fondazione Volta*, 21.1.1939, p. 640: *nobile compito di mostrare col prossimo Convegno che la matematica non è una provincia d'Israele.*

[79] Capristo (2006), p. 166: *costituisce un'ulteriore prova del coinvolgimento e dell'attiva partecipazione alla politica antisemita varata dal regime da parte dell'istituzione guida del fascismo nel settore culturale.*

Bompiani, well before the approval of the racial laws, who pressed for meticulous racial controls and for an inflexible application of ostracism, not only against Italians, but also against foreigners, and even carried out a censorship of citations and bibliographies.[80] He was doing the same, after all, within the UMI Scientific Commission.

Bompiani's zeal, although shared by Severi, caused serious difficulties in the organisation of the Conference, so much so that when Severi intervened at the meeting of the Volta Foundation in March 1939, he began to be really worried. He had had to remove the American J. Douglas ('Jew, and an angry Jew to boot'[81]), the Czech E. Čech, who was erroneously denounced as Jewish by Bompiani,[82] and the Hungarian B. Kerékjártó. Other names he put forward in their place, Bompiani rejected or kept in abeyance. In the meantime, Severi noted the repercussions that the Italian racial turn had had abroad and the contrasts with foreign scholars aroused by the criteria adopted for the choice of participants. His concerns about the progress of the preparatory work were such that he consulted the cabinet of the Presidency of the Council of Ministers on the advisability of inviting at least some foreign Jews to the congress.

An equally delicate issue derived from the fact that, before being issued, invitations to foreigners had to be first examined and authorised by the Ministry of Foreign Affairs and diplomatic institutions, which were responsible for preliminary investigations of a political reliability and racial purity. Severi and Bompiani were caught off guard by the reactions of these institutions. In addition to the racial exclusions, others arrived that they did not expect: the Danish H. Bohr (son of a Scandinavian father and a Jewish mother, who, like his brother N. Bohr, 'has taken an attitude of public hostility to racist principles'[83]), the Belgian T. de Donder ('anti-fascist, socialist militant and free-mason'[84]), the Dutch B.L. van der Waerden, the German C. Siegel, the French P. Montel ('great dignitary of Freemasonry'[85]) and the Russian B. Kagan ('communist'[86]).

In addition to this, since March the Academy received many refusals to participate as an implicit or explicit form of protest against the exclusion of their Jewish colleagues: J. A. Schouten, J. L. Synge, É. Picard and H. Lebesgue, E. T. Whittaker and A. Eddington, C. J. de la Vallée Poussin, L. Godeaux, O. Veblen, and M. Morse. On March 11, 1939, Severi, furious at the way things were going on, asked

---

[80] ANL-AI: F. Severi to J.A. Schouten, Rome 27.3.1939: *piena responsabilità morale.*

[81] ANL-AI: note by F. Carli: *ebreo, ed ebreo arrabbiato per giunta.*

[82] ANL-AI: E. Bompiani to F. Carli, Rome 4.3.1939.

[83] ANL-AI: G. Sapuppo to L. Federzoni, Copenhagen 21.2.1939: *i due fratelli Bohr hanno assunto un atteggiamento di pubblica ostilità ai principi razzisti.*

[84] ANL-AI: MFA to AI, Rome 3.7.1939: *un militante antifascista, membro della sinistra liberale a tendenze radical-socialiste, massone.*

[85] ANL-AI: L. Federzoni to G. Ciano [MFA], Rome 16.2.1939: *esponente notissimo della massoneria.*

[86] ANL-AI: L. Federzoni to G. Ciano [MFA], Rome 16.2.1939: *russo comunista.*

Federzoni to solicit the support of MFA so that he could urgently replace the guests who refused to attend. Federzoni invited him not to dramatize and rather to organise a trip, at the expense of the Academy, to 'promote and facilitate the coming' of foreigners, to persuade the doubtful, in a word to seduce them.[87] The Cartan's jubilee (May 1939) appeared to be an opportunity to be exploited for this purpose. Severi participated to meet É. Picard, H. Lebesgue, A. Denjoy, and M. Fréchet and to convince them to accept the invitation, but there were embarrassing moments because some dismissed scholars attended the event: Enriques and Volterra's son, Edoardo.

Among the refusals, perhaps the most emblematic came from the American delegation. Veblen and Morse, in fact, informed by Schouten of his decision, 'said no' as an implicit sign of solidarity with their colleagues who had been banned, and in particular as a gesture of support for Levi-Civita. Birkhoff was the only one to accept at first (1.4.1939). Severi's disappointment was severe. American mathematicians are ungrateful—he voiced—who, after having 'watered' from the source of the Italian Masters, have turned their backs on them. Bompiani suspected that some outstanding refugees, such as Fubini, flamed hostility against fascist Italy. When also Birkhoff withdrew his participation, the situation escalated. After trying in vain to convince him to rethink his decision, and after discussing it with Bompiani, Severi asked the UMI for help. Until then, the Union had had nothing to do with the organisation of the Volta Conference and in general had peripheral relationships with the Academy of Italy, almost as if the two institutions had tacitly divided the spheres of mutual influence. The Birkhoff case was the first moment of alliance between the two bodies. To punish that filo-Jewish 'democratic burletta that American academia is',[88] Severi and Bompiani coined a retaliation plan: they menaced the withdrawal of the Italian delegation from the celebrations for the centenary of the University of Washington and from the ICM scheduled in Cambridge Mass. for 1940 (D2.8.4). Bompiani discussed this strategy with some UMI colleagues and with W. Blaschke, W. Süss, E. Sperner, H. Geppert, and G. Doetsch.[89] Severi liked the idea, and in mid-June 1939, he put it into practice.[90] Birkhoff immediately relented and, as a gesture of peace, proposed sending a text, so as to participate *in*

---

[87] ANL-AI: *Verbale del Consiglio della Fondazione Volta*, 11.3.1939, p. 647.

[88] ANL-AI: F. Severi to E. Bompiani and F. Carli, Chianciano (Siena) 14.6.1939: *che burletta la "democrazia americana"*!

[89] See D2.8.4 and in AS-UMI: E. Bompiani to E. Bortolotti and L. Berzolari, Roma 10.1.1939; W. Süss to E. Bompiani, Freiburg 12.1.1939; E. Bompiani to W. Süss, Roma 14.1.1939; E. Bompiani to W. Blaschke, Roma 18.1.1939; E. Bortolotti to E. Bompiani, Bologna 2.6.1939; E. Bompiani to E. Bortolotti, Roma 3.6.1939; L. Berzolari to E. Bompiani, Pavia 7.6.1939; E. Bompiani to L. Berzolari, [Roma] 23.6.1939; W. Süss to E. Bompiani, Freiburg 10.7.1939; E. Bompiani to L. Berzolari, Santa Marinella (Civitavecchia) 15.7.1939; E. Bompiani to W. Süss, Santa Marinella (Civitavecchia) 15.7.1939; F. Severi to E. Bompiani, Arezzo 18.7.1939; W. Süss to E. Bompiani, Freiburg 19.7.1939; E. Bompiani to L. Berzolari, Santa Marinella (Civitavecchia) 21.7.1939; L. Berzolari (E. Bompiani) to MFA, [21.7.1939]; H. Geppert to E. Bompiani, Giessen 25.7.1939.

[90] ANL-AI: F. Severi to A. Bruers, Chianciano (Siena) 14.6.1939.

absentia at the Volta Conference. Severi accepted but, in the reply, written in agreement with Bompiani, he did not fail in 'removing a few pebbles from his shoe': he felt sorry for Birkhoff who 'had been prevented'[91] from participating in the Volta Conference, deplored that in Cambridge plenary conferences be entrusted to only two Italians (L. Tonelli and F. P. Cantelli), and reiterated the warning to invite to ICM only scholars who were 'at present active members of Italian Academies and Universities',[92] i.e. only Aryans.

The racist orientation taken by the Italian government influenced the adhesion of foreign representatives to the Conference, especially those from democratic countries who reacted to the exclusion of their Jewish colleagues. There were a good number of adhesions (32), some variously motivated refusals and few explicit protests. Acceptance, like rejection, had a strictly individual character. In general terms, however, by guaranteeing their presence at a congress marked by conspicuous absences, the mathematicians who accepted the invitation collaborated with a regime committed to erasing the Jewish presence from the international circuit. The affair confirmed the compliance with the directives of the regime of many authoritative exponents of the scientific Italian and international community. Those who agreed to intervene were aware of the ostracism in force in Italy against Jewish scholars and knew that some of the first-rank representatives of their sectors would not be present. To accept the invitation meant to accept serious political conditioning and a blatant violation of internationally shared scientific values. The reactions of dissent were in the minority. The picture that emerges is that of a high culture that was not immune to anti-Semitism. The resurgence of anti-Semitism, like a karst river, was certainly not a very large or homogeneous phenomenon, but it ended up providing support for fascist state racism.

Documents

D2.8.1 J. A. Schouten to Reale Accademia d'Italia, Delft 6.4.1939

ANL-AI, fol. 1r
Dear Sirs, I could not immediately respond to the courteous invitation to participate in the Volta 1939 Conference addressed to me by the Royal Academy of Italy, as I wanted to first inquire whether Italian and foreign Jewish scholars should be invited or not. I have therefore sent a personal letter to Prof. Francesco Severi. Unfortunately, the answer I received leaves no doubt about that. I am therefore obliged to decline your invitation. However, please do not consider my refusal as a direct action against Italy: I have the utmost respect for Italian science,

---

[91] ANL-AI: draft of letter from F. Severi to D.G. Birkhoff, Fiuggi 14.8.1939: *Voi eravate il solo che avevate deciso di venire ed è un vero peccato che ne siate state stato impedito.*

[92] ANL-AI: F. Severi to D. G. Birkhoff, Rome 18.8.1939.

and I have feelings of deep friendship for my Italian colleagues. However, it is impossible for me to participate in a congress on Differential Geometry from which Italian and foreign scholars such as Tullio Levi-Civita, Guido Fubini, Beniamino Segre, D. van Dantzig and Ludwig Berwald are excluded due to racial prejudices.[93] I would like to point out that, for similar reasons, I also refused the invitation from the Canadian Government to participate in the 'international' congress of mathematicians held in Toronto[94], to which German scholars were excluded. With the utmost respect, yours truly J.A. Schouten

Meine Herren, Auf die freundliche Einladung der Reale Accademia d'Italia zur Beiwohnung des Convegno Volta 1939 habe ich nicht unmittelbar antworten können, da ich mich zunächst darnach erkundigen wollte, ob Italienische und ausländische Jüdische Gelehrte zur Tagung eingeladen werden sollten oder nicht. Ich wandte mich daher in einem persönlichen Schreiben an Prof. Francesco Severi. Seine erhaltene Antwort lässt leider in dieser Beziehung keine Zweifel bestehen. Leider befinde ich mich daher in der Lage Ihre freundliche Einladung abschlagen zu müssen. Ich möchte Sie bitten in dieser Weigerung keine gegen Italien gerichtete Aktion zu sehen, ich habe die grösste Achtung für die Italienische Wissenschaft und hege die freundschaftlichsten Gefühle für meine Italienischen Kollegen. Es ist mir aber unmöglich einen Kongress über Differentialgeometrie mitzumachen von dem Italienische und ausländische Gelehrte wie Tullio Levi-Civita, Guido Fubini, Benjamino [sic!] Segre, D. van Dantzig und Ludwig Berwald auf Grund von Rassenvorurteilen ausgeschlossen werden. In derselben Weise habe ich damals die Einladung der Canadischen Regierung zur Beiwohnung des „internationalen" Mathematikerkongresses in Toronto abgeschlagen, da man sich weigerte die Deutschen Gelehrten einzuladen. Mit vorzüglichen Hochachtung Ihr sehr ergebener J.A. Schouten

D2.8.2 J. A. Schouten to O. Veblen, Delft 6.4.1939

OVP, *JAS*, fol. 1r

Dear Sir and Colleague, included I send you copy of a letter to the Reale Accademia d'Italia concerning the *Convegno Volta* 1939 on differential geometry, algebra and topology. I hope that my American Colleagues will agree with

---

[93] Levi-Civita and Segre had been dismissed from Rome and Bologna, respectively, in the autumn of 1938. David van Dantzig (1900–1959), well known for works in differential geometry and topology, would be dismissed from his chair in Delft in the spring of 1940 when the Germans occupied Holland. Ludwig Berwald (1883–1942), from the German University of Prague, was deported in 1941 and died in the Łódź Ghetto in 1942.

[94] The International Congress of Mathematicians was held in Toronto, Canada, from 11 August to 16 August 1924.

my point of view. Perhaps you will be so kind as to show the copy to the other mathematicians in Princeton. With my most heartily greetings, sincerely yours J.A. Schouten

D2.8.3 O. Veblen to J. A. Schouten, [Princeton] 15.4.1939

OVP, *JAS*, fol. 1r
My dear Schouten: I congratulate you on your letter of the 6th of April to the Reale Accademia d'Italia about the Volta conference. I also declined the invitation, but only with a statement that I was unable to accept in the present circumstances. I shall be very glad to show the copy of your letter to the other mathematicians here. I am unable to do as much as I should like to help to find places for those who are still being forced out of their positions in Europe, but it still turns out to be possible to do something at rather rare intervals. With best greetings, yours sincerely, Oswald Veblen

D2.8.4 E. Bompiani to S. Visco, Santa Marinella (Civitavecchia) 21.7.1939

AS-UMI, *correspondence 1939*, fol. 1r
Dear Visco,[95] I respond to your circular concerning the fiftieth anniversary of the University of Washington. I have no opportunity to go to the U.S. in November; but I believe, having been there three times, to have a fair knowledge of the language and the environment, and therefore I could be conveniently sent (not at my expense, of course) to represent the University of Rome.[96] I would not have written this if I did not see the opportunity to go there from a point of view that we must deal with. The USA has become the refuge of many former Jewish professors at our universities, who naturally carry out activity contrary to our country. So it is that (keep this news strictly confidential) none of the American speakers invited to the next Volta Conference accepted. Now the International Congress of Mathematicians is being prepared, which will take place at Cambridge Mass. (U.S.) in 1940.[97] Invitations to hold general conferences have already been issued and Italian geometers have been completely excluded, although they undoubtedly count for something. It is not a risky assumption

---

[95] Sabato Visco (1888–1971), Dean of the Faculty of Sciences of the University of Rome from 1938 to 1944.

[96] Enrico Bompiani had spent three sojourns as visiting professor in the United States in 1929, 1930, and 1931.

[97] The congress, planned for September 4–12, 1940, would not take place.

that this is due to the influence of our former Jews. If I had the opportunity to go to the United States, taking advantage of my acquaintances there, I could either try to change the situation regarding the invitations, or, if this prove impossible, inform the Italian Government about decisions concerning our participation in the Congress. If you believe, take it into account; but keep in mind [...] that mathematics is one of the sectors most conquered by Judaism; and that to defend it, our will is not enough, but also funds are needed (and they are desperately limited). Many kind regards [E. Bompiani]

Carissimo Visco, Rispondo alla tua circolare riguardante il cinquantenario dell'Università di Washington. Io non ho occasione di andare in novembre negli S.U.; però credo, essendoci stato tre volte, di avere una discreta conoscenza della lingua e dell'ambiente e perciò potrei esservi convenientemente inviato (non a mie spese, s'intende) a rappresentarvi l'Università di Roma. Non avrei scritto questo se non vedessi l'opportunità di andarvi da un punto di vista del quale bisogna che ci occupiamo. Gli S.U.A. sono diventati il rifugio di parecchi ex-professori ebrei delle nostre Università e vi esplicano naturalmente un'attività a noi contraria. Così è che (tieni questa notizia rigorosamente per te) nessuno degli Americani invitati al prossimo Convegno Volta ha accettato di venire. Ora si sta preparando il Congresso Internazionale dei Matematici che avrà luogo a Cambridge Mass. (U.S.A.) nel 1940. Sono già stati diramati gli inviti per tenere le conferenze generali e da essi sono stati completamente esclusi i geometri Italiani che pure senza dubbio contano qualche cosa. Non è un'ipotesi azzardata che ciò sia dovuto all'influenza degli ex-nostri ebrei. Se avessi occasione di andare negli S.U., approfittando delle mie conoscenze colà, potrei o cercare di modificare la situazione riguardo agli inviti, o, se ciò riuscisse impossibile, informare di essa il Governo italiano per le decisioni riguardanti la nostra partecipazione al Congresso. Tieni quel conto che credi di quanto precede; ma tieni presente [...] che la Matematica è uno dei settori più battuti dall'ebraismo; e che a difenderlo non basta soltanto la nostra volontà, ma occorrono anche i mezzi (del tutto limitati). Molti cordiali saluti [E. Bompiani]

2.9   Disbelief and Disappointment

The news of the promulgation of anti-Jewish measures reached mathematicians 'in the last days of vacation of a mild and sweet summer.'[98] It was a bolt from the blue that upset individual destinies. Everyone reacted differently, according to character, but bewilderment, disbelief and disappointment were the prevailing reactions to a

---

[98]Corti (1987): *una fine d'estate mite e dolcissima.*

series of measures by which the State transformed a group of its citizens into a pariah caste:

> We said goodbye to each other when the measures against the professors seemed probable, but not yet certain. ... Many others of those affected leave a deep void in academia, but very few have honoured Italian science in the world quite like you.[99]

> We are all a bit dazed because nobody certainly expected so much. [...] The bill concerning children, well, I just did not expect it ...[100]

> 13 September 1938. Back from the mountains I pick up this notebook. Strange! After all that has happened [...] it seems to me that I am the same that I was two months ago. [...] At that time I was like someone who, seeing stones fall in front of his house, strives, and struggles to remove them. Now in front of my house there is a mountain. I don't deal with it anymore, or, if I take care of it, I don't think about pushing it away. I regret that it blocks the light for me.[101]

Dismay derived from certain factors, including the extent of the discriminatory decrees, the immediacy of execution of the purge and the fact that children had also been affected, through the expulsion from schools.

Litotes is perhaps the form of understatement that best describes the disoriented, problematic interpretation of the racial laws as given by Italian Jewish mathematicians. Most of them were more likely to say what racial laws were not, rather than what they were. First, they were *not a political* persecution.[102] Many people, in fact, had remained substantially detached from the political life of the country, so they could not conceive of being prosecuted for any kind of political activity or militancy whatsoever, especially if anti-fascist. That of 1938 was *not a religious* persecution.

---

[99] G. Castelnuovo to T. Levi-Civita, Rome 7.9.1938, in Nastasi and Tazzioli (2000), p. 187: *Ci siamo lasciati quando apparivano probabili, ma non ancora sicuri i provvedimenti che furono presi nei riguardi dei professori. ... Molti altri dei colpiti lasciano nell'insegnamento un vuoto profondo, pochissimi però hanno onorato come te, nel mondo, la scienza italiana.*

[100] A. and G. Terracini to B. Terracini, [Turin] 5.9.1938, in Terracini L. (1990), p. 445: *Siamo tutti un po' intontiti perché non si aspettava certamente tanto. [...] Quello riguardante i bambini, poi, proprio non me lo aspettavo ...*

[101] Emilio Artom (1938), in Treves (1954), pp. 45–46: *13 settembre 1938. Di ritorno dalla montagna riprendo questo quaderno. Strano! Dopo tutto quello che è avvenuto [...] mi pare di essere quello di due mesi fa. [...] Allora ero come chi, vedendo cadere davanti alla sua casa dei sassi, s'ingegna e si affanna per rimuoverli. Ora davanti alla mia casa c'è un monte. Non me ne occupo più, o, se me ne occupo, non penso a allontanarlo. Rimpiango che mi tolga la luce.*

[102] A testimony by Renato Treves (Treves 1990, pp. 27–28) appears as antithesis: 'Expulsion from school and university was experienced more *as a political than as a racial persecution*, and those who emigrated often ended up establishing bonds of friendship more with the community of Italians and children of Italian anti-fascists linked to the cult of Mazzini and Garibaldi, than with the Jewish communities very numerous other there, that were perceived as distant for language and tradition'. (*L'espulsione dalla scuola e dall'Università fu vissuta* come una persecuzione più politica che razziale *e, chi emigrava, finiva spesso per stabilire vincoli di amicizia più con la comunità di italiani e di figli di italiani antifascisti legati al culto di Mazzini e di Garibaldi che non con le comunità ebraiche assai numerose ... lontane per lingua e per tradizione.*) The adjective 'political' should not be taken, however, in a philological sense. Treves means that the expulsion was experienced as a 'rights-related persecution'.

For men born and raised in secular and integrated families, for people who had entered a synagogue only a handful of times in their life, the idea of being considered as Jews in terms of practitioners of a certain cult was inconceivable. Finally, the racial laws were *not* perceived as a *biological-racial* persecution. The idea of belonging to a pure Italic race which would generate the 'new fascist man' and that of a genetical inferiority of the Jews appeared devoid of scientific foundations, even in the eyes of those who plainly rode the wave of German racism.

2.10 Fix Things Up, and then Leave

Despite the defamatory propaganda against the Jews that had been launched in late 1937, and then more vigorously during the summer of 1938, the introduction of the measures for the defence of the race surprised Italian society. It is true that the average Italian could not ignore the anti-Semitic climate that had been growing over the last year nor the increasingly insistent and pervasive propaganda. The archive of the General Office of Public Security, Police Division, however, shows that the preparatory phase that should have been a necessary precondition for the intended racial policy and the forging of a national racial awareness had failed. Because of the regime's contradictory and shifting behaviour towards the Jewish community, racial legislation was generally perceived not only as incomprehensible but also as completely disproportionate considering the size of the Italian Jewish community which was one of the smallest in Europe.

Anonymous informants stigmatised how the Italians and the Vatican, with their traditional pietism, nourished Jewish defeatism. Jews denounced harassment, humiliations, and mass suicides resulting from the racial laws, and the Italian people believed them, pitied them, and stood with them in solidarity, rather than backing the government in the purge.[103] Compliant priests released false baptismal certificates, backdated to before 1 October 1938. Aryans lent themselves to the game of the wealthy Jews and, instead of reviling them, helped them to liquidate their real estates and commercial enterprises, without worrying about the consequences for the national economy and even setting up anonymous firms with the Jews, or acting as a nameplate. The hoteliers, far from reporting the Jewish tourists, protested at the damage caused to them by racial politics and greatly profited by systematically hosting Italian and foreign Jews registered under a false name.

It was not only the general population that was caught on the back foot. High culture too was taken by surprise, even more so because the racial census and the first provisions of law had been passed in the traditional holiday period of the academic world. The swift coordination among propaganda and legislative action, in addition

---

[103] However, there was no single response from the Catholic Church to the racial laws, instead revealing a series of infinitively varied responses according to the time, place, individual, and the provision in question. Neither Pius XI nor Pius XII took a clear stance in condemning the measures.

to a speedy enforcement of persecutory measures convinced even the most sceptical or indifferent Italians to admit that the regime was not joking. Rapid countermeasures had to be taken. Universities mostly resorted to temporary contracts to cover the teaching positions left vacant. In Turin, Bologna, and Rome, the question arose as to whether open national competitions should be held.[104] The leaders of UMI asked Mussolini for an extraordinary recruitment plan, so that the chairs of mathematics did not go to naturalists or physicists. Moreover, there was the urgency of replacing Levi-Civita, Fubini, Segre, Levi, and Enriques on the boards of *Annali di Matematica pura ed applicata*, *Periodico di Matematiche* and *Bollettino dell'UMI*, to ensure continuity of the publications.

In the Jewish world, it was a race against time to 'sort things out' (D2.10.1). Firstly, there were assets and properties to be liquidated. Fubini, Fano, and Volterra were aware that it was necessary to act quickly, especially because saving the patrimonies, either by registering them in the name of Aryan figureheads or by illegally exporting them abroad, was essential in order to ensure any sustainable migration project.

As well as things, there were people. Castelnuovo, Volterra, Fubini, Fano, Terracini, and Levi all had large family circles, which included not only wives and children but also elderly parents, brothers-in-law, uncles, cousins, distant relatives, in addition to domestic staff (chauffeurs, gardeners, cooks, nannies, maids). Racial laws prevented Jews from having Aryan domestic servants. Hence the need to hire them under false contracts, in order not to throw into the street those people who for decades had been part of their family, who had raised them and their children, and who had no other income or savings.

The most delicate and painful issue, however, remained the children, who were given priority in all arrangements. Volterra, Fubini, Castelnuovo, and Enriques, who had adult children, needed to help them find an occupation, or leave the country. Some of them, in fact, had lost any source of income. Apart from Edoardo Volterra, who in 1938 was already full professor of Roman law in Bologna, Enrico Volterra, Ugo Fano, Eugenio and Gino Fubini, Gina Castelnuovo, and many others were assistants or lecturers, who had been dismissed without any redundancy settlement. Emma Castelnuovo and Adriana Enriques had been successful in national competitions for teachers of mathematics and physics in middle and secondary schools, but this made no difference because all Jews had been removed from the rankings, as

---

[104] See e.g. AS-UMI: R. Einaudi to L. Berzolari, Turin 21.12.1938: 'The definitive fate of the chair of Geometry at the University of Turin does not seem very clear. Some decisions will be taken on this at the sitting of 9 January 1939. It would be very appropriate for the group of mathematicians to be aware of the decisions of other universities about a possible competition in Geometry at that time. I would be very grateful if you would inform me promptly of any information you may have in this regard'. (*La sorte definitiva della cattedra di Geometria dell'Università di Torino non sembra molto chiara. Nella seduta del 9 Gennaio si prenderà qualche decisione in proposito. Sarebbe molto opportuno che il gruppo dei matematici fosse per quell'epoca a conoscenza delle decisioni delle altre Università circa un eventuale concorso di Geometria. Le sarei molto grato se Ella mi comunicasse tempestivamente le informazioni che potrà avere al riguardo.*)

under racial legislation they could not work, not even in private schools. For those who had young children, such as Terracini, Segre, and Mortara, their plight was even more complicated. It was a matter of explaining the persecution to babies, young children and teenagers who (from one day to the next) had seen their world crumble, lost friends and schoolmates, and were obliged to suspend all their usual activity. Regarding school children, there was also the problem of how to let them continue their studies, hence the decision to enroll them in Jewish schools, or have them study privately while waiting for departure.

Finally, there were the immaterial patrimonies to be liquidated. In the case of mathematicians, these were cultural and publishing enterprises, mathematical journals, book series, study centres, etc. At stake was the survival of 'beloved creatures': for Enriques the journals *Periodico di Matematiche* and *Scientia*, the National Institute for the History of Sciences, the Mathematical Seminar of the University of Rome, and the series of scientific texts published by Zanichelli; for Mortara the *Giornale degli Economisti*[105]; for Levi-Civita the *Annali di Matematica pura ed applicata*; for Levi and Segre the UMI *Bollettino* and the Mathematical Institute of Bologna.

At this moment of frenetic activity, friendship made the difference, especially among the members of the School of Geometry: they supported each other, the old reassured the young, they discussed how to react and what to do, while some encouraging news began to spread: the Ministry of the Interior was planning to start a procedure of counter-discrimination, aimed at exempting from racial measures those Italian Jews who had distinguished military, political, or cultural merits.

Documents

D2.10.1 Manuscript memo re. Volterra, Rome, first days of November

ANL-*Volterra*, fol. 1r-2r

Submit the application to the Minister of the Interior [Mussolini] to obtain the inverted discrimination of Senator Prof. Volterra pursuant to art. 10-11 and 13b. Ask that the benefit of counter-discrimination be extended to members of the family of Sen. Volterra [...]. Attach evidentiary documents (Almagia's service certificates of war work in original). For the best result: submit an independent application for all the persons indicated below.

Regulate the housekeeper of Ariccia and Enrico [chauffeur] with a regular rental contract.

---

[105]Zanni (1977).

(a) Fire all domestic staff (including driver and gardener) within 15 days after 17 Nov.
(b) Hire Jewish servants *or*
(c) Use Gustavo's wife's servants.[106]
(d) Fire the chauffeur *for now* and temporarily hire a Jewish one. Subsequently, take back Enrico as a car owner and pay him a lump sum.

Domanda al Ministro interni per ottenere dichiarazione discriminazione Senat. prof. Volterra ai sensi dell'art. 10-11 e 13b. Chiedere che il beneficio della discriminazione sia esteso ai membri della famiglia del Sen. Volterra [...]. Allegare documenti probatori (attestati servizio Almagià durante guerra in originali). Ad ogni buon fine: presentare domanda autonoma per tutte le persone avanti indicate.

Regolare il custode di Ariccia ed Enrico con regolare contratto di affitto

(a) *Licenziare tutto personale domestico (anche autista e giardiniere) entro giorni 15 dal 17 Nov.*
(b) *Assumere domestici ebrei* oppure
(c) *Utilizzare domestici moglie di Gustavo*
(d) *Licenziare* per ora *l'autista e assumerne per ora uno ebreo. Successivamente riprendere Enrico intestatario di macchina sua e retribuirlo a forfait.*

2.11 Practices of 'Counter-Discrimination'

Rumours of an exemption of certain categories of Jews from the application of racial measures began to circulate in September 1938 but were confirmed only on November 17. Based on an article of the decree issued, the Ministry of the Interior could 'discriminate' case by case, i.e. exempting from the application of racial measures those Italian Jews who had proven exceptional merit in the political, military, or cultural fields. Senators, members of the PNF from 1919 to 1922, legionaries from Fiume and recipients of the Croix de Guerre in the four wars of the century (Libyan War, First Word War, Ethiopian War, and Spanish War) were automatically exempted. Other people were required to scrupulously document their honours, detailing their applications with letters, certificates, and any other element proving the exceptionality of their cases. Applications were to be sent to the Ministry of the Interior (to the MEN for faculty members only) and were examined by special committees. If accepted, this would lead to the immediate reinstatement of the individual and recovery of his full civil and political rights (although this process never actually happened).

---

[106] Gustavo Volterra (1909–2001), the last son of Vito, had married the Aryan Emilia Cosattini. She could maintain her servants.

The procedure was extremely humiliating,[107] but everyone capitulated to it: Volterra submitted the application in his quality of Senator of the Kingdom (D2.11.5); Segre and Fubini for distinct scientific merits (D2.11.2 and D2.6.2); Ascoli, Terracini, Fano, and Del Vecchio for military honours (D2.11.1, D2.11.3, and D2.11.4); Enriques, Castelnuovo, Levi-Civita for scientific and family merits (D2.11.6).

The gathering of the documents required for counter-discrimination was at the origin of unspeakable episodes of bad bureaucracy, especially targeting university professors, who from mid-October onwards were urged to collect two sets of documents: those required to obtain the pension and those to submit for the reverse discrimination practices. The first were standard administrative certificates (birth certificates, service status, etc.). A counter-discrimination dossier, on the other hand, was significantly more difficult to construct. It constituted a painful exercise of post-rewriting of his/her own political and professional trajectory.

As the applications show, the nature of the documents that were produced was the most varied imaginable, even though a tendency to synthesis and neutral prose prevailed among mathematicians. In other words, in mathematics, there were no analogous cases to that of Carlo Foà, son of the illustrious Pio Foà, professor of Physiology in Milan, who presented a 21-page memoir, comprehensive of large chapters dedicated to his political activity as a fervent fascist and to active combat of Zionism that he had practiced in Milan.[108] Volterra, who applied not for himself but for his children, to whom the privilege of counter-discrimination would have been extended ex officio, merely highlighted his military merits and the Senatorship (D2.11.5). Levi-Civita submitted his application very late and played the card of his father's political path, a Garibaldian patriot and a senator (D2.11.6 and Fig. 2.3). Fubini limited himself, with evident indignity, to drafting a few lines underlining his scientific reputation (D2.6.2). Ascoli, in addition to documenting his military and scientific merits, outlined two proposals for his alternative use (D2.11.1). The idea was clear: taking for granted the fact that Jews would not be reintegrated into teaching (even if counter-discriminated), he proposed other forms of activity where he could continue to offer his contribution to the country.

The scrutiny of applications gave rise to a system of corruption and indecent clientelism, perpetuated even by the most eminent fascist hierarchs, such as Roberto Farinacci. Finally, Castelnuovo and Enriques obtained reverse discrimination status through the intercession of Gentile; Levi-Civita and Fubini with the intervention of MEN; Volterra through direct intercession of the King. Many dossiers, however, including those of Terracini and Segre, were rejected.

---

[107] Palumbo (2019).

[108] APICE: *Prof. Carlo Foà, Stato di servizio militare; Stato di servizio nel P.N.F.; Rapporti con l'Ebraismo*, fols. 1–10 and *Prof. Carlo Foà, Pubblicazioni e discorsi di argomento politico; Pubblicazioni su argomenti di politica universitaria e culturale, Pubblicazioni di argomento filosofico-religioso con riflessi politici; Pubblicazioni su argomenti di medicina sociale e corporativa*, fols. 1–11.

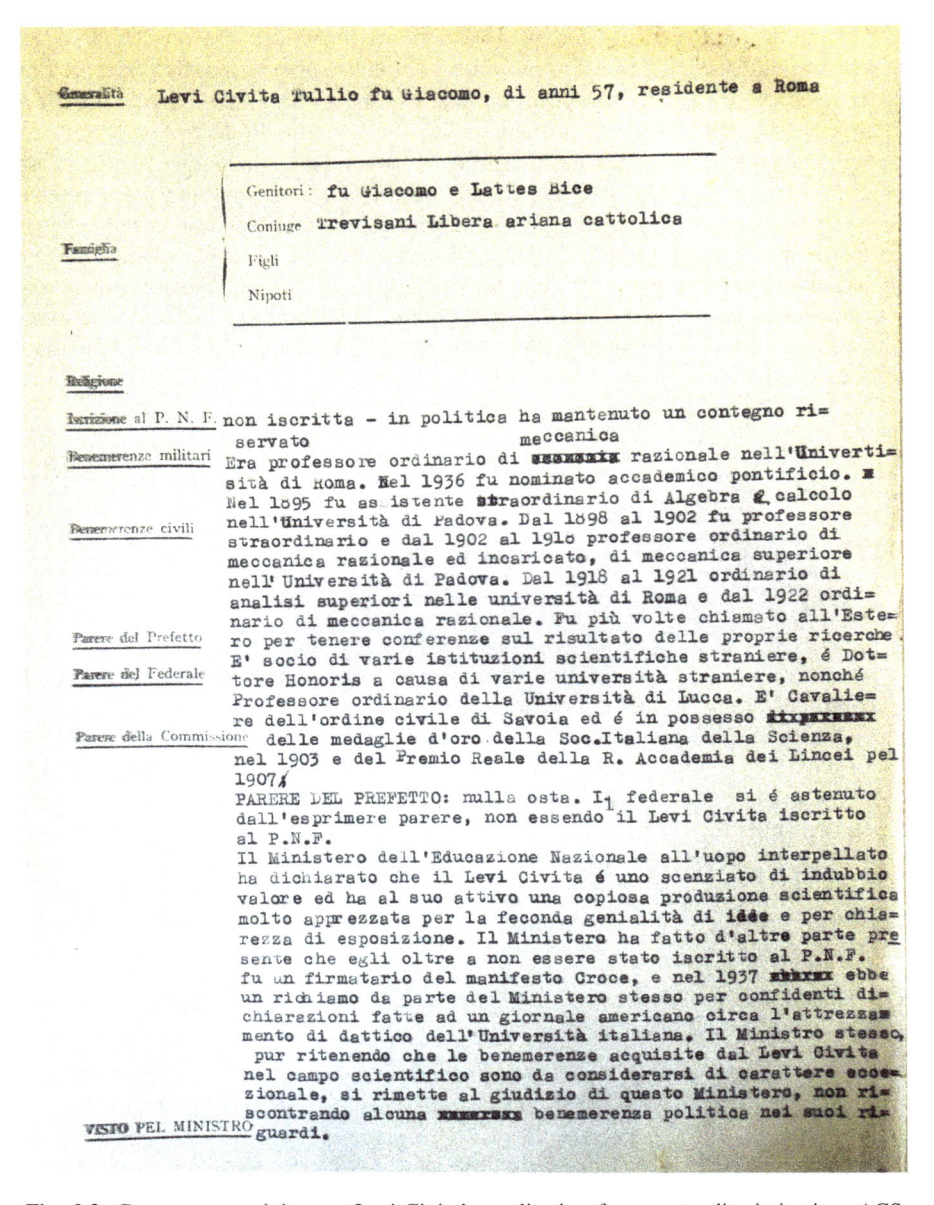

Fig. 2.3 Government opinion *re.* Levi-Civita's application for counter-discrimination, ACS-*Ebrei*

Communication of the outcomes of the scrutiny procedure was very late in most cases. Suffice it to say that the application of Volterra and his family, presented in November 1938 and not subject to the exam of the Commission (due to Volterra being a Senator) was accepted on January 28, 1939, but the Prefecture of Rome did not inform Volterra until the following May 17! Meanwhile, the purged scholars lost precious months. The hope of being counter-discriminated delayed in particular their first steps towards emigration, sometimes with dramatic effects, as in the case of Ascoli. In the chaos of the administrative offices, many had to decide whether to stay or leave without knowing what would happen in the event of a successful application for counter-discrimination status. Fano, Fubini, and Mortara, in fact, would be informed of having obtained counter-discrimination status only after they had already left the country.

Documents

D2.11.1 G. Ascoli to A. Pepere, Turin 21.10.1938

APICE-*Ascoli*, fols. 1r, 2r-v
Respected Rector,[109] by adhering to your kind invitation, I send you a reminder about my possible use in fields other than teaching. Given my previous verbal communications, it does not seem necessary to bother you with a new interview; if, however, in my interest, you see the opportunity, I am ready to come to Milan on any day of your choosing. I forgot to mention the statement quoted in the memorandum, and which must exist in the official documents [D1.6.5]; it was written under the friendly advice of Professors Foà and Picone on the occasion of my call to Milan, to reassure the Faculty about my real feelings towards the Regime.[110] I take this opportunity to thank you for your very nice and cordial interest in our painful circumstances, and to profess myself respectably yours, Guido Ascoli

Memorandum. Ascoli Guido, Professor of Mathematical Analysis at the R. University of Milan.

He participated in the World War, Artillery division, achieving the rank of Lieutenant of Complement; he was in military operations (Karst and Piave) from March 1917 to the end of the war (Campaigns 1917-1918); he was awarded the War Merit Cross. He suffered a light wound from the explosion of an enemy grenade.

---

[109] Alberto Pepere (1873–1940), Rector of the University of Milan from 1935 to 1940.

[110] Carlo Foà (1880–1971), professor of Physiology at the University of Milan from 1925 to 1938 and Mauro Picone (1885–1977), professor of Analysis at the University of Rome from 1932. Picone was an old friend of Ascoli. Both had been trained at the Higher Normal School in Pisa.

He is registered with the Fascist School Association; he is not enrolled in the P.N. F., but a statement directed in 1934 to the Trustee of the A.F.S. [Associazione Fascista della Scuola], a copy of which was sent to the Rector of the R. University of Milan testifies to his feelings of devotion to the fascist regime.

He has been married since 1925 to an Aryan and has two children aged 6 and 11, of Catholic religion. His wife is a primary school teacher in Turin. Both are devoid of private property and live modestly from their work income.

While dealing with various branches of Mathematical Analysis, Ascoli has devoted himself mostly, in memoirs and academic courses, to issues having greater relevance to physical and technical applications. He believes that this special activity can usefully be framed in that of the National Research Council and the Institute for the Applications of Calculus. He points out that the Council itself supervised the drafting of a monograph, still in progress, on *Funzioni speciali interessanti le applicazioni.*[111]

The modalities of his collaboration should be determined by the Research Council. Ascoli could:

(a) Systematically attend to the compilation of monographs in Applied Mathematics, according to the programme already approved by the Council, whose development would thus be greatly accelerated.
(b) Study a collection of numerical tables, particularly suitable for applied research, which, as well as serving as a natural and necessary complement to the afore-mentioned monographs, would also mark a significant step towards Italian autonomy in the cultural field.
(c) Take up the study of issues that could be indicated to him from time to time by the Council or the Institute for the Applications of Calculus.

He points out that for the family conditions it would be of great interest to him to continue to reside in Turin. Turin, 21 October 1938—XVI

Magnifico Rettore, aderendo al Vostro gentile invito vi trasmetto un promemoria relativo alla mia possibile utilizzazione in campi diversi dall'insegnamento. Date le mie precedenti comunicazioni verbali non mi sembra necessario importunarvi con un nuovo colloquio; se però, nel mio interesse, ne vedeste l'opportunità, sono pronto a venire a Milano in qualunque giorno ed ora vogliate indicarmi. Dimenticai di accennare alla dichiarazione citata nel promemoria, e che deve esistere in atti; essa fu redatta dietro amichevole consiglio del prof. Foà e del prof. Picone in occasione della mia chiamata a Milano, per tranquillizzare la Facoltà circa i miei reali sentimenti nei riguardi del Regime. Colgo l'occasione per ringraziarvi del Vostro così simpatico e cordiale interessamento per le nostre dolorose circostanze, e per dirmi con ossequio rev.^{mo} Guido Ascoli

---

[111] The book was not published.

Promemoria. Ascoli Guido, ordinario di Analisi matematica nella R. Università di Milano. *Partecipò alla guerra mondiale in Artiglieria da campagna, conseguendo il grado di Tenente di complemento; fu in zona di operazioni (Carso e Piave) dal Marzo 1917 alla fine della guerra (Campagne 1917-1918); conseguì la croce al merito di guerra. Ebbe una ferita leggera per scoppio di granata nemica. È iscritto all'A.F.S.; non è iscritto al P.N.F., ma una dichiarazione diretta nel 1934 al Fiduciario dell'A.F.S. e di cui fu inviata copia al Rettore della R. Università di Milano testimonia dei suoi sentimenti di devozione al Regime Fascista. È coniugato dal 1925 con una ariana e ha due figli di anni 6 e 11, di religione cattolica. Sua moglie è maestra elementare in Torino. Ambedue sono privi di beni di fortuna e vivono modestamente del loro lavoro. Pur occupandosi di vari rami di Analisi matematica, si è dedicato per lo più, in memorie e corsi accademici, alle questioni aventi maggior attinenza con le applicazioni fisiche e tecniche. Ritiene che tale sua speciale attività possa venire utilmente inquadrata in quella del Consiglio delle Ricerche e dell'Istituto per le applicazioni del calcolo. Fa notare che dal Consiglio stesso ebbe incarico della redazione di una monografia, tuttora in corso di compilazione, sulle* Funzioni speciali interessanti le applicazioni. *Le modalità della sua collaborazione dovrebbero essere fissate dal Consiglio delle Ricerche. In particolare, egli potrebbe:*

(a) *Attendere sistematicamente alla compilazione di Monografie di Matematica applicata, secondo il programma già deliberato dal Consiglio, il cui svolgimento verrebbe così notevolmente accelerato;*
(b) *Studiare una collezione di tavole numeriche, adatta ai bisogni delle applicazioni, che, mentre servirebbe di naturale e necessario complemento alle Monografie, segnerebbe un passo sensibile verso l'autonomia italiana nel campo culturale;*
(c) *Attendere allo studio di questioni particolari che potrebbero essergli indicate volta per volta dal Consiglio o dall'Istituto per le applicazioni del Calcolo.*

Fa presente che per le su citate condizioni familiari sarebbe per lui di grande interesse continuare a tenere la propria residenza a Torino. Torino, 21 Ottobre 1938-XVI

D2.11.2 Memorandum *re*. Prof. Beniamino Segre, Bologna 22.10.1938, ACS-*contro-discriminazione*, fol. 1r, 2r, 3r

Family news. I was born in Turin on 16-11-1903. My father [Samuele Segre], a municipal employee, directed the Food Service of the City of Turin during the Great War, then became Chief Treasurer of the Municipality; he received praise and certificates of esteem from the Podestà H.E. Senator Thaon of Revel. I married

Fernanda Coen on 20 March 1932, and I had three children: Sergio (22 May 1933), Silvana (14 May 1934) and Ornella (15 June 1937).

Academic career. Doctor in Pure Mathematics (with full marks and honours) at the R. University of Turin, on 14-7-1923.

1923-24. Assistant to the chair of Rational Mechanics of the Royal University of Turin.

1925-26. Assistant to the chair of Analytic, Projective and Descriptive Geometry of the R. University of Turin.

1927. Lecturer in Analytic and Projective Geometry at the R. University of Rome.

1927–1931. Assistant to the chair of Infinitesimal Analysis of the R. University of Rome.

Ranked 1st in a national competition (in which two university professors participated as candidates, and none of the members of the Jury was Jewish), in 1931 I became extra-ordinary Professor at the R. University of Bologna. I was promoted to full professorship on 1st December 1934; starting from that year I was also Director of the Mathematical Institute.

Scientific and didactic activity. Up to now, my life has been entirely dedicated to the School, i.e. to scientific research and the most complete enhancement of the Italian contribution to mathematics. This is clearly stated in my article on Analytic Geometry for the Treccani Encyclopedia [sic!][112] and in my *Prolusione* entitled *La Geometria in Italia dal Cremona ai giorni nostri* held in Bologna on 13 January 1932;[113] it is shown in the demanding reorganisation work which I carried out as Director of the Mathematical Institute of the Royal University of Bologna; for the Institute's benefit, I refused a very well-paid post in São Paulo (Brazil) which was offered to me in 1936 with great insistence.[114] The organisational work was judged highly meritorious by Prof. Wilhelm Blaschke of the University of Hamburg, who, last spring (1938), told me that our Mathematics Institute—after that of Rome—is the most prominent in Italy as far as he knows. I published 3 university courses (lithographed)[115] and over 120 original Notes and Memoirs.

---

[112] In fact, Segre had written for the Treccani Encyclopedia the entry *Coordinate* (vol. XI, 1931, pp. 294–303). The entry *Geometria analitica* had been prepared for the *Enciclopedia delle Matematiche Elementari* edited by L. Berzolari, D. Gigli, G. Vivanti (II_2, 1938, pp. 141–249).

[113] B. Segre, *La geometria in Italia, dal Cremona ai giorni nostri. Prolusione al Corso di Geometria Superiore, tenuta in Bologna il 13 gennaio 1932*, Annali di Matematica pura ed applicata, s. 4, 11, 1932, pp. 1–16.

[114] CA, *BSP*: M. Betti (Dean of the Faculty of Sciences of the University of Bologna) to B. Segre, Bologna 15.5.1936.

[115] F. Severi, *Conferenze di geometria algebrica, raccolte da B. Segre*, Roma, Litografie, 1927; B. Segre, *Esercizi e complementi di analisi algebrica*, Roma, Litografie, 1928 and 1930.

Distinctions and offices. Awarded by the R. Accademia Nazionale dei Lincei in 1926; in 1935, the Académie Royale de Belgique published one of my works in its *Mémoires couronnées*;[116] gold medal of the National Academy of the Sciences (called the XL), awarded in 1930. Honorary member of the Royal Academy of Sciences of the Institute of Bologna since 21 May 1933. Founding member of the Italian Mathematical Union. Lifetime fellow of S.I.P.S. Perpetual member of the Société Mathématique de France. Co-director of *Annali di Matematica pura ed applicata*.

Military news. In 1924-25 I did my military service first as an Officer Student, and then as an Officer Student Sergeant. In 1926 I served as Second Lieutenant of Campaign Artillery; I remained in the army—voluntarily and without stipend—a month longer than necessary, to meet specific needs of the Regiment (5th Campaign Artillery) where I was serving. Promoted to Reserve Lieutenant on 1 March 1935 (XIII).

Notizie sulla famiglia. *Nacqui a Torino il 16-II-1903. Mio padre, impiegato comunale, diresse durante la Grande Guerra il Servizio Annonario della Città di Torino, passò poi Tesoriere capo del Comune, ricevendo elogi ed attestazioni di stima dall'allora Podestà S.E. il sen. Thaon di Revel. Mi sposai con Fernanda Coen il 20-III-32 ed ebbi tre figli: Sergio (il 22-II-33), Silvana (il 14-V-34) ed Ornella (il 15-VI-37).*

Carriera accademica. *Dottore in matematica pura (con pieni voti e lode) presso la R. Università di Torino, il 14-VII-1923. 1923-24. Assistente incaricato presso la Cattedra in Meccanica Razionale della R. Università di Torino. 1925-26. Assistente di ruolo presso la Cattedra di Geometria analitica, proiettiva e descrittiva della R. Università di Torino. 1927. Libero docente in Geometria analitica e proiettiva presso la R. Università di Roma. 1927-1931. Assistente di ruolo presso la Cattedra di Analisi infinitesimale della R. Università di Roma. Essendo stato dichiarato 1° in un Concorso (al quale partecipavano come concorrenti due professori universitari, mentre nessuno dei membri della Commissione giudicatrice era Ebreo), passai nel 1931 Professore straordinario presso la R. Università di Bologna. Fui promosso ordinario il 1°-XII-1934; a partire da tale anno fui anche Direttore del relativo Istituto Matematico.*

Attività scientifica e didattica. *La mia vita fin qui fu interamente dedicata alla scuola, alla ricerca scientifica ed alla più completa valorizzazione del contributo italiano alle matematiche. Ne fanno fede il mio articolo su Geometria analitica nell'Enciclopedia Treccani, la mia* Prolusione *tenuta in Bologna il 13-I-1932 su La Geometria in Italia dal Cremona ai giorni nostri, e la complessa opera di riorganizzazione da me svolta quale Direttore dell'Istituto Matematico della*

---

[116]B. Segre, *Quelques résultats nouveaux dans la géométrie sur une V₃ algébrique*, Mémoires couronnés de l'Académie Royale des sciences, des lettres, et de beaux-arts de Belgique, (2), 16, 1936, pp. 3–99.

R. Università di Bologna, per amore della quale rifiutai un posto assai ben remunerato che nel 1936 mi veniva offerto con molta insistenza a S. Paulo nel Brasile. L'opera suddetta è stata giudicata altamente meritoria dal prof. Wilhelm Blaschke dell'Università di Amburgo, il quale, nella scorsa primavera (1938), mi dichiarava che il nostro Istituto Matematico—dopo quello grandioso di Roma—è il più ben organizzato fra quanti egli ne conosce in Italia. Ho pubblicato 3 corsi universitari (litografati) ed oltre 120 Note e Memorie originali.

Distinzioni e cariche. *Premiato dalla R. Accademia Nazionale dei Lincei nel 1926; dall'Académie Royale de Belgique nel 1935, la quale—l'anno successivo— pubblicò un mio lavoro nelle sue* Mémoires couronnées; *medaglia d'oro della Società Italiana delle Scienze (detta dei XL), conferitami nel 1930. Accademico onorario della R. Accademia delle Scienze dell'Istituto di Bologna, dal 21 maggio 1933. Socio fondatore della Unione Matematica Italiana. Socio vitalizio della S.I.P.S. Socio perpetuo della Société Mathématique de France. Condirettore degli Annali di Matematica pura ed applicata.*

Notizie militari. *Nel 1924-25 prestai servizio militare quale Allievo Ufficiale prima, e quale Sergente Allievo Ufficiale poi. Nel 1926 prestai servizio quale Sottotenente di Artiglieria da Campagna, e rimasi sotto le armi—volontariamente e senza assegni—un mese più del necessario, per sopperire a certe esigenze del Reggimento (il 5° Artiglieria da Campagna) presso cui prestavo servizio. Promosso Tenente di Complemento in data 1-III-1935 (XIII).*

D2.11.3 A. Terracini to MNE, Turin 31.10.1938

ACS-*contro-discriminazione*, fol. 1r-v
Following a similar invitation received by the Rector of the Royal University of Turin (letter no. 2665 cl. 2 fol. 1 on 19 October), and with reference to the declaration of the Grand Council of Fascism of 6 October, I, the undersigned Prof. Alessandro Terracini, full professor of Analytic Geometry with elements of Projective and Descriptive Geometry with Drawing at the Royal University of Turin, transmit to you the enclosed documents proving my belonging to a family of the category specified in the aforementioned Rector's letter. Specifically, I am the brother of a silver medal recipient for military valour in the 1915-18 war, who was moreover wounded in the war and receives a life pension for infirmity contracted in the war and attributable to the 7th category.[117] [...] Furthermore, wishing that my personal position as a soldier in the 1915–18 war be ascertained, I enclose a copy of my service record, which shows that I took part in the war, and that I have been recognised for the 1916, 1917 and 1918 campaigns. I am not aware, because I never asked about this, whether the War Merit Cross has been

---

[117]Benvenuto Aronne Terracini (1886–1968).

conferred on me, although I firmly believe that the service I rendered in the area of operations should have won me this honour. In recent days, I asked the Ministry of War for news about it. On this service, however, I enclose information in attachment no. 8. I ask this Ministry to proceed with the attached request if I should not have already been decorated. I ask the officers indicated therein, who are still alive and willing to do so, to testify on my war service, declaring that it was equivalent to the service commonly rewarded with the War Merit Cross. I further enclose, for all necessary purposes, my family status, proving that I am married with three children.[118] [. . .] I am a Reserve Captain in the Engineers Corps. I am registered with the P.N.F. With distinct regards, Prof. Alessandro Terracini

In seguito ad analogo invito ricevuto dal Rettore della R. Università di Torino con sua lettera n. di prot. 2665 cl. 2 fasc. 1 in data 19 corr., con richiamo alla dichiarazione del Gran Consiglio del Fascismo del 6 corr., io sottoscritto prof. Alessandro Terracini, ordinario di Geometria analitica con elementi di proiettiva e geometria descrittiva con disegno nella R. Università di Torino, vi trasmetto gli acclusi documenti comprovanti la mia appartenenza a famiglia di categoria specificata nella predetta lettera rettoriale. Precisamente, io risulto fratello di un decorato di medaglia d'argento al valor militare nella guerra 1915-18, il quale è per di più ferito di guerra e fruente di pensione vitalizia per infermità contratta in guerra e ascrivibile alla 7ª categoria. [. . .] Desiderando ulteriormente che sia accertata la mia personale posizione di combattente nella guerra 1915-18, accludo copia del mio stato di servizio, dal quale risulta che ho preso parte a quella guerra, e che mi sono state riconosciute le campagne 1916, 1917, 1918. Non mi consta, non avendo mai precedentemente fatte pratiche in merito, che mi sia stata conferita la croce al merito di guerra, sebbene io ritenga per fermo che il servizio da me allora prestato in zona di operazioni mi dovesse far riconoscere tale onorificenza. Ho rivolto in questi giorni domanda al Ministero della Guerra, per avere notizie in proposito. Su tale servizio, comunque, accludo indicazioni nell'allegato n. 8. Domando a cotesto Ministero che, ove io non risulti decorato di croce al merito di guerra, dia corso all'acclusa domanda. In essa domando che siano invitati a attestare sul servizio da me prestato in guerra gli ufficiali ivi indicati, i quali mi risultano tuttora viventi e disposti a attestare, se interrogati, nel senso da me richiesto: ciò allo scopo di stabilire che il servizio da me prestato in guerra equivale al servizio che veniva riconosciuto con la croce di guerra. Accludo ulteriormente, a ogni buon fine, il mio stato di famiglia, comprovante che sono ammogliato con tre figli. [. . .] Sono capitano di complemento nell'Arma del Genio. Sono iscritto al P.N.F. Con distinti saluti, Prof. Alessandro Terracini

---

[118] Alessandro Terracini had married Giulia Sacerdote in April of 1924. The couple had three children: Lore (born in 1925), Cesare (1927), and Benedetto (1931).

D2.11.4 E. Del Vecchio to H. E. the Minister of National Education [G. Bottai], Trieste 31.10.1938

ACS-*contro-discriminazione*, fol. 1r
In compliance with the invitation addressed by this Ministry to full-time university professors of Jewish race, I am honoured to send, with this letter, a copy of my brother's (Carlo Del Vecchio) state of service, certifying that he was awarded the War Merit Cross, and my consequent belonging to category c) of the October 6, 1938, provision of the Grand Council of Fascism. The degree of kinship is proven by the united paternal family status. I have the honour to attach a copy of my military state of service, which shows my status as a soldier in the Great War, while I wish to declare that procedures are still underway to obtain the recognition of the title of war volunteer and the granting of the Merit Cross. With the utmost observance, Prof. Ettore del Vecchio[119]

In ottemperanza all'invito rivolto da codesto Superiore Ministero ai professori universitari di ruolo di razza ebraica, mi onoro inviare, con la presente, copia del 'foglio matricolare' di mio fratello Carlo Del Vecchio, attestante, per essere egli insignito della croce al merito di guerra, la mia appartenenza alla categoria c) del disposto 6 ottobre c.a. del Gran Consiglio del Fascismo. Detto grado di parentela è provato dall'unito stato di famiglia paterna. Mi pregio allegare altresì copia del mio stato di servizio militare, dal quale risulta la mia qualità di combattente nella guerra mondiale, mentre desidero dichiarare che sono tuttora in corso pratiche per ottenere il riconoscimento del titolo di volontario di guerra e la concessione della croce al merito di guerra. Con la massima osservanza prof. Ettore del Vecchio

D2.11.5 V. Volterra to H.E. the Ministry of the Interior [B. Mussolini], Rome [18.11.1938]

ANL-*Volterra*: fasc. 7, *Discriminazione 1938*, fols. 1r, 2r-v, 3r-4r[120]
I am sorry that my health conditions prevent me from going down these days to the city to personally present to H.E. the present and annexed documents proving the work I performed as a volunteer Reserve Officer in the service of the Fatherland during the Great War. I therefore allow myself to send them with my son Gustavo Volterra. I would be very grateful to your H.E. if you would

---

[119]Ettore Del Vecchio (1891–1971) graduated in Mathematics from the University of Turin in 1914. *Libero docente* of Financial Mathematics from 1933, in 1935 he became extraordinary professor of General and Financial Mathematics at the University of Trieste.

[120]There are three versions of the application of Volterra: one signed by Vito Volterra with corrections, deletions, and additions, one type scripted with a few erasures, and a final draft, which is here reproduced.

kindly inform me whether these documents should be transmitted through the
Senate to the Ministry of the Interior, General Directorate for Demography and
Race, so that they are taken into consideration in accordance with the measures
deliberated by the Grand Council in the session of 6 Oct. If your H.E. would like
to offer me a hint of assurance in this regard, I would be extremely grateful. While
expressing my thanks to H.E., I have the honour of professing myself with the
greatest observance, yours, Vito Volterra

*Sono dolente che le mie condizioni di salute m'impediscano di scendere in questi
giorni in città per presentare di mia propria mano all'E.V. la presente e gli uniti
documenti comprovanti l'opera da me prestata quale volontario al servizio della
Patria durante la grande guerra in qualità di ufficiale di complemento. Mi
permetto perciò inviarli per mezzo di mio figlio Gustavo Volterra. Sarei
veramente grato all'E.V. se volesse farmi cortesemente noto se questi documenti
debbano essere trasmessi per mezzo del Senato al Ministero dell'Interno,
Direzione Generale Demografia e Razza, affinché vengano presi in
considerazione a norma delle disposizioni deliberate dal Gran Consiglio nella
Seduta del 6 corr. Se la E.V. volesse farmi avere un cenno di assicurazione a
questo riguardo, io ne sarei sommamente riconoscente. Mentre esprimo alla
E.V. i miei ringraziamenti ho l'onore di segnarmi con la maggiore osservanza.*

D2.11.6 Levi-Civita's Application for counter-discrimination, Rome 24.11.1938

ACS-*contro-discriminazione*, fol. 1r-v
Dear Commission for the Evaluation of Special Merits at the Ministry of the
Interior. I state that, in accordance with the decree law of 17 November 1938,
no. 1728, I belong to the Jewish race (despite having an Aryan wife.[121]) In
relation to art. 14, letter b, no. 6 of that law, I add the following merits:

I. Scientistic merits: specifically recognised through election as a member of the
main Italian and foreign Academies, including the Pontifical Academy of Sciences (see the attached curriculum with a list of publications), and the appointment as Knight of the Royal Civil Order of Savoy (documented appointment from
the attached participation).

II. Family merits: I am the son of the *quondam* Avv. Senatore Giacomo[122] who,
at 17 years of age, was with Garibaldi in Aspromonte, then in Bezzecca; an
eminent civilian, he enjoyed, especially in the Veneto, a wide reputation for
valour and honesty; a long-term President of the Council of the Bar Association;

---

[121] Libera Trevisani Levi-Civita (1890–1973).
[122] Giacomo Levi-Civita (1846–1922).

twice Mayor of Padua; Senator of the Kingdom, etc. I therefore ask to be counter-discriminated as provided for by the royal decree. Yours sincerely, Tullio Levi-Civita

Spett. Commissione per la valutazione di speciali benemerenze presso il Ministero dell'Interno. Premetto che, a termini del Decreto Legge 17 Novembre 1938, n. 1728, appartengo alla razza ebraica (avendo moglie di razza ariana). In relazione all'Art. 14, lettera b, n. 6 di detta legge, adduco le seguenti benemerenze:

I. scientifiche: *specificamente riconosciute colla elezione a socio delle principali Accademie italiane e straniere, tra cui la Pontificia Accademia delle Scienze (cfr. l'allegato curricolo con elenco delle pubblicazioni), e colla nomina a Cavaliere del Reale Ordine Civile di Savoia (nomina documentata dalla allegata partecipazione).*

II. famigliari*: sono figlio del* quondam *Avv. Senatore Giacomo, che fu a 17 anni con Garibaldi ad Aspromonte, poi a Bezzecca; civilista insigne, che godette, specialmente nel Veneto, larga rinomanza per valentia e probità; a lungo Presidente del Consiglio dell'Ordine degli Avvocati; due volte Sindaco di Padova; Senatore del Regno, ecc. Chiedo pertanto di essere discriminato a tutti gli effetti del citato R. Decreto. Con osservanza Tullio Levi-Civita*

2.12   The Jewish Schools

For some, discrimination involved not only an abrupt and unexpected awareness of personal identity, but also a sort of return to their own roots. This was the case of those who had school age children and needed to guarantee their education, and of the mathematicians hired at Jewish schools who, for the first time in their history, found themselves working in entirely Jewish teaching groups and in contact with an entirely Jewish student community.

In this regard, it should be remembered that the schools for children of the Jewish race, established in haste in September 1938 in areas where the Jewish presence was greatest, constitute one of the strongest and most courageous reactions of Italian Judaism and its institutions vis-à-vis persecution.[123] In many towns, there was no need to create *ex novo* such schools since they had operated since the pre-Risorgimento period. Throughout the post-unification phase, however, these schools had suffered a serious decline. Even where they were still open and in

---

[123] See, inter alia, Artom E.S. (1913); Colombo (1925); *Scuole Israelitiche* 1938; Ministero dell'Educazione Nazionale … 1941; Commissione Alleata in Italia, Sottocommissione dell'Educazione 1947; Piussi (1997); Minerbi (1998); Maida (1999); Capristo (2007).

good condition, it was necessary to renovate them thoroughly; on one hand, to welcome a flow of students never before experienced and on the other hand to integrate new members of the teaching staff, who had never taught previously in a school *of* and *for* Jews. Where possible, not only primary courses were created but also high schools, and even integrative courses for recent graduates who had been prohibited from enrolling at university.

In Turin, the Terracini brothers collaborated in the renovation of the premises of the ancient *Collegio Colonna Finzi* and organised courses of integrative culture for university freshmen. Bonaparte Colombo, Adelaide Diena, and Ugo Levi were recruited as teachers of mathematics and physics; Emilio Artom and his wife Amalia Segre as substitute teachers.[124] Emilio gave individual lessons to prepare young Jews to take the state exams as private students, advised those who intended pursuing the master's degree in rabbinic studies and taught spoken Hebrew to people preparing to leave for Palestine. Amalia was the school's Dean from 1942. Their sons, Ennio and Emanuele, managed the community library, conducted the youth conferences and the after-school programme, and raised funds for Zionism and refugees.

In Milan, in the autumn of 1938, just one Jewish kindergarten and primary school existed; hence, there was a strongly felt need to create a middle school and secondary study program.[125] The Jewish school began operating regularly on November 7, 1938, and comprised a complete curriculum in humanities and a section for each of the three main secondary schools (scientific lyceum, teacher training school, and technical institute). For scientific education, Augusto Levi and Guido Ascoli were approached as teachers, and immediately accepted, while Mario Attilio Levi instead refused, in the certainty (later disappointed) that he would have been counter-discriminated and reinstated in service.[126] In 1941, when the secondary school courses were almost consolidated, there arose the problem of the continuation of studies of a large number of students who had qualified. A para-university programme in General and Inorganic Chemistry was thus established. A special laboratory for chemistry and one for experimental physics were created and entrusted to Levi and Ascoli, respectively. As Ascoli wrote to Picone:

> A lot of work was assigned to me, as had never happened before, to which in November was added a private but advanced course which was awarded to me by the Jewish Community of Milan and for which I travel there three times a week, wasting a lot of time; work, which is, however, not unpleasant and to which I had agreed in September. On the other hand, these are times in which every resource must be put to good use.[127]

---

[124] Corinaldi (1988); Errera Foà (1984).

[125] Edallo (2018) and Edallo (2020).

[126] APICE: M. A. Levi to the Presidency of the Jewish Community in Turin, cc. A. Pepere, the Rector of the University of Milan, Turin 3.11.1938.

[127] G. Ascoli to M. Picone, 23.12.1941, in https://matematica.unibocconi.it/articoli/due-lettere-tra-mauro-picone-e-guido-ascoli: *mi si è rovesciata addosso una mole di lavoro come non mi era mai capitato, a cui in novembre si è aggiunto un insegnamento privato ma di carattere superiore che mi è stato offerto dalla Comunità Israelitica di Milano e per il quale mi reco là tre volte la settimana,*

In Trieste, it was decided that the primary school, founded in 1782, would be expanded to offer also a middle and secondary curriculum.[128] The premises were still those of the original school, and so some logistical ingenuity was required to make it feasible. Unlike primary school, the costs for running the middle school had to be covered entirely by the Jewish Community, at a time when the institution had to cope with the considerable losses caused by emigration (whether voluntary or due to the problem of citizenship) and conversions. Despite all these challenges, a protective and supportive environment was created where, for as long as it was possible, pupils received an excellent education. The teaching staff included two former university lecturers, namely the mathematician Bruno Tedeschi (who was appointed headteacher), and Guido Spiegel (German teacher). Among the others there were: Nedda Friberti (mathematics and physics), Adolfo Steindler (mathematics), Giorgio Schreiber and Ernesto Weiss (sciences).

Fausta Milla and Emma Senigaglia, grand-daughter of Salvatore Pincherle, both worked in Bologna; in Ferrara there was Emilio Teglio and, in Genoa, Vittorina Segre.[129] In Rome, Castelnuovo established the Clandestine University, in which Enriques held lectures in Higher Mathematics and History of Mathematics. Castelnuovo's youngest daughter, Emma, who had won the call in the summer of 1938, but who had not been able to take service due to the racial laws, taught mathematics at middle school level.[130]

The racial laws laid down that all aspects of the functioning and teaching in Jewish schools should be carefully supervised by an Aryan commissioner appointed by the Ministry. Thanks to the commitment of the individual teachers, however, these schools succeeded in offering their students a wide-ranging education, free from ideological conditioning, developing a teaching programme that an anonymous contributor to the newspaper *La Stampa* defined as 'shaped by that universalistic and internationalising character consisting of one of the most typical programmes and one of the most specific forms of pernicious dangers of the Jewish race'.[131]

---

con molta perdita di tempo; occupazione del resto non spiacevole e per cui avevo dato da settembre il mio assenso. D'altra parte, sono tempi in cui ogni risorsa va messa a frutto.

[128] Catalan et al. (2019) (eds.).

[129] Veziano (2007).

[130] Piperno Beer (2011); Castelnuovo E. (2001); Luciano (2013b).

[131] *L'epurazione razziale nelle scuole torinesi*, La Stampa, Turin 3.9.1938: *a carattere universalistico internazionaleggiante, in cui consiste uno dei tipici programmi ed una delle più specifiche forme di pericolosità della razza giudaica.*

Chapter 3
Fleeing from Italy

3.1 The Italian Racial Measures as Seen from Abroad

News of the racist turn of events in Italy rapidly spread abroad. Newspapers such as *The Times* and *The Guardian*, and periodicals such as *Nature*, informed readers of the anti-Jewish measures taken by the Italian government, and the effects they would have in the short and medium term.

'Aryans' in Italy. Nationalist doctrine in Italy hitherto has had the appearance of avoiding pronouncement upon theories of race. It has relied rather on fostering the totalitarian spirit upon imperial tradition with a consequent orientation to archeological research, for which the learned world is duly grateful to the personal interest of the Duce. Now, however, it is said to be time that the Italian frankly professed themselves to be "racist." A group of university professors, it is reported (*The Times*, July 15), working under the auspices of the Ministry of Popular Culture, has drawn up a pronouncement, appearing in the *Giornale d'Italia* of July 14, in which is stated what is to be regarded henceforth as the orthodox view of racial doctrine as applied to Italy. This statement has the merit that it recognizes that the idea of race is a purely biological concept, with which history, language and religion have nothing to do. It claims that the present population of Italy is in its majority Aryan, few elements of the pre-Aryan races remaining, and no immigration of populations capable of influencing the racial physiognomy of the nation having taken place since the Lombardic [Langobardic] invasion. The forty-four millions of inhabitants of Italy to-day, it is maintained, are for the most part descended from families which have been established there for at least a thousand years. While this may be conceded, but only so far as it goes, the inference that the Italian race – a term which in itself begs the question – can thus be regarded as a pure race is perhaps less readily to be accepted. The corollary that racism in Italy ought to be "essentially Italian with an Aryan-Nordic direction" is a hard saying, only partially intelligible, especially in view of the repudiation of linguistic evidence, in the light of the evident desire to dissociate the Italians from other members of the Mediterranean racial group by emphasizing its purely European characteristics as marking it out from all extra-European races. This view would ignore or deny any trace of kinship between the

E. Luciano, *The Jewish Mathematical Diaspora from Fascist Italy*, Science Networks. Historical Studies 64, https://doi.org/10.1007/978-3-031-64896-0_3

Mediterranean strain in the Italian and that in the peoples of North Africa, Arabia, and Palestine. This would seem to require a somewhat drastic re-interpretation of the facts.[1]

Racial studies in Italy. It is announced in the *Corriere della Sera* of October 5 that racial doctrine is to be made part of the curriculum of Italian universities. It is to be taught to students of natural science, medicine, and biology as well as to students of philosophy, educational theory, and literature. The innovation is made under a decree of the Minister of Education, Signor Bottai. A second course is to be devoted to study of the demographic problem. Italy has been slow in following Germany in the adoption of racial doctrine, nor, as was shown by the report of the Italian men of science on the racial question, has the doctrine been adopted entirely in the form, nor with enthusiasm, which it has aroused in the country of its origin. The physical characters of the main element in the Italian population would naturally call for some modification, and account for the stress, which in Italy is laid on continuity of cultural history.[2]

Alarmed by the information that appeared in the journals and by rumours circulating through correspondence, foreign mathematicians addressed their Italian colleagues to ask for confirmation and to express their solidarity. Cartan, for example, wrote to Levi-Civita: 'Mr. Fubini, whom I saw recently, gave me news of several of our Italian mathematician friends. There's no point telling you what our feelings are'.[3] Similarly, Picard to Volterra:

In recent months, we have often spoken here about you, your family, and the concerns that must have assailed you. We all live in sad times, where the world is close to madness, and where, under various attacks, what was good and noble in our civilisation risks perishing.[4]

Some *fuoriusciti* (i.e. political exiles) and Italians already living abroad, such as Max Ascoli, Gaetano Salvemini, and Aldo Mieli mobilised themselves.

In an initial phase, from September until the end of 1938, the situation appeared totally fluid, both in the eyes of Italians and of foreigners. Newspapers were vague and rumours were rife, for the most part unconfirmed. Italian Jews lived in the uncertainty of what the future may hold and, as Levi-Civita said to Cartan, there was nothing else but to wait: 'all my thanks for the sympathy you have expressed to me following recent anti-Semitic demonstrations. Until recently, I officially knew nothing, although I had learned enough, or rather too much, from the newspapers'.[5]

---

[1] *'Aryans' in Italy* (1938), p. 167.

[2] *Nature*, vol. 142, issue 3598, 15.10.1938, p. 142.

[3] ANL-*Levi-Civita*: É. Cartan to T. Levi-Civita, Paris 18.10.1938: *M. Fubini que j'ai vu récemment m'a donné des nouvelles de plusieurs de nos amis mathématiciens italiens. Il est inutile de vous dire quels sont nos sentiments.*

[4] ANL-*Volterra*: É. Picard to V. Volterra, Paris 30.11.1938. *Nous avons, ces derniers mois, souvent parlé ici de vous et de votre famille et des préoccupations qui ont dû vous assaillir. Nous vivons les uns et les autres dans de tristes temps, où le monde est près de folie, et où, sous des attaques diverses, ce qu'avait de bon et de généreux notre civilisation risque de périr.*

[5] ANL-*Levi-Civita*: T. Levi-Civita to É. Cartan, [Rome] 21.10.1938: *tous mes remercîments pour la sympathie que vous m'exprimez à la suite de récentes manifestations antisémitiques. Jusqu'à récent, je ne sais officiellement rien, tout en ayant appris assez, ou plutôt beaucoup trop, par les journaux.*

The opacity of the Duce's statements on the policy of excluding Jews led many to be confident in an act of retraction (*resipiscenza*) from the regime. Picard exhorted Volterra, for example, not to lose hope:

> In the sad times we are living in, we hardly dare to wish each other happiness. Let us tell ourselves, however, that very often things do not turn out as badly as we feared; bluffing does not always succeed, and common sense sometimes takes over. [...] The day before yesterday I had a visit from Enriques, who informed me of the situation of Italian mathematicians and of various Roman affairs. I hadn't seen him for a long time and, of course, the conversation turned for a moment to algebraic surfaces, with which we had both been concerned for so long. Dear friend, may the year 1939 be more benevolent to you than the one which now ends.[6]

3.2   The International Rescue Organisations

When the 'Italian question' arose, various international agencies to help victims of political and racial persecution were already up and running: the Society for the Protection of Science and Learning (SPSL) and the Emergency Committee in Aid of Displaced German Scholars (ECADGS) first and foremost, together with the American Friends Service Committee, the National Coordinating Committee for Aid to Refugees and Emigrants Coming from Germany, the *Comité International d'Aide aux Intellectuels*, the Central Committee for Refugees, and the Federation for Jewish Social Service. From the autumn of 1938, these became a new feature in everyday life for Italian intellectuals. Names never heard before—D.C. Thomson, E. Simpson, N. Searle, L. Farrand, S. Duggan, B. Drury—began to be pronounced daily.[7] Two addresses: 6 Gordon Square, London, W.C. 1, and 2 West 45th St., New York City, were noted for the first time in address books.

The two main agencies previously mentioned—SPSL and ECADGS—had been functioning since 1933 and had coordinated and financed the exodus of hundreds of scholars from Central Europe. Founded in London by W. Beveridge, L. Szilard, and Lord Rutherford under the name of Academic Assistant Council, the SPSL had registered about 2000 cases, 800 of which successfully handled, with a relocation of

---

[6] ANL-*Volterra*: É. Picard to V. Volterra, Paris 25.12.1938: *Par les tristes temps où nous vivons, on ose à peine s'adresser des souhaits de bonheur. Disons-nous cependant que bien souvent les choses ne tournent pas aussi mal qu'on le craignait; le bluff ne réussit pas toujours, et le bon sens reprend parfois ses droits. [...] J'ai eu avant-hier la visite de Enriques, qui m'a mis au courant de la situation des mathématiciens italiens et de diverses affaires romaines. Je ne l'avais pas vu depuis bien longtemps et, bien entendu, la conversation a dévié un moment sur les surfaces algébriques, dont nous nous sommes l'un et l'autre si longtemps occupés. Puisse, cher ami, l'année 1939 vous être plus clémente que celle qui finit.*

[7] David Cleghorn Thomson (1900–1980), General Secretary of the SPSL from 1938 to 1954; Esther Simpson (1903–1996), Assistant Secretary, later Secretary, of the SPSL, from 1933 to 1948; Nancy Searle, Secretary of the SPSL; Livingston Farrand (1867–1939), Chairman of ECADFS; Stephen Duggan (1870–1950), Secretary, then Chairman of ECADFS; Betty Drury (1907–1986), Executive Secretary of ECADFS from 1937 to 1944.

the refugee and family to the United Kingdom or the British Colonial Empire.[8]
D.C. Thomson served as General Secretary, flanked by two legendary Assistant
Secretaries: E. Simpson and N. Searle.

The Emergency Committee in Aid of Displaced German Scholars had been
created in New York on the initiative of A.E. Cohn, B. Flexner, F.M. Stein, and
S. Duggan, the Director of the Institute of International Education in May of 1933.
F. Warburg and A. Gregg served as advisors to the organisers and helped gain
support from other refugee and philanthropic organisations. L. Farrand became
Chairman of the Committee; Duggan its Secretary, and after the death of Farrand,
its Chairman; Stein became its Treasurer. E.P. Murrow served as Assistant Secretary
until 1935, followed by J.H. Whyte (1935–1937), B. Drury (1937–1944) and finally,
by F.F. Park (1945). The Committee brought 335 scientists out of Nazi-fascist
territories, out of about 6,000 dossiers submitted by applicants.[9]

The regulations of both the SPSL and ECADGS were designed for German-
speaking refugees and were rather strict. Neither of the two organisations could
directly approach a university, research institute, or college to present a candidate. It
was the individual institutions which, during recruitment, could contact these agencies, who would then suggest a shortlist of names. If the proposed individual was
judged of interest by the institution, SPSL and ECADGS provided economic and
logistical support, including help in obtaining the documents necessary for
expatriation.

In addition to being 'the latecomers' in the field of international assistance to the
victims of racial persecution, Italians had trouble in understanding the structure of
these agencies. The difficulty arose above all with the ECADGS, whose procedure
was applied inflexibly to all applicants, regardless of their reputation or number of
solicitations and intercessions received in support. This explains why, from
mid-September 1938, self-nomination candidatures and letters of recommendation
from Italian and foreign colleagues accumulated in its offices, invariably followed by
courteous but firm answers of rejection from the secretaries:

> You know, perhaps, about the work of the Emergency Committee in Aid of Displaced
> German Scholars which was formed in May 1933 to assist German professors and
> *Privatdozenten* who had lost university posts under the National Socialist regime because
> of racial, religious or political reasons. Today, because of the spread of totalitarian doctrines,
> it has seemed necessary to broaden the scope of the Committee's work, and to change its
> name to The Emergency Committee in Aid of Displaced Foreign Scholars. When you – or
> anyone else – writes to some individual for assistance for a displaced scholar, the letter is
> almost invariably forwarded to me as Secretary of the Emergency Committee in Aid of
> Displaced Foreign Scholars. Upon receipt of your letter, Professor Chamberlain at once
> asked me to do whatever was possible. I personally wrote the regulations of this Committee.
> The very first of these regulations was to the effect that we could not receive an application
> from an individual or the friends of an individual, but only from a university which requested

---

[8] Nossum (2012); Nossum and Kotulek (2015); Williams (2013).
[9] Duggan and Drury (1948).

that we make a grant for the support of that individual for at the most two years until the institution would be in a position to absorb the scholar permanently.[10]

In the autumn of 1938, after the *Anschluß Österreichs* (13 March 1938), the German annexation of the Sudetenland (30 September 1938) and the promulgation of anti-Semitic laws in various countries (Romania, Turkey, Italy, etc.), the United Kingdom and the United States prepared to face an even greater wave of migration. As Veblen explained to V. Hlavatý, at a time when rescue potentialities were nearly exhausted, they were receiving 'a flood of calls for help from Italy in addition to those from Austria'.[11]

Moreover, in the United States, xenophobia was gaining ground and many people wondered if, because of the growing seriousness of the situation, it might be necessary to rethink and revisit the overall system of reception of intellectuals 'in view of resentment arising among younger teachers in colleges who are finding the way to preferment obstructed by the placement of these foreigners ahead of them'.[12]

Faced with the Jewish intellectual emigration from racist Italy, which was beginning in this period, SPSL did not change its functioning regulation, instead extending it *de facto* to all non-German-speaking refugees. ECADGS, after a considerable tug-of-war, officially expanded its scope and from January 1939 changed its name from Emergency Committee in Aid of Displaced German Scholars to Emergency Committee in Aid of Displaced Foreign Scholars (ECADFS). However, such a process was far from linear (D3.2.1, D3.2.2, D3.2.3). A certain amount of scepticism towards Italians existed on the other side of the Atlantic, except in the case of important figures such as Levi-Civita or Fubini. The general impression was that:

> there is the slightest chance that that Committee can do anything in this situation. It has never taken on the Italian problem, even in the slightest way and with its depleted funds and the enormous increase in demands of German and Austrian victims, the situation is pretty desperate. I naturally hope something can be found for scholars of obvious eminence [...], but I am very sure that our Committee is not in a position to do anything definite. What a mess the whole situation is.[13]

SPSL and ECADFS were not alone in the immense task of assisting displaced scholars. They teamed up with various other international scientific bodies, such as the International Institute of Statistics and the International Astronomical Union; with religious institutions, such as the American Friends Service Committee;[14] and

---

[10] ECADFS Records, *GM*: S. Duggan [ECADFS] to L.I. Dublin, New York 3.1.1939.

[11] OVP, *AD*: O. Veblen to V. Hlavatý, [Princeton] 30.9.1938.

[12] ECADFS Records, *GM*: S. Duggan [ECADFS] to L.I. Dublin, New York 3.1.1939.

[13] ECADFS Records, *GM*: L. Farrand [ECADFS] to L.I. Dublin, New York 29.11.1938.

[14] The American Friends Service Committee was a religious society of friends (Quaker) founded organization working for peace and social justice in the United States and around the world. During the 1930s and through World War II, it helped refugees escape from Nazi Germany, aiding people who were not being helped by other organizations, primarily non-religious Jews and Jews married to non-Jews. They also provided relief for refugees in Vichy France and for individuals on both sides of the Spanish Civil War.

with philanthropic foundations. Among these, the Rockefeller Foundation stands out.[15] Since 1913, it had been funding research projects in five areas (medicine, natural sciences and agriculture, arts, international relations, and social sciences) and, as far as Italy is concerned, had financially supported the journal *Annali di Matematica pura ed applicata* in the first post-war period (1919) and the Institute of Public Health (*Istituto di Sanità Pubblica*) in 1928. In the mathematical field, however, only B. Segre had obtained a RKF grant in 1926, which he employed for a study stay in Paris. B. de Finetti and other young mathematicians, despite attempts, had never been selected as RKF fellows. Thus, it was no coincidence that when G. Demaria, professor at the Bocconi University in Milan and former consultant of RKF Economics Department, contacted T.B. Kittredge, assistant Director of the foundation, in order to inform him of the racial laws, the latter answered coldly:

> I have of course been informed of the action taken in connection with the appointments of professors of Jewish origin in the Italian universities. I fear there is not very much that the Foundation can do directly in dealing with such cases. If, however, there were institutions elsewhere that might wish to invite certain of the Italian scholars to join their staff, the Foundation might be disposed to collaborate with the institutions concerned by providing part of the salary over the first few years of scholars of recognized distinction.[16]

The role of RKF in supporting Italian émigrés would be generally limited and almost non-existent for mathematicians. In the report listing deposed student programmes under RKF up to December 1, 1939, there were no Italians. In the new programme for refugee scholars launched in August 1940, three fellowships for Natural Sciences were granted, one of which went to a student of Fermi settled in France since 1936: Bruno Pontecorvo. In the list of grant recipients from 1.6.1941 to 10.4.1943, once again there was no Italian.

Documents

D3.2.1 B. Drury [ECADFS] to G. Richter, New York 23.1.1939

ECADFS Records, *Italy*, fol. 1r
Dear Dr. Richter:[17] Let me acknowledge your letter of January 20 by saying that on November 9, 1938, because of the critical situation throughout the length and breadth of [war in] Europe, it seemed advisable to broaden the scope of this Committee's work and to change its name to the Emergency Committee in Aid of Displaced *Foreign* Scholars. We are therefore interested in receiving the curricula vitae of displaced professors and *Privatdozenten* from Italy, as well as from

---

[15] Siegmund-Schultze (2001). See also Gemelli (2000) and Lamberti (2006).

[16] SPSL, *GM*: T.B. Kittredge [RKF] to G. Demaria, Paris 20.9.1938.

[17] Gisela Richter (1882–1972), English archaeologist, Director of the Greco-Roman section of the Metropolitan Museum in New York (1925–1950).

Germany, Czecho-Slovakia, and Austria. Although the Committee's name has been changed, its regulations have not been materially altered. I believe you are familiar with the Committee's rules but am sending you herewith a copy of our last printed report so that you may have the latest available information about our program. Sincerely yours, Betty Drury - Assistant Secretary

D3.2.2 B. Drury [ECADFS] to J.P. Chamberlain, New York 8.2.1939

ECADFS Records, *Italy*, fol. 1r

Dear Professor Chamberlain:[18] May I write you unofficially about several matters which you brought to Dr. Duggan's attention some days before the Committee's executive meeting last week? In the agenda of the meeting Dr. Duggan included your letter of January 26 to him concerning the possibility of supporting professors in Cuba and also your letter of January 18 concerning the formation of a sub-committee of the Emergency Committee in Aid of Displaced Foreign Scholars to consider the cases of displaced Italian professors. Dr. Duggan will probably be writing you at greater length about both of these matters, but as you probably know he has left New York for a short rest, and I want to send you this brief note concerning the Committee's judgement on these two points. The Committee considered very carefully the material which you had sent from Mr. Edward Salinger of the Resettlement Division of the National Coordinating Committee about the opportunities in Cuba which Dr. Morton Kahn reported to him.[19] It was the opinion of the Committee that the matter of assisting refugee scholars to establish themselves in Cuba by means of annual renewable grants was outside the immediate work of the Emergency Committee. Concerning the formation of a sub-committee to take particular care of the cases of displaced Italian professors, the Committee seemed to feel that by and large the establishment of a special group of 'referees' might be unnecessary since the Committee had no special committee for Austrian refugees or for scholars from Czecho-Slovakia. They said, however, that they hoped they might continue to be able to call upon Professor Ascoli for information concerning specific cases as they had been doing during the past few months.[20] The members of the Committee had

---

[18] Joseph Perkins Chamberlain (1873–1951), professor of Public Law at Columbia University (1923–1950) and member of several organizations that provided aid to European Jewish refugees. He was a co-founder of the German Jewish Children's Aid.

[19] The National Coordinating Committee had been established in 1934 by the American Jewish Joint Distribution Committee at the suggestion of the State Department to coordinate the relief work of affiliated private refugee agencies in the United States. NCC's successor, the *National Refugee Service*, was founded in New York on 15 May 1939 to assist refugees from Europe fleeing Nazi persecution. Edgar Salinger (1887–1971) was the Director of this aid organization. Morton Kahn, executive Director of the Federation for Jewish Social Service.

[20] Max Ascoli (1898–1978). See Sect. 3.10.

found Professor Ascoli's advice most useful and would probably be turning to him for additional information from time to time. Sincerely yours, Betty Drury - Assistant Secretary

D3.2.3 J.P. Chamberlain to B. Drury [ECADFS], New York 16.2.1939

ECADFS Records, *Italy*, fol. 1r
Dear Miss Drury: I have received your letter of February 8. I am frank to say that I expected this answer, but thought it necessary to refer the material to you before going any further with it. I note also the action of the Committee in regard to the Italian professors. I think that the situation is substantially different from that of the Austrian refugees, who were really lumped in with the general German situation, and certainly the German scholars from Czechoslovakia, who could be classed with the same group. The Italians seem to me to be a different case because the Italians are not so well known and because they might have a special appeal to particular institutions. Furthermore, we all know that in their case there is difficulty in sending them to certain institutions in which a strong fascist sympathy exists on the part of present members of the Italian Department of the Faculty. However, I think that these difficulties can all be met in the way you suggest, by communicating with Dr. Ascoli and having him acting as consultant, which is probably a good way of doing it, as he can form an advisory committee of other Italians. Very sincerely yours, Joseph P. Chamberlain

3.3   Lack of Information

International rescue agencies were prompt in giving support to Italian professors removed from service. On 6 September 1938, just three days after the racial laws were enacted, the SPSL wrote to nine of its long-term consultants (R. Klibansky, F. Demuth, G.H. Hardy, F. Zeuner, F. Saxl, S. Brodetsky, C.S. Gutkind, C. Roth), asking for information on the fate of the Italians:

> We are anxious to obtain as quickly as possible complete and accurate information concerning the dismissal of Jewish university teachers and research workers in Italy. We are wondering whether you have names of colleagues who have already lost their positions. If so, we should be most grateful for any information you can give us. We understand that the university teachers liable to be dismissed include several eminent mathematicians.[21]

---

[21] SPSL, *Italy*: E. Simpson [SPSL] to S. Brodetsky, F. Demuth, C.S. Gutkind, G.H. Hardy, R. Klibansky, C. Roth, F. Saxl and F. Zeuner, London 6.9.1938. Selig Brodetsky (1888–1954), mathematician, Russian Jew, displaced from 1893 in London, and then to Bristol. Fritz Demuth (1876–1965), jurist and economist, emigrated to Switzerland in 1933, and later to England where he

A few days later, L.H. Engel, editor of the *Science Service*, was trying to locate a list of scholars and scientists now in danger of losing their positions through Italy's newly introduced anti-Semitic programme. Unable to do so, Engel contacted Duggan to enquire where to find a list of Italian academics who may be displaced as a result of Italy's new policies. The ECADGS redirected him to the SPSL which might have such a list or be in the act of drafting one. At the same time, Duggan asked SPSL and Science Service if they would be able to help in this matter.[22] However, the answers to both the agencies were all very generic. For example, S. Brodetsky replied:

> I am afraid that I cannot give you any names of Jewish university teachers who have been dismissed in Italy, as I have not obtained any definite information. I have, however, discussed the matter with one or two people who are well informed on recent events in Italy, and I think that it can be assumed that Jewish professors have been or will be dismissed. There certainly are a number of most eminent mathematicians among them, e.g., Professor Levi-Cürta [sic!], the famous Applied Mathematician. I shall give you any definite information that may come my way.[23]

The initial course of action of the associations was hampered by the legislative situation, which was still extremely confused, and by the chaos that reigned in the search for information. The decrees issued by the Italian government were ambiguous, the numbers of the dismissed Italian university teachers and the degree of rigour with which they would be applied equally unknown (D3.3.1, D3.3.2, D3.3.3). It was very difficult to acquire this information from the newspapers, especially now when events in Italy were overshadowed by the Czechoslovakian crisis. While waiting for the racial laws to be definitively laid down, SPSL and ECADGS could merely register the names of persecuted intellectuals that individual scholars and institutions signalled to them.[24]

became a highly regarded figure among Germans in exile. He was the driving force behind the Notgemeinschaft Deutscher Wissenschaftler im Ausland. Curt Sigmar Gutkind (1896–1940), German philologist of Jewish origin, he was expelled from Heidelberg University in 1933 and moved first to Paris and then to England. Godfrey Harold Hardy (1877–1947), number theorist, professor in Cambridge (1931–1942). Raymond Klibansky (1905–2005), historian of philosophy, German Jew, displaced in Italy, then in Brussels and finally in Oxford, where he was a professor at Oriel College. Cecil Roth (1899–1970), a British Jewish historian, was editor-in-chief of the *Encyclopaedia Judaica* and a reader in Post-Biblical Jewish Studies at the University of Oxford (1939–1964). Fritz Saxl (1890–1948), Austrian historian of art. In 1933 he moved to London where he became Director of the Warburg Institute. Frederick Everard Zeuner (1905–1963), geologist and paleontologist, German Jew, refugee in England since 1934, where he began working for the British Museum.

[22] ECADFS Records, *Italy*: B. Drury [ECADFS] to D.C. Thomson [SPSL], New York 16.9.1938 and B. Drury [ECADFS] to L.H. Engel [Science Service], New York 17.9.1938.

[23] SPSL, *Italy*: S. Brodetsky to E. Simpson [SPSL], London 20.9.1938.

[24] SPSL, *Italy*: E. Simpson [SPSL] to A. Calò, London 14.9.1938 and E. Simpson [SPSL] to M.L. Butters [The Warburg Institute], London 1.10.1938.

Distinguished Scholars Dismissed

Among the famous Italian Jews who are to be dismissed from their posts in accordance with this decree are Professor Artom, the specialist consultant to the Italian Royal Family, who also attended the Crown Princess when her daughter was born in 1934; Professor Tullio Levi-Civita, the famous mathematician, known as the Italian Einstein; Professor Giorgio del Vecchio, the well-known philosopher; Professor Guido Castelnuovo, the mathematician; Professor Gino Arias, the economist; Professor Dino Derossi, Professor Cesare Finzi and Professor Giorgio Tedesco.

Nor are we yet at the end of the anti-Jewish crusade. On the 1st October, it is officially announced, the Supreme Fascist Council will meet to consider further measures against the Jews. It is stated that the Council will decide to exclude all Jews from the Army, Navy and Air Force and introduce a strict *numerus clausus* against Jews in the free professions and in commerce, industry and finance.

In the meantime the Cabinet has set up a Superior Council for Racial Problems and Demography to formulate opinions on questions relating to race.

Problem of Emigration

The immediate effect of the two anti-Semitic decrees has been to cause a rush of the Jews to the foreign Consulates to inquire about the possibilities of emigration. The Jews have also begun to offer their businesses and property for sale and the rapidity with which the Jews have started to dispose of their shares on the Stock Exchange even caused the publication of an inspired assurance that no measures were contemplated against legitimate Jewish business activities.

Fig. 3.1 *The Zionist Review*, 8.9.1938, Archive of the Society for the Protection of Science and Learning

The first partial lists of dismissed scholars from Italy were thus put together piecemeal, on the basis of reports arriving in dribs and drabs, in a purely random manner (D3.3.4, D3.3.5, D3.3.6, and Fig. 3.1).

Also because of the very frequent cases of homonymy and the numerous kinships, scholars were often confused with each other: Giulio Bemporad was mistaken for his cousin Azeglio, Volterra and Castelnuovo seniors with their children. The names were often wrong: Volterosa is Volterra, Esiquez is Enriques (Fig. 3.2).

The only certain fact, for international organisations, was the fate of foreign Jews (even if naturalised Italians), for whom expulsion from the country was expected by

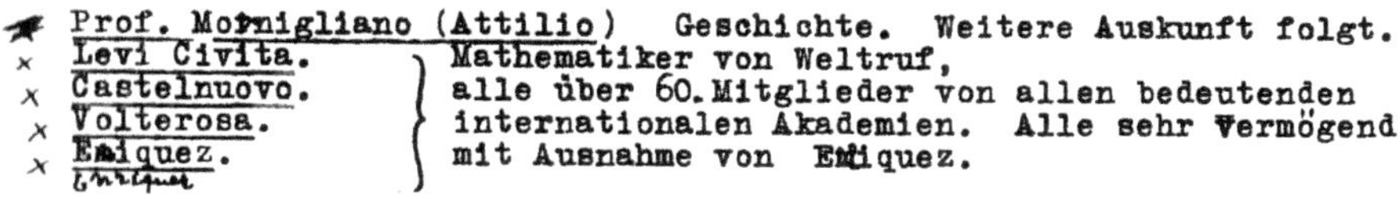

```
✒ Prof. Montigliano (Attilio)   Geschichte.  Weitere Auskunft folgt.
×  Levi Civita.         ⎫  Mathematiker von Weltruf,
×  Castelnuovo.         ⎪  alle über 60.Mitglieder von allen bedeutenden
×  Volterosa.           ⎬  internationalen Akademien.  Alle sehr vermögend
×  Enriquez.            ⎭  mit Ausnahme von Enriquez.
```

Fig. 3.2 List of Italian dismissed professors sent by the Warburg Institute to the SPSL, 30.9.1938, Archive of the Society for the Protection of Science and Learning

1 March 1939.[25] All other aspects and terms of the Italian persecution were obscure: did removal from service mean dismissal? Would people who fled from Italy continue to receive the pension? Would those who intended to emigrate be able to export money? (D3.3.7, D3.3.8) Furthermore, foreigners generally have difficulty within the jungle of Italian titles, in particular the difference between *assistente*, *aiuto*, *incaricato* and *ordinario* (i.e. full professor or regular assistant, or regular teacher of middle-secondary school, depending on the context) was obscure, so much so that on many occasions ECADGS had to ask Max Ascoli for explanation:

> I find so often in the curricula of the displaced scholars from Italy that the term "professor" is used much more frequently than it occurs in the German and even in the Austrian curricula for persons who taught in lyceums and gymnasia. Is the title 'professor' as used in Italy, synonymous with the German 'professor' (or *Privatdozent*) who belonged to the staff of a university or *Hochschule*?[26]

Finally, the practice of counter-discrimination was completely inexplicable and unconceivable. In this total confusion, a belief emerged that was destined to have a noticeable negative impact on the fate of aspiring emigrants: the idea that Italian Jewish mathematicians were old and rich, and therefore unwilling to leave (D3.3.9).[27] In fact, Castelnuovo, Volterra, and many others, already retired, had

---

[25] In the Memorandum drawn up by B. Drury to Duggan of 6 December 1938 (ECADFS, *GM: Scholars displaced from Italy*), for example, nine names were indicated and all except Mortara were foreign Jews who had sought refuge in Italy in the thirties and who were embarking on a second emigration: Sigmund Cohn (Political Economy); Heinrich Kahane (Romance Languages); Friedrich Friedmann (History, Philosophy, Germanic Literature); Helene Wieruszowski (Medieval History); Eduard Straus (Biochemist); Paul Oscar Kristeller (Philosophy); Arnold Reichenberg (German Language and Literature); Giorgio Mortara (Statistics); Peter Adalbert Silbermann (Education, Teaching of Languages); and Leonardo Olschki (Romance Languages).

[26] ECADFS Records, *Italy*: B. Drury [ECADFS] to M. Ascoli, New York 28.1.1939. In Nazi Germany academics were organized in three main categories: the permanent positions of *ordentlich Professor* (ordinary Professor) and *außerordentlich Professor* (extraordinary Professor), and the rank of *Privatdozenten*. A *Privatdozent* was a member of a Faculty who had passed the Habilitation. The title of *Privatdozent* had no equivalent in the Italian academic system.

[27] As an anonymous report stated (ACS, Pol. Pol., busta 219): "Padua, 13.1.1939. The opinion is spreading that except for those who will voluntarily abandon Italy, the remaining Jews will be tolerated and left undisturbed. To clarify this statement, someone maliciously added: The Government must be given time to skin them, without making them scream too much." (*Padova, 13.1.1939. Si va diffondendo l'opinione che eccettuati coloro che abbandoneranno volontariamente l'Italia, i rimanenti ebrei saranno tollerati e lasciati indisturbati. A chiarimento*

not been affected by the racial laws. Among the illustrious scholars in service, Levi-Civita and Enriques were supposed to remain in Italy.

Documents

D3.3.1 C. Roth to SPSL, London 6.9.1938

SPSL, *Italy*, fol. 481
Dear Madam [E. Simpson], I am scribbling a few words, before rushing off to France for ten days, to tell you that I have not had as yet any direct news from Italy. On the other hand, from Mussolini's decree it is pretty obvious who will be dispossessed. You will find some indication of it in a monograph by Eliezer ben David, *Gli ebrei nella vita culturale italiana, 1848-1928* (Città di Castello, 1931), which enumerates the majority of the extraordinarily distinguished roll of contemporary Italian Jewish savants. I am leaving my copy on my desk, and should you desire to consult it you might perhaps ring up my Secretary, Miss M. Benjamin, of 28 Compayne Gardens (Maida Vale 1567), who is authorized hereby to lend it to you in my absence. Yours very truly, Cecil Roth

P.S.: Some extremely illustrious Italian Jews are listed in my *Jewish Contribution to Civilisation*.[28]

D3.3.2 G. Bing [The Warburg Institute] to E. Simpson [SPSL], London 7.9.1938

SPSL, *Italy*, fol. 463r-v
Dear Miss Simpson, I do not know whether you keep the *Neue Zürcher Zeitung*.[29] In case you do not, the following information will interest you as an addition to the particulars about the dismissals in Italy which Dr. Saxl is trying to get. The NZZ is on the whole very accurate. The numbers of dismissed university teachers are up till now as follows: 50 at Turin

40 at Florence

20 at Milano

(date 5th of Sept.)

---

di tale affermazione qualcuno ha malignamente soggiunto: Bisogna ben dar tempo al Governo di pelarli, senza farli strillar troppo.)
[28] C. Roth, *The Jewish Contribution to Civilization*, London, Macmillan and Co., 1938.
[29] *Neue Zürcher Zeitung*, 5.9.1938, p. 2.

There have been new dismissals at Vienna University by which 20 persons of the Medical Faculty only are affected! These are no Jews, but Aryans, mostly Catholic, all of them "unbearable" for the new regime. Yours sincerely Gertrud Bing[30]

D3.3.3 C.S. Gutkind to E. Simpson [SPSL], Italy 8.9.1938

SPSL, *Italy*, fol. 483r-v
Dear Miss Simpson, unfortunately, I don't know more about the distressing dismissal of Jewish university teachers and research workers in Italy than what the unreliable newspapers tell. The *Giornale d'Italia* of to-day speaks of altogether 174 university professors (I imagine that in this number assistants and lecturers are more or less included in so far as these too had fixed salaries) who have not yet been dismissed, in the proper sense of the word, but, as the decree runs, temporarily suspended from work.[31] Momigliani's [sic!] excellent history of Ital. lit., introduced in Italian Schools as textbook as late as January 1938 (!) (I believe this date is exact), has been put on the *Index librorum prohibitorum* of the Italian State-schools;[32] Momigliani [sic!] himself is suspended from work, accordingly to the *Giornale d'Italia*. I am very very sorry, but for important personal reasons I am unable to give you further information. In about a fortnight, I shall be back at London, and Bedford College will be the best place for getting immediately into touch with you. Yours sincerely G.

P.S. Everybody here is bewildered of the quite unexpected suddenness of this un-understandable Jew-baiting, and people, although absolutely unable, for the moment at least, to react, behave in the most touching and humane way, which is a certain consolation. Italy, anyhow, is not Germany, or, Italians are not Germans.

D3.3.4 A. Calò to E. Simpson [SPSL], London 12.9.1938

SPSL, *Italy*, fol. 487
Dear Miss Simpson, I send you a list of Italian Jewish professors, whom I *personally* know. I have written to my friends in Rome to send me more exact information and the addresses of some of the people. I will then write you again. I am waiting now for the solution of my pecuniary situation. When I will know

---

[30] Gertrud Bing (1892–1964), German historian, worked as a librarian for the Kulturwissenschaftlichen Bibliothek Warburg, the nucleus of the future Warburg Institute.

[31] *I professori giudei nell'Università italiane*, Il Giornale d'Italia, XXXVIII, n. 211, 6.9.1938, p. 1.

[32] Attilio Momigliano, *Storia della letteratura italiana*, Messina, Principato, 1932.

exactly, what my condition will be here I will send you back my completed information forms. With best regards, Yours faithfully, Aldo Calò[33]

D3.3.5 F. Saxl to E. Simpson [SPSL], London 14.9.1938

SPSL, *Italy*, fol. 465

Dear Miss Simpson, Information from Italy comes very slowly, I am sorry to say. These are the few names which I got to-day: Limentani (Ordinarius für Philosophie); Momigliano (Ordinarius für italienische Literaturgeschichte); Bonaventura (Außerordentlicher Prof. für experimentelle Psychologie); Rubinstein (Assistent für mittelalterliche Geschichte – er ist ungarischer Nationalität); Frau Kahane (Lehrauftrag für Neugriechisch – Griechin, mit einem österreichischen Juden verheiratet).[34] Professor Olschki wrote in addition that there are more than 100, and that he is willing to find out from Italy (he himself is at Geneva).[35] When I hear from him, I will let you know. Yours very sincerely, F. Saxl

D3.3.6 M.L. Butters [The Warburg Institute] to E. Simpson [SPSL], London 30.9.1938

SPSL, *Italy*, fol. 470

Dear Miss Simpson, Dr. Bing asks me to send you the enclosed list of dismissed Italian scholars which we have received. The list is mainly of the Faculty of Literature and Philosophy, and even for that Faculty is not complete. It appears to refer to scholars all over Italy, and we are informed that not only are those concerned forbidden to teach or publish, they are also deprived of right to take any anonymous paid employment. In some cases, however, help is not urgently needed, either because those dismissed are elderly and would not leave Italy, or because they are fairly well off. Yours sincerely, M.L. Butters

---

[33] Aldo Calò, assistant in the Institute of General Surgical Clinic and Surgical Therapy of the Faculty of Medicine and Surgery of the University of Rome.

[34] They were all scholars from the University of Florence: Ludovico Limentani (1884–1940), Moral Philosophy; Attilio Momigliano (1883–1952), Italian Literature; Enzo Bonaventura (1891–1948), Experimental Psychology; Nicolai Rubinstein (1911–2002), Medieval History and Renée 'O Toole-Kahane (1907–2002), Romance Philology.

[35] Leo Samuele Olschki (1861–1940) was a renowned Italian publisher of Prussian Jewish origin. Accused of espionage at the outbreak of WW1, he had been driven into exile in Geneva, where he had established a branch of the Florentine firm. In 1938, he was forced again into exile in Geneva, where he died.

D3.3.7 H. Shapley to B. Drury [ECADFS], Cambridge Mass. 24.12.1938

ECADFS Records, *Italy*, fol. 1r

Dear Miss Drury: The material you sent by telegram the other day was exactly what we needed here. I wonder if you could easily get some information concerning the Italian situation. If I understand correctly the date of March 1 was set as the time when Jews must leave Italy, but I believe that referred to those who had come into Italy and were not actual residents.[36] But since scores of Jewish professors have been dropped from the Universities, I have wondered if their dropping is also an exile. In other words, must men like Professor Levi-Civita leave Italy, or may they remain in Italy without positions [?]. Do you have any information as to whether these discharged Italian professors receive pensions?[37] There is no immediate hurry for answers to these questions. It would be useful in some of my work if I could in a week or so know how these things stand. Possibly you can get in touch with an appropriate information agency and at least get partial answers. Very sincerely yours Harlow Shapley[38]

D3.3.8 B. Drury [ECADFS] to H. Shapley, New York 4.1.1939

ECADFS Records, *Italy*, fol. 1r

Dear Professor Shapley: Let me thank you for your letter of December 24 concerning the difficulties faced by the Jewish professors in Italy. The situation is still very far from being definitely settled and any information we can get at present concerning their future may be entirely wrong next week. Even the Italian consulate is unable to give us definite information. We do know, however, that all Jews who have gone to Italy since 1919, even though they have become Italian citizens, must leave by March 1. Concerning the Italian professors of Jewish race, we are fairly sure that they will have to give up their positions. Whether they will receive pensions or whether they will have to leave Italy will probably depend on how many of them there are and how strong the feeling of anti-Semitism becomes. At least they are not required to leave the country by March 1. The measures taken against the Jews in Italy are not yet nearly so drastic as those taken in Germany and Austria. We have obtained the

---

[36] The words "understand correctly … not actual residents" are marked. Next to them is added by hand, on the left margin: "residents or those who had come from other countries since 1919".

[37] The words "exile … without positions" are marked. Next to them is added, by hand, in the left margin: "exile in retirement." In the right margin is added: "Probably retired. May have to leave, may not".

[38] Harlow Shapley (1885–1972), Head of the Harvard College Observatory (1921–1952), political activist and humanitarian. From the late 1930s he headed a nationwide effort to bring refugees from Germany to the United States and was personally involved in the resettlement of at least 1000 refugee families.

above information from the National Coordinating Committee which has as its purpose the assistance of Jewish refugees. When the situation becomes clarified we shall be glad to obtain further information from them for you. Sincerely yours, Betty Drury – Assistance Secretary

D3.3.9 B. Drury [ECADFS] to H. Shapley, New York 24.1.1939

ECADFS Records, *Italy*, fol. 1r
Dear Professor Shapley: We have obtained [some further] information concerning the situation of Italian professors of Jewish race from Professor Max Ascoli of the Graduate Faculty of the New School for Social Research. Professor Ascoli tells us that all Italian teachers of Jewish race will be displaced from their positions. They will, however, receive a pension which will be in proportion to the number of years they have taught. The pension of those who have taught for a considerable length of time will be sufficient for them to live on. This will not be the case, however, where the scholar in question has only been teaching a few years. Also, as in Germany, there is no security for these scholars, as pensions may be revoked by the government at will. Professor Ascoli spoke of Professor Levi-Civita as a case in point. He will receive a pension which is large enough to meet his needs. Although he has as yet made no attempt to find a position outside of Italy, he has some intention of leaving should a position be offered him. We shall be glad to let you know if any further information concerning displaced Italian scholars comes to our attention. Sincerely yours, Betty Drury - Assistance Secretary

3.4   The Bureaucracy of Emigration

The racial laws provided for the forced expulsion only of foreign Jews. Unofficially, however, the regime also encouraged the exodus of Italian Jews, relying on rabbis and community Presidents for assistance (D3.4.7).[39] From the autumn of 1938 to February 1940, there were several interventions by the leaders of Italian Judaism who invited the co-religionists to move to Palestine or to phantom colonies that the regime seemed willing to create in the colonial domains (D3.4.6, D3.4.8, D3.4.11, D3.4.12). Again in February 1940, Mussolini informed D. Almansi, the newly-elected President of the Union of the Italian Jewish Communities (UCII), that Italian Jews must gradually (but absolutely) leave the country, and between the autumn of

---

[39] See also, in ACS-*Razzismo* 1939: notes by anonymous informants of the Secret Political Police from Turin 15.1.1939, Rome 5.2.1939, Turin 12.12.1939, Rome 17.12.1939; Rome 2.2.1942.

1940 and the summer of 1941 a project for the solution of the Israelite question was carried out, culminating with the idea of creating a Jewish enclave in the Italian colonial empire.[40]

Contrary to the extreme precision with which the lists of persecuted scholars were compiled and integrated, the sources concerning emigration in the archives of the General Office of Demography and Race are tremendously incomplete. However, according to the incomplete data collected in these documents, it seems plausible to outline the following frame: approximately 4,000 Italian Jews left the country between September 1938 and June 1940; a further 4,000 took refuge in Switzerland after the armistice (8 September 1943) and following Nazi occupation of Italy. Among them there were 11 mathematicians, 10 physicists, 2 astronomers, and 3 statisticians with their families.[41]

While the bureaucracy of discrimination was efficient, that of expatriation was terribly 'Italian style'. The Ministry of Foreign Affairs exerted no action of any kind whatsoever in relation to the intellectual diaspora, although occasionally urging the police to facilitate the issuance of passports. Faced with a riddle of contradictory indications from the regime (D3.4.1, D3.4.10)—facilitate the emigration of Jews or discourage it?—and in the absence of shared protocols, embassies and consulates gave different answers from place to place. Often, they were not able to give any information; in other cases, they provided incorrect or incomplete information. As in the case of counter-discrimination procedures, a system of corruption flourished, which officials and even fascist hierarchs wickedly took advantage of.

University professors, from this point of view, were advantaged because they had always annually requested the renewal of passports to participate in conferences, meetings of editorial boards, in order to go abroad as visiting professors, etc. However, the revocation of visas decided by Bottai in the summer of 1938 cancelled the validity of their documents. Moreover, the countries for which passports were requested, and which had to be specified, were usually the European States, except for Russia and Spain, so in 1938 very few people, irrespective of the revocation situation, would have a passport or visa valid for expatriation in the United States or South America. In the third place, although an extension of the passport was often requested for the wives, who accompanied the scholars on work-related missions abroad, the document had practically never been extended to children.

Gathering the documents for expatriation (passports, tourist visas, etc.) thus became an eternal process, which exhausted the aspiring exiles and which was resolved only when the official channels were abandoned in favour of personal

---

[40] Almansi (1945) and ACS-*Razzismo* 1940, 1941: notes by anonymous informants of the Secret Political Police from Rome 21.1.1939, Bordighera 2.1.1940; Rome 14.3.1940. See also *The position of the Jews* ..., 1941.

[41] According to a census by the Ministry of Interior on 28 October 1941, 5,966 Italian Jews had definitively abandoned the country, out of a population of 45,410 individuals in August 1938. See ACS-*Razzismo* 1941: Ministry of the Interior, Report *Al Duce*, 14.10.1940, fols. 1–5; Ministry of the Interior, *Situazione ebraica al 28 ottobre 1941*, fols. 1–3. See also ACS-*Razzismo* 1941: Ministry of the Interior. Report dated 19.7.1942, fols. 1–10.

networks and footholds so as to reach the top of the bureaucracy chain (D3.4.2, D3.4.3, D3.4.4, D3.4.5). The help of foreign colleagues was often essential. For example, Lefschetz and Borel personally accompanied Fubini to the Italian embassy and the American consulate in Paris. Veblen sent Edoardo and Enrico Volterra the *vademecum* of documents needed to obtain a visa for the United States, a document that paradoxically seemed impossible to obtain in the Italian offices. Volterra and Levi-Civita mobilised their contacts in the Holy See, A. Flexner and Veblen those in the American embassies in New York. The situation of individuals who emigrated with the support of the SPSL was slightly simpler, because this organisation helped in obtaining permits and documents to be shown to the Home Office (D3.4.9).

Documents

D3.4.1 MFA to Emigration Offices of Genoa, Trieste, Naples and Palermo, 17.11.1938

ACS-*Ebrei*, fol. 1r
Expatriation of Italian Jews. The question was posed to this Ministry whether, following the provisions recently issued for the defence of the race, the requests of Jewish individuals with Italian citizenship and resident in the Kingdom from the period prior to 1919, aimed at obtaining expatriation in transoceanic countries and especially in the Americas, are to be satisfied. It should be noted that, in principle, there should be no obstacles to the expatriation of the aforementioned elements, for whom the same treatment must be adopted as that in force for non-residents or for citizens of the redeemed lands. Therefore, they will not be obliged to present an act of call or employment contract, as it is sufficient for them to demonstrate that they can be admitted to the countries to which they intend to go. However, it is understood that this treatment cannot, for obvious reasons, be applied to those who ask to go to the United States permanently. With regard to them, this Office will have to limit itself to reporting to this Ministry the cases which, in its opinion, are worthy of very special consideration. As for the others, they must be told that there is no possibility of accepting their requests, due to the noted American restrictions on expatriate quota numbers.

Espatrio di ebrei italiani. *È stato posto il quesito a questo Ministero se, in seguito alle disposizioni recentemente emanate per la difesa della razza, siano da assecondare le domande di elementi ebraici, di cittadinanza italiana, residenti nel Regno da epoca antecedente al 1919 – dirette ad ottenere l'espatrio in paesi transoceanici e soprattutto nelle Americhe. Per opportuna norma, si fa presente che, in linea di massima, non debbono essere frapporti ostacoli all'espatrio dei suddetti elementi, nei riguardi dei quali dovrà essere adottato un trattamento eguale a quello in vigore per gli allogeni delle terre redente. Non sarà perciò fatto loro obbligo di presentare atto di chiamata o contratto di lavoro, essendo sufficiente da parte loro la dimostrazione che essi potranno essere ammessi nei paesi verso i quali hanno intenzione di dirigersi. Resta, tuttavia, inteso che tale trattamento non potrà, per ovvie considerazioni, essere usato a coloro che chiedono di recarsi negli Stati Uniti in via definitiva. Nei loro confronti, codesto Ufficio dovrà limitarsi a segnalare a questo Ministero i casi che, a suo giudizio, siano meritevoli di specialissima considerazione. Quanto agli altri, dovrà essere loro significato che non sussiste la possibilità di accogliere le loro eventuali richieste di espatrio in quota a causa delle note restrizioni americane.*

D3.4.2 G. Mortara to L.I. Dublin, Milan 29.11.1938

ECADFS Records, *GM*, fol. 1r

Dear Dr. Dublin:[42] [. . .] Many of my American correspondents are insisting on the point that my coming to the United States would be a help to the solution of my problem. But, as I have already told you, in the present situation I cannot leave my family here; on the contrary my most urgent wish is that of bringing them abroad where they can find again regard and tranquillity. The American consular authority, in order to give on our passports their visa for immigration, are requesting a declaration from a well-known firm or person in the United States 'that Professor Giorgio Mortara and his family, consisting of his wife and their four children, will not become public charges'. Am I too indiscreet to beg you to consider the possibility of delivering yourself or to obtain from the Metropolitan Life Insurance Company a declaration in these terms? I should be extremely grateful to you for it. Being already in possess of all the other documents requested for the visa, if I presented such a declaration, I could obtain the visa within a few days, and I might undertake the preparation for our prompt transfer to America. Of course, we should leave *only* after some opportunity of work for me were found; therefore, the declaration would not involve any financial responsibility for the person or firm that would deliver it. On this point I engage myself by my word of honor. I have also exposed my situation to Professor Hotelling,[43] writing to him by the same post. I hope that it will be possible to obtain this indispensable document, which will remove the last obstacle to the granting of the visa. Yours sincerely Giorgio Mortara

---

[42] Louis Israel Dublin (1882–1969), Jewish Lithuanian statistician, was Vice-President of the Metropolitan Life Insurance Company.

[43] Harold Hotelling (1895–1973), American statistician, professor of Economics at Columbia University (1931–1946).

D3.4.3 G. Fubini to N. Vigna, Paris [1.12.1938]

ASPoliTo-*Fubini*, fol. 1r
Dear friend,[44] I am being called to Princeton University; I should be in New York on January 15th. I would therefore need *immediately* and *urgently* a new copy (not legalised) of an act that proves that I have been full professor in the R. Polytechnic for the last two years. Another copy, legalised in Rome, should be sent to me as *soon as possible*. Of course, I will immediately pay not only the legal costs, but also the travel (and ancillary) expenses, if any of your employees would kindly volunteer to go to Rome to collect the aforesaid documents as soon as possible. I am of course ready to pay the expenses in advance as soon as I am informed of the sum owing. Forgive the nuisance, and please accept, with my thanks, my most affectionate and cordial greetings. Yours G. Fubini

Caro amico, Mi si preannuncia la mia chiamata alla Princeton University; dovrei essere a New-York il 15 Gennaio. Avrei perciò bisogno immediato e urgentissimo *di una nuova copia (non legalizzata) di un atto che dimostri essere io stato negli ultimi due anni prof. ordinario nel R. Politecnico. Un'altra copia, legalizzata a Roma, dovrebbe essermi spedita* nel termine più breve. *Naturalmente io pagherò immediatamente non solo le spese legali, ma anche le spese di viaggio (e accessorie), se qualche loro impiegato volesse cortesemente recarsi a Roma per riavere in tempo quanto più breve possibile le pratiche relative. Sono naturalmente pronto a pagare le spese in anticipo, appena preavvisato della somma a mio debito. Perdoni la noia, e coi miei ringraziamenti gradisca il mio saluto più affettuoso e cordiale, aff. G. Fubini*

D3.4.4 L.I. Dublin to G. Mortara, New York 23.12.1938

ECADFS Records, *GM*, fol. 1r
Dear Professor Mortara: Your letter of November 29th has aroused my every interest and I have increased my activity to find something for you. In this connection, I have written as par the enclosed to Mr. Dollard of the Carnegie Corporation.[45] Apparently, something is developing in that quarter. I have also gotten in touch with Professor Chamberlain of Columbia University, who is chairman of a Committee to Help Refugees, and he has many resources at his disposal to enable men like yourself to arrive in the country. I am sending you a copy of his letter to me dated December 22nd. Will you please write to him directly, sending me a copy, giving him the information which he needs with regard to the visa. I have already clarified the matter as to your age, as you will see

---

[44] Nicola Vigna, General Secretary of the Polytechnic of Turin.
[45] Charles Dollard (1907–1977) member of the Carnegie Corporation of New York from 1938.

from my reply. It is extremely hard in these times to find those who are willing to make affidavits. I have reached the limit of my own ability to do this, which I, of course, regret exceedingly in your case. One grant chance is to find a teaching position for you and to have one of the universities ask for you. I hope with the New Year to be able to write you something encouraging along these lines. Sincerely yours Third Vice-President and Statistician [L.I. Dublin]

D3.4.5 H.R. Marraro to E.S. Bailey 27.12.1938

IAS, *GF*, fol. 1r
My dear Miss Bailey:[46] The case of Professor Guido Fubini is not clear to me. Does he wish to come to the United States to settle permanently here? If so, I do not think there is any possibility as the Italian immigration quota, I have been told, has been filled up to 1941. If, on the other hand, he wishes to come for the duration of the course he has been invited to give at your Institute, all he needs is to have his passport made valid for the United States. If you will be good enough to clarify this matter for me and give me additional information as to his present status and all other aspects of the case that may facilitate my task, I will be glad to take the necessary steps to bring about the desired effect. With cordial wishes for a happy and prosperous New Year, I remain, sincerely yours, Howard R. Marraro[47]

D3.4.6 [Note by an Anonymous Informant of the Secret Political Police], Turin 15.1.1939

ACS-*Ebrei*, fol. 1r
In these days, Italian Jews are very anxious about the discussions that the English newspapers have predicted between the Premier and the Duce on the Jewish question.[48] The mention made in the aforesaid newspapers of a project to

---

[46] Esther S. Bailey, Secretary of the Institute for Advanced Study.

[47] Howard Rosario Marraro (1897–1972), an Italian-American historian, Director of the Italian Interuniversity Bureau at Columbia University.

[48] See, e.g., *Chamberlain to Discuss Refugees with Duce; U.S. Embassy Assures Jews on Ethiopia*, The Jewish Telegraphic Agency, Bulletin of January 11, 1939: "On the eve of Prime Minister Neville Chamberlain's visit here, diplomatic circles expressed confidence today that the British statesman, in his discussions with Premier Benito Mussolini, would continue to develop the subject of a general solution of the refugee question along the lines advanced recently by United States Ambassador William Phillips. The British and American governments, working in close cooperation, are hopeful of obtaining cooperation of the Rome-Berlin axis toward a feasible solution. It is believed here that a role similar to that which Il Duce played in the Munich peace-making should

establish a region in the Harar as a refuge for Jews who will have to leave Italy has raised many concerns on the grounds that the climate of that region is not suitable. Many people wonder if all Italian Jews would be obliged to go there or if it will be optional for everyone to transfer to another country. As a matter of fact, very few are suited to a life dedicated exclusively to agriculture, and very few possess the agricultural culture necessary for colonisation of that land. Everyone hopes that the talks of these days may also result in an international agreement on this question and that at least the anti-Semitic campaign that is still raging in all newspapers may ease off, and that the measures taken against their race will be applied more benevolently.

In questi giorni gli ebrei italiani sono in molta ansia per le discussioni che i giornali inglesi hanno prognosticato, essere tenute fra il Premier ed il Duce sulla quistione ebraica. L'accenno fatto nei detti giornali di un progetto di costituzione di una regione nell'Harar quale rifugio per gli ebrei che dovranno lasciare l'Italia ha destato molte preoccupazioni per tema che il clima di quella regione non sia adatto. Molti si domandano se tutti gli ebrei italiani sarebbero obbligati a recarvisi o se sarà facoltativo ad ognuno il trasferimento in altra Nazione. E ciò perché ben pochi sono quelli adatti ad una vita dedicata esclusivamente all'agricoltura, e pochissimi sono quelli che posseggono quella cultura agraria necessaria per la colonizzazione di quel paese. Tutti sperano che dai colloqui di questi giorni possa derivare un accordo internazionale anche sulla loro quistione e che almeno sia attenuata la campagna antisemita che infuria ancora su tutti i giornali, e che i provvedimenti presi contro la loro razza siano applicati più benignamente.

---

not be overlooked. This belief has its genesis in Mussolini's well-known ambition to be regarded as an arbitrator of international problems. Any United States Government proposal for Jewish colonization in Ethiopia will preclude the possibility of racial persecution, the American embassy advised foreign Jews who expressed opposition to plans for settlement in Italian possessions on the ground that Jews would be under the domination of an anti-Semitic government. Even though a habitable section of Ethiopia was designated for Jewish colonization, there could be no sense of permanence or tranquility under the anti-Jewish Rome regime, said Jewish circles in commenting on the suggestion for Ethiopian colonization reported to have been made to Premier Mussolini by Ambassador Phillips on behalf of President Roosevelt. It was authoritatively learned that Mr. Phillips proposed to Premier Mussolini the settlement of Jews in the fertile hills 300 miles south of Addis Ababa, near the frontier of the British colony of Kenya. He was said to have suggested that foreign Jews, who face expulsion from Italy under a Fascist decree, be allowed to bring their capital from Italy to Ethiopia since they would not be leaving Italian territory. Premier Mussolini's refusal of the plan was prompted, it was reported, not so much by opposition to the colonization as to concessions which Mr. Phillips asked along with it, since development of Ethiopia by Jews is thought to be within the scope of the Premier's plans."

D3.4.7 The Prefect of Mantua to the Ministry of the Interior, Mantua 16.1.1939

ACS-*Razzismo* 1939, fol. 1r-v

Treatment of Jews. The measures adopted to protect the race have now been fully implemented in this Province which counts, in the capital alone, about 500 Jews. In addition to the measures of mere execution of the R.D.L. 17 November 1938-XVI, no. 1728, administrative and police measures were taken to prevent the circumvention of restrictive rules established by the aforementioned royal decree law. However, it should be noted that the Jewish element, which in this Province has rooted traditions since the time of the Gonzagas and represents the richest class in the field of business, does not desist from devious action, which goes from messianic threats against the local hierarchies, guilty (according to them) of having applied the restrictive measures with extreme rigour, to the disintegrating action of public opinion, which is not easy to pursue with police means, given the traditional cunning of this race. It cannot be ruled out that the Jews of the Province of Mantua, who in the past identified themselves with the dominant Freemasonry in the Red Period [1919-1920], are able to find, even in the Aryan field, ties or help or, at the very least, pietism in non-Mantuan environments. It is certain that the affected Jews, even if not blatantly, are determined enemies of the Regime and, therefore, resolved not to miss an opportunity to carry out their hidden work of dissolution. It is therefore necessary to find a way to facilitate the exodus of the Jews and in this sense, under verbal authorisation given to me, I spoke to the chief rabbi of this community [Gustavo Calò] to induce him to persuade the acolytes of his jurisdiction to abandon this country. However, the difficulties that lie ahead are not slight because:

1. although abroad the condition of the Jews in Italy and Germany is verbally pitied, nevertheless, no country seems willing to grant them hospitality.
2. the restrictive provisions on currencies do not allow those who wish to leave to take money or national securities abroad to an extent that allows them to survive.
3. in many cases, the Jews residing here have contracted family ties with Aryan elements who, by blood, somehow manage to make the new plight of Jews less serious.

The rabbi pointed out to me the possibility of mass emigration to some island in the Aegean or to some other Italian colony, which would overcome at least one of the difficulties, namely, that of the transport of securities. I thought it is necessary to highlight the above points because while for the moment the Jews, faced with the decisive and firm attitude of the hierarchies of the Regime, refrain from any public or private action, nevertheless, as historical experience teaches, they do not restrain from sharpening in the shadows their weapons for the day of rescue, when the slackening of rules (which time necessarily brings to every resolve) will give them the opportunity to re-weave new networks to the detriment of national interests. The Prefect Raffaele Montuori

Trattamento dei giudei. *Le misure adottate a tutela della razza hanno avuto ormai piena attuazione in questa Provincia che conta, nel solo Capoluogo, circa 500 giudei. Oltre ai provvedimenti di mera esecuzione del R.D.L. 17 novembre 1938-XVI, n° 1728, sono state prese misure amministrative e di polizia per impedire l'elusione delle norme restrittive impartite dal precitato Regio Decreto Legge. Senonché occorre far presente che tutt'ora l'elemento ebraico, che in questa Provincia ha radicate tradizioni sin dal tempo dei Gonzaga e rappresenta il ceto più ricco nel campo degli affari, non desiste da un'azione subdola, che va dalle minacce messianiche contro le gerarchie locali, colpevoli, a loro dire, di avere applicato con estremo rigore i provvedimenti restrittivi, all'azione disgregatrice della pubblica opinione, che non è facile perseguire con i mezzi di polizia, data l'astuzia tradizionale della razza. Non è da escludere che i giudei della Provincia di Mantova, che si sono identificati in passato con la massoneria dominante nel periodo rosso, riescano a trovare, anche nel campo ariano, legami od aiuti o, quanto meno, pietismo in ambienti non mantovani. Certo è che i giudei testé colpiti, anche se non palesemente, costituiscono nemici decisi del Regime e, perciò, risoluti a non tralasciare occasione per condurre a termine la loro occulta opera di dissolvimento. S'impone, pertanto, di trovar modo di agevolare l'esodo dei giudei ed in tal senso, giusta autorizzazione verbalmente datami, ho parlato al rabbino capo di questa comunità per indurlo a fare opera di persuasione presso gli accoliti della sua giurisdizione affinché abbandonino il territorio nazionale. Le difficoltà che si prospettano sono però non lievi:*

1. *perché, sebbene all'estero si compianga verbalmente la condizione dei giudei in Italia e in Germania, tuttavia, nessun paese sembra disposto ad accordare loro ospitalità;*
2. *perché le disposizioni limitative sulle valute non consentono, a coloro che siano ben intenzionati a partire, di recare moneta o titoli all'estero in misura tale da consentire di affrontare le prime esigenze della vita;*
3. *perché in molti casi i giudei qui residenti hanno contratto vincoli familiari con elementi ariani i quali, per i legami di sangue, in qualche modo riescono a rendere meno gravi le nuove condizioni fatte agli ebrei.*

Il rabbino mi faceva presente la possibilità di una emigrazione in massa in qualche isola dell'Egeo o in qualche altro possedimento italiano, ciò che supererebbe almeno una delle difficoltà e, cioè, quella del trasporto di valori mobiliari. Ho creduto necessario far presente quanto sopra perché, se pel momento i giudei, di fronte al deciso e fermo atteggiamento delle gerarchie del Regime, si astengono da qualsiasi azione pubblica o privata, tuttavia, come la esperienza storica insegna, essi non ristanno dall'affilare nell'ombra le loro armi per il giorno della riscossa, quando l'allentamento, che il tempo reca necessariamente in ogni atteggiamento di forza, darà loro modo di ritessere nuove reti a danno degli interessi nazionali. Il Prefetto Raffaele Montuori

D3.4.8 [Note by an Anonymous Informant of the Secret Political Police], Geneva 31.1.1939

ACS-*Ebrei*, fol. 1r

The Head of the World Jewish Congress, Goldmann,[49] declared that the only possible colonisation option for Jews is that of Palestine, later adding in confidence that the Duce's alleged promises to Chamberlain to put Abyssinian territories at the disposal of Jews is a false trail to attract foreign Jews into Abyssinia, with the sneaky intention of laying his hands on their capital as soon as they are Italians.

Il capo del congresso ebraico mondiale Goldmann ha dichiarato che la sola colonizzazione possibile per gli ebrei è quella della Palestina, aggiungendo poi in confidenza che le pretese promesse del Duce a Chamberlain di mettere territori abissini a disposizione degli ebrei è un trucco per attirare gli ebrei stranieri in Abissinia solo per poter porre man bassa sui loro capitali appena saranno italiani colà.

D3.4.9 D.C. Thomson [SPSL] to the Under Secretary of State, Home Office - Aliens Department, London 17.3.1939

SPSL, *BS*, fol. 214

Dear Mr. Grant, I beg to make application for the granting of a permit to Professor Benjamino [sic!] Segre, Via Alamandini 103, Bologna, Italy, to come to this country and take up work with Professor Semple of King's College, London.[50] He is a brilliant mathematician, and this Society has agreed to make a grant of £122 a year to supplement the amount contributed by his mathematical colleagues in this country, making a total of £250 a year. The willingness of his colleagues in this country to co-operate with this Society on his behalf proves beyond doubt that Professor Segre is a person of great eminence, and we hope that a permit may

---

[49]Nahum Goldmann (1895–1982), educated at German universities where he studied philosophy and law, during the Mandate period was involved in a range of Zionist causes, including negotiations with the British, aimed at realizing the idea of Jewish statehood. In 1936, he helped organize the World Jewish Congress, and was the first chairman of its executive board; he later served as its President for many years.

[50]John Greenless Semple (1904–1985), professor of Mathematics at Queen's University, Belfast (1930–1936), then at King's College, London (1937–1969).

be granted to him accordingly. I am, Yours sincerely, David Cleghorn Thomson –
General Secretary

D3.4.10 The Minister for Exchange and Currencies [Ministero per gli scambi e le valute] to MFA, Rome [4.5.1939]

ACS-*Ebrei*, fol. 1r

Reference is made to cable no. 209108 of 24 March 1939, with which, for due
examination, a copy of a cable by the royal consulate general in Jerusalem was
sent, about the repercussions in Palestine of our laws on the defence of the race.
To this end, concerning the considerations and reports here contained, it is noted:
[. . .]

2. As regards the proposed possibility of exports of goods from Italy, with
   partial release of the capital of Jews subject to the obligation of expatriation,
   it should be noted that this act is currently permitted pursuant to circular
   no. 210 of 24 January 1939 of the National Institute for Exchange and
   Currencies.
3. As regards the request for travel concessions to Jewish scientists who will
   have to travel from Italy to Palestine, as well as regarding the export to the
   aforementioned country of volumes and libraries of Jewish scholars destined
   for that University, please note that the matter will be examined when
   concrete requests are made to this Ministry, case by case.[51]

*Si fa riferimento al telespresso n. 209108 del 24 marzo u.s. col quale è stata
trasmessa, per l'esame di competenza, copia di un documentato telespresso del
R. Consolato Generale in Gerusalemme circa le ripercussioni in Palestina delle
nostre leggi sulla difesa della razza. Al riguardo sulle considerazioni e
segnalazioni in detto telespresso contenute che più frequentemente ricorrono o
che più da vicino interessino la competenza di questo Ministero si osserva:* [. . .]

---

[51] This point refers to the library of Umberto Cassuto (1883–1951), professor of Hebrew and
Semitic Languages in Rome, who emigrated to Palestine in August 1939. Cassuto would ask
Cecil Roth for help in organizing the shipment of his books (Wiener Library London, 507/II,
Italian Jews. Cecil Roth Original Correspondence: U. Cassuto to C. Roth, Rome, 21.7.1939 and
24.7.1939). Roth (1965, p. 206) would recall: "Another scholar of great merit who had received an
academic appointment in Jerusalem, addressed me a touching letter telling me that he did not have
the possibility of carrying with him the books that he needed on a day-to-day basis, and begging me
to help him. It was a moment in which my mother had placed a small fund at my disposal, so that I
could use it as I saw fit. In order not to offend my friend's sensitivity, I made him believe that it was
a public fund and not a private charity." The transport of personal libraries also involved Giorgio
Mortara. On 7 March 1939 the commercial attaché of the Royal embassy of Rio de Janeiro
communicated to Rome that Mortara had brought his personal library with him to Brazil and had
donated it to the *Instituto Brasileiro de Geografia e Estatística*.

2. *Per quanto riguarda la possibilità prospettata di esportazioni di merci dall'Italia con regolamento parziale mediante sbloccamento di capitali di ebrei soggetti all'obbligo dell'espatrio, si fa osservare che la cosa già a suo tempo presa in considerazione è presentemente attuabile ai sensi della circolare n. 210 del 24 gennaio u.s. dell'Istituto Nazionale per i Cambi con l'Estero.*

3. *Per quanto riguarda infine la richiesta di agevolazioni di viaggio a scienziati ebrei che dall'Italia dovranno recarsi in Palestina, come pure per quanto riguarda l'esportazione verso il predetto paese di volumi e Biblioteche di letterati ebrei destinati a quella Università si fa presente che la questione verrà esaminata quando saranno fatte a questo Ministero richieste concrete e per singoli casi.*

D3.4.11 [Note by an Anonymous Informant of the Secret Political Police], Genoa 23.8.1939

ACS-*Ebrei*, fol. 1r
Establishing colonies in Abyssinia means:

1. finding a way to make large areas attractive for foreign investment, to the advantage of our empire.
2. arranging many Italian Jews who, also for humanitarian reasons, must be put in a position to work and live, and you should keep in mind that Italian Jews are better than their co-religionists of other nationalities.
3. mitigating and perhaps completely eliminating the hostility that the Jewish world has for Italy following the anti-Semitic action carried out, an action that everyone, however, recognises as much more humane than that carried out in Germany: indeed, many attribute responsibility to Hitler for the Italian initiative.

I affirm that hostility would be mitigated and perhaps eliminated because Jewish solidarity is much greater than it may appear. I read, I no longer remember where, a story intended to deny this profound solidarity: after a Sardanapalesco banquet in the Rothschild house in Paris, the Chief Rabbi [Israel Lévi] stood up and said «Now let's turn our sad thoughts to brothers who suffer rejection and persecution throughout the world and ... let us raise a cup to their health!» Solidarity, according to the moral of the story, would be reduced to a pathetic memory. This is not the case: in Egypt, for example, there is a relief committee for the Jews, which has unlimited powers and helps the co-religionists in all ways, as broadly as possible and without excessive formalities. By arranging the Jewish colonies, Italy would do its own thing and at the same time would give new proof of its humane character that always shines through all its action.

Fare stabilire dalle Colonie in Abissinia significa:

1. *trovare il modo di mettere delle vaste zone in grado di rendere attirandovi il capitale straniero, concorrere a mettere in valore il nostro impero;*
2. *sistemare molti elementi israeliti italiani, che, anche per ragioni umanitarie, vanno messi in condizione di lavorare e di vivere, e tenete presente che gli ebrei italiani sono migliori e di parecchio dei loro correligionari di altre nazionalità;*
3. *attenuare e forse eliminare completamente l'ostilità che il mondo ebraico ha per l'Italia a seguito dell'azione antisemita svolta sin oggi, azione che tutti, però, riconoscono molto più umana di quella svolta in Germania, anzi proprio ad Hitler molti attribuiscono la responsabilità della iniziativa italiana. Affermo che verrebbe attenuata e forse eliminata l'ostilità perché la solidarietà israelita è molto maggiore di quella che può apparire alla superficie. Lessi, non ricordo più dove, una storiella che vorrebbe negare questa profonda solidarietà e cioè che, dopo un banchetto sardanapalesco che aveva avuto luogo in casa Rotschild a Parigi, il rabbino maggiore presente, si alzò e disse «Ora rivolgiamo il nostro mesto pensiero ai fratelli che soffrono in tutto il mondo reietti e perseguitati e . . . leviamo il calice alla loro salute!» La solidarietà, secondo la morale della storiella, si ridurrebbe ad un patetico ricordo. Non è così, in Egitto, per esempio, funziona un comitato di soccorso per gli ebrei, il quale ha poteri illimitati e sovviene i correligionari in tutti i modi, con larghezza come ogni singolo caso può richiedere e senza eccessive formalità. Sistemando le colonie ebraiche l'Italia farebbe il proprio interesse e nel contempo darebbe novella prova dei sentimenti di umanità che non sono mai disgiunti da ogni sua provvidenza.*

D3.4.12 [Note by an Anonymous Informant of the Secret Political Police], Rome 21.8.1940

ACS-*Ebrei*, fol. 1r

Jewish problem. For some time, *La Stampa* of Turin has pursued a campaign aimed at convincing the world that all Jews must be relocated to Madagascar.[52] For now, I have not collected sufficient judgments to be able to report on the matter and I can only say that the evolution of the problem, as far as the Italian Jews are concerned, continues to be followed with particular interest and little understanding since Italian Jews in the Kingdom (both for their low numbers and for their activity) have never enjoyed any serious influence. The purpose of this report is, however, to underline the unfortunate ending of the last article published on Sunday by Carlo Barduzzi in the aforementioned newspaper:[53] "The Palestinian solution - concluded the author - has always limped, due to the reasonable

---

[52] *La questione giudaica nella nuova Europa. La necessità dell'allontanamento e dell'isolamento; il Madagascar è la soluzione più adatta*, La Stampa, 4.8.1940, p. 2.

[53] C. Barduzzi, *La questione giudaica e la soluzione del Madagascar. Clamore inopportuno*, La Stampa, 18.8.1940, p. 4.

opposition of Italy." In the *Nota* published in *Informazione Diplomatica* (17 [sic!] February 1938)[54] it was said that the Jews could have an autonomous territory, but not in Palestine. That Note, so famous and inappropriately remembered, whose aims were opposed to those that today's journalist would have us believe, began by saying *verbatim*: "Recent journalistic controversies have aroused in some foreign circles the impression that the fascist government is about to inaugurate an anti-Semitic policy. In responsible Roman spheres, it is possible to affirm that this impression is completely wrong." After having declared that "a specifically Jewish problem does not exist in Italy given the irrelevance of proportions," the *Nota* assured that "the fascist government has never thought, or thinks, of adopting political, economic and moral measures contrary to Jews as such." The great Turin newspaper, in the manner of many of its confreres, committed - as everyone sees - a gaffe by digging up a document that, after the severe denials it received from the facts, would have been better to let sink in oblivion, at least out of due regard for its Author, an oblivion assigned it by all those who subsequently wrote about the delicate subject.

Problema Ebraico. Da qualche tempo La Stampa di Torino ha iniziato una campagna tendente a convincere il mondo che tutti gli ebrei che lo abitano debbono venire relegati nel Madagascar. Non ho raccolto, per ora, giudizi sufficienti per poter riferire in proposito e solo posso dire che l'evoluzione del problema, per quanto concerne gli italiani ebrei, continua ad essere seguita con particolare interesse e scarsa comprensione poiché costoro, nel Regno, sia per il minimo numero, sia per l'attività svolta, non godettero mai di alcuna seria influenza. Lo scopo del presente rapporto è, invece, quello di segnalare l'infelice chiusa dell'ultimo articolo apparso domenica a firma Carlo Barduzzi nel predetto quotidiano. "La soluzione palestinese – concludeva l'Autore – ha zoppicato sempre per l'opposizione ragionevole dell'Italia." Nella Nota dell'Informazione Diplomatica del 17 Febbraio 1938 era detto che un territorio autonomo i giudei lo potranno avere, ma non in Palestina. Quella Nota, famosa e così inopportunamente ricordata, i cui scopi erano opposti a quelli che oggi si vorrebbe far credere, esordiva infatti dicendo testualmente: "Recenti polemiche giornalistiche hanno potuto suscitare in taluni ambienti stranieri l'impressione che il Governo Fascista sia in procinto di inaugurare una politica antisemita. Nei circoli responsabili romani si è in grado di affermare che tale impressione è completamente errata." E, dopo avere dichiarato che "non esiste in Italia un problema specificatamente italiano data la irrilevanza delle proporzioni", la Nota assicurava che "il Governo Fascista non ha mai pensato né pensa di adottare misure politiche, economiche e morali contrarie agli ebrei in quanto tali." Il grande quotidiano torinese, alla maniera di molti suoi confratelli, ha commesso – come ognuno vede – una "topica" rinvangando un documento che, dopo le severe smentite che ebbe dai fatti, era meglio lasciare nell'oblio, almeno

---

[54] L'Informazione Diplomatica n. 14 was dated February 16, 1938.

per doveroso riguardo verso il suo altissimo Estensore, oblio in cui lo abbandonarono tutti coloro i quali scrissero, in seguito, a torno il delicato argomento.

3.5   The Expulsion of Foreign Jews from Italian Soil

The plight of foreign Jews who were naturalised Italians was particularly serious (D3.5.6). In 1942, A. Dresden published a survey entitled *The Migration of Mathematicians*, giving the list of mathematicians who had managed to land in the United States from 1933 to 1942.[55] Four Italians were here mentioned: Fubini and three foreign Jews (Izaac Opatowski, Robert Frucht, and Wolfgang R. Wasow) who had found initial refuge in Italy and who in the autumn of 1938 had been forced to move on to other shores. The situation was not unusual. As it was their second move, they were quicker to react, more organised; they knew what procedure to follow in applying, and above all they had no choice, being forced by law to leave Italy by 1 March 1939. Who were these second time refugees?

Born in Warsaw in 1905, Izaac Opatowski graduated in Engineering from Turin Polytechnic in 1929 and in Mathematics in 1932 (Fig. 3.3a and b).

Between 1934 and 1936, he was a substitute teacher at the technical institute in Turin and from 1936 to 1938 worked at the Fiat automobile company. Fully integrated into the Italian mathematical community, in just six years after graduation, he published a dozen works and attended important conferences, such as the first UMI national Congress in 1937.

In the summer of 1938, he wrote to Levi-Civita:

> Some recent events and my being a foreigner and an Israelite push me to ask you for advice. [...] The recent government measures make it clear to me that the class of people to which I belong is unfortunately no longer desired in Italy and make me seriously fear that I will be faced with an abrupt change in my situation. Just yesterday I heard of a person who, being in similar conditions to mine, was ordered to leave Italy within eight days. So, I take the liberty of disturbing you in the hope that you could kindly give me some advice. [...] I know several foreign languages; could I hope to have some chance abroad? And where? I've heard of fellowships that are available in the United States also to foreign people.[56]

---

[55] Dresden (1942).

[56] I. Opatowski to T. Levi-Civita, Turin 5.8.1938, in Nastasi and Tazzioli (2003): pp. 569–570: *Alcuni avvenimenti dei giorni scorsi e le mie qualità di straniero e di israelita mi suggeriscono di pregarla di un consiglio. [...] I recenti provvedimenti del Governo, mi fanno apparire in modo chiaro, che la classe di persone a cui appartengo è purtroppo non desiderata in Italia e mi fanno temere seriamente di trovarmi di fronte ad un brusco cambiamento della propria situazione. Proprio ieri ho ricevuto notizia di una persona che trovandosi in analoghe condizioni alle mie, è stata intimata di lasciare l'Italia entro otto giorni. Mi sono permesso perciò di disturbarla, nella speranza che forse Ella potrebbe gentilmente darmi qualche consiglio. [...] Conosco parecchie*

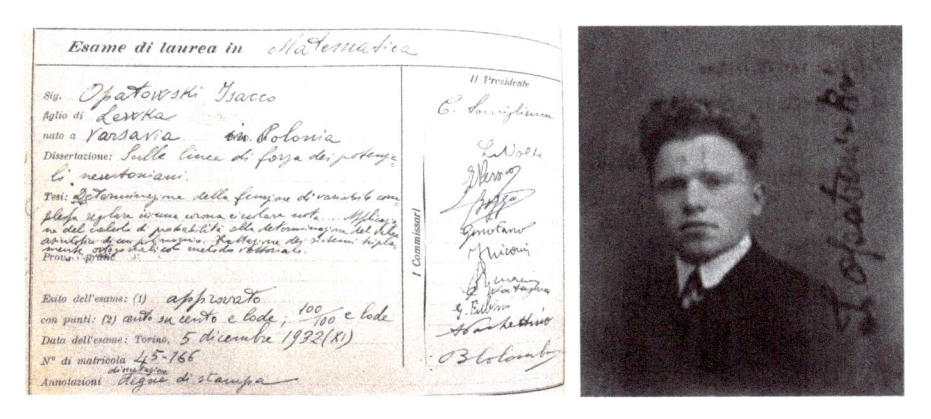

Fig. 3.3 (**a**) (left) and (**b**) (right) Minutes of Opatowski's graduation exam, Turin 1932, ASUT

> I presented my case to a former compatriot of mine, Prof. O. Zariski, at Johns Hopkins University, who suggested I apply for a fellowship by courtesy, i.e., without salary, attaching letters of recommendation from Italian professors.[57]

Fubini, Levi-Civita, and Colonnetti helped him to flee. Opatowski was a mathematician of versatile talent. First, his early mentors recommended him to J. Rey Pastor for a position in Argentina; at the same time, they suggested that he look for an opening in the States. Thanks to the endorsement and reference letters signed by 4 Italians (Fubini, Levi-Civita, Tricomi, and Somigliana) and by two former compatriots (S. Mandelbrojt and O. Zariski), Opatowski was given a job at the Institute of Technology of the University of Minnesota. He remained there until 1945, when he moved to the University of Chicago.

Robert Frucht (1906–1997), born in Brno, a Ph.D. obtained *cum laude* from the University of Berlin, had been displaced in Trieste since 1930. Here, he married, and his daughter was born in 1933. He spoke five languages (German, English, French, Italian, Czechoslovak) and was able to stenograph in two (German and Italian). Frucht had worked as an actuary in the mathematical office of the *Riunione Adriatica di Sicurtà* (1930–1935), then in the reinsurance office of the same company (1935–1937) and in 1938 he was appointed assistant manager of this office. However, his activity did not interrupt his studies and in eight years he had published 24 works. Frucht was an algebraist and a projective-differential geometer, highly appreciated in Italy and abroad despite his young age. He addressed the American

---

lingue estere; potrei sperare in qualche possibilità all'estero? E dove, secondo il suo parere? Ho sentito dei "fellowships" che negli Stati Uniti sono accessibili anche agli stranieri.

[57] I. Opatowski to T. Levi-Civita, Turin 11.11.1938, in Nastasi and Tazzioli (2003): p. 572: *Ho interessato nella questione un mio ex-connazionale, prof. Zariski della Johns Hopkins, il quale m'ha consigliato di fare una domanda per una "fellowship by courtesy" che sarebbe una fellowship gratuita, allegando lettere di raccomandazione di professori italiani.*

Friends Service Committee, which transmitted his file to the ECADFS.[58] His application, in which he highlighted three references (J. von Neumann, S.S. Wilks, and G. Pólya) was received on 10 March 1939 and was examined by H. Weyl, who noted:

> External circumstances forced Frucht, immediately after he had gotten his Ph.D. in Berlin to go into the insurance business. A number of valuable papers on theory of groups and differential geometry bear witness to the tenacity and success with which he has kept up his mathematical research. G. Pólya, and his teacher, I. Schur, have a very high opinion of his mathematical ability. Another group of his papers deals with mathematical actuarial problems. In a short time, he rose to the responsible position of vice manager of the Reinsurance Office of the *Riunione Adriatica di Sicurtà*. He is an exact and hard worker, and I am sure he would make a first-rate actuarian or instructor in mathematics in a college or secondary school. His plight is most serious. Born in Brno as an Austrian citizen, later (on account of his birthplace) a Czech citizen, after 1930 he acquired the Italian citizenship and married an Italian. Now he will lose his Italian citizenship, must leave Italy before March 1939, and almost all countries are closed to him because he is stateless.[59]

Recommended by Veblen, von Neumann, and Weyl for a position in the Department of Mathematics at the University of Michigan in Ann Arbor, Frucht was denied an immigration visa to the United States because as an insurance mathematician he could not claim to have taught. He arrived in Argentina and then in Chile, in May 1939 (D3.5.1, D3.5.2, D3.5.7).

Wolfgang Wasow (1909–1993) studied mathematics in Göttingen. He passed his *Staatsexamen* (a government licensing examination for future teachers) in January 1933, two days before Hitler became chancellor; he then applied for a teaching position but was refused because of his Jewish ethnicity. Wasow left Germany in 1933 and spent time in Paris and Cambridge before taking a job as a teacher at boarding schools for children of German emigrants in Italy, first in Florence (1935–1937) and then in Lana in Alto Adige (1937–1938).

In his *Memories of Seventy Years,* he recalled his time in Italy:

> There is one aspect of my life in Italy I have not yet touched upon. That is my peculiar situation as a refugee from one fascist country living in another fascist country. One reason is that Italian fascism, while its manifestations were omnipresent, did not impinge on my personal life. . . . Let me add here a good political joke which I remember from those years. In his efforts to increase the birth-rate in Italy, Mussolini sent a handsome gift of money to all parents of triplet babies. In one such case, after hearing that the mother had called the children, patriotically, Vittorio-Emanuele (the name of the reigning king), Benito

---

[58] ECADFS Records, *RF*: B. Drury [ECADFS] to C.S. Salmon [Refugee Service American Friends Service Committee], New York 28.3.1939. Founded in 1917 to support wounded civilians and victims of World War I, the American Friends Service Committee was a Quaker organisation dedicated to supporting peace and development programmes around the world.

[59] ECADFS Records, *RF*: Frucht's application, received on March 10, 1939. Hermann Weyl (1885–1955), mathematician, theoretical physicist, logician, and philosopher, had left Göttingen in 1933, particularly as his wife was Jewish. The third faculty member elected to the School of Mathematics at the Institute for Advanced Study, Weyl collaborated intensively with Veblen on the emigration of mathematicians to America and in 1934 founded the German Mathematicians' Relief Fund. See also Frucht (1982).

(Mussolini's first name), and Italia, he telephoned the mother and asked her what the babies were doing. She answered: 'Vittorio dorme, Benito succhia, Italia piange' (Vittorio sleeps, Benito sucks, Italy weeps). [...] Compared to the spontaneous mass hysteria in Germany at the time, the public enthusiasm for the Mussolini regime in Italy, when I was there, gave the impression of being a spectacle staged by the authorities with only lacklustre support from the masses. [...] Only after the Hitler-Mussolini Pact in the summer of 1938 did our daily lives begin to be seriously affected by the political system.[60]

After the introduction of the racial laws, Wasow emigrated to England and then to the United States in 1939. He spent a period at Goddard College in Vermont (1939–1941), before taking a fellowship at the New York University, arranged by R. Courant. Wasow's wife, a talented designer, continued to teach art at Goddard, before moving to the Connecticut College for Women (D3.5.3, D3.5.4, D3.5.5). Wasow completed his PhD studies in 1942 under the supervision of K.O. Friedrichs while teaching at the Connecticut College. A specialist in singular perturbation theory, he became full professor of mathematics at the University of Wisconsin where he remained till his retirement in 1980.[61]

Documents

D3.5.1 H. Weyl to J.D. Branson [American Friends Service Committee], Princeton 27.1.1939

OVP, *RefGen*, fol. 1r.
Dear Miss Branson:[62] We have now received additional information concerning refugees whose records we have already sent you, which I send you on separate sheets for convenience in filing. The Italian decree which affects Frucht may be of general interest, i.e. that because he is married to an Italian he can stay in Italy, although he will lose his job. Yours sincerely, Hermann Weyl

D3.5.2 O. Veblen to G.Y. Rainich, Princeton 6.2.1939

OVP, *BS*, fol. 1r
Dear Rainich,[63] I have talked with Weyl and von Neumann about the inquiry in your letter of February 4, and what I have to say is based on this conversation. We

---

[60] Wasow (1986).

[61] O' Malley (2010), pp. 12–13.

[62] Julia D. Branson, secretary of the American Friends Service Committee.

[63] George Yuri Rainich (1886–1968) Ukrainian mathematical physicist, emigrated to the United States in 1922, taught first at Johns Hopkins University (in Maryland) and then at the University of Michigan, where he remained until his retirement (1956).

are interpreting the word "Germany" broadly, to include Czechoslovakia and Italy. On this basis we arrive at three names, Frucht, Löwner, Segre. Of these three Frucht probably comes the closest to the specifications in your letter, for he has had extensive experience with an insurance company in Italy, is thirty-two years old, and is active mathematically. On the other hand, Segre is more eminent mathematically, is thirty-six years old, but is further from applied mathematics. Löwner is forty-five years old, is held in extremely high esteem by the analysts, and has done work in hydromechanics.[64] We have had so many appeals from refugees that we have recently get in touch with two agencies which are handling the problem systematically. In order not to cross wires, I am asking the American Friends Service Committee (attention of Miss Julia D. Branson), 20 South 12th Street, Philadelphia, Pa., to send you the facts about these men which we have placed in their possession. With regard to Löwner I might say that he is known personally to von Neumann, who regards him as having an agreeable personality although somewhat shy. Of course, you realize that there are at least two or three very good men who are already in this country, to whom the opportunity which you are writing about might be very welcome. With best greetings, yours sincerely Oswald Veblen

D3.5.3 C.S. Salmon [American Friends Service Committee] to H. Weyl, Philadelphia 9.3.1939

OVP, *RefGen*, **fol. 1r.**

Dear Dr. Weyl: I wish to send out some information about the following professors whom you referred to us. 1) George Jaffe 2) Wilhelm Gross 3) Fritz Reiche 4) Wolfang Wasow.[65] Can you please tell me if Professor Wasow came to the United States and what prospects the others have for coming here soon? Sincerely yours Charlotte Salmon

---

[64] Karl Löwner (1893–1968), had been dismissed from the Charles University of Prague and put in jail. After paying the 'emigration tax' twice over he was allowed to leave the country with his family. von Neumann arranged a position for him at Louisville University.

[65] George Cecil Jaffe (1880–1965), full professor of Theoretical Physics at the University of Gießen (1926–1933). In 1939 he would emigrate to the United States, where he would be visiting lecturer until 1942, then associate professor at Louisiana State University in Baton Rouge. Wilhelm Gross (1883–1944), professor of Mining Science at the Wroclaw University of Technology, emigrated to the Netherlands. Arrested in Holland, he died in Auschwitz. Fritz Reiche (1883–1969), a student of M. Planck, had been dismissed from the University of Breslau in 1933. With the help of R. Ladenburg, A. Einstein, and the ECADFS Reiche would emigrate with his family to the United States in 1941 and went on to work with NASA and the United States Navy on projects related to supersonic flow. Charlotte Salmon was the executive secretary of the American Friends Service Committee.

D3.5.4 American Friends Service Committee to H. Weyl, Philadelphia 23.3.1939

OVP, *RefGen*, fols. 1r–2r.
Dear Dr. Weyl: I had not received your note when I talked with you yesterday, over the phone, or I might have been a little more intelligent on the subject. After talking with you, however, I called the Shipley School and spoke with Mr. Lynes, Assistant Principal. He felt that Dr. Wasow would fit better in a boys' school than in a girls' school. He also felt that they should have a teacher with experience in American schools for teaching advanced Physics, in preparation for college board examinations. However, he said that he would be glad to talk with Dr. Wasow any time he was in Philadelphia. I was considering whether Dr. Wasow should come on down to Bryn Mawr right away, but it seems to me that it would be better if he and his wife [Gabrielle 'Gabi' Wasow-Brill] could go together. Although I have not talked to Dr. Lynes about it, it seems to me that she, too, might well fit into a school for younger girls. I wish they would invite the Wasows for a weekend, but I don't think that is likely. However, I do think it might be worth their while to come to Bryn Mawr. If they let me know the date, I'll be glad to make an appointment. [. . .] Sincerely yours Charlotte Salmon

D3.5.5 H. Weyl to C.S. Salmon [American Friends Service Committee], Princeton 25.3.1939

OVP, *RefGen*, fol. 1r.
Dear Miss Salmon: *Re Wolfang Wasow.* The Wasows returned to New York yesterday. Mrs. Wasow must practice English for at least half a year before she will be fit for a school job; in the meantime, it is better for her to try to find a commercial job as designer. In the field of designing, she has much talent and experience. Mr. Wasow's English is good, but he also needs practice. It would be preferable if he could for two or three months assist in and be introduced to the work in an American school under an au pair arrangement (free room and board but no salary). I have heard that some schools in New England make such arrangements for refugees. I suppose a summer camp would also be a very useful experience for him. If he comes to Philadelphia in the near future, he will certainly visit you, and may ask for an appointment with Dr. Lynes. Both the Wasows are healthy, honest, and candid people, and very charming. They deserve your special attention. [...] Sincerely yours, Hermann Weyl

D3.5.6 Jewish Central Information Office to SPSL, Amsterdam 14.4.1939

SPSL, *Italy*, fols. 512–517.
Dear Sir, This, a special report on *The Foreign Jews of Italy* has been placed at our disposal by a trained economist who had spent a number of years in Italy, until, like so many other non-Italian Jews, he had now to leave the country. Copies of this report, which we here-with present, may be had at this office, at the price of hfl. 0.50. You are free to use as material any data here given, without as usual naming us as your source. We shall be obliged if, in view of the personal nature of many of those statements, you will refrain from reproducing them literally. Yours faithfully, Jewish Central Information Office – Public Service Institute[66]

Jewish Central Information Office – Public Service Institute
The Foreign Jews of Italy

The 12th March, 1939, and after

Government Legislation and Popular Feeling

The legal measures which the Italian Government has been enacting against the Jews since last autumn, may be divided into two big groups. On the one hand, Italy too has, to an ever-increasing extent, eliminated the Jews from public life. The decree of the 5th September, 1938, forbade them to teach at public schools, and the decree of the 17th November, 1938, promulgating as it did very drastic provisions, forms an equivalent of the Nuremberg Laws. It already contains regulations aiming at the activities of Jews in industry, and these have since been considerably tightened up by new decrees and measures. All these measures are directed against the Jews of Italian nationality, and it is as yet far from certain that there is not going to be a worsening in the special regulations designed to outlaw and disgrace Italian Jewry on the German model.

The Position of Foreign Jews

The position is somewhat different with regard to legislation concerning foreign Jews, which has virtually come to an end. Its purpose, however, is plain from the measures hitherto adopted. The decree of the 7th September, 1938, which constitutes the basis for these measures, deprives foreign-born Jews of Italian citizenship if acquired after January 1st, 1939. In addition, it orders at six months' notice which expired on the 12th March, 1939, the expulsion from Italy and the Italian possessions of all foreign Jews that have settled there after January 1st, 1919. There is one odd exception granted in these provisions: in the event of a marriage being recognised between an Italian national of the Christian faith

---

[66]The Jewish Central Information Office was founded by Alfred Wiener (1885–1964) in Amsterdam in 1933. It collected and shared information about the Third Reich and the Nazis' persecution of Jews. See also *Jewish Central Information Office* (1938).

and a Jewish post-1919 immigrant, the children, if baptised, are considered Christians. On the other hand, where the Italian partner is Jewish, both parents and children are subject to the Italian Jewish enactments.

Six Months' Permit for Study and Business Purposes

Of the pre-1919 Jewish immigrants, a large number was engaged in Italian economic life, and it was they who were to be hit exclusively by the decrees. About the middle of February 1939, a new regulation was published, permitting persons of the Jewish persuasion to remain in Italy, after the 12th March, 1939, for study and business purposes, for a period not exceeding six months. To judge by previous experience, permission will be granted at once for two or three months. In contradistinction to the practice hitherto observed, foreign Jews will be unable to establish permanent residence in Italy, nor will they be permitted in future to engage in any active business in Italy. There is no objection, however, to a foreign Jewish partner to an Italian firm spending a few months in Italy every year, and supervising business in this manner. Since Italy is a Roman country, where formalities occupy a prominent place, it is only necessary carefully to arrange for the Jew not to appear officially as owner, representative, or otherwise. He may of course, upon written agreement, issue instructions to the owner, or director of the firm.

About 7.000 Foreign Jewish Businessmen Concerned

A large number of Jewish immigrants since 1919 had acquired Italian citizenship, or established permanent residence without changing their nationality. Though little noticed during the last few years they have been living in Italy in retirement to spend there the interest of their foreign investments. Others again were establishing, with imported capital, some kind of partnership in industry, purchasing or founding businesses or factories, or acting as agents for Italian or non-Italian firms. During the last few years, there has been a marked decrease in the numbers of labour permits issued to foreign Jewish employees. These Jews were the principal target of legislation. Since the number of foreign Jews in Italy is generally believed to be between 15 and 18,000, it may be assumed that some 6 to 7000 Jews in business were actually hit by the decrees. These foreign Jews were compelled, without exception, to dispose of their business engagements no later than March 12, 1939. Extensions were granted only in cases where evidence could be produced that a firm was in process of liquidation. Foreign Jews appointed liquidators were granted an extension of one or two months, so as to enable them to wind up the current affairs. Thus, before the 12th March, 1939, a large number of Jewish businesses, factories, also agencies and small estates have been sold and transferred into Italian "Aryan" hands. Up to the 12th March, 1939, no pressure appears to have been exerted. Owners of businesses and factories concerned were only informed that on March 12 there would be an automatic withdrawal of the licence which is required for every business establishment in Italy. It is, however, almost certain that now, and during the next few months, a searching control will be instituted as to what has become of those businesses.

Moreover, as early as September 1938, the foreign exchange authorities have imposed a tight control on many firms owned by foreign Jews and closely connected with the import and export trade, and thus subject to the foreign exchange regulations. There are, however, a number of firms which the banks know were Jew-owned, and which yet, up to the 12th March, have had no interference from the foreign exchange control. Since November 1938, the banks have been making systematic enquiries of all their customers, known to them to be foreigners, asking them for their exact nationality, and also whether or not they were Jews. Such questionnaires were circulated among all foreigners, irrespective of nationality, and indiscriminately among private individuals and businessmen.

2.500 Lire per head Permitted for Emigration

The banking accounts of foreign Jews were blocked about the middle of November, i.e. they were transferred from the various safe deposits and branches to the head offices, the accounts boing entered under the heading "Lira Vecchio". The employment of "Lira Vecchio" has been subject to certain regulations already since 1935, when the foreign exchange restrictions came into force. Of these "Lira Vecchio" accounts, the depositor was free to withdraw 5000 Lire per month; any payment above this amount was subject to the approval of the foreign exchange control boards ("ISTCAMBI", i.e. *Istituto per i cambi con l'estero*). Since Italy still recognises the formal law, it was possible for husband and wife having separate accounts, to withdraw 5000 Lire per month each. Besides there has never been a disinclination to appreciate special requests, e.g. for new clothes or furniture, for the payment of doctors' or solicitors' bills. Though originally it had been laid down that 'Lira Vecchio' must be used only to cover expenses on behalf of parents or children, these regulations too were considerably relaxed. Several cases are known of payments, up to 30,000 Lire, being authorised for the support of more remote relations – to enable them, for example, to purchase the tickets for emigration, and the equivalent of the foreign exchange allowed for emigration purposes. Jewish emigrants have been allotted, on application, a sum of 2150 Lire per head in foreign exchange, while 350 Lire could be taken out in cash. These 2500 Lire represent the sum which the foreign exchange regulations permit Italians to take abroad three times a year, for one month each. Railway tickets, boat passages on Italian liners, may be paid for in Lire, also the freight for furniture, free house of destination. The sums left on the accounts are blocked. A new decree of January 1939 provides for these sums to be used for export purposes, which however amount to a scheme of dubious advantage, as this export will be open only to goods with a small selling chance. Besides no more than 40% of the export may be paid in Lire, while the rest must be surrendered in foreign exchange. It is expected that in the course of the year new regulations will be issued, admitting of a more profitable use to be made of the blocked accounts, although the Lira share may be lowered considerably below 40%.

Hardships of Emigration

Of the Jews aimed at by these regulations, no more than one half has up to the 12th March been able to quit the country, because only this half has been in a position to secure a transit visa, or even a permanent visa, for another country. Worst of all is the lot of the Polish Jews in Italy, as they have to contend with the Polish passport regulations. German and Austrian Jews, on the other hand, as far as they are still left in the possession of their citizenship, have encountered no difficulty with the German passport authorities. A very large section of the foreign Jews in Italy is unable to leave the country, not only on account of the passport difficulties but also for want of financial means. Part of them again was unable to leave because the visas had not arrived yet, or because the boat only leaves later in the year, in May or June. These Jews have made applications for an extension of their stay, though none of these applications had been answered up to the 12th March. It appears, however, that no difficulties will be placed in the way of these people.

Anti-Jewish Legislation Unpopular in Italy

Whereas hitherto, as far as possible, I have been relating such facts only as I have been able to verify personally, the following must be regarded as my own estimate. There can be no doubt that the anti-Jewish legislation has failed to find favour either among the common people or in industrial quarters. Altogether its principles are little understood, for notwithstanding all press asseverations to the contrary, the race idea is clearly abandoned, since baptised children of a mixed marriage are considered full-blooded Italians. As far as the religious aspect is concerned, very limited appreciation can be expected in a country which statistics show to be 98.5% Roman Catholic, the Jews forming ½% of the rest. In contradistinction to other countries, in Italy the Jewish businessman is not feared by the Gentile. As a matter of fact, I should say that in business the Italian rather outwits the Jew. In industry, naturally, the sudden dismissal of many Jews in executive positions has caused a certain amount of dislocation, since, in contrast with Germany, no substitutes were readily available. It cannot be denied, though, that the anti-Jewish legislation was defended by such Italians as had reaped, or at least hoped to reap, some definite economic advantage from these new enactments.

No Demand for Officially Sponsored Race Periodical

The discrepancy between legislation and popular feeling, in spite of press propaganda, is also evidenced by the following fact: The new race periodical *Difesa della Razza*, which has a large sale because all State officials, big business establishments, Party members, etc., are compelled to subscribe to it, had been rather in vogue shortly after its first appearance; next to the fashion journals at the kiosks, it then was this periodical that struck the eye. During the last few weeks, I have noticed many kiosks which no longer displayed it at all, and others where it had taken a back seat. News agents have assured me that the private sale has greatly decreased. It must also be pointed out that there has been no change, on the part of Italian officials, always considerate in dealing with foreigners, in their courtesy, during the last few months, towards those even whom they knew to be Jews. There has not come to my knowledge, or to the knowledge of my numerous acquaintances in Milan, one single instance of an anti-Jewish outrage in public, either in the trams, or in the streets, or in public bars, etc. In Milan, at the beginning of October 1938, about five or six shops had fixed up boards, stating that these were Christian shops. Upon enquiry I found that these boards were removed after two or three days, at the instance of the authorities, i.e. the official Shopkeepers Union. About the middle of October, a decree was issued, albeit only for internal use, forbidding retail traders to exhibit that sort of boards. Emigration papers have been supplied by the authorities to foreign Jews without any kind of chicanery, although it did take some time until everything was ready. But this delay is known to be due to the fact that the chief amendments to the relevant decrees were not published until the beginning of January, and consequently, the central offices at Rome had to deal with some thirty thousand applications collected within four weeks.

Expediency rather than Conviction

My own personal impression is that the Italian anti-Jewish legislation is of a purely political nature. In view of the tremendous distrust of Italy prevalent in Germany, the Italian Government had to provide some proof of its good will which was to impress the outside world that there did exist a cordial intimacy between the two nations. Such proof could only be supplied through the visible conclusion of a military and political alliance which would have had to be published, and this, of course, Signor Mussolini did not desire. The result, therefore, mechanic rather than reasoned, was the conclusion of a cultural alliance which, in the German view, implies, as a matter of course, the Jewish question. Thus, it is not anti-Jewish conviction but political expediency which makes the Jew the innocent victim of Mussolinian opportunism.

D3.5.7 B. Drury [ECADFS] to G.Y. Rainich, New York 22.5.1939

ECADFS Records, *RF*, fol. 1r

Dear Professor Rainich: Thank you for your letter of May 17 letting us know that Dr. Robert Frucht has been invited to lecture at the University of Michigan on actuarial mathematics. As it happens, we are not able to give you the necessary information which would be helpful to him in securing visas for himself and his family to enter the United States. I am taking the liberty of forwarding your letter to the National Council of Jewish Women at 165 West 46th Street and asking them to get in touch with you.[67] It is their office which handles matter of this sort which come to the attention of various committees assisting refugees from Germany and Central Europe. Sincerely yours, Betty Drury - Assistant Secretary

3.6 To Stay or Leave?

Whether to leave or to stay was a very difficult decision to make, and it depended on many factors. The first, on which there was no room for manoeuvre, were the immigration quotas established by each country and the increasingly selective policies adopted since the twenties, both in the United States and in Latin America.

A more general consideration also weighed heavy: fascist Italy was not nazi Germany. Before 1943, nobody feared pogroms and acts of violent anti-Semitism such as those from which the Jews of Central and Eastern Europe were escaping (D3.6.2). The choice of leaving or staying was therefore, at least to some extent, a decision. The first emigration wave (1938–1941) was not an escape, nor an expulsion.

Beyond the normative framework that regulated racial persecution and migratory flows until 1943, several largely subjective variables conditioned the choice: the fact of possessing or not a certain mentality, that is, the strength and courage to reinvent one's life and career abroad, possibly in distant and unknown lands (D3.6.3); whether or not the individual had savings and adequate financial means to transplant the family abroad; the fact of being able to count on a network of contacts and, last but not least, the fact of having or not an internationally established reputation. Very few Italian mathematicians possessed all these requisites.

Age was the first crucial factor. In 1938, Volterra was almost 80 years old, Enriques, Levi, and Fano approaching seventy, Levi-Civita was 65, and Fubini 60 (Figs. 3.4 and 3.5). Mathematicians who were elderly and already retired like Volterra, Castelnuovo, Vivanti, Rimini, Almansi, and Loria, or people who felt as though they were at the end of their career like Fubini and Maroni did not consider the idea of leaving. Or, if they did so, they did it very hesitantly. As a matter of fact, however, they would also have fewer opportunities. Foreign universities did not willingly hire an aged scholar, except in the case of real excellence. In some countries, it was legally impossible to do so. Borel, for example, informed Edoardo

---

[67]ECADFS Records, *RF*: B. Drury [ECADFS] to A. Mayerson [Director, New York Section National Council of Jewish Women], New York 22.5.1939. The National Council of Jewish Women grew out of the 1893 World's Columbian Exposition in Chicago with two purposes. First, this new organisation would keep Judaism alive by informing women of their religious duties and their role in rearing Jewish children. Secondly, its members would "take part as a large group in all that concerns the welfare of mankind," solving together growing social problems that were impossible for the individual to combat alone. From the 1930s onward the organization maintained an active interest in European refugees.

Fig. 3.4 Volterra and
Levi-Civita, late thirties,
Ceccherini Silberstein
Family Archive

Fig. 3.5 Levi-Civita (second from the left) and Volterra (first on the right), Argentière, late thirties,
Ceccherini Silberstein Family Archive

Volterra that his father, having passed the age of 65, could never have been assigned a stable position in any French university, although the *Institut Henri Poincaré* and many other institutes would be happy to have him as a guest lecturer and temporary fellow.

Among the senior mathematicians, Fano accepted expatriation only as a last resort and *sub condicione*, only on condition of going to Switzerland because he did not want to move to a country that might enter a war with his homeland. Mortara and Fubini fled not for themselves, but for their children. Levi-Civita evaluated the idea of emigrating, but only if he could find a position in South America. However, that Castelnuovo, Enriques or Levi-Civita did not think about leaving is less strange than may seem at first glance. All of them, in fact, had been abroad but for short periods; none of them was the 'wandering Jew' with the suitcase always in hand as described in literature.

Emigrating was, in fact, the only alternative for the youngest, such as Segre, Colombo, Jacchia, Bedarida, Tedeschi, who would have the right mentality to do so but who were the most economically penalised by the racial laws, having been discharged from service with very poor pensions, or no pension at all (D3.6.1). Their careers were just beginning. Segre and Jacchia had some opportunities, Segre as one of the best algebraic geometers of the young generation and Jacchia as a specialist in the avant-garde astronomical sector of magnitude sequences. Colombo, Bedarida, Tedeschi, on the other hand, were not known abroad, their works were only published locally (nationally, at best), and their teaching experience was limited.

Finally, for the generation of forty and fifty-year-olds, such as Terracini, Ascoli, Maroni, Colombo, and Mortara, a serious obstacle was represented by families, particularly by young children, babies, or adolescents who had been kicked out of primary, middle, and high schools (Fig. 3.6, 3.7, and 3.8).

The 'humble' schoolteachers, and even more so women, had no choice. Emblematic, in this sense, is the case of Artom, a name already mentioned several times in the previous pages. A former pupil of Segre and Enriques, *ordinario* (regular teacher) of Mathematics and Physics at the scientific lyceum in Turin, he was a typical Jew of the post-Risorgimento generation, who saw a career as a route to social standing. Educated in patriotic values and the cult of the State, he considered the Great War as the epilogue of the Risorgimento struggles and volunteered for the Army. Fascism, in which he had initially placed his trust, betrayed him with the racist drift of 1938. Faced with the racial laws, which condemned him to professional marginalisation and social downgrading, Artom evaluated whether to emigrate or not, concluding:

> Sometimes I fantasised about possible emigration of my family, but the idea of abandoning my Italy wiped out all the attraction of these dreams. And I looked with sadness at my books, purchased one by one for decades, each of which had a story, now made dearer by the use that my children made of them and that perhaps I would need to partly abandon; I looked at

Fig. 3.6 Benedetto, Cesare
and Lore Terracini, summer
of 1938, Terracini Coll.

the small objects and furniture of the house of which I was so fond and I saw the unravelling
of a home built with so much love during my humble life.[68]

Talking about a 'choice between leaving and staying' is quite improper for
women also. There were few female researchers at the time and their visibility was
much lower than that of their male colleagues. There was no female full professor,
for example, in mathematical and physical disciplines. This explains why almost all
the Jewish women of science remained in Italy, sometimes to take care of the parents
and elderly relatives after the departure of the family's men. Among those who
remained at home were the daughters of Castelnuovo, Enriques, and Volterra: Emma
Castelnuovo and Adriana Enriques, both mathematicians, and Luisa Volterra, a
naturalist, who did not want to interrupt the research in biology that she was

---

[68]Emilio Artom (1940–41), in Treves (1954), p. 68: *A volte fantasticavo su possibili emigrazioni
della mia famiglia, ma l'abbandono della mia Italia toglieva ogni attrattiva a questi sogni. E
guardavo con mestizia i miei libri, acquistati uno ad uno da decenni, ognuno dei quali aveva una
storia, resi ormai più cari dall'uso che ne facevano i miei figli e che forse avrei dovuto in parte
abbandonare; guardavo i piccoli oggetti e i mobili della casa a cui ero tanto affezionato e vedevo il
disfacimento di un edificio costrutto con tanto amore durante la mia umile vita.*

Fig. 3.7 Irma Levi Colombo and her granddaughter Marina, 1935, Colombo Coll.

Fig. 3.8 Giorgio Mortara and his family, *A Noite*, Rio de Janeiro 19.1.1939

Fig. 3.9   Guido Fubini with his wife, Annetta Ghiron Fubini, late thirties, Fubini Coll.

conducting in collaboration with her father Vito and her (Aryan) husband Umberto D'Ancona. Only Nedda Friberti, Nella Mortara, sister of Giorgio Mortara, and Gina Castelnuovo, another daughter of Guido Castelnuovo attempted the path of emigration, Friberti and Castelnuovo with the United States as their destination, Mortara headed towards South America. The applications of Friberti and Mortara were not retained by ECADFS. Gina Castelnuovo was to succeed, after painful hardships, in finding job in Missouri. Mortara joined her brother Giorgio in Rio de Janeiro at her own expense but returned to Italy in 1940 after the rupture of diplomatic relations between Italy and Brazil.

Finally, wives and mothers also played a driving role in the decision to stay or leave (Fig. 3.9). Rosa Cassin Fano, Anna Ghiron Fubini, Laura Ottolenghi Mortara, Giulia Sacerdote Terracini, and Adriana Treves Colombo were elegant and educated ladies, typical products of the education that late nineteenth century Italian upper middle classes imparted to girls; they knew languages, they had studied and travelled.

Although they did not have a clear awareness of political events, they reacted to the trauma of the racial laws even more readily than their husbands and were ready before the men to take into consideration the idea of fleeing, to keep the families together, and to ensure a future for their children. Even at the cost of embarking on a collision course with their husbands (as in the case of Fano) they managed the logistics of departure and, having arrived at destination, quickly reorganised houses, lives, and family routines.

Documents

D3.6.1 G. Castelnuovo to B. Segre, Rome 11.9.1938

CA, *BSP*, fol. 1r-v

Dear Professor, you are among the hardest hit by the recent measures; having arrived just a few years ago at the chair, you see abruptly cut short a career that had begun so brilliantly and offered so much hope for the future! I therefore wish to bestow a word of esteem and friendship in this tempestuous moment. We old people who have spent most of our lives in happy times, feel how painful it must be for young people to see their hopes cut short and ideals shattered. Certainly, one day, under another sky, you will be able to resume your mission as a teacher and lover of science. But the difficulties and uncertainties of the present moment are certainly not favourable to mathematical research. I am sure, however, that neither difficulties nor uncertainties will diminish the enthusiasm that you have brought to our austere studies. With the hope of seeing you soon settled in some new place, I assure you of my cordial friendship. Yours affectionate G. Castelnuovo

Caro Professore, Ella è tra i più duramente colpiti dal recente provvedimento; arrivato da pochi anni alla cattedra, vede bruscamente troncata una carriera che aveva avuto inizio così brillante e dava tante speranze per l'avvenire! Voglio dirigerle perciò una parola di stima e di amicizia in questo procelloso momento. Noi vecchi che abbiamo trascorso la maggior parte della vita in un periodo felice, sentiamo quanto debba esser doloroso per loro giovani veder troncate le speranze ed infranti gli ideali. Certo Ella potrà un giorno sotto altro cielo riprendere la missione di insegnante e cultore della scienza. Ma le difficoltà e le incertezze del momento non sono certo propizie alla ricerca matematica. Son sicuro però che né difficoltà né incertezze varranno ad attenuare l'entusiasmo che Ella ha portato ai nostri studi austeri. Con la speranza di saperla presto sistemato in qualche nuovo posto Le invio le espressioni della mia cordiale amicizia; suo aff.^{mo} G. Castelnuovo

D3.6.2 [Note by an Anonymous Informant of the Secret Political Police], Milan 22.5.1939

ACS-*Ebrei*, fol. 1r

Anti-Semitism. It is claimed that many foreign Jews are still in Italy and indeed that many have recently come here with six-month residence permits. Others, Italian Jews, who had gone abroad in desperation faced with what they felt as a grave injustice towards them, not resisting the call of the Italian homeland, returned. In this painful affair which has brought ruin and troubles to so many families, one generally gets the impression that a considerable softening approach is going to be pursued, and that Italy, although it can no longer behave as before the so-called racial policy, wants to avoid reaching those points that have marked the "furor teutonicus" against the Jews; above all, a broad interpretation of the discrimination, allows us to alleviate the plight of these involuntary and, to a large extent, innocent martyrs of political necessity. Such are the rumours that run rife and that are marked by trust in the heart and mind of the Duce.

Antisemitismo. *Si afferma che ancora parecchi ebrei stranieri siano in Italia ed anzi che molti vi siano venuti in questi ultimi tempi, con permessi di soggiorno di sei mesi. Altri, nazionali, che si erano recati all'estero in preda a disperazione per quella che sentivano essere una grave ingiustizia nei loro confronti, non resistendo al richiamo della Patria italiana, sono tornati. In questa penosa faccenda che ha recato rovina e guai a tante famiglie, si ha in genere la impressione che un notevole temperamento si vada apportando, e che l'Italia pur non potendolo più per le premesse adottate per la così detta politica razziale, voglia evitare di giungere a quelle punte che hanno contrassegnato il "furor teutonicus" contro gli ebrei colà, e, sovrattutto, che una larga interpretazione delle discriminazioni, consenta di rimettere un po' in sesto questi involontari e, in gran parte, innocentissimi martiri di una necessità politica. Tali sono le voci che corrono e che sono improntate a fiducia nel cuore e nella mente del Duce.*

D3.6.3 [Note by an Anonymous Informant of the Secret Political Police], n.p., [October 1939]

ACS-*Ebrei*, fol. 1r

Concern of feeling persecuted, and offence at being charged with the most serious accuses against the Italian homeland that they had served until then. Others - and it seems there are not many by now – stay, because they do not have the strength to leave the homeland that they served so devotedly, and they are not capable of emigrating. There are people abroad who left in a moment of despair, are infinitely sorry for having left Italy and, despite everything, would like to return. Now you would think that these people would have such proof of Italian feelings that the entire matter could be shut down by saying 'what it was, was', now let's not talk about it anymore and go back to how it was before. And the horizon that internally is becoming clearer every day, due to the ingenious honesty of the new Secretary of the P.N.F. [Ettore Muti], could leave room for such hopes by making them feasible.

Preoccupazione di sentirsi dei perseguitati, e offesa di sentirsi addebitare le cose più gravi contro la Patria Italiana che avevano fino ad allora servito. Altri – e pare non siano ormai molti – sono rimasti non sentendosi la forza di lasciare la Patria che avevano servito fino allora devotamente e non potendola lasciare. Ve ne ha di quelli all'estero che, partiti in un momento di disperazione, si dolgono infinitamente di avere lasciata l'Italia e malgrado tutto vorrebbero tornare. Ora si pensa che costoro, avrebbero una tale prova di sentimenti italiani che potrebbe nei loro confronti essere smobilitata la faccenda, e dire quello che è stato è stato, ora non se ne parli più e si torni come prima. E l'orizzonte che all'interno va chiarendosi ogni giorno di più per la geniale onestà del neo Segretario del P.N.F. potrebbe lasciare adito a siffatte speranze rendendole attuabili.

3.7   A Family Affair

The choice to leave for the children's sake and in their interest, the urgency of ensuring their future are leitmotifs of letters and a peculiarity of Jewish Italian mathematical emigration. Some mathematicians, such as Terracini and Mortara, had young children, who could not be left behind (D3.7.1, D3.7.2); others had adult children, who had to continue their university studies (D3.7.5). Gino and Eugenio Fubini, Ugo Fano, Enrico and Edoardo Volterra, Giulio Levi, and Gina Castelnuovo were researchers or university professors in turn and had been dismissed together with their fathers (D3.7.3). From this point of view, the Jewish intellectual diaspora after 1938 is analogous to the Great Italian emigration (D3.7.4). It was a brain drain, surely, but it was also a family affair, involving the entire family nucleus and relying on family networks.

Those who had friends or relatives already abroad were in an advantageous position. Giulio Racah, theoretical physicist, and nephew of Fano, came to the London offices of the SPSL, to ask for information on how to help his uncle Gino and cousins Roberto (student of Engineering) and Ugo (physicist) to leave Italy. Guido Segre, who until 1938 had been consul general for the Italian government in Boston, went to the office of the ECADFS to ask what chances there were for his cousin Beniamino who was eligible for support, having been a full professor in Bologna. Less than two weeks later, ECADFS would send the curricula of the two Segres, Guido and Beniamino, to H.L. Bevis, President of Ohio State University and professor of Law and Political Science at Harvard and the Graduate School of Business Administration and Public Administration.[69]

A mathematical School is also a sort of family, and emigration was therefore also a School affair. The Masters supported their protégés and often helped them financially. Sociability was a crucial factor at this juncture. While the Jewish circles in Turin or Rome existed only in the darkened minds of anti-Semites, it is also true that the families of Castelnuovo, Volterra, Enriques, Levi-Civita, Fubini, Fano, Fermi, Amaldi were friendly and interconnected. Fathers and children had worked and lived side by side for decades, and therefore many were not left alone in the face of

---

[69] ECADFS Records, *BS*: B. Drury [ECADFS] to H.L. Bevis, New York 31.3.1939.

expatriation. The correspondence of Castelnuovo, Volterra, and Levi-Civita vividly reflects this aspect:

> Rasetti accepted the role of Director of the Physical Institute of Quebec in Canada. It is another serious blow to the Physics School in Rome. Terracini leaves with his family on the 24th for Argentina, Tucumán; it seems that in the latter University they have money and initiative because they have managed to quickly complete the paperwork for the landing permit. And did your Enrico get it? And where is Edoardo?[70]

Generational boundaries were crossed: hundreds of people were set in motion, in the hope that some assistance would emerge. Levi-Civita, for example, helped Enrico Volterra, his former pupil and son of his friend and colleague Vito, in the procedures to obtain a visa for Argentina. Faced with the difficulties he encountered, he turned to Levi and Terracini, by then already resettled in Argentina, who had been helped in turn by Levi-Civita to find a position in Rosario and Tucumán, respectively.

Disciplinary boundaries were also crossed. This passage from a letter by Levi-Civita to Volterra is emblematic: 'Castelnuovo wrote to me that his daughter Gina reports that Veblen has received my letter, but at the present time he can do nothing for Edoardo'.[71] In those days, Levi-Civita and Veblen were in fact doing their utmost to help a biologist (Gina Castelnuovo), a jurist (Edoardo Volterra) and two engineers (Gino Castelnuovo and Enrico Volterra). Analogously, A. Reichenberger, a romance philologue, pointed out to ECADFS the case of the two Terracini brothers: Benvenuto Aaron, a linguist and a glottologist, and Alessandro, a mathematician, although he obviously knew the production of Benvenuto only.[72]

Documents

D3.7.1 A. Terracini to O. Veblen, Turin 16.9.1938

OVP, *AT*, fol. 1r

My dear Professor Veblen, As you probably know, the Italian Government has decided to suspend for the moment from the service the Jewish professors. I don't know yet which will be the latest decisions upon this matter. But - although it

---

[70] ANL-*Volterra*: G. Castelnuovo to V. Volterra, Bolzano 19.8.1939: *Rasetti ha accettato il posto di direttore dell'Istituto fisico di Québec nel Canada. È un altro grave colpo per la Scuola fisica di Roma. Terracini parte con la famiglia il 24 per l'Argentina e Tucumán; pare che in quest'ultima Università abbiano quattrini e iniziativa perché sono riusciti a sbrigare rapidamente le pratiche per il permesso di sbarco. E il tuo Enrico lo ha ottenuto? Ed Edoardo dov'è?*

[71] T. Levi-Civita to V. Volterra, Vallombrosa-Saltino (Florence) 26.7.1940, in Nastasi and Tazzioli (2000), p. 192: *mi ha scritto Castelnuovo che sua figlia Gina [. . .] riferisce che il Veblen ha ricevuto la mia lettera, ma nel momento attuale non può far nulla [per Edoardo].*

[72] ECADFS Records, *AT*: Interview memorandum April 6, 1939, and B. Drury [ECADFS] to A. Reichenberger, New York, undated but April 1939.

would be very painful for me to leave my country - I must keep seriously up my mind about the possibilities concerning my future, not only because I wish to continue my work, but also as I must think about my wife and three children. Therefore, I ask you if you think there should be some chance for me to work in an American University or College. I permit me to remind you I always worked, and continue to work, in different fields of Geometry, and especially in projective differential Geometry. In any case I add a curriculum, which I am always disposed to complete with further informations or testimonials. During my work in Italian Universities, I always endeavored to reach didactic efficacy, and I think I should soon be able to get accustomed to the particularities of the American Universities. I hope that in consideration of your positions it will not be too difficult for you to help me in any way, and I thank you sincerely for the trouble. I beg you to agree my best regards. Yours truly Alessandro Terracini

D3.7.2 G. Mortara to SPSL, Milan 16.11.1938

SPSL, *GM*, fol. 352

To the Society for the Protection of Science and Learning - 6 Gordon Square London W.C.1. Having heard that you are generously taking care of searching for possibilities of employment in favour of the university professors, whose careers have been cut short by "racial" measures, I am taking the liberty of exposing you my case and of appealing to your help in the difficult task to find an opening abroad. From 1910 till a few weeks ago, I was professors of statistics, demography and economics in the Royal Italian universities and was unanimously appreciated and honoured in my mother-country. Today, in consequence of the measures against Italian citizens of Jewish descent, I have lost my chair, and whatever suitable occupation is precluded to me and to my children.[73] I am 53 years old and in good health and energetic; therefore, I hope to may continue for several years my scientific activity abroad, if I can find a starting point for the recommencement of it. Meanwhile, my four children (of 18, 16, 15, 7 years of age) might find abroad the possibility of completing their education and to obtain some employment. The present situation makes indispensable and urgent my emigration and that of my family; *I cannot leave them behind*. I am not unknown in the international scientific world: I am a member of the International Statistical Institute, a 'Fellow' of the Econometric Society and in 1920 I was elected an Honorary Fellow of the Royal Statistical Society (London). I have contributed to the Economic and Social History of the World War, edited by the Carnegie Endowment for International Peace, with a volume on "Public Health in Italy during and after the war" and have also contributed to several inquiries conducted by international bodies, such as the League of Nations, the International Chamber

---

[73] Alberto (born in 1920), Marcella (1922), Guido (1923) and Valerio (1931).

of Commerce, etc. A complete information about my scientific life and work is contained in the printed notice, in English, of which I allow me to send you six copies under separate cover (I shall send you other copies if useful).[74] I allow me to enclose the copies of two letters of the Rockefeller Foundation, referring to the possibility of a contribution from the Foundation to the institution which would offer me an appointment abroad. I am enclosing also a memorandum about my work as consulting statistician and economist: I should willingly collaborate in such quality with some banking or industrial firm if I could not find an academic occupation, or as supplementary job. I shall receive with gratitude any hint from you about the possibility of my occupation abroad. Of course I should prefer to find an opportunity in the British Empire or in the United States, but in the present situation I should accept with gratitude whatever occasion of decorous work in whatever civilized country. In the hope of your kind answer, I remain Yours sincerely Giorgio Mortara

D3.7.3 Guido Fubini to O. Veblen, Paris 30.11.1938

IAS, *GF*, fol. 1r-v
Dear Sir, From a letter of prof. Levi-Civita I know that he has written to you about my position.[75] Considering that I have no answer from him, I fear his letter lost, and I dare write directly to you. I was since more than 30 years Professor of Calculus and of higher analysis in the high school [sic!] and in the University of Turin; but the new racial laws of the *fascismo* have obliged me to leave Italy with my wife and two sons (25 and 27 years old).[76] A very brief account of my scientifical work is here included. I beg you to let me know if you think that some possibility exist for me of living and teaching in the U.S. An invitation from an University makes very easy the emigration, which is now strictly limited *only* by the Italian government. Money is immediately not important for me; my wish is now to obtain the tranquility for all my family after the very bad past days. I beg you to accept my best thanks and regards of your faithfully Guido Fubini

---

[74]*Notes of the scientific work of Prof. Giorgio Mortara*, Città di Castello, Società Anonima Tipografica 'Leonardo Da Vinci', 1938-XVI, 18 pp. The paper is kept in ECADFS, *GM*.
[75]FUBINI.1.
[76]Fubini had married Anna 'Annetta' Ghiron (1892–1973) in 1910 and the couple had two sons: Eugenio (born in 1910) and Gino (1913).

D3.7.4 [Note by an Anonymous Informant of the Secret Political Police], Chiusa Pesio (Cuneo) 19.3.1939

ACS-*Ebrei*, fol. 1r–2r

The Jewish problem in Piedmont continues to fascinate everyone, due to the possible developments to come and to the measures taken in this regard that are rumoured to be still in the pipeline. There is absolutely no anti-Semitism in this region, and everyone deplores the anti-Jewish attitude of recent months. The Catholic clergy, which in Piedmont had always found assistance, understanding and financial aid from the Jews, today is clearly placed on their side, tries to support them morally, without influencing religious belief. If it were not for the racial laws and the exclusion from schools of the Jewish element, here one would not even feel the discomfort that the enacted measures have caused. Especially in small towns, Jews are very well-liked and respected; traders and professionals have not yet felt any difference in treatment; on the contrary, in general they have received testimonies of sympathy and pity from the entire population. This also happens in Turin, except for a few exceptional examples. For some time, it has been verified that Jewish homes are visited by distinguished Catholic priests, who bring them the comfort of a good word urging them to be patient, that time will give them justice. This fact, noted by the population, is much discussed, as have been various sermons given in churches about the anti-Semitic struggle. However, pessimism persists in the Jewish environment, and many, worried about the future of their children, are preparing to leave the country by making use of commercial ties with neighbouring countries. Jews of foreign origin, who settled in Italy after 1919, have almost entirely left; only a few cases, for special family or health reasons, await a resolution that cannot be delayed. Whole families have left for both America and Palestine, and the Zionist Committee in Zurich is providing for those who need their help. I am aware that large sums have been raised both in France and in England and America for this purpose, and that North America is willing to welcome a large contingent of Italian Jews, professionals of value, for example in Mexico which has invited medical specialists and industrialists. Many Piedmont Jews went to the Côte d'Azur and settled there temporarily, so much so that in Nice a colony that is quite important for the personalities who compose it, has been formed. In recent days, the problem of Aryan service personnel in Jewish families has become topical, as the R. Prefecture has revoked almost all the temporary permits previously granted. Many have filed an appeal with the competent Ministry of the Interior and are now awaiting the decisions. But the Prefecture of Turin does not seem to intend to grant extensions, except in very exceptional cases, while those who submitted a request for discrimination hoped to be favoured, at least pending the outcome of the practice. Everyone is wondering, however, why an appeal which has been presented and motivated does not halt the Prefectural measure, pending examination of their position.

Il problema ebraico in Piemonte continua ad appassionare tutti, per i possibili sviluppi avvenire e per i provvedimenti che si vocifera saranno ancora a breve scadenza, presi a tal riguardo. Antisemitismo non ne esiste assolutamente in questa Regione e tutti deplorano l'atteggiamento antigiudaico di questi ultimi mesi. Il Clero Cattolico, che in Piemonte aveva sempre trovato assistenza, comprensione ed aiuti finanziari dagli ebrei, oggi si è chiaramente posto al loro lato, cercando di appoggiarli moralmente, e senza cercare di influire nella credenza religiosa. Se non fosse delle leggi razziali e della esclusione dalle scuole dell'elemento ebraico, qui non si sentirebbe nemmeno il disagio che i provvedimenti emanati hanno provocato. Nei piccoli centri specialmente, gli ebrei sono molto ben voluti e rispettati; i commercianti ed i professionisti non hanno sentito a tutt'oggi alcuna differenza di trattamento, anzi, in linea generale hanno avuto attestazioni di simpatia e di commiserazione da tutta la popolazione. E ciò accade anche in Torino stessa, salvo pochi esempi eccezionali. Da qualche tempo si verifica che le case degli ebrei sono visitate da esimi sacerdoti cattolici, che portano loro il conforto di una buona parola esortandoli a pazientare, che il tempo darà loro giustizia. Questo fatto notato dalla popolazione è molto commentato, come sono state commentate varie prediche tenute nelle chiese sull'argomento della lotta antisemita. Con tutto questo, nell'ambiente Ebraico perdura il pessimismo, e moltissimi, preoccupati dell'avvenire dei loro figli, si preparano ad espatriare valendosi di conoscenze commerciali coi paesi vicini. Gli Ebrei di origine estera, che si erano stabiliti in Italia dopo il 1919, sono quasi interamente partiti; solo pochi casi ed in speciali condizioni di famiglia o di salute, attendono risoluzione che non può tardare. Intere famiglie sono partite sia per l'America che per la Palestina ed il Comitato Sionista di Zurigo sta provvedendo per quelle che ne richiedono l'aiuto. Mi consta che forti somme sono state raccolte sia in Francia che in Inghilterra ed America allo scopo, e che l'America del Nord è disposta ad accogliere un forte contingente di ebrei italiani, professionisti di valore; così il Messico che ha invitato medici specialisti ed industriali. Molti ebrei piemontesi si sono poi recati nella Costa Azzurra ed ivi si sono stabiliti temporaneamente, tanto che a Nizza si è formata una colonia abbastanza importante per le personalità che la compongono. In questi ultimi giorni il problema del personale di servizio ariano presso famiglie ebree è diventato di attualità, avendo la R. Prefettura revocato quasi tutti i permessi temporanei prima concessi. Molti hanno inoltrato ricorso al competente Ministero degli Interni ed attendono ora le decisioni. Ma la Prefettura di Torino sembra non intenda concedere proroghe, se non in casi eccezionalissimi, mentre quanti hanno inoltrato domanda di discriminazione speravano di poter essere favoriti, in attesa almeno dell'esito della pratica. Tutti si chiedono però perché il ricorso presentato e motivato, non sospende, in attesa dell'esame della loro posizione, il provvedimento Prefettizio.

D3.7.5 [Note by an Anonymous Informant of the Secret Political Police], Rome 20.6.1939

ACS-*Ebrei*, fol. 1r
Among the Jews of Italian nationality, the slowness is denounced with which the commission responsible for examining the documentation produced by those who have requested discrimination is proceeding. It is said that among the applicants are persons who have presented numerous documents from which the civil and patriotic merits they have acquired are very evident; for them a quick examination would be enough to obtain discrimination, but the fact that even for these cases the commission proceeds slowly, suggests that the commission itself either does not want to take responsibility for its decisions, or has instructions to postpone such decisions for as long as possible. Since discrimination can mean life and non-discrimination can mean death, these Jews would like to proceed swiftly at least for those cases that can be resolved quickly, with a sure conscience, based on the detailed information in the attached documentation. It is also said that some well-to-do Jews, having already initiated their children to higher education which they could no longer continue in Italy, have now sent them abroad to institutional boarding schools, etc. to which they must pay the fees in foreign currency that they draw from our banks. There are those who think that this creates a new exodus of foreign currency and could serve as a pretext for sending money (converted into foreign currency) abroad.

Roma, 20.6.1939. Fra gli ebrei di nazionalità italiana si deplora la lentezza con la quale procede la commissione incaricata di esaminare la documentazione esibita da coloro che hanno chiesta la discriminazione. Si dice che fra i discriminanti vi sono di quelli che hanno presentato numerosissimi documenti dai quali risultano evidentissime le benemerenze civili e patriottiche da essi acquistate e per costoro basterebbe un rapido esame per giungere alla discriminazione, ma il fatto che anche per questi casi la commissione procede a rilento, fa supporre che la commissione stessa o non vuole assumersi la responsabilità delle proprie decisioni, oppure ha istruzioni di rimandare tali decisioni il più a lungo possibile. Poiché per taluni la discriminazione può significare la vita e la non discriminazione può significare la morte, si vorrebbe da questi ebrei che la commissione procedesse speditamente almeno per quei casi che possono essere per la larghezza della documentazione allegata risolti rapidamente, con sicura coscienza. Si dice poi che alcuni ebrei agiati avendo già i figli avviati a studi superiori che non potevano più continuare in Italia, li hanno inviati all'estero, presso convitti istitutivi ecc. ai quali debbono corrispondere le rette in valuta estera che attingono alle nostre banche. C'è chi pensa che questo dia motivo ad un nuovo esodo di valuta estera e potrebbe servire di pretesto per inviare all'estero denaro convertito in valuta estera.

3.8 'Many People Looking for *Giobbi*, and *Giobbi* Few and Ugly'[77]

Very few were able to move abroad without the need to find paid employment. Among the mathematicians, only Fano and Fubini had sufficient assets, the others needed to find a job if they wanted to leave Italy. Hence what Fubini would have called, with painful irony, 'the hunt for *giobbi*', commenced.

The would-be emigrants from racist Italy started with a great disadvantage. In the first place, they were the latecomers and were destined to beg for the crumbs of an international scientific solidarity long since exhausted in favour of refugees from the Third Reich (D3.8.2, D3.8.4). This was especially true for those who played the American card. Looking for a position in the United States in the autumn of 1938 meant facing ruthless competition, in an intellectual market close to saturation point.

Emigration had looked like an attractive option well before 1933, but as the Nazification of the European continent advanced, the pace of emigration had greatly quickened. At the same time, the effects of the Great Depression were manifest. As Rider (1984, p. 123) pointed out the worst year for American college budgets and staff rosters was 1933–34, just after the promulgation of the German Civil Service Law. The impact of the Depression on faculties had varied according to the size, nature, and location of the institution and the suffering had not been equally apportioned among faculty members of different ranks. In response to the crisis, most institutions had used salary cuts to help bring expenditures in line with income. Courses had been dropped whenever possible. New PhD graduates had seen their employment options and job prospects dim. The normal retention rate (i.e. the appointment of new PhDs as teaching or research assistants) had doubled.

By 1936, economic conditions had improved yet, in both the United States and the United Kingdom, proposals to absorb displaced scholars did not meet with unqualified approval. Precisely in the days when the Italian problem leapt into the headlines, clippings of comments about the situation in the field of mathematics were accumulating on the desk of the ECADFS office (D3.8.1). As L.W. Cohen recalled:

> Young American mathematicians were finding it hard to get appointments, and the question of whether to bring in foreign mathematicians to occupy positions which would then not be available to American mathematicians was debated. Veblen took what I would call the broader view. I hesitate to attribute views to Veblen, but the considerations that seem to have actuated him were two: a concern for the welfare of mathematics itself, and a humane concern for certain individuals who had talent. Veblen was a grand man, and the people for whom he made it possible to come to the United States made a great contribution to mathematics. G.D. Birkhoff opposed him on this. [. . .] Birkhoff said: «If these distinguished people come and take the positions, the young American mathematicians will become hewers of wood and drawers of water.»[78]

---

[77] F. Rasetti to E. Amaldi, Roma 8.5.1940 in Battimelli and De Maria (1997), pp. 127–128: *gente in cerca di giobbi molti, e giobbi pochi. Gina ha un giobbo temporaneo e racchio a Philadelphia.* See also Goodstein (2001).

[78] Coen (1985), pp. 8–9.

Opposition was generally couched in terms of academic protectionism where colleges and universities were exhorted to 'buy American (or British)' (D3.8.5 and D3.8.6). A reluctance to favour young foreigners over American or British students was also, for some mathematicians like Volterra's old friends G.C. Evans and G.A. Bliss, a consequence of the sense of responsibility that established scholars felt for their own students. Some others, like the Jewish Russian-born S. Lefschetz, feared that the prolonged crisis would result in them being overrun with foreigners, however well qualified. Talking about Fine Hall in the early fifties, G.C. Rota would recall:

> No one who talked to Lefschetz failed to be struck by his rudeness. [...]. He was rude to everyone, even to the people who doled out funds in Washington and to mathematicians who were his equals. I recall, Lefschetz meeting Zariski, probably in 1957 [...]. After exchanging with Zariski warm and loud Jewish greetings (in Russian), he proceeded to proclaim loudly (in English) his scepticism on the possibility of resolving singularities for all algebraic varieties. [...] His colleagues must have been surprised when Lefschetz himself started to develop anti-Semitic feelings which were still lingering when I was there. One of the first questions he asked me after I met him was whether I was Jewish. In the late thirties and forties, he refused to admit any Jewish graduate students in mathematics. He claimed that, because of the Depression, it was too difficult to get them jobs after they earned their Ph.D.s. He liked and favoured red-blooded American boyish Wasp types (like Ralph Gomory), especially those who came from the sticks, from the Midwest, or from the South.[79]

As a matter of fact, anti-Semitism meandered in certain American and Italo-American circles. American scientists explicitly took position against the racial theories promulgated in Germany and Italy in a manifesto published by the *New York Times* and *Washington Post* on 11 December 1938, which collected more than 1,200 signatures in the American academic environment. However, anti-Semitic prejudices were quite widespread to the point that, in his defence against charges of Communism, Veblen would have described his efforts to place refugees as having seriously encountered the opposition of 'American anti-Semites' (Sect. 3.11). Anti-Semitic sentiments emerge from the excerpts of various correspondences, whose hostile comments about foreigners often served as code words to mask anti-Semitism.[80] To cite a single example, Birkhoff opposition to the election of Lefschetz as President of the American Mathematical Society had interesting overtones:

> I have a feeling that Lefschetz will be likely to be less pleasant even than he had been, in that from now he will try to work strongly and positively for his own race. They are exceedingly confident of their own power and influence in the good old USA. [...] He will get very cocky, very racial and use the *Annals* as a good deal of racial perquisite. The racial interests will get deeper as Einstein's and all of them do.[81]

---

[79] Rota (1989), pp. 230, 232.

[80] See Siegmund-Schultze (2009); Solow (1942); Reingold (1981), pp. 319–327; Phillips (1994), pp. 6–8; Mac Lane (1994), pp. 9–10 and Parshall (2022).

[81] G. Birkhoff to R.G.D Richardson, 18.5.1934, in Reingold (1981), p. 321.

In autumn of 1938, both the SPSL and the ECADFS were noticeably sensitive about the perils of nationalism and anti-Semitism, and both clearly feared nationalistic and anti-Semitic opposition to their efforts:

> behind the reluctance to hire foreign scholars on grounds of nationalism or protectionism lay a complex of attitudes, not always announced openly. Their geographical distance from events in Europe permitted Americans the luxury of isolationism in academia as well as foreign policy, and contributed to a sort of xenophobia, however muted its expression. An undertone of anti-Semitism ran through many of the nationalist assertions made by British and American scholars. Private utterances were likely to be more blatantly anti-Semitic than public statements, but enough of the latter survive in the documentary record to suggest the scope of prejudice.[82]

Alongside these anti-Semitic accents in high culture, there was also what some Italian emigrants who took refuge in America called 'mild' or 'calm' anti-Semitism, so different from the pathological and virulent anti-Semitism of Nazi Germany, yet quite pervasive and ubiquitous.[83] It was a tacitly accepted kind of antagonism, a superstition from which neither the intellectual nor the common man were immune; a prejudice cemented by old stereotypes about morphometric and cultural characteristics: the physiognomy, on the one hand, and the customs, cultural identity, religion, attitude and social behaviour of the Jews, on the other.

Stereotypes with which American and Italo-American societies were impregnated often emerge in a priori unsuspected contexts. Thus, for example, M. Litzinger, one of the first American researchers in abstract algebra, while visiting the Volterras, wrote to her mother that 'Volterra's wife looks Jewish'. Analogously, V. Snyder specified to B. Drury, the secretary of the ECADFS: 'he [Segre] is strikingly fair, has blue eyes, and no trace of appearance or manner that would suggest a Jew' (SEGRE.3). The traits of this gentle anti-Semitism are fully portrayed in some articles by G. Prezzolini, Director of the *Casa Italiana* at Columbia University from 1929, and New York correspondent for the newspaper *Tempo* (D3.8.7 and D3.8.8). In the face of these testimonies, it should be noted that neither Fermi, nor Segré, nor Fubini, nor Fano's children mentioned repercussions of this kind of anti-Semitism at work, nor did they declare having suffered serious emotional distress because of it. After all, without a caftan, without Kosher food, without Yiddish speech, they were hardly recognisable. On the whole, they seem to have felt more the anti-Italian prejudice than anti-Semitic discrimination: familism, the emigrant who gets away with it in ways that are not always legitimate, who circumvent the rules and teach others to circumvent them, a certain snobbery on the part of intellectuals, etc.

Finally, three other circumstances must be taken into account in understanding the challenges faced by Italian mathematicians looking for a job abroad in autumn of 1938. The first: unlike their colleagues fleeing Central and Eastern Europe, Italians

---

[82] Rider (1984), p. 133.

[83] As Reingold writes (1981, p. 337): "Certainly, the American university with all its peculiarities appears innocently idyllic in contrast to the German university at the end of the Weimar era. The Nazi era had the ironic effect of ending the idyll by disclosing the dangers of anti-Semitism." On exclusionary practices in Harvard, Yale, and Princeton, see Synnott (2010).

were not in immediate danger of life and, therefore, in some ways it seems natural that their applications should be postponed in favour of those of scholars for whom the prospect of deportation was tragically imminent. The assessment of academic merits was sometimes arbitrary, due to the fear for the safety of scholars' lives. For example, from 1938, several German scientists who had remained in Germany were sent to concentration camps. For a time, rescue agencies could get prisoners out of concentration camps and out of Germany if emigration was guaranteed. The events of Kristallnacht, when Jewish men over sixty were arrested and incarcerated, made it imperative that refuge also be provided for older, even retired, scholars. As Hitler's net widened to include Austria, Czechoslovakia, and Poland, mathematicians from these countries suffered the same plight. Nothing similar happened in Italy. No imminent personal danger from violence and possibly starvation, no risk of arrest and incarceration.

Secondly it must be considered that most of the Italian mathematicians who tried to leave the country were pure mathematicians and did not consider any other prospect of employment than that in academia. This fact, of course, limited their possibilities. A restricted minority, even among young people, accepted other options. The astronomer Jacchia was ready to accept any job, even as an interpreter or tour guide (D3.8.3). The statisticians Bachi and Mortara were willing to work in banks, insurance companies, or commercial enterprises. Mortara was ready to accept cumulative solutions: 'I think it will be difficult to find within a few weeks a permanent academic occupation for me, but I hope it will be relatively easy to find several contemporary or successive opportunities of temporary work: a *cumulative* solution of my problem instead of a *unique* one'.[84]

Thirdly, the debate over unemployment and enforced emigration took place in the context of discord, especially in the English-speaking mathematics community, about the relative professional merits of teaching and research. In somewhat brutal terms the prejudice was: if people are brought in who are primarily interested in research work, they will either object to elementary teaching or will do it in a less perfunctory way. The ECADFS in particular wanted to place scholars in research settings but ones hopefully leading to a permanent position, perhaps involving an occasional graduate course. This was to reduce the peril of xenophobia and anti-Semitism, but the effect was often to the contrary: 'some faculty members greatly resented giving special privileges to foreigners at a time when money was hard to get for research and when others were forced to carry heavy teaching loads'.[85] In connection with this debate, Italian émigrés were doubly disadvantaged: they had been trained within a vigorous tradition of both teaching and teachers training, and they were willing to teach, but the linguistic bias prevented them from being charged with both basic and advanced courses.

---

[84] ECADFS Records, *GM*: G. Mortara to L.I. Dublin, Milan 29.11.1938.
[85] Reingold (1981), p. 316.

Documents

D3.8.1 Job Scarcity in Field of Mathematics Because of Foreign Influx, 16.9.1938

ECADFS Records, *Italy*, fol. 1r
Mr. Flexner[86] sends the attached clipping, with the comment that he believes the members of the Committee will be interested. It expresses a theory that he has advanced so many, many times in Committee meetings – that various fields in academic life, particularly the discipline of mathematics – are saturated, and that to continue to introduce foreign scholars will mean lack of opportunity for young Americans. He spoke with a good deal of feeling on this point over the telephone yesterday. Will you not bring this up at the next meeting? Or give him an opportunity to do so?

D3.8.2 O. Veblen to A. Terracini, Princeton 4.10.1938

OVP, *AT*, fol. 1r
Dear Professor Terracini: I am making inquiries to find out whether there is any place in this country where there would be an academic opening for you. You will easily understand the difficulties, namely that we have absorbed so many of the scholars who were displaced from Germany that we are dangerously near the saturation point. I am sure that what has already been done has been a great advantage to this country and that we could benefit by further absorption of European scientists. I will do everything I can, but it would not be right for me to hold out any expectation of success. In addition to the saturation effect which I have just mentioned, there is also the fact that recent political events are strengthening the reactionary influences in this country as well as elsewhere. If and when I got any information which might be helpful to you I will write again. In the meantime, please accept my thanks for the reprints which I have received from you both recently and in former years. With cordial good wishes, Yours sincerely, Oswald Veblen

---

[86] Bernard Flexner (1865–1945), a New York lawyer and a prominent member of the Zionist Organization of America, was one of the founders of the ECADFS.

D3.8.3 L. Jacchia to S.A. Mitchell, Bologna 18.11.1938

ECADFS Records, *LJ*, fol. 1r–2r

Dear prof. Mitchell:[87] I do not know if my name is known to you. I have been till now assistant in the Bologna Observatory and a member of Commission 27 (variable stars) of the I.A.U. [International Astronomical Union] I apply to you for a personal affair, and I hope you will be so kind and excuse my boldness. In effect I am at present in such a hopeless situation, that I feel obliged to try every way of escape. As a Jew – though I have only a quarter of Jewish blood, 3 of my 4 grandfathers being Arian – I am chased from my place without any practical possibility of earning my livelihood in this country. I have not a cent of personal fortune and I must care for my mother [Elisabetta Carpi]. Foreseeing the present state of things, I had planned to go to Stockholm past August, to the meeting of the U.A.I., in order to try to find a settlement abroad by personal speaking with foreign astronomers.[88] The Ministry did not allow my journey, even in private form, always for racial reasons. I charged then one of the Italian delegates to take upon himself this task, but the mission did not yeald to anything. Now, having considered the matter thoroughly, I have decided to desist from any attempt to pursue elsewhere my scientific career and to leave off the luxurious science that is Astronomy. It is a terrible renouncing for me, but in an epoche like the present one it is indispensable to be in first place realistic. The matter is now to find in some part of the world the means of supporting myself. Happily, I have not limited myself, during my life, to the astronomical studies and I think that the other knowledges I have acquired may be of some use to me at present. I am 28 years old, bachelor, have the degree of doctor for physics obtained at the Bologna University in 1932. I speak and write nearly as well as my mother-language, Italian, the following languages: French, German, English, Spanish, Portuguese, Swedish, Danish, Norwegian, Netherlandish, Romanian, Russian, Hungarian. Besides I write well and speak sufficiently well Polish, Czech, Slovenian, Croatian, Greek, Turkish, Arabic, Esperanto. Some knowledge of Finnic and Bulgarian. I think that with this knowledge of languages it shall not be impossible for me to find an employment in some school, industry or commercial firm. Europe offers very few possibilities and therefore my aims are directed toward America. Unfortunately, I have no acquaintances there and so I have thought of you as a person who could help me by means of his wide influence and his numerous acquaintances. Would you charge yourself with this nuisance and try and help me from my critical situation? I repeat you that I am ready to do anything. I have even proposed my services as interpreter to travelling

---

[87] Samuel Alfred Mitchell (1874–1960), a Canadian-American astronomer, President of Commission on Solar Eclipses and Commission on Stellar Parallaxes and Proper Motions, and Director of the Leander McCormick Observatory at the University of Virginia.

[88] The triennial conference of the International Astronomical Union, representing 27 countries, had been held at Stockholm during the week August 3–10, 1938.

companies. Please excuse me again, dear professor, you shall consider it a great boldness from me that I so ply for things like these to a person I do not know personally, but I ask you to consider the condition of anxiety in which I find myself, seeing stopped so suddenly my scientific career and thinking of my uncertain future. I hope that it will not cost you a too great loss of time to look after whatsoever an employment which could suit to a polyglot, acquainted with physics, chemistry, and mathematics. Thanking you beforehand for all you will do for my sake, I beg you to agree my best regards. Yours faithfully, Jacchia

D3.8.4 D.C. Thomson

America and Academic Exiles. Progress in and obstacles to placement, The Scotsman 17.8.1939, p. 11

The inquirer into the question of America's reception and absorption of refugee scholars and scientists is in the first place struck by two main factors in the situation. First of all, the problem is being faced without the background of dynamic optimism which was so familiar a feature of opinion in the States prior to the depression; and, secondly, the amount of sympathetic, constructive, and incompletely co-ordinated work being undertaken by individuals is very considerable when compared with the plans and accomplishment of the several national committees. The first factor to which I have referred of necessity plays a large part in the formation of official policy. Thousands of able young graduates formerly used to a sense of living in an expanding universe are now constantly made aware of the closing of doors against them in academic centres, and the contraction of the market for their services. It is perfectly natural that the problem of refugee placement should be considered in all academic centres in close relation to the policy of the graduate schools and their experiences in seeking careers for their students, even when it is widely realised that many universities still have some distance to go before they can claim that all their disciplines are fully staffed. Financial problems loom large in the minds of college Presidents, whether of State or private institutions, and the situation has not been made more easy in this respect by the fact that professors in the former category are no longer exempt from federal Income-tax, following a recent decision of the Supreme Court, or the insecurity with regard to the yield of those investments upon which the private institutions depend.

Individual Efforts in Salvage. I have recently visited between twenty and thirty academic institutions in America, and interviewed also the officials of the best-known learned and scientific societies and educational trust funds. Everywhere one came across individuals engaged in attempts to salvage some colleague or friend with whom they had come into contact in Europe, and in a great number of cases all this energy and money were being expended without any co-operation with national co-ordinating activity in the field, with the consequent dangers of overlapping and failure. Whatever may be said in criticism of this isolated endeavour, it was among enthusiasts of this group that I constantly met the finest type of attitude with regard to the problem as a whole. In most instances these individuals were working not merely on account of simple emotional humanitarian feeling, but because of a realisation of the basic principle involved, the challenge to intellectual and academic freedom. It was only comparatively rarely that I met with an official in an academic institution who was to any degree kindled or apprehensive on this issue.

Anti-Semitic Feeling. While the question of anti-Semitism or anti-alien feeling undoubtedly cropped up from time to time, it was rarely a matter of first-class importance except in the minds of the official committees. So far as the official attitude of the universities is concerned, the outlook of the more enlightened Faculties is seen in the words of the report recently issued by a Harvard Committee: "In the United States anti-Semitic feeling has operated within the universities themselves, in the form of a prejudice which is difficult to prove and never officially proclaimed. Though the prejudice has not, as in totalitarian countries, caused the wholesale dismissal of professors, it has made it difficult for Jews otherwise eligible to obtain initial appointments, and, there is reason to believe, has retarded their advancement to higher rank when appointed. No graver reflection could be cast on the academic profession than that any of its members should be willing to compromise time-honoured educational and scholarly standards by racial or religious discrimination." My own impression is that the undoubted fact of a rising tide of anti-Semitism unduly affects the minds of those national committees who work in New York where the tide is highest as yet, while it rarely exercises much influence in the reaction to the problem of exile scholar placement in the most progressive institutions.

Effect on American Scholars. As an instance of the spreading realisation that the steady infiltration of skilled foreigners is prejudicing the status of American scholars, I quote the following from the publication of the American Mathematical Society; it is taken from a paper by Dean Birkhoff of Columbia, who has been discussing the influx of eminent non-American mathematicians: «With this eminent group among us, there inevitably arises a sense of increased duty toward our own promising younger American mathematicians. In fact, most of the newcomers hold research positions, sometimes with modest stipend, but nevertheless with ample opportunity for their own investigations, and not burdened with the usual heavy round of teaching duties. In this way the number of similar positions available for young American mathematicians is certain to be lessened, with the attendant probability that some of them will be forced to become 'hewers of wood and drawers of water'. I believe that we have reached a point of saturation, where we must definitely avoid this danger.» What is true of mathematics is in some degree also true in other disciplines. It is important, I think, to stress the pioneering imagination which has been displayed in some of the smaller Colleges and in at least one of the negro institutions. Carlton College in Minnesota and Blackmountain College in North Carolina maintain quite a number of refugees on their budget without reference, so far as I know, to any of the great educational or refugee funds. Howard University at Washington and Notre Dame University at South Bend, Indiana, have shown considerable enlightenment in strengthening their Faculties by bringing in eminent teachers from Europe.

A Saturation Point. There can be little doubt that in most of the large academic centres, for one reason or another, the situation in this matter resembles that of the saturated solution in chemistry. It may well be argued that very little further can be accomplished by the statesmanlike methods of the Emergency Committee in Aid of Displaced Foreign Scholars which has in the past five years co-operated in financing the re-establishment of 125 exiles. Although this Committee has circularised all listed American Colleges on more than one occasion, nevertheless instances have occurred where institutions of standing, even within a hundred miles of New York, have insisted that they have never heard of such facilities and that they would otherwise have taken advantage of them, which only demonstrates how ineffective letters are and how essential personal and continuous contact will always be in this matter. In addition to the carefully planned activities of this body, tribute must be paid to the successful placement accomplished in very many instances by Dr Wilbur Thomas of the Oberlander Trust, and of the untiring endeavours of Dr Alvin Johnson of the New School of Social Research. Mention must also be made of the quiet collecting of information by the American Friends Service Committee and the effective propaganda and relief of the Catholic Refugee Committee in New York.

Contribution to American Culture. The success or otherwise of this exile placement work must ultimately be judged not on the basis of the number of unfortunate people provided with sanctuary but from the acknowledged enrichment of faculty strength achieved and the permanent integration of skill in the fabric of American culture. After all, this is of greater moment than the successful wheedling of apprehensive Deans or the subtle circumvention of local obstacles or prejudices which can only find a temporary solution to the problem. No more constructive and imaginative thinking has been contributed to this problem than that untiringly given by a small group of individuals acting sometimes alone and sometimes in co-operation with others, among whom the foremost names are those of Alvin Johnson, Eduard Heimann, Paul Tillich, Robert Ulich, Oswald Veblen, Harlow Shapley, and Elias Avery Lowe, while a large number of scholars seeking refuge in the States have particular reason to be grateful for the tireless energy and sage counsel of Waldo Leland, of the American Council of Learned Societies, and his deputy, Mortimer Graves. Two most promising developments have unfortunately been held up in recent months. The first was a plan set in motion by a group of the most prominent College Presidents, under the chairmanship of James Conant,[89] for the raising of independent funds to establish "asylum fellowships" in all the main institutions. While between one and two

---

[89] James Conant (1893–1978), President of Harvard University from 1933 to 1953.

hundred academic leaders have now given their support to this project, it hangs fire through lack of a leader for its money-raising activities. Little, it is feared, can be hoped from this source now until 1940, although the testifying of these prominent figures to the existence of an undeniable obligation on the part of the university community was a matter of great moment.

Pending Developments. In the second place, some delay is being experienced in the setting up of horizontal discipline committees throughout the country on the model of the Philosophers' and Psychologists' Groups which are already in being. This development looked like taking the burden off the shoulders of individual faculty members and providing the Deans in the various centres with expert advice on available exile teachers. The extension of this project was welcomed in every place where I heard it discussed. One final development looks like coming to fruition before the autumn – namely, the formation of an international co-operating group of academic leaders who will seek to prevent the threatened dislodgment of refugee scholars who have successfully found asylum in countries such as Turkey and Japan, and who are now jeopardised by economic pressure from Germany. Great Britain must be regarded as one of the clearing house countries and America as the great terminal country in this crisis, and in most cases the methods they are employing bear a striking resemblance to each other. Great Britain has no reason to be ashamed of the proportion of scholars for whom she has successfully found places, still less of the great sum of money which is annually subscribed for the maintenance and extension of this work by her own academic institutions and teachers, and the frank and open propaganda for this cause which they foster in their midst.

D3.8.5 V. Snyder to O. Veblen, Ithaca 12.12.1939

OVP, *BS*, fol. 1r
Dear Veblen: Thank you for sending me a copy of your recent letter to B. Segre. I am sorry, but must feel that it is not the final word. That is, if there are funds when the awards for 1940–41 are made, his case will still be considered. About the same time that Segre wrote you he told me that on account of the war his chances for an extension of his stipendium were very small. Of course, this is easy to understand, under the circumstances. It is evident that we cannot go on indefinitely in forcing things to create places for Jews. In too many cases the recipients have taken the appointments not only as a matter of course, but in some instances, they have the attitude that we should feel grateful that they would deign to accept them. In the case of Segre, I do not know how that phase of the situation would work out; I have the feeling that his own attitude would be entirely proper, but I

have nothing to go on except his very cordial and kindly relations with me.[90] Perhaps I am over jealous; in the little part of mathematics that I know something about, his work is just about the last word. [...] Sincerely yours, Virgil Snyder

D3.8.6 G. Salvemini to H. Shapley, Cambridge Mass. 29.6.1940

ECADFS Records, *EdV*, fol. 1r

Dear Professor Shapley: Columbia University has never helped any Italian exiles or refugees. Nicholas Murray Butler does not like to displease Mussolini.[91] Edward Volterra is a man of great ability and fine moral character.[92] If you help him, you will help a worthy man. During the present chaos it will be impossible for him to get out of Rome. But if he gets an appointment somewhere in this country, he will be enabled to come when things have begun to settle down there. Sincerely yours, G. Salvemini[93]

D3.8.7 G. Prezzolini

L'emigrazione ebraica italiana negli Stati Uniti, pp. 248–250

In the years 1938–40, Italian Jewish emigration was limited compared to the figures reached by Italian emigration in the years before the end of emigration. The (approximately) 2,000 Italian Jews who came here are a drop of water compared to the peak of 1920 with its (approximately) 350,000 emigrants. But there was a significant contextual difference: the emigration of 1920 was proletarian and southern, that of 1938–40 was bourgeois and central or northern. It was also a minor exodus compared to the European Jewish emigration in the years before the war from countries dominated by the Nazis; but, for Italy, it was a new phenomenon because, up to that point, Italian Jews had never been pushed or forced to leave in significant numbers. The Italian Jews who came to the United States were, almost without exception, all well-to-do, members of the middle

---

[90] Virgil Snyder (1869–1950), professor at Cornell University (1910–1938) had been in close contact with Italian geometers. He had studied in Turin with Corrado Segre in 1922 and with Severi in 1929. On that occasion he had met Beniamino Segre as a student. See SEGRE.3.

[91] Nicholas Murray Butler (1862–1947), President of Columbia University from 1901 to 1945.

[92] Edoardo Volterra (1904–1984), first son of Vito Volterra, was professor of Roman Law at the University of Bologna. Dismissed in 1938, he was looking for a position in the States. See Sect. 5.2.

[93] Gaetano Salvemini (1873–1957), professor of Modern History at the University of Florence, was dismissed for political reasons, and his Italian citizenship was revoked in 1926. As a *fuoruscito* (i.e. a political exile) he continued to actively organize resistance against Mussolini in France, England, and finally the United States. In 1934, Salvemini accepted a position created especially for him, to teach Italian civilization at Harvard University, where he would remain until 1948.

classes, neither important capitalists nor poor devils (contrary to the legend that makes every Jew a rich man, there are several thousand Jews in Italy who have no property). They were merchants, small industrialists, teachers, lawyers, doctors, minor bankers, insurers, entrepreneurs, and almost all of them were able to transport here at least a part of their assets, for in spite of the laws which set out to impoverish them, the practical application of said laws by the officials of the Italian State was generally benevolent, and tended to alleviate, as far as they could, what they had to apply reluctantly. In some cases, the officials themselves indicated to the Jews how they could circumvent the law. Some of these emigrants already had, by reason of their trades, money deposited in New York (not only Jews, but also fascist Italians) and others easily found non-Jewish families who would exchange goods and money to their greater or lesser advantage. But no 'big capitalists' or 'big industrialists' came. For the first time, Ellis Island saw the arrival of troops of Italians who presented themselves as emigrants but were different from the Italian emigrants from the South, in terms of appearance, clothing, manners, and linguistic ability. Many knew a little English, had a few dollars to get through the first few months, and did not belong to rural classes. But while this emigration was so distinct from the mass of ancient Italian emigrants, described to us in De Amicis' *Sull'Oceano*[94] and by Commissioner Corsi in his book on Ellis Island,[95] it was equally distinguished from that of the Jewish emigrants who came from Eastern Europe; the rabbis with caftans and unkempt black beards, were not to be seen among Italians; there was no misery and rags, no dejection and hunger, no numerous families, no harsh guttural accent of Yiddish, the curious language spoken by the Jews of Poland, Russia, and parts of Germany and Austria. There was almost never a need for them on Ellis Island to prepare kosher food (which excludes religiously impure animals such as pigs, or animals not killed according to Synagogue rules). Not only were the Americans astonished not to see the type of Italian who passes here for characteristic, with the black moustache protruding from the upper lip, short in stature and earthy colour, but also the Jews from here who had not travelled in Italy were astonished to see such different co-religionists. [...] Without a rabbi to accompany them, those who practiced the faith were aggregated to the Portuguese Spanish Synagogue, which is of the same Sephardic cult as the Italian synagogues (while the majority of New York's Jews belonged to the Ashkenazi cult).

---

[94] Edmondo De Amicis, *Sull'Oceano*, Milano, Treves, 1889. It is a well-known Italian novel that deals with the theme of Italian emigration between the nineteenth and twentieth centuries. In describing his journey from Genoa to Buenos Aires, the author illustrates the misery and tenacity of the emigrants, forced by inhuman living conditions to abandon their homeland.

[95] Edoardo Corsi, *In the Shadow of Liberty: The chronicle of Ellis Island*, New York, The Macmillan Company, 1935. Edoardo Corsi (1896–1965) was an Italian-American association executive and government official, particularly active in the areas of immigration, labor relations, and social welfare. In 1933 he had been appointed Commissioner of Immigration and Naturalization at Ellis Island, where he humanized the entry procedures.

L'emigrazione ebraica italiana degli anni 1938–40 fu piccola rispetto alle cifre raggiunte dall'emigrazione italiana degli anni precedenti la chiusura dell'emigrazione. I (circa) 2000 Ebrei italiani che vennero qui son una goccia d'acqua rispetto alla cataratta del 1920 con i suoi (circa) 350000 emigranti. Ma come qualità ci fu questa grossa differenza: che l'emigrazione del 1920 fu proletaria e meridionale, quella del 1938–40 fu borghese e centrale o settentrionale. Fu pure un fatto di piccola importanza rispetto all'emigrazione ebraica europea negli anni precedenti la guerra dai paesi dove dominarono i nazi; ma però, rispetto all'Italia, nuovo, perché prima d'allora gli Ebrei italiani non eran mai stati spinti o costretti a partire in rilevante numero. E gli Ebrei italiani che vennero negli Stati Uniti furon, quasi senza eccezioni, tutti benestanti, appartenenti ai medi ceti, né grandi capitalisti né poveri diavoli (contrariamente alla leggenda che fa d'ogni Ebreo un ricco, ci son in Italia varie migliaia di Ebrei nullatenenti). Furon mercanti, piccoli industriali, insegnanti, avvocati, medici, minori banchieri, assicuratori, imprenditori e quasi tutti poteron trasportare qui una parte almeno delle loro sostanze, perché nonostante le leggi che li spogliavano, l'applicazione pratica fattane dai funzionari dello Stato italiano fu in generale benevola, e tendente a lenire, in quanto essi potevano, quello che dovevan fare a malincuore. In taluni casi i funzionari stessi indicarono agli Ebrei in quali modi avrebbero potuto sfuggire alla legge. Alcuni di questi avevano già, per ragion del loro commercio, denaro depositato a New York (non soltanto Ebrei ce l'avevano ma anche degli Italiani fascisti) ed altri trovaron facilmente famiglie non ebree che cambiaron beni e denaro con vantaggio più o meno grande di quest'ultime. Non vennero però dei 'grandi capitalisti' o dei 'grandi industriali'. Per la prima volta Ellis Island vide sbarcare degli Italiani in quantità ragguardevole che si presentavano come emigranti ma si distinguevano dall'emigrante italiano proveniente dal Mezzogiorno per il loro aspetto, vestiario, modi, capacità linguistica. Molti sapevano un po' di inglese, avevano un po' di dollari per passare i primi mesi, e non appartenevano a classi rurali. Ma mentre questa emigrazione si distingueva tanto dalla massa degli antichi emigranti italiani, descrittaci nel Sull'Oceano *di De Amicis e dal Commissario Corsi nel suo libro su Ellis Island, essa si distingueva altresì, altrettanto, da quella degli emigrati ebrei che proveniva dall'Europa orientale; non si vedevan fra loro i rabbini col caftano, e la barba nera incolta, non si vedevano miserie e stracci, non si vedeva avvilimento e fame, non si vedevano numerose famiglie, non si sentiva l'aspro gutturale accento dello yiddish, la curiosa lingua parlata dagli Ebrei della Polonia, della Russia e di parte della Germania e dell'Austria. Non c'era quasi mai bisogno per loro a Ellis Island di preparare il cibo kosher (che esclude gli animali impuri religiosamente come il porco, o non uccisi secondo le regole della Sinagoga). Non soltanto gli Americani eran meravigliati di non veder il tipo dell'Italiano che passa qui per caratteristico, con i baffoni neri sporgenti dal labbro superiore, basso di statura e di color terragno, ma anche gli Ebrei di qui che non avevan viaggiato in Italia eran meravigliati di veder dei correligionari così differenti. [. . .] Non avendo con sé un rabbino, quelli che frequentavano il culto furon aggregati alla Sinagoga*

*portoghese spagnola, che è dello stesso culto Sefardita delle Sinagoghe italiane
(mentre la maggioranza degli ebrei di New York appartiene al culto Askenazi).*

D3.8.8 G. Prezzolini

Gli Italo-Americani sono antisemiti?, pp. 260–263

In Long Island, there is a small town called R. that is completely inhabited by a
Jewish population from Eastern Europe. In an apartment of a two-storey house, a
family of Italian Jews who had arrived here in 1939 found accommodation on the
upper floor. They were not converted, but they did not attend the Synagogue.
They considered themselves 'freethinkers' or something like that. But they had no
difficulty in following Italian social customs. On Christmas Day it happened that
the little girl remembered what she and her companions used to do in Italy and
bought a fir tree, covered it with candles, lit them and displayed it on the terrace.
The hostess of the house, an Israelite from here, and a strict observant, flew into
a rage: «I did not know that you were Gentiles. I wouldn't have rented the
apartment to you . . . » The girl's mother made a scornful response. It was a bad
time. This anecdote, which was told to me with a smile by an Italian Jew of old
acquaintance, depicts the situation of many Italian Jewish families when they
encountered the Jewish population from here. They felt bewildered. Another
source of marvel for these families was finding themselves excluded from many
hotels, health resorts, and sea baths simply because they were 'Jewish'. For a
family that came from Italy, even if as a result of the racial laws, the idea that free
America allowed these social offences to those people who had obtained entry
into the United States precisely because of the democratic principles of equality,
seemed a great contradiction, and was a source of bitterness. They did not know
that, in the United States, a distinction is made between political and social
equality. Political equality exists very widely, except for blacks in certain southern states, but social equality is less strong than in Europe. Classes, races, sections
of the population feel separated and closed off from each other much more than in
certain areas of Europe, such as France and Tuscany. Thus, it came about that the
Italian Jews also felt separated from the Italo-Americans, with whom they did not
merge at all (of course, with some exceptions). A couple of Italian Jews, now
American citizens, clearly stated to me that Italo-Americans are 'anti-Semites'.
What is the truth's rate of this statement? If you were to ask the spokesmen of the
Italo-Americans, you would hear them unanimously answer that it is false. Even
if you were to ask Italo-Americans one by one, I think almost all of them would
answer in the same way: they are all American citizens and they make no
distinction of race or religion. On the contrary, someone will mention the
common interest that Italo-Americans and Jews have in fighting religious or
racial discrimination, and the campaign of Italian journalism goes in this sense.
These are two minorities – they will say – which both have a common interest in
eliminating differences in treatment due to colour, race or religion. All this is true,

in a way, but the sentiment of those two Jews who openly spoke to me of an anti-Semitic attitude by Italo-Americans also expresses some truth. First of all, the Italo-American feels that the Italian Jew is a foreigner because the Jew is an Italian different from him, i.e., he is a social product of modern Italy, which was formed and developed when the Italo-American was no longer in Italy. That same impossibility that ninety-nine out of a hundred Italo-Americans have of reading a modern Italian novel (that is, something later than Carolina Invernizio and Matilde Serao,[96]) whose language they do not understand, whose feelings they do not share, whose customs they do not appreciate, makes the Italo-American look at the new Italian generations with a mixed feeling of distrust and superiority. Now, the Italian Jews who came here all arrived from modern, middle-class Italy, and not from the Italy that the emigrants have left. Their culture, their language, and their customs, placed them outside the common Italo-American type, generally of peasant and southern origin. Secondly, Italo-Americans are Catholic, by family heritage and by force of social bond. Abandoned by the Italian government and exploited by their American masters, they found social refuge in the churches, comfort in the churches, social communion in the organisations that grew up around the churches. What were these Italians who came from Italy if not Catholics? For Italo-Americans, the existence of Italian Jews was, in most cases, something unheard of, because the South of Italy has known no Jews for centuries. The Jew is for them, at best, a kind of infidel. If the Italo-American is ignorant and superstitious, the Jew may be Christ's torturer. Third, in some areas of economic life, Italo-Americans are at odds with Jews (not Italians, but Jews). For example, in the wine industry, it can be said that, with a few exceptions, Italo-Americans have been good creators of wine industries, but the wine trade made by Italians is all, in general, in the hands of Jewish shopkeepers. Now, it is well known that this kind of trade relationships inevitably generates friction and resentment. He who produces the wine, and has an almost paternal relationship with the product, one who sees something born from his hands that has an individual character, feels a certain jealousy at the profits made by another, who is not able or less able than he is in producing the same thing, but who knows how to trade it, to make it popular and profitable. Finally, during the unfortunate period in which Italy found itself at war with America, the Jews from here regarded (with due exceptions) the Italians as allies of the Germans in the struggle against them. I saw, to make just one observation, the disappearance of advertisements for New York Jewish firms from Italian-language periodicals. It's too natural and human. We would be in the world of angels if it were otherwise. But this explains why many Italo-Americans reciprocated with a certain antipathy toward the Jews (and therefore also those who came from Italy, and who in many respects were Italians like them, or even more so). In short, if I ever were to say that Italo-Americans were anti-Semitic, I would not deny that some of them were,

---

[96]Carolina Invernizio (1851–1916) and Matilde Serao (1856–1927) were two of the best-known women writers of the time.

and that the unfortunate conditions of recent years have produced in Italo-Americans a certain feeling of distrust and suspicion of Jews, far greater than ever before in Italy.

Nell'isola di Long Island, c'è una cittadina che si chiama R. completamente abitata da una popolazione ebraica proveniente dall'Europa orientale. In un appartamento d'una casa a due piani trovò alloggio al piano superiore una famiglia di Ebrei italiani rifugiatisi qui nel 1939. Questi non eran convertiti, ma non frequentavano la Sinagoga. Si consideravano 'liberi pensatori' o qualche cosa di simile. Ma non avevano nessuna difficoltà a seguire i costumi sociali italiani. Il giorno di Natale avvenne che la bambina si ricordò di quel che facevano lei e le sue compagne in Italia e comprato un abetino lo riempì di candelette, le accese e lo espose sul terrazzino. La padrona della casa, israelita di qui, e stretta osservante, salì su tutte le furie: "Non sapevo che foste 'gentili'. Non vi avrei affittato l'appartamento ....". La mamma della bambina rispose per le rime. Fu un brutto momento. Quest'aneddoto, che m'è stato raccontato sorridendo da un israelita italiano di vecchia conoscenza, dipinge la situazione di molte famiglie ebraiche italiane, quando vennero in relazione con la popolazione ebraica di qui. Si sentiron spaesate. Un'altra meraviglia di queste famiglie fu il fatto di trovarsi escluse da molti alberghi, luoghi di cura, bagni di mare soltanto per il fatto che esse erano 'ebree'. Per una famiglia che veniva dall'Italia, sia pure per causa delle leggi razziali, l'idea che la libera America permettesse queste offese sociali a quelle persone che avevano ottenuto d'entrare negli Stati Uniti proprio per via dei principi democratici di eguaglianza, pareva una grossa contraddizione, ed era sorgente di amarezza. Non sapevano che negli Stati Uniti si fa distinzione fra eguaglianza politica e eguaglianza sociale. L'eguaglianza politica esiste molto largamente, salvo che in certi Stati del sud per i negri, ma l'eguaglianza sociale è meno forte che in Europa. Classi, razze, sezioni della popolazione si sentono separate e si chiudono una dall'altra molto più che non in certi paesi d'Europa, come la Francia e la Toscana. Così avvenne che gli Ebrei italiani si sentiron separati anche dagli Italo-Americani, con i quali non si son punto fusi (salvo sempre eccezioni personali). Un paio di ebrei italiani, diventati oramai cittadini americani, mi hanno nettamente affermato che gli Italo-Americani sono 'antisemiti'. Che cosa c'è di vero in questa affermazione? Se interrogaste i portavoce degli Italo-Americani li sentireste rispondere unanimi che si tratta d'una bubbola. Anche se interrogaste uno per uno gli Italo-Americani, credo che quasi tutti risponderebbero nello stesso modo: che son tutti cittadini americani e che non fanno distinzione di razza o di religione. Anzi, qualcuno accennerà al comune interesse che Italo-Americani ed Ebrei hanno nel combattere la 'discriminazione' religiosa o razziale, ed a campagna del giornalismo italiano in questo senso. Si tratta di due minoranze – diranno – che hanno ambedue interesse comune a far scomparire le differenze di trattamento dovute a colore, razze o religione. Tutto questo è vero, in un certo senso, ma il sentimento di quei due Ebrei che mi hanno apertamente parlato di un 'antisemitismo' degli Italo-Americani esprime pure qualche cosa di vero. Prima di tutto l'Italo-Americano sente straniero l'Ebreo italiano perché questo Ebreo è un italiano diverso da lui, cioè a dire è un prodotto sociale dell'Italia moderna, che si è formata e sviluppata quando l'Italo-Americano non era più in Italia. Quella stessa impossibilità che hanno novantanove su cento degli Italo-Americani di leggere un romanzo moderno italiano (che sia cioè posteriore a Carolina Invernizio e a Matilde Serao), di cui non capiscon la lingua, non condividono i sentimenti, non apprezzano i costumi, fa sì che l'Italo-Americano guardi alle nuove generazioni italiane con un sentimento misto di diffidenza e di superiorità. Ora gli Ebrei italiani che vennero qui, provenivano tutti dall'Italia moderna e borghese, e non dall'Italia che gli emigrati avevano lasciato. La loro cultura, la loro lingua, il loro costume li poneva fuori dall'Italo-Americano comune, in generale di origine paesana e meridionale. Secondariamente, gli Italo-Americani sono cattolici, per eredità familiare e per forza di vincolo sociale. Abbandonati dal Governo italiano e sfruttati dai padroni americani, socialmente trovaron rifugio nelle chiese, conforto nelle chiese, comunione sociale nelle organizzazioni che crebbero intorno alle chiese. Che cos'eran questi Italiani che venivano dall'Italia se non eran cattolici? Per gl'Italo-Americani l'esistenza di Ebrei italiani era, nella maggioranza dei casi, qualche cosa d'inaudito, perché il Mezzogiorno d'Italia non conosce Ebrei da secoli. L'Ebreo è per loro, tutt'al più, una specie d'infedele. Se l'Italo-Americano è ignorante e superstizioso, l'Ebreo sarà magari il torturatore di Cristo. Terzo punto, in alcuni settori della vita economica gli Italo-Americani si trovano in contrasto con gli Ebrei (non Italiani, ma Ebrei). Per esempio nell'industria del vino, si può dire che, salvo poche eccezioni, gli Italo-Americani son stati dei buoni creatori d'industrie vinicole, ma che il commercio del vino fatto dagli italiani sia tutto, in generale, nelle mani di negozianti ebrei. Ora è noto che questo genere di relazioni commerciali genera inevitabilmente delle frizioni e dei rancori. Chi produce, ed ha col prodotto quella relazione quasi paterna di chi vede nascere dalle proprie mani qualche cosa che ha una individualità, prova una certa gelosia ai guadagni che fa un altro, che non è capace o è meno capace di lui nel produrre la stessa cosa, ma sa commerciarla, farla popolare e redditizia. Infine, durante il disgraziato periodo in cui l'Italia si trovò ad essere in conflitto con l'America, gli Ebrei di qui considerarono (sempre salvo le debite eccezioni) gli Italiani come alleati dei Tedeschi nella lotta contro di loro. Io vidi, per fare una sola osservazione, scomparire la pubblicità di ditte ebraiche di New York dai periodici in lingua italiana. È una cosa troppo naturale ed umana. Saremmo nel mondo degli angeli se fosse diversamente. Ma ciò spiega come molti Italo-Americani abbiano ricambiato gli Ebrei (e quindi anche quelli che venivano dall'Italia, e che sotto tanti aspetti eran italiani come loro o più di loro) con una certa antipatia. Insomma, se io non dicessi mai che gli Italo-Americani son stati antisemiti, non negherei nemmeno che alcuni di essi lo sian stati, e che le disgraziate condizioni di questi anni hanno prodotto negli Italo-Americani in generale un certo sentimento di diffidenza e di sospetto per gli Ebrei molto maggiore di quel che mai ci sia stato in Italia.

3.9 Solidarity Chains: Tullio Levi-Civita

Because of the difficult reality facing them, Italians wishing to expatriate immediately turned to the colleagues with whom they were in contact, as well as to international organisations. This is a second similarity between Jewish intellectual emigration from fascist Italy and the Italian historic proletarian emigration: both were supported by chains of spontaneous solidarity, which were created already in the days immediately following the introduction of the racial laws. At the top of these chains were, for the Italian mathematical diaspora, Tullio Levi-Civita, Max Ascoli, and Oswald Veblen.

Levi-Civita was a figure of extraordinary scientific and moral stature. Brought up in a progressive liberal family,[97] educated in secular values, he had obtained fundamental results in the field of absolute differential calculus and was the main spokesman of the theory of general relativity in Italy, both from the point of view of diffusion and from that of active research, with studies on the integration of equations in the static case, relativistic optics, and relativistic celestial mechanics.

Socialist, pacifist, internationalist, he fought to mend relations with German speaking mathematicians interrupted by the First World War. In the twenties and thirties, he was a true ambassador of Italian mathematics in the world and contributed enormously to the spread of Italian culture in dozens of countries (Fig. 3.10).

In 1936 and 1937, he made two long tours of the United States and South America, receiving celebrity-like receptions, and leaving an indelible mark with his teaching.[98] In Peru, Argentina, Brazil and Bolivia, as well as in Poland, Russia, Turkey, Japan, an 'unforgettable recollection' of his trips was preserved.[99] The network of correspondence he had established with colleagues, scientific societies, academies, institutions, research centres and universities spanned the whole world. His foreign students and colleagues, many of whom had spent a study sojourn in Rome to work under his supervision, had remained in contact with him (Fig. 3.11).

Levi-Civita's international profile clearly explains why he was the most important and effective node of the solidarity network created in favour of mathematicians

---

[97] Davi and Simone (2015).

[98] Levi-Civita's correspondence with the IAS governance (in ANL-*Levi-Civita*) is particularly interesting from the point of view of the history of international relations. Veblen invited Levi-Civita to the IAS in April 1936, having learned that he was in the United States to attend the feast at Harvard (O. Veblen to T. Levi-Civita, 2.4.1936). Levi-Civita accepted the invitation but was worried about his English (T. Levi-Civita to O. Veblen, 16.4.1936). The invitation was renewed the following May, first by A. Flexner, who recalled how important his visit to Rome had been and how useful Levi-Civita's advice had been to him in organising the IAS, and then by H. Weyl who assured Levi-Civita that everyone was waiting for him eagerly (A. Flexner to Levi-Civita, 4.5.1936; H. Weyl to Levi-Civita, 7.5.1936). At this point, Levi-Civita formally accepted the invitation and congratulated Flexner on the success of his efforts to organise the IAS (T. Levi-Civita to A. Flexner, 18.5.1936).

[99] ANL-*Levi-Civita*: G. García to T. Levi-Civita, Lima 27.8.1938: *inolvidable recuerdo*. See also ANL-*Levi-Civita*: G. García to T. Levi-Civita, Lima 2.3.1939; R.A. Marotta [President of the Instituto Argentino de Cultura Italica] to T. Levi-Civita, Buenos Aires 19.12.1938.

Fig. 3.10 From left to right: Leonida Tonelli, Rosetta Cassin Fano, Gino Fano, Annetta Ghiron Fubini, Guido Fubini, Tullio Levi-Civita, Libera Trevisani Levi-Civita and Francesco Severi, border station with Russia, September 1925, Ceccherini Silberstein Family Archive

Fig. 3.11 From left to right: Tullio Levi-Civita and the Grausteins, Ceccherini Silberstein Family Archive

displaced from racist Italy. In 1938, Levi-Civita was at the apex of his fame. For him, emigration would be a sensible and sustainable choice: he had been dismissed with a good pension, having many years of seniority of service, and had better chances because his prestige would assure him a stable position in any university or research centre in the world. From the point of view of family ties, having no children or parents alive, he needed to take care only of his wife Libera Trevisani, who would follow him everywhere, as she had always done for scientific trips. Quite astonishingly, however, Levi-Civita barely examined the offers made to him by the South America and from the moment of dismissal to that of his death (29 December 1941), devoted himself entirely to helping those who intended to leave.[100]

After the announcement of the racial laws, everyone (not only mathematicians, but also several physicists and engineers) turned to him for advice on how and where to find a position abroad. His correspondence was dominated by requests for help from Italian and foreign scholars, such as Hilda Geiringer and Myron Mathisson,[101] requests which Levi-Civita resolved mainly alone with the sporadic contribution of Castelnuovo and Volterra. Levi-Civita monitored the activity of international organisations in support of refugees and the Rockefeller and Guggenheim Foundations, inquiring about competitions for mathematics chairs announced around the world, soliciting individuals and institutions, and writing testimonials, letters of introduction and recommendations.[102] Above all, he mobilised the network of international relations that he had built up in 40 years of scientific activity (Fig. 3.12).

It was a network of truly global dimensions, photographed by a very interesting document kept in his archive. Less than a week after the promulgation of the racial laws, Levi-Civita had to provide E.O. Lovett with a list of foreigners to whom the excerpts of the lectures he held at the Rice Institute during his last visit should be sent (D3.9.1). He named 85 scholars or institutes, 43 of whom would help him in accommodating fleeing colleagues. Among these were the Argentinian, Peruvian, and Brazilian mathematicians whom he had met during his trip to Latin America in 1937. From that stay, the memories of which were for Levi-Civita 'among the most precious and joyful of his entire life',[103] a web of scientific and friendly

---

[100] Retired to a sort of self-imposed house arrest, he underwent a strong physical decline that led him to die in less than 3 years.

[101] See D2.2.2, VOLTERRA.5, and M. Mathisson to T. Levi-Civita, Cambridge 25.8.1939, in Nastasi and Tazzioli (2003), pp. 184–185. Myron Mathisson, an expert in mathematical physics, had enjoyed a correspondence with Levi-Civita since 1932. In the autumn semester of 1937, Levi-Civita invited him to give a series of seminars in Rome about Hadamard's problem relating to the diffusion of waves. In August 1939 he left Warsaw for Cambridge. Levi-Civita, J. Hadamard, and Volterra wrote the reference letters that would allow Mathisson to obtain a chair of Theoretical Physics at the University of Jerusalem in summer of 1939.

[102] Levi-Civita had been advisor of the RKF since 1923 and occasional advisor of the John Simon Guggenheim Memorial Foundation. See, e.g., in ANL-*Levi-Civita*: John Simon Guggenheim Memorial Foundation to T. Levi-Civita, New York 31.7.1939.

[103] ANL-*Levi-Civita*: T. Levi-Civita to G. García, [July 1939]: *fra i più preziosi e lieti della mia vita*. See also ANL-*Levi-Civita*: T. Levi-Civita to R.A. Marotta, 7.1.1939 and G. García to T. Levi-Civita, 7.2.1940.

Fig. 3.12 Solidarity chains that Levi-Civita mobilized to help Italian mathematicians, courtesy of Elena Scalambro

contributions was woven. The election as *academico asociado* of the *Academia de Ciencias Exactas, Fisicas y Naturales* in Lima, 'as a sincere tribute of sympathy to his work of intellectual rapprochement'[104] between Italy and Peru, the exchange of publications, the request to Levi-Civita to send his own and his students' works for South American periodicals all testified to strong relationships that led to concrete offers of help in the moment of persecution. In the imminence of the racial laws, J. Rey Pastor, C. Pla, G. García, A. Rosenblatt, I.M. Azevedo do Amaral not only proved willing to welcome Levi-Civita, but also tried to find positions for Fubini, Terracini, Jacchia, Ascoli, Levi, and Edoardo Volterra (D3.9.2, D3.9.3, D3.9.4).

Documents

D3.9.1 T. Levi-Civita to E.O. Lovett, 8.9.1938

ANL-*Levi-Civita*, fol. 1r
Dear President Lovett,[105] I have been once more delighted by, and am deeply obliged to you, for your extremely kind letter of August 10. Gratefully accepting your cordial suggestion, I am enclosing a list of 85 foreign (I mean non-Italian) addresses to which I beg the October (1938) Pamphlet to be send through the

---

[104]ANL-*Levi-Civita*: G. García to T. Levi-Civita, 7.7.1939: *como un sincero tributo de simpatía a su labor de acercamiento intelectual entre nuestros país.*

[105]Edgar Odell Lovett (1871–1957), mathematics professor and President of Rice Institute (1912–1946).

gracious care of the Rice Institute, the President, and the other distinguished scientists of the Institute itself being of course implicitly included among the recipients of the booklet.[106] Furthermore, I shall be very glad to receive in Rome some 65 copies (the total of my rather indiscreet demands being accordingly 150) for friends and eventually Italian scientific Institutions. It seems however very likely that in the meantime I shall be expelled from these last, as a Jew. But it was not my intention to entertain you also with racial news. I wish only to thank you sincerely for your so benevolent reply to my definitive approval of the material to be printed. With the kindest regards, also on behalf of my wife, I am Very faithfully yours [T. Levi-Civita]

1. C. Raymond Adams – Brown University – Providence R.I.
2. Raimond Clare Archibald – Brown University – Providence, R.I.
3. Harry Bateman – Cal. Inst. of Technology – Pasadena, Cal.
4. G.D. Birkhoff – Cambridge Mass.
5. V. Bjerkness – Institute for Theor. Astrophysics – Blindern, V. Aker (Oslo), Norway
6. W. Blaschke – Hamburg
7. H. Bohr – Charlottenlund (Denmark)
8. É. Borel – Paris
9. M. Born – Edinburgh (England)
10. L.E.J. Brouwer – Laren N.H. (Holland)
11. A. Buhl – Toulouse (France)
12. C. Carathéodory – München (Germany)
13. É. Cartan – Paris (France)
14. J. Chazy – Paris (France)
15. A.W. Conway – Dublin (Ireland)
16. J.L. Coolidge – Cambridge Mass.
17. Th. De Donder – Brussels (Belgium)
18. A.S. Eddington – Cambridge (England)
19. A. Einstein – Princeton N.J.
20. L.P. Eisenhart – Princeton, N.J.
21. F. Engel – Giessen (Germany)
22. H. Favre – Zürich (Switzerland)
23. H. Fehr – Geneva (Switzerland)
24. S. Finikoff – Moscow (U.S.S.R.)
25. G. García – Lima (Peru)
26. S. Goldstein – Cambridge (England)
27. W.C. Graustein – Cambridge Mass.
28. J. Hadamard – Paris (France)
29. Fr. Humbert – Montpellier (France)
30. G. Julia – Versailles (Seine et Oise France)
31. B. Kagan – Moscow (U.S.S.R.)
32. N. Kryloff – Kiev – (Ukraine U.S.S.R.)

(continued)

[106] Levi-Civita had visited Houston on his trip to America in 1936. He had given three lectures at the Rice Institute, published in 1938: T. Levi-Civita, *Three lectures on mathematical subjects, delivered at the Rice Institute in September, 1936*, The Rice Institute Pamphlet, Vol. XXV, October, 1938, No. 4. A nice account of Levi-Civita's trip to the USA is kept in Ceccherini Silberstein Family Archive: [Travel diary of Libera Trevisani Levi-Civita], 27 August–19 December 1936.

33. J. Larmor – St. John's College, Cambridge (England)
34. S. Lefschetz – Princeton N.J.
35. C. de Losada y Puga – Lima (Peru)
36. A.E.H. Love – Oxford (England)
37. N. Lusin – Moscow (U.S.S.R.)
38. A.J. McConnell – Trinity College Dublin (Ireland)
39. N. Moïsseiev – Moscow (U.S.S.R.)
40. Marston Morse – Fine Hall – Princeton N.J.
41. F.D. Murnaghan – The Johns Hopkins University – Baltimore Md.
42. O. Neugebauer – Copenhagen (Denmark)
43. J. von Neumann – Fine Hall – Princeton N.J.
44. N.E. Nörlund – Copenhagen (Denmark)
45. C.W. Oseen – Stockholm (Sweden)
46. J. Pérès – Paris (France)
47. É. Picard – Paris (France)
48. M. Planck – Berlin (Germany)
49. L. Prandtl – Göttingen (Germany)
50. J. Rey Pastor – Buenos Aires (Argentina)
51. Dean R.G.D. Richardson – Brown University – Providence R.I.
52. A. Rosenblatt – Lima (Peru)
53. J.A. Schouten – Delft (Holland)
54. J.A. Shohat – University of Pennsylvania, The College – Philadelphia, Pa
55. A. Speiser – Zürich (Switzerland)
56. W. Stepanow – Moscow (U.S.S.R.)
57. C. Störmer – The Institute of theor. Astrophysics – Blindern, V. Aker, Oslo (Norway)
58. D.J. Struik – Belmont, Mass.
59. J.L. Synge – Dep. of Applied Math. – University of Toronto (Canada)
60. D.J. Tamarkin – Brown University – Providence R.I.
61. E. Terradas – Observatorio de la Universidad Nacional de la Plata (Argentina)
62. T.Y. Thomas – Princeton, N.J.
63. O. Veblen – Princeton, N.J.
64. H. Villat – Paris (France)
65. G.N. Watson – University of Birmingham (England)
66. R. Wavre – Geneva (Switzerland)
67. H. Weyl – IAS – Princeton, N.J.
68. E.T. Whittaker – Edinburgh (England)
69. A. Wintner – Johns Hopkins University – Baltimore Md.
70. K.P. Williams – Indiana University – Bloomington, Indiana
71. Th. V. Kármán – California Institute of Technology – Pasadena, Cal.
72. J. Kampé de Fériet – Université de Lille (France)
73. Fr. Lasareff – Moscow (U.S.S.R.)
74. L.N. Srettensky – Moscow (U.S.S.R.)
75. Library of the Kon. Ak. Van Wetenschappen – Amsterdam (Holland)
76. Library of the Indian Academy of Science – Bangalore (India)
77. Bibliothèque de l'Académie Royale des Sciences, des Lettres et des Beaux Arts de Belgique – Brussels (Belgium)
78. Bibliothek der Leop.-Carolinischen Akademie der Naturforsher – Halle a.S (Germany)
79. Library of the Royal Society – Burlington House – London (England)
80. Library of the National Academy of Sciences – Washington D.C.
81. Seminario Matematico de la Universidad – Buenos Aires (Argentina)
82. Mathematical Seminar – Rua Tres Rios – São Paulo (Brazìl)
83. To the Editor of Nature – St. Martin's Street – London W.C.2. (England)

(continued)

84. Zentralblatt für Mathematik – Copenhagen (Denmark)
85. Library of the Mathematical Institute – Fine Hall – Princeton N.

D3.9.2 J. Rey Pastor to T. Levi-Civita, [31.12.1938]

ANL-*Levi-Civita*, p.c.

Dear friend: I send you and your wife a cordial New Year's greeting, wishing you resignation in the face of current difficulties. If you dare to sail again, it might be possible to obtain a new invitation from the Argentine Universities.[107] I hope that this time the behaviour is more correct, without repetition of last time's ugly events (which luckily, I managed to mitigate at the last minute, thanks to the discovery of private information). If in principle you are willing to work this summer, I will speak to some influential people. Terradas will help me.[108] I try to help the people you recommend, although the situation is difficult due to the large numbers. I believe it is a professional duty and even an act of human solidarity to do all that I can. I have managed to bring Mieli, who is in a difficult situation, to Paris.[109] Very cordial greetings and best wishes for the new year, J. Rey Pastor[110]

Please compare Mr. Vignaux's note on polygenous functions presented to the Academy with my report *Funciones complejas de variable binaria* 1936.[111]

---

[107] Levi-Civita had travelled across Latin America in 1937. A nice account of the trip is kept in Ceccherini Silberstein Family Archive: [Travel diary of Libera Trevisani Levi-Civita], Lima 6–18 August 1937.

[108] Esteban Terradas i Illa (1883–1950) professor of Analysis at the University of Madrid. He had left Spain at the end of October 1936 and moved to Argentina, where he remained till 1941.

[109] Aldo Mieli (1879–1950), one of the most outstanding Italian historians of science, founder, and first perpetual secretary of the Académie internationale d'histoire des sciences, editor in chief of the journal *Archeion*. With the fascist regime in full force, Mieli was barely tolerated for being non-conforming, his socialist and pacifist militance and his sexual orientation. After a decade spent in Paris (1928–1939), he moved to Argentina in February 1939, when he took the direction of Instituto de Historia y Filosofía de la Ciencia founded by the Universidad Nacional del Litoral in Santa Fe.

[110] Julio Rey Pastor (1888–1962) played a significant role in aiding Italian refugees looking for a position in Latin America. Formerly professor of Geometry at the University of Madrid, he had moved to Argentina in 1927. In Buenos Aires he had founded an influential mathematical Seminar (1928) and had invited important foreign mathematicians to give short courses, including Federigo Enriques (1925), Francesco Severi (1930), and Tullio Levi-Civita (1937).

[111] J.C. Vignaux, *Sulle funzioni poligene di una variabile bicomplessa duale*, Atti della Reale Accademia Nazionale dei Lincei. Rendiconti. Classe di Scienze fisiche, matematiche e naturali, 6, XXVII, 4, 1938, pp. 641–645; J. Rey Pastor, *Funciones complejas de variable binaria*, Boletín del Seminario Matemático, IV, 19, 1936, pp. 101–116.

Querido amigo: Le envío a Ud. y su esposa un cordial saludo de año nuevo deseándoles resignación ante las dificultades actuales. Si Ud. se animara a navegar nuevamente sería quizás posible lograr una nueva invitación de las universidades argentinas. Espero que esta vez el comportamiento sea más correcto, no repitiéndose la fea acción de la vez pasada (que por suerte logré atenuar a última hora, gracias a la indiscreción de averiguar cosas privadas). Conteste si en principio está dispuesto, para trabajar este verano ante los personajes influyentes. Terradas me ayudará. Procuro ayudar a sus recomendados, aunque la situación es difícil por la gran afluencia. Creo un deber de solidaridad profesional y hasta humana hacer cuanto puedo. He logrado traer a Mieli que está en difícil situación en Paris. Saludos muy cordiales y feliz año nuevo les desea J. Rey Pastor

Ruégole compare la nota del Sr. Vignaux presentada a la Academia sobre funciones polígenas con mi memoria Funciones complejas de variable binaria 1936.

D3.9.3 R. Treves to T. Levi-Civita, Montevideo 16.1.1939

ANL-*Levi-Civita*, fol. 1r-v
Illustrious Professor, Please excuse me if, due to the worries and emotions of these first months of American life, I have delayed so much in thanking you for your cordial welcome and the good, paternal advice you gave me before leaving Italy. Prof. Rey Pastor, to whom you introduced me, was very kind to me. I found in him an authoritative and confident adviser. Having the opportunity to write to him, if you would like to tell him all the gratitude I feel to him, I will be very, very grateful. Last November, I gave some lectures in Montevideo, I will give some others in Buenos Aires when I go next April. I hope to be able to start again the career path already undertaken in Italy. If you see Prof. Sandro Levi, please greet him on my behalf and tell him that I always think of him with great affection.[112] To you, a heartfelt thank and a devoted greeting from your Renato Treves[113]

---

[112] Alessandro Levi (1881–1953), philosopher of law, was Levi-Civita's cousin. Dismissed from the University of Parma in 1938, he fled to Switzerland in autumn of 1943, and he held courses in the Camps for refugee students in Geneva. See Levi A. (1957), pp. 391–418.

[113] Renato Samuele Treves (1907–1992), philosopher and sociologist from Turin, taught Philosophy of Law at the University of Urbino from 1935. In 1938 he emigrated to Argentina, where he taught General Legal Theory, Philosophy of Law and Sociology at the University of Tucumán. As Treves recalled (Treves 1985, p. 91): *"Mi imbarcai a Napoli verso la fine dell'ottobre del 1938 e, dopo diciotto giorni di navigazione (allora non si parlava di viaggi aerei), sbarcai a Montevideo. Per un complesso di circostanze fortunate, dopo aver tenuto due conferenze in quella città ed aver partecipato successivamente ad un convegno dell'Associazione argentina su indicata, con l'appoggio del Presidente della medesima, Carlos Cossio, che era professore a La Plata, ma*

Illustre Professore, La prego scusarmi se, per le preoccupazioni e le emozioni di questi primi mesi di vita americana, ho ritardato tanto nel ringraziarla della cordiale sua accoglienza e dei buoni, paterni consigli che ha voluto darmi prima di partire dall'Italia. Il prof. Rey Pastor, a cui Ella ha voluto presentarmi, fu con me gentilissimo. Ho trovato in lui un consigliere autorevole e sicuro. Avendo occasione di scrivergli, se Ella vorrà dirgli tutta la gratitudine che provo per lui, le sarò molto, molto riconoscente. Nel novembre scorso, ho tenuto alcune conferenze a Montevideo, altre ne terrò a Buenos Aires quando mi vi recherò nel prossimo aprile. Spero così poter ricominciare la carriera già percorsa in Italia. Se vede il prof. Sandro Levi, la prego salutarlo da parte mia e dirgli che lo penso sempre con molto affetto. A Lei, un grazie di cuore e un saluto devoto dal suo Renato Treves

D3.9.4 C. Pla [Universidad Nacional del Litoral] to T. Levi-Civita, Rosario 14.4.1939

ANL-*Levi-Civita*, fol. 1r

Distinguished Professor, I have delayed my response to your kind letter of 24/1/39, because the board of Directors of the Faculty was in recess, and we also lacked a budget. Even though, at this time and for various reasons, the Faculty has not definitively assigned the available resources, at its meeting on the 10th of this month, the board resolved to approve the report of the Commission of professors that studied the organisation of the Institute of Stability sanctioning the ordinance that accompanied it. At the same time, they authorised me to recruit Eng. Enrico Volterra, proposed by you, to take over directorship of the Institute.[114] The remuneration will be at most eight hundred pesos – with the possibility of a reduction in the current year, given the global reduction made in the budget that will affect various items. However, in any case, I will try not to give less than the above amount. I appreciate your generous advice regarding the organisation of the Institute of Mathematics – which will be taken over by Prof. Beppo Levi, who you indicated to me – and I was pleased to meet your suggestion regarding Stability.[115] I hope that the presence of these professors will provide us with

---

molto influente anche a Tucumán, sua città natale, ottenni un contratto annuale per insegnare Introduzione al diritto in quella Università, insegnamento che iniziai nel maggio del 1939."

[114] Enrico Volterra (1905–1973), son of Vito Volterra, had been Levi-Civita's assistant. In February 1939 he had left Italy for Cambridge. When Italy entered the war, Enrico was imprisoned in an internment camp on the Isle of Man, where he learned of his father's death. Released through the efforts of well-placed British scientists, notably Archibald V. Hill, Enrico spent the rest of the war years in England, earning a PhD in engineering and working on plastic and rubber materials under G. I. Taylor and for the British Admiralty. See Sect. 5.2.

[115] See Chap. 12.

much satisfaction, and since I feel you are linked to the work I am doing, I dare to ask you to send me an original work of yours, to be published by our Faculty. We would feel honoured by the collaboration of such an eminent professor. I would also be grateful if you could make any suggestions that you deem convenient, for the best organisation of the two Institutes, in the full conviction that your advice will be carefully listened to by the undersigned. Cultivating the hope of counting on your collaboration and of having you at some point among us, even if temporarily, believe me, distinguished professor, your admirer and friend, Cortés Pla

Distinguido profesor: He demorado la respuesta a su amable carta del 24/1/39, por cuanto el Consejo Directivo de la Facultad se encontraba en receso y además carecíamos de Presupuesto. Aún cuando en este momento, por razones diversas, la Facultad no tiene definitivamente asignado los recursos de que dispondrá, el Consejo Directivo de la Facultad en su reunión del 10 corriente, resolvió aprobar el informe de la Comisión de profesores que estudió la organización del Instituto de Estabilidad sancionando la Ordenanza que le acompaño. Al mismo tiempo resolvió autorizarme para contratar al Ing. Enrico Volterra, propuesto por Ud., para ocupar la Dirección del Instituto. La retribución será a lo sumo de Ochocientos pesos – con posibilidad de disminución en el corriente año – dada la rebaja hecha en el Presupuesto en forma global y que incidirá sobre las diversas partidas. Sin embargo, en toda forma, trataré de que no sea inferior a la suma antedicha. Agradezco su generoso concepto acerca la organización del Instituto de Matemáticas – para cuja Dirección traeremos al prof. Beppo Levi que Ud. me indicó – y me agradaría conocerla referente al de Estabilidad. Espero que la presencia de esos señores Profesores, ha de proporcionarnos múltiples satisfacciones, y como lo siento a Ud. vinculado a la obra que estoy realizando, me atrevo a pedirle tuviera a bien, enviarme un trabajo suyo original, para ser publicado por nuestra Facultad, que se sentirá honrada con la colaboración de tan eminente profesor. Le agradecería también me hiciera las sugestiones que estimare conveniente, para la mejor organización de los Institutos, en el pleno convencimiento que su palabra será escuchada por el suscripto con toda atención. Abrigando la esperanza de contar con su colaboración y de tenerlo alguna vez entre nosotros, aun cuando fuere temporariamente; créame, distinguido profesor su admirador y amigo Cortés Pla

3.10 Solidarity Chains: Max Ascoli

On the other side of the Atlantic, the 'patriarchs of the Italian Jewish intellectuals eager for the embrace of Uncle Sam'[116] were Max Ascoli and Oswald Veblen. The first, a graduate in Law from Ferrara with a thesis on Philosophy of Law supervised by A. Levi, was with G. Salvemini and the brothers C. and N. Rosselli, one of the main voices of the anti-fascist press from the twenties onwards.[117] Arrested in 1928 as part of an investigation into the group *Giovane Italia*, he saw his academic career compromised in 1930 and began to take his first steps towards a move to the United States. This plan came to fruition in 1931, when Ascoli obtained a scholarship from the RKF. To develop a research project on the crisis of democracy in America, Ascoli travelled across the United States, visiting Yale, Columbia, Chicago, and Washington DC, before finally landing at Harvard where he became better acquainted with influential figures such as F. Frankfurter, Supreme Court justice and adviser to F.D. Roosevelt. At the end of the RKF grant, A. Johnson, Director of The New School for Social Research and an exceptional talent-scout, managed to insert him into the Graduate Faculty of Social and Political Sciences.[118] The arrival at the New School represented a decisive turning point in his intellectual and political biography. Ascoli found himself projected into the centre of a network of scientific relations of extraordinary level and shifted the axis of his interests towards political and economic disciplines. In a few years, thanks to three books—*Intelligence in politics* (1936), *Political and Economic Democracy* (1937) and *Fascism for Whom?* (1938)—he succeeded in carving out a prestigious position within the so-called University in Exile. At the same time, he integrated perfectly into the cultural, journalistic, political, and diplomatic circuits of New York. His marriage with Marion Rosenwald Stern, daughter of the philanthropist J. Rosenwald, assisted his process of American-isation, granting him various contacts with the highest levels of the political-administrative establishment and heads of institutions, including the American Institute of Pacific Relations, the Bureau of Latin America

---

[116]The description refers to Max Ascoli and appears in a letter from P. Milano to M. Ascoli, Zurich 21.11.1938 (in Camurri 2009, p. 60): "All are preparing to leave, and I believe that before long you will become the young patriarch of the Italian Jewish intellectuals eager for the embrace of Uncle Sam". (*Tutti si preparano ad andarsene, e credo che tu diventerai fra non molto il giovane patriarca degli intellettuali giudeo-italiani anelanti l'abbraccio con lo zio Sam*).

[117]Ascoli was not related with the mathematicians Giulio Ascoli and Guido Ascoli.

[118]Alvin Johnson (1874–1971), economist, co-founder, and first Director of The New School (1922–1945), helped to save numerous central European scholars from persecution by the Nazis and brought them to a specially created division of the New School which became known as the "University in Exile". There, among others, he worked with Max Ascoli. For further details, see Camurri (2009), p. 55 and ECADFS Records, *MA*: Stacy May [RKF] to E.B. Murrow [ECADFS], New York 15.8.1933; E.B. Murrow [ECADFS] to Stacy May [RKF], New York 17.8.1933; E.B. Murrow [ECADFS] to Dean C.E. Clark [School of Law, Yale University], New York 17.8.1933. The dossier of Ascoli included recommendations from L. Einaudi, F. Ruffini, and B. Croce and a curriculum vitae in English, dated 3.6.1933, with a list of 11 American scholars and other prominent Italo-Americans who could endorse him.

Relations, the *Comité International d'Aide aux Intellectuels*, and the Council for Foreign Relations.

From October 1934 onwards, Ascoli used these links to come to the aid of intellectuals and scientists, but also of ordinary citizens trying to leave Europe for the United States.[119] As of the autumn of 1938, he articulated his commitment in four areas of action: 1) raising awareness among his American colleagues about the problems faced by the new wave of refugees from Italy (Ascoli explained the racial laws, the meaning of the titles, and intervened personally and with his own means to solve delicate situations, etc.); 2) the creation—unfortunately without success—of a sub-committee under the umbrella of ECADFS, specifically devoted to the rescue of Italian dispossessed scholars (D3.10.1, D3.10.2); 3) the Mazzini Society, the most significant attempt made on American soil to give a unified direction to anti-fascist activities; 4) the support to individual scholars, especially in the area of humanities and social sciences, but also in rescue of physicists, psychologists, etc., often acting as a guarantor (D3.10.3). It was Ascoli, for example, who indicated the names of P. Treves, L. Ginzburg, E. Tagliacozzo and E. Gianturco to W.E. Knickerbocker for the Department of Romance Languages of the City College of New York.[120] The RKF and ECADFS regularly sent Ascoli lists of aspiring exiles and asked for information and opinions.[121] Thus, Ascoli reviewed the dossiers of T. Ascarelli, E. Volterra, E. Colorni, M. Einaudi, F. Liuzzi, L.J. Wollemborg, E. Levi d'Ancona, G. Levi della Vida, A. Pekelis, and many others. Ascoli was mainly interested in scholars of his own area of expertise but, thanks to his friendship with the Fermis and Fubinis, he often intervened also in favour of scientists. In January 1941, for example, he recommended (together with the astronomer H. Shapley) Giorgio de Santillana, a physicist and historian of science, and a former student of Enriques, who had been a visiting lecturer in History of Science at Harvard University since 1936.[122] Ascoli's recommendation, together with those of D. Canfield Fisher and G. Prezzolini, led to the confirmation of the grant for de Santillana, who would remain in the United States until his death. Finally, in 1942, Ascoli sponsored the candidacy of chemist R. Calabresi and, although he was acting outside his field of expertise, his statement and high opinion of the candidate were however considered the 'most helpful to the Committee in considering her case'.[123]

---

[119] In October 1934, Ascoli informed A. Nussbaum (Columbia University, School of Law) that ECADFS had extended its work to former assistants and young scholars who had not yet been teachers. This was the first of countless cases that he would follow. ECADFS Records, *MA*: A. Nussbaum to E.B. Murrow [ECADFS], New York 18.10.1934.

[120] ECADFS Records, *MA*: M. Ascoli to W.E. Knickerbocker, 25.2.1939. To these names the ECADFS added that of Attilio Momigliano (see ECADFS Records, *MA*: B. Drury [ECADFS] to M. Ascoli, New York 27.2.1939).

[121] See e.g. ECADFS Records, *MA*: B. Drury [ECADFS] to M. Ascoli, New York 14.12.1939; B. Drury [ECADFS] to M. Ascoli, New York 19.9.1940 and B. Drury [ECADFS] to M. Ascoli, New York 3.10.1940.

[122] ECADFS Records, *MA*: B. Drury [ECADFS] to M. Ascoli, New York 29.1.1941.

[123] ECADFS Records, *MA*: B. Drury [ECADFS] to M. Ascoli, New York 1.10.1942.

Documents

D3.10.1 J.P. Chamberlain to S. Duggan [ECADFS], New York 30.12.1938

ECADFS Records, *GM*, fol. 1r

Dear Duggan: [...] I had a talk with Professor Ascoli last week. He is concerned over the Italian professors and suggested that he try to form a committee of Italians here who would be sympathetic, both university men and outsiders. We agreed that this committee would have to be very carefully formed, since there are so many *Fascisti* about, and businessmen, in particular, would scarcely dare to help in this situation for fear of consequences to themselves. I think Professor Ascoli is taking the matter up with you, and it seems to me that if he does form such a committee, which I think might be useful as meeting the needs of the Italians in particular, he should have some relationship, even though informal, with you and your committee. Professor Ascoli suggested as members Professor Borgese at Chicago, who is now living at the Beaux Arts Apartments in New York; Professor W.Y. Elliott of Harvard; Professor Max Ascoli of New York; Professor Ferrando of Vassar; and Professor Spingarn, formerly of Columbia.[124] Very sincerely yours, Joseph P. Chamberlain

D3.10.2 J.P. Chamberlain to S. Duggan [ECADFS], New York 18.1.1939

ECADFS Records, *MA*, fol. 1r

Dear Duggan: Edward Warburg has written me about a Professor Otto Rosenthal, an Austrian-born Italian citizen teaching at Rome.[125] He has lately lost his job in

---

[124]Giuseppe Antonio Borgese (1882–1952) taught German Literature and Aesthetics at the Universities of Turin, Rome, and Milan until 1931 when, due to his opposition of the fascist regime, he was forced to move to the United States. Here he declared himself a political exile and became an American citizen in 1938. When the Italian-American antifascist Mazzini Society was founded in 1939, Borgese joined it. He was professor in the Universities of Chicago and California until the end of World War II. William Yandell Elliott (1896–1979), professor of History and Government at Harvard University (1925–1963), served as an adviser to numerous government agencies. During the war, he was an official of the Office of Production Management and later vice chairman of the War Production Board. In the late 1940s, he was staff Director of the House committee on foreign affairs and foreign aid. In a famous discussion at the Liberal Club in 1926 Elliott had condemned fascism and expressed his doubts as to the permanence of the fascist regime. Guido Ferrando (1883–1969), professor of English Literature at the University of Florence since 1920, he had also served as Vice Director and Headmaster of the British Institute of Florence. In 1932 he emigrated to the United States, where he continued his anti-fascist activity. He became professor of Italian and Philosophy at Vassar College and was appointed chairman of the Department of Italian in 1933. Joel Elias Spingarn (1875–1939), professor of Comparative Literature at Columbia University (1899–1911). Spingarn died a few months later, on July 26, 1939.

[125]Edward Mortimer Morris Warburg (1908–1992) was an American philanthropist and a founder of the Harvard Society for Contemporary Art (1928). Otto Rosenthal (1890–1944), lecturer in

the university and wants, if possible, to come here. I think Mr. Warburg will write you about him. This brings up the question of whether your committee would like to form a sub-committee of Italian professors and persons interested in Italy, such as Professor Ascoli suggested to me. I wrote you about this matter, and suggest that if you feel it would be advisable, you get in touch with Professor Ascoli, and we can make an appointment. Very sincerely yours, Joseph P. Chamberlain

D3.10.3 M. Ascoli to J.P. Chamberlain, New York 21.1.1939

ECADFS Records, *EdV*, fol. 1r
Dear Professor Chamberlain: Since I had the pleasure of seeing you a few weeks ago, I have been out of New York and lost in an ocean of troubles. May I now take the liberty of sending you some memoranda about Italian scholars in distress. You already know about Ascarelli, who is going to come here, with a wife and three children, to do, God knows what.[126] Volterra [Edoardo] is a first-rate Romanist. He will be willing to go anywhere – Puerto Rico, South America, any place where he will have a chance to make a living and go on with his work. A particularly painful case is that of Professor Tagliacozzo,[127] a younger man, and perhaps more of the type in which your committee is interested. He has already got a visa and will land here in a few days, full of good intentions and with ten dollars in his pocket. I have already written to the Catholic committee about the Jewish scholars who happen to be somewhat baptized. Let us hope in the grace of God. With my best regards, Sincerely Max Ascoli

---

German at the University of Rome from 1936 to 1938. Arrested in Florence on 11 January 1944 and deported to Auschwitz, he did not survive the Shoah.

[126] Tullio Ascarelli (1903–1959), professor of Commercial Law at the University of Bologna. Despite the support of Ascoli and the requests to the ECADFS, Ascarelli would not have managed to reach the United States. He would first flee to England, where he obtained a modest position at the London School of Economics and Political Science. Disappointed by the lack of recognition of his merits, he reached France where he obtained a doctorate in commercial law from the University of Paris. In June 1940, he was forced to flee again due to the German occupation. He crossed Spain and Portugal before arriving at the Universidade de São Paulo, where he will remain until 1952. See Ascarelli (2017).

[127] Enzo Tagliacozzo (1909–1999), professor of History, Philosophy, and Political Economy at the lyceum U. Dini in Pisa. Anti-fascist, student, and assistant of G. Salvemini, was in the editorial staff of the journal *Quaderni di Giustizia e Libertà*. He fled to the United States and, thanks to Salvemini, got a position at Harvard University. In collaboration with a circle of Italian political intellectuals forced into exile—Bruno Zevi (1918–2000) and his wife Tullia Calabi (1919–2011), Aldo Garosci (1907–2000), and Renato Poggioli (1907–1963)—Tagliacozzo founded and edited the antifascist political review *Quaderni Italiani*.

3.11 Solidarity Chains: Oswald Veblen

Nephew of the Norwegian-American sociologist T. Veblen, awarded a first A.B. at the University of Iowa and a second at Harvard, in 1900, Veblen received most of his mathematical training at the University of Chicago from an inspiring trio: O. Bolza, H. Maschke, and E.H. Moore (supervisor of his doctoral dissertation).[128] Under their guidance, he laid the basis for the important work he was later to achieve in the fields of foundations of geometry, projective geometry, topology, differential geometry and its application in relativity theory, and atomic physics. Leaving Chicago, Veblen taught mathematics at Princeton University from 1905 to 1932, except for the period of the First World War, when he was appointed to the Office of Ballistics Research at the new Aberdeen Proving Ground in Maryland, and for the academic year 1928–1929 when he taught at Oxford as part of an exchange programme with G.H. Hardy. In 1929, funds were supplied for Fine Hall at Princeton, and Veblen provided most of the ideas that went into its design. He wanted the mathematicians in Fine Hall to be able to 'group themselves for mutual encouragement and support. [It had to be a place where] the young recruit and the old campaigner [could have] those informal and easy contacts that are so important to each of them'.[129] Veblen helped organise the Institute for Advanced Study, resigning from the H.B. Fine professorship to become the first professor at the Institute in 1932. Together with J.W. Alexander, A. Einstein, J. von Neumann, and H. Weyl, he went on to establish Princeton as one of the leading centres in the world for mathematical research.[130]

In 1932, Veblen spent time in Germany and lectured at Göttingen, Berlin, and Hamburg. That experience 'gave him a first-hand glimpse of the approaching turbulence in Germany, and he subsequently worked tirelessly to help place refugees who came to the United States'.[131] Since that date he started an unflagging activity of refugees placing. His effort in aiding displaced European mathematicians represented indeed a break in his mathematical career, as Veblen declared in 1953:

> The fact is that I am the opposite of a communist, namely, an individualist of the old-fashioned American type that grew up in the 19th century. My life has been devoted to my science (Mathematics), to my students and to my institutions of learning. This career has been interrupted by the following episodes:
>
> 1. Voluntary service, after I was 37 years old, in the American Army in the First World War.
> 2. Much activity, against the opposition of American anti-Semites, in finding academic positions for refugees from Nazi Germany and Fascist Italy.
> 3. Service as a civilian, after the age of 61, with the Armed Forces of the United States in the Second World War.

---

[128] On Veblen's biography see Despeaux, Dumbaugh and Lorenat (2022); Mac Lane (1964); Montgomery (1963).

[129] Leitch (2015), p. 489.

[130] Batterson (2007) and Parshall (2022), pp. 171–180.

[131] Zund (1999), pp. 307–308.

4. Efforts to find academic posts for refugees from China and from behind the Iron Curtain in Europe.[132]

For Italian mathematicians who wanted to leave, Veblen was a reference figure, both as a consultant of the ECADFS and The American Friends Service Committee and as an individual colleague sensitive to their plight (D3.11.2). Veblen had been to Italy several times; he had corresponded with Volterra and Levi-Civita since the late nineteenth century and had consulted them about the organisation of mathematical life in Rome before shaping the Institute for Advanced Study. Almost all Italian would-be refugees turned to him: mathematicians (Fubini, Terracini, Segre, Ascoli), astronomers (Bemporad, Jacchia), engineers (Enrico Volterra and Gino Castelnuovo), a full professor in Roman Law (Edoardo Volterra), a lecturer in Embryology (Gina Castelnuovo), many physicists, sometimes directly, sometimes through Levi-Civita.

Veblen always responded to their appeals, without fostering illusions but also without cutting off hope. He collected information on openings all over the world, circulated the requests for help, carefully examined and ranked the applications, before submitting them to universities, research institutions, laboratories, archives, libraries, or schools worldwide. Not infrequently, he also performed mundane and precise tasks, typing out curricula written in illegible handwriting, correcting English, and reviewing the files before proposing a candidate for a position. Often, he and his wife Elizabeth hosted the Italian emigrants and helped them to fit into a country like America with a lifestyle and issues very different from Italy. The sons of Volterra, Castelnuovo, Fano, and Fubini, who had known Veblen as children, ended up viewing him as a sort of Master.

Veblen also intervened publicly in at least two occasions: when he resigned from the *Zentralblatt* editorial board in protest against Levi-Civita's dismissal, and when he refused to participate in the Volta Conference of 1939 in public protest against the exclusion of Jewish scholars (Sects. 2.8 and 3.12). For these gestures, as well as for his scientific contributions, Veblen would be appointed a foreign member of the Lincei Academy in 1947, nominated by Castelnuovo.

Veblen inserted Italians into the solidarity network he had built for refugees from Nazi Germany, circulating their requests for help and involving colleagues from all over the world.[133] Thanks to him, many mathematicians mobilised in favour of Italians and teamed up with agencies to find positions and bring them overseas. International alumni like Zariski, Snyder, Rosenblatt, took advantage of their historic connections to Italy and served as intermediaries for collaboration, recruitment,

---

[132]OVP: Subject File, 1918–1960, box 22, Communist charges against Veblen, 1950–55, 25.5.1953, fol. 1r.

[133]Veblen was, from 1933 onwards through and beyond the war, the nexus of a web of personal connections providing aid to distressed mathematicians. Curiously, he primarily provided this aid from no official position; rather, he simply exploited his knowledge of the community and his extensive personal relationships. This institution-man was, paradoxically, a hyper-individualistic source of succour. See Aubin, Hollings, Kennedy, Kent, and Luciano (2022).

and mentoring. The engagement of these scholars represents a fundamental lesson in academic rescue and highlights the political role mathematicians can play through scientific diplomacy. Nevertheless, their action did not always suffice. On many occasions, Veblen's efforts clashed, especially within the American mathematical world, as was in the cases of Segre and Terracini. However, it was thanks to Veblen that Fubini was invited to the Institute as a temporary member in December 1938 (D3.11.1, D3.11.2).

More research should be needed to understand Veblen's prioritisation schemes of ranking mathematicians from Italy who needed placement and to understand the opposition to his efforts mounted by American anti-Semites. However, his correspondence with the Italian exiles restitutes an intensely human iconography of individual and institutional choices, of scientific and ethic responsibilities in evaluating and recruiting colleagues. Veblen's letters prove that throughout the crisis he remained a gentleman and a gentle man who combined a concern for the welfare of mathematics itself with a sympathetic and humane concern for mathematicians. A scholar who, in aiding the Italian refugees, acted as a statesman, a diplomat, a good person, and a grand man, working not merely on account of simple scientific realisations, but also of emotional involvement and humanitarian feeling.

Documents

D3.11.1 O. Veblen to T. Levi-Civita, [Princeton] 1.12.1938

IAS, *GF*, fol. 1r
Dear Professor Levi-Civita: I have had to wait a long time before answering your letter of October 30, but today Dr. Flexner is sending a formal invitation to Fubini to become a member of our Institute for the second term of the present year.[134] We did not have the funds necessary for offering a really suitable stipend, but are hoping that the invitation will at least be of help to him in getting a visa for a trip to this country. There is a great deal of sympathy here for our scientific colleagues who are having such grave trouble, and various measures are being proposed to help them, but you of course understand the nature of the difficulties. I am planning in a day or two to send to South America some information about displaced Italian mathematicians in the hope that something may be done in that part of the world. Many of your friends have been inquiring about you, and the fact that I had had a letter from you was the only information which I was able to give them. I hope you will not fail to write again from time to time. My wife joins

---

[134]Levi-Civita had presented to Veblen the case of Fubini on October 30, 1938 (FUBINI.1). Abraham Flexner (1866–1959), the Director of the Institute for Advanced Study in Princeton from September 1930 to June 1939, invited Fubini to be a member of the Institute in the second term of the academic year 1938–39 on December 1, 1938 (IAS, *GF*: A. Flexner to G. Fubini, Princeton 1.12.1938). A copy of the formal letter of invitation was sent to Levi-Civita.

me in cordial greetings to Mrs. Levi-Civita and yourself. Yours sincerely Oswald
Veblen

D3.11.2 T. Levi-Civita to O. Veblen, [Rome] 12.12.1938

ANL-Levi-Civita, fol. 1r

Dear Professor Veblen, I am highly indebted to you for the substantial interest and
open-hearty sympathy, manifested by your friendly letter of December 1,[135] and,
over all, for having accepted as soon as possible my warm request concerning
Professor Fubini. I trust that not only he, but the Institute too, will be quite
satisfied. I learn with particular interest your generous initiative, at home and
abroad, to help the displaced Italian mathematicians. My wife and I cordially
thank Mrs. Veblen, you and other kind friends for your special interest about our
persons. I may assure that we are, fortunately, both in good health, though, of
course, troubled by the actual situation with the unpleasant disabilities implied,
not only in cultural, social and economic regards, but also in every day life. Please
to accept our best wishes and the renewed expression of my grateful thoughts.
Sincerely yours [T. Levi-Civita]

3.12 Case Study. An Episode of International Censorship: Levi-Civita, *Zentralblatt* and Tricomi's Announcement of *Mathematical Reviews*[136]

When W. Süss took over the chair of the *Deutsche Mathematiker-Vereinigung* in
October 1937, he had three tasks ahead: to reconcile his former teacher and mentor
L. Bieberbach with the association, to carry out the purge of Jewish members
(*Judenfrage in der DMV*) by proposing new guidelines on their expulsion and on
the management of relations with members who had emigrated abroad, and to
strengthen the position of the society in the publishing market.[137] Regarding the
latter task, Süss's action began in March 1938 when he drew the attention of
Dr. Dames, chief of the Mathematics Section at the Office for Scientific Affairs
(*Amt für Wissenschaft*) to the Jewish influence in publishing. In the crosshairs were
the editorial boards of journals such as the *Mathematische Annalen* and the
Mathematische Zeitschrift, on which the Jews O. Blumenthal and I. Schur sat, the
'yellow series' of Springer, in which the English-speaking domain had an émigré

---

[135] D3.11.1.

[136] See Luciano (2022b).

[137] On the topic, see Remmert (1999); Siegmund-Schultze (1993a and 1993b).

(R. Courant) as series editor, and the international reviews journal *Zentralblatt für Mathematik und ihre Grenzgebiete*, directed by another Jewish émigré, O. Neugebauer. F. Springer himself—Süss venomously pointed out—did not have completely Aryan blood. In addition to the Jewish question, another factor had to be taken into account: political circumstances and the reduction of human and economic resources required cooperation between the two German review journals which had previous always been rivals: the *Jahrbuch über die Fortschritte der Mathematik*, published since 1868 in Berlin by de Gruyter and under the aegis of the Prussian Academy of Sciences, and Springer's *Zentralblatt*, published in Göttingen since 1931. In short, a comprehensive restructuring of the German bibliographic system was urgent and mandatory.

In the autumn of 1938, two events brought about a drastic acceleration to the situation: the expulsion of Levi-Civita from the board of the *Zentralblatt* and the rumours about the creation of a new international review journal in America. This, in brief, is the timeline of events. On 11 October 1938, Neugebauer, receiving in Copenhagen the volume of the *Zentralblatt*, noticed an anomaly: the name Levi-Civita (the only Italian mathematician on the editorial board) had disappeared from the cover. He immediately wrote to Springer and, receiving no reply, addressed Levi-Civita:

> I would be very grateful if you would be so kind as to write to me explaining why suddenly you are no longer among the editors of the *Zentralblatt*. In the last drafts of the previous issue, your name was still listed. In the printed copy, your name was no longer there. Despite my prompt question, I have not yet received a reply from the publisher. That is why I would like to ask you personally to write down in detail for me what really happened.[138]

Levi-Civita, bewildered, replied. He had not received the volume, nor any communication that he had lost, or was about to lose, his position on the editorial board. 'I could explain the matter to myself', he added, 'by relating it to the recent anti-Semitic measures adopted in Italy [...]. In that case, it would be a matter of an immediate extension to Germany of the abovementioned laws, even before the interested persons had any inkling. I don't know enough to say'.[139] On 27 October, Neugebauer received the publisher's reply. With cold cynicism, Springer explained to Neugebauer that the removal of Levi-Civita had come about when, because of the racial laws, he had lost his chair at the university. This 'seemed to be enough to cancel his name without first waiting for a corresponding request from the competent authorities'. Springer took this opportunity to introduce a further request:

---

[138] ANL-*Levi-Civita*: O. Neugebauer to T. Levi-Civita, Kopenhagen-Charlottenlund 23.10.1938: *Le sarei molto grato se fosse così gentile da scrivermi qual è il motivo per cui Lei improvvisamente non è più tra i redattori del Zentralblatt. Nelle bozze dell'ultimo fascicolo il suo nome era ancora riportato. Nell'esemplare a stampa il suo nome non c'era più. Malgrado la mia domanda tempestiva non ho ancora ricevuto risposta dall'editore. Perciò vorrei pregarla personalmente di scrivermi dettagliatamente che cosa sia realmente avvenuto.* In Levi-Civita's archive it is conserved only the Italian translation of the letters exchanged with Neugebauer.

[139] ANL-*Levi-Civita*: T. Levi-Civita to O. Neugebauer, [Rome] 26.10.1938.

Our *Zentralblatt* is an internationally recognised journal that is published by a German publisher and pursues purely scientific objectives. The preservation of its international character is considered recommendable by the bodies involved in the publication of the *Zentralblatt* and which have been particularly concerned with the problem of scientific journals published in Germany. The need for clear preservation of its scientific character is also recognised. Nevertheless, the German mathematicians, the competent bodies and, together with them, the publisher, share the opinion that scientific rigour and international character would not be impaired in any way if the editorial staff, in principle, prevent emigrants from reviewing works by German authors. I believe that you would be doing a service to the cause and to the publisher if you gave your unconditional and binding assent to the future management of the journal.[140]

The request was naturally considered inadmissible by Neugebauer, who immediately resigned from the editorship of the *Zentralblatt*. The printed circular with which he announced the move, dated 15 November 1938, contained an explicit allusion to the Levi-Civita case:

Since one of the editors has been removed from the *Zentralblatt für Mathematik*, without informing himself, me or any other of the editors, and having also been required to take into account, in the distribution of reviews, all points of view except those that are purely objective, I have resigned from the editorial staff of the *Zentralblatt*. I would like to warmly thank all my collaborators for their excellent work over many years, and above all for the understanding with which they adapted to the (not always easy) demands that have had to be made.

As a result of the events, H. Bohr, R. Courant, G.H. Hardy, J.D. Tamarkin, Veblen and a large number of collaborators in all countries resigned from the editorial office (D3.12.1, D3.12.2).[141] Neugebauer's successor was E. Ullrich, an analyst from Giessen who was one of the signatories of the *Bekenntnis der Professoren an den deutschen Universitäten und Hochschulen zu Adolf Hitler und dem nationalsozialistischen Staat* (1933). Ullrich immediately began to contact new potential authors, so as to fill the positions left empty by the resignations. For Italy, Bompiani and Severi agreed to join the editorial board. In Severi's case, this was a cowardly act against Levi-Civita, a colleague to whom he had been linked by a 10-year friendship and connections, first in Padua and then in Rome. At the same time, Süss and the Deutsche Mathematiker-Vereinigung put pressure on de Gruyter and Springer to complete the reorganisation of the German review system, with the merger of the *Jahrbuch* and the *Zentralblatt*. The negotiations were led by H. Geppert, professor in Giessen from 1935 and director of the *Jahrbuch* from 1939. The Reich Ministry of Education entrusted to him the plan for the reorganisation of German review journals and transferred him to the University of Berlin, where he held the chair of mathematics from 1940 to the time when he committed suicide, four days before Germany's surrender.

Springer offered strong resistance to Geppert's and DMV's projects. The publisher had other plans, namely to discuss the situation of the *Zentralblatt* with the

[140] ANL-*Levi-Civita*: F. Springer to O. Neugebauer, Berlin 27.10.1938.

[141] R. Courant to T. Levi-Civita, New York 2.12.1938 and New York 22.12.1938 in Nastasi and Tazzioli (2000), pp. 283–284.

Americans and, if possible, to build a joint venture with the group (Courant, Veblen, etc.) that was creating the *Mathematical Reviews* (D3.12.3).[142] To this end, in April 1939, Springer's chief adviser for mathematics publications, F.K. Schmidt, was sent to New York on a mission that Süss tried to oppose in every way. Schmidt's trip, however, did not have the desired effect. Firstly, Courant's emigration to the United States accelerated the scientific movement that would lead to the birth of the *Mathematical Reviews*. Secondly, thanks to the connections that they enjoyed in ministerial circles, Süss and Geppert now had in their hands the unified *Generalredaktion* of the two journals: *Jahrbuch* and *Zentralblatt*.

As soon as he took office, Geppert immediately began to contact foreign colleagues, tried to convince the French mathematicians to collaborate[143] and, via Severi and Bompiani, asked several young Italian mathematicians to join the team. F.G. Tricomi was among the first to be contacted. It was the spring of 1939.

Meanwhile, the Germans strove to thwart the plan of the *Mathematical Reviews*. The friendly relations between Süss, Geppert, W. Blaschke, H. Hasse, and the Italians led the Deutsche Mathematiker-Vereinigung to seek the support of the UMI to oppose the American initiative. After all, at that time Bompiani was Vice-President of UMI, Severi needed German cooperation for the organisation of the 1939 Volta Conference, and both sought revenge for the non-invitation of Italian geometers as plenary speakers in the ICM to be held in Cambridge Massachusetts. The three issues—*Mathematical Reviews*, the participation of the German delegation in the Volta Conference and the withdrawal of the Italian delegation in the 1940 ICM—thus ended up intertwining and producing a vicious circle of exchanges of favours. Berzolari, President of the UMI, stated this expressly in one of the numerous letters to the Ministry of Foreign Affairs:

> Finally, I would like to add that Germany intends to act in agreement with Italy and that it is interested in attending the Congress [1940 ICM] in order to try to prevent the creation in the U.S. (as already aired) of a periodical of bibliographic information in competition with the *Zentralblatt für Mathematik* which has been printed in Germany for many years and in whose editorial committee Italy is represented by H.E. the Academic of Italy F. Severi and Prof. E. Bompiani. This is also in our interest to ensure a fair presentation of the Italian mathematical contribution.[144]

Geppert thus explicitly asked UMI to report a shortlist of possible reviewers for the *Zentralblatt*, and Bompiani did so. However, the UMI bureau did not have the

---

[142]Reingold (1981), pp. 327–333. On Courant's role in this project, see also Reid (1976), pp. 210–221.

[143]Eckes (2018).

[144]AS-UMI: L. Berzolari (but E. Bompiani) to MFA, [Pavia-Roma] [21].7.1939: *Infine aggiungo che la Germania si propone di agire d'accordo con l'Italia e che ha interesse ad intervenire al Congresso per cercare di evitare che venga creato negli S.U.d'A. (com'è stato già ventilato) un periodico d'informazione bibliografica in concorrenza col* Zentralblatt für Mathematik *che da anni si stampa in Germania e nel cui Comitato di Redazione l'Italia è rappresentata da S.E. l'Accademico d'Italia F. Severi e dal Prof. E. Bompiani. È questo un interesse anche nostro per garantire un'equa presentazione del contributo matematico italiano.*

means to keep the general situation under control. For every Sansone who asked Bompiani what to do, many others responded independently.[145] Some accepted the proposal to join the *Zentralblatt*, others like Tricomi dropped it and accepted the invitation to pass to the 'enemy' *Mathematical Reviews*. The behind the scenes that preceded the choice of reviewers for the *Zentralblatt für Mathematik* and the lack of involvement of some 'big names' reflect the crisis of Italian mathematics in the last phase of the fascist regime.

That Tricomi was an irreducible anti-fascist is well known. The testimonies of Guido and Emma Castelnuovo, Colonnetti and various others about how he helped his colleagues escape deportation during the Nazi occupation of Rome have long since been acquired.[146] What was however unknown was Tricomi's anti-fascist contribution, which preceded his experience of partisan struggle and the months he spent in hiding in Rome as a representative of the Action Party. This was an anti-fascist contribution of cultural character, exercised through publications. In the period 1938–1942, Tricomi published many works containing evident political statements: the text of a lecture held at the Mathematical Seminar of the University and Polytechnic of Turin after the racial laws dedicated to the mastership of Corrado Segre (a Jew) at the Teacher Training School; some articles on ethical-political topics published in the Waldensian periodical *La Luce*, in the Catholic monthly *La Festa*, and some others in the journal *Il Saggiatore*.[147] In particular, Tricomi assumed with enthusiasm the co-editorship of the last journal, founded by G. Einaudi in March 1940, and published in it a report entitled *Una nuova rivista di bibliografia matematica* dedicated to the founding of the *Mathematical Reviews*.[148]

Tricomi did not enjoy cordial relations with UMI. He had not hidden his hostility to the Aryan-isation of the Union, he had criticised the management of the *Bollettino* by Bompiani and the organisation of the first two national UMI conferences (Florence 1937 and Bologna 1940). The coolness was reciprocated. Berzolari attributed to Tricomi some character defects and a deplorable internationalist tendency, which led him to appreciate foreign mathematical production more than Italian contributions. With Bompiani there were old grudges linked to the 1938 Royal Prize for

---

[145] AS-UMI: G. Sansone to E. Bompiani, Monguelfo (Bolzano) 17.8.1939: "O. Neugebauer writes to me inviting me to collaborate on the new bibliographic journal *Mathematical Reviews*; what do you advise me to do?" (*O. Neugebauer mi scrive invitandomi a collaborare alla nuova rivista bibliografica* Mathematical Reviews; *che cosa mi consigli di fare?*) Bompiani evidently advised Sansone to drop the invitation because on the following 20th Sansone wrote again (AS-UMI: G. Sansone to E. Bompiani, Monguelfo (Bolzano) 20.8.1939): "I thank you for your letter. I see it was appropriate for me to write to you about the *Mathematical Reviews*; so, I know how necessary it is to answer that it is impossible for me to cooperate." (*Ti ringrazio della tua lettera. Vedo che è stato opportuno ti abbia scritto in merito alle* Mathematical Reviews; *così so quanto è necessario per rispondere che mi è impossibile collaborare.*) Sansone was the treasurer of UMI.

[146] See, for example, Natalini and Mattaliano (2004), pp. 4–7.

[147] Tricomi (1938–1940, 1940a, 1940b, 1940d, 1942, 1943a, 1943b, 1943c).

[148] Tricomi (1940c).

Mathematics, which had seen them compete as rivals, and which had finally been awarded *ex aequo* to Bompiani and Picone. The Tricomi-Severi relationship fared better. Tricomi had begun his career as assistant to Severi in Padua in 1921–1922 and had followed him to Rome in 1922. He esteemed Severi as a mathematician and this respect was reciprocated. On the other hand, Tricomi considered Levi-Civita a true Master. After the frequent exchanges in Rome, in the early twenties, the two had kept in touch by correspondence and had met on numerous occasions, both in Italy and abroad.

Among the few Italian mathematicians who had continued to maintain a wide network of international partnerships even in times of cultural autarky, Tricomi had immediately learned of Levi-Civita's expulsion from the *Zentralblatt*. Neugebauer, who he met in Denmark in the summer of 1939, had also told Tricomi about the *Mathematical Reviews*. Moreover, as a contributor to the *Jahrbuch* since 1920, for which he had written over a thousand reviews, mainly of Italian works on applied mathematics, Tricomi had been informed of the project to reorganise the German review system. Geppert first, followed by Ullrich, Blaschke and G. Doetsch, had contacted him to join the editorial staff of the *Zentralblatt*. In short, in March 1940 Tricomi had all the elements to understand that dedicating a paper to the *Mathematical Reviews* would have been considered an inappropriate, if not hostile, step.[149] Despite all of this, Tricomi decided to publish the following text in the first issue of *Il Saggiatore*:

A *new bibliographic journal*. A scheme has recently been published, with a sample issue, announcing a new journal of mathematical bibliography, *Mathematical Reviews*, which will be first published in the U.S. early next year under the umbrella of the American Mathematical Society and the Math. Association of America. It is a journal which, like the former *Jahrbuch über die Fortschritte der Mathematik* (from 1868 onwards), *Zentralblatt für Mathematik* (from 1931) and *Zentralblatt für Mechanik* (from 1934), aims to publish short summaries, edited by specialists, of all the mathematical works that appear in the main periodicals throughout the world, as well as reviews of new books etc. In terms of internal organisation and other features, the new journal is closer to the *Zentralblatt* which was founded and directed until last year by one of the two directors of the *Reviews*: the distinguished historian of mathematics, Prof. O. Neugebauer. (The other director is the well-known analyst Prof. J.D. Tamarkin of Brown University in Providence, R.I.). One novelty is the fact that the new journal is accompanied by a service of microphotographic copies of mathematical papers, which will enable all subscribers, but especially those who live far from the major centres of study, to obtain (at a reasonable cost) a faithful copy of any article reviewed in the *Mathematical Reviews*. It is also worth noting that, thanks to the ample financial resources made available by the Carnegie Corporation, the Rockefeller Foundation, etc., the subscription at the new journal is quite cheap, i.e., $13 per year, reduced to $6.50 for members of the two associations mentioned above. For this reason, too, as well as because of the high reputation of the directors and the first collaborators whose names are listed, it is easy to foresee that the *Mathematical Reviews* will soon become

---

[149] See AS-UMI: H. Geppert to E. Bompiani, Giessen 14.1.1939; E. Bompiani to H. Geppert, Rome 15.1.1939; H. Geppert to E. Bompiani, Berlin 3.2.1940; E. Bompiani to H. Geppert, [Rome] 19.2.1940.

widely distributed and will perhaps make the two *Zentralblatt* superfluous, which, in some respects, would not be a great misfortune![150]

That Tricomi was able to publish such a text to print is surprising. That he sent the offprint to UMI and Blaschke is even more so. Tricomi, with his typically edgy style, recalled how the events unfolded in the book *La mia vita di matematico attraverso la cronistoria dei miei lavori*:

> In the first issue of *Il Saggiatore* I announced, as was my duty, the publication – which had just started – of the *Mathematical Reviews*, adding the easy prophecy that this periodical would soon become an indispensable tool for every mathematician. To my surprise, this provoked a fierce rebuke from my colleague W. Blaschke, who accused me of the serious crime of not having passed over in silence a publication edited by a German émigré! In fact, O. Neugebauer, *Mathematical Reviews* first editor, was a German Jew, who had emigrated to the United States after the advent of Nazism in Germany, certainly not by his own free will! I did not reply to such silly letter (to say the least) from Blaschke. After some time, I was warned by friends who had become aware of it, that my telephone was being monitored and that I was sometimes being tailed, "as a result of a complaint by a certain German professor." As was confirmed to me after liberation, it was Blaschke. But fascist Italy was not Hitler's Germany (above all it did not have the same diabolical efficiency) and it all ended there, without further trouble for me. I saw Blaschke again in 1952, at a congress in Salzburg, but I turned a blind eye.[151]

---

[150]Tricomi (1940c): *Nuova rivista bibliografica. È stato di recente pubblicato un prospetto, con alcune pagine di saggio, annunciante una nuova rivista di bibliografia matematica:* Mathematical Reviews, *che comincerà a pubblicarsi negli S.U., dal principio del prossimo anno, sotto gli auspici della American Mathematical Society e della Math. Association of America. Si tratta di una rivista che, analogamente a quel che fanno i già esistenti* Jahrbuch über die Fortschritte der Mathematik *(dal 1868 in poi),* Zentralblatt *(dal 19**) e* Zentralblatt für Mechanik *(dal 193**) si propone di pubblicare dei brevi riassunti, a cura di studiosi specificamente competenti, di tutti i lavori di matematica che compariranno nei principali periodici del mondo intero, nonché recensioni di nuovi libri ecc. Per l'organizzazione interna e altri caratteri, la nuova rivista si avvicina di più al* Zentralblatt *che fu, del resto, fondato e fino all'anno scorso diretto da uno dei due Direttori delle* Reviews: *l'illustre storico della matematica, Prof. O. Neugebauer. (L'altro direttore è il ben noto Analista Prof. J.D. Tamarkin della Brown University in Providence R.I.). Una novità è costituita dal fatto che alla nuova rivista è unito un servizio di copie microfotografiche di pubblicazioni matematiche, che darà modo a tutti gli abbonati, ma specialmente poi a quelli che vivono lontano dai grandi centri di studio, di procurarsi, con una spesa che si assicura modesta, una copia fedele di qualsiasi articolo recensito nelle* Mathematical Reviews. *È inoltre degno di esser fatto rilevare che, grazie ai larghi mezzi finanziari messi a disposizione dalla Carnegie Corporation, dalla Rockefeller Foundation, ecc., la nuova rivista potrà essere data in abbonamento ad un prezzo relativamente assai modesto, e cioè $13 annui, ridotti a soli $6,50 per i membri delle due associazioni ricordate in principio. Anche per questo motivo, oltre che per la chiara fama dei direttori e dei primi collaboratori di cui si conoscono i nomi, è facile prevedere che le* Mathematical Reviews *prenderanno presto larga diffusione e renderanno forse superflui i due* Zentralblatt, *ciò che, sotto certi riguardi, non sarebbe poi una gran disgrazia!*

[151]Tricomi (1967), pp. 64–65: *Nello stesso primo fascicolo del* Saggiatore *annunciai, come era mio dovere, l'inizio – avvenuto giusto allora – della pubblicazione delle* Mathematical Reviews, *aggiungendo la facile profezia che questo periodico sarebbe presto divenuto uno strumento indispensabile ad ogni matematico. Con mia sorpresa, questo mi procurò una fiera rampogna da parte del collega W. Blaschke, che mi accusava del grave reato di non aver passato sotto silenzio una pubblicazione diretta da un emigrato tedesco! Invero O. Neugebauer, il primo direttore di M.R.*

Shortly after the publication on *Il Saggiatore*, Blaschke wrote to Tricomi to express his disappointment (D3.12.5). At the same time, Blaschke informed Geppert, Bompiani, and Severi (D3.12.4). All of them knew that it could not have been a misunderstanding and that Tricomi knew perfectly well the situation of German publishing and the events that had led to the founding of *Mathematical Reviews* (D3.12.6, D3.12.7). The documents do not allow us to say whether and what measures were taken against Tricomi. Bompiani alluded to having informed '*chi di dovere*' (the authority involved), but no traces of this intervention have been found in the archives (D3.12.8, D3.12.9). Undoubtedly, by refusing to correct or attenuate his statements, by refusing to privately apologise to Blaschke and Geppert, Tricomi made himself the protagonist of an act of intellectual courage, placing himself opposite the many who fell in line, kept silent, and turned away.

Documents

D3.12.1 O. Veblen to F. Springer, [Princeton] 5.12.1938. Copy for the Information of prof. Levi-Civita

ANL-*Levi-Civita*, fol. 1r

Dear Doctor Springer: I have your friendly letter of November 21 which Professor Courant transmitted to me in accordance with your request.[152] In the meantime, you will have received the resignation from the editorial board of the *Zentralblatt* which I sent in jointly with Professors Tamarkin and Courant.[153] I fear that so far as I am concerned personally, this decision is irrevocable. It would be quite impossible for me to remain on a board of editors from which so eminent a mathematician and so unimpeachable a personality as Professor Levi-Civita had been dismissed. In addition, I am convinced that the restrictions which you have felt yourself obliged to impose upon Professor Neugebauer make it unreasonable to expect that the *Zentralblatt* can continue to be regarded as a useful scientific enterprise.[154] It is most particularly in the case of an abstract journal that full editorial freedom is essential. I am writing you this letter before I

[Mathematical Reviews], *era effettivamente un ebreo tedesco, emigrato negli Stati Uniti, non certo per sua libera decisione, dopo l'avvento del nazismo in Germania! Poiché io non diedi alcuna risposta a tale sciocca lettera (per non dir peggio) del Blaschke, dopo qualche tempo fui avvertito da amici che ne erano venuti a conoscenza, che il mio telefono era controllato, che venivo talvolta pedinato ecc. "in seguito ad una denuncia di un certo professore tedesco" che – come mi fu confermato dopo la liberazione – era appunto il Blaschke. Ma l'Italia fascista non era la Germania hitleriana (soprattutto non ne aveva la diabolica efficienza) e tutto finì lì, senza ulteriori guai per me. Rividi poi il Blaschke nel 1952, ad un congresso a Salisburgo, ma feci finta di niente.*

[152] OVP, *General Correspondence, 1902-1960*, box 13: Springer Verlag to O. Veblen, 21.11.1938.

[153] Jacob David Tamarkin (1888–1945) and Veblen were members of the journal's Editorial Board since 1932 and 1931, respectively.

[154] Otto Neugebauer (1899–1990) was the first editor-in-chief of *Zentralblatt* (1931–1938).

have consulted with any of my colleagues, because I realize how much your firm has in the past done for the cause of science, and I wish to make it quite clear that my own reaction is completely uninfluenced by the opinion of anyone else. For the same reasons I think I should let you know that I favor the establishment of a mathematical abstract journal in the United States. This is because it seems to me that at the present time and probably for a considerable period in the future it is only in this country that such a journal will have the requisite freedom from political influence. Also I think that it may be possible to obtain for such an enterprise something like the complete international cooperation which the *Zentralblatt* had a few years ago. There is as yet little if any nationalistic bias to be seen in this country, and there are indeed many who, like myself, refrain even from the use of such an expression as "American mathematics". I suppose that under the existing political conditions it would be difficult for you to enter into an arrangement by which the existing *Zentralblatt* would be transferred to this country. If that were possible, however, I have no doubt that it would be interpreted as a very significant gesture in the direction of international scientific solidarity. Sincerely yours Oswald Veblen

D3.12.2 J.L. Synge to O. Veblen, [Toronto] 19.1.1939

OVP, *JLS*, fols. 1r–2r
Dear Veblen, I thought you might like to see how my correspondence with Springer ended. Yours sincerely John L. Synge[155]

Dear Dr. Springer, I thank you for your courteous and prompt reply to my letter of December 19. If the *Zentralblatt* were a purely German publication the situation would be quite different. But such a journal cannot be conducted successfully by one country alone and so very wisely international cooperation was sought and obtained. Having established its reputation by international cooperation, it would seem natural that the publishers of the *Zentralblatt* should avoid any action likely to alienate foreign collaborators. In my personal opinion Professor Levi-Civita is one of the greatest of living mathematicians, and I can see no reason why his dismissal by the Italian authorities should lead to his dismissal by the *Zentralblatt*. It appears to me a clear case of the intrusion of politics into science, and by this act the publishers have alienated such a body of foreign opinion that it will be a physical impossibility to secure quick and efficient service abroad by means of which the *Zentralblatt* has hitherto operated so successfully. As for the regulation prohibiting the reviewing of papers of German authorship by emigrants, it seems to me that such action was uncalled for until cause for complaint arose. Any reviewer of any nationality who allowed his personal feelings to sway

---

[155]John Lighton Synge (1897–1995), professor of Applied Mathematics at the University of Toronto, reviewer for the *Zentralblatt* (1935–1939).

his judgment is not a fit person to act as a reviewer. But from my personal acquaintance with the emigrants, I have no reason to suppose that they would belong to this category. They might justly claim that the prohibition in question should have been accompanied by another, namely, that the papers written by emigrants should not be reviewed in Germany. So far from being tactful, as your letter suggests, the prohibition introduced by the publishers of the *Zentralblatt* appears to me insulting to a body of mathematicians for whose academic eminence and personal integrity I have a high regard. It is with great regret that I sever my connection with the *Zentralblatt*. But I would only remain a collaborator on following conditions:

1. That Professor Levi-Civita be reinstated
2. That the regulation regarding the reviewing of German papers by emigrants be rescinded
3. That all those who have resigned on account of recent events be invited to return.

Since there is nothing to indicate any possibility that your firm would reconsider its policy in this way, I beg you to accept my resignation as a collaborator of the *Zentralblatt*. Yours faithfully, J.L. Synge

D3.12.3 O. Veblen to A. Flexner, [Princeton] 9.2.1939

OVP, *AF*, fols. 1r–4r
Dear Doctor Flexner: The *Zentralblatt für Mathematik* was founded by Otto Neugebauer, a professor in Göttingen, in 1931, and from the beginning has been published by Julius Springer of Berlin. It was one of a group of coordinating journals published by Springer in various fields of science. From the beginning it has served a useful purpose in bringing together in convenient form and with suitable criticism all the published mathematical work of the world. It has had an eminent board of Associate Editors, and a very large number of collaborators and reviewers, most of whom are themselves mathematicians of some distinction. After the National Socialists came into power in Germany, Neugebauer, who is perhaps the leading historian of mathematics, moved to Copenhagen and continued to edit the *Zentralblatt*. The Springer publishing house has an excellent record for cooperation with scientific men, but Neugebauer and his friends have recognized for some time that the situation would probably change by the time that the political authorities were ready to apply pressure to Springer. This happened sooner than we expected, perhaps in consequence of Munich. Springer dropped the distinguished Italian mathematician, Levi-Civita, from the list of associate editors without consulting Neugebauer. In response to a letter of inquiry from Neugebauer, Springer stated that this was a consequence of the anti-Semitic decrees in Italy, and added that Neugebauer must give him before the 1st of December a formal engagement that no article by a German author should be

reviewed by an emigre. Neugebauer resigned as of the 1st of December, and was followed in this action by a number of the associate editors. A very large number of the collaborators (the exact number is of course not available to me at present) have also resigned. Springer has replaced Neugebauer by an obscure mathematician named Ullrich and added a few German and Italian and one Polish associate editors. As soon as I had resigned from the board, I brought the situation to the attention of Mr. Keppel of the Carnegie Corporation and Mr. Weaver of the Rockefeller Foundation.[156] Both were interested in Neugebauer as well as in the problem of the *Zentralblatt*. As a result of collaboration from one or both of the foundations, Neugebauer has been appointed to a chair of the history of mathematics at Brown University, where it is hoped that he will be able to continue his researches in this field under favorable auspices. The Carnegie Corporation has authorised its President "to commit the Corporation to a grant of not more than $66,000 to permit the establishment in this country of an international journal of mathematics, on presentation of a satisfactory plan of operation." [. . .] Yours sincerely Oswald Veblen

D3.12.4 W. Blaschke to E. Bompiani, 8.5.1940

AS-UMI, *Bompiani correspondence*, fol. 1r
Dear friend, I am sending you a copy of a letter that I wrote to Tricomi. I find Tricomi's paper quite unpleasant in political terms. With warm greetings from our family to yours, your devoted Blaschke

Lieber Freund, beiliegend sende ich Dir den Durchschlag eines Briefes, den ich an Tricomi geschrieben habe. Ich finde Tricomis Veröffentlichung in politischer Beziehung ziemlich unerfreulich. Mit herzlichen Grüßen von Haus zu Haus Dein ergebener Blaschke

D3.12.5 W. Blaschke to F. Tricomi, 8.5.1940

AS-UMI, *Bompiani correspondence*, fol. 1r
Dear Mr. Tricomi, you were kind enough to send me several reprints of your work. I noticed a report from you in *Saggiatore* March 1940 p. 26: *Una nuova revista di bibliografia matematica*. It should be clear to you that the *Mathematical Reviews* is a company founded essentially by Jewish emigrants in America with

---

[156]Frederick Paul Keppel (1875–1943), President of the Carnegie Corporation (1922–1941); Warren Weaver (1894–1978), Director of the Division of Natural Sciences at the Rockefeller Foundation (1932–1955).

the purpose of replacing the *Jahrbuch für Fortschritte der Mathematik* and the *Zentralblatt für Mathematik*. I find it regrettable that you consider the present time to be appropriate to promote these efforts. Yours sincerely, Blaschke

Sehr geehrter Herr Tricomi, Sie waren so freundlich, mir wieder mehrere Sonderdrucke Ihrer Arbeiten zu schicken. Dabei fällt mir ein Bericht von Ihnen auf im Saggiatore *marzo 1940 S. 26:* Una nuova revista di bibliografia matematica. *Es dürfte Ihnen klar sein, daß es sich bei den* Mathematical Reviews *um ein Unternehmen handelt, das im wesentlichen von jüdischen Emigranten in Amerika mit dem Zweck gegründet wurde, das* Jahrbuch für Fortschritte der Mathematik *und das* Zentralblatt für Mathematik *zu verdrängen. Daß Sie den gegenwärtigen Zeitpunkt für geeignet halten, diese Bestrebungen zu fördern, finde ich bedauerlich. Ihr ergebener Blaschke*

D3.12.6 H. Geppert to F. Tricomi, [Berlin] 15.5.1940

AS-UMI, *Bompiani correspondence*, **fols. 1r–2r**
Dear Mr. Tricomi, among the reprints that you sent me a few days ago and for which I thank you heartily, there is an article in the *Saggiatore* that refers to the new-born *Mathematical Reviews*. I see from this article that you are not properly informed about the new development of our German bibliographic system, and I take this opportunity to inform you of the exact situation. As you know, in November 1938 Mr. Neugebauer resigned from the editorial board of the *Zentralblatt*. His departure was mainly determined by political considerations, since it did not seem tenable, in the long run, for a German emigrant to publish a leading German magazine from abroad, with a German publisher, and this situation also led to difficulties in the actual management. The editorial staff was led by Professor Dr. Ullrich in Giessen, one of our best function theorists, and at the same time our Italian colleagues agreed to take a decisive part in the design of the *Zentralblatt*. In the course of the negotiations that I conducted myself, Mr. Bompiani and Excellency Severi, as well as myself, joined the editorial board of the *Zentralblatt*. It was a great regret to me that you refused Mr. Ullrich's request to cooperate. Objectively, it cannot be said that the scientific level of the *Zentralblatt* has been affected by the change of management; in my opinion, it has even improved considerably, both in terms of the content of the reviews, which are in some cases much more thorough and detailed than before, and in terms of the completeness of the literature covered. The current group of reviewers consists of a considerable number of outstanding mathematicians who are in no way inferior to their colleagues working in the *Reviews*. The fact that the *Zentralblatt* was in the process of being redesigned made it desirable that this periodical, originally founded as a competitor of the *Jahrbuch*, should be associated with it in some way. Very long negotiations, in which I myself played a decisive part, showed that such a project would be welcomed by the German

Association of Mathematicians, but in view of the present circumstances of the war and in view of the competing journal founded by the Americans, it did not seem appropriate to go ahead with it for the time being. On the other hand, an agreement has been reached between the publishers and the journals involved so that, from the beginning of this year, competition between the two journals will be eliminated and cooperation between their editorial offices will begin. Both the *Jahrbuch* and the *Zentralblatt* are managed with the decisive participation of the German Mathematical Association, which can be seen in the title, and both periodicals are currently under my supervision. This collaboration between editorial staff is intended to prevent people working in both journals from being subjected to excessive stress or being forced to undergo it, which would mean the death of any scientific journalistic initiative. We are compelled, in order to maintain our scientific journals, to use all legitimate means to stem the process of proliferation of periodicals. Incidentally, it may interest you that American, French and English publishers, as well as American librarians, have raised a strong objection to this process, which contradicts every moral law, and which will sooner or later lead to our desired success. For all these reasons, I deeply regret that your statements in the *Saggiatore* do not do justice to the scientific and political concerns of the moment. I assume, however, that you were not aware of some information that has now been communicated to you. Notwithstanding the above remarks, which I have been obliged to make as director of the German review system, I am very pleased that you are one of the most vibrant members of the *Jahrbuch* editorial staff and I commend you for the great work you do every year and for taking on the commitment, especially in today's difficult times. Thank you very much. I hope that the necessary decisions I have had to make will not jeopardize your relationship with the *Jahrbuch*. I also share the contents of this letter with Mr. Blaschke, who sent me a carbon copy of his letter to you. With best regards, I remain yours sincerely, Harald Geppert

Lieber Herr Tricomi, unter den Sonderdrucken, die Sie mir vor einigen Tagen zugesandt haben und für die ich Ihnen herzlich danke, befindet sich ein Aufsatz im Saggiatore, *der sich auf die neu erscheinenden* Mathematical Reviews *bezieht. Ich ersehe aus diesem Aufsatz, das Sie über die neue Entwicklung unseres deutschen Referatenwesens nicht richtig orientiert sind und benutze die Gelegenheit, um Ihnen die genaue Sachlage mitzuteilen. Wie Sie Wissen, ist im November 1938 Herr Neugebauer aus der Redaktion des Zentralblattes ausgeschieden; sein Ausscheiden war hauptsächlich durch politische Erwägungen bestimmt, da es auf die Dauer nicht tragbar schien, das sein deutscher Emigrant vom Auslande au seine führende deutsche Zeitschrift bei einem deutschen Verlag herausgab und diese Sachlage auch Mißlichkeiten in der eigentlichen Berichterstattung zur Folge hatte. Die Redaktion wurde von Herrn Professor Dr. Ullrich in Giessen, einem unserer besten Funktionen-theoretiker, übernommen und gleichzeitig erklärten sich unsere italienischen Kollegen bereit, in entscheidendem Masse an der Gestaltung des Zentralblattes teilzunehmen. Im Verlaufe der Verhandlungen, die ich selbst geführt habe, sind die Herren*

Bompiani und Exzellenz Severi, wie auch ich selbst, dem Herausgeberkreise des Zentralblatts *beigetreten. Dass die damals von Herrn Ullrich an Sie angetragene Bitte zur Mitarbeit von Ihnen abgelehnt wurde, habe ich selbst am meisten bedauert. Bei objektiver Beurteilung kann man nicht sagen, dass das* Zentralblatt *durch diese Umstellung in der Redaktion in seinem wissenschaftlichen Niveau gelitten hat, meiner Auffassung nach hat es sich sogar wesentlich gebessert, sowohl was den Inhalt der Besprechungen selbst anlangt, die zum Teil wesentlich gründlicher und ausführlicher sind als früher, als auch was die Vollständigkeit der erfaßten Literatur anbetrifft; auch der jetzige Kreis der Referenten besteht aus einer stattlichen Zahl von hervorragenden Mathematikern, die den bei den* Reviews *tätigen Kollegen in keiner Weise nachstehen. Die Tatsache, der Neugestaltung des* Zentralblatts *ließ es in immer dringenderem Masse erwünscht erscheinen, daß das ursprünglich als Konkurrenzunternehmen zum* Jahrbuch *gegründete* Zentralblatt *in irgendeiner Form mit dem* Jahrbuch *vereinigt würde. Sehr langwierige Verhandlungen, an denen ich selbst entscheidend teilgenommen habe, haben ergeben, daß eine solche Einigung von der Deutschen Mathematiker-Vereinigung von Herzen begrüßt würde, aber unter den jetzigen Kriegsumständen und im Hin blick auf das von den Amerikanern gegründete Konkurrenzunternehmen im Augenblick nicht opportun ist. Hingegen ist eine Einigung zwischen den Verlegern und den beteiligten Stellen dahin erzielt worden, daß seit Beginn dieses Jahres die Konkurrenz zwischen den beiden Organen beseitigt und eine Zusammenarbeit zwischen ihren Organisationen eingeleitet ist. Sowohl das* Jahrbuch *über die Fortschritte der Mathematik als auch das* Zentralblatt *für Mathematik werden unter entscheidender Anteilnahme der Deutschen Mathematiker-Vereinigung, die auf dem Titel vermerkt ist, geleitet, und zwar unterstehen zur Zeit beide Blätter meiner Aufsicht. Diese Zusammenarbeit der Organisationen soll einmal verhindern, daß die bei beiden Blättern tätigen Referenten übermäßig beansprucht oder gezwungen eines derartigen Verfahrens den Tod jeder wissenschaftlichen publizistischen Initiative bedeuten würde und daß wir im Interesse der Aufrechterhaltung unserer wissenschaftlichen Zeitschriften gezwungen sind, mit allen Mitteln auf eine rechtmäßige Beschränkung dieses Reproduktionsverfahrens zu dringen. Im Übrigen dürfte es Sie interessieren, daß von Seiten der amerikanischen, französischen und englischen Vorleger, wie auch der amerikanischen Bibliothekare, ein energischer Einspruch gegen diese, jedem moralischen Gesetz widersprechende Anwendung der Reproduktionsverfahrens erhoben worden ist, der wohl über kurz oder lang auch zu dem gewünschten Erfolg führen wird. Aus allen diesen Gründen bedaure ich außerordentlich, daß Ihre Ausführungen im* Saggiatore *den augenblicklichen geltenden wissenschaftlichen und politischen Belangen durchaus nicht gerecht werden; ich nehme aber an, dass Ihnen ein Teil der mitgeteilten Patsachen unbekannt war. Unbeschadet der obigen Ausführungen, die ich als Redakteur der deutschen Referatenblätter zu machen schuldig war, bin ich sehr erfreut darüber, dass Sie eines der am lebhaftesten Mitglieder des Referentenstabes des* Jahrbuches *sind, und ich bin Ihnen für die Große Arbeitsmenge, die Sie jährlich für das* Jahrbuch *auf Ihre Schultern nehmen, gerade in den heutigen schwierigen Zeiten sehr zu Dank verbunden. Ich hoffe, daß die notwendigen Feststellungen, die ich treffen mußte, Ihr Verhältnis zum* Jahrbuch *nicht belasten werden. Herrn Blaschke, der mit von seinem Brief an Sie einen Durchschlag übersandt hat, teile ich den Inhalt dieses Schreibens ebenfalls mit. Mit vielen freundlichen Grüßen verbleibe ich Stets Ihr ergebener Harald Geppert*

D3.12.7 H. Geppert to E. Bompiani, Berlin 16.5.1940

AS-UMI, *Bompiani correspondence***, fol. 1r-v**

Dear Mr. Bompiani, thank you very much for your last letter.[157] Since I have been charged by the German Association of Mathematicians with establishing contact with the Union, I would like to inform you, as the chairman, of an incident caused by Mr. Tricomi. Tricomi published an article in the first issue of *Saggiatore* last month, under the title: *Ein neues mathematisches Referatenorgan.* In this essay, he expresses his enthusiasm about the American founding of the *Mathematical Reviews* and makes a rather disparaging assessment of the *Zentralblatt für Mathematik*, which, in his opinion, is completely superfluous. I assume that Mr. Tricomi was not informed about the question of the *Zentralblatt*, the essential cooperation between his Italian colleagues and the D.M.V., as well as about the cooperation with the *Jahrbuch über die Fortschritte der Mathematik*, and I therefore wrote him a detailed letter, a copy of which I am sending you for use. Mr. Blaschke sent Tricomi a much sharper letter in which he expressed his surprise that Tricomi, at this very moment in time, thought it appropriate to unduly praise an organ set up by German emigrants and Jews as deliberate competition against Germany. Perhaps you will have the opportunity to enlighten Mr. Tricomi about the Italians' position on our German bibliographical journals. With many kind regards from our family to yours, I remain respectfully yours, Harald Geppert

Lieber Herr Bompiani, vielen Dank für Ihren letzten Brief. Da ich von der Deutschen Mathematiker Vereinigung damit beauftragt bin, die Verbindung mit der Unione herzustellen, möchte ich Ihnen als dem Vorsitzenden von einem Zwischenfall Kenntnis geben, den Herr Tricomi verursacht hat. Tricomi hat im ersten Heft des Saggiatore*, das im vorigen Monat erschienen ist, einen Aufsatz veröffentlicht mit der Überschrift:* Ein neues mathematisches Referatenorgan. *In diesem Aufsatz äußert sich begeistert über die amerikanische Gründung der* Mathematical Reviews *und urteilt recht abfällig über das* Zentralblatt für Mathematik, *das nach seiner Meinung ganz überflüssig ist. Ich nehme an, daß Herr Tricomi über die Frage des* Zentralblattes, *die wesentliche Mitarbeit der*

---

[157] AS-UMI: E. Bompiani to H. Geppert, [Rome] 19.2.1940.

italienischen Kollegen und der D.M.V., sowie über die Zusammenarbeit mit dem Jahrbuch über die Fortschritte der Mathematik nicht orientiert war, und habe ihm deswegen einen ausführlichen Brief geschrieben, dessen Durchschlag ich Ihnen zur gefl. Benutzung beifüge. Herr Blaschke hat an Tricomi ein wesentlich schärfer abgefaßtes Schreiben gerichtet, in dem er Ihn sein Befremden darüber ausdrückt, daß Tricomi gerade im jetzigen Augenblick es für gut befindet, ein von deutschen Emigranten und Juden als absichtliche Konkurrenz gegen Deutschland gerichtetes Organ in ungebührender Weise zu loben. Vielleicht haben Sie einmal Gelegenheit, Herr Tricomi über die Stellung der Italiener zu unseren deutschen Referatenblättern aufzuklären. Mit vielen freundlichen Grüßen von Haus zu Haus verbleibe ich stets Ihr ergebener Harald Geppert.

D3.12.8 E. Bompiani to W. Blaschke, Rome 16.5.1940

AS-UMI, *Bompiani correspondence*, fol. 1r
Dearest Blaschke, I will respond to your letter immediately. You are definitely right. The matter had not escaped Severi and me (also as co-directors of *Zbl.f. Math.*) and who could put Tr[icomi] in place was immediately informed. So he will learn to do his job. Note that in the same sending of reprints he sent a lecture in which, without any need, he makes a dead man speak because he is a Jew (the living ones are not enough for him);[158] and he adds unpleasant comments regarding U.M.I.[159] He gives himself the airs of a moralistic superman [...]. Naturally, in that first issue of *Mathem. Reviews* that he appreciates so much, there is not a single Italian article reviewed: and perhaps this is why he appreciates it! [E. Bompiani]

Carissimo Blaschke, Rispondo immediatamente alla Tua. Hai perfettamente ragione. La cosa non era sfuggita a Severi ed a me (anche come condirettori del Zbl.f.Math.*) ed è stato subito avvertito chi può metter Tr[icomi] a posto. Così imparerà a fare il suo mestiere. Nota che nello stesso invio di estratti ha mandato una conferenza in cui, senza nessun bisogno, fa parlare un morto perché ebreo (non gli bastano quelli vivi); e aggiunge commenti antipatici nei riguardi dell'U.*

---

[158] Tricomi (1938–1940).

[159] Bompiani is referring to a phrase of the report about the UMI conference, which Tricomi also published in the *Saggiatore* (*Congresso nazionale di matematica*, Il Saggiatore, I, 1, 1940, p. 51): "In this second conference, a marked preference is shown for Applied Mathematics (even more so that at the first event; for example, 8 sections out of 11 have been reserved! This is exaggerating already! You almost get the feeling that someone wants to pass off true science as contraband." (*In questo secondo congresso viene, ancora più che nel primo, data una spiccata preferenza alle matematiche applicate cui, per esempio, sono riservate 8 sezioni su 11! Qui c'è già un po' d'esagerazione! Si ha quasi l'impressione che si voglia far passare come di contrabbando la scienza pura, coprendola col mantellino delle applicazioni*).

M.I. Si dà arie da superuomo moralista [...]. *Naturalmente in quel primo fasc. della* Mathem. Reviews *ch'egli apprezza tanto non c'è recensito un articolo italiano: e forse per questo l'apprezza!* [*E. Bompiani*]

D3.12.9 E. Bompiani to H. Geppert, Rome 20.5.1940

AS-UMI, *Bompiani correspondence*, **fol. 1r**
Dear Professor Geppert, thank you for your letter and for the copy of the letter sent to Tricomi. As I wrote to my friend Blaschke, Severi and I had already noticed the inappropriateness and lack of consistency of Tricomi's comments; and the matter has already been reported to whom it concerns. Severi, on whose behalf I also reply, asks me to tell you that he himself will write to you soon (he is very busy at the moment). Best wishes for the continuation of military successes and special greetings for you, your wife and child. My most cordial greetings [E. Bompiani]

Caro prof. Geppert, Grazie della Vostra lettera e della copia della lettera mandata al Tricomi. Come ho scritto all'amico Blaschke, Severi ed io avevamo già notato l'inopportunità e la non consistenza dei rilievi del Tricomi; e la cosa è già stata segnalata a chi spetta provvedere. Severi, a nome del quale anche rispondo, m'incarica di dirVi che presto Vi scriverà Egli stesso (ora è occupatissimo). I migliori auguri per la continuazione dei successi militari e particolari per Voi, la Vostra Signora e il bambino. Molto cordialmente. [*E. Bompiani*]

3.13 The 'Choice' of the Destination: The United States

The choice of destination can be seen as a fictitious choice just as much as that between leaving or staying. While, in fact, the difficulties of finding money and documents could in some terms be remedied (and, in this sense, the letters document examples of considerable generosity within the mathematical community), to identify the destination it was necessary to already have a network of contacts ready for prompt intervention. The first who moved about, such as Fubini, Terracini and Segre, began to write to foreign colleagues on 5 September 1938, just two days after the publication of the first *Provvedimento per la difesa della razza*.

The network of contacts needed to be spread across the territory, so as to be able to present the candidate to various institutions (universities, colleges, libraries, research centres), even of minor importance or in remote locations. It also needed to include determined and influential personalities, capable of advising the candidate in the preparation of the dossier, to support him with authority, enhancing the main strengths of the curriculum in a situation of strong competitiveness with other refugees, and finally to hire him or make others hire him, defeating the reticence, xenophobia, and subliminal anti-Semitism of native-born colleagues.

The options also depended largely on the type of mathematics that the displaced scholar actually did. At a time of increasingly pronounced disciplinarisation, when different national styles had emerged, being an Italian geometer or an Italian analyst took on a very specific meaning. In the international mathematical arena, one explicitly spoke of an Italian School in Geometry, of an Italian School in Logic and of an Italian School in Mathematical Physics. Moreover, the status and characters of certain sub-disciplines depended on the national setting: in Italy, for example, statisticians and demographers did not have skills in probability calculus, actuaries did not have competence in economics. An academic labour market like that of the United States, therefore, could not easily absorb figures like Bachi, Mortara, B. Tedeschi, and M. Pugliese. Similarly, the Italian tradition of studies in history of science had no equivalent in the Anglo-Saxon countries, as de Santillana realised when, in order to remain in the United States, he was forced to revise the approach he had learned from Enriques. In short, people did not choose where to go, but went where they could.

The first thought of Italian emigrants generally went to France, where they had woven a capillary web of scientific relations for over 40 years. Fubini and his sons Gino and Eugenio, Volterra and his sons Edoardo and Gustavo, Enriques and his children Giovanni, Ugo Fano, and many others took shelter in Paris, in the first period. France, however, was only a temporary solution, an intermediate step towards other destinations. The most coveted terminus were the United States and the United Kingdom. In particular, young people like Segre had grown up with the image of America as a free, democratic country, the new mecca for scientific studies. The lure of the New World was the most attractive. They had long dreamed of going there, and some would have done so, if they had not been discouraged by their mentors. Moreover, for at least ten years, the fascist cultural propaganda had passed off America as a pro-Jewish country and the American mathematical community as a Jewish lobby par excellence.

Beyond intentions and hopes, finding an opening overseas was extremely difficult. Italians relied on various old friends in the United States, people who had formed and maintained relations of friendship and scientific collaboration with Italian mathematicians since the end of the nineteenth century. Lovett and Birkhoff had met and respected Volterra since 1897; Coolidge had encounter Segre at the ICM of Heidelberg (1904) when, impressed by his plenary lecture, he had asked him to publish it in the *Bulletin of the American Mathematical Society* (the translation by J.W. Young actually appeared in 1905). Others trusted in the gratitude of Evans, Wilczynski, Lane, Snyder, Zariski, Lefschetz, perceived as pupils who 'came out' of the Italian mathematical Schools. It was plausible that Americans felt a strong obligation to counter the racism of the Italian government with a full expression of the internationalist ideal of science. It made sense that, out of human empathy and humanitarian considerations, they would help their old Masters. It was reasonable that Italian mathematicians were singled out for rescue for their recognised intellectual importance and influential role in the scientific arena.

However, there were several problems: some of these American scholars were men of an advanced age, at the end of their careers or already retired; others were too young and did not yet have the credit and power to enforce the appointment of a refugee. It must also be borne in mind that ties were loosened considerably in the twenties and thirties. Furthermore, many of the personal and institutional connections crucial in the placement of émigré mathematicians resulted from the wide-ranging philanthropy of the Rockefeller Foundation, as transmitted by the International Education Board. Work within the Board, and the international friendships and collegial connections fostered when reviewing the fellowship applications in mathematics, added an intensely personal sense of responsibility for the careers and safety of European colleagues deprived of their positions. Italy had remained marginal with respect to the two main programmes to improve education on an international scale: the IEB system of capital grants for selected mathematical institutes and the world-wide programme of awarding travelling research fellowships to promising young scientists. Consequently, Italian mathematicians missed out on the professional and personal payoffs these programmes had, as well as the related dynamic web of visits, connections, and correspondence. It was therefore not surprising that, even if the quick response of American mathematicians to the Italian racial laws bespoke of a lively international network of contacts, the applications of Italians were almost all rejected.

At first, in the spring of 1933, American rescue agencies expected that the need for their assistance would be temporary and certainly they did not anticipate having to cope with subsequent waves of Austrian, Italian, Czechoslovakian dismissals. However, as Hitler's pressure tactics increased and expanded, still more became potential emigrants. The crisis in Nazi Europe presented the rest of the academic world with both obligations and opportunities. On the one hand, there was the chance to secure the services of renowned scholars, profiting by the stupidity and brutality of the racist governments. On the other hand, a feeling of responsibility often accompanied the realisation of opportunity.

Conscious of the effects of the Depression, the American organisations concerned over the fate of displaced scholars had wanted to avoid a nationalistic reaction to the coming of foreigners and the danger of arousing anti-Semitism. Consequently, they had decided that they would aid scholarship, rather than provide relief to suffering. In other terms, selection of individuals would be based on merit only; what can be done should be reserved to those of high scientific value. They had wanted to save the scholar for the sake of scholarship. In late 1938, however, the general situation worsened. The careers and lives of more and more scholars were threatened; a panic flight from persecution and danger was imminent while the unemployment situation remained discouraging. Many thought that America had done all it could to assimilate refugees. Under the impact of the huge waves of people who continued to come across the ocean, many thought that humanitarian considerations should be definitively laid aside.

Not everyone adopted that stance. So, Veblen and Weyl, the two principal agents of the mathematicians in aiding Italian émigrés, ran at the Institute for Advanced Study an informal placement bureau for displaced mathematicians, open also to the 'hopeless cases'. Both Veblen and Weyl had previous experience in the field. In May 1933 Veblen had gone to the RKF about Nazi-induced transfers and became a member of ECADFS. At the same time, Weyl had established a German Mathematicians' Relief Fund and had spearheaded attempts to incorporate more refugees in the mathematics program at the Institute. But, by late 1938, the Institute could no longer make invitations to displaced scholars seeking to emigrate, since funds for stipends were exhausted. Exasperated at the inactivity of the ECADFS, Veblen and Weyl felt that the power of assimilation in the country was not yet exhausted. Hence, they came up with the idea of compiling dossiers of displaced scholars. One copy of the refugee files was to be sent to H. Shapley, Head of the Harvard College Observatory, one to the American Friends Service Committee, one to the SPSL. The ECADFS and RKF were also kept informed. Not only were Veblen and Weyl aiding the non-eminent, but they also stopped the restriction of placements to institutions with research capabilities. Veblen and Weyl were now placing refugees in any willing 4-year college or even in junior colleges.

Precisely in this period, on 30 October 1938, Veblen was informed first by Levi-Civita and then by Zariski of the promulgation of the Italian racial laws. He immediately spoke about it with Lefschetz, who on 1 December 1938 wrote (at Veblen's suggestion) to Father John Francis O'Hara, President of the University of Notre Dame. The fourth of ten children of the United States consul to Uruguay, O'Hara had studied at the Catholic University in Montevideo and served as private secretary to E.C. O'Brien, the United States ambassador to Uruguay. He had travelled in Argentina and Brazil and had a substantial knowledge of the Latin American world. For this reason, Roosevelt appointed him as a delegate to the 1938 Pan-American Conference in Lima. The occasion could be beneficial in providing effective help to Italian refugees, inviting South American institutions to accept them. Lefschetz and Zariski knew the Italian mathematical situation better than Veblen, and he placed absolute faith in their opinions. Lefschetz, in particular, identified six cases of potential exiles: Enriques, Fano, Fubini, Levi-Civita, Segre, and Terracini (D3.13.1). Lefschetz's assessments came too late in order to be presented by O'Hara at the Lima conference, however his evaluations and ranking (Fano, Fubini, Levi-Civita, Enriques, Segre, and Terracini) were to have an essential role in determining individual destinies, both because Fano, Levi-Civita, and Enriques had no intention of emigrating, and because Segre and Terracini, who instead had the intention and need to do so, were presented as distinguished mathematicians, but of a lower level than the first four. Veblen would try to correct Segre and Terracini's assessments, but without success (D3.13.2).

A few days later, the American Friends Service Committee of Philadelphia, a Quaker organisation that was establishing a refugee service staff and committee, contacted Veblen to ask him to cooperate with them (D3.13.3). At that moment the ECADGS had not yet opened its sphere of action to the Italians, which is why Veblen promptly took advantage of the new partner. Lefschetz's assessments of

Fano, Fubini, Levi-Civita, Enriques, Segre, and Terracini were sent to the American Friends Service Committee in mid-December. At the same time, Veblen drafted two lists of emigrants.

Refugees
Miss Bösse Käthe – Egyptologist
Prof. Eduard Helly
~~Alt~~
Kottler
Thirring
~~Mann~~
Frucht (von N[eumann])
Helly
Terracini (Lane Graustein)
Duschek
~~Frucht~~
Schrodinger
Baule
Lubelski (Acta Arithmetica)
Refugees December 14/38
Dr. Eugene Lukacs
Dr. Alfred T. Brauer
Prof. Guido Fubini
Prof. Levi-Civita?
Prof. Fano
Prof. Enriques
Prof. Segre
Prof. Terracini
~~Prof. I. Schur?~~
~~Prof. O. Neugebauer?~~
Prof. Čech?
~~Prof. E. Hellinger~~
~~Prof. Hans Hamburger~~
~~Prof. Artur Rosenthal~~
~~Prof. Dehn~~
~~Prof. Carl L. Siegel?~~
~~Prof. Hecke~~
~~Dr. Wolfang Wasow~~
~~Prof. F. Ehrenhaft~~
~~Bernard Baule~~
~~Prof. Hans Thirring?~~
~~Dr. Stefan Stefan Piotrowski?~~ (in N.Y.C.)
~~Prof. Helly~~
~~Toeplitz?~~
~~Dr. E. Rothe?~~
Castelnuovo's son [Gino]
Kottler
Schrodinger
Helly

In mid-December Weyl too drafted an annotated list of refugees.

Refugees From Prof. Weyl Dec. 14, 1938 Still in Germany
47 [years old] Bernhard Baule (Graz) catholic
62 O. Blumenthal (Aachen)
44 A. Brauer (Berlin)
60 M. Dehn (Frankfurt a.m.)
51 F. Ehrenhaft (Wien)
32 Robert Frucht (Trieste)
Hans Hamburger (Berlin)
55 E. Hellinger (Frankfurt a.M.)
Ludwig Hopf (Aachen) (Applied Math. (Einstein interested) Popular books in physics)
35 Anton E. Mayer (Wien) Engineering + Geometry
Max Reiss (Wien) Physicist in Vienna. Pupil of Ehrenhaft
Robert Remak (Berlin)
Artur Rosenthal (Heidelberg)
Isaac Schur (Berlin)
50 Hans Thirring (Wien)
O. Toeplitz
Wolfang Wasow (Lana, Bolzano) teacher 27–28**
In Europe outside of Germany
36 S. Lubelski (Warsaw) Acta Arithmetica
Gottfried and Hermann Noether (Göteborg)
Fritz Noether (Tomsk)
Hans Schwerdtfeger (Prag)
45 Artur Winternitz (Prag)
In neutral countries
P. Bernays (Zürich)
55 Eduard Helly (New York)**
H. Müntz (Stockholm) (was in Russia)
Erwin Schrödinger (Oxford)
23 Noether Gottfried** school M.I.T.
24 Noether Hermann

The cases, discussed by Veblen and Weyl, were sent to the American Friends Service Committee on 17 December 1938 (D3.13.4). Veblen's understanding was that an attempt had to be made through the various connections of the Committee to find colleges and private schools in which suitably qualified refugees could find employment. His hope was that it might be possible to send some representative who can visit institutions in parts of the country which have not yet been reached, to seek for openings.

In addition to the lists, a typed record was produced by Veblen and Weyl for each aspiring refugee, with the following entries: Date; Name in full; Present address; Place of birth; Nationality; Date of birth; Religion (if this had bearing on the type of relief agency which might be asked for assistance); Married or unmarried? Children or other dependents? What languages do you command? Education (with names of universities and dates of degrees); Field of research; References (with addresses); Positions held (rank, place, dates); and Publications.

Records were compiled for Fubini, Segre, Terracini and for two naturalised Italian foreign Jews (Frucht and Wasow) counted in the Italian quota because they arrived from Trieste and Bolzano respectively. For Fubini, passages from Levi-

Civita's letter of introduction were transcribed. (FUBINI.1) For Segre and Terracini, Lefschetz's assessments were copied, to which further few comments by Veblen, Zariski, and Snyder were added. Files were sent in subsequent slots to the American Friends Service Committee and updated over time. The moment of being added to the records can be considered the entry point of the aspiring exile into the circuit of the solidarity network managed by Veblen and Weyl.

In early 1939, Shapley asked the ECADFS for information about the refugee situation in Italy. The ECADGS, which had not yet opened its agenda to Italians, redirected him to Max Ascoli and asked Ascoli to give Shapley any data which would be useful to him in his project (D3.13.5). Shapley intended to establish the 'Harvard National Research Associates'. These consisted of a small group of displaced scholars of an age beyond that at which places might be readily found in a college or university. The facilities of the library and laboratories were to be placed at their disposal for personal research, but they were not required to give lessons.[160] This did not mean life tenure for any scholars, but rather that a modest yearly grant would be available to them so long as the capital and interest of the fund raised by the efforts of Shapley should last.[161] Among the Italians, Fano (67 years old), Enriques (67), Levi-Civita (65), Fubini (59), and Maroni (60) could have benefited from this plan, but none of them were lucky enough to do so.

The first batch of dockets concerning the refugees in the field of mathematics and physics was mailed by Veblen and Weyl to Shapley and the American Friends Service Committee on 11 January 1939 (D3.13.6 e D3.13.7). The first 20 records prepared all refer to names included in the lists drawn up in mid-December. There were 3 Italians (Fubini, Segre, Terracini), a foreign Jew from Italy (Frucht), 7 mathematicians leaving Austria (Baule, Duschek, Ehrenhaft, Gross, Helly, Kottler, Mayer), 7 from Germany (Bernays, Brauer, Dehn, Hamburger, Hellinger, Jaffe, Reiche), one from Czechoslovakia (Löwner), and one from Poland (Lubelski). A second batch of 18 records would be sent on 17 January, and a third (5 records) on 21 January.

The occupation of Austria and the Sudetenland was evidently at the root of this new wave of exiles. Veblen and Weyl perceived that the organisation established by their School of Mathematics would soon no longer be sufficient. As Weyl explained to H. Freudenthal:

> Unfortunately our funds for stipends are at present exhausted, and if our friends succeed in coming over here with students' or professors' visas, to join the Institute temporally, we are of course aware of the responsibilities that will fall upon us, although I cannot at the moment

---

[160]Money which had been raised by the efforts of Shapley would be subsequently (by arrangement with the donors and with the full consent of Harvard University) turned over to the ECADFS. Five of the original group at Harvard would be taken over by ECADFS and six other distinguished scholars, 48 years of age or older, would be added from among the Emergency Committee's grant recipients at other institutions. The universities agreed that the Associates should be enabled to carry out the same type of scholarly pursuit as members of their permanent professional staff. In almost all cases, the National Research Associates remained dependent on the particular funds in this account.

[161]On Shapley's plan, see, e.g., Duggan and Drury (1948), pp. 84–85.

see clearly how to meet them, and although we had to decline formally all further commitments. We hope for the support of several relief committees that will begin to operate in the near future and which are providing with material concerning exiled mathematicians (about 35 cases altogether). [...] *We must try to unite forces.*[162]

At a time when America was facing a very intense wave of migration, after three years of smaller flows (D3.13.8), the ECADFS was stalling, so Weyl suggested that the American Friends Service Committee contact the SPSL. The convergence of forces hoped for by Weyl, however, immediately emerged as complicated. Shapley's movement for rather large-scale assistance for the refugees was taken over, at least tentatively, by the Association of American Universities under the presidency of J.B. Conant (Harvard), which slowed down the procedures. At the end of January 1939, the situation was unchanged. Shapley lamented: 'Things go too slowly with the refugee programme to suit me or the refugees. But we are doing things. Here at the Observatory, we are managing to rescue at least four of the Central European astronomical exiles'. One of the four was the Italian Luigi Jacchia. Weyl and Veblen shared Shapley's irritation at the slow pace of rescue missions:

> Your help and energy are a great comfort to me. [...] If one could only pump some life blood into the Duggan Committee [ECADFS]! I understand well enough that the slowness of all these operations gets on your nerves. Einstein says "Committees are set up to prevent things from being done.[163]

Shapley replied in a scornful way:

> The Duggan Committee – if you mean the Emergency Committee for the Aid of Foreign Scholars – had at last accounts only six thousand dollars at hand and I am asking for a part of that for Dr. Freundlich (for whom I got a place at Tufts College and the Harward Observatory) and for Dr. Jacchia of Bologna who has got as far as London. But I believe the Committee could get more money if there were more need. But its determined policy is not to do anything except upon appeal from universities or colleges, which can afford positions and guarantees. It is the proposed new committee with large ambitions that needs to have not only life pumped into it, but driving, uncomplicated ambitions. [...] I despair of immediate action on my large plan and think things best handled individually where we can do it. It is rather unfortunate that hard pressed University Presidents will undertake large movement and then go off on mid-winter vacations.[164]

By February 1939, Weyl too had lost hope of a quick implementation of any large-scale plan and only by force of inertia he continued to pass on 'bits of details about European scholars for whom the soil of their own countries burns under the soles of their feet'. Veblen, however, had not yet thrown in the towel and returned to

---

[162] OVP, *RefGen.*: H. Weyl to H. Freudenthal, [Princeton] 11.1.1939.

[163] OVP, *RefGen.*: H. Weyl to H. Shapley, [Princeton] 30.1.1939.

[164] OVP, *RefGen.*: H. Shapley to H. Weyl, Cambridge Mass. 31.1.1939. See also H. Shapley to H. Weyl, Cambridge Mass. 23.2.1939: "I have sent along to Marshall Stone some of the material you have sent me and he has replied; but to my disappointment, his replies are not tinged with glowing enthusiasm. Possibly he is merely wise and realizes the general hopelessness, or perhaps he would prefer to have you or me perform some magic. [...] Meanwhile the powerful plutocrats of New York City are suffering exile at Miami Beach, Palm Springer, and Santa Barbara, California. It's a hard and bitter winter, you know."

the American Friends Service Committee, proposing the idea he had already advanced at the end of 1938: sending some sort of an agent around to the less distinguished colleges, with a view to discovering places where refugees would be valuable (D3.13.9, D3.13.10).

At the end of March 1939, copies of the 47 files concerning mathematicians and physicists collected by Veblen and Weyl were transmitted by the American Friends Service Committee to the SPSL. Among these were the records of Fubini, Segre, Terracini, and Frucht. At the end of April 1939, the Italians were included in a new rescue plan drawn up by Conant. It had been Shapley who had suggested to Veblen and his colleagues that they send him a list of the names, former academic positions, ages and rough rankings of mathematicians and mathematical physicists yet to be rescued from Central Europe or in the process of being rescued.[165] The more impressive the list—Shapley wrote—the more Conant would be able to make much more headway in New York in getting the large movement under way. The list was sent by Veblen on 29 April (D3.13.11).

	Former academic position	Born	
Issai Schur	Full Prof., Univ. Berlin	1875	
Guido Fubini	Full Prof., Polytechnic Sch. and Univ. Turin	1879	
Felix Hausdorff	Full Prof., Univ. Bonn	1868	
Karl Löwner	Full Prof., German Univ. Prague	1893	
Hans L. Hamburger	Full Prof., Univ. Cologne	1889	
Max Dehn	Full Prof., Univ. Frankfurt a.M.	1878	Probably has 1yr. position in Trondheim, Norway
Otto Blumenthal	Full Prof., Tech. Hochsch., Aachen	1876	
Beniamino Segre	Full Prof., Univ. Bologna	1903	
Ludwig Berwald	Full Prof., German Univ. Prague	1883	
Alessandro Terracini	Full Prof., Univ. Turin	1889	
Hermann Muentz	Prof. Univ. Leningrad	c. 1893	
Robert E. Remak	Privatdozent, Univ. Berlin	1888	
Paul I. Bernays	Ausserord. Prof., Univ. Göttingen	1888	
Arthur Rosenthal	Full Prof., Univ. Heidelberg	1887	
Ernst E. Jacobsthal	Ausserord. Prof., Tech. Hochsch. Berlin	1882	
Alfred T. Brauer	Privatdozent, Univ. Berlin	1894	1939–40 Inst. Adv. Study 1 yr
Fritz Reiche	Full Prof., Univ. Breslau	1883	
George Jaffe	Full Prof., Univ. Giessen	1880	
Ludwig Hopf	Full Prof., Tech. Hochsch. Aachen	1884	
Hans Thirring	Full Prof., Univ.Vienna	1888	

(continued)

---

[165] Meanwhile, non-Aryans in Czechoslovakia were prohibited from leaving the country.

	Former academic position	Born	
Ernst Hellinger	Full Prof., Univ. Frankfurt	1883	1939–40 Northwestern Univ. 1 yr
Felix Ehrenhaft	Full Prof., Univ. Vienna	1879	
Friedrich Kottler	Asso. Prof. Univ. Vienna	1887	
Edward Helly	Privatdozent, Univ. Vienna	1884	
Paul Funk	Full Prof., German Inst. Tech., Prague	1886	
Bernard Baule	Privatdozent, Tech. Hochsch., Graz	1891	
Anton E. Mayer	Privatdozent, Tech. Hochsch., Vienna (accepted by Faculty / 38)	1903	Teaching experience for some years
Artur Winternitz	Ausserord. Prof., German Univ., Prague		

As for the Italians, who were in excellent positions, there are some peculiarities to note. Fubini, second in the ranking, was not actually looking for a position, because he had already joined the IAS. Segre, in turn, ranked eighth, had landed in the United Kingdom ten days previously and the funds raised to help him by his English colleagues ensured his livelihood until the end of the year. The most urgent case would be that of Terracini, in tenth position. The ranking was explained by Veblen and Weyl on the following May 2nd. The names had been arranged very roughly in order of scientific importance. The tabular statement had been constructed incorporating Veblen's and Weyl's attempts to rank the various mathematical and physical refugees according to scholarship, personality, teaching ability, and adaptability. Below is the definitive document produced by Veblen and Weyl.

Mathematicians and physicists whose records have been sent to American Friends Service Committee

May 2, 1939

In the column 'Acad. Position' means: 'held no academic position proper'. Privatdozentur *is*, Assistantship *is not*, counted as such. For exact positions, see individual records. The significance of scientific output is the basis for rating under column 'Scholarship'. In judging Scholarship age has to be taken into account. 'B' indicated successful research work of considerable originality.

	Scholarship	Personality	Teaching ability Adaptability	Acad. Position	Birth date	Remarks
Bernard Baule	C	A	A		1891	Catholic, in jail and concentration camp for more than a year
Gustav Bergmann	C+	-	-		1906	Interesting for combining math., psychology, philosophy, and law practice
Paul Bernays	B+	B	B		1888	
Ludwig Berwald	B+	A	B		1883	(Record will follow)
Otto Blumenthal	B	B	A		1876	
Alfred T. Brauer	B	B	A		1894	
Max Dehn	B	A	A		1878	
Felix Ehrenhaft	B	C	C		1879	
Robert Frucht	B-				1906	Good actuarial work
Guido Fubini	A	A	B		1879	
Paul Funk	B-	B	B		1886	
Hilda Geiringer	B+	B	A		1893	
Wilhelm Gross	-	-	-		1883	Prof. of mining engineering
Hans L. Hamburger	A	A	B		1889	
Felix Hausdorff	A	A	—		1868	
Ernst Hellinger	C+	B	A		1883	
Edward Helly	C	B	B		1884	
Paul Hertz	B	C	C-		1881	Impossible as a teacher for undergraduates. It is added: Died March 24, 1940
Ludwig Hopf	B	B	A		1884	It is added: Died Dec. 1939
Ernst E. Jacobsthal	B	B	A		1882	
George Jaffe	B	A	A		1880	
Stanislaus Jolles					1857	Too old to be placed
Friedrich Kottler	B	B	B		1887	

	Scholarship	Personality	Teaching ability Adaptability	Acad. Position	Birth date	Remarks
Heinrich Loewig	C	-	-		1904	
Karl Löwner	A	B	A		1893	
S. Lubelski	B-	C	C	?	1902	
Anton E. Mayer	C	B	A		1903	As assistant had considerable teaching experience
Hermann Muentz	A	C	B	-	c. 1893?	
Fritz Noether	B	A	B		1884	
Max Pinl	C+	-	-		1897	Actuarial experience
Felix Pollaczek	B (?)	B	B	-		Combination of pure mathematics and physics (telegraph and telephone)
Fritz Reiche	B	B	A		1883	
Max Reiss	C	B	-	-	1903	
Robert E. Remak	A-	B	C		1888	
Arthur Rosenthal	B	B	A		1887	
Peter Scherk	C	B	B	-	1910	
Issai Schur	A	A	A		1875	As scholar, personality and teacher, superior to all others. It is added: Died Jan. 10/41
Stefan Schwarz	C	B (? or A)	-	-	1914	Fellowship would be suitable
Hans Schwerdtfeger	C	B	B	-	c. 1904 ?	
Beniamino Segre	A	A	B		1903	
Wolfgang Sternberg	B	B	B		1887	
Alessandro Terracini	B	B	B		1889	
Hans Thirring	B	A	A		1888	
Otto Toeplitz	B	B	A		1881	It is added: Died Feb. 1940
Artur Winternitz	C	B	C		1893	
Walter Froelich	B-	B	B			
[Guido] Ascoli	A-	A	B			
Gotthard Günther	B-	A	B			

The ranking, shared with Shapley, the American Friends Service and the SPSL, was used to indicate suitable names to fill in the few positions still open in the United States, such as one in Massachusetts, one in Louisville, for which Löwner and Segre were considered, and one in Kansas City (for which Baule, Rosenthal and Thirring were recommended).[166] Given the positive results obtained in these three institutions, Shapley relaunched the idea of a general plan, aimed at medium-small colleges or universities:

> Two of the three direct approaches I have made to the Presidents of small colleges or universities – namely Tufts, University of Louisville, and University of Kansas City – have resulted in useful interest in the assimilation of well-chosen exiles. [. . .] But this apparent success of the direct approach worries me to this extent. Should we not deliberately solicit attention at a selected group of smaller institutions in America? Perhaps you have been doing that and it is only an accident that these three that I have had the opportunity of dealing with directly have responded so favorably. I am beginning to lose hope in the program for general help. Mr. Conant seems to find continual difficulty in getting a National Committee formed. The summer is coming, interest is waning in a generalized project, I am afraid.[167]

Hence the proposal to write a formal letter to a selected group of smaller colleges outlining the general situation, listing the names of men available with a very brief comment on each, and stating what possibilities there might be for temporary help from the various committees and foundations. Birkhoff and Shapley 'agreed that the distribution of these first-rate and high second-rate scholars among smaller American institutions would in the long run be very advantageous, providing at the same time they defended not too feebly the inherent rights of their own graduate students'.[168] Veblen favoured the idea and wrote about it to the ECADFS. However, in Veblen's opinion, such a plan would be successful only if the ECADFS 'stirs, and if they find a man of the same efficiency as, for instance, their former Secretary Mr. Murrow'.[169]

By the end of June, the shortlist of refugees (20 names out of 60) was ready for sending to 150 smaller colleges. On the Italian front, among the outstanding middle-aged mathematicians, only Terracini appeared (in fourth position). Fubini had rightly disappeared as Veblen had managed to get a direct call for him to the IAS. However, Segre also had disappeared, his name being absent both from the list of middle-aged candidates and from that of young promising mathematicians. Ascoli, included in the records under the recommendation of Veblen and Fubini, and the astronomer Bemporad also did not make the selected list. At this stage, it was Shapley who

---

[166] See D3.13.9, SEGRE.5 and OVP, *RefGen.*: H. Shapley to H. Weyl, Cambridge Mass. 24.5.1939. Hans Thirring (1888–1976), Austrian theoretical physicist, and Head of the institute for Theoretical Physics of the University of Vienna. In 1938, was accused of dealing with the "Jewish" theory of relativity, his friendship with Albert Einstein and Sigmund Freud as well as his pacifist attitude. After his forced leave, he worked as a consultant for various companies such as Elin AG and Siemens.

[167] OVP, *RefGen.*: H. Shapley to H. Weyl, Cambridge Mass. 24.5.1939.

[168] OVP, *RefGen.*: H. Shapley to H. Weyl, Cambridge Mass. 24.5.1939.

[169] OVP, *RefGen.*: H. Weyl to H. Shapley, [Princeton] 30.5.1939.

halted procedures, given that he did not write the formal letter accompanying the list, as per the agreements with Veblen and Weyl, because he had not received any assurance from the committees and foundations on the partial financial coverage of the invitations. As he wrote to Weyl on the day of the invasion of Poland: 'some weeks ago I decided to diminish very much my activities for the next few months on refugee problems. But the war news of today probably more effectively releases me from work on refugees than any strong resolution on my part or advice from the university or the doctor or friends'.

A new list of refugee mathematicians and physicists, classed as 'especially urgent', was drawn up by Veblen and Weyl on 18 July 1940. 17 names, including only one Italian: Bernays, Berwald, Born, Fenchel, Fried, Fürth, Geiringer, Hamburger, Pollaczek, Pólya, Pringsheim, Scherk, Schur, Segre, Tarski, Weinstein, and Mahler. These names, with the exception of Schur, entered the mass-rescue plan, proposed by the Director of the New School for Social Research, A. Johnson. In fact, faced with the new emergency caused by the war upon the countries of Western Europe, the RKF had come forwards to offer additional assistance to the cause of the displaced scholars. Johnson had already organised the University in Exile for the benefit of displaced German scholars and now offered to receive refugees from other countries and to find chairs for them in the New School for Social Research, provided that funds were forthcoming. Johnson presented Veblen with his plan and asked him for information on the academic merits of Pringsheim and Segre (D3.13.12). However, Segre's nomination was unsuccessful.

Even before the organisation of the ECADFS, the RKF had undertaken its extensive programme in aid of the displaced scholars. From the very beginning of the Emergency Committee's activities, the Foundation rendered great service with its advice, the result of the long and continuous experience in the field of international scholarship and scientific collaboration. By 1940, the RKF made a grant to the Committee of $10,000 a year for two years to enable its assistance in the administration of additional activity involved in the organisation of a placement programme. According to the plan, the New School, the Institute of International Education, and the ECADFS were all authorised to recommend scholars to the Foundation. The final decision in all cases, however, rested with the Foundation. The Segre case was proposed again, together with that of Ascoli, on 8 October 1940. For the Italians, that was the last chance.

As the sphere of Nazi control expanded (France had capitulated on June 25, 1940), the rescuers became concerned for their colleagues trapped in Belgium and France. On 14 October 1940, Weyl presented to W. Weaver 'a number of French mathematicians who may be in need of help from America and who could make valuable contributions to our mathematical life over here'.[170] These were A. Denjoy, H. Cartan, J. Leray, R. de Possel, J. Delsarte, C. Chabauty, C. Ehresmann, C. Pisot, J. Dieudonné, and E. Kogbetliantz. Seven of them were characterised as Bourbakist. To them, Lefschetz added É. Cartan, J. Hadamard, H. Lebesgue, É. Picard, and A. Weil (Fig. 3.13).

---

[170] OVP, *RefGen.*: H. Weyl to W. Weaver [RKF], [Princeton] 14.10.1940.

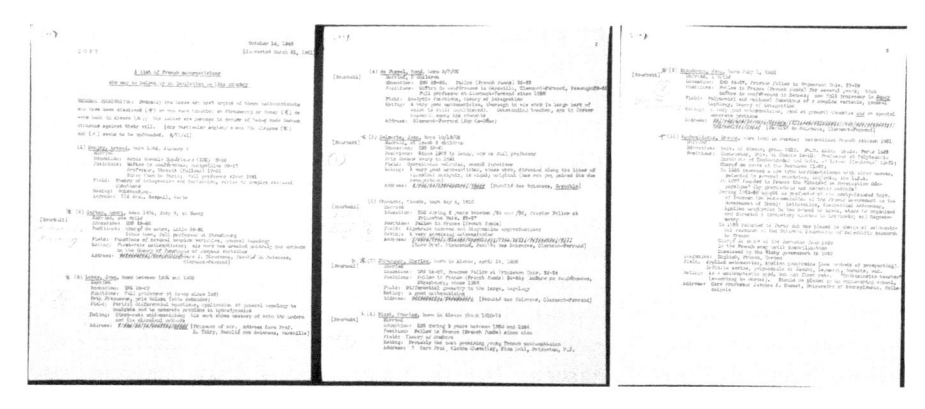

Fig. 3.13 Plan to aid the Bourbakist group to emigrate in United States, 14.10.1940, OVP, *RefGen*

At the same time, Lefschetz reproposed various names that, over time, he had recommended to Johnson: L. Ahlfors, K. Borsuk, E. Čech, H. Freudenthal, H. Hopf, C. Kuratowski, K. Mahler, R. Nevanlinna, and W. Sierpiński. As is evident, there was no Italian.

In early 1941, the RKF and Alvin Johnson's organisation jointly undertook a final rescue action (D3.13.13). Out of thousands of applications, a hundred were selected, with a quarter of them for positions at the New School for Social Research (D3.13.14, D3.13.15). Neither Segre nor Ascoli succeeded in obtaining a fellowship.

By July 1941, the rescue agenda of the Italians can be considered concluded. At this junction, an urgent need was felt for a single organisation which would serve as the centre for exchange of information and advice about the whole problem of the displaced foreign scholars. Upon the advice of the RKF, the ECADFS was selected for that function. After all, following the summer of 1940, the Committee had already invited Johnson (University in Exile), Shapley (National Research Associates), F. Aydelotte (IAS), H. Kraus (American Friends Service Committee), H.A. Moe (Oberlaender Trust), and C.A. Riegelman (National Refugee Service) to join it as representatives of their organisations.

On 2 July 1941, Weyl advised Shapley and the American Friends Service Committee that it was time to stop sending them every tiny detail that he and Veblen received about refugees. The work of placing refugee scholars—explained Weyl— seemed by then to be concentrated chiefly in the ECADFS, which cooperated directly with the Rockefeller Foundation and the New School for Social Research. All the records were passed on to L.H. Seelye, engaged by the ECADFS as field secretary to visit colleges and universities in various parts of the United States in order to stimulate interest in the rescue programme.[171] At the same time, an advisory committee was organised, composed of scholars in the different fields of learning

---

[171]Laurens Hickok Seelye (1889–1960) former President of St. Lawrence University, Canton, New York (1935–1940), was member of the Executive Committee of ECADFS from 1940 to 1944.

who were eminently qualified to give advice on ways and means of using the scholars to advantage.

The overall balance of the last years of efforts is traced by Weyl in the letter to Shapley dated 18 July 1941 (D3.13.15). Despite the joint efforts of Veblen, Weyl, Shapley, the American Friends Committee, the ECADFS, the RKF and Alvin Johnson's organisation, Italians did not have good results in terms of recruitment. Between 1939 and 1942, according to the report drawn up by A. Dresden, 66 mathematicians arrived in the United States, categorised as follows by nationality: Austria (11); Czechoslovakia (5); France (8); Germany (30); Italy (4); Poland (5); Hungary (2). The Italians were Fubini, Frucht (German-born naturalised Italian), Opatowski (Polish-born naturalised Italian), and Wasow (German-born naturalised Italian). Of the 288 displaced scholars assisted through the ECADFS, there were 9 Italians: Enzo Bonaventura (Psychology), Umberto Moshe David Cassuto (Philology), Luigi Jacchia (Astronomy), Leonardo Olschki (Philology), Alexander Pekelis (Law), Giulio Racah (Physics), and Bruno Rossi (Physics) to which two Rosenwald fellows would be added: Renata Calabresi (Psychology) and Dario Viterbo (Sculpture). No mathematician was included in the grants awarded by the Rockefeller Foundation.

Documents

D3.13.1 S. Lefschetz to Father J.F. O'Hara, [Princeton] 1.12.1938

OVP, *RefGen*, fol. 1r, 2r
My dear Father O'Hara:[172] Upon the suggestion of Professor Veblen, I am taking the liberty of mentioning to you the names of the following outstanding mathematicians, Professors in Italian universities, who have just lost their positions because of the racial policy of the Italian government.

Fano, Gino. 67 years old. Married. Professor of descriptive geometry at the University of Torino. A first-rate mathematician and a very attractive personality. His speciality is algebraic geometry, and he is one of the most outstanding men in that field.

Fubini, Guido. 59 years old. Married. I do not know him personally but according to all reports he is a very interesting and forceful personality. He is an excellent mathematician having profound influence in several fields of mathematics. He is a Professor in the School of Engineering of the University of

---

[172] John Francis O'Hara (1888–1960), a member of the Congregation of Holy Cross and prelate of the Catholic Church, was the President of the University of Notre Dame (1934–1939). As a member of the U.S. delegation at the Eighth International Conference of American States (convened at Lima, Peru, on December 9, 1938), O'Hara was to travel to South America and could be strategic in presenting there the cases of displaced Italian mathematicians.

Torino and has made important contributions to engineering research. He is one of the most prominent mathematicians of Italy.

Levi-Civita, Tullio. 65 years old. Married. Professor at the University of Rome. He is perhaps the best known of all Italian mathematicians. Suffice it to say that he is a member of the Papal Academy[173] and member of the outstanding world academies. In recent years he has spent one term at the Institute for Advanced Study and also lectured in South America.[174] He has been the recipient of numerous scientific honors. He comes from an outstanding family, closely associated with the life of Venice.

Enriques, Federigo. Age 67. Married. One of the two founders of the modern Italian school of algebraic geometry, a school whose renown it would be difficult to exaggerate. He is also known for his contributions to the philosophy of science and general history of science. He is a very cultured man. He is a Professor at the University of Rome and has also lectured in Spain and South America.[175]

The four gentlemen mentioned above are world authorities on mathematics, their names being known to any mathematical scholar worthy of the name and their mere connection with any institution anywhere would greatly stimulate its scientific atmosphere. In addition, I wish to mention the following who are also men of considerable distinction.

Segre, Beniamino. Age 36 [sic!].[176] Married. Has three small children. His economic situation may be serious soon because his length of service has been comparatively short. Until this year he was a Professor at Bologna where he held an important chair in geometry in which he was successor to a line of distinguished geometers (among them Cremona and Enriques). He is a specialist in algebraic geometry and differential geometry. He is very productive mathematically and I believe he is destined to have a long and active scientific career. He is a very pleasant fellow and comes from a good Italian family. In fact, he is a nephew of Carro [Corrado] Segre whose fame as a geometer was world wide.

Terracini, Alessandro. Age 39 [sic!].[177] Married and has several small children. He is a Professor at the University of Torino. His speciality is differential geometry. I believe that he would be a good acquisition as he has been active for a good many years and promises to continue so.

I am only giving you very summary information regarding these scientists, but it is hardly necessary to say that if more complete data is desired, we would eagerly

---

[173] The Pontificia Academia Scientiarum renewed and reconstituted in 1936 by Pope Pius XI.

[174] Levi-Civita had been invited to give a series of lectures in Princeton in 1936 (27 August– 21 December) and in South America in 1937 (13 July–12 October).

[175] Enriques had been visiting professor in Spain in 1920 and in Latin America in 1928.

[176] Segre was 35 years old.

[177] Terracini was 49 years old.

give it to you. In fact, we could easily provide detailed reports regarding each of these men. I was very sorry not to have had the opportunity of seeing you before your departure for South America and hope that your trip will prove to be most pleasant. I also believe that it will prove most useful for all of us. It is indeed an important mission on which you have been sent by the President of the United States. With best regards, I am, Sincerely yours, S. Lefschetz

D3.13.2 O. Veblen to Father J.F. O'Hara, [Princeton] 2.12.1938

OVP, *RefGen*, fol. 1r
Dear Father O'Hara: Because of a rush of other affairs, I was unable to send you the promised letter about the Italian mathematicians before your departure from New York. When I finally got around to it, I was lucky enough to be able to get Lefschetz to discuss the matter with Professor Zariski of Johns Hopkins and write you directly.[178] These men know the Italian situation much more closely at first hand than I do, and I think that you can place absolute reliance on the information Lefschetz has sent you.[179] I may add that I think he has been a little too restrained in what he says in favor of the two younger men, Segre and Terracini. In my opinion they are both mathematicians of a very high order. The Institute for Advanced Study is inviting Professor Fubini, who is now in Paris, to come to Princeton for the second term of the present academic year.[180] The money we have available for the purpose is a very small amount indeed, but we hope that it may at least help him to get across the ocean with his family. I need not add that I shall be glad to cooperate in any way possible in trying to find new opportunities for any of these colleagues. I was disappointed that you were not able to visit Princeton before you sailed, but hope that we may see you on your return trip. Yours sincerely, Oswald Veblen

---

[178] Oscar Zariski (1889–1996) was probably the best man to assess the Italian refugees because he had spent in Rome six years (1921–1926) studying with Castelnuovo, Enriques, and Severi, before moving to the United States in 1927.

[179] Salomon Lefschetz (1884–1972) had been to Italy several times, the last in Rome in 1931, when he had presented the conference *Ricerche recenti in topologia* at the Mathematical Seminar directed by Castelnuovo and Enriques.

[180] See Chap. 8.

D3.13.3 C.E. Pickett [American Friends Service Committee] to O. Veblen, Philadelphia 7.12.1938

OVP, *RefGen*, fol. 1r
Dear Dr. Veblen: Dr. Aydelotte[181] has written to us that you are concerned about the refugee problem and would like to work in cooperation with us, finding places for refugees, especially in teaching. The American Friends Service Committee is organizing a Refugee Service staff [The Home Service Field Secretary] and a Refugee Committee to take care of the appeals that come to us, in a more adequate way than heretofore. As many of our cases are professional people, your help would be of value to us. Could you attend committee meetings? Or would you prefer to cooperate with us in the educational field? I should be very glad to see you to talk of this matter, if you could come to Philadelphia. Very sincerely, Clarence E. Pickett[182]

D3.13.4 O. Veblen to J. Branson [American Friends Service Committee], [Princeton] 17.12.1938

OVP, *RefGen*, fols. 1r, 2r
My dear Miss Branson: May I, first of all, thank you for your consideration and the large amount of time and patience which you devoted to hearing my story about the refugees. I mean to send you a considerable number of cases and have talked with Professor Weyl, who will do the same. My understanding of the general plan is that an attempt will be made through the various connections of your organization to find colleges and private schools in which suitably qualified refugees could find employment. I hope that it may be possible, when you have had time to develop your organization, actually to send some representative who can visit institutions in parts of the country which have not yet been reached, to seek for openings. I have already written to Professor Carl Menger, Mathematics Department, University of Notre Dame, Indiana, to ask him to send directly to you any information which he has with regard to the whereabouts of Professor Thirring.[183] I left in your hands a brief record of Professor Thirring's scientific life and standing, and we thought that the first step would be to find what his actual circumstances are and whether anything can be done to help him. Another case

---

[181] Franklin Ridgetway Aydelotte (1880–1956), Director of the Institute for Advanced Study in Princeton (1940–1947).

[182] Clarence Evan Pickett (1884–1965) a Quaker leader and professor in Biblical Literature at Earlham College, was executive secretary of the American Friends Service Committee (1929–1950).

[183] Karl Menger (1902–1985), the son of the economist Carl Menger, taught Mathematics at the University of Notre Dame from 1937. Menger had attended physics lectures by Thirring at the University of Vienna, in 1920–21.

which I think is a subject for inquiry before anything else is done, is that of Professor Friedrich Kottler, Streichergasse 4, Vienna III, Germany, for whom I enclose a curriculum vitae.[184] According to his letters he is desperately anxious to get away from the Nazis and would be willing to do any kind of respectable work in order to support himself. Nevertheless, there remains in my mind a question whether he would be better off in a new and untried environment than where he is. The answer to this question could perhaps be obtained by a discreet investigation on the spot. I am also enclosing statements about two Italians, Professors Terracini and Segre, both scientists of very high rank, and very active at the present time. I am asking Professor Zariski of Johns Hopkins University to write you about them. I am leaving Princeton next Tuesday for a month's vacation, but I think that you may receive accounts of some of the other scientists whose difficulties have come to our attention, from Professor Weyl and from Miss Blake, the secretary of the School of Mathematics. Yours sincerely, Oswald Veblen

P.S. A letter from Professor Ladenburg to Professor Thirring has just been returned, with the single word "Zurück."[185]

P.P.S. I understand from Miss Blake that the records referred to as enclosed will have to wait and follow this letter.

D3.13.5 B. Drury [ECADFS] to M. Ascoli, New York 6.1.1939

ECADFS Records, *MA*, fol. 1r

Dear Professor Ascoli: Professor Harlow Shapley asked us the other day for information about the refugee situation in Italy. I felt that we could give him very little information other than that which we obtained from the National Coordinating Committee.[186] If he has not already been in touch with you about this problem, could you give him any data which would be useful to him in his project? We wrote him that we were turning to you for information. I am enclosing for your information a copy of Professor Shapley's letter. Sincerely yours, Betty Drury - Assistant Secretary

---

[184] Friedrich Kottler (1886–1965), Austrian theoretical physicist. Dismissed in 1938, immigrated to America, settling in Rochester, New York, where he worked at the Eastman Kodak Research Laboratory.

[185] Rudolf Walter Ladenburg (1882–1952) was a German atomic physicist. He emigrated from Germany as early as 1932 and became a Brackett research professor at Princeton University. He was the principal coordinator for job placement of exiled physicists in the United States.

[186] D3.3.8.

D3.13.6 H. Weyl to H. Shapley, [Princeton] 11.1.1939

OVP, *RefGen,* **fol. 1r**

Dear Professor Shapley: Here is the first batch of dockets concerning the refugees in the field of mathematics and physics, which we have compiled in our office. The rest will soon follow. It takes us much more time than we had expected to collect all the material. We are sending the same information to the American Friends Service Committee (Miss Julia Branson), 20 South 12th Street, Philadelphia, Pa. The opinions given were meant primarily as information for the Friends Service Committee. To institutions interested in the men and desiring their services they should be passed on, as you will see, *with discretion.* If the men invited by our Institute without stipend – namely, Hamburger, Hellinger, and Rosenthal[187] – succeed in getting their visas for entering the country, you may soon expect us to apply for your assistance, as our own funds for stipends are at present exhausted. With best regards, yours sincerely Hermann Weyl

P.S. Following is a list of the records enclosed: Bernard Baule, Paul I. Bernays, Alfred T. Brauer, Max Dehn, Adalbert Duschek, Felix Ehrenhaft, Robert Frucht, Guido Fubini, Wilhelm Gross, Hans L. Hamburger, Ernst Hellinger, Edward Helly, George Jaffe, Friedrich Kottler, Karl Löwner, S. Lubelski, Anton E. Mayer, Fritz Reiche, Beniamino Segre, Alessandro Terracini.[188]

---

[187] Hans Hamburger (1889–1956), professor in Cologne. Left Germany in 1939, and from 1941 to 1946 he was lecturer at the University of Southampton. Ernst Hellinger (1883–1950), professor at Frankfurt a/M (1914–1936). He was arrested on November 13, 1938, but his friends were able to arrange a temporary job for him at North-western University at Evanston, Illinois. He was released from the Dachau camp after six weeks, on condition that he emigrated immediately. He joined the Faculty at North-western University in 1939. Arthur Rosenthal (1887–1959), professor at Munich and at Heidelberg, moved to the Netherlands in 1936, then to the United States in 1939. He was appointed lecturer and research fellow at the University of Michigan in 1940 with a promotion to assistant professor in 1943. He also had a visiting professorship at the University of New Mexico (1941–1942).

[188] Bernhard Baule (1891–1976), professor of mathematics at the Technische Hochschule, was charged with treason, because of his Catholic and pacifist activities. To gain his release from the Dachau camp, Weyl arranged for a job at Trinity University (Texas). Paul Bernays (1888–1977), dismissed from the University of Göttingen in 1933, moved to Switzerland, whose nationality he had inherited from his father, and got a position at the ETH in Zurich. Alfred Brauer (1894–1985), assistant to I. Schur at the University of Berlin (1926–1935), left Germany in 1939 and went to the Institute for Advanced Study as Weyl's assistant. He built up the first mathematics library at the Institute for Advanced Study. Max Dehn (1878–1952), professor at Kiel, Breslau, and Frankfurt a/M, was forced to retire in 1935. He spent the year 1939–1940 as professor at the Norwegian Institute of Technology in Trondheim. In October 1940 came to the United States by way of Siberia and Japan. He obtained a professorship of mathematics and philosophy at the University of Idaho, in Pocatello. Adalbert Duschek (1895–1957), *Privatdozent* at the Technical University in Vienna. On April 22, 1938, shortly after the Anschluss, he was given leave of absence with immediate effect and shortly afterwards he was dismissed. From 1940 he worked as a technical employee and consultant for the company Elin AG. Felix Ehrenhaft (1879–1952), professor of Experimental Physics at the University of Vienna, emigrated first to England, then to the United States. Eduard Helly (1884–1943), lecturer at the University of Vienna, held important positions in the actuarial

D3.13.7 H. Weyl to J. Branson [American Friends Service Committee], [Princeton] 11.1.1939

OVP, *RefGen*, fols. 1r, 2r

Dear Miss Branson: Here is another batch of dockets concerning refugees in the field of mathematics and physics: Bernard Baule, Paul I. Bernays, Alfred T. Brauer, Max Dehn, Adalbert Duschek, Felix Ehrenhaft, Robert Frucht, Guido Fubini, Wilhelm Gross, Hans L. Hamburger, Ernst Hellinger, Edward Helly, George Jaffe, Karl Löwner, S. Lubelski, Anton E. Mayer, Fritz Reiche. I am sorry that this is not yet the end; the rest will soon follow. Compiling all the material takes considerably more time than we had expected. We are sending the same to Professor Harlow Shapley of Harvard, who also requested such information. The opinions given are meant primarily as information for your committee. To institutions interested in the men and desiring to secure their services they should be passed on *with discretion*. I am afraid too much emphasis has been laid on research; however, our information as to teaching and linguistic abilities is somewhat scanty. It seems to me it would be highly important for you to make contact and exchange information with the Society for the Protection of Science and Learning, 6 Gordon Square, London W.C. 1, England, which has done the most efficient work for academic refugees in Europe. Their fourth annual report is just out. If you wish, I could send it to you. With regard to the three men who have been invited to the Institute for Advanced Study without stipend – Hamburger, Hellinger, Rosenthal – I have to remark that our funds for stipends are at present exhausted. So we may soon be in the position of applying to you for succor in their behalf. Dr. Hans Freudenthal, Molenbeekstr. 6^1, Amsterdam Z, Holland, advised us under date of December 27 last that they all have obtained, or are about to obtain, temporary visas to Holland.[189] Personally I wish to thank you for the list of requirements imposed by the several states on the practice of medicine. A dismal prospect. Yours sincerely, Hermann Weyl

P.S. You may wish to note the time limit in the case of Robert Frucht. [see Chap. 3.5]

---

field and as consultant with financial institutions. In 1938, he escaped to America where he got a position at the Paterson Junior College and later at Monmouth Junior College (Long Branch, N.J.). Salomon Lubelski (1902–1941), Polish number theorist, died in Majdanek concentration camp. Anton E. Mayer (1903–1942), assistant at the chair of Descriptive Geometry at the Vienna University of Technology, (1930–1938). After the annexation of Austria, he emigrated to England.

[189] OVP, *RefGen*.: H. Freudenthal to J. von Neumann, O. Veblen and H. Weyl, Amsterdam 27.12.1938.

D3.13.8 H. Shapley to H. Weyl, Cambridge Mass. 17.1.1939

OVP, *RefGen*, fol. 1r
Dear Dr. Weyl: I acknowledge the receipt of the dockets for twenty refugees in the fields of mathematics and physics. Since my movement for rather large-scale assistance for the refugees has been taken over, at least tentatively, by the Association of American Universities, progress is and will be naturally slow. But I think the progress will continue to be blessed with a positive sign. As you intimate, a very serious difficulty ahead of us is the problem of getting the men released from Germany and Austria. I believe it is still possible, but not without trouble and complications, for the increasing number of exiles in Czechoslovakia and in Italy. Sincerely yours, Harlow Shapley

D3.13.9 H. Shapley to H. Weyl, Cambridge Mass. 11.3.1939

OVP, *RefGen*, fols. 1r
Dear Dr. Weyl: [...] I have awakened the interest of the University of Louisville in the possibility of an exiled mathematician, but must await discussion with a representative of the place before I can be sure what is the next step. When I have had this conference I shall probably ask you and Veblen to suggest an appropriate person to call, if we can fix the matter up financially. For instance, I should like to have an estimate, by return mail if it is possible, of the minimum decent stipend one could offer Beniamino Segre. But there are probably more distinguished mathematicians on the list – men more worth rescuing. Very sincerely yours Harlow Shapley

D3.13.10 O. Veblen to J.B. Conant cc. H. Shapley, [Princeton] 29.4.1939

OVP, *RefGen*, fols. 1r
Dear President Conant: At the suggestion of Professor Shapley, I am taking the liberty of sending you a list of mathematicians and mathematical physicists from Central Europe who have lost their positions and have not yet found any sure haven elsewhere. The names are arranged very roughly in order of scientific importance. They are taken from a still longer list of refugees whom Professor Weyl and I have been trying to help in one way or another. If there is any way in which I can help with the refugee problem, either in its general aspects or in special cases, please do not hesitate to call on me. Yours sincerely, Oswald Veblen

D3.13.11 A. Johnson [The New School for Social Research] to O. Veblen, New York 6.9.1940

OVP, *RefGen*, fols. 1r

Dear Professor Veblen: we are contemplating the addition of a considerable number of refugee scholars to our staff. Among the persons we are considering are Peter Pringsheim,[190] the former Professor of Physics at the University of Berlin, and Beniamino Segre, the Italian Mathematician. It occurs to me that you might be able to give us information about them and thus to facilitate eventual action. May I ask you, therefore, to provide us with such data as you may have concerning these scholars? Above all, will you give us brief but concrete appraisals of their scholarly and teaching careers? Thank you very much for any assistance you can lend. Sincerely, Alvin Johnson Director

D3.13.12 S. Duggan [ECADFS] to H. Miller [RKF], New York 23.12.1940

ECADFS Records, *BS*, fol. 1r

Dear Dr. Miller:[191] Dr. Frank Aydelotte has written me about the cases of four mathematicians who were still in Europe and who had been suggested at one time as possible candidates for assistance by the Rockefeller Foundation. These men were the following: Professor Beniamino Segre of the University of Bologna; Professor Hans Hamburger, University of Cologne; Dr. Felix Pollaczek, has worked in statistics and communication problems; Dr. Kurt Mahler, temporary assistant lecturer at Manchester University.[192] Dr. Aydelotte has asked me whether any action was taken by the Foundation on any of these men, and since I do not know, I am writing to you for information. I should greatly appreciate a word from you at your convenience. Sincerely yours, Chairman [S. Duggan]

---

[190] Peter Pringsheim (1881–1963), dismissed in 1933, moved to Belgium, where he was arrested in May 1940. With the help of high-ranking political figures (Pringsheim's sister Katia Mann was the wife of Thomas Mann) he was released on 6 December 1940. His former colleague James Franck campaigned for him, and finally a third-party and limited job was managed at the University of California.

[191] Harry Milton Miller Jr. (1895–1980) served the Natural Sciences Division of the Rockefeller Foundation as a fellowship administrator (1932–1934), then as assistant Director (1934–1946).

[192] Felix Pollaczek (1892–1981), Austrian-French engineer and mathematician. Dismissed from the German Postal, Telephone, and Telegraph Services in 1933, he reached Paris, where in 1939 he was appointed as Maître de Recherches at the Centre National de la Recherche Scientifique. Kurt Mahler (1903–1988), number theorist, left the University of Königsberg in 1933 and accepted an invitation by L. Mordell to go to Manchester. At the start of World War II he was interned as an enemy alien in Central Camp in Douglas, Isle of Man, where he met Segre.

D3.13.13 S. Duggan [ECADFS] to F. Aydelotte, 29.1.1941

ECADFS Records, *BS*, fol. 1r
Dear Aydelotte: Some weeks ago you wrote asking whether any further action had been taken by The Rockefeller Foundation on the cases of Professor Beniamino Segre, Professor Hans Hamburger, Dr. Felix Pollaczek and Dr. Kurt Mahler. I have had a reply from Mr. Harry M. Miller, Assistant Director of the Natural Sciences, in which he made the following statement: "at the present time we have under consideration the possibility of a type of local assistance to the more distinguished scholars who have taken refuge in England, and until it is determined whether this plan can go forward, we are holding in abeyance all requests for individuals." I shall let you know if I learn of any change in the situation. Sincerely yours, Chairman [S. Duggan]

D3.13.14 A. Johnson [The New School for Social Research] to O. Veblen, New York 23.5.1941

OVP, *BS*, fol. 1r-v
Dear Mr. Veblen: Since last August about a thousand names have been submitted for consideration under our project for enriching American scholarship and science by rescuing European refugee scholars. We have invited almost 100 scholars to the United States, of whom 25 are now here and at work. Thus nearly the entire fund available is earmarked. Perhaps a few of those invited will be unable to get here. In that case, some money would again be available, but it would be a small amount and the scope of the project in the future will be extremely limited.

I want to use this moment to communicate with colleagues and friends who have been in touch with us concerning refugee scholars for whom, unfortunately, we will be unable to do anything within the limits of the present project. We have been restricted in our selections first of all by a legal factor. We could appoint only scholars eligible for a non-quota visa by virtue of having done sufficient teaching in a university. Moreover, we could get financial support only for persons young enough so that, during the initial two-year period at the New School, they may hope to win a permanent post in some American institution. The general rule was to appoint nobody over fifty-five. Naturally, many able persons were ruled out because of these restrictions. Additional reasons why persons whose names were suggested, or concerning whom we made inquiries, could not be selected, included the following:

1. The candidate was reported to be in no particular danger or distress.
2. The candidate could not get permission to leave his country, or a change in circumstances caused a change in his plan to leave.

3. The candidate was a refugee scholar established at an institution whose work was continuing, and which had given hospitality to a group whose status would have been damaged by the premature departure of any one individual.
4. The candidate was in a relatively safe country, e.g., Brazil.
5. Our efforts to gather the necessary data about the candidate were more or less fruitless.
6. The candidate was, in a general way, eligible and available, but proved to be, according to the consensus of opinions of colleagues consulted, not the first choice among the many in his field.

You will understand that we were able to select fewer than one in ten of the names submitted, and that the person indicated above [B. Segre], concerning whom we had correspondence with you, was omitted from our list only after careful consideration. I am sending you this information primarily for the sake of this scholar, since you may want to seek some other way of helping him. We would be glad to provide you with whatever further data we have. May I thank you again for the help you have given us, without which it would have been impossible to administer this project successfully. And may I add the suggestion that, if you learn of additional refugee scholars who meet the technical requirements, you continue to send me their names. Should some now trying to come over fail to do so, we would, again with the advice of our friends and colleagues, appoint others to their places. To that limited extent, our project continues to function, and I hope for your further cooperation. Sincerely yours, Alvin Johnson Director

D3.13.15 H. Weyl to H. Shapley, [Princeton] 18.7.1941

OVP, *RefGen*, fols. 1r, 2r
Dear Professor Shapley: In reply to your kind letter of July 8, I am sending you herewith copies of letters dated June 18, 1941, to Dr. Laurens H. Seelye of the Emergency Committee for Displaced Foreign Scholars; and dated March 22, 1941, to Alvin Johnson; and copy of a list of French mathematicians which I drew up in October last after a conversation with Warren Weaver, at his express desire. The letter to Dr. Seelye contains the names of those refugee mathematicians in this country whom I knew to be unplaced at that time [Arthur Rosenthal, Wolfgang Sternberg, Stefan Bergmann, Alexander Weinstein, Peter Scherk, Alfred Basch].[193] Since then A. Weistein has got a one-year job at the University

---

[193]Wolfgang Sternberg (1887–1957), former professor at Bresalu (1927–1933), moved to Prague, then to the United Kingdom before fleeing to the United States, in February 1939. He got a position at the Cornell University and worked at the Ballistic Research Laboratory in Lakehurst. Stefan Bergman (1895–1977) taught at Berlin and at the Technological Institute of Tomsk. He emigrated first to Russia, then to Paris and finally, in 1939, to the United States, where he lectured at MIT, at Yeshiva College in New York City, and at Brown University in Providence, Rhode Island.

of Toronto, and I could add to the list the names of two physicists, Felix Ehrenhaft and F.R. Kottler, and one mathematicians, Hans Fried.[194] I am not absolutely sure whether Kottler is still unplaced. Here is a list of refugee mathematicians in this country who are only temporarily placed (I cannot guarantee that it is complete): Alfred T. Brauer, Max Dehn, Guido Fubini, Hilda Geiringer, Kurt Gödel, Ernst Hellinger, Edward Helly, R. von Mises (?),Wolfgang Pauli (physicist), George Pólya, Fritz Reiche (physicist), Carl L. Siegel, Alfred Tarski, André Weil, Alexander Weinstein.[195] About most of these men you will find information in the records which I have sent you. As far as I remember this is not the case for the following men, who are all of the very first rank: Gödel, Pauli, Siegel, who are temporarily placed at our Institute; R. von Mises, about whom you know better than I; and André Weil, now residing at 1 Chambers Terrace, Princeton, who was recently rescued from France by the Alvin Johnson organization. I enclose Pólya's curriculum vitae. The letter to Alvin Johnson about the French "Bourbaki" group is self-explanatory. Since writing that, I have received a form letter from Alvin Johnson advising me that nothing can be done for them.

Alexander Weinstein (1897–1979) lectured at the University of Breslau (1928–1933). He spent the years 1933–1940 in Paris, before escaping to the United States through Portugal. Weinstein taught at the Free French University in New York, did war work in the research group of Harvard University chaired by G. Birkhoff, worked at the Carnegie Institute of Technology of Pittsburgh and the Naval Ordnance Laboratory in Maryland. He also received a temporary appointment at the University of Toronto in 1941. Peter Scherk (1910–1985), dismissed from the German University of Prague, emigrated to the United States in 1939. Worked at the Taft school, and at the Universities of Yale, Indiana, and Saskatchewan. Alfred Basch (1882–1958), obtained the doctorate from the Vienna Technical University and combined teaching at technical schools in Dresden, Prague, and Vienna with engineering practice. In 1938 he went to the USA, where he taught at Holy Cross College (Worcester, Mss.), in the Paterson Junior College and in the Harvard Summer School of Engineering (1942).

[194] Hans Fried (1893–1945), a teacher in Vienna from 1927 to 1938, emigrated to England, then to the United States. After some months in New York, a half year at the Haveford Workshop and the summer of 1941 at Brown University, he became a research assistant at the Sproul Observatory in the autumn of 1941.

[195] Richard von Mises (1883–1953), professor of Applied Mathematics at Berlin (1920–1933), later at Istanbul (1934–1939), in 1939 moved to the United States, where he was visiting professor at the Graduate School of Engineering of Harvard University, lecturer at MIT, and visiting professor at Brown University. In 1944 he was appointed as Gordon McKay professor of Aerodynamics and Applied Mathematics at Harvard University. Wolfgang Pauli (1900–1958), professor of Theoretical Physics in Zürich (1928–1940), in 1940 moved to the United States, where he was employed as a professor of Theoretical Physics at the Institute for Advanced Study. Georg Pólya (1887–1985), professor at Zürich (1928–1940), in 1940 was called to a permanent appointment at Stanford University. Carl Siegel (1896–1981), professor at Frankfurt a/M, 1922–1937, and at Göttingen, 1938–1940, emigrated in 1940 via Norway to the United States, where he joined the Institute for Advanced Study. André Weil (1906–1998), lectured at Aligarh University (Delhi) and at Strasburg. Arrested in Finland at the outbreak of the Winter War on suspicion of spying, he was imprisoned in Le Havre and then Rouen. After the fall of France in June 1940, he met up with his family in Marseille, whence they sailed to New York in January 1941. He spent the remainder of the war in the United States, at Haverford College, and at Lehigh University, supported by the Rockefeller Foundation and the Guggenheim Foundation.

Nevertheless I know that the project is not yet dead and that the Rockefeller Foundation is still interested in it. In the above-mentioned conversation with Warren Weaver, Veblen and I recommended in the first place the following mathematicians whom it would be desirable to bring over from Europe: Beniamino Segre and Hans L. Hamburger, both living in England, and Felix Pollaczek living in unoccupied France. Veblen also took up the case of Segre directly with Alvin Johnson and has been informed by the same form letter as mine that nothing could be done for him. On January 3, 1941, the Rockefeller Foundation informed me "we have under consideration a plan which, if approved, will afford a type of local assistance to scientists of eminence in England, and until it is determined whether we shall be able to forward such a plan, requests for assistance to scholars in England to come to this country will be held in abeyance." Veblen's suggestion was for the Rockefeller Foundation to give financial assistance to the English "Society for the Protection of Science and Learning" so as to enable them to increase their aid to these refugees, but we have never heard whether anything has been done about it. When we first set up our records for the Friends Service Committee, we expected very much from their help in placing people in smaller colleges. But two recent letters from that Committee seem to indicate that the whole arrangement had fallen into oblivion, and as a matter of fact there is not a single case on our list where they have been successful in placing a man, though they have been helpful in a number of other ways. That is why I thought that we should have to revamp our procedure. What I have done is simply this, - communicated directly about the most urgent cases with Dr. Seelye and with Dean Richardson of Brown University.[196] But the Friends Service Committee also wants us to continue sending them news of the cases already in their hands. You have probably heard that Issai Schur, about whom I negotiated with you several times, died in Palestine early this year.[197] Thanking you for your continued interest, Sincerely yours, Hermann Weyl

P.S. Our Institute has taken steps to invite Langevin, but it seems a delicate matter to get in touch with him, and even more so to get him out of France.

---

[196] Roland George Dwight Richardson (1878–1949), Head of the Department of Mathematics of Brown University (1915–1942) and Dean of the Graduate School (1927–1948).

[197] Issai Schur (1875–1941), outstanding mathematician known for his fundamental work on the representation theory of groups, in number theory and analysis, had been dismissed in 1935. He had received invitations to go to the United States and to United Kingdom, but he had declined them all, unable to understand his persecution. Schur had left Germany for Palestine in 1939, broken in mind and body, having the final humiliation of being forced to find a sponsor to pay the 'Reich flight tax'. Without sufficient funds to live in Palestine he had been forced to sell his library to the Institute for Advanced Study.

3.14   The 'Choice' of the Destination: United Kingdom

Soon after the Nazi takeover in Germany, the Academic Assistance Council was founded on the initiative of W. Beveridge, then Director of the London School of Economics. Nobel laureates E. Rutherford, then President of the Royal Society, and A.V. Hill were appointed President and vice-President, respectively (22 May 1933). The American counterpart of the Academic Assistance Council, the ECADGS, was to be founded a few weeks later. The Council, reorganised as SPSL in 1936, had a threefold purpose: to coordinate the attempts by individual British scholars and institutions to place refugees in appropriate research or teaching positions; to create a fund for financial support of displaced scholars, and to act as a placement service putting academics in touch with institutions. In regular contact with scholars established in the United Kingdom and the United States, like Bohr, Weyl, and Courant, the Society requested their confidential opinions about the merits of each would-be refugee (D3.14.1, D3.14.2). In many cases, an initial arrival in Great Britain was followed by a later re-establishment in the United States.

Frequent referents for mathematics were S. Brodetsky (University of Leeds), a Russian-born applied mathematician, member of the World Zionist Executive and the President of the Board of Deputies of British Jews; the number theorists Godfrey H. Hardy and John E. Littlewood (both in Cambridge), Louis Mordell (Manchester, field of expertise: the geometry of numbers), and the algebraic topologist pupil of Veblen and Lefschetz, John H.C. Whitehead (Oxford).

None of these had thematic, political, philosophical, and cultural affinities with the Italians. The Italian mathematicians were not Zionists; number theory was not cultivated in Italy and after the death of Bianchi its followers were a very tiny group; geometry of numbers and algebraic topology were not among the Italian highlights. The Italians had contact with other British mathematicians. Baker, for example, had been on friendly terms with Segre since the late 19th. Baker and Roth had met Castelnuovo and Terracini at the ICM of Cambridge in 1912. Hodge, Lowndean professor of Astronomy and Geometry at Cambridge during the 1930s, had spent time in Rome working with Severi and Enriques.

The lack of direct relationships with SPSL mathematics consultants had a notable impact on the (failed) emigration of Italian mathematicians to Great Britain and the colonies of the British Empire. The SPSL, in fact, was never a general emergency relief organisation and always operated as an academic placement service. On initial contact, refugees were asked to complete a questionnaire. Then, the Society had a network of scholars from whom confidential opinions were solicited about the applicants; as was to be expected, opinions sometimes diverged, complicating the task of prioritising requests. Personal contacts and recommendations played an essential role, even for refugee mathematicians with strong academic resumés.

The financial coverage of the SPSL in turn had an impact on determining the fate of the Italians. At the time leading up the outbreak of the Second World War, the SPSL was financially on a firm footing and in early 1939 the going rate for a maintenance grant to a single scholar placed in Great Britain was £182 p.a. However, in December 1939 the Society Grants Committee had reluctantly come to the conclusion that all grants should be considerably reduced, and in many cases discontinued, in spite of several successful fundraising events.

In light of these facts, it is understandable that the SPSL had a limited role in helping the Italians. In its Archive 14 'Italian files' are stored. Seven of these belonged to mathematicians: Castelnuovo, Fano, Maroni, Segre, Terracini, Giulio Supino, and Scipione Treves. However, only Fano, Segre, and Terracini had direct contact with the Society. Other files contain merely the dates of dismissal and reinstatement in their university positions, and very little other information.

As far as Fano is concerned, the SPSL acted as a middleman. Fano's family had succeeded in exporting its fortune into Switzerland. The Society agreed to act as a conduit for monthly payments to Fano from his own family funds. At the start, SPSL sent Fano £10 cheques drawn at a British bank; later a Swiss bank account in the name of SPSL was set up in Lausanne for this purpose. In a farcical twist, the Society ran into trouble with the Foreign Exchange Control for violating the British Defence Finance Regulations, and the arrangement was terminated in 1941.

In Segre's case, the SPSL was invited to complement and channel financial support established by a group of British mathematicians which collected funds for an earmarked grant. The initiative was coordinated by J.G. Semple and W.V.D. Hodge. The Society assisted Segre and his family also in the application for a permit to enter and work in England. Moreover, when the comprehensive internment of Italian citizens in Great Britain was set off—' Collar the lot!' Churchill is quoted as having said in response to the Italian declaration of war—the Society appealed for his release.

Terracini, despite multiple moves, never managed to obtain a grant from the SPSL.

For Segre, and the other Italian academics who went to the United Kingdom (Enrico Volterra, Leo Pincherle, etc.), however, Great Britain was supposed to be a temporary destination, en-route towards the United States or Latin America. In regular contact with the ECADFS, the Society frequently forwarded information about vacancies in Great Britain and its dominions, and vice-versa asked the Emergency Committee to circulate through the American universities the files in its lists. Neither for Segre nor for Volterra or Terracini did mediation end up in an appointment overseas.

Documents

D3.14.1 H. Weyl to E. Simpson [SPSL], 1.7.1940

OVP, *RefGen*, fol. 1r
Dear Miss Simpson: I am deeply impressed by the way in which your Society, in the teeth of the present grim circumstances, is carrying on its work on behalf of the Central European refugees. Unfortunately, so far no opportunity whatsoever

has turned up for Dr. Hamburger in America. Professor Veblen recommended him in Melbourne, but very likely his application fell through just as did that of Beniamino Segre. There is still a list of first-rate mathematicians with only temporary provisions over here, including for instance Alfred Brauer, Kurt Gödel and Carl L. Siegel, who are in Princeton. In recent months a number of Polish mathematicians have been absorbed, and we have worked hard on the Copenhagen group.[198] The train of Hitler's victims in other lands detracts from the interest in the remaining German refugees who, besides, share in the public eye, to whatever slight degree, the suspicion and the taint of the German name. It is a pity that there materialized no large-scale plan for bringing over to this country before, or right after, the outbreak of the war, Central European refugee intellectuals who are now so hopelessly caught in the whirlpool of events. I have the greatest interest in Dr. Hamburger, and Professor Veblen and I will continue our efforts, but at the present moment I can hold out but little hope. Yours sincerely, Hermann Weyl

P.S. I have learned this moment that Dr. Max Dehn, up to now in Trondheim, Norway, got a position at the University of Idaho, Southern Branch, Pocatello, Idaho, U.S.A.

D3.14.2 E. Simpson [SPSL] to H. Weyl, London 8.7.1940

OVP, *RefGen.*, fol. 1r
Dear Professor Weyl, Thank you for your letter of July 1st. It is of course a great disappointment to us that you are at present unable to help Dr. Hamburger. Our only hope for scholars now lies in U.S.A., as there is no chance in Europe now and will not be for a long time to come. We appreciate very much the efforts already made by colleagues in U.S.A., especially the mathematicians who have co-operated so magnificently. We understand the difficulties you are up against. Nevertheless, we beg you to persist and try and achieve what seems almost impossible, because here it *is impossible*, and if hope in USA is taken away, our scholars must despair. For seven years we have striven to help them build up a new existence in freer surroundings, which could benefit from their exceptional knowledge and experience. Considering the forces arrayed against such work, we had up to the outbreak of war been singularly successful. Now the war has swept this away, and it will be very long before there can be hope again of building up what has been destroyed. A few individuals will win through, no doubt; for the rest we must *turn to the New World*. Yours sincerely, Esther Simpson - Secretary

---

[198]Weyl refers to the group of physicists working at the Institute for Theoretical Physics of the University of Copenhagen, under the guidance of N. Bohr.

P.S. We are very relieved to hear that Dr. Dehn has managed to get to U.S.A. from Norway.

3.15   The 'Choice' of the Destination: Latin America

South America was the most popular destination for Jews leaving racist Italy, who joined the political exiles of the first wave (the Italian anti-fascists who arrived in the twenties and thirties) and preceded the so-called *Los de después*, i.e. the fascists and Nazis who would take shelter there in the post-Second World War period. As the chances of obtaining a post in the English-speaking countries faded, the aspiring immigrants began to take into consideration Argentina, Brazil, Peru, Uruguay, Venezuela, Bolivia, Paraguay, Colombia, and Cuba, which were among the few routes they could still glimpse (D3.15.1, D3.15.2, D3.15.4)[199]

There was no lack of favourable conditions, starting from a linguistic and cultural affinity between Italy and Latin America, certainly greater than that with America or England.[200] In the first place, there were many contacts with scholars such as Rey Pastor, the eclectic pioneer of mathematics in Latin America, and with young scholars such as Rosenblatt, García, Vignaux, Pla, J. Babini, J. Blaquier, and F. La Menza, who had attended the international congress of mathematicians in Bologna in 1928 and had published their works in Italian mathematical journals, thanks to the help of Volterra, Levi-Civita, Castelnuovo, Enriques, Fubini, Levi, and Segre. In addition to this, there was the imprint that the teaching of Enriques (1908, 1928), Volterra (1910, 1921), Severi (1930), and Levi-Civita (1923, 1937) had left in Argentina, Peru, and Brazil.

Two other decisive and partly overlapping circumstances played in favour of emigration to South America in the eyes of Jewish mathematicians fleeing from racist Italy: the presence of the Italian mathematical mission in Brazil and the fact that three newly formed universities needed to appoint their staff. The so-called Italian mission (*missione italiana*) arrived in São Paulo in 1934, the year in which the Faculty of Philosophy, Sciences, and Literature was founded.[201] Pressed by the need to cover twenty-two chairs, the Faculty had to recruit foreign scholars, particularly German, French, and Italian. The appointment of Italians was clearly a political act, carried out in the name of the 'affinity of Latin blood' between Italy and Brazil, and facilitated by the good relations that fascism entertained with the

---

[199] In this sense, it is not surprising that Laurence Duggan, son of the Chairman of ECADFS, was entrusted with bringing the cases of displaced Italian scholars to the attention of South American delegates at the 8th International American Conference in Lima (December 1938). See D3.15.1, D3.15.2, and D3.15.4.

[200] For an Italian, Castilian was incommensurably easier to learn than English, in a short time.

[201] On the Italian mission see, for example, Vieira Souza da Silva and Monteiro de Siqueira (2018); Vieira Souza da Silva and Bontempi (2020), and Wataghin (1992).

Estado Novo led by G. Vargas and with the local para-fascist movement *Ação integralista*. The fascist regime sent five scholars to São Paulo, with the dual task of transmitting their discipline and spreading fascist ideology: G. Wataghin (Theoretical Physics), E. Onorato (Mineralogy and Petrography), G. Albanese (Geometry), L. Galvani (Statistics), and L. Fantappié (Analysis). Fantappié and Albanese, who were to remain in Brazil until the outbreak of the war, laid the foundations of the Department of Mathematics of São Paulo, also creating (in 1936) the first Brazilian periodical, the *Jornal da Matemática pura e aplicada*, which published works by Segre, S. Cinquini, and the texts of the conferences held by Levi-Civita during his visit to Brazil.

At the time of the promulgation of the racial laws, Fantappié, a notoriously partisan intellectual, suggested calling Mortara (Jewish but considered to be a good fascist). The suggestion was accepted. Mortara left Italy for Brazil in January 1939, where he stayed till his death. Practically in the same period, the Italian and Brazilian authorities began to talk about a second Italian Mission, this time in Rio de Janeiro. The national Faculty of Philosophy, founded in April 1939, needed to recruit its staff and looked for high-skill and expert Masters, capable of training a new generation of Brazilian researchers. Minister Capanema addressed Fantappié and Wataghin, asking for suggestions. Fantappié proposed A. Bassi; Wataghin named Segre and Terracini (D3.15.3). In the list of candidates drawn up by the Dean Raúl Leitão da Cunha and dated April 17, 1939, Levi-Civita also appeared, suggested for the chair of Higher Analysis. After several tug-of-wars between the Italian and Brazilian Ministries of Education, and various unpleasant manoeuvres within the UMI, finally, three successful candidates were invited to Rio de Janeiro: A. Terracini for Geometry, G. Mammana (Analysis), and L. Sobrero (Mathematical Physics and Theoretical Physics). The three were expected in Brazil by the end of July. Terracini, in the meantime, had received an invitation to transfer to the University of Tucumán and opted for Argentina at the last minute. In his place, Bassi was hired and remained in Brazil for the rest of his life, while Mammana returned to Italy in 1942, just before Brazil declared war on the Axis powers (21 August 1942).

In 1938–1939, also some newly created Argentine Departments had an immediate need to hire staff and so they implemented a targeted recruitment campaign for persecuted professors, recruitment that would lead to the contracts proposed to A. Herlitzka, L. Lattes, B. Levi, R. Treves, G. Arias, M. Finzi, R. Mondolfo, the Terracini brothers, and many others. Unfortunately, such a trend was greatly slowed down from 1941 onwards until its eventual end, when visas were no longer released to the so-called *indeseables*, i.e. individuals belonging to ethnic or religious groups which were considered scarcely integrable into the Argentine racial pot or who did not fit the characteristics of Argentinian-ness.

In the mathematical field, in the early thirties, an interesting development had been achieved in Buenos Aires and La Plata, largely thanks to the impulse supplied by Rey Pastor. However, in the inland cities, the teaching of mathematics was relatively recent and was basically aimed at the preparation of students in the field of engineering and other applied specialties. In 1937, a group of professors from the

Universidad Nacional del Litoral, including J. Olguín, S. Rubinstein, C. Dieulefait and J. Babini, proposed the creation of an Institute of Mathematics and an Institute of Stability. On 10 December 1938, some months after the promulgation of the racial laws in Italy, a commission was appointed to examine the proposal, headed by the Dean of the Faculty C. Pla and formed by professors C.F. Dieulefait, E. Lacal, S. Rubinstein, J. Olguin, M. Erlichman, F.L. Gaspar, and J. Babini and delegate student Néstor Ulivi. On 23 December, the constitution of the two Institutes was approved. The discussion opened on the appointment of their directors and teachers. At this juncture Pla decided to avail himself of the counselling of Rey Pastor and Vignaux who, aware of the difficult plight of Italians, forwarded the request to Levi-Civita. The latter immediately proposed Beppo Levi as Director of the Institute of Mathematics and Enrico Volterra as Director of the Institute of Stability. To cover the chairs of Analysis and Geometry Ascoli, Segre, and G. Supino were all suggested. Trusting in the advice of Levi-Civita, who acted as guarantor of the scientific talent and soft skills of Levi and Volterra, Pla approved their appointment on 10 April 1939. After endless difficulties related to the release of a tourist visa for entry into Argentina, Levi landed in Rosario on 6 November 1939, assuming the leadership of the Institute two days after his arrival. Volterra, meanwhile, was forced to stay in England, having failed to obtain the passport.

Historically remarkable in itself, the moment when Terracini joined the ranks of Universidad Nacional represents the beginning of its golden years. Founded in 1914, this institution included the faculties of Engineering, Pharmacy, Law, and Philosophy. Under the umbrella of the latter, the *Profesorado en Matemática* (i.e. the class designed for teacher training) had been established in April 1937. In June 1939, the third year should be implemented and the existing academic personnel, which comprised just two professors (F. Cernuschi, an engineer who taught Probability, Statistics and Theoretical Physics and J. Würschmidt, a physicist from Cologne, expatriated for racial reasons, who held the courses of Experimental Physics), did not suffice. Hence the invitation to Terracini, a mathematician of genuine distinction, whose arrival would be hailed by students, colleagues, and local authorities as the 'starting point of a new period in the evolution of mathematical studies in the north-west of Argentina'.[202]

---

[202] *Evolución de las ciencias en la República Argentina ...,* 1979, p. 201: *punto de partida de un nuevo período en la evolución de los estudios matemáticos en el noroeste argentino.*

Documents

D3.15.1 C.L. Melton [Counselor at Law] to J.W. Alexander, New York 5.1.1939

OVP, *RefGen*, fol. 1r

Dear Doctor:[203] Shortly before the new year, I had occasion to visit the Cuban embassy in Washington, D.C., and there had a conversation with one of the Cuban representatives concerning the status of immigration to Cuba. I was informed, at that time, that the regulations for entry to Cuba had just been changed involving no change in the temporary admission for those who had quota numbers, providing these quota numbers could be reached within a period of six months, or so, but that for others, there was a very steep advance in the necessary amount of capital guaranteeing their support in Cuba. On the other hand, I ascertained that entry into Trinidad could be accomplished without very much difficulty for those under the age of 40 years. If you still need further information, I would only be too glad to supply it. Very sincerely yours, Charles L. Melton

D3.15.2 Interview with Laurence Duggan, 11.3.1939

ECADFS Records, *GM*, fol. 1r

Laurence Duggan came in this morning to return the papers of displaced scholars which we sent him in November and December when he was attending the conference in Lima.[204] He said he was unsuccessful in finding positions for any of those candidates. He had an opportunity, however, to speak about the scholars to the ministers of education of both Columbia and Venezuela. In general, throughout South America opportunities for refugees are quite limited. There is little money with which to pay university salaries. Interest exists, however, as well as good will in academic circles. Mr. Duggan doubts whether many more can be placed. Those few scholars already taken on have been most successful. Papers returned concerned: Sigmund Cohn, Paul Oskar Kristeller, Heinrich Kahane, Arnold Reichenberger, Friedrich Friedmann, Giorgio Mortara, Eduard Strauss, Adalbert Silbermann, Leonardo Olschki[205]

---

[203] James Waddell Alexander II (1888–1971), topologist, one of the first members of the Institute for Advanced Study (1933–1951), and a professor at Princeton University (1920–1951).

[204] Laurence Duggan (1905–1948), the son of Stephen Duggan, was Head of the Latin American Division (1930–1939). Duggan assisted Secretary of State Cordell Hull at major conferences in Lima, Peru and Havana, Cuba. Positions he held included Chief of the Division of the American Republics as well as Political Adviser and Director of the Office of the American Republics.

[205] Sigmund Cohn (1898–1997), a lawyer and then a judge in Germany's Weimar Republic, in 1934 immigrated to Italy, then to the United States, where he was hired by the Georgia School of Law in Athens. Paul Oskar Kristeller (1905–1999), reader of German at the *Scuola Normale* of Pisa

D3.15.3 A. Terracini to E. Bompiani, Turin 5.6.1939

Terracini Coll., fol. 1r

Dear Bompiani, first, I would like you to immediately receive the expression of my deep satisfaction and my sincere congratulations on the very high and well-deserved recognition of your work.[206] And now I would like to talk to you about a topic that is very close to my heart. Among others, my name was mentioned for the new Faculty in Rio de Janeiro. The name of Segre was also raised but, according to what he wrote to me, probably he will not follow up on the matter on his part, having other positions in his sight; and indeed, he himself has again designated my name. I hear now that the Comm. Castruccio, R. consul general in S. Paolo, has also proposed you.[207] As you can imagine, succeeding in finding accommodation is of fundamental importance for the present and the future of myself and my children; and one of the very few possibilities that I still glimpse is precisely that of Brazil. So, I would be very grateful if you could tell me whether, as far as you are concerned, it is an initiative of the comm. Castruccio or not; and – in this second case – if you would be willing not to insist on your own behalf, and eventually to support my call. I know this is no small favour; but as I told you before this is a matter of vital importance to me. I thank you already with all my heart, and I send you a cordial greeting. Yours, Alessandro Terracini

Caro Bompiani, in primo luogo desidero ti giunga subito l'espressione del mio vivo compiacimento e delle mie congratulazioni sincere per l'altissimo e meritato

---

(1935–1939), in 1939, thanks to the help of G. Gentile, managed to expatriate to the United States and first find a job at Yale University, then at Columbia University in New York. Heinrich Kahane (1902–1992), a Romance philologist and linguist, moved to Florence in 1933, then to Cephalonia (Greece) and finally managed to emigrate to the United States in 1939, where he worked as a research assistant in Comparative Literature at the University of Southern California, Los Angeles (1939–1941). Arnold Gottfried Reichenberger (1903–1977), reader in German at the University of Milan (1934–1938). In 1939–1940 taught at the New School for Social Research in New York City. Friedrich Georg Friedmann (1912–2008), German historian, graduated in Rome in 1937. Dismissed from the Giulio Cesare classical lyceum, in 1939 he emigrated first to London and then to the United States where he taught in various high schools. Eduard Strauss (1876–1952) was a chemist and a philosopher. Dismissed due to the racial laws from the Georg-Speyer-Haus, a chemotherapeutical institute in Frankfurt am Main, he immigrated with his family to Italy in 1938. In December of the same year, he had to leave Italy and immigrated to the United States via Havana, Cuba. Peter Adalbert Silbermann (1878–1944), professor and founder of the evening high school in Berlin, in 1933 emigrated first to Italy, where he worked at the University of Rome, then to the United States. Leo Samuele Olschki (1861–1940), a renowned Italian publisher of Prussian Jewish origin, emigrated in Geneva in 1938. See also Chap. 3.3, p. 118.

[206] Bompiani had won the Royal Prize for Mathematics awarded by the Academy of Lincei for the year 1938.

[207] Giuseppe Castruccio (1887–1985), consul general in San Paulo (1928–1940).

riconoscimento della tua opera. E ora vorrei parlarti di un argomento che mi sta molto a cuore. È stato fatto, tra altri, il mio nome per la nuova Facoltà di Rio de Janeiro. Anche quello di Segre è stato fatto; ma, a quanto egli mi ha scritto, probabilmente egli non darà per parte sua seguito alla cosa, avendo altre sistemazioni in vista; e anzi egli stesso ha nuovamente designato il mio nome. Sento ora che il comm. Castruccio, R. Console generale a S. Paolo, ha anche proposto il nome tuo. Come puoi pensare, il poter trovare per me una sistemazione ha un'importanza fondamentale per il presente e l'avvenire mio e dei miei figli; e una delle pochissime possibilità che intravedo è proprio quella del Brasile. Quindi io ti sarei molto grato se tu mi dicessi se nei tuoi riguardi si tratta di una iniziativa del comm. Castruccio o no; e – in questo secondo caso – se tu saresti disposto a non insistere nei tuoi riguardi, e eventualmente a appoggiare una mia chiamata. So che non ti chiedo un favore da poco; ma come ti ho detto si tratta per me di un problema di importanza vitale. Ti ringrazio sin d'ora di tutto cuore, e ti mando un cordiale saluto. Tuo aff. Alessandro Terracini

D3.15.4 B. Drury [ECADFS] to J.P. Chamberlain, New York 24.7.1939

ECADFS Records, *Italy*, fol. 1r
Dear Professor Chamberlain: Dr. Duggan's secretary has asked me to tell you about a conversation I had with Dr. Duggan's son, Laurence Duggan, who is, as you know, in the State Department. He came into the office shortly after his visit to South America at the time of the Lima Conference. He made a number of attempts during the Conference to bring the cases of displaced Italian scholars to the attention of the South American delegates who might be in a position to do something for them. (We had, before he left, made up a file containing the papers of seven of these scholars). Unfortunately Mr. Duggan said absolutely nothing whatsoever came of these attempts. Sincerely yours, Betty Drury - Assistant Secretary

3.16   Case Study. A Solidarity in Two Verses: Alfred Rosenblatt in Lima

This is quite a strange case of solidarity in two directions: a Pole who had been helped by Italians in the late twenties to emigrate to Peru and who helped his Masters from many years earlier, in the dramatic period of racial persecution. This was Alfred Rosenblatt. Born on June 22, 1880, in Krakow, the son of a professor of Law at the Jagiellonian University, he became a Doctor of Philosophy himself there in 1908; his formal supervisor was S. Zaremba. In the years 1908–1910, Rosenblatt did research training in Göttingen, where his teachers included F. Klein, D. Hilbert, and E. Landau. *Privatdozent* in 1919, he was proposed by S. Dickstein as candidate for chairs in Poznan and Lvov in 1923, 1926, and 1928, but he was always rejected, also due to opposition from Zaremba and Sierpiński. In 1926 and 1928, Rosenblatt spent periods of study in Italy, during which he worked in contact with Fano, Fubini, Enriques, Severi, and Levi-Civita. He attended the International Congress of Mathematicians in Bologna, giving (in Italian) talks in the geometry section and in the section of mathematical physics.[208] Desperately trying to find a permanent position, in May 1930 Rosenblatt informed Dickstein:

> Simultaneously with a letter from you, I received a letter from Professor J. Rey Pastor from Buenos Aires in which he reports on my appointment in La Plata (Argentina, near Buenos Aires, not Lima in Peru!). We will soon receive official notification of the terms, etc. I am being awarded the chair of Mathematics here with the help of Enriques, Severi, Levi-Civita, and Einstein, to whom, in the last instance, the Dean from La Plata turned for an opinion about me. I enjoy the recognition that I have abroad, evidence of which, moreover, I have just had in Liege and Paris. Einstein also told me: I am surprised that Poland allows so many to go, they have so few people.[209]

Rosenblatt obtained paid leave from the Jagiellonian University and a business passport for Argentina, but because of the military coup in September 1930 he could not go there. In the meantime, García, who was Dean of the Faculty of Exact Sciences at the University of San Marcos in Lima, visited Poland.[210] It was there that he established contacts with Sierpiński and Rosenblatt, who soon after wrote to him about the risks many Poles of Jewish origin, including himself and his family, faced due to anti-Semitism. García promptly invited Rosenblatt to work as a lecturer in Lima. His contract included taking over the chair of Astronomy and Geodesy. Rosenblatt accepted the invitation, and in 1936 arrived in Peru.[211] He took with him from Europe the latest news on functional analysis, topology, and other mathematical areas, algebraic geometry included, which was not yet available in Peru at that time.

Shortly after his arrival in Lima, Rosenblatt invited Levi-Civita to go to Peru as a visiting professor. Levi-Civita arrived on 4 August 1937, gave two talks at the local Seminar, and held a course of nine lectures on *The relativistic two-body problem, its solution in the first approximation and possible astronomical control*. A speech

---

[208] A. Rosenblatt, *Varietà algebriche a tre e più dimensioni*, in *Atti del Congresso Internazionale dei Matematici Bologna 3–10 settembre 1928*, Bologna, Zanichelli, vol. 4, 1929, pp. 93–114; *Sopra le varietà algebriche a tre dimensioni fra i cui caratteri intercedono certe disuguaglianze*, vol. 4, 1929, pp. 123–128; *Sopra certi moti permanenti dei liquidi viscosi incompressibili*, vol. 5, 1929, pp. 165–174.

[209] A. Rosenblatt to S. Dickstein in Ciesielska and Maligranda (2019), pp. 60–61.

[210] Godofredo García (1888–1970) was professor of Rational Mechanics in the Faculty of Sciences of the University of San Marcos beginning in 1919 and later served as Dean (1928–1940) and Rector (1941–1943).

[211] Velásquez López (1990).

preceding the lectures on Levi-Civita's scholarly achievements was given by García, which was later translated into Polish by Rosenblatt and published in *Wiadomości Matematyczne*. Shortly afterwards, Rosenblatt addressed Castelnuovo, Fano, and Fubini, asking them to contribute to the *Revista de Ciencias de la Facultad de Ciencias Biológicas, Físicas y Matemáticas de la Universidad Mayor de San Marcos* and to exchange it with Italian journals.[212] The mathematical libraries of Turin and Rome subscribed and received the journal from 1937 to 1939. Finally, in 1938, Rosenblatt (together with other mathematicians from USM) founded the *Academia Nacional de Ciencias Exactas, Físicas y Naturales del Perú*. Fubini and Terracini were immediately elected correspondent members.

These circumstances were fundamental. When, in 1938, Italians were looking for a host country, Levi-Civita directed them to García and Rosenblatt. In the questionnaires of these mathematicians, preserved in the archives of the SPSL, Peru is expressly indicated as one of the favourite destinations, along with Argentina and Brazil. Positions were offered to Fubini, Terracini, and Jacchia (the later as assistant astronomer in the Lima Observatory) (D3.16.1, D3.16.2). Recruitment was not entirely successful, but it still had some positive results. Thanks to Rosenblatt, Levi-Civita, Fubini, Terracini, and Levi joined the Academy of Sciences of Lima and published in the *Revista* and in the *Actas de la Academia*. The figure of Volterra was solemnly commemorated in Lima.[213]

Documents

D3.16.1 A. Rosenblatt to G. Fubini, Lima 4.11.1938

Fubini Coll., fol. 1r
Distinguished Professor Guido Fubini, I had the pleasure of receiving your kind letter [. . .]. I deplore the attitude of the most distinguished scientists in this regard: unfortunately, Europe is now going through a critical situation, in which the high value of the men who made an epoch in the scientific progress of civilised nations is not recognised and venerated. I have taken note of your desire to come to Peru, and when the University Board discuss the plan of conferences and invitations for the next year, I will inform them of your wish, which I will support in the most decisive way. If my efforts are successful, I will be very pleased to communicate it to you immediately. Please accept, distinguished Professor, my high esteem and consideration. [A. Rosenblatt]

Illustre Professor Guido Fubini, ho avuto il piacere di ricevere la sua gentile lettera [. . .]. *Deploro l'attitudine al riguardo dei più illustri scienziati:*

---

[212]BSM-*Fano*: A. Rosenblatt to G. Fano, Lima 21.4.1937.
[213]Rosenblatt (1942).

disgraziatamente ora l'Europa attraversa una situazione critica, nella quale non si riconosce e non si venera l'alto valore degli uomini che fanno epoca nel progresso scientifico dei popoli civili. Ho preso nota del suo desiderio di venire in Perù, e quando discuteremo al Congresso Universitario il piano di conferenze e proiezioni che deve aver luogo l'anno prossimo, farò presente il suo desiderio, che appoggerò nel modo più deciso. Se i miei sforzi avranno esito favorevole, mi farà un vero piacere di fargliene parte immediatamente. La prego gradire, illustre professore, l'espressione della mia alta stima e considerazione. [A. Rosenblatt]

D3.16.2 G. García to G. Fubini, Lima 16.12.1938

Fubini Coll., fol. 1r
Dr. Guido Fubini, very illustrious Professor, and friend: Now I have the pleasure of greeting you and informing you that I have arranged for you to enter the *Cuerpo de Catedráticos* de la Universidad de 'La Libertad'. This is one of the three minor universities of Peru, in the city of Trujillo located on the coast, and equipped with all kinds of comforts and a beautiful climate. If it is suitable for you, you could obtain there the chairs of General Mathematics and Physics; for each of the two assignments you will be given the corresponding credit of five hundred gold soles per month; this would be from 1 April; later, you could occupy the chair of Atomic Physics also. If you find the proposal convenient, you can immediately write to the Rector of the University of La Libertad, Department of La Libertad - Trujillo, Dr. Ignacio Meave Seminario, indicating such conditions.[214] It would be opportune if you could send me a duplicate of the letter, in case the Rector comes to Lima, as is very possible. In that case, I will deliver it personally. Given the living conditions that the city of Trujillo offers, the salary is enough to live comfortably; the population is highly cultured, and the town is equipped with all kinds of transport to easily travel to the capital, Lima, at any time.

It is advisable to hurry with your response, as the holiday period is approaching. Your most respectful and reverent friend, Godofredo García

N.B. The exact address to which you should write is Dr. J. Ignacio Meave Seminario, Rector of the University of 'La Libertad', Trujillo – Peru, South-America

Señor Doctor Guido Fubini. Muy ilustre Profesor y amigo: Por la presente tengo el agrado de saludarlo y comunicarle, que he gestionado su ingreso al Cuerpo de

---

[214]Ignacio Meave Seminario (1877–1954), Rector of the Universidad Nacional de Trujillo (1931–1944).

Catedráticos de la Universidad de "La Libertad". Es ésta una de las tres Universidades Menores del Perú, que funciona en la Ciudad de Trujillo, situada en la Costa y dotada de toda clase de comodidades y de un hermoso clima. Si así le conviniera, podría obtener allí las Cátedras de Matemáticas Generales y de Física, por cuyas dos asignaturas dan el haber correspondiente de quinientos soles oro (soles oro 500) al mes; esto sería a partir del 1° de Abril próximo, también posteriormente puede Ud. ocupar la Cátedra de Físico - Atómica. Si Ud. encontrara conveniente la propuesta, inmediatamente puede escribir al señor Rector de la Universidad de "La Libertad", Departamento de La Libertad – Trujillo, Dr. Ignacio Meave Seminario, indicando sus condiciones. Sería oportuno me enviara un duplicado de la carta para el caso en que, el señor Rector viniera a Lima, como es muy posible, en ese caso le entregare personalmente. Dadas las condiciones de vida que posee la Ciudad de Trujillo, el sueldo sin ser crecido es suficiente para vivir con comodidad; su población es muy culta y está dotada de toda clase de comunicaciones para transportarse con facilidad en cualquier momento a la Capital, Lima. Conviene apresure su respuesta, pues el periodo de vacaciones se aproxima. Su amigo de su mayor consideración y estima, Godofredo García

N.B. *La dirección exacta a la cual debe dirigirse es: Dr. J. Ignacio Meave Seminario, Rector de la Universidad de La Libertad, Trujillo – Perú Sud-América*

3.17 Applications

A total of six Italian Jewish mathematicians (Ascoli, Fano, Friberti, Segre, Terracini, Tedeschi), two astronomers (Jacchia and Bemporad), one statistician (Mortara), and four engineers/mathematicians (Ugo Fano, Enrico Volterra, Scipione Treves and Giulio Supino) applied to SPSL and/or ECADFS. Just three applications (Fano, Jacchia, and Segre) were retained, and two out of three only formally. Fano's grant was in fact fictitious, as he was self-financed. In other words, the SPSL merely provided the formal cover for Fano necessary to regain access to the assets that he had illegally exported to Switzerland. Segre was awarded a fellowship thanks to a group of British geometers, who contributed to a fund to secure his maintenance in Cambridge, London, and successively in Manchester. The other nine applications were rejected.

A combination of various factors stands behind such failures. In addition to those already mentioned (labour market close to saturation point, high competition, anti-Semitism, absence of immediate danger to life for Italian mathematicians) it should be remembered that, regardless of the reputation enjoyed, for both organisations it was necessary to follow a procedure, i.e. to fill in a questionnaire declaring biographical and personal data, and to submit a curriculum vitae, better if accompanied by testimonials, i.e. letters of reference written by mathematicians residing in the country for which the scholar applied.

The difficulties encountered by Italian mathematicians in completing these documents were countless. Having had no previous experience of filing a dossier, nor contacts with the SPSL and the ECADFS before the autumn of 1938, the first communication between Italians and these organisations generally took place by letter, written in fairly stunted English. At that point, the secretaries explained which documents to produce. From there, a long back and forth of letters and documents commenced. Sometimes Italians sent files in several tranches or applications that were not completely compiled. Mortara, for example, submitted to the SPSL his curriculum vitae and the list of publications in six copies on 16 November, and the next day mailed the completed questionnaire. Mortara was however the only one who, wisely, wrote the curriculum vitae in English and had it printed.[215] In the meantime, in the wake of more than understandable anguish, he also sent a part of these documents to various colleagues, including A.L. Bowley, L. Dublin, etc. without telling them that he had presented them to both the SPSL and the ECADFS. The foreign colleagues then also forwarded Mortara's file to the two agencies, so clogging up their offices. The secretary of SPSL had to explain A.L. Bowley, for the umpteenth time, the procedure to follow.[216]

Moreover, the questionnaires were requested by British and American friends and colleagues.[217] Sometimes they did not reach destination. For example, G.B. Jeffrey, a physicist at the University College London and former President of the London Mathematical Society, asked for the questionnaire for Segre on 18 October 1938. When Segre contacted the SPSL on 7 January, he was told: 'last October we gave our questionnaire to Professor Jeffrey for you, but it does not seem to have reached you'.[218] The completed questionnaire arrived at the SPSL at the end of January. Segre had lost three and a half months, precious time in this situation.

The model itself, moreover, had been designed for refugees of Central and Eastern Europe and included fields that were truly incomprehensible to Italians, such as the question regarding religious identity: 'Jewish orthodox or reformed?'. Italian academic titles (*assistente ordinario, incaricato, straordinario, libero docente, aiuto*) were not included in multiple choice fields. Other entries, such as PhD or habilitation, did not have an equivalent in Italy.

Some refugees also left empty the fields that embarrassed them, for example, the question regarding the financial means at their disposal. Others committed strong naivety in indicating their preferred destinations. Mortara, for example, excluded

---

[215] See footnote 74, p. 156, D3.7.2 and SPSL, *GM*: G. Mortara to SPSL, Milan 16.11.1938.

[216] SPSL, *GM*: D.C. Thomson [SPSL] to A.L. Bowley, London 18.11.1938. On 23 November, the SPSL registered receipt of the questionnaire, duly filled in.

[217] For example, on 3 February 1939, Marcello Treves asked the SPSL to send the questionnaires to Mario Giacomo Levi, formerly chief of the Laboratories for Chemistry in coal at the Milan University and to Giulio Supino, former professor of Mathematics and Hydraulics at the University of Bologna (see SPSL, *Italy*: M. Treves to E. Simpson [SPSL], London 3.2.1939).

[218] SPSL, *BS*: E. Simpson [SPSL] to B. Segre, London 10.1.1939.

Russia 'for political reasons', Terracini the tropical countries for the climate, Segre
the 'countries too far away' because he had left his elderly parents in Italy. Fano
indicated a single destination (not a country, but a single town): Lausanne.

3.18   Rejections, Failures, Missteps

Once the questionnaires had been completed, the time came for preparing the
dossier. The missteps committed by Italians were at this stage even more harmful.
Here, too, documents were often sent in several instalments, causing considerable
annoyance to the secretaries of the SPSL and ECADFS who, having hundreds of
files to manage, did not tolerate this behaviour. Terracini, for example, sent to the
SPSL and ECADFS a very short curriculum at the beginning of December 1938,
ten days after the list of his publications, three days later another list of his papers
which he begged the Society to add to other papers, and a month later a commented
list of his works. In the latter document, he referenced the publications by numbers
with which he had indicated them in the list sent in January, but perhaps he no longer
had the list at hand, because some did not correspond. The testimonials of
Castelnuovo, Fano, and Levi-Civita were addressed to the SPSL even later, from
January to March 1939, after they had been returned to Terracini from the offices of
the Universities in Aberdeen and Durham where he had unsuccessfully applied for
two vacant positions.[219] The same event occurred with ECADFS. Terracini sent a
short resume of his scientific activity in autumn of 1938 and the annotated list of
publications in mid-May 1939, when the ECADFS had already corresponded about
his case with various American colleagues. Also in this situation, the documents
were gradually added to the dossier:

> Let me acknowledge your letter of May and say that we have had correspondence with both
> Professor Sperry and Dr. Reichenberger about your case. [...] However, we are glad to add
> the papers you sent to your dossier, and to keep you in mind. Although we do have copies of
> a curriculum for you, it is a very brief one, and we would like to have a more amplified
> account of your career for our files.[220]

Curricula written by hand, with often incomprehensible graphs and in approximate English, did not present an impressive image, even of excellent mathematicians
such as Fubini or Fano. Sending an identical file to both the United States and the
United Kingdom was not appropriate, just as it was not appropriate to send documents such as the extract from the MEN *Bollettino* containing the result of a
competition committee for a university chair in Italy, instead of a description in
English of his research activity (D3.18.1, D3.18.2, D3.18.3).

---

[219] See TERRACINI.4 and SPSL, *AT*: A. Terracini to SPSL, Turin 1.1.1939; E. Simpson [SPSL] to
A. Terracini, London 4.1.1939; E. Simpson [SPSL] to A. Terracini, London 12.5.1939. The last
letter closed with a rather dry comment: "You will hear from us as soon as we have definite news to
report."
[220] ECADFS Papers: B. Drury [ECADFS] to A. Terracini, New York 19.5.1939.

The choice of referees was often not very sensible. Italians indicated old mathematicians as referees, like Coolidge and Baker (D3.18.4), colleagues who were themselves refugees like Fano and Fubini, mathematicians whose name was compromised at international level for political reasons, and even people like Severi and Bompiani, who never provided references for dispossessed Jews. Some testimonials were not properly structured. Apart from Levi-Civita, who had produced reference letters since the late twenties, Italians had no experience in this matter. In Italy, recommendation letters had always had an eminently confidential character. For this reason, Castelnuovo and Fano wrote two testimonials for Terracini that were not very useful, because they focused on his teaching skills, which were the least expendable in university systems such as the American and English ones. The distinction between research universities and colleges escaped them completely and led them to think that the strength of Terracini's curriculum was the fact that, from 1925 to 1938, he had taught 13 courses in Higher Geometry on different topics, for many of which he had also carried out research.

Documents

D3.18.1 B. Segre to ECADFS, Bologna 27.12.1938

ECADFS Records, *BS*, fol. 1r–3v

My dear Sir, I permit myself to apply to you owing to expose you my case. I was full professor of mathematics (and precisely of Analytics [sic!], Projective and Descriptive geometry) in the University of Bologna, and Director of the relative Mathematical Institute but recently I was dismissed from these and the other academical charges I held, on account of my being a Jew, so that at present I have no means of subsistence. I would be very grateful to you, if you could procure me a situation fit for my capacity. To this aim I enclose a copy of the report of the Commission which judged me when I was advanced as full professor, and a short curriculum vitae of myself.[221] As for languages, (besides Italian) I know well French, pretty well German and a little less English: but I am sure to be able to command also this language in a short time. I will accept joyfully every place which permit a life, also if modest. I can dispatch to you, if that will be necessary, further informations on myself, copies of my scientific publications (surpassing a hundred), and others documents and justifications which you are willing to request from me. If you wish to get directly something more on my account, you can apply for that to the following university professors: prof. Solomon

---

[221] Segre sent the ECADFS and the SPSL the same cv, with two attachments: the *Chronological List of scientific publications of Dr. Prof. Beniamino Segre from 1923 to 1932* and the *Relazione della Commissione giudicatrice per la promozione del professore Beniamino Segre a ordinario di geometria proiettiva e descrittiva nella R. Università di Bologna* (eds. F. Severi, M. De Franchis, G. Scorza), Bollettino Ufficiale del Ministero dell'Educazione Nazionale, p. II, 23.5.1935, n. 21.

Lefschetz, Princeton University, Princeton Mass.; prof. Virgil Snyder, Cornell University, Ithaca (N. J.); Prof. Oscar Zariski, Johns Hopkins University, Baltimore (Ml).[222] I thank you in advance and send you the most distinguished regards. Your Beniamino Segre

Prof. B. Segre Curriculum Vitae

I was born in Turin (Italy) on 16th of February 1903, from Samuele and Leonilda Segré (both Jews). I took the doctorship in Mathematics (with the maximum votes and the commendation) on 14th July 1923, at the University of Turin. I kept the place of Assistant Professor from 1923 to 1931 in the Universities of Turin and Rome, excepted the year 1924–25 in which I made the military service as an artillery-officer, and the year 1926–27 in which I was in Paris with a Rockefeller prize.

Precisely I was:

In 1923–24 Assistant of Rational Mechanics in Turin with Prof. Somigliana.

In 1925–26 Assistant of Analytics Geometry in Turin with Prof. Terracini; and during 1927–1931 Assistant of Infinitesimal Analysis in Rome with Prof. Severi.

In 1931, having gained the Concourse proclaimed by the University of Bologna, I was called as an Associate Professor (Prof. straordinario) of Analytics, Projective, and Descriptive Geometry in the University of Bologna.

From 1934 I was full Professor (Prof. ordinario) of the same instruction in the same University, and also Director of the relative Mathematical Institute. But I was dismissed from these charges on 14 December 1938, in consequence of the precautions taken by the Italian government against Jews.

I obtained the following prizes:

in 1926, Rockefeller – prize for improvement abroad;

in 1927, prize of the R. Accademia dei Lincei in Rome;

in 1927, Corrado Segre Prize, of the Turin University;

in 1930, gold medal of the Società Italiana delle Scienze;

in 1935, prize of the Académie Royale de Belgique in Brussels.

I was: member of the R. Accademia delle Scienze dell'Istituto di Bologna, one of the four directors of the *Annali di Matematica pura ed applicata*, founder member

---

[222]Lefschetz had met Segre in Rome in 1931 but the two scholars had not entered into a friend relationship. Snyder had met Segre in Rome in 1929 and Segre had reviewed the book by V. Snyder, A.H. Black, L.A. Dye, *Selected topics in algebraic geometry II*, Bollettino dell'UMI, II, 14, 1935, p. 194. Zariski and Segre were close friends. In the years 1926–1927 they had met regularly at the Caffè Greco in order to gossip and play chess. Segre had reviewed the book by O. Zariski, *Algebraic surfaces*, Bollettino dell'UMI, 14, 1935, pp. 115–117.

of the Unione Matematica Italiana, perpetual member of the Société Mathématique de France, member of Mathesis, of the Circolo Matematico di Palermo, of the Società Italiana per il Progresso delle Scienze.

The number of my scientific publications surpasses till now a hundred.

I got married with Fernanda Coen in 1932, and I had three sons: Sergio in 1933, Silvana in 1934, and Ornella in 1937.

D3.18.2 B. Segre to SPSL, Bologna 7.1.1939

SPSL, *BS*, fols. 21–22
My dear Sir, I permit myself to apply to you owing to expose you my case. I was full professor of Mathematics (and precisely of Analytics [sic!], Projective, and Descriptive Geometry) in the University of Bologna: but recently I was dismissed from this and the other academical charges I held, on account of my being a Jew, so that at present I have no means of subsistence. I would be very grateful, if your Society could procure me a situation fit for my capacity. To this aim I enclose the chronological catalogue of my scientific publications, a copy of the Report of the Commission which judged me when I was advanced to full professor, and a short Curriculum Vitae of myself. As for languages, (besides Italian) I know well French, pretty well German, and a little less English: but I am sure to be able to command also this language in a short time. I will accept joyfully every place which may permit me a life also if modest, continuing my mathematical researches. I can dispatch to you further informations on myself, copies of my scientific publications, and others documents and justifications which you are willing to request from me. If you wish to get directly something more on my account, you can apply for that to the following university professors: Mr. W.V.-D. Hodge, Pembroke College, Cambridge; Mr. Leonard Roth, 21 Brycedale Crescent Southgate – London N14; Mr. J.G. Semple, 4 Ruskin Close London N.W. 11.[223] I thank you in advance and send you the most distinguished regards. Yours Beniamino Segre

---

[223] William Vallance Douglas Hodge (1903–1975), Lowndean professor of Astronomy and Geometry at Cambridge (1936–1970), had been in contact with Severi since the late 1920s, and with Segre since the publication of his article *Intorno ad un teorema di Hodge sulla teoria della base per le curve di una superficie algebrica*, Annali di Matematica pura ed applicata, s. 4, 16, 1937, pp. 157–163. Leonard Roth (1904–1968), lecturer in Mathematics at Imperial College in London, had spent the academic year 1930–1931 in Rome with a Rockefeller fellowship, working in close contact with Castelnuovo, Enriques, and Severi. Segre, Roth, and their families had become close friends. John Greenless Semple (1904–1985), professor of Mathematics at King's College, London (1937–1969), worked in algebraic geometry according to the Italian approach.

D3.18.3 B. Drury [ECADFS] to B. Segre, New York 28.1.1939

ECADFS Records, *BS*, fol. 1r
Dear Professor Segre: Let me acknowledge your letter of December 27 with which you enclose curriculum vitae and a report of your work by the *Ministero dell'Educazione Nazionale*. I very much wish there were something we could do to assist you. As it happens the Emergency Committee cannot because of its regulations bring the availability of displaced scholars to the notice of colleges and universities unless asked to do so by such institutions. Upon receiving such a request from a college or university the Committee will suggest candidates suitable for the position to be filled. Because of the number of demands made upon it, the Committee has been obliged to limit its activities to the cases of scholars who have been displaced from university posts where they held the rank of professor or Privatdozent. I feel I must tell you that because of unfavorable economic conditions there are almost no openings in this country at the present time. Nevertheless, we shall be glad to place your papers on file and to keep you in mind even though there is but a slight chance of our learning of a suitable opportunity. Sincerely yours, Betty Drury - Assistant Secretary

D3.18.4 H.F. Baker to B. Segre, Cambridge 12.3.1940

CA, *BSP*, fols. 1r-v, 2r
Dear Professor Segre, I hear from Professor Hodge that you are applying for a post in Melbourne; I am sending a testimonial, via Hodge, with this. You would not be *very* far from Sidney, where the Professor of Mathematics, T.G. Room, is an old pupil of mine.[224] I am *very* sorry circumstances seem to make it difficult to keep you here; and I had looked forward to our meeting often with the return of warm weather. But I have been in the house for 6 weeks. Would you be so kind as to tell me the best postal address of Castelnuovo – and Enriques – and Severi; I may wish to write to them in the near future. With our kind respects to yourself and Signorina Segre, and kind regards to your children, yours truly H.F. Baker

St. John's College, Cambridge (Engl.) 12 March 1940

---

[224] On the lectureship at Melbourne see D3.14.1 and Chapt. 10, pp. 563–564. The position will be given to Eric Russell Love (1912–2001). The interview was conducted by Hodge (CA, *BSP*: W.V.D. Hodge to B. Segre, Cambridge 28.3.1940). Thomas Gerald Room (1902–1986) had been a lecturer in Cambridge from 1928 to 1935, before moving to the University of Sydney. Henry Frederick Baker (1866–1956) Lowndean professor of Astronomy and Geometry at St John's College, Cambridge from 1914 to 1936, had been in close contact with C. Segre and the Italian School of geometry since the late 19th century and made their work the subject of his 1911 London Mathematical Society presidential address.

I have known personally Professor Beniamino Segre, and his wife and children, since September 1940 [sic!]; and I have known his eminence and position in the mathematical world of Italy for a very long time. His remarkable work and discoveries in Geometry make me believe that he would be a great accession to university work in Australia, and indeed here in Cambridge, if present circumstances allowed of this. I recommend his claims to the best of my ability, without any reservations. H.F. Baker Lowndean Professor of Astronomy and Geometry in the University of Cambridge 1914–1936

3.19   The Handicap of English

The issue of language skills has already been mentioned. The documents that we have transcribed faithfully reproduce the level of written English of Italian mathematicians. From this point of view, they were in a position of inferiority vis-à-vis the refugees from Nazi Germany. English, unlike French and German, was not part of the training of Italian mathematicians. It should be borne in mind that, on top of this fact, there was also a political reason: the Italy of 1938 was that of cultural autarchy, of the indefinite defence of Latinity, the country where the Royal Academy of Italy had appointed a specific commission charged with the task of identifying foreign terms and suggesting Italian equivalents. For the mathematical vocabulary, Severi had worked on this commission.

In somewhat brutal terms, it can be said that in 1938 all aspiring refugees from Italy read English, and almost all wrote it (many badly or very badly), but no one spoke it. Women (wives, daughters, etc.) were better placed because modern languages were a typical element of the education of young women from respectable families. Amusing testimony is provided by the young mathematician Marie Litzinger who took tea at the Volterras' house and told her parents about the poor linguistic competences of Edoardo Volterra and Gino Castelnuovo (D3.19.1).[225]

From the point of view of reading, English was a minority language in the material patrimonies of Italian mathematicians. The percentage of volumes and books in English in the libraries of Fano and Terracini constituted about 6%. The fact that people read so little in English was also related to the experiences of publishing in English or American journals which had been generally disappointing, if not humiliating, for Italians. The practice of peer review, the limit in page number, the very structure of the papers (with few footnotes, limited references, synthetic style, etc., the vary-typed composition) were aspects that had left Fano, Terracini,

---

[225] Marie Litzinger (1899–1952) was an American mathematician known for her research in number theory, homogeneous polynomials, and modular arithmetic. Litzinger earned her bachelor's degree and master's degree at Bryn Mawr College in 1920 and 1924, respectively. While at Bryn Mawr, Litzinger received a European travel fellowship which allowed her to study at the University of Rome in 1923 and 1924.

and Segre puzzled (D3.19.3, D3.19.4). As a consequence, in 1938, very few Italian mathematicians had written a work in English, and even when this did occur, it was the translation into English of books and articles which originally appeared in Italian.

Correspondence, even with American and British colleagues, was all in French or Italian. Volterra had his wife Virginia who translated the drafts of the letters that he wrote in Italian to Birkhoff and Lovett, but to friends like Evans he wrote directly in Italian. Castelnuovo did the same in his correspondence with Veblen. Fubini admired Levi-Civita for the 'really excellent' English recommendation letter that he had addressed to IAS in his support.

The most critical point, however, was that of speaking. None of the Italian mathematicians was fluent in English. In all their tours across England and America, Volterra, Levi-Civita, Fano had expressed to the colleagues their concern with giving lectures and conferences in a language that they did not master, and ultimately, they spoke in French or had the texts of their talks translated and revised by native English-speaking colleagues, friends, and relatives. G. Sutton, a physicist who followed the course held by Fano at Aberystwyth in 1923, commented with amusement:

> In my third year, Fano came from Italy to give lectures on geometry, but as far as I was concerned, these were less successful. Fano's English was eccentric, and we had insufficient preparation for the task.[226]

The handicap of English was one of the determining factors behind the rejection of applications by Italian mathematicians (D3.19.2). Both in addressing individuals and presenting themselves to international agencies like SPSL and ECADFS, Italians focused on their teaching skills. As a matter of fact, Fano, Fubini, and Levi had almost forty years of experience teaching courses, both basic and advanced programmes. Terracini had a 20-year teaching career. They were universally praised in Italy as inspiring lecturers of extraordinary clarity and effectiveness. However, in addition to not being considered suitable for the Anglo-American university system, the fact of not being able to lecture in English evidently penalised them.

Documents

D3.19.1 M. Litzinger to Her Parents, 29.11.1923

TriCollege Libraries Digital Collections (Bryn Mawr College, Haverford College, Swarthmore College) Marie Litzinger Papers, fols. 1r–2v
Dearest Papà and Mamma, [. . .] On Monday Signorina Eugenia was invited to take us to tea at the Volterra's. All the Mathematicians here are Jewish, Signorina says, and Volterra's wife looks Jewish.[227] She is quite lovely, with beautiful hair

---

[226] Sutton (1965), p. 19. See also Hoyle (1997).
[227] Virginia Almagià Volterra (1875–1968).

like yours, Mamma, but hers is a little coarsest. They have three sons and a daughter.[228] The oldest one (about twenty-two) spent last summer in England, so he very bashfully spoke a few English sentences. Castelnuovo's son, of the same age and skinny-ness and bashful disposition was also present.[229] I gather they were there for our benefit, but they looked most uncomfortable. People were always entering or leaving the room, and they had to stand up, looking all arms and legs and pining for some place in which to hide them. They stood it as long as they could and then fled after tea was served. I liked the Volterra one. [...] A professor of astronomy, his wife, an old man, and his daughter were among those present. Volterra was most entertaining. [...] I didn't say ten words the whole time, and understood only part of what was said, but it was fun. [...] Heaps of love, Marie

D3.19.2 O. Veblen to A.B. Coble, [Princeton] 6.5.1939

OVP, *BS*, fol. 1r

Dear Coble:[230] I must ask your pardon for not having written sooner about Beniamino Segre. Since I do not know him personally I cannot be too definite about his command of English, but my understanding is that he has been studying the language, and there is every probability, in view of his general intelligence and resourcefulness, that he will be far enough along to meet all demands by the time that he reaches Urbana. Yours sincerely, Oswald Veblen

D3.19.3 G.H. Hardy to B. Segre, 29.8.1939

CA, *BSP*, fols. 1r–2v

Dear Dr. Segre, I have just run through your long Lond. Math. Soc. Paper (the one of 50 pp.) 'secretarially'.[231] I have made on the MS a considerable number of small corrections of a linguistic or typographical character. The 'English' is good for a foreigner, but repeatedly 'just not' right. You will of course be able to reconsider anything in the proofs. It is difficult for me to be sure that I am always right. There are one or two ways in which you disregard English idiom rather

---

[228] Edoardo (1904–1984), Enrico (1905–1973), Gustavo (1909–2001) and Luisa (1902–1983).

[229] Gino Castelnuovo (1903–1995), third son of Guido Castelnuovo and Elbina Enriques.

[230] Arthur Byron Coble (1878–1966), Head of the Department of Mathematics at the University of Illinois at Urbana-Champaign (1934–1947).

[231] B. Segre, *On limits of algebraic varieties, in particular of their intersections and tangential forms*, Proceedings of the London Mathematical Society, s. 2, 47, 1942, pp. 351–403. At time, Godfrey Harold Hardy (1877–1947), Sadleirian professor at Cambridge, was President of the London Mathematical Society.

systematically, and I would suggest that, when you see the proofs (which may, I fear, not be for a considerable time), you should observe carefully where they occur. The points which I noticed particularly were

1. insertion of commas when there is no logical justification: e.g. 'The trans-formations which we made in §20, are ...' [I *invent* a sentence]. If you put a comma after *20*, you *must* also put one after *made*. Possibly Italian usage is different. *No* comma is *best* in such cases.
2. Putting qualifying adverbs or phrases *before* the main verb or statement: 'we rationally transform the equations (20)'. The 'rationally' should follow (20). These are cases of definite error. There are many others in which one cannot say that what you write is *wrong*, but where the perusal balance of your sentence is obviously 'un-English'. In such cases it is difficult for a secretary to know what to do, and it is quite likely that I am inconsistent. After all, a secretary's time is limited.

May I also suggest that, if you make long typewritten copies, you should, if possible, use rather better paper? This MS has to pass through at least 3 hands before it gets to the printer, and the flimsy paper gets gradually into a state of disarray which makes it extremely troublesome to handle. I realize that geometry is more difficult for a foreigner to express than analysis, since the sentences are much less dominated by formulae and tend inevitably to become rather long and complex. Yours sincerely G.H. Hardy

D3.19.4 G.H. Hardy to B. Segre, 5.9.1939

CA, *BSP*, fol. 1r-v

Dear Dr. Segre, Your result is quite new to me. I have looked into Koksma, but can find nothing like it. But I would advise you to consult Davenport.[232] Your English is, on the whole and in the circumstances, remarkably good: there are very few definite 'mistakes', and such 'un-Englishisms' as there are, are almost entirely of a few definite types. It is for this reason that I thought it worth writing. It is, I suppose, true that the worst English writer (and we have some pretty bad ones) writes in a way which is at bottom more 'English' than the best foreign writer. Yours sincerely G.H. Hardy

---

[232] Hardy refers to the manuscript of the paper by B. Segre, *A complete parametric solution of certain homogeneous Diophantine equations of degree n in n + 1 variables*, Journal of the London Mathematical Society, 19, 1944, pp. 46–55. The two papers mentioned are: J.F. Koksma, *Diophantische Approximationen*, Berlin, Springer, 1936 and H. Davenport, *Note on an identity connected with Diophantine approximation*, Proceedings of the Cambridge Philosophical Society, 34, 1938, pp. 109–110.

3.20 A Matter of Style

Of the five mathematicians who managed to leave Italy in the first migration wave of 1938–1939 (Fubini, Fano, Terracini, Segre, and Levi), four belonged to the Italian School of Algebraic Geometry. Fano, Levi, and Segre were classical algebraic geometers, Terracini and Fubini projective-differential geometers.[233] Inserting themselves into a clearly defined geometric tradition, and the fact of making no mystery of it, did not play in favour of would-be refugees.

The image of Italian mathematics that had been conveyed abroad from the end of the nineteenth century to 1938, in dozens of different contexts, was that of a discipline and a community divided into research schools each of which was distinguished by its own style. This was not just a rhetorical artifice but corresponded to a deep conviction. To attribute to the research project a national connotation was neither incoherent nor improper for a scholar like Volterra, Castelnuovo, or Enriques, who were however convinced that, in order for a School not to deteriorate, continuous exchange was necessary, a supranational circulation of individuals and ideas which should not be hindered either by political power or by forms of parochialism or cultural autarchy. In the field of geometry, however, the academic leadership of Severi and Bompiani had led in the thirties to a sort of isolationism which was counterproductive and, at times, hysterical. Generally, it was the young scholars who paid the price—the topologising geometers (Segre, Bassi, P. Buzano, ...)—and people who worked in abstract algebra, such as F. Conforto, who had abandoned these studies or had been marginalised.

Considering this context, it is understandable why the applications of Italian geometers systematically encountered a difficult reception, especially in the American geometrical environment dominated by Lefschetz who had a vision of algebraic geometry decidedly different from that of the Italian School. The peculiar character of the latter was marked by its purely geometric approach and the fundamental role attributed to intuition. Synthetic procedures, of a purely projective nature, were skilfully mixed by the scholars belonging to this team to arrive at results that had been 'seen' through geometric intuition, more than rigorously proved.[234] In his studies, Lefschetz, despite having received the legacy of the Italian Masters, had gone further, believing that new methods were needed to address the classical problems of algebraic geometry. Lefschetz himself described his goal with an evocative metaphor: 'to plant the harpoon of topology in the whale body of algebraic geometry'.[235] Also for this reason, he did not believe that collaboration with Italian geometers, still too anchored to classical—in a certain sense, outdated—research

---

[233] As a matter of fact, until 1916, Fubini had not been 'a pure geometer' and had had to his credit excellent output both in the field of analysis and in that of applied mathematics.

[234] According to Fano, for example, intuition was conceived as "the faculty of observing, of perceiving facts directly, without reasoning." (*la facoltà di osservare, di percepire fatti direttamente, senza ragionamento.*)

[235] Lefschetz (1968), p. 854.

methodology and style, would be stimulating or profitable for the development of the 'new' algebraic geometry in the United States.

Claiming membership of the Italian School had a boomerang effect, as in the cases of Terracini and Segre. At first, when presenting themselves in England and America, both explicitly placed themselves in the Italian tradition, remarking that they belonged to that community and that they had taken their lead from C. Segre. Terracini, for example, wrote:

> Let me add that, according to a tradition that traces its origins back to the mid-19th century, geometry was particularly cultivated in England and Italy. If opinion cannot be accused of self-referentiality, I dare to think that perhaps a contact between those two geometric schools would be far from useless. (D3.20.1)

In the United Kingdom, some people, such as Baker, Semple, and Milne, appreciated this kind of statement. Indeed, Italian geometers were considered excellent precisely because they represented this research approach. Semple, for example, defined Segre as 'the foremost of the younger school of Italian geometers, and that means that he is one of the best of the younger generation of geometers in the world'.[236]

In contrast, in the United States thinking in terms of membership of the Italian School was far from profitable. Despite the long and friendly connection with Italy, in the eyes of Lefschetz or Zariski or Snyder, the Italian geometric tradition was in sharp decline and stagnation, a de-evolution due to two factors: insufficient topological and algebraic culture, and lack of interpenetration between these two branches and classical algebraic geometry (D3.20.2). This explains the rejection of all applications submitted by Italian geometers. The most positive evaluation obtained would be that of Snyder: 'if Segre makes good (which includes being able to see and assume the American point of view) in a two-year probation, I feel that he can land something'.[237] The fact of no longer being able to perform algebraic geometry in a non-Italian way, after a decade of autarkic politics in the mathematical field, meant that Segre and Terracini were unable to get any opening. As Veblen commented in 1941:

> I think that he [Segre] would have found a place in this country if it were not for the opposition of Lefschetz. I cannot help thinking that this opposition is due to the fact that Segre does not go wholeheartedly in the Lefschetz direction for algebraic geometry, as Zariski is going. (D3.20.3. See also D3.20.4)

---

[236] SEGRE.9.

[237] OVP, *BS*: V. Snyder to O. Veblen, Ithaca 28.2.1939.

Documents

D3.20.1 A. Terracini to SPSL, Turin 11.12.1938

SPSL, *AT*, fol. 355

The Society for the Protection of Science and Learning – London. As you know, according to the decisions of the Italian Government, Jewish Professors are no more allowed to continue their service in the Italian Universities. On these conditions I must keep seriously up my mind about the possibilities concerning my future, not only because I wish to continue my scientific work, but also as I must think about my wife and three children.[238] Therefore, I ask you to help me in order to find some opportunity to work. I always worked and continue to work in different fields of Geometry, and especially in projective (algebraic and differential) Geometry. In any case I add a curriculum, which I am always disposed to complete with further informations or testimonials. I think that Professor H.F. Baker (Sc. D., LL.D, F.R.S., formerly Lowndean Professor and Fellow of St. John's College, in the University, Cambridge, now in retirement) is also disposed to give informations about my scientific work and position. I permit me to add that, according to a tradition which traces its origin back to the middle of the nineteenth century, Geometry has been particularly cultivated both in England and Italy. Geometry schools flourished and flourish in both countries, also when the impulse towards geometrical researches has failed in some other countries. If it could not be taxed with self-conceit, I should dare think that perhaps a contact of those two geometrical schools should not be quite useless. I thank you very sincerely for your help and for your trouble. I beg you to agree my best regards, yours truly Prof. Alessandro Terracini

D3.20.2 O. Veblen to W. Weaver, [Princeton] 8.10.1940

OVP, *BS*, fol. 1r, 2r

Dear Weaver: Although you did not ask me to do so, I think it might be well to put down in black and white the definite recommendations which Weyl and I made in our conversation yesterday: that Segre, Hamburger, [Felix] Pollaczek and Mahler should be invited through the New School. With regard to Segre, the point is that he has done work of very considerable value in two principal geometrical fields, - differential geometry and algebraic geometry. I should say that, as compared with some of his contemporaries, he tends to be very geometrical in the classical sense of the word. This tendency has fallen somewhat into decay in the United States, and I think that it would be very desirable to revive it. Since there are several other

---

[238] Giulia Sacerdote (1899–1974) had married Terracini on April 16, 1924. They had three children: Lore (1925–1995), Cesare (1927–1972), and Benedetto (1931).

people who feel the same way, I think it will be possible to find a place for Segre within a reasonable time. He has already made progress with the language during his stay in England, where he has held a position which was, if I am not mistaken, supported by subscriptions made by English geometers. [...] If you would like me to send you more complete data about any or all of these men, I shall be very glad to do so. You will doubtless recall that we are keeping such data here rather systematically. Yours sincerely, Oswald Veblen

D3.20.3 O. Veblen to T.Y. Thomas, [Princeton] 25.6.1941

OVP, *BS*, fol. 1r

Dear Tracy: I was of course delighted at your election to the National Academy, and circulated a telegram for signatures which got on more or less at random.[239] [...] This reminds me to pass on to you an idea which I have had on my mind for some time, - namely that it might be a good move for you to call in Beniamino Segre as a junior colleague. I don't know whether you have looked at any of his work in differential geometry. Some of it is, I think, quite creditable. He has also done a considerable amount of work in algebraic geometry, and perhaps also a little in the direction of foundations. Anyway he seems to be interested in a variety of geometrical fields. I think that he would have found a place in this country before now if it were not for the opposition of Lefschetz. I cannot help thinking that this opposition is due to the fact that Segre does not go wholeheartedly in the Lefschetz direction for algebraic geometry, as Zariski is going. Whatever the merits of the case may be, it seems to me that it would be a healthy thing to have a competing point of view in algebraic geometry maintained in this country. For at least two years Segre has been in England, where he has been supported by a fellowship contributed largely by English mathematicians out of their own pockets. He has a fairly large family, and is probably not of much direct use in the war. It would therefore, I think, be a good deed to relieve the English mathematicians of his support. I think if an application came from the University of California to the Emergency Committee in Aid of Displaced Foreign Scholars, there might be a chance of financial help from this committee. If you or Hedrick[240] would let me know when the application was being sent in, I could write to one or two members of the committee who could increase its chances of the decision being favorable. I enclose herewith a curriculum vitae for Segre. I am leaving today for Brooklin, Maine, and therefore shall probably not be here to

---

[239]Tracy Yerkes Thomas (1899–1983), Ph.D. in 1923 from Princeton University, professor at the University of California, Los Angeles (1938–1944). He had just been elected a member of the National Academy of Sciences.

[240]Earle Raymond Hedrick (1876–1943), provost and Vice-President of the University of California (1937–1942).

sign this letter when it is typed. With best greetings to Mrs. Thomas, and also to your father and mother when you write to them, Sincerely, Oswald Veblen

D3.20.4 T.Y. Thomas to O. Veblen, Los Angeles 1.10.1941

OVP, *TYT*, fol. 1r
Dear Veblen, I received your letter written just before you left Princeton for Maine but did not answer it promptly, as it seemed very difficult to do so at that time. You recall that you were concerned with the case of Beniamino Segre and the difficulty in question was due to the fact that funds were not allowed this year even for that appointment of an instructor which was requested. It seems impossible here to do anything for Segre or anyone else for that matter. Possibly the situation may change but this is the present situation at least. In a letter from my mother the other day she mentioned that Mrs. Veblen had written her a nice letter and that you have undertaken the good work of taking care of an English family. It must be very hard on such people when they realize what is taking place in England no matter how pleasant conditions are in this country. [...] Sincerely yours, Tracy Thomas

3.21 Months of Anxiety

Despite the efficiency of the SPSL and ECADFS offices, and despite the promptness and speed with which colleagues responded to aspiring refugees, assuring them that they did everything possible to help them, the last three months of 1938 and early 1939 were filled with anguish. Only Fano and Fubini succeeded in leaving Italy by the end of the year. The others continued for many months to look for a space of intellectual survival.

Colleagues did not hide the difficulties they were encountering in welcoming the new waves of displaced scholars from Italy, Austria, and Czechoslovakia. Lefschetz, for example, openly suggested to Segre that he abandon the English-speaking route and pointed instead to Latin America:

> Dear Mr. Segre, Your various letters have reached me (they still reach us in America), but if I did not answer you, it is because I had nothing very specific to tell you. I am one of those who have tried to do something for you from various sides, and I hope that one day one of these efforts will succeed. In the meantime, you must have both patience and courage, and perhaps start studying Spanish rather than English, with a view to South America. I am not

without hope that your great friend and Master, *Sua Eccellenza* Severi, will manage to do something for you as he promised me.[241]

Thanking him for the list of publications, Veblen wrote to Terracini: 'I hope to make good use of it. Up to the present I have not heard of any suitable opening for you on this side of the Atlantic'.[242]

In turn, Italians understood, but did not give up. Addressing the SPSL at the end of November, Mortara still hoped: 'Certainly there are great difficulties, but I hope they will not be insurmountable!'[243] The same goes for Segre. In addition to anxiety, the prevailing feeling was truly sincere gratitude to individuals and organisations, as testified by Terracini's correspondence (D3.21.1) and by the fact that Segre continued to send money donations to the SPSL until 1950, a tangible sign of recognition of the help he had received.

The first months of 1939 saw the departure of all foreign Jewish mathematicians who had found initial refuge in Italy: Opatowski, Wasow, and Frucht. Among the Italian-born mathematicians, Mortara left for Brazil in January, Segre in April for England. Those who, like Ascoli, Tedeschi and Friberti, had postponed the decision to flee while awaiting the outcome of the counter-discrimination procedure were too late. Terracini, Levi, and Ascoli, young people such as Enrico Volterra, Luigi Jacchia, and Ugo Fano, as well as the astronomer Giulio Bemporad who had searched for a position abroad late or unsuccessfully, continued to wait for an offer (D3.21.3, D3.21.4). The United States appeared increasingly unattainable, while hopes were raised on the South American front. In June 1939, Brazil and Argentina relieved Terracini, Levi, and Enrico Volterra of their anguish, calling them to Tucumán and Rosario, respectively. Preparations had to be made quickly, because international observers agreed that the winds of war were arriving.

The outbreak of the war resulted in a desperate acceleration of departures. Fortunately, Terracini and Levi managed to leave Italy just after the Nazi invasion of Poland, but with the beginning of the war visas became invalid. Volterra Jr., who did not want to travel to Argentina with a tourist visa, remained blocked in London. Fano's youngest son, Roberto, who had remained in Turin to complete his university exams (fourth year of Engineering Studies at the Polytechnic) was unable to obtain a passport for France, so he travelled to Switzerland and then to Lisbon, where he embarked for the United States.

---

[241] CA, *BSP*: S. Lefschetz to B. Segre, Paris 2.1.1939. *Cher M. Segre, Vos diverses lettres me sont bien parvenues (elles nous parviennent toujours en Amérique), mais si je ne vous ai pas répondu, c'est que je n'avais rien de bien précis à vous dire. Je suis de ceux qui ont essayé de faire quelque chose pour vous de divers côtés, et j'espère qu'un jour une de ces démarches aboutira. En attendant il vous faut avoir tant patience que courage et vous mettre peut-être plutôt à l'étude de l'espagnol, que de l'anglais, en vue de l'Amérique du Sud. Je ne suis d'ailleurs pas sans espoir que votre grand ami et maitre,* Sua Eccellenza *Severi, arrive à faire quelque chose pour vous comme il me l'a promis.*

[242] SPSL, *AT*: O. Veblen to A. Terracini, [Princeton 18.4.1939].

[243] SPSL, *GM*: G. Mortara to SPSL, Milan 25.11.1938. See also D3.21.2.

Documents

D3.21.1 A. Terracini to O. Veblen, Turin 16.10.1938

OVP, *AT*, fol. 1r

My dear Professor Veblen, I have just received your letter, and I am very sincerely grateful to you. I understand quite well that the problem to find some academical position in your country for foreign mathematicians is a very difficult one. Therefore, my gratitude for you and the others American Professors who strive in order to find a solution in still greater. Moreover, as perhaps you know, since I sent my first letter to you the Jewish Professors, without any exception, have been definitively forbidden from teaching in the Italian Universities. I still hope to receive some further favourable news from you. You can imagine how glad I should be. Renewing my best thanks, I remain yours truly Alessandro Terracini

D3.21.2 G. Mortara to L.I. Dublin, Milan 11.11.1938

ECADFS Records, *GM*, fol. 1r

Dear Dr. Dublin, Your letter of warm sympathy has reached me in a day of greater affliction caused by a new and harder blow. The hope inspired by its concluding sentence has relieved my pain. With all my heart I hope, as you do, that there is still for me a country to live and to work in. I wish to express you my deepest feelings of gratitude for the action you have undertaken on my behalf.[244] If you would 'unearth' even an opportunity of minor importance, but sufficient to ensure a modest living for my family, I should take it willingly. What is essential for us is to find a starting point in America. I think it may be useful for your action in my favour to know the possibility to which alludes the enclosed letter of the Rockefeller Foundation.[245] Maybe this possibility could be realized even in the case my occupation were not a strictly 'academic' one. I allow me to send you some other copies of the printed notice about my scientific work. With thankful and friendly greetings, I remain Yours sincerely Giorgio Mortara

---

[244]ECADFS, *GM*: L.I. Dublin to L. Farrand, New York 23.11.1938; L. Farrand to L.I. Dublin, New York 29.11.1938 with a *Memorandum on Professor George Mortara* and D3.4.4.

[245]SPSL, *GM*: Tracy B. Kittredge [RKF] to G. Demaria, Paris 7.10.1938.

D3.21.3 L. Jacchia to H. Shapley, 7.2.1939

OVP, *LJ*, fol. 1r, 2r

Dear Professor Shapley: I thank you for your letter of Jan. 16 and for all the efforts you are making in trying to come to my rescue.[246] In the meantime you will have received my letter of three weeks ago. My situation is unchanged and all I can say you is that my last hope is pointed toward you. As I told you in the letter, a statement by you that a position would be vacant for me at Cambridge within a presumable epoch would procure me a temporary subsidy fund by the Society for the Protection of Science and Learning. I hope that you will not have any difficulty in addressing to me or to the Society (6, Gordon Square, London W.C. 1) such a formal statement, while the negotiations with the I.A.U. [International Astronomical Union] are going on. My only possibility of material life in the future weeks depends on it. The little money I had taken with me is now turning to the end and all my desperate endeavours to get work of whatever a kind in this country have implacable failed. My London friends, who are German refugees, have still their money in Germany: they had to get it partially transferred through some commercial trick, but after the recent drastic measures of the German Government, this business has been cut off and so they are unable to offer me any help. Besides, I am extremely preoccupied for my mother, who is still in Italy. The political tension between France and Italy is growing more and more threatening and I fear that, if it will go on this way, she will not be allowed – as a common Italian citizen – to enter France and England. At present, till I have no means for supporting her, I am unable to call her to London. Turning to the kind of work you speak me of in your letter, of course I would be very glad to perform it. I consider the question of exact magnitude sequences down to very faint stars as a fundamental one and, as it is directly connected with my previous works, I would put the greatest enthusiasm in its most radical solution. I understand very well your obligation toward young American astronomers and the difficulties you have to worry about because of the narrowness of your budget, and I am only too sorry to be a cause of troubles. To my excuse I have only the fact that I have no guilt in my distress and that my mother's existence compels me to continue my struggle for life. You write that I should not be too optimistic – unfortunately I am a refugee, and the consciousness of this fact is sufficient today to strip off every kind of optimism. At any rate I can assure you that all my efforts in America will be aimed to find a speedy solution to my further existence, so that my presence there may not be a handicap to other

---

[246] The case of Jacchia was presented by Shapley in person during his visit at the ECADFS Office. In an Office Memorandum dated January 16, 1939, Drury noted: "Miss Lisowski tells me that Harlow Shapley is going to speak to Dr. Duggan about the case of Luigi Jacchia, presumably astronomer. We have nothing in file about Dr. Jacchia. He is not known either to Professor Max Ascoli or to Professor Fermi (who arrived several weeks ago to take up work at Columbia). He is the celebrated physicist." See also ECADFS, *LJ*: B. Drury [ECADFS] to H. Shapley, New York 25.1.1939.

scientific workers. My London address is that written on the heading of this letter – please write me no longer to Bologna. Yours very truly, L. Jacchia

D3.21.4 A. Terracini to O. Veblen, Turin 29.3.1939

OVP, *AT*, fol. 1r
Dear Professor Veblen, Referring to your kind letter of last October, I take the liberty to remember me to you, and beg you again *not to forget my name* in the case that any academical opening might be found in your country. Of course, it would not matter in what University: but I am longing to retake my academical work in any way. I send you, for any case, a list of my printed papers. I beg you to excuse me for the trouble, and to agree my thanks and best regards. Sincerely yours Alessandro Terracini

3.22 The War Years

Those who had not been able to leave Italy by the end of 1939 had almost no more options left to them. The outbreak of war marked a drastic setback in the migration flow. Enrico Volterra, as mentioned, was no longer able to leave for Rosario, where he was supposed to take over the Institute of Stability. Segre was unable to land in the United States, as had seemed possible in April 1939 when he departed from Italy. The stay in the United Kingdom, which was not supposed to exceed a semester, and which should have been an intermediary step, lasted for seven years. Meanwhile, journeys were becoming much more difficult. Fano and Enriques, who in September 1939 had been shuttling between Lausanne, Paris, and Italy, no longer moved around. Edoardo Volterra, who had settled in Paris with his wife Nella Mortera and their daughter Laura, waiting for a call to the Americas that would never come, returned to Rome, and stayed there.

On 10 June 1940, Italy entered the war on the side of Germany. International relations with many countries were broken. Refugees in England, such as Segre and Enrico Volterra, were interned as enemy aliens. The contracts of the mathematicians of the Italian Missions in São Paulo and Rio were terminated or suspended. Italians living abroad were to be repatriated, although all the Jewish mathematicians, and some Aryan colleagues, like Bassi and the physicist G. Occhialini managed to stay abroad.

From 1941 onwards, acts of violent anti-Semitism multiplied. Jews were accused of spreading false news about the fate of the war and of making defeatist propaganda. The first terrible Allied bombings increased hostility towards them. The *Minculpop* archive records dozens of anonymous reports and denunciations against Jews who 'profited and prospered', while Italians fought at the front. In mid-October 1941 the

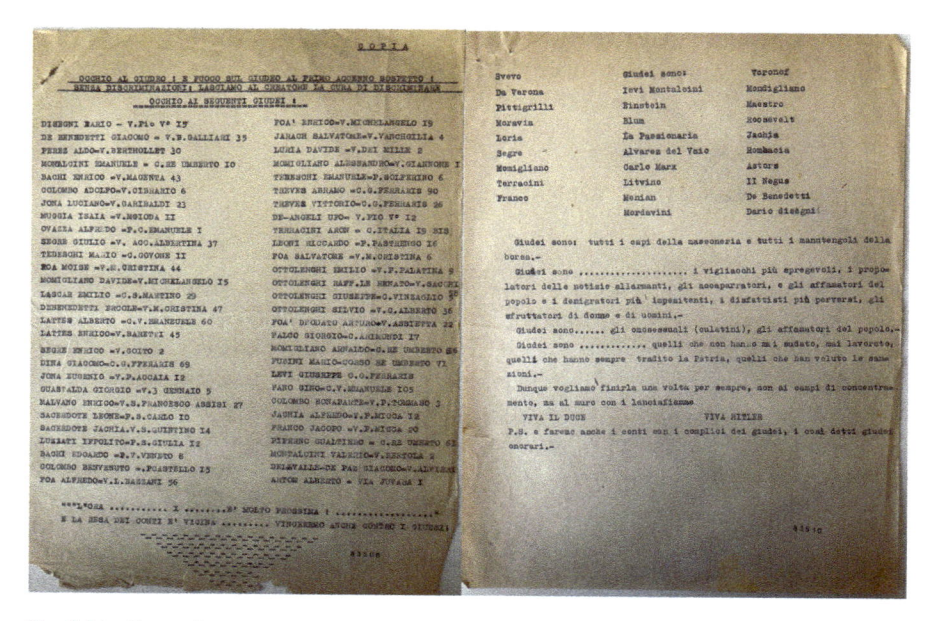

Fig. 3.14 Copy of towns posters appeared in Turin's streets, October 1941, ACS-*Razzismo*

first fire-bombings of synagogues occurred, while in the streets of various towns, posters appeared with lists of Jews and slogan such as: 'We want Jews in concentration camps!! Death to the Jews! We do not want Jews in concentration camps, but against the wall with flamethrowers'[247] (Fig. 3.14).

As Emanuele Artom recalled:

> The Synagogue [of Turin] burns on the night between November 20 and 21, 1942. Some passers-by said: «It is good for the Jews who wanted war.» [...] I'm going to see. The interior is all destroyed and covered with rubble. All around, almost intact, the walls with the four towers. The Community is also on fire and the school is unsafe. [...] Back home I discuss with my parents about what to do, because we could be killed. [...] This feeling that you can lose everything, even life, a feeling already emerging with the anti-Semitic campaign and now stronger with the increasing probability of loss, is very educational, because it teaches us that we are negligible particles of the world and that after our death everything will proceed as before.[248]

---

[247] ACS-*Razzismo*: *Vogliamo gli Ebrei in campo di concentramento! Morte a Giuda!, Morte agli Ebrei! Non vogliamo gli Ebrei in campo di concentramento, ma bensì al muro coi lanciafiamme.*

[248] Emanuele Artom (1940–44), in Schwarz (2008), p. 34: *Il Tempio brucia nella notte fra il 20 e il 21 novembre 1942. Qualche passante diceva: «Sta bene agli Ebrei che hanno voluto la guerra.» [...] Vado a vedere. L'interno è tutto distrutto e coperto di calcinacci. Tutt'intorno quasi intatte le mura con le quattro torri. Anche la Comunità è incendiata e la scuola pericolante. [...] Tornato a casa discuto con papà e la mamma sul testamento da fare, perché potremo venire uccisi. [...] Questa sensazione che si può perdere tutto, anche la vita, già cominciata con la campagna antisemita e ora divenuta più forte con l'aumentare delle probabilità, è molto educativa, perché*

The years between 1942 and 1945 were painful even for those who had managed to leave the country. The Italian government stopped paying pensions to Jews dispossessed in 1938. Even for those who did not suffer internment, even those who did not have their hard-won contracts terminated, these were 'years of anxious searching for news, through radio and newspapers, with their sirens and blackboards, in the persistent longing that Fascism and Nazism would collapse'.[249] Correspondence with those who had fled from Italy ceased or became very slow and irregular. No more news of children, parents, siblings, relatives, and friends arrived. Castelnuovo systematically lamented with Volterra and Levi-Civita that he knew nothing about friends outside nor about his daughter Gina, because from the United States of America he received air mail, but only irregularly.[250] Segre was in no way able to get in touch with his parents, who remained in Turin. In four years, Terracini received fewer than a dozen letters from his brother-in-law Aldo Sacerdote, in Rome.

The scarce information that did manage to arrive from colleagues told a frightening reality: mass deportations, liquidations of ghettos, forced shifts of Jews to 'colonies' in Central Europe. People were shocked by the news of the atrocities committed by the Nazis in Poland, Czechoslovakia, and Hungary. Fubini and Terracini felt the grief of their impotence when they received the plea of the Czechoslovak geometer Ludwig Berwald who begged to be called to Latin America. Despite Veblen's support, Berwald had always been ranked second in the lists of names presented for openings in Argentina. Neither the ECADFS nor the SPSL had managed to help him, even though he had tried to emigrate in early 1938. In the autumn of 1941, Terracini received a postcard from the ghetto of Lodz, in which Berwald said

> he knew he was going to be deported - nor could I say today if he understood the euphemistic use of this verb - and pleaded for a call to Tucumán. Perhaps never as at that moment did, I feel the grief of my impotence: I could only report the request to some friends, who unfortunately immediately came to the same conclusion: the impossibility, despite best intentions, of any attempt to save Berwald. [...] And a greater sorrow stemmed from the fact that he felt so disoriented, that in his postcard he added: «*I profess evangelical religion.*»[251]

---

insegna che siamo delle particelle trascurabili del mondo e che dopo la nostra morte il tutto procederà come prima.

[249] Terracini L. (1989), pp. 362–363: *años de ansiosa búsqueda de noticias dadas por la radio y los periódicos, con sus sirenas y sus pizarrones, siempre con el anhelo persistente de que el fascismo y el nazismo se derrumbaran.*

[250] ANL-*Castelnuovo*: G. Castelnuovo to V. Volterra, Cortina 29.8.1940: *dagli Stati Uniti di America, ci arrivano lettere per via aerea, ma anche queste irregolarmente.*

[251] Terracini A. (1968), p. 139: *mi diceva di aver saputo che stava per essere deportato – né oggi saprei dire se egli si rendeva conto dell'uso eufemistico di questo verbo – e supplicava per avere una chiamata a Tucumán. Mai forse come in quel momento sentii l'angoscia della mia impotenza: non ho potuto che riferirne ad alcuni amici, che purtroppo giunsero subito alla conclusione dell'impossibilità, nonostante le migliori intenzioni, di ogni tentativo per salvare il Berwald. [...] E una maggiore tristezza derivava dal fatto che egli si sentiva così poco parte in causa, che nella sua cartolina mi aveva aggiunto: «Ich bin evangelischer Religion.»*

On 2 December 1941 Veblen wrote to Terracini:

Professor Fubini has told me what you wrote to him about Professor Berwald. I have the very highest esteem for Professor Berwald personally, and also much admiration for his scientific work. In spite of all this, I have thus far been unable to think of any way of helping him. The only means that I can see of getting him away from Europe would be an invitation to some university. This has already been done in so many of the North American universities that all the possibilities seem to be exhausted. However, the matter is very much on my mind and I will speak of it to other colleagues whenever I have an opportunity. At present it looks as if it will very soon be too late.[252]

Berwald and his wife had already been deported to the Lodz ghetto. Berwald died there on 27 March 1942.

Finally, a last question to be taken into consideration is the war effort: the willingness or not to take part in military projects, devoting oneself to studies of applied mathematics. In the United Kingdom and United States, there was a certain reluctance to involve Italians in this type of projects and, before doing so, references were requested from colleagues who had welcomed refugees and worked with them. On the part of the refugees themselves, there was generally a sort of reluctance to abandon theoretical studies to devote themselves to a type of activity that they felt far removed from their fields of expertise and to which they did not feel like making a quality contribution. The failure of Italian refugee mathematicians to be included in military research contributed to their isolation. Among Italian mathematicians (or scholars with a solid training in mathematics) who were involved in war work, we ought to mention Eugenio Fubini, Ugo and Roberto Fano, and Luigi Jacchia (D3.22.1, D3.22.2).

Documents

D3.22.1 H. Shapley to S. Duggan [ECADFS], Cambridge Mass. 22.3.1943

ECADFS Records, *LJ*, fol. 1r
Dear Dr. Duggan: I have had to decide that I can't come down for the meeting of the Emergency Committee tomorrow. On Wednesday afternoon I must be in New Haven; all day Thursday in Brooklyn; and Friday in New York. Another special trip would be out of the question because of the tremendous amount of pressure here. I am sorry to miss this meeting, because I know it will be important. From Miss Drury you will have learned that we have made some attempts, without much luck because of the confusion in the government, to get some of our mathematical people into special Government work. The matter is still pending, and we may place one or two. Just now from the Massachusetts Institute of Technology, where several of the Observatory computers are now working on Government work of high importance, comes the appeal for more calculators.

---

[252] OVP, *LB*: O. Veblen to A. Terracini, [Princeton] 2.12.1941.

The new job that has come requires that the computers, who must be somewhat experienced with ordinary computing machines, and with the handling of simple mathematics and large arrays of numbers, must also be citizens, either of the naturalized or the born sort. They are paid from one hundred and twenty-five to one hundred and fifty dollars a month, depending on experience. They are needed at once. Have some of our refugee scholars got their second papers – scholars who are qualified for the Government computing? Possibly on Friday I shall have time to drop in at the office and get a direct report of the discussion and decisions of tomorrow's luncheon. Three of the special calculating-bureau members at the Massachusetts Institute of Technology are of the refugee class, Miss Mayer, who has been doing some computing for me during the past year here at the Observatory, Dr. Kopal, who is in charge of much of the work, and Dr. Luigi Jacchia.[253] Kopal and Jacchia were formerly recipients of assistance through the Emergency Committee. Both are doing excellent work. Very sincerely yours, Harlow Shapley

D3.22.2 L. Jacchia to S. Duggan [ECADFS], Cambridge Mass. 12.3.1944

ECADFS Records, *LJ*, fol. 1r

Dear Dr. Duggan: In response to your letter of February 29th I am glad to provide you with the following informations concerning my present status. Since August 1941 I have been working full time in fields other than Astronomy and connected with the war effort; I have published no scientific papers or books since October 1941. From August 1941 to the end of 1942, I was active as foreign-language expert and Director of the Italian broadcasts at the World Wide Broadcasting Foundation (WRUL [World Radio University Listeners]) in Boston, whose broadcasting facilities had been turned almost exclusively to anti-fascist propaganda. Since the beginning of 1943 (after the OWI [Office of War Information] took over WRUL) I have been working in mathematical capacity for Navy projects at the Massachusetts Institute of Technology, where I have the position of a D.I.C. staff member. My citizenship status has not undergone any alteration since my last communication to your Committee. Sincerely yours, Luigi Jacchia

---

[253] Ursula Mayer, observer (possibly Volunteer) at Harvard before 1952. Zdenek Kopal (1914–93), a Ph.D. *summa cum laude* in physics and mathematics at the Charles University of Prague, took an appointment at Harvard College Observatory at the end of 1938, supported by the asylum fellowship plan at Harvard. He worked also on ballistics for the US Navy at the MIT during World War II and after the Second World War.

3.23   The Last Frontier of Hope: Who Sought Refuge in Switzerland

On 3 September 1943, general G. Castellano signed the armistice at Cassibile, in Sicily. It was made public five days later by marshal P. Badoglio, the Head of the Italian government. From that moment on, the darkest phase began for Italy: Nazi occupation. The first executions of Jewish civilians were recorded on 8 September in Turin, Milan, Bologna, Trieste, Rome, and Naples. Roundups began a few days later. The persecution of rights had become the persecution of lives.

In those moments of absolute confusion, people were called on to make decisions of vital importance. It was their last chance to leave. This latest wave of migration was a completely different phenomenon from the one described so far. From a quantitative point of view, it was decidedly more intense and consistent, involving in less than a month 20,000 Italians. These were not only Jews, but also anti-fascists hunted for political reasons, disbanded soldiers, people who had not responded to the call to arms of the RSI. It was no longer an emigration but an escape, with one single destination: Switzerland.[254] It affected almost exclusively people from Northern Italy, bourgeoisie, high and humble social classes indistinctly. Nor was it a brain drain, but an escape by individuals hunted by the Nazi-Fascists. It was a very risky emigration because it was totally illegal, so refugees had to be truly committed and invest all the assets they could muster to recruit a good *passeur* and to bribe the border guards.

Those who could not or did not seize this last opportunity were lost. Two years of life in hiding, in the terror of allegations and roundups, between continuous movements, search for false documents and safe shelters, awaited them. In this last wave of migration, three mathematicians managed to escape: Bonaparte Colombo, former professor of Descriptive Geometry at the Military Academy of Turin and in charge of the teaching of Complementary Mathematics at the University; Bruno Tedeschi, lecturer of Actuarial Mathematics at the University of Trieste and two secondary school teachers: Nedda Friberti and Bianca Ottolenghi. Colombo and Friberti would later become involved in an extraordinary teaching experience, with the Italian University Camps.

3.24   Case Study. The 'University in Exile' of G. Colonnetti

A valuable scientist, renowned for his contributions to engineering, construction statics and mathematical theory of elasticity, Gustavo Colonnetti belonged to that 'spiritual and moral collective that kept alive another Italy alongside the official

---

[254] On this last emigration see, e.g., Sarfatti (1981); Broggini (1993); Broggini (1999); Lasserre (1995) and Wisard (1998).

one'.[255] Anti-fascist due to his nature as a man of culture, a free spirit and a fervent Catholic, Colonnetti refused to compromise with the regime and was forced into exile in Switzerland in the autumn of 1943. For him, it was the beginning of a new stage of academic activity in the service of victims of political and racial persecution. Sheltered in Lausanne, he created 'a strip of Italian university in a foreign land'[256] by establishing six Camps for interned military students located in Freiburg, Geneva, Huttwil, Lausanne, Mürren, and Neuchâtel.

This experience aroused historiographical interest already in the 80s and 90s,[257] but through the examination of unpublished sources held in Italian and Swiss archives, including the Colonnetti Archive in Turin and that of the State University of Milan, it is now possible to outline precisely the contours of what was not only a beautiful story of solidarity and hope but also a cultural adventure of extraordinary intensity, aimed at re-instilling the love for study and free thought in those young people to whom the material and moral reconstruction of the country should be entrusted.

Born in Turin on 8 November 1886, Colonnetti graduated with honours in Civil Engineering in 1908 and in Mathematics three years later, while he was already assistant to C. Guidi for the chair of Construction Science. Mathematical studies gave him the opportunity to weave a wide network of relationships with Segre (his thesis supervisor), Peano, Somigliana, Fano, as well as with Terracini, with whom he graduated in the same session and to whom he would be bound by bonds of friendship for the rest of his life.

Ranked third in the competition for the chair of Mechanics Applied to Construction and Machines at the School of Naval Engineering in Genoa, Colonnetti moved to the Ligurian capital in 1910. He remained there until 1914 when he moved to the University of Pisa, as Director of a laboratory of resistance tests of materials. Despite the proximity to neutralist circles, during the Great War Volterra entrusted him with management of the laboratory for the testing of steels of grenades, within the Inventions Office he had created in 1917. His expertise was not slow to manifest itself also in this sector: Colonnetti invented an electro-magnetic apparatus for steel tests, which was more efficient than those used in the Libyan campaign. Returning to Turin in 1919, Colonnetti first held the chair of Higher Mechanics and, from 1928, that of Construction Science. In 1922, he was appointed Director of the Polytechnic and from January 1923 he sat on the UMI Scientific Commission.

The Turin season was a period of central importance in his biography, both for the publication of the famous volumes *Fondamenti della statica* and *Principii di dinamica* (1927, 1929), and for the ideological choices he made. The twenties were in fact a time of growing tension and the Polytechnic was no stranger to the

---

[255] Antonicelli (1934), in *Testimonianze in memoria di Gustavo Colonnetti* (1973), p. 23: *collettivo spirituale e morale che teneva viva 'un'altra Italia' accanto a quella ufficiale.*

[256] Colonnetti (1973), p. 35: *questo lembo di università italiana in terra straniera.*

[257] Signori (1983); Twardzik (2006).

struggle for human rights and fundamental freedoms. E. di Rovasenda recalled—
among others—the first protests and

> when we all resisted together (there were three hundred of us in the first course) the violence
> of the elderly freshmen [...] and how one of those, in a black shirt and with a revolver,
> threatened to put down the rebellion in Prof. Fubini's class of mathematical analysis. And I
> remember the uproar, the screams for and against the Director Colonnetti.[258]

Colonnetti was among the few, in the Italy of those years, not to exploit his office
or the prestige he enjoyed for political purposes, and indeed to claim the autonomy
of men of science and the value of collaboration vis-à-vis personal political orien-
tations. All Turin knew, however, that he had refused to enter the PNF—in 1946, his
name would be listed on the wall posters among people who never joined the party—
and his failed *ralliement* began to cause sensation. In the same period, Colonnetti
(who in 1919 had joined the Popular Party) gradually approached the circles of
Catholic university action and in particular the Italian Catholic Federation of University Students (FUCI), where he met intellectuals sympathetic with him in the
defence of the core values of Christian socialism, such as P.G. Frassati, A. Severi,
and I. Bonini. The refusal of the Turin FUCI adherents to register with the Fascist
University Groups (GUF) and the fact of having publicly condemned the alleged
obligatory nature of this registration cost him his first appearance at the political
police office and the risk of being sent to confinement. Among his countercurrent
choices, the organisation of group exams stands out. Colonnetti had always criticised
the system of university exams, hoping that they would be replaced by a global
interview in which the student had the opportunity to deal with the entire teaching
staff, and vice versa the teaching staff could appreciate collegially the maturity of the
candidate. Rovasenda recalled how things went:

> There were a dozen of us taking that exam, on the entire programmes of the two-year study
> plan, which was called the 'group exam'. Our adhesion [...] was an act of trust in Colonnetti
> [...]. The group exam was an expression of scholarly freedom in a world increasingly
> suffocated by the totalitarianism of the regime, for us young people it was a manifestation of
> autonomy, an act of individual courage.[259]

Following various pressures, Colonnetti was forced to resign in 1926 from the
position of Director of the Polytechnic without having completed his 3-year term.
Relieved of most of his institutional commitments, in 1930 he bought Villa Ricci in
Pollone near Biella, which became the summer residence of his family and their

---

[258] di Rovasenda (1923), in *Testimonianze in memoria di Gustavo Colonnetti* (1973), p. 18: *quando
resistemmo tutti insieme, eravamo trecento in primo corso, alle violenze degli anziani [...] e come
uno di quelli, in camicia nera, minacciò con la rivoltella la nostra ribellione, nell'aula di analisi
matematica del Prof. Fubini. E rammento i parapiglia, gli urli pro e contro il Direttore Colonnetti.*

[259] di Rovasenda (1923), in *Testimonianze in memoria di Gustavo Colonnetti* (1973), pp. 18–19:
*fummo una dozzina a sostenere quell'esame, comprensivo di tutto il biennio, che fu detto 'esame di
gruppo'. La nostra adesione [...] nacque da un atto di fiducia in Colonnetti [...]. L'esame di
gruppo era un'espressione di libertà universitaria in un mondo sempre più soffocato dal
totalitarismo del regime, ed era da parte di noi giovani una manifestazione di autonomia, un
atto di coraggio individuale.*

permanent home from the beginning of the Second World War. Here, dozens of men of culture, journalists, and intellectuals were hosted, including B. Croce, A. Germano, and F. Antonicelli. In the thirties, Colonnetti progressively strengthened his contacts with the circles of anti-fascism and clandestine political action and one of his most iconic acts dates to 1932: the refusal to wear the black shirt in solemn demonstration of homage to the Duce who was visiting the Royal School of Engineering; he agreed to receive Mussolini in his laboratory only briefly (*in rapido passaggio*) and wearing a white coat. The election to the Pontifical Academy of Sciences in 1936 sanctioned the prestige he enjoyed in the Catholic environment, on the other hand it also led him to take an interest in three problems that would become, with the passing of time, leitmotifs of his intellectual commitment: social housing, the modernisation of the Italian school and university, and the responsibility of men of science.

In the autumn of 1938, Colonnetti witnessed the discrimination of colleagues and long-time friends, such as Levi-Civita, Supino, and Volterra. Fubini and Fano left the Polytechnic of Turin, together with one of his favourite pupils: Franco Levi.[260] Faced with the shame of racial policy, he was among the few who were not silent and addressed the Pope asking him to explicitly show solidarity with the Jews. When Pius XII replied that he doubted the appropriateness of the initiative, Colonnetti did not hesitate to express his disappointment. Shortly after this, he went to Paris for the last time as a professor of the Polytechnic of Turin, to give a series of lectures at the Sorbonne. It was in those days of bitterness and distrust that he settled on moving to Switzerland, where he knew he could rely on a network of scientific contacts. He was in Lausanne at least three times in 1940 and, in April 1941 asked the Ministry of National Education if he could go there for a few months as a visiting professor. Despite concerns, Bottai could not refuse authorisation.

As Colonnetti's travels in Switzerland intensified, the Polytechnic management began to doubt the real motivations that kept him away from university life. On 23 July 1943, the Director A. Bibolini informed the Ministry that Colonnetti had not regularly taught for some time and denounced the scientist's intention to transfer the Study Centre on Building Materials to Pollone. Two days later, the regime fell and Colonnetti again took over as President of the Faculty Council of the Polytechnic, a post which he held until the following autumn. As his first act, he decided to publicly commemorate Fubini, who had passed away in New York in the previous June. This was, however, a brief pause, an 'explosion of joy'[261] that preceded the return to an even more brutal regime. Immediately after the armistice, Colonnetti was suspended again from his post at the Polytechnic. Afraid for the safety of himself and his family, he began to think of fleeing to Switzerland. His fears were well-founded: in the

---

[260] Levi F. (2002–2003); *Testimonianze: 75° compleanno di Franco Levi.*
[261] Szegö (1943), in *Testimonianze in memoria di Gustavo Colonnetti* (1973), p. 28: *l'esplosione di gioia.*

autumn of 1943, he was suspended from his post at the Polytechnic, tried as a deserter and sentenced *in absentia* for political crimes in March 1944.[262]

Colonnetti fled Italy on 18 September 1943 and, from Pollone, reached Lugano on foot with his daughter Elena. His wife Laura and the other four children joined him later, crossing the border on 1 November. Less than a month after his arrival in Switzerland, Colonnetti had already obtained a position at the *École d'Ingénieurs* in Lausanne. However, this position did not enable him to provide for the needs of his family and forced him to separate from his children. Only his wife Laura stayed with him in Lausanne, where she became almost a legend among refugees. 'Aunt Lalla'—as everyone affectionately called her—did her best to provide material aid and books, with tenacity and fortitude, combined with a healthy dose of common sense: 'Saving principles is a beautiful thing but helping others is also just as beautiful', she often jokingly reminded her husband. [263]

In the years spent in Switzerland, Colonnetti did not dedicate himself only to his studies or teaching, indeed it may be said that these are two marginal aspects of his life as an exile. Believing that he had to do something concrete for his country, he immediately began to devote himself to three projects. The first was the creation of a Study Centre for Italian Reconstruction, an eminently apolitical organ involving architects and engineers such as M. Mazzocchi, A. Roth, A. Olivetti, and E. Nathan Rogers.[264] The second was the activity of columnist for various Swiss periodicals, in which he published several pieces under the pseudonym of *Etegonon*, crasis of the motto *Etiam si omnes et ego non*. The crisis of the Italian university and the theme of freedom of thought which he had begun to develop in the thirties were here addressed with renewed verve, and it is speaking of the purge of people who had sold themselves to the fascist regime that Colonnetti wrote one of his perhaps most intense pieces:

> There is another reason that will allow a thorough review of the frameworks of university teaching in Italy; and it is to purge the University of all those who have been direct accomplices or profiteers of the regime, or who, in order to obey the regime, sacrificed the dignity of the School and betrayed their educational mission. They are guilty of a new kind of crime: the crime of prostitution of science. They must be inexorably driven out of the University, whiplashed, like the merchants from the Temple.[265]

---

[262] *Condamnations dans le nord*, Courrier de Génève, 28.3.1944.

[263] ACT: L. Badini Confalonieri to G. Colonnetti, [September-October 1944]: *Salvare i principii è una bella cosa ma aiutare il prossimo è anche altrettanto bello.*

[264] ACT: *Centro Studi in Svizzera per la ricostruzione italiana*, [1944].

[265] Etegonon (alias G. Colonnetti), *L'Università*, Gazzetta Ticinese, a. I, 19, Saturday 2.9.1944: *un'altra ragione v'è che permetterà di rivedere a fondo i quadri dell'insegnamento universitario in Italia; ed è quella di epurare l'Università da tutti coloro che sono stati complici diretti o profittatori del regime, o che, per obbedire al regime, hanno sacrificato la dignità della scuola e tradita la propria missione educatrice. È di costoro un nuovo genere di reato: il reato di prostituzione della scienza. Essi vanno inesorabilmente cacciati dall'Università, a colpi di frusta, come i mercanti dal Tempio.*

The work that 'best characterises Colonnetti as a Master and as a Man',[266] however, is the creation and organisation of six University Camps for interned military students located in Freiburg, Geneva, Huttwil, Lausanne, Mürren, and Neuchâtel. To reconstruct and understand this experience, it must be remembered that in less than a month between September 8 and October 1, 1943, Switzerland saved about 20,000 Italians fleeing German occupation and the Italian Social Republic. This number rises to 40,000 units if we consider the entrances from 1939 to 1945. In this last wave of migration, various intellectuals and scientists managed to escape: U. Terracini, C. Marchesi, A. Fanfani, G. del Vecchio, G. Strehler, A. Lanzillo, L. Szegö, M. Donati, etc. Most refugees were shunted off first to transit camps (*campi di smistamento*), where civilians were separated from the military, then to quarantine hubs, and finally to labour camps, where they remained until repatriation. Until 1943, Switzerland institutionalised only one form of internment, that for French, Italian, and Polish soldiers.[267] The life of the inmates, although different in various camps, was generally miserable. Refugees, especially simple soldiers, were crammed into stables and sheepfolds and forced to engage into penal labour. Only in a few cases did some officers manage to obtain paid employment on farms or as manual labourers (Fig. 3.15a and b).

To improve this situation, in September 1943 the *Fond Européen de Secours aux Étudiants* (FESE) involved the *Eidgenössisches Kommissariat für Internierung und Hospitalisierung* (EKIH) in the formulation of a support programme aimed at providing refugees with that intellectual and moral help that represents 'an imperative necessity and constitutes an indispensable complement to material aid'.[268]

The secretary of the FESE, A. de Blonay, circulated a questionnaire in almost all 150 internment camps located in Switzerland, to conduct a census on university students. As of 13 November, 1,140 duly completed questionnaires had already been received by FESE. Among these, 1,015 came from students at universities and institutes of higher studies, 120 from graduates and five from teachers, principally in Northern Italy. The students of Economics were the most numerous, followed by those of Engineering, Literature, Law, Mathematics, and Physics. Seeing the census results, Colonnetti, Monseigneur A. Jelmini and P. Bolla, the Vice-President of the Federal Court, then addressed the President of the Swiss Confederation:

> Among the civil and military refugees are many young graduates or university students, who, in the camps, necessarily devoid of means of study and mentorship, experience a spiritual discomfort much more serious than any other suffering. I do not ask for them freedom, nor material comforts greater than those that Switzerland can give and generously has given. On the contrary, I am myself willing, renouncing the concession that has been

---

[266] Supino (1969), p. 9: *più lo caratterizza come Maestro e come Uomo*. See also Badini Confalonieri (1973); Badini Confalonieri (1978) and *Gustavo Colonnetti per chi lo conobbe* (1973).

[267] Military internment was in fact governed by international conventions that did not apply to civilian refugees, although in 1945 a camp for civilians would be established in Lausanne-Pully, directed by A. Fanfani. See Levi A. (1947); Einaudi (1997); Fanfani (2012).

[268] AFB: *Les universitaires italiens internés en Suisse*, November 1943, memorandum signed by de Blonay: *un'imperiosa necessità e costituisce un complemento indispensabile all'aiuto materiale*.

Fig. 3.15 (**a**) (above) and (**b**) (below) Internment camps in Switzerland, 1944, ACT

made to me, to share with them the life of the camp, in order to be able to continue even in exile that work of intellectual and moral education, which has always been for me the highest and dearest mission. To this end, it would be enough to bring young people together in a single camp, in which myself and some colleagues be also allowed to live.[269]

Despite the opposition of the military hierarchies, the Federal Council accepted the request and allocated 200,000 francs to the creation of internment camps for university students. A *Comité d'aide aux universitaires italiens en Suisse* was then set up in Lausanne, chaired by Colonnetti and Bolla, who soon sent a working protocol to EKIH. Their first attempt was to demand, for Italian refugees, free registration in Swiss faculties and maximum freedom of movement, while continuing to be subject to the control of military authorities. The proposal was rejected but the large number of prominent personalities who supported it pushed EKIH to agree to gather part of the young Italian internees in special camps under para-military regime, located in the surroundings of various French universities. Thanks to the efforts of Swiss and Italian intellectuals such as A. Stucky, R. Secrétan, and Colonnetti, the organisational structure was quickly set in motion. To select students from the applications received, an examination board was convened in Mürren, Lyss, and Olten, consisting of four subcommittees, one for each macro-disciplinary area: Engineering, Economics, Literature, and Science. The exam, in front of Swiss and Italian teachers, consisted of an oral interview of a few minutes on each subject included in the study plan for which the internee had applied for admission to the camps. Colonnetti immediately ascertained the uselessness of this test and, years later, would recall the sense of human participation experienced during the interviews:

Talking to those young people about mathematics or physics, literature, or history, was obviously useless. [. . .] I decided to make contact with them on a more humane level. I asked them where they came from, what miseries they had witnessed, what news they had of their distant families. And I saw their gaze rekindled, and their souls opened to hope; and I listened to the heartfelt voices that supplicated me to welcome them into the university camps, to help them to resume their interrupted studies; and I judged them from man to man, without worrying about what they knew or what they had forgotten [. . .].[270]

---

[269]Letter published by Colonnetti, under the pseudonym of Minimo, in the periodical *In attesa*, n. 12, 25.8.1944, within the title *Università Italiana in esilio*: *Sono tra i rifugiati civili e militari molti giovani laureati o studenti universitari, i quali, nei campi necessariamente privi di mezzi di studio e di guida, accusano un disagio spirituale assai più grave di ogni altra sofferenza. Io non domando per essi la libertà, né comodità materiali maggiori di quelle che la Svizzera può dare e generosamente ha dato. Al contrario sono io stesso disposto, rinunciando alla concessione che mi è stata fatta, a sottomettermi volenterosamente alla vita del campo, pur di poter continuare anche nell'esilio quell'opera di educazione intellettuale e morale, che è sempre stata per me la più alta e la più cara. Basterebbe a tal fine riunire i giovani in un campo solo, in cui fosse concesso abitare anche a me e qualche collega.*

[270]Colonnetti (1945), p. 218: *Parlare a quei giovani di matematica o di fisica, di letteratura o di storia, era evidentemente inutile impresa. [. . .] Decisi di prender con essi contatto su di un terreno più umano. Chiesi loro di dove venissero, di quali miserie fossero stati testimoni, quali notizie avessero delle loro famiglie lontane. E vidi riaccendersi il loro sguardo, ed aprirsi il loro animo alla speranza; ed ascoltai le voci accorate che mi scongiuravano di accoglierli nei campi universitarii, di aiutarli a riprendere gli studi interrotti; e li giudicai da uomo a uomo, senza preoccuparmi di quel che sapevano o di quel che avevano dimenticato [. . .].*

Due to the limited funding granted, only 540 of the 1,140 applications were selected. Candidates admitted were divided into four camps (Lausanne, Freiburg, Neuchâtel, and Geneva) in which parallel university courses were organised, allowing interned soldiers to enrol in Swiss institutions and finish their studies.[271] Additional programmes for officers and sub-officers called University Studia (*Studi Universitari*) were created in Mürren and Huttwil afterwards. For a short time, a camp was also active in Herzogenbuchsee. The camps attracted:

> young people of all conditions and backgrounds; young people who had left the University on the eve of graduation; young people that the University had not even seen, having been called to arms immediately after the diploma. Restless and deeply troubled consciences, in which the sudden collapse of every hierarchy, of every military discipline, had dug a furrow not yet filled by the rise of the sense of personality that for too long had been debased and compromised.[272]

The camps, which reported to the Swiss authorities and General Inspectorate, had the pyramid structure typical of military bodies. Lieutenant colonel M. Zeller, a former professor of Photogrammetry at ETH Zurich and almost caricature figure, was their general Director. He identified, for each camp, a military commander, while each University proposed a Rector to coordinate the teaching. Then there was the *chef des études* who acted as a link between the Rector and students. Despite the strict regulations established by the inspectorate, each reality still had its own characteristics, partly related to the composition of the teaching staff and partly due to the interactions between this and the local context. Thus, for example, Huttwil constituted a vigorous model of effective collaboration between military and civilians, thanks to the intelligent guidance of A. Montel and R. Dellea, but was sadly marked by the manifestations of anti-Semitism against Jewish teachers (G. Fano, A. Levi and P. D'Ancona).[273]

For Bolla and Colonnetti, Lausanne immediately represented the ideal place to bring together Italian refugee students and teachers of the technical-scientific area. Firstly, the city had given asylum to a large number of 'special' refugees, such as A. Olivetti, and could therefore offer numerous guarantors. Moreover, the city boasted prestigious cultural institutions, such as the *École d'ingénieurs* and the *Cercle mathématique*. For these reasons, Lausanne was chosen as the seat of the main Italian University Campus (IUC) for Engineering and Sciences, inaugurated on 26 January 1944 and operating until May 1945, hosting about two hundred students.

---

[271] The destination was often casual, except for students of Engineering and Architecture who were mainly sent to the Lausanne camp and those of Law, who were admitted mainly in Geneva.

[272] Colonnetti (1945), p. 217: *giovani di tutte le condizioni e di tutte le provenienze; giovani che avevano lasciato l'Università dopo di averla per più anni frequentata, ed esser giunti alla vigilia di una laurea; giovani che l'Università non avevano neppur vista, essendo stati chiamati alle armi subito dopo il conseguimento, spesso affrettato, di una maturità classica o scientifica. Coscienze inquiete e profondamente turbate, in cui il crollo improvviso di ogni gerarchia, di ogni disciplina militare, aveva scavato un solco non ancora colmato dal sorgere del senso della personalità che per troppo tempo era stata avvilita e compromessa.* See also Colonnetti (1946).

[273] See Montel (1945) and in ACT: G. De Marchi to G. Colonnetti, Vevey 27.2.1945.

From November 1943 Colonnetti, who was appointed Rector of IUC, faced a period of frenetic activity. To overcome a series of obstacles posed by the federal and cantonal offices, 'all his patience, his tact and his power of conviction were needed'.[274] The first essential issue to tackle was the recruitment of teachers. It was Colonnetti who selected them personally, beginning by contacting former colleagues and alumni. The fruits of these interviews were not long in coming. The Lausanne Camp was to count as professors 29 prominent scholars including L. Einaudi (Economic Policy), A. Fanfani (Economic History of Italy), G. Tedeschi (International Trade and Economic Policy), G. Fano (Analytic and Descriptive Geometry), M.G. Levi (Industrial Chemistry), F. Levi (Resistance of Materials), L. Szegö (General and Inorganic Chemistry) and 22 assistants, including B. Colombo (Special Mathematics) and M. Dedò (General Mathematics).[275]

In addition to defining the teaching staff, Colonnetti examined the files of candidates who, for one reason or another, had not been able to sit for the interview or had not been admitted to the camps in the first round of selection. The *Hôtel des Étrangers*, where the Colonnettis' lodged, thus became a free port through which 'a gallery of humanity'[276] transited, dozens of men who identified in Colonnetti a spiritual guide and 'venerated him like a father'.[277] Many other refugees collaborated, highlighting situations of greatest urgency. For example, Fubini and Fano provided their contacts with the international Jewish committees of rescue, while the engineers R. De Benedetti and V. Consolo suggested that Colonnetti consider, exceptionally, the request of Nedda Friberti, a mathematics teacher and lecturer in Trieste, who had been interned in Oberhelfeuschill (St. Gallen) a few days after the deadline for entrance exams to the camps.[278]

One particularly complicated question regarded the fragmentary character and heterogeneity of the candidates' *curricula*, and recognition of their qualifications, of which the internees were often unable to provide any documentation.

The most critical issue, however, was linked to the plurality of purposes which, from the very beginning, Colonnetti intended to award the camps: to offer the possibility of recovering the years of study lost, but also (and above all) to offer young people who had known only the discipline of fear and weapons an intellectual experience that would project them towards reconstruction, re-educating them to free and critical thought after 20 years of fascist indoctrination. It is from this point of view that the *Grande Campo* in Lausanne stands out from the others: for its cultural life, of a liveliness and dynamism as singular as extraordinary, taking into account the historical moment.

---

[274] Szegö (1923), in *Testimonianze in memoria di Gustavo Colonnetti* (1973), p. 29: *ci voleva tutta la pazienza, il tatto ed il potere di convinzione del Prof. Colonnetti.*

[275] ACT: *Quadro degli assistenti e capi-gruppo della Facoltà di Ingegneria*, fols. 1–3.

[276] Badini Confalonieri and Colonnetti (2006), p. 6: *un campione di varia umanità.*

[277] ACT: G. Carloni to G. Colonnetti, Lützelflüh 21.4.1944: *lo venera come un Papà.*

[278] ACT: V. Consolo to G. Colonnetti, Lausanne 29.2.1944, in Luciano (2017b), p. 160.

At first, to smooth out the cultural inequalities between students, Colonnetti inaugurated the publication of the handouts (*dispense*). The editorial production was intense, and, between December 1944 and May 1945, 66 issues appeared with the editorial mark 'Campo Universitario Italiano Lausanne' cyclostyled and distributed by a specific office. Among these texts were the lecture notes on Analytic and Descriptive Geometry by Fano, exemplified in his masterful lectures given at the Turin Polytechnic (1910, 1935), and those of the course of Construction Science taught by Colonnetti himself. These texts also reached (through the FESE) Italian prisoners of war in France, Germany, and Poland and, after the closure of the camps, were sent to Rome to be made available to veteran students and teachers.

The most interesting element of the activity conducted in the IUC was, however, constituted by the conferences, entrusted both to exiles (Marchesi, Del Vecchio, B. Caizzi, Einaudi, Fanfani, ...) and to Swiss intellectuals (A. Vodoz, G. Wyss), which took place every Wednesday in a classroom of the Palais de Rumine and which gained immense success. The plurality of ideas about Europe's future reflected the different leanings of the speakers and their mutual respect for each other's opinions. For many internees, these lectures represented the only opportunity to overcome the anguish caused by the dramatic news on the progress of the war that reached the camp.

Among the first to speak was Colonnetti, on the theme *The spiritual premises of reconstruction* (*Le premesse spirituali della ricostruzione*, April 17, 1944). It was a text of great emotional appeal, in which he illustrated to his students the moral and deontological principles that had guided (and would continue to drive) his career as a scientist and his conduct as a man: Catholic faith, struggle against materialism, love for research for its own sake, defence of freedoms and rights, the overcoming of all nationalism, in the name of a higher and more universal conception of contemporary reality. Reconstruction—said Colonnetti in that circumstance—cannot be reduced to the material rebuilding of houses, bridges, and roads, but must be preceded by an analysis of the sources of democratic immaturity that have caused the clouding of consciences and that have allowed the affirmation of totalitarianism, leading to the immense tragedy of war. Even before the technical and scientific challenges posed by reconstruction, it is therefore necessary to discuss as a priority its spiritual premises. For Colonnetti, the equation holds whereby scientific research corresponds to technological progress, which in turn forms the basis of all lasting economic and social advancement. To regain the highest values of European civilisation, exported to the world through the circulation of men and ideas that had constituted the only positive aspect of colonialism, Colonnetti argued for the need to overcome nationalist ideologies and class egoisms, building a united Europe that should know how to give 'fair resolution to what it had now been agreed to call the social question'.[279]

On 12 June and 10 July 1944, Colonnetti returned to speak to his students, this time about *The problems of university life* (*Problemi della vita universitaria*). In the

---

[279]Colonnetti (1944a), handout no. 107, republished in Colonnetti (1973), p. 15: *equa risoluzione di quella che si è ormai convenuto di chiamare la questione sociale.*

modern era—he stated—the university institution has lost its vocation, and the sinister utilitarian end has consequently prevailed. The solution proposed is radical: to adopt separate systems for those who want to be trained for professional life and for those who want to devote themselves exclusively to research; open the way of studies to all those who deserve it, however disadvantaged their starting conditions are, and close it 'to the inept, even if largely endowed with personal goods, because studying is not a luxury or a pastime, but a social service'.[280] The closing of the conference was yet another appeal to the young people of that strip of Italian university in a foreign land, to personally participate in the spiritual reconstruction of the Italian university:

> You know that your Rector must scrupulously refrain from talking to you about politics. Here, then, my discourse stops, and I will be careful not to tell you through which aberrations of thought, through which degenerations of customs, those who have dragged Italy into the current abyss have undermined and corrupted the university life. There were [...] teachers who, out of cowardice or lust for honours, betrayed their mission by putting themselves at the service of the most shady interests of the regime; there were institutions which, unmindful of their traditions, did not disdain being turned into barracks; there were crowds of young people who, in an orgy of turmoil, agreed to renounce the most sacred of all freedoms, the freedom of thought.[281]

The organisation of the public conferences and the publication of their texts gave rise to strong disagreements between Colonnetti and the Swiss military authorities. The officers in charge of the camp in Lausanne did not look favourably on the political content of some of these speeches, considering them incompatible with the strictly scientific character that should distinguish the teaching in the camps. The federal authority, through Zeller, intervened to stop these meetings, branding them as contrary to the functioning rules established by the EKIH and went so far as to envisage the suspension of the courses and the dissolution of the IUC if they were not immediately interrupted. Colonnetti's opinion was clear:

> In the young internees who had witnessed the collapse of the Italian army, the sense of military discipline was irreparably destroyed. Reconnected with the university world where we think, reason, discuss, they had to feel the irresistible need to think, to reason, to discuss.

---

[280]Colonnetti (1944b), handout no. 116, republished in Colonnetti (1973), p. 42: *agli inetti, anche se largamente dotati di beni di fortuna, perché studiare non è un lusso od un passatempo, ma è un servizio sociale.*

[281]Colonnetti (1944b), handout no. 116, republished in Colonnetti (1973), pp. 34–35: *Voi sapete che il vostro Rettore deve scrupolosamente astenersi di parlarvi di politica. Qui s'arresta quindi il mio discorso, ed io mi guarderò bene dal dirvi attraverso quali aberrazioni del pensiero, attraverso quali degenerazioni del costume, coloro che han trascinata l'Italia nel baratro attuale abbiano minata e corrotta anche la vita universitaria. Sta di fatto [. . .] che vi furon maestri che per viltà o per brama di onori, tradirono la loro missione mettendosi al servizio dei più loschi interessi del regime imperante; che vi furono istituti che, immemori delle loro tradizioni, non sdegnarono di trasformarsi in caserme; che vi furono folle di giovani che, in un'orgia di clamori, accettarono di rinunciare alla più sacra di tutte le libertà, alla libertà del pensiero.*

No other discipline could be imposed on them other than that resulting from a recognised superiority in the field of thought.[282]

Faced with the threat that his students would be dispersed, Colonnetti decided to abolish the conferences, regardless of their theme. This is how he commented on the epilogue of the story with Bolla:

> You know, dear President, that I was particularly concerned with leading the students to study and discuss the great social and political problems of the post-war period; and I had of course chosen the simplest way: to talk to them about these problems and to offer them the freedom to discuss them. But there is obviously also another way to achieve the same result, and that is to forbid them any discussion.[283]

Meanwhile, the end of lessons and the opening of the exam session approached. Despite the shortness of the school period, all the teachers had managed to complete the programmes, and some had even set up additional lectures. The assessments were largely satisfactory: very few fails, a few passes with the minimum mark, but many with high marks or with honours. Colonnetti thanked his colleagues individually (D3.24.1 and D3.24.2).

Satisfaction was, however, marred by a new phase of tension with the military authorities. F. Bellia, consul of Italy in Bern, asked Colonnetti to share with the students of IUC the text of the proclamation launched by the Liberation Committee for Upper Italy to celebrate the anniversary of the fall of the fascist regime. At the same time, E. Nathan Rogers, G. Pozzi, G.E. Sessa, and C. Bianchi asked him to collect the names of students in the Lausanne-Vevey camp who had entered the CLN (National Committee for the Liberation of Italy). The act of having authorised these initiatives put Colonnetti in the spotlight once again with the authorities.[284]

The IUC closed its doors in the last week of July 1944 and students and assistants were assigned to forced agricultural labour. While the Rector left Lausanne to spend a few days with his children, discontent was spreading in some sectors. It was Colonnetti who had drawn up the list of assistants who could remain in Lausanne, deciding to exempt from forced labour those who needed to complete the drafting of the handouts.

---

[282] G. Colonnetti, *Relazione al Ministro dell'Educazione Nazionale* [Guido De Ruggiero], 2.8.1944, republished in Colonnetti (1973), p. 77: *Nei giovani internati che avevano assistito allo sfacelo dell'esercito italiano, il senso della disciplina militare era irrimediabilmente distrutto. Rimessi a contatto col mondo universitario dove si pensa, si ragiona, si discute, essi dovevano sentir sorgere irresistibile il bisogno di pensare, di ragionare, e di discutere. Nessun'altra disciplina poteva loro venire imposta che non fosse quella derivante da una riconosciuta superiorità nel campo del pensiero.*

[283] ACT: G. Colonnetti to P. Bolla, Château d'Oex 15.5.1944: *Ella sa, caro Presidente, che io mi preoccupavo soprattutto di condurre gli allievi verso lo studio e la discussione dei grandi problemi sociali e politici del dopoguerra; ed avevo naturalmente scelta la via più semplice: quella di parlar loro di questi problemi e di offrir loro libertà di discuterne. Ma v'è evidentemente anche un'altra via per raggiunger lo stesso risultato, ed è quella di proibirglieli.*

[284] ACT: G. Colonnetti to CLN, Delegation of Lugano, Lausanne 25.8.1944.

Fig. 3.16   Colonnetti and his wife Laura Badini Confalonieri, ACT

Inevitably, the decision was criticised, and he was accused of favouritism. Meanwhile, in August, Colonnetti drew up a detailed report on the functioning of the camps and pointed out to the Federal Delegation some aspects that could be improved. In particular, he proposed increasing the number of places available, so that deserving people could access them; he asked to divide people according to the faculties closest to the various camps, and above all hoped that the 'relations between academic and military authorities would be defined in a very different way'.[285]

During the summer holidays at Château d'Oeux, Colonnetti and his wife Laura also intensified their commitment to the collection and transmission of news on displaced Italians, deportees, or prisoners of war (Figs. 3.16 and 3.17).

In an Italy divided in two by the Gothic Line, being able to obtain information was very difficult. Thanks to their contacts with the Holy See and with the legation of Italy in Bern, Gustavo and Laura managed to help many families desperate for news on the fates of their nearest and dearest. Among the requests for help evaded was that of mathematician G. Cassinis, whose son Roberto, an aspiring lieutenant of the Naval Engineers, had no longer sent news of himself after the armistice.

Throughout the summer, there were no announcements either about the reopening of the camp or about whether its rectorate still pertained to Colonnetti, although dozens of applications from aspiring students continued to arrive. The critical political circumstances and the stalled war situation finally pushed the federal

---

[285] ACT: *Per una migliore organizzazione dei campi universitari in Svizzera*, fols. 1–4: *vengano in ben altro modo definiti i rapporti tra autorità accademiche e militari.*

Fig. 3.17 Colonnetti's wife and children, Château-d'Oex, 1944, ACT

authorities to decide to resume the activity of the IUC for the following academic
year. About 140 students were admitted to the winter semester, under the guidance
of 25 teachers and assistants. In addition to the lessons, also the conferences resumed
but the experience of Colonnetti at the head of IUC was clearly coming to an end
because of the irreconcilable contrast between his ideals as a scientist and the rigidity
of the military Swiss institutions. In a secret report, inspired by Zeller, the military
commander of the camp (A. Tommasi) accused Colonnetti of 'frequently staying
with communist agitators', of undermining the authority of Italian officers and of
granting 'full tolerance to demonstrations of a political nature'. At the same time he
proposed the opening of an investigation, in order to re-establish the discipline in the
IUC, greatly compromised by the management of Colonnetti, 'who is not a military
man and who has never kept silent about his feelings completely contrary to
everything and everybody military'.[286]

---

[286] AFB: *Personaldossier* G. Colonnetti E4264, 195/196; E5791, 1, 7/53–56: *trattenersi
frequentemente presso agitatori comunisti; piena tolleranza a manifestazioni di carattere politico;
che non è un militare e che non ha mai taciuto i suoi sentimenti del tutto contrari a quanto sia
militare.*

Colonnetti reacted by leaving the position of Rector of the IUC, comforted by the support of most of his students and staff. The countless farewell letters he received echoed a similar sentiment: 'We will always remember you as the first one who spoke to us about Italy and freedom with the words we were waiting for' some internees wrote to him.[287] Colonnetti took his leave on 30 November 1944, giving one of his most passionate speeches, *Transient and Permanent Phenomena* (*Fenomeni transeunti e permanenti*), almost a sort of spiritual testament, in which he drew a suggestive comparison between the principles of statics that are at the foundation of a building and those on which the economic, social, and political edifice of a nation should stand. Following his resignation, two Swiss Rectors were appointed: A. Stucky and R. Secrétan. Meanwhile, on 2 December 1944, the EKIH established a Faculty of Engineering in Huttwil and appointed Tommasi *chef d'études*. Tommasi left the *Grande Campo* in Lausanne, replaced by P. Malinverni, former professor of General Mathematics in Huttwil. The IUC was definitively terminated on 14 May 1945.

Meanwhile, at the end of November 1944, a group of eleven refugees met in Geneva, for whom the Italian government (installed under the presidency of I. Bonomi) urgently asked for repatriation to Rome in order to entrust them with some delicate government tasks. After frantic preparations, A. Alessandrini, G. Boeri, F. Carnelutti, G. Colonnetti, L. Einaudi, C. Facchinetti, T. Gallarati Scotti, L. Gasparotto, S. Jacini, C. Marchesi, and A. Orlando left Switzerland on December 4 and, following a stop in Lyon, landed in Ciampino on December 10. A few days later, Colonnetti presented to the Italian authorities the work carried out in Swiss university camps and asked for the recognition of credits for exams taken by internees. The government authorised him to officially communicate the news of examination credit approval on Radio London on December 30.[288]

Moved from the awareness that 'Italy in ruins asked for dedication and initiative for commitments that were even greater and more urgent than those of families'[289] the Colonnettis left their children in Switzerland and went through 12 years of intense work. Chair of the CNR and elected member of the High Council of Education, Gustavo dedicated himself to school, university, and the responsibility of scientists towards technological progress. Laura committed to a recovery programme designed for veterans. However, Colonnetti's political career stopped at the Constituent Assembly due to his battle in favour of the annulment of all university appointments made from 1923 onwards without a competition procedure, as a gift in exchange for political merits. Needless to say: Colonnetti would have lost

---

[287] ACT: The students of the IUC to G. Colonnetti, Lausanne 23.11.1944: *La ricorderemo sempre come il primo che ci parlò dell'Italia e della libertà con le parole che tanto attendevamo.*

[288] ACT: *Communication aux étudiants italiens*, fol. 1. The text of the radio message, transmitted on 30.12.1944, is reproduced in Colonnetti (1973), pp. 97–98.

[289] Badini Confalonieri and Colonnetti (2006), p. 8: *un'Italia in frantumi chiedeva dedizione e iniziativa per necessità che erano ancora più grandi e impellenti di quelle loro famigliari.* See also *Laura e Gustavo Colonnetti . . . 2000.*

this battle because, as A.C. Jemolo frankly affirmed, there soon appeared the contrast between:

> the men of austerity, who refused any compromise, and those who believed it was necessary instead to heal the wounds, to reconcile, to all shake their hands together, having no repugnance to shaking even scarcely clean hands, letting oblivion cloak the past, collective faults and individual faults, and above all relying on the principle of the consolidated statement: everyone keeps what he has, it does not matter whether it was rightfully or wrongly acquired.[290]

Documents

D3.24.1 G. Fano to G. Colonnetti, Lausanne 16.11.1943

ACT, fol. 1r

Distinguished Professor, Prof. Carletti,[291] whose name I have already given you, and who must have written to you directly, sends me the attached postcard from which some interesting news appear that he had not, it seems, communicated to you. You will see whether, and to what extent, to take this news into account, for the patriotic work that has begun. With the most cordial greetings, Yours affectionate Gino Fano

15-XI-1943

Distinguished Professor, Thank you very much for your interest in me. I wrote, as you yourself suggested me, directly to Prof. Colonnetti: if I can be useful to you, I will work humbly, but with passion. Please tell Prof. Colonnetti that in Mürren, among the officers, there are many students, including the son of one of the best professors of the Politecnico in Milan, whose name escapes me, and there are also some university professors. If Prof. Colonnetti has influence on the Swiss authorities, he could do great good to all students, civilian and military: the Poles have managed to concentrate all their students in a university city, thanks to the support of the YMCA.[292] In this sense, Prof. Dedò, Chisini's assistant, is working as an

---

[290] Jemolo (1945), in *Testimonianze in memoria di Gustavo Colonnetti* (1973), p. 42: *i personaggi dell'austerità, del rifiuto di ogni compromesso e chi riteneva occorresse invece sanar le ferite, riconciliare, darsi tutti la mano, non avendo ripugnanza a stringere anche mani scarsamente pulite, facendo scendere l'oblio sul passato, colpe collettive e colpe individuali, e soprattutto facendo gran leva sul principio della situazione consolidata: ciascuno conservi quel che ha, non importa se bene o male acquisito.*

[291] Ernesto Carletti (1897–1971), graduated in Mathematics in Bologna, was one of the few teachers who refused the fascist loyalty oath. After the armistice (8.9.1943) he had to flee to Switzerland.

[292] The Young Men's Christian Association, a non-governmental federation founded in London in 184, was involved in supporting millions of prisoners of war and in helping refugees, particularly displaced Jews. YMCA co-organized and worked in camps for Polish civilian refugees and prisoners of war in Fribourg, Feldbach, Losone, etc.

organiser at the YMKA in Münchenbuchsee (Bern).[293] Yes, the farmers here all come from border areas: Como and Varese. In these days other Italians have arrived, Jews from Milan (Levy and Friedmann). Forgive me for bothering you and, please, always count on my gratitude, Ernesto Carletti

P.S. I just now remember the name of the university professor I was talking about before: he is Prof. *Danusso* of the Politecnico di Milano.[294]

Egregio Professore, Il Prof. Carletti, del quale già le ho dato il nome, e che deve avere scritto a lei direttamente, mi dirige l'unita cartolina, dalla quale appare qualche notizia interessante che non aveva, pare, comunicata a lei. Vedrà lei quale conto se ne possa tenere, per l'opera patriottica iniziata. Coi più cordiali saluti, suo aff. Gino Fano

15-XI-1943
La ringrazio vivamente dell'interesse che dimostra per me. Ho scritto, come lei stesso mi consigliava, direttamente al Prof. Colonnetti: se potrò esservi utile, lavorerò umilmente, ma con passione. Voglia dire al Prof. Colonnetti che a Mürren, tra gli ufficiali, ci sono moltissimi studenti, tra gli altri uno figlio d'uno tra i migliori professori del Politecnico di Milano, di cui mi sfugge il nome, e ci sono anche alcuni professori d'Università. Se il prof. Colonnetti ha influenza sulle Autorità Svizzere, potrebbe far del gran bene a tutti gli studenti, civili e militari: i polacchi erano riusciti a far concentrare tutti i loro studenti in una città universitaria, grazie all'appoggio dell'YMCA. In questo senso lavora il prof. Dedò, assistente di Chisini, assunto come organizzatore dall'YMKA di Münchenbuchsee (Bern). Sì, i contadini di qui sono tutti di zone di frontiera: di Como, di Varese. Sono giunti in questi giorni altri italiani, israeliti di Milano (Levy e Friedmann). Perdoni se l'ho importunata e voglia contare sempre sulla mia riconoscenza Ernesto Carletti

P.S. Ricordo il nome del prof. universitario di cui parlavo prima: si tratta del prof. Danusso del Politecnico di Milano.

---

[293] Modesto Dedò (1914–1991), former assistant of Oscar Chisini in Milan, in 1943 refused the induction into army and fled to Switzerland where he was interned in the Münchenbuchsee camp. He served as assistant of G. Fano for the course of Analytic, Projective, and Descriptive Geometry at the Lausanne camp. With the support of YMCA Dedò organized both literacy and university level courses at the Universities of Lausanne and Neuchâtel.

[294] Arturo Danusso (1880–1968), professor at the Polytechnic of Milan (1915–1950). His son Ferdinando (1921–2006) had enrolled at the Polytechnic of Milan but with the outbreak of the Second World War he was called to arms as an artillery officer. After the armistice, he reached his family in Cernobbio and escaped to Switzerland, where he remained interned until the liberation.

D3.24.2 G. Colonnetti to G. Fano, Lausanne 24.7.1944

ACT, fol. 1r

Distinguished colleague, At a time when the University Camp of Lausanne is
about to close, I would like to express to you all my gratitude for the work you
have carried out as a teacher of Projective, Descriptive, and Analytic Geometry,
with such a high spirit of patriotism and solidarity for the young soldiers interned.
I am sorry that circumstances have not allowed me to obtain adequate compensation for your service. However, I have signalled your work to the Governance
of the École d'Ingénieurs, which has recognised the possibility of attributing to
you a part of the Finances de Cours that is shared among Swiss teachers; this
quota would amount (for the winter semester) to fr. 275, which will be paid to you
if the competent authorities give the necessary consent. In any case, I would like
to declare that I make it my duty to report your work to the Italian Government as
soon as I can do so. For the moment I can only confirm my gratitude for your
precious collaboration and ask you to accept my thanks and most cordial greetings. [G. Colonnetti]

*Egregio Collega, Nel momento in cui sta per chiudersi il Campo Universitario di
Losanna io desidero esprimerle tutta la mia riconoscenza per l'opera da Lei
prestata in qualità di docente di Geometria Analitica Proiettiva e Descrittiva, con
così alto spirito di patriottismo e di solidarietà per i giovani militari internati.
Sono dolente che le circostanze non mi abbiano permesso di ottenere per questa
sua prestazione un adeguato compenso. Ho però segnalata l'opera sua alla
Direzione dell'École d'Ingénieurs, la quale ha riconosciuta la possibilità di
attribuirle una quota di quella parte delle Finances de cours che viene ripartita
fra i docenti svizzeri; tale quota ammonterebbe (per il semestre d'inverno) a
fr. 275 che le verranno versati se le competenti autorità daranno il necessario
consenso. Tengo in ogni caso a dichiararle che mi farò un dovere di segnalare
l'opera sua al Governo Italiano non appena mi sarà possibile farlo. Per il
momento non posso che confermarle la mia riconoscenza per la sua preziosa
collaborazione, e pregarla di gradire coi miei ringraziamenti i più cordiali saluti.
[G. Colonnetti]*

Chapter 4
Gallery: Those Who Failed to Leave

The word absence is strongly polysemic. It indicates a state or condition in which something expected, wanted, or looked for is not present or does not exist; a failure to be present at a usual or expected place but also an inattention to present surroundings or occurrences. The multiplicity of meanings lends itself well to reflect the plurality of forms of absence that distinguishes, in the long run, the collective history of mathematicians who fled for political and racial reasons. To understand when and why a name is missing from a list of aspiring exiles, it is rarely enough to invoke a purely scientific motivation. Other factors come into play: painful choices, forced decisions, human participation, distress, etc. There were several reasons behind failed emigrations: some scholars moved too late, or made errors of judgement, for example by declaring in their curriculum their fascist merits; others were not internationally renowned or dealt with mathematical disciplines that were considered marginal or of little interest. Many would-be refugees were early career scholars. Age and gender were two discriminating factors.

Despite their efforts, four mathematicians (Ascoli, Maroni, Friberti, and Tedeschi) and two astronomers (Bemporad and Horn d'Arturo) missed the opportunity to emigrate. In actual fact, Maroni and Horn d'Arturo made only a modest attempt: a letter to the SPSL, to report their cases, without following up on this action. No trace of an attempt at emigration on the part of Del Vecchio, extraordinary professor of actuarial mathematics in Trieste has been found.

4.1 Guido Ascoli: The (Im)possibility of Escape

Closest to reaching the goal of expatriation was Guido Ascoli, full professor of Analysis at the University of Milan. Ascoli was a Jew of mixed blood, born from a Jewish father and an Aryan mother. Just over fifty years old (he was born in Livorno on 12 December 1887), graduated in Pisa in 1907, with a thesis on projective-differential geometry supervised by Bianchi, Ascoli had had the talent to pursue an

E. Luciano, *The Jewish Mathematical Diaspora from Fascist Italy*, Science Networks. Historical Studies 64, https://doi.org/10.1007/978-3-031-64896-0_4

Fig. 4.1 The Ascoli family, 1933, courtesy of Cristina Ascoli, Davide Ascoli, Irene Ascoli, Maria Bolgiani, Guido Bolgiani, Paola and Marco Salbol

academic career. For family reasons, however, after the postdoctoral specialisation in Pisa, he had been forced to return in Livorno, where he had found a job in a company producing electrical conductors. Subsequently, he had entered the teaching profession, first in Spoleto, then in Genoa, Cagliari Caserta, and Florence. In May 1916, despite his weak constitution, he was called to arms and sent to the front as a soldier of the 44° Field Artillery Regiment. Injured in the explosion of a grenade in 1918, he ended his military service as a reserve artillery lieutenant by being decorated with the Cross of Merit. Once back to civil life, Ascoli resumed teaching, first in Parma and, from 1920, in Turin, at the scientific lyceum Galileo Ferraris.

Arrival in Turin marked a watershed in his private and working trajectory. On the personal side, in 1925 Ascoli married Mauriziana Sossi, a non-Jewish primary school teacher; in 1927 and 1932 their two children were born: Renato and Gigliola. Both were baptised (Fig. 4.1).

The arrival in Turin also marked a turning point in Ascoli's career, thanks to meeting Fubini. Under his impulse and encouragement, Ascoli published twenty works in ten years, allowing him to rank third in the first national competition in which he participated, for the chair of Algebraic Analysis at the University of Cagliari (1930). With the publication of the book *Lezioni elementari di Analisi Matematica per i licei scientifici* (1924), he also gained a reputation as a specialist in the teaching of this subject and in 1931 the text earned him the Ministerial Prize for Mathematics.

Called by Tonelli to the University of Pisa in 1932, in the summer of 1934 Ascoli relocated in Milan: he was the second full professor of the newborn State University.

His shift from Pisa to Milan was far from peaceful. The story began on 14 July 1934, when the Milan Faculty of Sciences decided to nominate Ascoli to fill the chair of Algebraic and Infinitesimal Analysis, which had remained vacant since the retirement of G. Vivanti. On 1 August, the Rector A. Pepere communicated the news to Ascoli, who willingly accepted the transfer and sent the programme of the course to be delivered in the a.y. 1934-1935. No one, until then, had cared about his political position. It was taken for granted that Ascoli had taken the loyalty oath (as a schoolteacher) and that he was a member of the fascist party. That was not the case. On 27 August, spurred by some colleagues, the Rector asked Ascoli to document his political position. Ascoli replied: 'In response to your card, protocol no. 4976 dated 27 of this month, I am honoured to inform you that I am not registered with the P.N.F., and that, as of today, I have not applied for registration'.[1] What happened next is not clear but, either the communication did not reach the office or was passed over in silence, because on 3 October the decree of Ascoli's transfer arrived from the Ministry of National Education. Three Milan colleagues (L. Cambi, E. Grill, and A. Desio) remonstrated and asked Pepere to urgently convene the Faculty Council to discuss a very important issue related to the nomination of Ascoli. The Council met twice, and finally ratified the transfer, although Pepere specified to the minister:

> As far as I am concerned, I inform Your Excellency that when the Dean of the Faculty of Sciences informed me that the Faculty itself was orienting towards Prof. Ascoli for the succession of Prof. Vivanti, I immediately asked if Ascoli was enrolled in the P.N.F. and since I had formal assurance, I did not raise objections to his transfer.[2]

Everything appeared to be solved, if not for the fact that shortly afterwards two mathematicians, O. Chisini and B. Finzi, proposed entrusting Ascoli with the teaching of Higher Analysis (23.11.1934). The Faculty, the Council and the academic senate of the University of Milan approved, but the Ministry opposed the appointment:

> This Ministry has examined the proposal of this Faculty of Sciences for assignment of the course of Higher Analysis to Prof. Guido Ascoli, who holds the chair of Algebraic Analysis in the same Faculty. In this regard, it should be noted that only exceptionally tenured professors may be appointed to paid teaching positions in their own Faculty, when there is an absolute lack of the possibility of providing otherwise. In the present case, it does not appear that that condition can be regarded as existing, since it is not plausible that, for the abovementioned discipline, there be no lecturers or assistants or other competent persons in Milan. On the other hand, it should be borne in mind that the role of professor in charge is a separate office from that of a tenured professor, and that, under the provisions of art. 275 of

---

[1] APICE-*Ascoli*: G. Ascoli to A. Pepere, Lemie (Torino) 29.8.1934: *In risposta al Suo foglio n° 4976 in data 27 corr. mi onoro comunicarle che non sono iscritto al P.N.F., e che, a tutt'oggi, non ho fatto domanda di iscrizione.*

[2] APICE-*Ascoli*: A. Pepere to MEN, Milan 5.11.1934: *Per quanto mi riguarda rendo noto a V.E. che quando il Preside della Facoltà di Scienze mi informò che la Facoltà stessa stava orientandosi verso il prof. Ascoli per la successione del prof. Vivanti io chiesi subito se l'Ascoli fosse iscritto al P.N.F. e poiché ne ebbi assicurazione formale, non sollevai obiezioni al suo trasferimento.*

the unique text of laws on higher education, the requirement of enrolment in the P.N.F. is required for professors in charge, a requirement that Prof. Ascoli does not possess. For the above reasons, this Ministry does not believe it can allow the aforesaid appointment of Prof. Ascoli.[3]

Ascoli naturally cared about this post because he 'yielded' and wrote to the Rector:

I am honoured to inform you that I have recently submitted to the Trustees of the university Section of the Fascist Association of the School a written application for admission to the Section itself, confirming the verbal request that I submitted more than a month ago; the application is here attached, in addition to other documents, including a detailed declaration of my stand with regard to ideals and the fascist regime [D1.6.5]. Unable, in the present moment, to translate my position into a solemn act of adhesion and commitment, and as current provisions mean an application for registration with the P.N.F. could not be taken into consideration, I would be grateful if you would take note of the aforementioned declaration; and I therefore send you a signed copy, begging you to keep it among those documents written in my regard.[4]

The act was judged very positively, so much so that on 19 December 1935, the three colleagues who had denounced Ascoli's failure to undersign the PNF would join with O. Chisini, G. Polvani, G. R. Levi, B. Finzi, and R. Monti Stella in signing an enthusiastic report on Ascoli's work, praising his 'excellent teaching qualities, the modernity of the course and clarity of presentation, his enthusiastic love for the school and zeal in fulfilment of academic duties',[5] along with the effective

---

[3] APICE-*Ascoli*: MEN to A. Pepere, Rome 13.12.1934: *Questo Ministero ha esaminato la proposta di codesta Facoltà di Scienze per il conferimento dell'incarico di Analisi Superiore al prof. Guido Ascoli, titolare di Analisi Algebrica nella Facoltà medesima. Al riguardo deve osservare che ai professori di ruolo possono conferirsi incarichi di insegnamento retribuiti nella loro stessa Facoltà solo eccezionalmente, quando manchi in modo assoluto la possibilità di provvedere altrimenti. Ora, nella specie, non sembra che tale condizione possa considerarsi esistente, non essendo ammissibile che, per la disciplina su accennata, manchino, in codesta Città, liberi docenti o persone per altro titolo competenti. D'altra parte, è da tener presente che l'ufficio di professore incaricato è ufficio distinto da quello di professore di ruolo, sicché, giusta il disposto dell'art. 275 del T.U. delle leggi sulla istruzione superiore è da richiedersi, per coloro i quali vengano assunti all'ufficio stesso, il requisito della iscrizione al P.N.F.; requisito che il prof. Ascoli non possiede. Per le su esposte ragioni, questo Ministero non ritiene di poter consentire al conferimento dell'incarico su accennato al prof. Ascoli.* Ascoli would be assigned the teaching of Higher Analysis following the death of Vladimiro Bernstein in 1936.

[4] APICE-*Ascoli*: G. Ascoli to A. Pepere, Turin 20.12.1934: *Mi onoro informarla che ho in questi giorni presentato al Fiduciario della Sez. Universitaria dell'Associazione Fascista della Scuola domanda scritta di ammissione alla Sezione medesima, a conferma della richiesta verbale inoltrata da più di un mese, e che alla domanda ho allegato, oltre ad altri documenti, una circostanziata dichiarazione della mia condizione spirituale nei riguardi delle idealità e del Regime fascista. Nella impossibilità in cui per il momento mi trovo di tradurre in un solenne atto di adesione e di impegno questa mia posizione, presentando domanda di iscrizione al P.N.F., domanda che per le vigenti disposizioni non potrebbe esser presa in considerazione, mi sarà grato se Ella vorrà prendere conoscenza della suddetta dichiarazione; e Gliene invio pertanto una copia firmata, con viva preghiera di volerla conservare tra i documenti che mi riguardano.*

[5] APICE-*Ascoli*: Minutes of the meeting of the Council of the Faculty of Sciences of 19.12.1935: *qualità didattiche ottime, per modernità del corso e chiarezza dell'esposizione, vivo amore alla scuola e zelo nel compimento dei doveri accademici.*

contribution he had made to the activities of the Mathematical and Physical Seminar and the specialisation course in Mathematics and Physics; in short, a series of activities that had greatly satisfied the Faculty. In the following years, Ascoli no longer encountered any problem of a political nature, indeed his career took off at national level, as shown in the achievement of a prestigious award from the *Scuola Normale Superiore* of Pisa for the book *Le equazioni alle derivate parziali dei tipi ellittico e parabolico* (1935).[6]

At the time of his dismissal from service, Ascoli had just four years of seniority under his belt in the role of university professor. However, he applied for counter-discrimination for military merits (D2.11.1). The file, which the Milan university offices sent by mistake to the Ministry of National Education instead of to the Ministry of the Interior, was returned to Ascoli, and arrived at the commission only in mid-January 1939.[7] While waiting for his case to be examined, Ascoli began to look around. He left Milan and moved to Turin with his family, devoting himself to private teaching and collaborating with the Milan Israelite community. Picone tried to help him, both by stimulating him not to abandon research, and by procuring him some paid research work at the INAC.

When Ascoli started thinking about emigration, it was decidedly late. Ascoli was a full professor, but he had never attended an international congress, nor he had ever been abroad, or published a work in a non-Italian journal. He had limited family savings and, with a wife and two small children who depended on him, it was difficult to take such a step. Without contacts with foreign scholars, Ascoli turned to his former mentor Fubini, who had already resettled in Princeton. Fubini immediately took action. First, Fubini asked Ascoli for his curriculum vitae, which he then circulated in Princeton. This was the version that came to Veblen, with his own annotations:

Curriculum Vitae Guido Ascoli [received through Fubini, June 30, 1939]

Address: via Bianze 21, Turin, Italy

Born Dec. 12, 1887, Leghorn, Italy–English

Married; 2 children, born 1927 and 1932

Religion (if this has a bearing on the kind of relief organization which might be called on for help): ###

Languages mastered: Italian, French, German, English

Ph.D. 4 July 1907, Univ. of Pisa

---

[6] Only on one occasion, in May 1938, the University of Milan asked the University of Pisa to communicate the date of Ascoli's loyalty oath. The University of Pisa replied (31.5.1938): 'I inform you that Prof. Guido Ascoli was extraordinary professor of Algebraic Analysis at this University in the year 1932-33, but the oath report was sent to the Ministry of National Education and therefore I am not able to communicate the date of the oath itself'.

[7] APICE-*Ascoli*: MEN to A. Pepere, Rome 31.12.1938; A. Pepere to G. Ascoli, Milan 2.1.1939; G. Ascoli, [*Dichiarazione*], 7.1.1939.

Positions held: Professor, Univ. of Milan 1934-38 [Veblen notes: earlier positions?]

References [name and addresses of 2-3 prominent people, especially in the United States, who are acquainted with the man's work, character, and personality]: Prof. Guido Fubini-Ghiron, Institute for Advanced Study, Princeton, N.J.; Prof. S. Lefschetz, Princeton University, Princeton, N.J.; Prof. Oswald Veblen, Institute for Advanced Study, Princeton, N.J.

Field: Analysis; differential equations and theory of functions of a real variable

Publications: list of 49 available

Note: Professor Fubini regards him as the most distinguished analyst in Italy after Tonelli. His work in differential equations and theory of functions of a real variable is of a very high order, and Prof. Tonelli is of the same opinion. Prof. Ascoli is of a most "sympathetic" personality. We cannot, however, speak with complete confidence of his powers of organization, not because we think he is a poor organizer but because we do not know of his experience.

31.8.1939: Rating of Guido Ascoli: Scholarship A**, Personality A, Teaching ability, Adaptability B

The American route immediately appeared unfeasible, which is why Fubini opportunely abandoned it and suggested Ascoli to look at the Universidad Nacional del Litoral, which in those same days was contacting mathematicians from all over the world to recruit staff for its newborn Institute of Mathematics. On 16 April, Fubini, Lefschetz, and Veblen mailed the following cable to Rey Pastor: 'We believe excellent for Rosario Professors Ascoli of Milan, Berwald of Prague, Segre of Bologna, Terracini of Turin. Letter follows'.[8] The next day, as promised, they added:

The telegram which we sent you yesterday was the result of a conference following the receipt of your kind invitation to Professor Fubini. After discussing the problem, we thought that we would take the liberty of sending you two or three suggestions based on our knowledge of certain important mathematicians who have been separated from their bases of operation by the events in Europe.

We mentioned first the name of **Professor Guido Ascoli** of the University of Milan, now at Via Bianze 21, Torino, Italy, because Professor Fubini regards him as the most distinguished analyst in Italy after Tonelli. His work in differential equations and theory of functions of a real variable is of a very high order, and Professor Tonelli is of the same opinion. Professor Ascoli is of a most "sympathetic" personality. We cannot, however, speak with complete confidence of his powers of organization, not because we think he is a poor organizer but because we do not know of his experience. He is about forty years old.

Professor Ludwig Berwald, Ovenecká 43, Prague XIX, Moravia, has done many and important researches in the higher differential geometry. He has touched on the most modern branches of this geometry, as for example on Finsler and Cartan spaces, and has also made contributions to the formal side of the calculus of variations. He is a geometer of very considerable distinction, and a man of high character, who would be an excellent colleague. He is fifty-five years old. He was Ordentlicher Professor at the German University of Prague from 1927 until January 1939.

---

[8]ECADFS Records, GA: G. Fubini, S. Lefschetz, O. Veblen to J. Rey Pastor, [Princeton] 16.4.1939: *Crediamo ottimi per Rosario Professori Ascoli di Milano, Berwald di Praga, Segre di Bologna, Terracini di Torino. Segue lettera.* The words "Terracini di Torino" are cancelled in the cable.

Professor Beniamino Segre is a man of thirty-six years, who has published extensively in differential geometry, algebraic geometry and the theory of convex bodies. Segre possesses in a very high degree the power of organization, and he is well liked by all his colleagues, not only in Italy but also in other countries. This is proved by the fact that we have had letters about his present plight from many of the countries of Europe. He has been full professor, and Director of the Mathematical Institute of the University of Bologna from 1934 until December 1938. His address is Via Alamandini 10^3, Bologna, Italy.

We beg you to pardon our presumption in offering this advice upon which we have ventured because of our intense interest in the progress of science. We beg you to accept our most distinguished greeting. Sincerely yours.[9]

Despite the support of the three outstanding mathematicians in Princeton, Rosario's position went to Beppo Levi, as suggested by Levi-Civita. Ascoli made no other attempts and resigned himself to living a segregated life. After September 8, 1943, he moved to San Michele d'Asti (in the country village of Dusino). His whole family were saved from deportation to the extermination camps thanks to protection from the local population.

In May 1945, considering 'imminent, if not already occurred, his readmission to service',[10] Ascoli contacted the Allied Commissioner of Milan, and communicated his real current address where any further provisions on his case could be sent. At least on paper, Milan was one of the fastest Universities in processing the reintegration practices and in welcoming back Jewish teachers. The Rector F. Perussia wrote to Ascoli on 1 June:

Distinguished Professor, I am pleased to greet you on behalf of myself and the University as a whole, which with deep pleasure sees a teacher who so highly honoured the chair and science return to its family. I can assure you that the measures to reinstate in their offices the university professors affected by the racial laws have already been issued by the government of Rome and all that remains is to carry out the individual bureaucratic procedures. I would be grateful, therefore, to let me know if you can, as of now, return to your office, in which case I would think of asking you to accept some academic assignment in the interest of the University. I renew to you, distinguished Professor, my warmest and most sincere congratulations together with those of all my colleagues.[11]

---

[9]ECADFS Records, *GA*: G. Fubini, S. Lefschetz, O. Veblen to J. Rey Pastor, [Princeton] 17.4.1939.

[10]APICE-*Ascoli*: G. Ascoli to the University Commissioner of Milan, Turin 21.5.1945: *Ritenendo imminente, se non già avvenuta, la mia riammissione in servizio.*

[11]APICE-*Ascoli*: F. Perussia to G. Ascoli, Milan 1.6.1945: *Illustre Professore, Sono lieto di salutarla a nome mio personale e della Università tutta che con vivo compiacimento vede rientrare in seno alla sua famiglia un maestro che così altamente onorò la cattedra e la scienza. Posso assicurarla che il provvedimento che reintegra nel loro ufficio i professori universitari colpiti dalle leggi razziali è già stato emanato dal governo di Roma e non rimangono che da espletare le pratiche burocratiche individuali. Le sarei grato, quindi, di farmi conoscere se Ella può, fin d'ora, rientrare in sede, nel qual caso penserei di pregarla di accettare qualche incarico di carattere accademico nell'interesse della Università. Le rinnovo, illustre Professore, le mie più vive e sincere congratulazioni unitamente a quelle dei Colleghi tutti.* The official letter of reinstatement is dated 2.6.1945.

The return to Milan devastated by bombings, and where Ascoli was forced to live in makeshift housing,[12] but above all the experience in the university purge commission of which he was vice-President (the presidency went to A. Galletti) disappointed him terribly. On 18 October 1945, he vented his dissatisfaction to Picone:

> In these times, in which so many relationships have necessarily loosened or been destroyed, your constant remembrance and your faithful friendship, which remained faithful even in the worst times - and I want to proclaim it on every occasion, in honour of your courage and your right conduct - are of great comfort to me. In recent years, many things, strengthening my natural optimism, have created in me a great loss of esteem of men, and especially of those who by the height of their mind should also be an example of integrity and character; and the participation in the purge of the University of Milan, putting me in contact with so many small and big cowards, represented the coup de grace. So, it is a real consolation to find among the former university fascists a good person, who used fascism only to implement a good and beautiful idea that has honoured and continues to bring honour to Italy, above and beyond any sectarianism. I was pleased to know that your institution has also found in the new political environment the consideration that it deserves [. . .]. You are absolutely right in thinking that in order to remove us from political isolation and to re-evaluate ourselves in front of the world, the scientific contribution that each of us will be able to provide will be highly necessary; and for this reason it would be urgent first of all to fully re-establish cultural exchanges with other countries, and especially with America, where surely in recent years a lot of work has been done, also by the men of value who emigrated there. On my behalf, as far as my very modest strength and unfavourable working conditions allow me, I will not fail to collaborate in this very effective task together with my protégés.[13]

Reinstated in service in June 1945, Ascoli moved to the University of Turin in 1949, and remained there until his death in 1957.

---

[12] APICE-*Ascoli*: G. Ascoli to F. Perussia, Milan 4.4.1946 and F. Perussia to G. Ascoli, Milan 12.4.1946.

[13] G. Ascoli to M. Picone, Turin 18.10.1945, in http://matematica.unibocconi.it: *In questi tempi, in cui tanti rapporti si sono per forza di cose allentati o distrutti, il tuo costante ricordo e la tua fedele amicizia, che fedele rimase anche nei tempi peggiori-e ci tengo a proclamare in ogni occasione, ad onore del tuo coraggio e del tuo retto sentire-sono per me di gran conforto. Molte cose negli ultimi anni, facendo forza al mio naturale ottimismo, hanno creato in me una grande disistima degli uomini, e specialmente di quelli che per altezza di ingegno dovrebbero essere anche esempio di dirittura e di carattere; e la partecipazione all'epurazione dell'Università di Milano, mettendomi a contatto con tante piccole e grosse vigliaccherie, mi ha dato il colpo di grazia. Per cui trovare tra gli ex-fascisti universitari una brava persona, che del fascismo in fondo, si è servito solo per mettere in opera una buona e bella idea che ha fatto e fa grande onore all'Italia, al di fuori e al di sopra di ogni settarismo, è una vera consolazione. Io sono stato lieto di sapere che la tua istituzione ha trovato anche nel nuovo ambiente politico quella considerazione che merita [. . .]. Hai perfettamente ragione pensando che a toglierci dall'isolamento politico ed a rivalutarci di fronte al mondo varrà moltissimo il contributo scientifico che noi sapremo fornire; e per questo sarebbe urgente anzitutto ristabilire in pieno gli scambi culturali con gli altri paesi, e specialmente con l'America, dove certo in questi anni si è lavorato molto, anche per opera degli uomini di valore che vi sono emigrati. Per mio conto per quanto mi permetteranno le mie modestissime forze e le sfavorevoli condizioni di lavoro, non mancherò di collaborare insieme con i miei scolari a questo utilissimo compito.*

4.2   Arturo Maroni: A Secluded Scholar

Born in 1878, Arturo Maroni was full professor of Analytic and Projective Geometry at the University of Pavia. Graduating in 1901 in Pisa with E. Bertini as advisor, and a fellow student of Fubini, Maroni began his career as a mathematics teacher in high schools and technical institutes of Padua, Perugia, Naples, and Florence, before winning a national competition in 1934 for a chair in Geometry at the University of Cagliari.[14] From there he obtained the transfer to Modena and finally to Pavia in 1937. At the time of discrimination, Maroni therefore had the same length of service as Ascoli, despite being almost ten years older.

His research, not unanimously appreciated in Italy, concerned plane algebraic curves, families of twisted curves and ruled surfaces, systems of curves belonging to algebraic surfaces and group theory. Maroni also devoted himself to the problem of determining the maximum number of double points of an algebraic surface immersed in ordinary ambient space, but without solving it completely. He was the author of many geometry entries in the Italian *Enciclopedia Treccani*.

Considered a teacher of great clarity, but cold in manner and unable to involve young people in research, Maroni was detached from the Italian mathematical community and even isolated within the Italian School of Geometry to which he felt to belong.[15] Agnostic and secularised, in the census form he answered negatively to all questions about his religious status and vehemently rejected the loophole of conversion.[16]

In autumn of 1938, the Rector of the University of Pavia, P. Vinassa, wrote a brief note to the Ministry of National Education about the Jewish professors who left the University, summarising (with an attitude of benevolent, albeit generic protection) their respective positions and merits as well as suggesting alternative employment for each of them. Maroni was the only one for whom the Rector did not advance any possibility of a work that allowed him to use his academic preparation, even with a drastic *diminutio* of his professional status. Maroni was the only full professor in Pavia who did not apply for counter-discrimination. The documents kept in the archive of *Demorazza* provide shocking evidence of the opinions held of Maroni by his Rector and colleagues:

> Maroni, Dr. Arturo, full professor of Geometry. Of modest economic means, recently entered university teaching after secondary school. He does not hold any special titles. He

---

[14] Zappa (2004).

[15] According to Ciliberto and Sallent Del Colombo (2018, pp. 8–9), when Bertini asked Enriques in 1902 to give him Severi as an assistant, in the place of Maroni, Enriques replied: 'I don't feel that I can agree to this. It does not seem right to me to favour another mediocre individual over our own mediocre individuals, one who may not be lacking in education but is ill-equipped to teach due to his excessively cold character' (*non mi par giusto di preferire ai nostri mediocri, un altro mediocre, non privo forse di studio, ma mal dotato per l'insegnamento stante il suo carattere eccessivamente freddo*).

[16] Signori (1997), p. 457. See also Signori (2002).

is in the curious position of being an anti-Semitic Jew; and therefore 'despised by God and his enemies'. He awaits the arrival of his pension and nothing else.[17]

At the time of persecution, Maroni was sixty years old; he had very few, and weak, relationships with Italian colleagues, none with foreigners. The chagrins and mortifications suffered by his children and wife made him evaluate the idea of emigrating. However, this was a plan that Maroni did not pursue with any tenacity. His case was presented to the SPSL, but he did not enter in direct contact with the agency. He did not attempt the American emigration route by addressing the ECADFS; nor did he turn to any Italian colleague, except for Severi, not even for a discussion on how to move. After receiving the standard response letter that the SPSL sent to all would-be emigrants, Maroni surrendered. He started working in the Jewish school of Florence as a mathematics teacher and, following the liberation of the city, he taught provisionally at the University of Florence. His own family was saved, but his sisters were deported and murdered in Auschwitz. Reinstated in service on 16 June 1945, Maroni moved to the University of Florence in 1948, and remained there definitively until his retirement in 1953.[18]

4.3   Bruno Tedeschi: A Hyper-Fascist Jew

Born in Trieste in 1898, Bruno Tedeschi graduated in Genoa in 1920. Having immediately entered the teaching profession, in 1922 he won the competition for chairs of mathematics in technical institutes and worked in Gorizia and then in Trieste. From 1923 he held various precarious positions at the University of Trieste: honorary assistant, effective assistant, and *aiuto* to the chair of Financial and Actuarial Mathematics (1923–1933); assistant professor of Statistics (1927–1929), responsible for the course in Financial and Actuarial Mathematics (1929–30), substitute teacher of Special Applications of Insurance Mathematics in the School of Specialisation for graduates of the University of Trieste (1933–34). In 1934, he qualified as a *libero docente* of Financial and Actuarial Mathematics. A volunteer in the First World War when he was still a university student, Tedeschi was a model soldier and received various medals and honours. A fervent fascist, he was a member of several party organisations, and a senior member of the National Security Militia.

As a result of the racial laws, Tedeschi lost his position in the school and his status as lecturer was revoked. As a temp in academia, he did not receive any pension. Tedeschi did not lose heart. He confided that he was known and appreciated in the

---

[17] ACS-*Contro-discriminazione*: P. Vinassa to MEN, Pavia 3.11.1938: *Maroni Dr. Arturo, ordinario di Geometria. Di modeste condizioni economiche, entrato da poco tempo nell'insegnamento universitario dopo quello secondario. Non presenta titoli speciali. Si trova nella curiosa posizione di essere un ebreo antisemita; e perciò "A Dio spiacente ed a' nemici sui." Attende la liquidazione della pensione e nulla altro.*

[18] P. Fraccaro to A. Maroni, Pavia 16.6.1945 and A. Maroni to P. Fraccaro, Florence 1.7.1945, in Torchiani (2010), p. 213.

fascist circles of Trieste and therefore hopefully applied for counter-discrimination for military and political merits (9.3.1939).[19] His request (and that of his father Erminio) was accepted on 16 January 1940, not without having met strong reservations from E. Grazioli, federal secretary of the PNF in Trieste.[20] At the same time, Tedeschi was looking at emigration. His age played in his favour (just forty years old), as did the fact of travelling alone, i.e. without parents, a spouse, or children to take with him, and a good academic output. Tedeschi was an applied mathematician, specialised in actuarial science, with a portfolio of 29 publications on probability calculus, statistics, and finance mathematics. Against him, there was his lack of an international reputation: his publications, all in Italian, had only appeared in Italian journals.[21] Tedeschi had participated in just one international congress, in Paris in 1937, and his incursions in the international scientific parterre were limited to some reviews of foreign texts on statistics and financial mathematics for the *Bollettino dell'UMI*. To his detriment was, however, the fact that he made no secret of his political faith. The curriculum vitae which Tedeschi sent to ECADFS in February 1939 is constructed in a frankly astonishing way (TEDESCHI.1). In addition to declaring that he was a representative of the Opera Balilla, a Centurion of the Voluntary Militia for National Security, and a qualified senior commander of the avant-garde legions, Tedeschi adopted a language typical of the worst fascist rhetoric. The record is written in Italian and is untranslatable in English, not for the terms but for certain expressions, such as 'healthy and resistant physique, necessary to be employed in the American States and in their colonies', which seem like something from a film by the *Istituto Luce*. The ECADFS recorded receipt of the curriculum on March 16, 1939, and mailed Tedeschi the standard letter.[22] Tedeschi gave up his attempt and did not take any other action. In the meantime, he became head of the Jewish School of Trieste, a post he held until the school closed in 1943 (Fig. 4.2).

Purged because of his fascist past, after the war Tedeschi became an inspector at the Ministry of Public Instruction. He returned to his studies at the beginning of the fifties, first in Trieste and then in Rome, at the Faculty of Statistical Sciences, where

---

[19] The Trieste Central Police Station carried out a series of checks on Bruno Tedeschi but without any result. See the report drafted on 16.2.1939 in Catalan et al. (2019), p. 132.

[20] See Positive opinion with reservations of the federal secretary of the National Fascist Party in Trieste, in response to Tedeschi's application for discrimination, Trieste 9.3.1939 and Discrimination issued to Tedeschi and his father Erminio, Trieste 16.1.1940, in Catalan et al. (2019), pp. 133–134.

[21] *Giornale di Matematica finanziaria, Annali della R. Università di Trieste, Bollettino dell'Istituto statistico economico della R. Università di Trieste, Rivista Italiana di Ragioneria, Bollettino dell'UMI* and *Giornale dell'Istituto Italiano degli Attuari*. Tedeschi also had to his credit some works of mathematics didactics published in the *Bollettino di Matematica* and three textbooks (Geometry, Physics, Arithmetic) for middle schools.

[22] ECADFS Records, *BT*: B. Drury [ECADFS] to B. Tedeschi, New York 16.3.1939.

Fig. 4.2 Bruno Tedeschi in the courtyard of the Trieste Jewish school, courtesy of Museo della Comunità ebraica di Trieste 'Carlo e Vera Wagner', in Catalan, Di Fant, Perissinotto eds. (2019), p. 132

he taught Social Insurance.[23] He remained in Rome until his death in 1979, without ever denying his 'turbulent fascist past'.[24]

Documents

Tedeschi.1 Curriculum vitae of prof. Bruno Tedeschi, February 1939

ECADFS Records, *BT*, fols. 1r, 2r, 3r, 4r
Son of Erminio Tedeschi and Giuseppina Ulman, born in Trieste on 17.10.1898, single, domiciled *in Trieste, via Ginnastica 54.*

Born in Trieste, in the Kingdom of Italy during the World War, volunteer in the R. Italian Army, later professor of Mathematics and Physics in the R. Higher Technical Institutes and in charge of Financial and Actuarial Mathematics at the R. University of Trieste. Excluded from teaching by racial laws because he is a Jew.

Qualifications. Until 1915 he carried out his studies at the civic Royal High School of Trieste; 1915 graduated from the Physics-Mathematics section of the

---

[23] Marbach and Rizzi (2011), p. 56.

[24] Guerraggio et al. (2016), p. 4: *turbolento passato fascista.*

R. Technical Institute 'Sarpi' of Venice; 1920 graduated in Mathematics (R. University of Genoa, adv. prof. Gino Loria[25]); Graduated in Mathematics (R. University of Genoa); Graduated in Accounting (R. Technical Institute of Genoa); 1935 qualification (unanimous votes) for lecturing in Financial and Actuarial Mathematics (Ministry for National Education, Rome)

Languages spoken. Italian, French, German and he is studying English

Extracurricular service. 1915–1917 employed at the Cassa di Risparmio di Genova

High school. 1919-1938 regular professor of Mathematics and Physics and Mercantile Calculus in the advanced course of the R. Technical Institute Carli of Trieste. Secretary of the Council of professors

1935-1938 included in the list of people eligible to serve as Heads (Official Bulletin of the Ministry of National Education of 26.9.1935/XIII, Part II, no. 39 and following)

University Teaching. 1923–1927 volunteer assistant to the chair of Financial and Actuarial Mathematics at the R. University of Trieste, being prof. Filippo Sibirani the holder of the chair;[26] 1927–1933 tenured assistant to the same chair; 1933–1938 help to the same chair; 1927–1929 in charge of the course in Methodological Statistics; 1929–1930 he taught the course of Financial and Actuarial Mathematics; 1933–1935 in charge of the teaching of "Special applications of mathematics to insurance"; 1935–1938 he held courses on the following topics: Calculation of mathematical provisions on groups of policyholders; Mathematical reserve also from the point of view of the technical budget; Calculation of the rate in investment problems

Assignments. 1933 during the summer holidays he replaced the Head; 1934-1938 Director of the Physics Department of the R. Technical Institute representing the Opera Balilla[27] within the Scholastic Fund; 1934 in charge of the presidency of the R. Technical Institute Carli; 1935 member of the State Exams Board in Zara; 1936 commissioner for exams at the Toppo-Wasserman Institute in Udine; 1936-1937 member of the board for the State qualifying exams

---

[25] Gino Loria (1862–1954), professor of Higher Geometry in the University of Genoa (1886–1935).

[26] Filippo Sibirani (1880–1957), professor of Financial Mathematics at the University of Trieste (1922–1935).

[27] The Opera Nazionale Balilla for the assistance and for the physical and moral education of youth (*Opera Nazionale Balilla per l'assistenza e per l'educazione fisica e morale della gioventù*) was a youth organization established in 1926 and placed under the umbrella of the Ministry of National Education. Complementary to the school, the Opera Nazionale Balilla aimed not only to spiritual, cultural, and religious education but also to pre-military, gymnastic-sports, professional and technical education according to fascist ideology. It admitted young people from 6 to 18 years old, divided into three subgroups: *figli della lupa* (6–8 years old), *balilla* (8–14 years old), and *avanguardisti* (14–18 years old).

to practise the profession of Economics and Commerce; 1937 proposed as a representative of the Opera Balilla within the Disciplinary Council of the R. Provveditorato agli studi (State Education Department) in Trieste; 1937 commissioner at the Technical Institute of S. Donà di Piave; 1937 President of the Commission for technical qualification in Vasto and Foggia; 1937–1938 leader of the Local Unit of the Italian Youth Red Cross; 1938 Chairman of the Commission for Technical Qualification in Pola and Fiume.

Scientific societies of which he is a member. Italian Mathematical Union; Society of the Scuola Normale di Pisa; Italian Institute of Actuaries; Italian Society for the Progress of Sciences; Mathesis; Adriatic Society of Natural Sciences[28]

Congresses which he attended. 1932 National Congress of Actuaries, Trieste; 1934 International Congress of Actuaries in Rome and Congress of the Italian Society for the Progress of Sciences in Naples; 1935 Congress of the same Society in Palermo; 1936 Congress of the same Society in Tripoli and International Congress of Teaching in Rome; 1937 International Congress of Actuaries in Paris and Congress of the Italian Society for the Progress of Sciences in Venice

Military titles. Volunteer in the R. Italian Army during the World War, now Reserve Artillery Captain, approved for position of major; Centurion of the Voluntary Militia for National Security, approved as senior Centurion; 1926–1938 commander of avant-garde legion, he was in charge of youth organisations.

Honours received. 1918 medal for honours awarded to the Trieste freedom fighters; 1921 authorised to bear the medal established in memory of the 1915–1918 war, the medal established in memory of the Unification of Italy, and the inter-allied medal of Victory; 1932 bronze medal of the Ministry of National Education; 1934 appointed Knight of the Crown of Italy; 1936 Silver Medal of the Ministry of National Education; 1938 Seniority Cross of the Militia for National Security. Various letters of commendation.

Papers. 29 publications as per attached list.

He wishes to be relocated at: a university, an institute of mathematical, actuarial or statistical studies, an astronomical observatory, a middle school, a scientific society, an insurance company, a state or municipal or private statistical office, possibly as an Italian correspondent, or accountant in the American States and in

---

[28] The Italian Institute of Actuaries had been established in 1929. The promoters and first organizers of the Institute had been Paolo Medolaghi (first president), Francesco Paolo Cantelli, Guido Castelnuovo, and Bruno De Mori. The Italian Society for the Progress of Sciences was a scientific association originated during the Meetings of Italian Scientists of the pre-unification period. It had been reconstituted in 1906 on the initiative of Vito Volterra, Arturo Issel, and Pietro Romualdo Pirotta. The Adriatic Society of Natural Sciences had been created in Trieste in 1874.

the colonies (for which he has the necessary health conditions and resistant physique).[29]

Figlio di Erminio Tedeschi e della fu Giuseppina Ulman, nato a Trieste il 17.10.1898, celibe, domiciliato a Trieste in via Ginnastica 54. Triestino, nel Regno durante la guerra mondiale, volontario nel R. Esercito Italiano, successivamente professore di matematica e fisica nei R. Istituti Tecnici superiori e incaricato di matematica finanziaria ed attuariale alla R. Università di Trieste. Escluso dall'insegnamento in seguito alle leggi razziali, perché ebreo.

Titoli di studio. *Fino al 1915 svolse i suoi studi alla civica Scuola Reale superiore di Trieste*

1915 licenziato dalla sezione fisico-matematica del R. Istituto tecnico 'Sarpi' di Venezia; 1920 laureato in matematica (R. Università di Genova, prof. Gino Loria); Diplomato in magistero di Matematica (R. Università di Genova); Diplomato in ragioneria (R. Istituto tecnico di Genova); 1935 abilitazione (unanimità di voti) alla libera docenza in matematica finanziaria ed attuariale (Ministero per l'Educazione Nazionale, Roma)

Lingue conosciute. *Italiana, francese, tedesca e sta studiando l'inglese*

Servizio extra scolastico. *1915-1917 impiegato alla Cassa di Risparmio di Genova*

Scuola Media. *1919-1938 ordinario di matematica e fisica e calcolo mercantile nel corso superiore del R. Istituto Tecnico Carli di Trieste. Segretario del Consiglio dei professori; 1935-1938 nell'elenco degli idonei a Preside (Bollettino Ufficiale del Ministero dell'Educazione Nazionale del 26.9.1935/ XIII, parte II, n° 39 e successivi)*

Insegnamento Universitario. *1923-1927 assistente volontario alla cattedra di matematica finanziaria ed attuariale alla R. Università di Trieste essendo titolare della cattedra il prof. Filippo Sibirani; 1927-1933 assistente di ruolo alla stessa cattedra; 1933-1938 aiuto alla stessa cattedra; 1927-1929 incaricato di statistica metodologica; 1929-1930 tenne il corso di matematica finanziaria ed attuariale; 1933-1935 incaricato di "applicazioni speciali della matematica alle assicurazioni"; 1935-1938 tenne corsi sui seguenti argomenti: Il calcolo delle riserve matematiche su gruppi di assicurati; La riserva matematica anche dal punto di vista del bilancio tecnico; Il calcolo del tasso nei problemi riguardanti gli investimenti*

Incarichi avuti. *1933 durante le ferie estive sostituì il Preside; 1934-1938 Direttore del gabinetto di fisica del R. Istituto tecnico rappresentante dell'Opera Balilla in seno alla Cassa Scolastica; 1934 incaricato della presidenza del*

---

[29] The words 'in the American States and in the colonies' are added by hand.

R. Istituto tecnico Carli; 1935 in commissione di esame di Stato a Zara; 1936 commissario unico agli esami dell'istituto Toppo-Wasserman di Udine; 1936-1937 fece parte della commissione per gli esami di Stato di abilitazione all'esercizio della professione in economia e commercio; 1937 proposto quale rappresentante dell'Opera Balilla in seno al Consiglio di disciplina del R. Provveditorato agli studi in Trieste; 1937 commissario unico all'Istituto tecnico di S. Donà di Piave; 1937 Presidente della Commissione per l'abilitazione tecnica a Vasto e Foggia; 1937-1938 dirigente dell'Unità locale della Croce Rossa Italiana Giovanile; 1938 Presidente della Commissione per l'abilitazione tecnica nelle sedi di Pola e di Fiume

Società scientifiche delle quali fa parte. *Unione Matematica Italiana; Società della Scuola normale di Pisa; Istituto Italiano degli Attuari; Società Italiana per il progresso delle scienze; Mathesis; Società Adriatica di Scienze Naturali*

Congressi ai quali prese parte. *1932 Congresso nazionale degli attuari Trieste; 1934 Congresso internazionale degli attuari Roma e Congresso della Società Italiana per il progresso delle Scienze Napoli; 1935 Congresso della stessa Società a Palermo; 1936 Congresso della stessa Società a Tripoli e Congresso internazionale dell'insegnamento Roma; 1937 Congresso internazionale degli attuari Parigi; Congresso della Società Italiana per il progresso delle Scienze Venezia*

Titoli militari. *Volontario nel R. Esercito Italiano durante la guerra mondiale, ora capitano di complemento di artiglieria, idoneo a maggiore; Centurione della Milizia volontaria per la sicurezza nazionale, idoneo a seniore; 1926-1938 quale comandante di una legione di avanguardisti, si occupò delle organizzazioni giovanili*

Onorificenze avute. *1918 medaglia istituita dal comitato per le onoranze ai triestini combattenti; 1921 autorizzato a fregiarsi della medaglia istituita a ricordo della guerra 1915-1918, della medaglia istituita a ricordo dell'Unità d'Italia, della medaglia interalleata della Vittoria; 1932 medaglia di bronzo del Ministero per l'Educazione Nazionale; 1934 nominato Cavaliere della Corona d'Italia; 1936 medaglia d'argento del Ministero per l'Educazione Nazionale; 1938 Croce di anzianità della Milizia Nazionale per la Sicurezza Nazionale. Varie lettere di lode.*

Pubblicazioni. *29 pubblicazioni come da elenco allegato.*

Desidera collocarsi presso: una università, un istituto di studi matematici, attuariali o statistici, un osservatorio astronomico, una scuola media, una società scientifica, una compagnia di assicurazioni, un ufficio statistico statale o comunale o privato eventualmente quale corrispondente italiano, contabile o ragioniere negli Stati Americani e nelle colonie, per recarsi nelle quali ha il necessario fisico sano e resistente.

4.4 Nedda Friberti: An Aspiring Female Refugee

The story of Nedda Friberti is that of a career that was never born. Born in Trieste on 15 September 1913, to Oscarre Freiberger and Sara Corinna Luzzatto, Nedda's family Italian-ised its surname as Friberti in 1928.[30] After attending the scientific lyceum Guglielmo Oberdan in Trieste, Nedda enrolled at the University of Padua, where she graduated in Mathematics in 1935 with a thesis on algebraic geometry, supervised by A. Comessatti. In 1938, she received a second degree, in Physics. Assistant professor of Higher Geometry in 1936–1937, winner of two competitions for chairs for middle and high schools in 1937 and 1938, at the time of expulsion Friberti had just one year of teaching in the Technical Institute of Trieste to her name. For her there was to be no pension or liquidation, with no title or merit to be asserted for the purposes of counter-discrimination.

A young woman, dynamic and tenacious, Friberti presented her case to ECADFS on 24 February 1939 (FRIBERTI.1). She declared that she was an Italian citizen of Jewish religion, she had a valid passport and an *affidavit* for the United States, that she spoke French and German, and she was studying English. She was ready for any kind of employment: in scientific institutes, astronomical and meteorological observatories, as a college teacher, even as a manual worker or an employee in some industries. The outcome of her application was obvious: despite her professional qualifications, Friberti 'appeared to lack sufficient relational capital to activate the migration process'.[31] The Committee's answer was in fact categorical in rejecting her application. ECADFS could and must deal exclusively with scholars with the rank of professors or *Privatdozenten*, because in light of unfavourable economic conditions there were almost no openings in the country.[32]

After this refusal, Friberti was taken on as a teacher of mathematics in the Jewish school of Trieste, where she worked with Bruno Tedeschi until 1943[33] (Fig. 4.3).

Friberti miraculously managed to escape to Switzerland, with her family, after the Nazi occupation of Trieste. Interned in Oberhelfenschwil (St. Gallen)[34] she asked to be employed in the IUC of Lausanne as an assistant, secretary, or typist, but her application was submitted too late and could not be accepted (FRIBERTI.3). Back in Italy, she taught mathematics and physics in technical, agricultural, industrial, commercial, and surveyor institutes, from 1951 until her retirement.

---

[30] The application for a modification in Italian form of the German surname Freiberger (or Freiliergër) was submitted by Nedda's father on 6.3.1928. The simplified Friberti moniker was naturally extended to Oscarre's family: his wife Sara Corinna and their children Enzo (born in 1911) and Nedda.

[31] Gissi (2016), p. 71. Friberti's case has also been rediscovered by the Refugee Scholars project, Northeastern University https://historycollection.com/scientists-history-abandoned-cruel-reality-scholarly-woman-refugee-fleeing-holocaust-america/2/.

[32] FRIBERTI.2.

[33] Catalan et al. (2019), p. 157.

[34] No. 288 of Jewish Refugee Records, 1936–1946, RG-58.001M.

Fig. 4.3 Some of the teaching staff of the Israelite school in Trieste: B. Tedeschi (kneeling in the photo), N. Friberti (third on the right), courtesy of Museo della Comunità ebraica di Trieste 'Carlo e Vera Wagner', in Catalan, Di Fant, Perissinotto eds. (2019), p. 157

Documents

Friberti.1 N. Friberti to B. Drury [ECADFS], Trieste, received on 24.2.1939

ECADFS Records, *NF*, fol. 1r

To Miss Drury-Secretary of the Emergency Committee in aid of displaced foreign scholars. I permit to address myself to you in order to beg some informations as to the possibility to find a situation there. I am 25 years old, am an Italian citizen, was born in Trieste and profess Jewish religion. I studied at the Padua University, where I was graduated in Mathematics (D. Sc.) in 1935 (Thesis: Bi-rational transformation of a curve in itself interpreted as hyperbolic plane's rotation) with Professor A. Comessatti (votation 110/110) and in Physics in 1938 (Thesis: Global Sun radiation at Trieste in 1937) with Professor F. Vercelli (votation 110/110).[35]

1936-37: I have been Assistant to the chair of Mathematics, having deserved a scholarship in superior geometry;

1937: I have been qualified to the teaching in Public undergraduate Schools (Sc. Medie Inferiori);

1938: id. id. (for the Scuole Medie Superiori);

---

[35] Annibale Comessatti (1886–1945), Severi's assistant for twelve years (1908–1920), was professor of Descriptive Geometry and its applications at the University of Padua from 1922 to 1945. Francesco Vercelli (1883–1952), geophysicist, directed the Geophysical Institute of Trieste (1919–1952).

1937-38: I teached [sic!] Mathematics in the Technical Institute Leonardo da Vinci in Trieste.

Now, as all forms of activity in public teaching is forbidden to me in my country, I wish to get out of it. I have my passport, and also an "Affidavit" for the U.S.A., and the Italian Government will not, I am sure, hind my departure from Italy. I know French, German and am learning also English. I will be very grateful to you, if you would let me know whether I have any possibility to find a situation in some scientific Institute, Astronomical Observatory, or in the Meteorological Service (I have already worked in the Geophysical Institute of the R. Comitato Talassografico Italiano.[36]) In the case that this were not possible, I should be glad to obtain whatever situation in the undergraduate teaching as well as in some industry. Thanking you in advance for all you will do for me, I beg, dear Miss Drury, to remain your very faithfully Dott. Nedda Friberti

Friberti.2 B. Drury [ECADFS] to N. Friberti, New York 25.2.1939

ECADFS Records, *NF*, fol. 1r

Dear Dr. Friberti: Let me acknowledge your letter received yesterday. I very much wish there were something we could do to assist you. As it happens the Emergency Committee cannot because of its regulations bring the availability of displaced scholars to the notice of colleges and universities unless asked to do so by such institutions. Upon receiving such a request from a college or university the Committee will suggest candidates suitable for the position to be filled. Because of the number of demands made upon it, the Emergency Committee has been obliged to limit its activities to the cases of scholars who have been displaced from university posts where they held the rank of professor or *Privat-dozent*. I feel I must tell you that because of unfavorable economic conditions there are almost no openings in this country at the present time. Nevertheless we are glad to place your letter, with its brief summary of your career, in our files and to keep you in mind even though there is but a slight chance of our learning of a suitable opportunity. Sincerely yours, Betty Drury-Assistant Secretary

---

[36]The Geophysical Institute of Trieste had been created under the umbrella of the Royal Thalassographic Committee, established by V. Volterra in 1910, and charged with 'the physical-chemical and biological study of the Italian seas, mainly in relation to the navigation and fishing industry, and the exploration of the atmosphere with regard to air navigation'.

Friberti.3 V. Consolo to G. Colonnetti, Lausanne 29.2.1944

ACT, fol. 1r

Dear Professor, Eng. De Benedetti[37] sends me the attached application of Dr. Nedda Friberti, currently interned in the camp of Oberhelfeuschill (St. Gallen), begging me to pass it on to you, in order to see if you can do anything to help. I think that, given the measures in force, it would be difficult. I am, however, sure that, with your enormous benevolence and understanding, you will consider the case. Please greet Mrs. Colonnetti for me and accept my regards as yours faithfully, Valerio Consolo

Egregio Professore, L'ing. De Benedetti mi manda l'acclusa domanda della D.ssa Nedda Friberti, attualmente internata a Oberhelfeuschill (St. Gallen), con la preghiera di passarla a Lei per vedere se può fare qualche cosa. Penso che, date le disposizioni vigenti, possa essere difficile. Sono, ad ogni modo, sicuro che, con la Sua enorme bontà e comprensione, prenderà la domanda in considerazione. La prego di ricordarmi alla Signora Colonnetti e di credermi Suo dev.mo Valerio Consolo

4.5   Giulio Bemporad: An Astronomer of the Old Guard

Giulio Bemporad was born in Florence on 3 January 1888, and studied first at the University of Pisa and then in Catania, where he graduated in Mathematics in 1910. Appointed at the Catania Observatory from 1909, in 1911 he was called to cover the role of assistant astronomer at the Astronomical Station of Carloforte, on the island of San Pietro, in the south-west region of Sardinia. In 1920 he became Director, of the Station. After fourteen years on the island, he managed to obtain a transfer to the Capodimonte Observatory in Naples, directed by his cousin Azeglio Bemporad. The work experience of the two cousins—the free thinker, even rebellion, and Zionist Giulio, and the fascist authoritarian man Azeglio—was bitter. The contrast between the two culminated with the evaluation of 'mediocre' (*mediocre*) that Azeglio gave on Giulio's work relating to the year 1931 and with the request for his immediate transfer.

Relocated at the Pino Torinese Observatory in March 1933, in the autumn of 1938 Giulio Bemporad was a fifty-year-old gentleman, with almost twenty years of career behind him, and with only one dependent: his sister Giselda, who had always

---

[37]Rodolfo De Benedetti (1892–1991), engineer and entrepreneur, escaped to Switzerland with his family in November 1943. After a period in the Bellinzona refugee camp, he was assigned to confinement in Lucerne.

lived with him. He had no qualifications to apply for counter-discrimination, and emigration was therefore a sensible, although difficult, path to undertake.

The first obstacle was the strongly anti-Semitic tendency at the Turin Observatory. Its Director, L. Volta, was a despicable figure, who immediately removed the Bemporads' allocated living quarters at the Observatory and asked the pensions department of the Ministry how to reduce Bemporad's December salary, given that in that month he had worked only 14 days before the discrimination ruling arrived. This was a man who, when the *Cassa Depositi e Prestiti* asked him for the address of Prof. Bemporad for payment of the severance sum, replied: 'he does not have the title of Professor, having been deprived of his qualification'.[38] From Volta, Bemporad could not expect any help, not even a reference letter.

Other two circumstances played against him: the fact of being continually confused with his cousin Azeglio (much older and notoriously fascist) and the fact that he was no longer able to perform field work in the cold and at night, and therefore could accept only desk activities, such as calculations of stellar motions and errors.

Despite this, Bemporad enjoyed the esteem and appreciation of mathematicians such as Hadamard (who would present his case to Veblen), and astronomers such as F. Schlesinger, Director of the Yale Observatory and H. N. Russell of the Princeton Observatory. Fubini too considered him 'a man of very considerable scientific value'.[39] Their support, however, was not enough. Neither the American Friends Service Committee, to which Veblen submitted Bemporad's application, nor the SPSL, to which Bemporad himself turned, succeeded in finding a place for him. On the outcome of his file weighed also the negative evaluation by Shapley who did not consider Bemporad's case a priority and suggested that he try his luck in South America or Turkey, instead of in the United States.[40]

After expulsion, Bemporad devoted himself with great commitment to assisting the Jewish community and from January 1939 he was one of the most active members of the Committee of Assistance for Jews in Italy (COMASEBIT), then Delegation for the Assistance to Jewish Immigrants (DELASEM) of which he was also general secretary. On the publication of the police order of November 30, 1943, which provided for the sending of all Jews to concentration camps and the confiscation of their assets, Giulio and his sister took refuge in the Turin countryside. In the autumn of 1944, Giulio reached Rome, where he was called on to coordinate the Central Palestinian Office, which was responsible for organising the emigration of Italian Jews to Palestine. He participated in the first conference of the representatives of the Zionist groups of liberated Italy (Rome, January 12–15, 1945) and was a member of the executive committee of the Italian Zionist Federation which proposed the spread of Jewish culture among Italian Jews by establishing an Association for

---

[38] L. Volta to Cassa depositi e prestiti, [Turin] 3.9.1939 in Schiavone 2015, p. 30: *Non gli spetta il titolo di Professore essendogli stata tolta la libera docenza.*

[39] BEMPORAD.1.

[40] BEMPORAD.2–5.

Jewish Language and Culture (*Agudath ha-lasciòn*). He died of lung cancer on 9 July 1945, before he could be reinstated at the Pino Torinese Observatory. On his death, the journal *Israel* published a moving obituary:

> Italian Judaism loses one of its best men, one of its most educated and active men, one of the most prepared to perform managerial functions, most determined to fight strenuously when necessary. In Turin where he resided, and in Rome where the events of the war had temporarily led him, his activity in the Jewish field was continuous and tireless.[41]

Documents

Bemporad.1 O. Veblen to H. N. Russell, [Princeton] 17.4.1939

OVP, *GB*, fol. 1r[42]

Dear Henry:[43] Here is some information about an astronomer, Giulio Bemporad, who was the Head of the Observatory of Turin, and has been dismissed under the anti-Semitic regulations. The information comes from Professor Hadamard. Would you mind looking it over and letting me know your impressions of the case? About all that I know to do would be to send on this information, together with whatever remarks you would care to add, to the Friends Service Committee in Philadelphia. Is there, however, something more effective to do in case the man is really a good one? Professor Fubini regards Bemporad as a man of very considerable scientific value.[44] Yours sincerely, Oswald Veblen

Giulio Bemporad [typed on May 9, 1939]

Forwarding address–Care Professor Jacques Hadamard, 12 rue Emile Faguet, Paris (14°), France

Born Jan. 3, 1888, Florence, Italy–Italian, Jew

Unmarried; dependent sister, Giselda Bemporad[45]

Languages: English, French, modern Hebrew; some German and English

---

[41]Obituary of Giulio Bemporad published in *Israel* 30, n. 31–32, 12.7.1945, p. 1: *L'Ebraismo italiano perde uno dei suoi uomini migliori, più colti, più attivi, più preparati a svolgere funzioni direttive, più decisi a combattere strenuamente quando occorresse. A Torino ove risiedeva, a Roma ove lo avevano condotto temporaneamente le vicende della guerra, la sua attività in campo ebraico è stata continua, instancabile.*

[42]On April 20, a copy of the letter is sent to F. Schlesinger (1871–1943), Director of the Yale University Observatory (1920–1941).

[43]Henry Norris Russell (1877–1957), Director of the Princeton University Observatory (1912–1947).

[44]Fubini had been Bemporad's colleague in Turin for six years (1932–1938).

[45]Giselda Bemporad (1890–1978).

References: Gen. J. Perrier, Rue Auber 9, Paris, France; Prof. Frank Schlesinger, Director, Yale Observatory, New Haven, Conn.; Prof. H. Kimura, International Latitude Observatory, Mizusawa, Ivate-Ken, Japan; Prof. F.P. Cantelli, Istituto di Matematica Attuariale, Univ. of Rome, Italy; Sir H. Spencer Jones, Royal Astronomer, Greenwich Observatory, Greenwich, England[46]

Field: Positional, theoretical, and geodesic astronomy, numerical computation, meteorology

Dr. of Math. 1910 University of Catania

Positions held: Royal Astronomical Station, Carloforte: assistant 1911-20, Director 1920-25; Royal Astronomical Observatory, Capodimonte, Naples 1925-32; Royal Astrophysical Observatory, Catania, and Privatdozent in Astronomy at University of Naples, 1932-33; Royal Astronomical Observatory, Turin, and Privatdozent in Astronomy at Univ. of Turin, 1933-

Publications: list of 43 available

Notes. Giulio Bemporad is younger brother of astronomer who was Head of the Observatory of Turin.[47] During 14 years (1911-25) he was in charge first as assistant, then as Director, of the systematic observations of the variations of latitude at the Station of Carloforte, Sardinia (about 20,000 determinations). Prof. B. Wanach of the Geodätisches Institut, Potsdam, praises his utmost competence and industry in this work; the results of Carloforte are "unsurpassed in accuracy and freedom from systematic errors."[48] During the subsequent 8 years he participated in the reduction of photographic plates for the Astrographic Catalogue at Catania and Naples. [...]

---

[46] George Perrier (1872–1946), major general, geodesist, and professor at the École Polytechnique. He had been vice-president of the Geographical Society (in 1938). Hisashi Kimura (1870–1943), Director of the International Latitude Observatory at Mizusawa, Japan, since 1899. Francesco Paolo Cantelli (1875–1966), professor of Financial and Actuarial Mathematics at the University of Rome (1931–1951). He had graduated in Mathematics in 1899 at the University of Palermo with a thesis on celestial mechanics and he had worked at the Astronomical Observatory of Palermo until 1903. Harold Spencer Jones (1890–1960), Astronomer Royal at the Royal Observatory in Greenwich (1933–1956) and president of the Royal Astronomical Society (1937–1939).

[47] In fact, Azeglio Bemporad (1875–1945) was not Giulio's brother but Giulio's cousin. Azeglio was never Director of the Turin Observatory. He directed the Specola of Capodimonte, in Naples (1912–1934), then the Catania Observatory (1934–1938).

[48] Bernhard Wanach (1867–1928) was the head of the Astronomical Section at the Prussian Geodetic Institute in Potsdam (1922–1928).

Bemporad.2 H. N. Russell to O. Veblen, Princeton 18.4.1939

OVP, *GB*, fol. 1r

Dear Oswald: Thank you for sending me this dossier about Bemporad, which I return. He is a good man. The work which he records is of the best quality and of notable amount. His record is definitely a high one and he certainly merits any aid which can be given him. I would suggest that you communicate with Schlesinger, at Yale, as he probably knows him personally (which I do not) and is an expert in his field of work. I have no hesitation, however, both on the enclosed showing and from my general knowledge of his professional work and reputation, in endorsing Bemporad as a fully worthy candidate for help. With best wishes, Very sincerely yours, Henry Norris Russell

Bemporad.3 F. Schlesinger to O. Veblen, New Haven Connecticut 1.5.1939

OVP, *GB*, fol. 1r

Dear Veblen: I have your letter of April 20 relating to G. Bemporad (I have been away for a few days, hence my delay in replying). I know Bemporad and his scientific record well. He is a very competent observer and an adept practical astronomer. I have referred him to the new Observatory at Istanbul and to the two observatories in the Argentine, but I doubt very much if he will succeed in making connections at these places since his age is against him. He is nearly sixty years of age. By all means refer his case to the Friends Committee as being a very worthy one. In the meantime his colleagues will keep his needs in mind and will be on the lookout for a possible opening. It is not he but an elder brother who was the Head of the Observatory at Turin. I have not heard directly of this elder brother's case, but he too must surely have been dismissed; he gets the benefit of a pension.[49] Giulio also receives a pension, but it is much smaller and is not at all adequate for his needs. Sincerely yours, Frank Schlesinger

---

[49] Azeglio Bemporad (1875–1945), cousin (not brother) of Giulio, was Director of the Catania (not Turin) Observatory from 1934 to 1938. Although he was Catholic and had joined the PNF, was dismissed in 1938. Azeglio remained in Catania, forced to live in hardship, and in 1943 he was reinstated, even if he did not resume his activity.

Bemporad.4 O. Veblen to F. Schlesinger, [Princeton] 10.5.1939

OVP, *GB*, fol. 1r

Dear Schlesinger: We have prepared a statement about Giulio Bemporad and sent it to the American Friends Service Committee. I fear, however, that they may not know how to do much for this sort of case. Therefore I hope that you and other astronomers will keep him in mind. You remarked in your letter that Bemporad is "nearly sixty years of age." Actually, however, he is only fifty-one (born January 3, 1888). I mention this because it may make it easier to find an opening for him. Yours sincerely, Oswald Veblen

Bemporad.5 H. Shapley to H. Weyl, Cambridge Mass. 23.6.1939

OVP, *GB*, fol. 1r

Dear Dr. Weyl: For your records please note that Professor Schlesinger of Yale and I have conferred on the problem of Dr. G. Bemporad. We have written Bemporad to try to find a place at Istanbul and also gave him some addresses in South America to which to write. We agree that there is really no place for him in America. Also Giulio Bemporad, as his older brother in Naples, has some pension and therefore is not starving. If the situation improves for the Jews in Italy, he is very likely to be reinstated. Very sincerely yours, Harlow Shapley

Chapter 5
Gallery: Dispersed Families

5.1 The Castelnuovo–Enriques Families

Castelnuovo and Enriques were both from families of high cultural status and strong civil values. Guido Castelnuovo's maternal grandmother, Adele Levi della Vida, an educator, opened the first Froebelian kindergarten in Venice in 1869; his father Enrico was a patriot, a novelist and Director of the *Scuola Superiore per il Commercio* in Venice; his maternal uncle Luigi Luzzatti was an outstanding economist and politician.[1] Similarly, Federigo Enriques' father, Giacomo, was a wealthy merchant and his mother Matilde Coriat came from one of the most prominent families in the community of Livorno.[2] Both families were non-practising Jews. Guido and his sister, Federigo and his sisters and brothers were all registered as Jews but they did not follow any religion. Enriques had an attitude somewhat analogous to that of C. Segre, a religious sense of mathematical research, which he conceived—in some sense, a legacy of the Talmudic tradition—as an acute logical criticism, a creative drive, and a tension towards the discovery of truth.[3] Castelnuovo could have

---

[1] On Castelnuovo's biography see e.g. Gario (2016); Rogora (2016).

[2] On Enriques' biography see e.g. Bottazzini et al. (2001); Scarantino (2004) and Bussotti (2008).

[3] Chisini (1947), p. 122 recalled: 'He was not an Aryan, but he never used any difference in favor of those who had the same origins as him: the students in whom he was particularly interested in were all Aryans and some fervent Catholics [. . .]. The only observation he made once to me, linked to the merits of his lineage, was that, in his opinion, the undeniable logical-mathematical attitude of the Jews depended on the tradition of Talmud studies, whose comments are exercises of the most acute and pushed logical construction and critique'. (*Non era ariano, ma nessuna differenza usò mai in favore di coloro che avevano le sue stesse origini: gli allievi suoi di cui particolarmente si interessò sono tutti ariani e qualcuno cattolico fervente [. . .]. L'unica osservazione che mi faceva, legata ai pregi della sua stirpe, è che, secondo lui, la innegabile attitudine logico-matematica degli ebrei dipende dalla tradizione degli studi sul Talmud, i cui commenti sono esercitazioni della più acuta e spinta costruzione e critica logica.*)

E. Luciano, *The Jewish Mathematical Diaspora from Fascist Italy*, Science Networks. Historical Studies 64, https://doi.org/10.1007/978-3-031-64896-0_5

written what his uncle Luzzatti, a theorist of freedom of science and conscience, had written to Bishop Bonomelli:

> I was born Jewish, and I return proudly to these origins every time that I am reproached for being an Israelite and that being so exposes me to danger. There is a dignity in bearing the brunt of persecution and it would be cowardly to escape it. But outside of this, my education, my aspirations lead to Christianity, in the wider sense of the word.[4]

Despite the six years that separated them, the academic careers of Guido and Federigo have strong similarities. Both treasured the mentorship of Cremona in Rome and of Segre in Turin. Castelnuovo graduated from Padua, then he obtained a one-year training position in Rome where he attended Cremona's lectures, and from 1887 to 1891 he was given the position of D'Ovidio's assistant in Turin. As for Enriques, after his graduation from the *Scuola Normale Superiore* in Pisa, in 1892 he too obtained a one-year training position in Rome, then in November 1893 he settled in Turin to study at Segre's School but a couple of months later he left for a teaching position at the University of Bologna. They both won a chair: Castelnuovo in Rome (1891) and Enriques in Bologna (1896). From 1922, when Enriques obtained a transfer, they were both in Rome, Enriques occupying the chair of Higher Geometry and Castelnuovo that of Analytic Geometry.

Their families were also born and grew up together. In July 1896, Castelnuovo married Elbina Marianna Enriques (1870–1960), Federigo's older sister, and from their union (a marriage of love, which caused a certain sensation among fellow mathematicians) five children were born: Mario (24.6.1897), Maria (15.7.1899), Gino (27.9.1903), Gina (21.4.1908) and Emma (12.12.1913). Three years later, in July 1899, Enriques also 'settled down', marrying Luisa Miranda Coen, daughter of Achille, full professor of History at the University of Florence. They had three children: Alma (1.1.1900), Adriana (2.8.1902), and Giovanni (24.1.1905). The parents were in constant contact, as were the cousins who were almost the same age. From the time that the Enriques arrived in Rome, there was never a Saturday evening when uncle Ghigo, as he was affectionately nicknamed, did not pop in to visit the Castelnuovos, together with other friends and colleagues, for a chat about mathematics, politics, culture, or any other topic.

From the point of view of research, Castelnuovo and Enriques were very close. The 'binomial' as they were jokingly named in the mathematical environment of the time, built from 1892 to 1906 one of the masterpieces of the Italian School of Geometry: the general theory of geometry over an algebraic surface and the classification of algebraic surfaces in relation to the genres. In about seventy works, the two brothers-in-law established an impressive series of results: the general theory of linear systems of curves on algebraic surfaces, the determination of invariants for

---

[4]L. Luzzatti to Bishop Geremia Bonomelli, 5.11.1899 in Luzzatti (1933), vol. II, pp. 553–554: *Io sono nato israelita e ci ritorno fieramente ogni volta che mi si rimprovera di esserlo e che l'esserlo mi espone a un pericolo. Vi è una dignità a sostenere il peso della persecuzione e sarebbe vile il cansarlo. Ma fuori di questo, la mia educazione, le mie aspirazioni intendono a un largo cristianesimo.*

birational transformations, the problems of surface classification, etc. For these contributions to the field, they both obtained the XL prize for mathematics of the Italian Society of Sciences (1893, 1896) and were awarded the royal prize of the Lincei Academy in 1905 and 1907, respectively.[5]

In addition to being two genius geometers, Castelnuovo and Enriques had a real awareness of the themes of mathematical education and instruction. Both were members of the Italian subcommittee of the International Commission for Mathematical Instruction (1909) and presidents of the Mathesis association of Italian teachers of mathematics (1911–1914, 1921–1933), as well as authors of highly appreciated textbooks. Enriques was editor-in-chief of the *Periodico di matematiche* (1921–1938), and editor of the celebrated book series for the teachers training *Questioni riguardanti la geometria elementare* (1900), then *Questioni riguardanti le matematiche elementari* (1912, several successive reprints) and *Per la storia e la filosofia delle matematiche* (1925).

From 1906, both combined geometric research with other study interests: for Castelnuovo, the calculus of probabilities, statistics, and actuarial mathematics; for Enriques, philosophy and history of mathematics and science.[6] Castelnuovo was moved by a strong sense of social and civil commitment, Enriques by a more philosophical mentality and by a constant desire to reach beyond disciplinary, cultural, linguistic, and national boundaries, a goal best fulfilled in the journal *Scientia*, which he directed from 1907 to 1916 and from 1921 to 1938.

Considered to be among the most authoritative voices of Italian mathematics and of the international philosophical-scientific debate, from the end of the nineteenth century, Castelnuovo and Enriques created a wide network of contacts. Enriques was one of the presidents of the International Congress of Philosophy in Geneva (1904) and, as president of the Italian Philosophical Society (1907–1913), organized the fourth International Congress of Philosophy in Bologna in 1911. Castelnuovo was secretary general of the ICM in Rome (1908). Due to his enthusiasm in establishing exchanges of ideas and interdisciplinary contacts, Enriques in particular encountered and maintained relationships with very different kinds of intellectuals, from mathematicians such as Klein, Picard, and Borel to philosophers such as X. Léon, from physicists such as Einstein, to historians of science such as G. Sarton. The living rooms of Enriques and Castelnuovo played host to Hadamard, Rey Pastor, P. Langevin, L. Godeaux, O. Fernandez, and countless others. A cosmopolitan intellectual, Enriques travelled several times in Egypt, Brazil, Greece, Tunisia, France, Belgium, Spain, England, Scotland, Ireland, Germany, Persia, South America, Turkey, and the Middle East.[7] He was a visiting professor in Geneva, Rio de Janeiro, Constantinople, Paris, Madrid, Edinburgh, Athens, Cairo, and Buenos Aires. Everywhere he went, he left a remarkable imprint, so much so that he was awarded honorary degrees from the Universities of St. Andrews, Liège, Buenos Aires, Montevideo, and Santiago de Chile. Bearers of 'a sense of Italian-ness,

---

[5] Bottazzini et al. (1998).

[6] See e.g. Menghini (2016); Giacardi (2023).

[7] De Benedetti (2001), p. 63.

declined in an open perspective, and at times fused with a broad sense of integration among Italian and Mediterranean cultures',[8] Castelnuovo and Enriques were not interventionists, but neither were they in favour of the ostracism of German mathematicians and indeed, in 1919, they were among the first to re-establish a dialogue with German-speaking mathematicians. The rise of fascism worried them but as men of science they limited themselves to intervening in those aspects on which they felt they could, and must, have their say, such as the Gentile reform. Together with Volterra, Castelnuovo drew up the report for the Commission established by Lincei regarding the problems of higher and middle education, a very strong *j'accuse* of the Gentile's reform project, and of its philosophical assumptions, which relegated scientific studies to a subordinate position compared to the humanities.[9] Even then, however, his interventions were marked by calm elegance:

> I have treated the Minister with great deference because he is a Minister and because, despite his extravagant ideas, he is an elevated person, but he is wrong in believing that any criticism is inspired by personal reasons. [. . .] A propos of our *Relazione*, I simply have to point out an article in *Il Popolo* (it must be the newspaper of Don Sturzo) dated August 21–22, in which, after reproaching the Academy for remaining too detached from the life of the nation, and after reporting various inaccuracies, it is said that this first attempt (the *Relazione*) attacking an interest in the problems of national culture was unlucky.[10]

Faced with the fascistisation of society and culture, and the denial of freedom of speech and the press, Castelnuovo and Enriques did not speak publicly. In the various communications of the Political Police, Castelnuovo is in fact invariably described in these terms:

> Confidential investigations carried out did not reveal that the aforementioned has ever pronounced inappropriate phrases towards the Regime and the Duce; he is considered incapable of doing so. Mr. Castelnuovo is of Jewish race but does not profess any religion. He is considered a correct person, honest in all respects and, according to appearances, he maintains indifferent demeanour towards racist politics, showing himself obsequious to laws and authorities.[11]

---

[8] Pompeo Faracovi in *Le città di mare e lo spirito scientifico* 2001, p. 32: *un senso di italianità non chiuso, fuso piuttosto col senso dell'integrazione fra cultura italiana e culture mediterranee.*

[9] G. Castelnuovo, *Sopra i problemi dell'insegnamento superiore e medio a proposito delle attuali riforme*, Relazione alla Acc. Naz. dei Lincei, Roma, Tip. della R. Acc. Nazionale dei Lincei, 1923.

[10] ANL-*Castelnuovo*: G. Castelnuovo to V. Volterra, Niederdorf 31.8.1923: *Ho trattato con molta deferenza il Ministro, perché è Ministro e perché, non ostante le sue idee stravaganti, è una persona elevata, che ha però il torto di ritenere che ogni critica sia ispirata da motivi personali. [. . .] Di nuovo intorno alla nostra* Relazione *ho da segnalarti soltanto un articolo del* Popolo *(deve essere il giornale di Don Sturzo) del 21-22 agosto, nel quale dopo aver rimproverato l'Accademia di tenersi troppo appartata dalla vita del paese, e dopo aver detto varie inesattezze, trova che questo primo tentativo (la* Relazione*) di interessarsi ai problemi della cultura nazionale non è stato fortunato.* Castelnuovo refers to the article *L'Accademia dei Lincei* appeared in Il Popolo, Rome 2.8.1923.

[11] Office of P.S. of Castropretorio at the Royal Police Headquarters, Rome 28.2.1939: *Da riservati accertamenti eseguiti non è risultato che predetto abbia mai proferito frasi sconvenienti nei riguardi del Regime e del Duce; ne è ritenuto capace di farlo. Il Castelnuovo è di razza ebrea, ma non professa alcuna religione. È ritenuto persona (retta ed) onesta (sotto tutti gli aspetti) e,*

The 1920s and 1930s were studded with personal successes for both Castelnuovo and Enriques. Thanks to an important allocation of funds from the Marco Besso Foundation, Castelnuovo founded the School of Statistical and Actuarial Sciences in Rome (1927–1935). Enriques was put in charge of editing the mathematics section of the *Enciclopedia Treccani* (1926–1937) and directed the National Institute of History of Science (1927–1938). In those years, Castelnuovo and Enriques dedicated the best of themselves to the School: as teachers and advisors, as charismatic Masters, they attracted dozens of Italian and foreign students to Rome and involved them in their initiatives. Indicative of this fact is the number of foreigners who participated in the Mathematical Seminar of Rome, organised by Castelnuovo and Enriques, and in the commemorative conference on Castelnuovo's retirement. On that occasion, Castelnuovo concluded his last lesson (28 May 1935) with these words:

> I cannot complain about the fate that has given me so much satisfaction in life, in my career, in science. I experienced the purest joys that come from scientific research and discovery. But I have always preferred the most human joys that derive from the affection of those around me. Therefore, this ceremony that has spontaneously gathered around me so many former disciples, fond and grateful to their old teacher, will be one of the dearest memories of my life and will be enough to comfort the silent months or years that remain to me.[12]

Castelnuovo and Enriques' families were strongly affected by the racial laws. Castelnuovo was already retired, while Enriques was removed from the chair of Higher Geometry after 42 years of a brilliant career. Beyond the purge, they faced multiple humiliations: the expulsion from all the Italian academies and scientific societies to which they belonged; the obligation to leave all positions of responsibility held up to that point; a ban on publishing or conducting any type of editorial activity. For Castelnuovo, one particularly painful event was his expulsion from the Lincei, of which he had been secretary for the class of Physical, Mathematical, and Natural Sciences from 1922 to 1925.

Their first reaction, as recalled by the women of the family, Luisa Cohen and Emma Castelnuovo, was disbelief. The hope of obtaining the counter-discrimination followed. Through the direct intercession of Gentile, both Castelnuovo and Enriques would eventually obtain it for their exceptional scientific merits, but the status was not extended to their children.[13] Waiting for the outcome of the process, the two families moved. Various issues were at stake, some even economic, such as the

---

stando alle apparenze, mantiene contegno indifferente nei riguardi della politica razzista, si mostra ossequiente alle leggi ed alle autorità.

[12] *Onoranze per il giubileo scientifico del prof. Guido Castelnuovo* 1937, p. 10: *Non posso lagnarmi della sorte che mi ha concesso molte soddisfazioni nella vita, nella carriera, nella scienza. Ho provato le gioie più pure che vengono dalla ricerca scientifica e dalla scoperta. Ma ad esse ho sempre preferito le gioie più umane che derivano dall'affetto di chi mi circonda. Perciò questa cerimonia che ha raccolto spontaneamente intorno a me tanti antichi discepoli affezionati e riconoscenti sarà uno dei ricordi più cari della mia vita e basterà a confortare i mesi o gli anni silenziosi che mi rimangono.*

[13] Guerraggio and Nastasi (1993), pp. 135–136, 166.

copyright on books published by Enriques with Zanichelli. Others were matters of both cultural and affective nature: if you do not want to kill something for which you are responsible, you must identify an appropriate successor, and do so in a short time. For the journals *Periodico di matematiche* and *Scientia*, this was in fact the case: both journals survived the war thanks to O. Chisini, P. Bonetti, and E. della Monica. However, the Institute for the History of Science, an authentic excellence in Italian humanistic research, was taken over by Severi and then cancelled by him in just a few months.

Emigration was just one of the things to think about. The older generation did not take it into account, while some of their children did. Helping them became the priority.

Enriques reached Paris in the autumn of 1938, and stayed there for a few weeks, meeting Fubini with his sons, who were leaving for America. Taking advantage of the many contacts he had with the intellectual French world, he tried to find accommodation for his son Giovanni, an engineer. His daughters Adriana and Alma did not feel like uprooting their families and moving abroad. Adriana, the girl who had photographed Einstein at the Bologna station in 1921,[14] had graduated *cum laude* in Mathematics in Rome in 1925 with a thesis on algebraic geometry supervised by Castelnuovo before becoming a middle and secondary school teacher. She had published several works on the history of mathematics, a volume of the series *Gli elementi di Euclide e la critica antica e moderna* (vol. III, Rome, Stock, 1927), an anthology of children's stories, five texts on arithmetic for primary school and a collection of books of arithmetic and geometry for middle and professional schools (1934–1936). In 1930 she married the engineer and industrialist U. De Benedetti, Director of a family-run factory in Turin. Her sons, Andrea and Federigo Jr., were still very young. After some strong indecision, Alma and Adriana decided not to leave.

Unlike his brother-in-law, Castelnuovo remained in Rome and began to mobilise his network of foreign partners in favour of aspiring emigrants such as Segre and Terracini. Two of his five children were planning to flee: Gino (1903–1995) and Gina (1908–2001). The first was an electronics engineer. Castelnuovo talked about him to Veblen who addressed F. B. Jewett, Director of the Bell Telephone Laboratories Inc. of New York (CASTELNUOVO.1). Jewett was not moved by the story. Pathetic cases such as that of this young very distinguished man, he replied, were now two a penny, which is why the energies of good Americans should be directed first of all towards finding a place for their compatriots, all the more so in a moment of recession (CASTELNUOVO.2-3). Veblen understood that there was no room for manoeuvre and limited himself to thanking Jewett and offering a generic recommendation:

> I am very grateful for the trouble which you have taken in the case of Dr. Castelnuovo. On account of the regard in which we hold his father, I hope that in case business conditions improve some place may be found for him.[15]

---

[14]Linguerri and Simili (2008), pp. 12–13.

[15]OVP, GC: O. Veblen to F. B. Jewett, [Princeton] 24.1.1939.

No further developments in Gino Castelnuovo's story are known, but he was unable to leave Italy and lived in hiding, mainly in Milan, with his wife Clara Mori and his son Enrico (1929–2014). After the war he would move to Turin, where he served as central technical Director of RAI since 1952.

By contrast, Gino's younger sister managed to emigrate.[16] Gina graduated in biology at the University of Rome in 1931, and was a fellow of the Corsi Foundation, and an assistant in the Zoological Laboratory and in the Hydrobiological Laboratory at the University of Rome. At the promulgation of racial laws, she was working in the *Pharmaco-therapeutisch Laboratory* of the University of Amsterdam with a fellowship of the National Council of Research. She had already published various works in Italian and French scientific journals, and had an international experience under her belt, having completed her specialisation in Amsterdam. She was acquainted with eminent scholars and spoke four languages: Italian, French, English (very badly, in truth), and German.

With great determination and promptness, Gina decided not to return to Rome; moved from Amsterdam to the *Marin Biologisk Laboratory* of Copenhagen and turned to the SPSL and the ECADFS, at the same time asking for help from many of her father's colleagues and other young people who, like her, were determined to leave. Gina and her cousin Adriana were childhood friends of Laura Capon Fermi. They grew up together in a compact, close-knit group, which Adriana would fondly recall in an interview with the *Corriere della Sera* on the threshold of her ninetieth birthday:

It was nice to be young in those summers [. . .]. Fermi was my great friend. We went skiing in Switzerland. [. . .] We went to the sea, we played chess. Fermi was a disaster as a dancer. Even Laura (Lalla) Capon, the beautiful daughter of the admiral he met in my house, before becoming his wife refused to dance with Enrico. He couldn't hear the music. Enrico clung to me in order to court her. Every Saturday we went to the Castelnuovos'; there came Fermi, Pontecorvo, Levi-Civita and once Ettore Majorana, although usually you could not see him.[17]

The first contacts with the SPSL and ECADFS were not encouraging but Gina did not allow herself to be discouraged. On 13 July 1939, with no job offer in hand, she was accompanied by her parents to the Rex and sailed to New York. Waiting for her were the Fermis, who hosted her at their home.[18] At the beginning of September, after a brief sojourn as guest researcher in a zoological laboratory near Boston,

---

[16] On Gina Castelnuovo's biography see Simili (2010), pp. 37, 41–49.

[17] M. Chierici, *Lessico famigliare. La figlia del matematico Federigo Enriques apre l'album dei ricordi*, La Stampa, 10.4.1992: *Era bello essere giovani in quelle estati. [. . .] Fermi era mio grande amico. Andavamo a sciare in Svizzera. [. . .] Andavamo al mare, giocavamo a scacchi. Fermi era un disastro come ballerino. Perfino Laura (Lalla) Capon, la bella figlia dell'ammiraglio conosciuta in casa mia, prima di diventare sua moglie rifiutava di ballare con Enrico. Non sentiva la musica. Enrico si aggrappava a me per farle la corte. Tutti i sabati si andava da loro [dai Castelnuovo]; veniva Fermi, venivano Pontecorvo, i Levi-Civita e una volta Ettore Majorana, anche se di solito non lo si vedeva.*

[18] ANL-*Castelnuovo*: G. Castelnuovo to V. Volterra, Sarentino 23.7.1939.

directed by Thomas H. Morgan, Gina joined Enrico and Laura in Ann Arbor and followed them in New York. 'From there – as her father explained to Volterra – she went to meet the most renowned biologists with the hope of finding some opening'.[19]

For six months things were not going well and Gina continued obtaining just temporary positions in New York (January 1940), Philadelphia (May–June 1940), Salisbury Cove (June 1940), and Bar Harbor (July 1940).[20] In the United States, however, she was not alone: Laura Capon Fermi and Annetta Ghiron Fubini supported her, both emotionally and economically. Levi-Civita also wrote to Veblen to present her case and Veblen's wife, Elizabeth, took the request to heart.[21] Veblen's role was essential in rescuing Gina Castelnuovo, even though biology fell outside the field of his competence. It was Veblen who tried to convince E. Witschi in offering a temporary position in Jowa;[22] it was Veblen who coordinated a fund, to which many mathematicians, admirers and friends of Guido Castelnuovo would contribute, thanks to which Gina was able to attend a summer school in Wolfeboro and to work at Mt. Desert Island Biological Laboratory at Salisbury Cove.[23] Finally, Gina would be hired in Missouri.[24]

From the moment of Gina's departure to Italy's entry into the war, the Castelnuovo and Enriques families lived in a sort of limbo, living a solitary life, and spending long periods of vacation in the mountains. Enriques published some works, under the pseudonym of Adriano Giovannini. On September 15, 1938, just a few days after the promulgation of the first racial laws, there appeared Castelnuovo's book *Le origini del calcolo infinitesimale nell'era moderna* (Bologna, Zanichelli). He would resume publishing scientific papers in 1945.

Meanwhile, the little girl of Castelnuovo's house, Emma, began working in the Jewish school in Rome. Emma had graduated in Mathematics in July 1936; soon after, she was hired as a librarian in the Institute of Mathematics and in August 1938 she won the competition for chairs of mathematics in the middle schools (the first in which she participated). She was supposed to take up service in September 1938, but this was prevented by the promulgation of the racial laws. Having lost her job as a librarian, she was appointed as a teacher in the Jewish middle school, inaugurated in November 1938, first in the premises of a former noble villa renovated in Via Celimontana, and then in the classrooms of the ancient Jewish kindergarten on the Lungo Tevere. Very young, but equally stubborn, Miss Emma was a tomboy capable of inciting a strike because of the absence of heating in the classrooms.[25] She was an

---

[19] ANL-*Castelnuovo*: G. Castelnuovo to V. Volterra, Rome 5.9.1939: *di là andrà a vedere i biologi più rinomati con la speranza di trovare qualche sistemazione.*

[20] Castelnuovo.6.

[21] Castelnuovo.4 and Castelnuovo.5.

[22] Castelnuovo.7 and Castelnuovo.15.

[23] Castelnuovo.8-14.

[24] Castelnuovo.16 and Castelnuovo.17.

[25] Della Seta (1996), p. 27.

exceptional teacher, with a natural talent for teaching and a real passion for this profession. So, managing to make the most of this first work experience, in less than a year Emma wrote a text of *Lezioni di Geometria Elementare* that she published under the name of Marcello Puma.[26] As she wrote:

> Marcello-a mathematician who had graduated with my father-managed that private school, Galileo Ferraris, which was first near Piazza di Spagna and then in Via Piave, at the corner of Via Flavia. This Marcello Puma called me to ask if I could write (under a false name) some textbooks for secondary school. So, we wrote a book for the publishing house Garzanti. He wrote the texts of Algebra and Trigonometry, I contributed with those of Geometry. These books, of which I still have a copy, disappeared altogether because one of the bombs dropped on Milan hit Garzanti and the printing press full-on.[27]

In reality, two copies of this handbook survived to the present day and suitably testify how Emma was capable of anticipating some didactic solutions for which she would become one of the most influential figures in mathematics education research in the post-war period. The efforts made by Emma and her colleagues to make the activity in the Jewish school 'normal' bore fruit. At the exams, the Jewish students obtained excellent results, although they were aware that they had no prospect of continuing their studies. At this point, Castelnuovo and Enriques did not resign themselves to the solution often displayed in Italian Judaism, namely that of aiming offer young people a practical and profession-oriented training, with an eye to eventual emigration to Palestine or South America.

Upon learning of the existence of the *Institut Technique Supérieur* in Fribourg, Switzerland, which allowed students to enrol without attending the courses, Guido Castelnuovo had the idea of creating a Clandestine University. Some colleagues and friends, both Jewish and Aryan, were involved: G. Coen, G. Bisconcini, R. Lucaroni, B. Cacciapuoti, as well as Enriques, of course (Fig. 5.1). The idea, welcomed with enthusiasm, was immediately put into action. In the autumn of 1941, the Clandestine University inaugurated courses in Analytic Geometry (Lucaroni), Algebraic and Infinitesimal Analysis (Bisconcini), Graphic Statics and Construction Science (G. Supino and V. Camis), and Projective Geometry (Castelnuovo and Enriques). Thanks to these 'courses of integrative mathematical culture', as they were prudently named, some of the best students of the Jewish school of Rome would be admitted to the third year of university studies in Engineering, after liberation.[28]

At the beginning of 1942, frightened by the course of the war, Adriana Enriques and her husband adopted false identities and went into hiding, first in Tuscany, and

---

[26] On the history of this text see Luciano (2013b), pp. 38–40.

[27] Natalini and Mattaliano (2004), p. 5. *Il fratello Marcello-un matematico che si era laureato con mio padre – dirigeva quella scuola privata, Galileo Ferraris, che prima era vicino a piazza di Spagna e dopo in via Piave, angolo via Flavia. Questo Marcello Puma mi aveva allora chiamato per domandare se potevo scrivere (sotto altro nome) qualche libro di testo per la scuola secondaria. Così l'abbiamo scritto con Garzanti. Lui aveva fatto quelli di Algebra e di Trigonometria, io quelli di Geometria. Questi libri, di cui ho ancora una copia, sparirono del tutto perché una delle bombe lanciate su Milano ha colpito in pieno Garzanti e la tipografia.*

[28] Castelnuovo E. (2001); Fiorentino (2003), pp. 107–110.

Fig. 5.1 Coen,
Castelnuovo and Bisconcini
in the courtyard of the
Jewish school, Rome, 1941,
in Castelnuovo E. 2001,
p. 67

then on the hills of Ivrea, as guests of the engineer and entrepreneur C. Olivetti. Maria Castelnuovo and Alma Enriques did the same. Their parents, on the other hand, were unable to understand that they were in danger and refused to leave Rome. The children of Castelnuovo and Enriques, their students and colleagues, all tried to convince them to take refuge in the Vatican City, but Enriques was adamant in his refusal. Both remained in Rome even during the occupation and, if they were saved, it was only thanks to Emma's colleague Marcello Puma, who phoned her two days before the roundup at the ghetto:

> my brother – the brother was Head of the police station here in Piazza Bologna – said that on 16 October [1943] there will be a raid: you must flee from home. My father absolutely did not believe it; however, my parents went as guests of the mathematician Tullio Viola, where they stayed for a month.[29]

---

[29] Natalini and Mattaliano (2004), p. 5: *mio fratello il fratello era Capo del commissariato qui di piazza Bologna – ha detto che il 16 ottobre ci sarà una razzia, fuggite da casa. Mio padre non ci credeva assolutamente e allora i miei genitori sono andati ospiti del matematico Tullio Viola dove sono stati un mese.*

Having gone into hiding in October 1943, Castelnuovo (under the false name of Guido Cafiero) and Enriques (as Adriano Giovannini) were helped by mathematician companions, such as Viola, Tricomi and A. Frajese, to hide in religious institutes and private boarding houses. Both, however, continued to go out and walk around Rome, at the risk of being recognised.

The Jewish school and the Clandestine University obviously closed in the autumn of 1943 but in those 'black days, of terrible anguish, between fear and pain, and countless discomforts',[30] people began to talk about the Italy to be rebuilt. As the German troops retreated, tragic news arrived: the shooting in Florence of Anna Maria Enriques, the illness of Enzo Enriques, brother of Anna Maria who was enlisted in the partisan gangs, the alleged deportation (which he would fortunately escape) of Mario Castelnuovo. However, it was from the lively exchange of ideas of that period that the activity of Castelnuovo and Enriques restarted, in a different realm.[31]

After the liberation of Rome (4 June 1944) a new season of civil commitment opened for both.[32] As extraordinary commissioner of the CNR and president of the Lincei, Castelnuovo directed the reconstitution of this Academy into a free society and the dismantling of the Academy of Italy and contributed substantially to the cultural reconstruction of the country. Enriques was given back the chair and put back in charge of *Scientia*, but he passed away less than two years later, on 14 June 1946.

Documents

Castelnuovo.1 O. Veblen to F. B. Jewett, [Princeton] 15.12.1938

OVP, *GC*, fol. 1r
Dear Jewett:[33] May I call your attention to the case of Dr. Gino Castelnuovo who has been cast adrift by the recent anti-Semitic measures in Italy? He is the son of one of the most distinguished of the Italian mathematicians, whom I have known for most of my scientific life, not only for his scientific eminence but also for his kindness and consideration for mathematicians in all parts of the world. I am told that the son is a very distinguished electrical engineer, and I am enclosing herewith a statement about him, which has been given me by one of my colleagues. Yours sincerely, Oswald Veblen

---

[30] Ragusa Gilli (2006), p. 119: *giorni neri, di terribile angoscia, tra paura e dolore, e innumerevoli disagi.*

[31] Castelnuovo E. (1997), pp. 232–234.

[32] *Prof. Guido Castelnuovo . . .*, 1945 and Roth (1952).

[33] Frank Baldwin Jewett (1879–1949), PhD in Physics in 1902 from the University of Chicago, was the first president of Bell Telephone Laboratories (1925–1940), and chairman of the Board of Directors of Bell Laboratories (1940–1944).

Castelnuovo.2 F. B. Jewett to O. Veblen, New York 16.12.1938

OVP, *GC*, fol. 1r
My dear Veblen: I have your letter of the 15th with its attached papers relative to Dr. Gino Castelnuovo which I am putting in line for proper consideration, although I doubt very much whether there is anything in the way of a position that we have to offer. The increasing tide of these pathetic cases which is rising is heartrending, particularly in the face of an inability to do very much by way of amelioration. With the best intentions and the most ardent desire to be of assistance, there is really not much that can be done, especially at a time of depressed industrial activity and with an urgent demand to take care of our own young people. However, I am referring this letter to the people at the Laboratories with the thought that possibly they may have need for some one of Dr. Castelnuovo's training and experience. Yours sincerely, F.B. Jewett

Castelnuovo.3 F. B. Jewett to O. Veblen, New York 19.1.1939

OVP, *GC*, fol. 1r
My dear Veblen: This is in further reply to your letter of December 15th regarding young Dr. Castelnuovo, who would like to come to the United States. Due to the fact that both you and Dr. [Abraham] Flexner have written me about him, we have gone to considerable pains to discover whether or not we have any possible opening. The answer, unfortunately, is in the negative. One can have the deepest sympathy for a person who, under present conditions, is anxious to get out of Italy and begin a free and independent life but, as you will readily appreciate, there is really not much that we in industry can do at times of depressed industrial activity such as the present. Our foremost obligation is to take care of as many young people of our own country as humanly possible. Thanking you, however, for having brought the case of Dr. Castelnuovo to my attention, I am yours sincerely F.B. Jewett President

Castelnuovo.4 Gina Castelnuovo to E. Richardson Veblen, 18.1.1940

OVP, *GC*, fol. 1r
Dear Mrs. Veblen,[34] thank you very much for your kind letter. I shall be enjoyed to see you, when you come to New York, and any time it should be all right for me; unfortunately I am not employed and I am always free. I thank very much

---

[34] Elizabeth Richardson (1882–1974), sister of the Nobel laureate Owen Willans, had married Oswald Veblen in 1908.

again Dr. Veblen, and I shall be very grateful if when he had the opportunity, he will remember me to the Rockefeller Institute. I hoped to get some money in my Laboratory, but I find that it is impossible; my Director is always very kind and gives me many promises, but unfortunately at the last moment he always says that he has not money for me.[35] I have heard that he has done the same thing with many refugees, who worked by him. I can't have a grant from the Committee of the Refugees because I have not assured a position for next year. Therefore, I am always in very great trouble; the only hope that I have now, it is the possibility to get a Guggenheim fellowship but I am afraid it would be impossible. I have received today a letter from Mrs. Levi-Civita[36] with many compliments for you, if I saw you. Hoping to see you very soon, will you accept my best regards. Yours very truly Gina Castelnuovo

Castelnuovo.5 Gina Castelnuovo to O. Veblen, New York 22.1.1940

OVP, *GC*, fol. 1r

Dear Dr. Veblen, I thank very much again you and your wife for your great kindness towards me. I have been really very glad to met you and I hope to have sometime the opportunity to visit you. Whenn [sic!] I came in your home I was really very discouraged, and I was left you I was not so pessimistic as I was before. I went to Bryn Mawr and the people of the College promised me to try to find a laboratory who called me and after to get a fellowship from the Committee of Refugies [sic!] in New York, and I hope that it would be possible.[37] I went also to talk with Dr. Ten Broek. He was very kind, he asked to me about my works and hearing that my specialised field of Biology was that of endocrinology, he said that this general field was interesting for the researches of his Laboratory, but he did not tell of any possible opening, only he said that if he for chance heard of any possible opening anywhere he would advise me. I know also that Dr. White of Botanical Dept. of the same Laboratory, to whom I have spoken, wishes have my help for tissue culture, field that I don't like so much and I don't know so well as this of endocrinology, but of course I should be too glad to enter in the Rockefeller Institute. I doubt that it would never possible, because to fine. I wished relate to you what I had done for your good help. Greateful [sic!] very much to you and to mrs. Veblen will you accept my best regards and my best thanks. Yours very sincerely Gina Castelnuovo

---

[35] In November 1939 Gina had obtained a temporary position in the Laboratory of Experimental Biology of the American Museum of Natural History. The Laboratory was directed by the zoologist Gladwyn Kingsley Noble (1894–1940) from 1928 to 1940.

[36] Libera Trevisani Levi-Civita (1890–1973).

[37] Anna Pell Wheeler, head of the Department of Mathematics at Bryn Mawr College presented Gina's case to the ECADFS (see SPSL, *GC*: A. Pell Wheeler to ECADFS, 28.11.1939).

Castelnuovo.6 Guido Castelnuovo to O. Veblen, Rome 22.1.1940

OVP, *GC*, fol. 1r-2r

Dearest friend, I was very pleased with your letter of 19 Dec. which arrived a few days ago (it spent almost a month on the road!). Thank you for what you did for that translation; the answer did not surprise me. I know that you live in the midst of many troubles; perhaps we experience less of this misery, but we hear about it all the time. It is very sad, after having spent the first part of life under a clear and propitious sky, to see in recent years such a cruel storm raging over a good part of humanity, and to live in the fear that the hurricane will hit us or our children. You, at least, and your family, are in a quiet area; I understand, however, that even overseas the possibility of pure scientific research may appear to be hopeless until humanity finds peace. Our mutual friends Enriques and Levi-Civita are fine; the Volterras' reasonably well. The School of Advanced Mathematics is running at full steam; ten conference courses! Students are missing, but professors and assistants make up the audience. Good news from Beniamino Segre who is in Cambridge (England), where Enrico Volterra also is at present; good news also from Terracini who had an excellent welcome in Tucumán, and from Beppo Levi, now in Rosario. I saw Racah two days ago, leaving for Jerusalem where he goes to cover the chair of Theoretical Physics.[38] Of my Gina you will have direct news; she works at the Museum of Natural Sciences of N.Y. She is happy with the work but would be happier if she could have a salary, grant or scholarship. Since she is highly esteemed and popular, I hope she will be able to find a solution. Until that day it would be good if your colleagues, who have known her and have influence, did not forget her. A thousand best wishes to you, your wife, and your children, from all of us. Yours affectionately G. Castelnuovo

Carissimo amico, Mi ha fatto molto piacere la tua lettera del 19 dic. arrivata pochi giorni fa (ha messo quasi un mese in viaggio!). Grazie di quello che hai fatto per quella traduzione; la risposta non mi sorprende. Sento che vivete in mezzo a molte miserie; noi forse ne vediamo meno, ma ne sentiamo parlare continuamente. È ben triste, dopo aver passato la prima parte della vita sotto un cielo sereno e propizio, veder negli ultimi anni infierire una così crudele tempesta sopra una buona parte dell'umanità, e viver nel timore che l'uragano investa noi o i nostri figli. Tu, almeno, con la tua famiglia, ti trovi in zona tranquilla; capisco però che anche al di là dell'Oceano possa apparire la vanità della pura ricerca scientifica, finché l'umanità non trova pace. Gli amici Enriques e Levi-Civita stanno bene; i Volt. discretamente. La Scuola di alta matematica funziona a tutto vapore; dieci corsi di conferenze! Mancano gli allievi, ma professori e assistenti costituiscono l'uditorio. Buone notizie da Ben. Se. che si trova a Cambridge

---

[38] Giulio Racah (1909–1965), a pupil of E. Fermi, former professor of Physics at the University of Pisa, emigrated to Palestine in 1939. He was appointed professor of Theoretical Physics at the Hebrew University of Jerusalem.

(Ingh.), dove è pure Enrico Vo.; da Terr. che ha avuto ottime accoglienze a Tucumán, e da B° Le. a Rosario. Ho visto due giorni fa Racah in partenza per Gerusalemme dove va a coprire la cattedra di Fisica teorica. Della mia Gina avrai notizie dirette; lavora al Museo di Sc. Natur. di N.Y., è contenta del lavoro, ma più contenta sarebbe se potesse avere una retribuzione, come stipendio o borsa di studio. Poiché è molto ben vista e ben voluta, spero potrà mettersi a posto. Fino a quel giorno sarà bene che i tuoi colleghi che l'hanno conosciuta ed hanno influenza non la dimentichino. Mille cose cordiali a te, alla Signora, ai tuoi figli, da parte di tutti noi. Tuo aff^{mo} G. Castelnuovo

Castelnuovo.7 O. Veblen to E. Witschi, [Princeton] 4.6.1940

OVP, *GC*, fol. 1r

Dear Professor Witschi:[39] A weak or so ago I saw Miss Castelnuovo, who is the daughter of an old friend of mine (Professor Castelnuovo is the 'Dean' of the Italian mathematicians). In answer to my inquiries about her progress in getting a foothold in this country she told me that what she would like best of all would be a chance to work with you in Iowa City. Yesterday I happened to be in New York, and took the occasion to talk with Miss Betty Drury, the Assistant Secretary of the Emergency Committee in Aid of Displaced Foreign Scholars, 2 West 45th Street. I learned from Miss Drury that what is most particularly needed is a formal application from the administration of the University, stating the amount of money which would be needed and the other relevant facts. The Emergency Committee used to require that the university should make some commitment in the direction of a future position for the applicant. At the present time, however, no university is able to do this, and consequently the Committee is being forced to revise its procedure; and I think the sort of statement which you made to the National Refugee Service, to the effect that you would take an interest in helping Miss Castelnuovo to find something more permanent in the future, would be sufficient.[40] Since the amount of money which would be required in this case is a modest one, I personally think there is a good chance that the application would be granted. Of course this is my own opinion, and does not come from Miss Drury or anyone connected with the Executive Committee. It might be of interest to you, since I gather you do not know Miss Castelnuovo personally, that she seems to me and to other people here who know her, to have an attractive personality and

---

[39] Emil Witschi (1890–1971), professor of Zoology at the University of Iowa (1927–1958).

[40] ECADFS, *GC*: E. Witschi to W. Haber [executive Director of the National Refugee Service], Iowa City 29.5.1940. The National Refugee Service was a aid organization founded in New York City in May 1939 to assist refugees from Europe fleeing Nazi persecution. It represented a reorganization of the National Coordinating Committee for Aid to Refugees and Emigrants Coming from Germany, which had been active since 1934 as an umbrella organization of refugee aid agencies.

to be likely to get along well in the American environment. She has a good command of English. Yours sincerely, Oswald Veblen

Castelnuovo.8 H. Kraus [American Friends Service Committee] to H. Weyl and S. Lefschetz, Philadelphia 4.6.1940

OVP, *GC*, fol. 1r

Dear Dr. Weyl: These two persons [Peter Scherk and Gina Castelnuovo] have been recommended to us as strong candidates for our Summer School for refugee scholars.[41] I believe they are both known to you and your committee. To what extent would you be able to help them meet the expenses of $100,00 each for room, board, and tuition for the eight weeks' period of the Summer School? Practically, each one of our 40 candidates wants a scholarship, and we can only help in a very few cases as far as cash grants are concerned since the funds of the Service Committee are already tied up in the whole project. I hope that the various professional committees will help their younger refugee colleagues to attend the Summer School which is likely to meet one of their very urgent needs. We are tentatively accepting both Dr. Scherk and Dr. Castelnuovo but the financial matter remains to be straightened out. May we have a word from you in this matter? Cordially, Hertha Kraus Consultant[42]

---

[41] During the summer of 1940 the American Seminar for Foreign Scholars was held at Brewster Free Academy, Wolfeboro, N.H. This was a Summer School for intensive study of the American Community, American Education, English. According to the report *Services for refugee scholars and teachers* published by the American Friends Service Committee (Philadelphia, 1941, pp. 8–9): 'It was [...] a co-operative undertaking in which foreign scholars, American volunteer tutors and professors, and foreign and American maintenance staff worked together, with the increasingly sympathetic help of the local community, toward the common goal of mutual understanding and adjustment of the newcomers to the American scene and the American educational system. The beautiful setting, on the shores of Lake Winnipesaukee, inspired release of tension and a genuine recreation'. In actual fact, only Peter Scherk, as a mathematician, was known to the administrators of the German Mathematicians' Relief Fund. On Scherk see footnote 193, p. 236.

[42] Hertha Kraus (1897–1968) had fled Cologne in the 1930s because she was a Jew and because of her involvement with the Democratic Socialist Party. At Bryn Mawr College, she was appointed as an associate professor of Social Economy and Social Research. She used her extensive contacts and experience in Germany to help form America's refugee programs and published a series of case relief records in International Relief in Action (1914–1943) so that critical helpers could learn principles and procedures for international relief. She was a restless collaborator and consultant of the American Friends Service Committee.

Castelnuovo.9 H. Weyl to H. Kraus, [Princeton] 6.6.1940

OVP, *RefGen*, fols. 1r
My dear Doctor Kraus: *Re Summer School: Peter Scherk and Gina Castelnuovo.*
Before this you will have received my letter of June 4 about Dr. Scherk and the
Summer School. As far as I know, Dr. Castelnuovo is a biologist, and I am not in
a position to help her financially out of my Mathematicians' Relief Fund. I know
that Professor Veblen (who has left Princeton for his summer vacation) is
interested in her, and I shall let him know about this matter. [. . .] Yours sincerely,
Hermann Weyl

Castelnuovo.10 G. Fubini to H. Weyl, [Princeton] 6.6.1940

OVP, *GC*, fol. 1r
My dear Professor Weyl, Miss Castelnuovo writes to me that you have received a
letter about Dr. Scherk and Dr. Castelnuovo.[43] She believes to be thoroughly
unknown to you, and asks me to give to you some informations. She is the
daughter of the Italian geometer Prof. Guido Castelnuovo (the man who, with
Prof. Enriques, constructed, for the first time, the theory of algebraic surfaces
under the point of view of birational transformations). She is Doctor of Biology,
and she has no money: her father, in *many* months, could send to her (I believe)
only 100 dollars. Now she had a job for two or three months in Philadelphia; and
now she hopes to find another job, since the job in Philad. is over.[44] My best
regards from yours Guido Fubini

In order to help my colleague's daughter, I am disposed to contribute with the
sum of $50 for the refugees (Amer. Serv. for Coll. and Univ. Teachers). G.F.

Castelnuovo.11 H. Weyl to G. Fubini, [Princeton] 13.6.1940

OVP, *GC*, fol. 1r
Dear Professor Fubini: Unfortunately, I am unable to contribute to Miss
Castelnuovo's expenses out of the fund which I administer, because that money
comes from mathematicians and is meant for mathematicians only. Dr. Kraus did
not know exactly about the nature of this fund. However, you will be interested in
the enclosed copy of a letter Veblen wrote to Miss Kraus. I am informed that

---

[43] OVP, *Ref. Gen.*: H. Kraus to H. Weyl, Philadelphia 28.5.1940.
[44] Gina had obtained a temporary job at Brin Mawr college. She would get another temporary work
at the Mt. Desert Island Biological Laboratory at Salisbury Cove for a few weeks in the summer
of 1940.

another $50 has been pledged for the same purpose, and Veblen thinks $150 would probably get Miss Castelnuovo through the summer. In these circumstances would it not be best for you to contribute your $50 also to meet Dr. Castelnuovo's expenses at Salisbury Cove? In that case you might make out your check to Professor Veblen. Or have you already promised it to Miss Kraus? I think she would be perfectly ready to transfer it. Yours sincerely, Hermann Weyl

Castelnuovo.12 G. Fubini to O. Veblen, Princeton 13.6.1940

OVP, *GC*, fol. 1r

My dear Prof. Veblen, From a letter of Prof. Weyl I learn that it can be very useful to Miss Castelnuovo to receive other $50 for her expenses at Salisbury Cove. If you agree, I will send to you immediately these $50; but obviously I wish that this offer remains secret and completely unknown. If it is not so, I fear that Miss Castelnuovo can refuse the sum. After the recent events of the war, we will follow the counsels of Mrs. Veblen, and I give up Seattle and the West. Perhaps we will spend our summer either in the Maine, or in the State of N.Y. I am certain that you will be so kind as to excuse me if I have taken the liberty to inconvenience you. With the best regards to Mrs. Veblen and to you, also in the name of my wife, yours sincerely Guido Fubini

Castelnuovo.13 G. Blake to O. Veblen, Princeton 13.6.1940

OVP, *GC*, fol. 1r

Dear Professor Veblen: Professor Alexander is working exclusively at home, so I called him up and at his request read him your letter to Miss Kraus about Miss Castelnuovo. He asked if I would not telephone Professor Lefschetz directly as he did not want to interrupt his work but said he would himself contribute $25 toward Miss Castelnuovo's expenses. I suggested that his check be made payable to you. Professor Lefschetz also promised $25 and gave me a considerable list of men to circularize with the material you had suggested. But when I spoke to Professor Weyl he told me of a letter he had from Professor Fubini referring to $50 which he was promising or paying to Dr. Kraus for Miss Castelnuovo's attendance at the Friends' "American Seminar". Professor Weyl sends you the enclosed copy of his reply to Professor Fubini. As this apparently makes up the required $150, Professor Weyl and I thought it best to postpone any further action at least until you had been informed, and Professor Lefschetz agreed. The pledges so far are as follows:

Prof. Veblen $50;

Prof. Fubini 50;

Prof. Alexander 25;

Prof. Lefschetz 25.

[Total] $150.

Professor Lefschetz suggested that his name might be used in approaching Professors Zariski, Lane of Chicago, H. Levy, Murnaghan, Schilling and E.B. Stouffer; and that I send information also to Profs. Einstein, Weyl, Fite, Ritt, Graustein, Coble and Virgil Snyder.[45] I notice also the attached memoranda by you in Miss Castelnuovo's file; does this mean that Dr. Louise Pearce, Dr. Florence Sabin and Mr. Harrison might also be approached in case of need?[46] Sincerely, Gwen Blake[47]

Castelnuovo.14 G. Fubini to O. Veblen, Princeton 18.6.1940

OVP, *GC*, fol. 1r
My dear Professor Veblen, I thank you for the news you gave me about Miss Castelnuovo, and I send you the enclosed check of $50. We all are struck by the European news; I endeavour not to think about and to spend my time by thinking only of mathematical problems. We will go, at the end of the month, to the village

---

[45] Ernest Preston Lane (1886–1969), chairman of the Department of Mathematics of the University of Chicago (1941–1946). Hans Lewy (1904–1988) came to the United States from Göttingen, spent the years 1933–1935 at Brown University and in 1935 he moved to the University of California, Berkeley. Lewy had met Gina Castelnuovo in 1929 when using a Rockefeller Fellowship to study algebraic geometry in Rome with Enriques. Francis Dominic Murnaghan (1893–1976), head of the Department of Mathematics of the Johns Hopkins University (1928–1948). Otto F. G. Schilling (1911–1973) completed his studies at Marburg in 1934, spent two years at the Institute for Advanced Study (1935–1937) and two years at Johns Hopkins (1937–1939), before moving to the University of Chicago. Ellis Bagley Stouffer (1884–1965), professor at the University of Kansas (in Lawrence) from 1914, was head of the Graduate School (1921–1955). Stouffer had come to Italy to meet the Italian geometers in 1926 and 1928. Joseph Fels Ritt (1893–1951) professor at Columbia University and head of the Department of Mathematics (1942–1945). William Caspar Graustein (1888–1941), chairman of the Division of Mathematics of Harvard University (1932–1937), and assistant Dean of the Faculty of Arts and Sciences (1939–1941). Arthur Byron Coble (1878–1966), head of the Department of Mathematics at the University of Illinois at Urbana-Champaign (1934–1947). There is probably a typo: William Benjamin Fite (1869–1932), instructor then assistant professor at Cornell until 1911, then at Columbia, had passed away in 1932.

[46] Louise Pearce (1885–1959) was a pathologist at the Rockefeller Institute from 1913 to 1951. Florence Rena Sabin (1871–1953) was the first woman to hold a full professorship at Johns Hopkins School of Medicine (1902–1925), the first woman elected to the National Academy of Sciences, and the first woman to head a department at the Rockefeller Institute for Medical Research (1925–1938).

[47] Gwen Blake was the secretary of the School of Mathematics at the Institute for Advanced Study.

of Placid Lake in the State of N.Y. and I will be very glad, if it will be possible, to make a trip as far as Brooklin. The best greetings of my wife and of my own to Mrs. Veblen and to you. Yours truly Guido Fubini

My address will be: The "William Hurley Cottage" 10 Pine Street—Village of Lake Placid Essex County (N.Y.). May I hope to see Mrs. and Mr. Veblen there?

Castelnuovo.15 E. Witschi to O. Veblen, Iowa City 14.7.1940

OVP, *GC*, fol. 1r-v

Dear Dr. Veblen, The deplorable refusal of our administration to make an application in behalf of a fellowship for Miss Castelnuovo was not based on any misunderstanding. Full information about the past and present [curriculum] vitae of the candidate had been submitted. It must rather be taken as an expression of the distrust harassed by so many Americans against "foreigners" in general and dislike of Jews in particular. I am putting this statement in non-diplomatic language since I consider this a purely personal letter to one who, as an alumnus, takes a special interest in matters of our university.[48] In the meanwhile, Dr. Gilmore has been replaced by Dr. Ch. Phillips, former Dean of the School of Commerce, now Acting President of our University.[49] This change is not likely to affect the present situation very much, since the deciding vote, probably, was that of Dr. J.H. Bodine, Head of our Zoology Department.[50] While I am directing a quite independent section of Embryology and Endocrinology, any administrative affairs have to go through Dr. Bodine's office and from there on to the Deans and the President. Along this bureaucratic course many well intended projects have found an inglorious end. I was interested to hear that miss Castelnuovo is at the present with Dr. Little at the Bar Harbor Laboratory.[51] I think it would be very fine if the Emergency Committee could decide to give her some support enabling

---

[48] Veblen was born in Decorah, Iowa, and did his undergraduate studies at the University of Iowa, where he received an AB in 1898.

[49] Eugene A. Gilmore (1871–1953), professor of Law, served as the 12th president of the University of Iowa (1934–1940). Chester A. Phillips (1882–1976), first Dean of the University of Iowa College of Commerce (1921–1950), served as *interim* president of the University of Iowa in 1940, between the administrations of Eugene Gilmore and Virgil Hancher.

[50] Joseph Hall Bodine (1893–1954), professor and chairman of the Department of Zoology at the State University of Iowa (1929–1954).

[51] Gina Castelnuovo had obtained a temporary job at the Jackson Laboratory for Mammalian Genetics in Bar Harbor, Maine. The Laboratory was directed by Clarence Cook Little (1888–1971), an American genetics, cancer, and tobacco researcher and a former University of Maine and University of Michigan president. In 1929 Little had founded the Jackson Laboratory in Bar Harbor, Maine, under the name Roscoe B. Jackson Memorial Laboratory with the purpose of discovering the causes of cancer and other diseases through research on mammals.

her to stay there for a year. I have written in this sense also to Mr. Haber.[52] I got first interested in Dr. C. through some of my Italian friends, esp. Dr. Montalenti, Director of Naples Station and Dr. D'Ancona, Padua (Mrs. D'Ancona is the daughter of Prof. Volterra, Rome, whom you may know).[53] Recently I met her at a meeting in New York and was very favorably impressed by her personality. With sincere regards, yours Emil Witschi

Castelnuovo.16 G. Castelnuovo to O. Veblen, Salisbury Cove 24.8.1940

OVP, *GC*, fol. 1r

Dear Dr. Veblen, I have just received a offer to go to work at the Missouri University – Columbia Missouri in the Dept of Zoology as researcher and technician. I have accepted at once, though not having an idea what kind of place it is, and what kind of people is there. I have to work for the Head of the Dept of Biology who is Dr. Mary Guthrie.[54] She wants that I work for her in some researches that she is doing, aiding her for the microscopical study and for preparation of the slides. It has been offered me the salary of Ds 1200,00 for a year. I have accepted at once, also if it is not a beautiful offer, in view of the present difficulties. I hope not to have been wrong in doing so. With Dr. Little in Bar Harbor, I had lost many hopes, because last time I talked with him, he changed very much for the first time, and he did not confirm at all the promises which he did the first time. Dr. Dahlgren advised me also to accept the offer of the Missouri University.[55] It seems for me now very strange to have to go so far away, in a place in which I don't know anybody. But anyway, I hope that in the future I can find something better somewhere. I hope not to be obliged leave here at once, as they wanted, and I asked if it was possible for me to go there about the middle of September in a way that I can finish before my works begun here. I hope that this will be possible. If you know anybody in Columbia let me know, because I should be very glad to know that some known person is there. I hope

---

[52] William Haber (1899–1988), professor of Economics at the University of Michigan in Ann Arbor from 1937, held important posts in U.S. government bodies, including that of chairman of the National Committee on Long Range Work and Relief Policy (1941).

[53] Giuseppe Montalenti (1904–1990), assistant of Zoology at the University of Bologna, 1933–1937. In those years he had won two competitions but could not be called because the fascist demographic laws prescribed that, to hold a professorship, it was necessary to be married. In 1939 he had moved to the Anton Dohrn Zoological Station in Naples. Umberto D'Ancona (1896–1964), professor of Zoology in Padua (1936–64). On 22 July 1926 he had married Luisa Volterra, daughter of the mathematician Vito and also a biologist, who collaborated with him for many years.

[54] Mary Jane Guthrie (1895–1975), professor of Zoology at the University of Missouri in Columbia (1939–1950).

[55] Ulric Dahlgren (1870–1946), professor of Biology at Princeton (1911–1939), trustee and president of the Mount Desert Island Biological Laboratory until 1937.

that you too will think that I have not taken a mistake accepting this offer. I hope really very much before leaving to be able to see you; going so far away I really don't know when I will be able to come to visit you in Princeton. With my best regards to you and to Mrs. Veblen, yours sincerely Gina Castelnuovo

Castelnuovo.17 O. Veblen to G. Castelnuovo, [Princeton] 3.10.1940

OVP, *GC*, fol. 1r
Dear Miss Castelnuovo: I was very glad to have your letter telling me about your arrival in Columbia. Since I myself grew up in Iowa, Columbia does not seem to me at all far away from the world, and I hope that you are going to like it. I think, with what can be described as a real job, you are making a better start than you would have been making in one financed by a philanthropic foundation. Even though your work is supposed to be exclusively mechanical, you will doubtless have opportunities to improve your English and to learn the ways of the indigenous student. My wife joins me in best greetings. Yours sincerely, Oswald Veblen

P.S. I have written to Professor Blumenthal of the Mathematics Department to mention your existence.[56]

5.2   The Volterra Family

Another dispersed family was that of Volterra. The only son of Abramo and Angelica Almagià, registered under the name of Samuel Giuseppe Vito, 'Mister Italian Science' (as he would come to be called abroad) was born into a community, Ancona, which in 1860 counted 1800 inhabitants, mostly residing in the ghetto.[57] His maternal grandfather, Vito, had also been a teacher in the local Jewish school while his maternal uncle Saul was a pillar of the community, so much so as to be elected deputy (the official representative of the Jewish community of the city). Volterra's mother was deeply religious, but her child grew up in the new era of emancipation. The Italo-Jewish roots of the name (although he would use only the name Vito), his record in the annals of the community, and a Jewish catechism, are the only concrete traces of Volterra's ties to the religious tradition of his ancestors.

Volterra's career was brilliant: full professor of Mathematical Physics at the age of 23 in Pisa, eight years in Turin on the chair of Rational Mechanics (1892–1900),

---

[56] Leonard Mascot Blumenthal (1901–1984), PhD in 1927 from Johns Hopkins University, a visiting scholar at the Institute for Advanced Study from 1933 to 1936, taught for the majority of his professional career at the University of Missouri.

[57] On Volterra's biography see Goodstein (2007); Guerraggio and Paoloni (2008); Coen (2008).

then the arrival in Rome in 1900, thanks to Castelnuovo. One of the most outstanding mathematicians of his time, with contributions ranging from analysis to mathematical physics, from celestial mechanics to mathematical biology, Volterra began very early to take an interest in the relationships between scientific research, economic development, and cultural and civil progress. It was this vision of science as a factor of growth of society and the economy which he identified as the motivation behind his launch into Italian political life. Named by the king Vittorio Emanuele III as a senator of the Kingdom of Italy in 1905 and Dean of the Faculty of Sciences of Rome in 1907, he promoted important teaching and research structures such as the Italian Society for the Progress of Sciences (1907) and the Italian Thalassographic Committee (1910).

An interventionist since the attack in Sarajevo, Volterra enlisted at the age of 55 as a volunteer in the Aeronautical Engineers in May 1915. For three years, he devoted himself entirely to war research, first dealing with shooting boards and then, at the invitation of G. A. Crocco, to studies connected with the air force and military airships. For the complexity of these activities, he obtained an Honourable Mention and the War Merit Cross. At the beginning of 1917, he set up the War Inventions Office and became its president, making several trips to France and England to foster scientific and technical collaboration with the Allies.

Since 1880, when as a student he had met G. Mittag-Leffler in Pisa, Volterra had carefully nurtured international partnerships. In 1899, he went to Paris for the first time, and in 1902 to the United Kingdom. In both countries, where he returned regularly over time, he made important friends: Poincaré, Picard, Borel, Painlevé, A. R. Forsyth, E.T. Whittaker, and many others. In 1909, he embarked on his first journey across the United States. He returned in 1912, and in 1919 for the last time, briefly visiting Clark University, the Rice Institute, the universities and research centres in Illinois, Chicago, Princeton, and Harvard, and pushing himself into South America. Also overseas, he had admirers and long-time friends, including Birkhoff, Evans, Veblen, Lovett, G. E. Hale, etc.

Married from 1900 to Virginia Almagià, Volterra was the father of four children: Luisa (1902–1983), Edoardo (1904–1984), Enrico (1905–1973), and Gustavo (1909–2001) (Fig. 5.2).

Casa Volterra, like those of Enriques and Castelnuovo, saw the elite of world science and culture pass through its living room: Mittag-Leffler, Hadamard, Klein, Birkhoff, and hundreds of others all preserved an indelible memory of Volterra's legendary hospitality in Rome and in his villa in Ariccia, and were linked to him by unfailing gratitude.

Volterra's prestige reached its peak in the early 1920s, when he was elected president of the *Società dei XL* (1920), the Lincei Academy (1923), and the newborn National Research Council (1923). Among the few who immediately recognised the threat to freedom of thought implicit in Mussolini's government already from the march on Rome, Volterra openly expressed his concerns. In 1923 he opposed the Gentile reform in his quality as president of Lincei and in 1925 he signed the *Manifesto of anti-fascist intellectuals*. From that moment on, inexorable blacklisting began against him, in order to deprive him of all institutional positions and to erase

Fig. 5.2 From left to right: Gustavo Volterra, Emilia (his wife), Vito Volterra, Virginia Volterra, Luisa Volterra, Silvia D'Ancona (her daughter) and Umberto D'Ancona (her husband), 1930, CA, Enrico Volterra Papers

his presence from public life. Between 1926 and 1928, Volterra was removed from the presidency of Lincei and the CNR; from 1930, when Parliament was abolished, he participated only sporadically in the activity of the Senate; in 1931, having refused to take the oath of allegiance to fascism, he was expelled from the chair.

The following year he was forced to resign from all Italian scientific academies to which he belonged. Volterra reacted to the 'invisibility' to which he was condemned by devoting himself to scientific work with renewed intensity. His fundamental contributions to biomathematics date back to this period, also stimulated by the collaboration with his son-in-law, Umberto D'Ancona, full professor of Zoology at the University of Siena, Pisa and Padua. His trips abroad became more and more frequent, and Volterra spent most of his time in Paris, where he held courses and conferences, and where he was elected President of the *Bureau International des Poids et Mesures*. In November 1936, he was appointed pontifical academician by the decision of Pius XI.

In 1938, anaesthetised by the marginalisation suffered, Volterra scarcely reacted to racial fascist politics. Officially, he was not affected by the racial laws, because he had already been dismissed in 1931. As a Jew, however, a series of supplementary humiliations was inflicted on him, for example by former students such as Fantappié, who turned their backs on him from one day to the next and who prevented Volterra from accessing the Institute of Physics and the annexed library. Compared to Levi-Civita, Enriques, or Castelnuovo, the professional downgrading and the obscuring of his own image were, however, less unexpected, and more forewarned, because

Volterra had already been cancelled *de facto* by the regime and had lived a good part of the years 1931–1937 in a kind of self-imposed exile in France.

Volterra's correspondence does not allow us to truly identify his motions and emotions in the dark days of racial persecution. Some allusions surface in his letters to J. Pérès regarding the 'very bad moments for the present and in view of the future' that the Volterra family was experiencing and to the negative consequences for his health that these troubles and sufferings caused.[58] Borel, Pérès, and Picard were the only ones to whom Volterra opened his heart, receiving in turn affectionate comfort. Borel, for example, wrote to him at the end of November: 'you can be sure that my heart is with you and with yours'.[59] Volterra, however, continued his work, giving lectures at the Sorbonne, at the Poincaré Institute, and completed (with his son Enrico) a volume of the *Collection de physique mathématique* edited by É. Borel and M. Brillouin.[60]

Emigration was not contemplated by a man approaching the age of eighty, but also in the global narrative of the Jewish mathematical emigration from Nazi-Fascist Europe, Volterra is objectively a marginal actor. Apparently, the only testimonial he signed in the years from 1933 to his death in 1940 was for H. Geiringer.[61] He also offered some modest assistance to some of his best students, 'pushed into every part of Europe by the gloomy shadows of the events that were pressing'.[62] But there was little, or nothing more: no traces of an endorsement in favour of any Italian mathematician, and even for the closest pupils like Supino, have emerged. There was no intervention in favour of the young physicists of via Panisperna. Perhaps Volterra's help was not requested, since he was considered a great old man, not to be bothered with indiscreet requests and mundane affairs. It may also be that a recommendation letter by Volterra was not considered really useful or effective. Maybe the phlebitis, from which he suffered from December 1938 and that made him lose the use of his legs, kept Volterra out of the dynamics of international solidarity. The fact is that, in the rescue chains that were mobilised for Italians, Volterra remained in the background.

Family, on the other hand, was at the centre of his concerns. The weeks following the promulgation of the racial laws were destined entirely towards two things: collecting the documents necessary to apply for counter-discrimination and taking

---

[58] ANL-*Volterra*: V. Volterra to J. Pérès, Ariccia 11.10.1938: *très mauvais moments pour le présent et en vue de l'avenir.* Joseph Pérès (1890–1962), a pupil of Borel, had worked for his doctorate under the guidance of Volterra in 1913–14.

[59] ANL-*Volterra*: É. Borel to V. Volterra, Paris 29.11.1938: *Vous pouvez être certain que je suis de cœur avec vous et les vôtres.* See also Volterra.1, Volterra.2, Volterra.3, and in ANL-*Volterra*: É. Borel to V. Volterra, 24.12.1938; V. Volterra to É. Picard, Rome 11.3.1939, É. Picard to V. Volterra, Paris 15.3.1939.

[60] See e.g. ANL-*Volterra*: V. Volterra to J. Pérès, Ariccia 11.10.1938, E. Volterra to J. Pérès, Ariccia 23.10.1938, and V. Volterra to J. Pérès, Ariccia 3.11.1938.

[61] Volterra.5.

[62] Krall (1955), p. 24: *sospinti in ogni parte d'Europa dalle cupe ombre degli avvenimenti che incalzavano.*

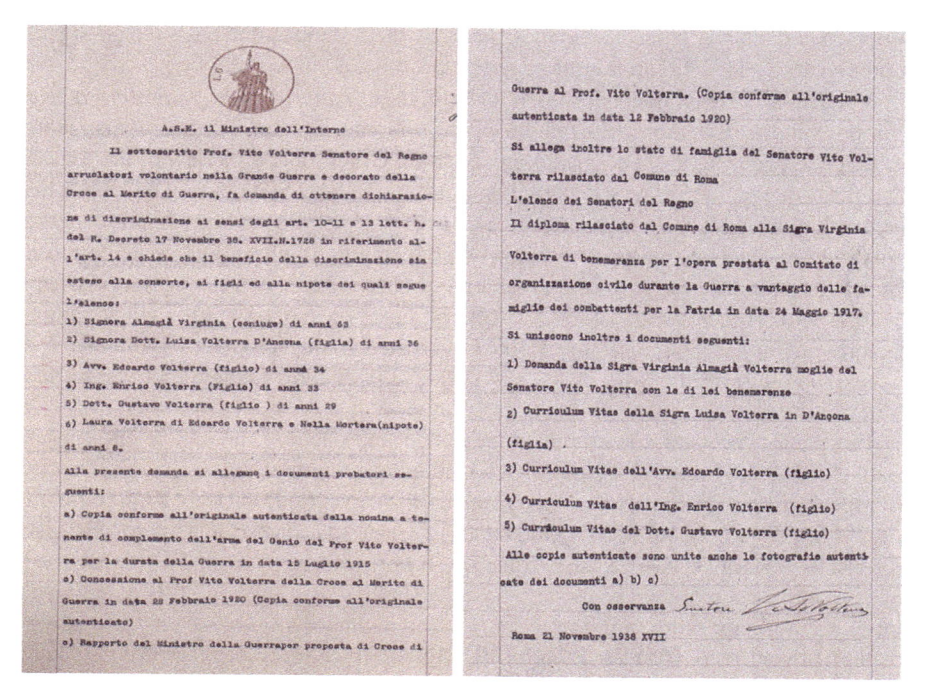

Fig. 5.3 Application for counter-discrimination by V. Volterra, ANL-*Volterra*

measures to safeguard the family heritage. Vito committed himself to the first task while his wife took care of the second, dismissing the servants and selling the properties, thereby in fact saving a major part of the assets of the Almagià–Volterra families. At the end of November 1938, Volterra applied for counter-discrimination for himself, wife, and children: Edoardo, full professor of Roman Law in Bologna, Luisa a researcher in biology, Enrico a physicist-mathematician and an engineer, and Gustavo an engineer (Fig. 5.3). He trusted in the positive outcome of the application, and rightly so, since the Senate informed him in January that the undersecretary of State for the Interior had ruled on the counter-discrimination of all senators belonging to the Jewish race and had also extended this recognition to their relatives.

For the Volterras, however, this last provision was of little significance because, by that time, the family had dispersed. Volterra's eldest son Edoardo had moved to Paris. Enrico reached London, from whence he intended to embark for America. Both were in extremely precarious conditions, which is why (with considerable effort) Volterra gathered his last remaining energy to help them. At first he wrote to W. H. Bragg in the hope that he would find an opening for Enrico in Scotland or at the Davy Faraday Research Laboratory of the Royal Institution in London.[63] He received no concrete proposals of help. He then turned to E. O. Lovett and O. Veblen, who involved L. P. Eisenhart and G. E. Beggs.[64] Also in this case, no success. Villat and Pérès proposed to invite Vito and Enrico in Paris to the *Institut de Mécanique*.[65] In mid-July hope appeared on the horizon: thanks to Levi-Civita, Enrico was called to direct the Institute of Stability at the University of Rosario. It seemed the definitive solution, but it was not. Due to a series of bureaucratic impediments and buck-passing games by the Italian and Argentine diplomatic authorities, Enrico would not manage to obtain the entry visa. The appeals to the Argentine academic authorities, to the Holy See, the consulates in Rome, Paris and London all failed, one after the other. The help of Beppo Levi, who had succeeded in entering Argentina with a tourist visa, was also ineffective, because Enrico refused to purchase it.[66] In the meanwhile, E. Sears invited him to the National Physical Laboratory of Teddington Middlesex, and promised to give him some advice and assistance to find a place in the United Kingdom.[67] Even this overture, however, did not lead to concrete results. In the end, thanks to C. E. Inglis, Enrico applied for a maintenance grant at the Engineering Department of Cambridge University. He would earn a PhD from King's College, in 1941.[68]

The older brother, Edoardo, at the time of the dismissal had better relocation prospects. Six months later he was still in Paris with his family, a displaced scholar for racial reasons in a group of political exiles that was thinning out day by day. Despite being considered one of the world's leading experts on Roman Law, he was more difficult to place, and all attempts made to find a position for him in the United States proved unsuccessful. Veblen, von Neumann and others then suggested to look towards Latin America:

> Concerning Volterra's son: don't you think that his chance is in South America? Reasons: 1) It seems that they are taking in there refugees from 'Latin' European countries. 2) Their legal systems are based on Roman, and not on English Common, Law. 3) As far as I know, jurisprudence plays a great role in their politico-academical lives.[69]

---

[63] VOLTERRA.7 and ANL–*Volterra*: W. H. Bragg to V. Volterra, London 5.4.1939. William Henry Bragg (1862–1942) was professor of Chemistry at the Royal Institution and Director of the Davy Faraday Research Laboratory from 1923.

[64] VOLTERRA.4, VOLTERRA.6 and OVP: O. Veblen to E. O. Lovett, [Princeton] 28.3.1939. Edgar Odell Lovett (1871–1957), president of Rice Institute (1912–1946), was a close friend of Volterra. In autumn of 1912 Volterra had held a series of lectures at the Rice Institute. L. P. Eisenhart (1876–1965) was head of the Mathematics Department at Princeton (1928–1945) and George Erle Beggs (1883–1939) chaired the Department of Civil Engineering at Princeton University (1914–1939).

[65] ANL-*Volterra*: J. Pérès to V. Volterra, Paris 5.3.1939.

[66] VOLTERRA.10–14.

[67] ANL-*Volterra*: J.E. Sears to V. Volterra, Teddington Middlesex 17.7.1939.

[68] VOLTERRA.8 and VOLTERRA.21. Charles Edward Inglis (1875–1952) was head of the Engineering Department of the Cambridge University from 1919 to 1943.

[69] OVP, *JN*: J. von Neumann to O. Veblen, Seattle 16.7.[1940].

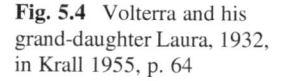

Fig. 5.4 Volterra and his
grand-daughter Laura, 1932,
in Krall 1955, p. 64

Even in this case, however, no success. Volterra's last desperate attempt to help
his son dates to shortly after Italy's entry into the war. Once again he turned to
Veblen and Birkhoff. Veblen contacted Rosenblatt, Father O'Hara, and the American consul in Naples.[70] These last rescue attempts were also unsuccessful. In the
summer of 1940, Edoardo returned to Italy with his wife and daughter Laura, and
from that moment he entered the anti-fascist resistance, taking considerable personal
risks (Fig. 5.4).

Arrested in Bologna in 1943, he was freed on 26 July, after the fall of Mussolini.
Already in connection with the anti-fascist forces, he moved to Rome to join the
partisan war, while his family hid in Roman convents. Reinstated in the chair in
1944, he became Rector of the University of Bologna in 1945 and was a leading
figure in the cultural reconstruction of the country.

The last year of Volterra's life was particularly painful. Bedridden, he could not
take part in Borel's Jubilee celebration. He was aware that he would no longer be
able to move, although his French colleagues reassured him that they would soon

---

[70] VOLTERRA.15-20 and OVP: O. Veblen to G. Blake, Princeton 15.7.1940.

meet again: 'perhaps the 1942 exhibition will provide the occasion; from here to then the horizon will undoubtedly have brightened'.[71] He had been unable to help his children and, meanwhile, one of them had been interned as an enemy alien on the Isle of Man.

Volterra died in Rome on 11 October 1940. According to his wishes, he was buried in the small cemetery of Ariccia, near the country villa that he had loved and where he had spent the most serene hours of his life. He was commemorated abroad, from France to the United States, from England to Argentina, as one of the greatest mathematicians that Italy had ever had. Yet, in Italy, he received no official commemoration. Indeed, in October 1943 the Nazi-Fascists knocked on the door of the Volterra house to deport him with his family, not knowing that he had passed away. Only in 1946 did Castelnuovo, as president of the reconstituted Lincei, decide to open the first session after the liberation with a moving dedication to Volterra:

> A weakness, a minimal gesture of consent to the dominant party, would have given him back influence and honours and would have brought him back to the leadership of the Italian scientific movement. But he did not bend. [. . .] He preferred to remain faithful to the ideals that had inspired his whole life, ideals of freedom, respect for the opinions of others, love for individuals and among peoples. Thus, ended in October 1940, between the indifference of the ruling class and the silence of our press, the life of one of the greatest sons of Italy, of whom we do not know whether to more admire the greatness of mind or the nobility of character.[72]

Documents

Volterra.1 V. Volterra to É. Borel, Ariccia 7.11.1938

ANL-*Volterra*, fol. 1r
My dear Friend, Thank you for your kind letter of 30 October and for all the kind words you addressed to me. The Duchess of Clermont Tonnerre was so kind to come and visit us here, and I was happy to see her again, to hear from you and to speak with her on so many subjects.[73] Among other things, I understood from her

---

[71] ANL-*Volterra*: É. Borel to V. Volterra, Paris 3.5.1940: *Peut-être votre exposition de 1942 ni en fournira-t-elle l'occasion; d'ici là, l'horizon se sera sans doute éclairci.* See also VOLTERRA.9 and ANL-*Volterra*: V. Volterra to É. Borel [first days of January 1940]; É. Borel to V. Volterra, Paris 18.1.1940.

[72] Castelnuovo G. (1946), p. 9: *Una Sua debolezza, un minimo gesto di consenso al partito dominante, Gli avrebbe ridato influenza ed onori e Lo avrebbe portato nuovamente alla direzione del movimento scientifico italiano. Ma Egli non si piegò. [. . .] Preferì rimaner fedele agli ideali che avevano ispirato tutta la Sua vita, ideali di libertà, di rispetto delle opinioni altrui, di amore tra gli individui e tra i popoli. Così si è chiusa nell'ottobre del 1940, tra l'indifferenza della classe dirigente e il silenzio della nostra stampa, la vita di uno dei maggiori figli d'Italia, del quale non sappiamo se più ammirare l'altezza dell'ingegno o la nobiltà del carattere.*

[73] Antoinette Corisande Élisabeth de Gramont, Duchess of Clermont-Tonnerre (1875–1954) was a French writer.

words that you are interested in my son Edoardo. How grateful I would be if you could do something for him at this difficult time! It would be a great relief for me to know that my son can count on your assistance. My health is not good; I am very weak and very tired. My wife does not want to take me away from the countryside because she fears it will increase my distress. For our trip to Paris, my wife's passport is not yet ready. I greatly regret the bereavements among our colleagues at the Institute. [. . .] I received a letter from Mr. Pérès. He told me that he had asked Mr. Thouzellier for the second volume of *Functionelles*.[74] I expect that the editor Gaut[hier] V[illars] will not delay sending us the first clichés. A word from you on this subject would be helpful. I very much wish not to miss seeing the end of this book which constituted a work in progress for so many years and formed the subject of the numerous conferences which I held in France thanks to your generous initiative. Please accept, my dear friend with Madame Borel the expression of my best memories and my affectionate devotion [V. Volterra]

Mon cher Ami, Merci de votre bonne lettre du 30 Octobre et de tous les aimables mots que vous m'adressez. La duchesse de Clermont Tonnerre a été si aimable de venir nous rendre visite ici et j'ai été heureux de la revoir, d'avoir de vos nouvelles et de m'entretenir avec elle sur tant de sujets. Entre autres j'ai compris par ses paroles que vous vous intéressez à mon fils Edoardo. Que je vous serais reconnaissant si vous pouviez faire quelque chose d'utile pour lui dans ce moment pénible ! Ce serait un grand soulagement pour moi de savoir que mon fils peut compter sur votre assistance. Ma santé n'est pas bonne ; je suis très faible et très fatigué. Ma femme ne veut pas m'éloigner de la campagne car elle craint augmenter mon agitation. Pour notre voyage à Paris le passeport de ma femme n'est pas encore prêt. Je regrette beaucoup les deuils parmi nos confrères de L'Institut. [. . .] J'ai reçu une lettre de Monsieur Pérès. Il me fait part d'avoir sollicité Monsieur Thouzellier pour le deuxième volume des Fonctionnelles. *Je compte bien que la Maison Gaut. V. ne voudra pas tarder à nous faire envoyer les premiers placards. Un mot de votre part à ce sujet serait utile. Je désire beaucoup ne pas manquer de voir la fin de cet ouvrage qui a constitué un travail poursuivi pendant de si longues années et a formé le sujet des nombreuses conférences que j'ai tenu en France grâce à votre généreuse initiative. Veuillez agréer, mon cher Ami avec Madame Borel l'expression de mes meilleures souvenirs et de mon affectueux dévouement* [V. Volterra]

---

[74]ANL-Volterra: J. Pérès to V. Volterra, Paris 16.10.1938. Étienne Thouzellier (1869–1946) was the *directeur général des éditions Gauthier-Villars* from 1920 to 1946. Gauthier-Villars had published the book by V. Volterra and J. Pérès, *Théorie générale des fonctionnelles. Tome 1: Généralités sur les fonctionnelles. Théorie des équations intégrales. Preface de Vito Volterra*, Collection de Monographies sur la Théorie des Fonctions publiée sous la Direction de M. Émile Borel, Paris: Gauthier-Villars, 1936. Other two volumes were in the pipeline: *Composition, Équations intégro-différentielles et aux dérivées fonctionnelles, Généralisations des Fonctions analytiques* and *Compléments et Applications*.

Volterra.2 V. Volterra to É. Borel, [Rome] 20.12.1938

ANL-*Volterra*, fols. 1r-2r

My dear Friend, in the end I received the passports with visas from the French consulate. A stay of three months is granted from the date of crossing the border and this can take place before 30 April. I really want to enjoy once again the sojourn, after the cold weather will have passed. But we also have a mass of issues to resolve in trying to save our assets from ruin. The only hope will be to get what is called discrimination, for myself and my family.[75] I would have the right to get it as a war volunteer and a man decorated with the War Merit Cross, but we have to wait. It was only today, through a letter forwarded to me by my son [Edoardo], that I was able to read a part of one of your letters (perhaps from last October) which in part concerned me. I very much regret having known it so late and not having thanked you yet for what you wrote. Your words touched me deeply. The proposals of help that you make, both through conferences and research activity, and the words with which you accompany them are such that I will accept them with gratitude if they became, as I believe, necessary. It is always in France, and through a friend like you, that I find effective help. Please extend best wishes from me and my wife, [V. Volterra]

Mon cher Ami, J'ai reçu enfin les passeports avec les visas du consulat français. Un séjour de trois mois est consenti à partir du passage de la frontière et celui-ci peut avoir lieu jusqu'au 30 Avril. Je désire beaucoup d'en profiter, le grand froid passé. Mais nous avons aussi une foule de questions à régler pour tâcher de sauver nos biens de la ruine. Le seul espoir sera d'obtenir ce qu'on appelle la discrimination de moi-même et de ma famille. J'en aurais le droit comme volontaire de guerre et décoré de la croix au mérite de guerre, mais il faut attendre. Ce n'est qu'aujourd'hui par une lettre qu'on m'a remis de la part de mon fils que j'ai pu lire un morceau d'une de vos lettres (peut-être de l'Octobre dernier) dont on avait fait un extrait qui me concernait. Je regrette beaucoup de l'avoir connu si tard et de ne vous avoir encore remercié de ce que vous avez écrit. Vos paroles m'ont profondément touché. Les propositions que vous faites soit pour les conférences que pour les recherches et les mots avec lesquels vous les accompagnez sont telles que je les accepterai avec reconnaissance si elles me seront, comme je crois, nécessaires. C'est toujours en France et par un ami tel que vous que je trouve un aide efficace. Veuillez présenter les meilleurs vœux de ma part et de la part de ma femme, [V. Volterra]

---

[75] See D2.11.5.

Volterra.3 V. Volterra to J. Pérès, Rome 10.3.1939

ANL-*Volterra*, fols. 1r-2v

My dear Friend, I am very grateful to you for your letters. Please forgive me if I have been so negligent in my answers. As you know I have been suffering from a very long and very painful illness. We try electric wave treatments, light baths, and all kinds of injections. I would like them to be effective so that I can get by. Thank you very much for all the kindness and the interest you have shown my son Enrico. Your proposals are very flattering and advantageous for him. He will answer you directly. Please accept my congratulations on your appointment to the editorial staff of the *Revue Générale des Sciences* and for your new position at the University.[76] I hope that you will be appointed as an examiner at the École Polytechnique leaving exams, which I have wanted for a long time. It is not necessary to tell you that I would be happy to give lectures at the Institute of Mechanics and I hope that my strength will allow me to do so. I will speak to Mr. Levi-Civita about your proposal. [. . .] I have good news from my family, but my family is dispersed. My son Edoardo is now in Paris where he works in libraries. I think you will have already seen or will see him. Forgive me if after such a long silence I am writing to you only to give you trouble. I send my best wishes for the recovery of your beautiful daughter. Please greet Madame Pérès for me and please accept, my dear Friend, my most affectionate attachment and all my esteem. [J. Pérès]

Mon cher Ami, Je vous suis très reconnaissant de vos lettres. Vous voudrez bien m'excuser si j'ai été si négligeant dans mes réponses. Comme vous savez je suis affligé d'une maladie très longue et très douloureuse. On essaie des traitements d'ondes électriques, bains de lumière et toute sorte d'injections. Je voudrais qu'ils fussent efficaces pour pouvoir m'en tirer. Merci beaucoup de toutes les amabilités que vous avez eu envers mon fils Enrico et de tout l'intérêt que vous lui avez démontré. Vos propositions sont bien flatteuses et avantageuses pour lui. Il vous aura répondu directement. Veuillez agréer mes félicitations pour votre nomination dans la rédaction de la Revue Générale des Sciences *et pour vos nouvelles fonctions à l'Université. J'espère que vous aurez la nomination d'examinateur de sortie à l'École Polytechnique, ce que je souhaitais depuis longtemps. Il n'est pas nécessaire de vous dire que je serais heureux de faire des conférences à l'Institut de Mécanique et je forme les vœux que mes forces puissent le consentir. Je parlerai à Monsieur Levi-Civita de votre proposition. [. . .] J'ai des bonnes nouvelles des miens mais ma famille est dispersée. Mon fils Edoardo est maintenant à Paris où il travaille dans les bibliothèques. Je pense que vous l'aurez déjà vu ou que vous allez le voir. Pardonnez-moi si après un si*

---

[76] In January 1939 Pérès had joined the Editorial Board of the *Revue Générale des Sciences* and had taken over the direction of a *Séminaire d'Aérodynamique appliquée*, in addition to his teaching at the University of Paris where he was Director of the Institute of Fluid Mechanics (1932–1954).

long silence je vous écris pour ne vous donner que des ennuis. Je forme les meilleurs vœux pour la guérison de votre belle fille. Veuillez me rappeler au bon souvenir de Madame Pérès et veuillez agréer, mon cher Ami, l'expression de mon plus affectueux attachement et de toute mon estime. [J. Pérès]

Volterra.4 E.O. Lovett to O. Veblen, Houston 23.3.1939

OVP, *EnV*, fol. 1r, 2r

My dear Professor Veblen: Professor Enrico Volterra, Dr. Eng. (Rome), second son of Senator Volterra, is seeking an academic appointment in the United States, and is prepared to start with a stipend that would meet his living expenses. He was born in Rome, June 11, 1905, is not married nor engaged to marry, has had wide experience in professional practice as a civil engineer, and until last September was holding a double appointment at the University of Rome, from which he was removed for racial reasons. He had been giving a course in construction in the School of Architecture, and one in graphical statics in association with Professor Levi-Civita of the chair of rational mechanics. These two courses of lectures are in published form, and I have copies of them.[77] I have also the author's account in English of his record, together with copies of nearly forty of his scientific papers, fourteen of which appeared in the *Rendiconti della R. Accademia Nazionale dei Lincei*, two in the *Comptes Rendus* of the Paris Academy of Sciences, and several in the proceedings of international congresses in which the author participated. Dr. Volterra left Italy some months ago, has been in Paris, where his address was the Hotel Lutetia, 43, Boulevard Raspail, and at present is in London, where his address is 30 Orsett Terrace, Hyde Park, London, W.2. I am enclosing herewith a copy of a recent signed photograph of the author, and I should be glad to place promptly at your disposal any or all of the above-mentioned documents. In my judgment young Volterra is a first-rate man, who would succeed in this country. I very much wish that I could arrange an appointment for him in this institution. With best wishes, I remain, very sincerely yours, Edgard O. Lovett

Volterra.5 H. Geiringer to V. Volterra, Istanbul 27.3.1939

ANL-*Volterra*, fol. 1r-v, 2r

Sir and revered Master, In Paris during the small Mathematical Congress of 1937, I had the honour of being introduced to you by Mr. de Mises. You were kind enough—a few weeks ago—to honour me by sending some of your important

---

[77]E. Volterra, *Corso di scienza delle costruzioni anno scolastico 1937-38*, Roma, Libreria Politecnica, 1938 and E. Volterra, *Nozioni di Statica Grafica*, Roma, Libreria Politecnica, 1937.

works. As I find myself in a rather precarious situation now, I dare to ask you for a favour. I have been in Turkey for five years as a mathematics teacher, alongside Mr. de Mises, and I was very happy to find the opportunity here to teach and continue my scientific research. Since my current contract with the Turkish government expires in October and I believe it will not be renewed—because of political developments in Central and Eastern Europe—I am forced to look for a new position which is now quite difficult. I thought about trying in the United States and especially in a College for Women and I now want to contact several institutions of this kind. It would be extremely useful to me if I could attach, to the file on which I will base my application, a few lines of recommendation from you. I am convinced that two words from your hand will make a decisive impression on the American authorities.[78] I do not dare hope that the separate offprints that I took the liberty of sending you over the last few years have attracted your attention, but I am counting on your kindness for victims of the current situation. I would like to attach to this letter a list of my publications and a curriculum vitae. I would be sorry, my dear and revered Master, to think that this request could annoy you. I hope that you will only find here a sign of my respect for the scientific and personal authority which you enjoy throughout the world, and of my confidence in your kindness. Expressing in advance to you my most sincere thanks, please accept my greatest respect, Hilda Geiringer

Monsieur et vénéré Maitre, A Paris pendant le petit Congrès Mathématique de 1937 j'ai eu l'honneur de vous être présentée par M. de Mises. Vous avez eu l'amabilité – il y a quelques semaines – de m'honorer par l'envoi de quelques-uns de vos importants travaux. Comme je me trouve pour le moment dans une situation assez précaire j'ose vous demander une faveur. Il a y cinq ans que je me trouve en Turquie comme professeur de mathématiques, a côté de M. de Mises, et j'étais très heureuse d'avoir trouvé ici la possibilité d'enseigner et de continuer mes recherches scientifiques. Étant donné que mon contrat actuel avec le gouvernement Turc expire en octobre et que je crois qu'il ne va pas être renouvelé – en conséquence de l'évolution politique en Europe centrale et orientale – je me vois forcée de chercher une nouvelle position, ce qui est maintenant assez difficile. J'ai pensé de l'essayer aux États Unis et surtout dans un College for Women et je veux m'adresser à présent à plusieurs institu-tions de ce genre. Il me serait extrêmement utile si je pouvais joindre au dossier sur lequel j'appuie ma candidature quelques lignes de recommandation provenant de vous. Je suis convaincue que deux mots de votre main feront une impression décisive sur les autorités américaines. Je n'ose pas espérer que les tirages à part que je me permettais de vous envoyer pendant les dernières années ont pu attirer votre attention mais je compte sur votre bonté pour les victimes de

---

[78]Volterra and Levi-Civita had met Geiringer at the *Réunion Internationale des Mathématiciens*, organized in Paris by the *Société Mathématique de France* in July 1937. Both wrote a testimonial for her. See also D2.2.2.

la situation actuelle. Je me permets de joindre à cette lettre une liste de mes publications et un curriculum vitae. Je serais désolée, mon cher et vénéré Maître, si je devais croire que cette demande vous dérange. J'espère que vous n'y trouverez que le signe de mon respect pour l'autorité scientifique et personnelle dont vous jouissez dans le monde entier, et de ma confiance en votre amabilité. En vous exprimant d'avance mes remercîments les plus sincères je vous prie d'agréer l'expression de ma plus haute considération Hilda Geiringer

Testimonial by V. Volterra

I am happy to be able to express my admiration for the mathematical talent of Ms. Hilda Geiringer. She worked in almost all areas of pure mathematics, and she also developed many interesting questions in applied mathematics. I will begin by citing a very important work on plasticity that she recently published in Monsieur Villat's *Mémorial des Sciences Mathématiques*.[79] It must be remembered that the theory of plasticity is one of the questions on the agenda in the world of engineers and scientists who deal with the science of construction. Ms. Geiringer, in recalling the old results of Boussinesq and others, does not fail to expose the most modern results and her original views. Many of Ms. Geiringer's works are devoted to classical mechanics, others to modern statistical theories. Nor did she neglect to carry out research into vibrations, which play such a great role today. Certain statistical theories developed by Ms. Geiringer have been used in the most popular textbooks. Several interesting works on the probability calculus were published by Ms. Geiringer, and she also dealt with mathematical philosophy, questions on education, high analysis and geometry. Her collaboration for several years with the *Jahrbuch über die Fortschritte der Mathematik* testifies to her versatility and the extent of her culture. I believe that the didactic work of Ms. Geiringer can always be treasured and to the benefit of any Institute of Mathematics.

Je suis heureux de pouvoir témoigner mon admiration pour le talent mathématique de Madame Hilda Geiringer. Elle a travaillé dans presque tous les domaines des mathématiques pures et elle a développé aussi beaucoup de questions intéressantes les mathématiques appliquées. Je commencerai par citer un ouvrage fort important sur la Plasticité qu'elle a publié récemment dans le Mémorial des Sciences Mathématiques *de Monsieur Villat. Il faut rappeler que la théorie de la Plasticité est une des questions à l'ordre du jour dans le monde des ingénieurs et des savants qui s'occupent de la science des constructions. Madame Geiringer en rappelant les anciens résultats de Boussinesq et d'autres ne manque pas d'exposer les résultats les plus modernes et ses vues originales. Beaucoup de travaux de Madame Geiringer sont consacrés à la Mécanique classique; d'autres aux théories modernes statistiques. Ella n'a pas non plus négligé de*

---

[79] H. Geiringer, *Fondements mathématiques de la théorie des corps plastiques isotropes*, Mémorial des sciences mathématiques dir. H. Villat, fasc. 86, Paris, Gauthier-Villars, 1937.

s'occuper de recherches sur les vibrations qui jouent aujourd'hui un rôle si grand. Certaines théories de statistique développées par Madame Geiringer ont été utilisées dans les ouvrages didactiques les plus répandus. Plusieurs travaux intéressant le calcul des probabilités furent publiés par Madame Geiringer, et elle s'occupa aussi de philosophie mathématique, des questions sur l'enseignement, de haute analyse et de géométrie. Sa collaboration pendant plusieurs années au Jahrbuch über die Fortschritte der Mathematik *témoignent de sa versatilité et de l'étendue de ses connaissances. Je crois que l'œuvre didactique de Madame Geiringer pourra être toujours utilisée avec profit dans tout Institut de Mathématique.*

Volterra.6 O. Veblen to G. E. Beggs, [Princeton] 28.3.1939

OVP, *EnV*, fol. 1r

Dear Beggs: I wonder if you can think of anything to do in response to this letter from Lovett? Vito Volterra, the father of the man in question, has been recognized during the last twenty years as the leading mathematician of Italy. He was a Senator and a member of all the important academies, etc. He is now very old, and also quite ill. I have doubtless met the son, but that was so long ago that I really know nothing about him except what is contained in this letter. Undoubtedly, however, from his record he must be a man of very great distinction. Any suggestion which you can make will be much appreciated. Yours sincerely, Oswald Veblen

Volterra.7 V. Volterra to W. H. Bragg, Rome 1.4.1939

ANL-*Volterra*, fol. 1r

Dear Sir William, I wish to express to you all my sincere gratitude for your great kindness to my son Dr. Enrico Volterra who is in London at present. I hear that you have taken an interest in him and that you have been of a great help with introducing him to many scientists and technicians. Your kindness has availed him a great deal. My health conditions have not been good these last few months, but I hope to be soon able to get up and take up again my active life. I renew all my best and sincerest thanks and remain yours truly Vito Volterra

Volterra.8 C. E. Inglis to V. Volterra, Cambridge 22.11.1939

ANL-*Volterra*, fol. 1r-v
Dear Dr. Volterra, It was most kind of you to write your thanks for the little I have done to help your son in Cambridge. I appreciated your letter very much indeed, and I shall certainly do what I can to further the interests of your son. I was very pleased to hear that his application for a maintenance grant had been successful and so far as this Department is concerned, we shall remit all fees which are ordinarily charged. Of course, I am familiar with your great contributions to scientific knowledge, and I regard it as a privilege to be of some assistance to your son. With kindest regards I am sincerely yours C.E. Inglis

Volterra.9 P. Montel to V. Almagià Volterra, Paris 29.12.1939

ANL-*Volterra*, fol. 1r-v
Dear Madam. Very often my thoughts went to you, Mr. Volterra and all of yours. I did not express them because they did not tell you anything new about the feelings of respectful affection that I feel towards you. I heard about you six weeks ago from Mr. Carcopino who sometimes visits us.[80] I was returning at that time from a three-month trip to America. It was in Argentina that the war surprised me, in Argentina where Universities, like Tucumán for example, would be happy to have foreign scholars as professors, but are not always followed by the government in their choices. After a short stay in Chile and Peru, I returned to France via New York. French scientific life continues with the modifications that circumstances have brought to it. A few days ago, we celebrated Mr. Picard's fiftieth academic anniversary. On January 14, we will have Mr. Borel's jubilee.[81] The great voice of Mr. Volterra will not be heard there but we all know that his thoughts will be there. "Majesty of the King, absent but present", as d'Annunzio said to that rock which is Quarto. News from you will be welcome. They will spread among your friends where they will be received with keen interest. Don't deprive us of them. Please accept, Madam, with my New Year wishes for yourself, your family, and your country, my deep respect and faithful affection. Paul Montel[82]

---

[80] Jérôme Carcopino (1881–1970), historian, directed the *École Normale Supérieure* from 1940 to 1942 and served as Rector of the University of Paris, after the dismissal of Gustave Roussy following the student manifestations of November 11, 1940.

[81] Institut de France, Académie des Sciences, séance du lundi 11 décembre 1939: *Commémoration du cinquantenaire académique de M. Émile Picard, Secrétaire perpétuel pour les sciences mathématiques*, Paris, Gauthier-Villars, 1939. The Scientific Jubilee of Émile Borel was celebrated at the Sorbonne on Sunday January 14, 1940.

[82] Paul Montel (1876–1975), a student of É. Borel, worked mostly on the theory of analytic functions of a complex variable and had been in contact with Volterra from the early 1920s.

Chère Madame, Bien souvent ma pensée est allée vers vous, M. Volterra et tous les vôtres. Je ne l'ai pas exprimée parce qu'elle ne vous avérait rien appris de nouveau sur les sentiments de respectueuse affection que j'éprouve à votre égard. J'ai eu de vos nouvelles, il a six semaines, par M. Carcopino qui nous rend parfois visite. Je rentrais à ce moment d'un voyage de trois mois en Amérique. C'est en Argentine que la guerre m'a surpris, en Argentine où les Universités, comme Tucumán par exemple, seraient heureuses d'avoir comme professeurs des savants étrangers, mais ne sont pas toujours suivies par le gouvernement dans leurs choix. Après de courts séjours au Chili et au Pérou, je suis rentré en France en passant par New-York. La vie scientifique française continue avec les modifications que les circonstances lui ont apportées. Il y a peu de jours, nous avons fêté le cinquantenaire académique de M. Picard. Le 14 janvier, nous aurons le jubilé de M. Borel. La grande voix de M. Volterra ne s'y fera pas entendre mais nous savons tous que sa pensée y sera. «Maestà del Re, assente ma presente», comme disait d'Annunzio au rocher de Quarto. Des nouvelles de vous seront les bienvenues. Elle se propageront dans le milieu de vos amis où elles seront reçues avec avidité. Ne nous en privez pas. Voulez-vous agréer, Madame, avec mes vœux de nouvel an pour vous-même, votre famille, votre pays, l'hommage de mon profond respect et de mon fidèle attachement. Paul Montel

Volterra.10 B. Levi to V. Volterra, Rosario 23.5.1940

ANL-*Volterra*, fol. 1r
Dear Professor, I send you greetings. Regarding the business of Prof. Enrico, I have no news. Since nothing came by the way of the nunciature, I wrote again to B[uenos] Aires at the address of Di Brocchetti, who had written to me that he would be returning to B[uenos] Aires by the end of April. I received no answer from him either. Out of prudence I did not mention to him that we had also tried another way, so I believe that at least a courtesy answer should have arrived. The lack of response therefore makes me think that these are issues of slowness and delay rather than insurmountable difficulties. So, I continue to hope. I would add that a little slackening in processes may also have been caused by the fact that the university authorities were concluding their term in these days: now we have had the elections, which have also gone very favourably, and this also allows us to hope. With affectionate regards B. Levi

Caro Professore, Le mando un saluto. Per l'affare del prof. Enrico non ho notizie. Visto che non giungeva nulla per la via della nunziatura, scrissi nuovamente a B. Aires all'indirizzo del Di Brocchetti, che mi aveva scritto che sarebbe stato di ritorno a B. Aires per la fine di aprile. Anche da lui non ricevetti risposta. Per prudenza non gli feci alcun cenno che si era tentata anche altra via, per cui debbo ritenere che per lo meno una risposta di cortesia debba arrivare.

La mancanza di risposta mi fa quindi pensare che si tratti piuttosto di lentezze che di difficoltà insuperabili. Continuo dunque a sperare. Aggiungo che un piccolo rilassamento nella pratica può anche essere stato causato dal fatto che le autorità universitarie scadevano di carica in questi giorni: ora abbiamo avuto le elezioni, che sono anche andate molto favorevolmente, ed anche questo permette di sperare. Con saluti affettuosi B. Levi

Volterra.11 T. Levi-Civita to A. Gemelli, Rome 29.5.1940

ANL-Levi-Civita, fol. 1r-v

Illustrious President,[83] Allow me to make recourse to your benevolent, long-standing support in the interests of my distinguished former assistant Prof. Enrico Volterra, son of our eminent colleague Sen. Vito, who naturally joins his prayers to mine. Shortly after the removal of all Jews from Italian universities, Enrico Volterra had submitted, under my advice, his candidacy for the newly instituted chair of Building Sciences, with the task of organising *ex-novo* a laboratory of Elasticity, Stability and Resistance of Materials, at the Universidad Nacional del Litoral in Rosario (Argentina). He was very well prepared for this task because of his valuable and numerous theoretical works, and for the special experimental expertise that he acquired in the most important laboratories of Europe. Indeed, undoubtedly on the basis of his excellent qualifications, he was chosen by the council of the Faculty of Sciences of Rosario and appointed unanimously on 14 April 1939. He was immediately informed by the Dean of the Faculty, Prof. Ing. Cortés Pla, of the appointment for the duration of three years, accompanied by the draft of the contract.[84] The essential point is that the Dean, in communicating the above, added that he would receive permission to disembark in Argentina as soon as possible. Since then, despite the official appointment and multiform solicitations exerted on the Dean and others, it has not been possible for Volterra to obtain the necessary permit to enter Argentina. Recently he turned for support and advised the same Faculty to turn to Monsignor Giuseppe Canovai, his childhood friend and auditor of the Nunciature in Buenos Aires.[85] Monsignor Canovai immediately took up this matter, as can be seen from the letter included, which was received yesterday by Mrs. Virginia Volterra, Enrico's mother. It

---

[83] Father Agostino Gemelli, born Edoardo Gemelli (1878–1959), Rector of the Catholic University of the Sacred Heart of Milan from 1924 to 1959 and president of the Pontifical Academy of Sciences from 1937 to his death. Between the autumn of 1938 and 1939 he tried to get the Church to accept the anti-Semitic policy, for example publishing with Ferruccio Banissoni the volume *Antropologia e Psicologia* (Milano, Bompiani, 1940).

[84] Cortés Pla (1898–1975), Dean of the Facultad de Ciencias Matemáticas of the Universidad Nacional del Litoral (1934–1943).

[85] Giuseppe Canovai (1904–1942) Auditor of the Nunciature (*Uditore di Nunziatura*) in Buenos Aires from May 1939 to January 1942. He had been Enrico Volterra's schoolmate.

appears from this letter that, despite the strong efforts of Canovai, the procedure is not progressing. Strong pressure should be exerted *in alto loco*. Dr. Giovanni Lampariello, now professor at the University of Messina, who was already a colleague of [Enrico] Volterra, when they were assistants together in Rome, one day spoke to Monsignor Montini, Adjunct to the Vatican Secretariat of State, regarding the anguished wait of Volterra: he said that he would gladly remember it on the occasion.[86] It is not known, however, whether he may have dealt with it, nor could any of us disturb him further. This being the case, I dare to appeal to you, even though I know how much you are absorbed by a thousand commitments, to ask if it is possible for you, by resorting to the great means which you have, also internationally, to fix a situation that should be smooth, but instead seems seriously compromised by the recent attitudes of the Argentine Government on immigration. In all cases, please accept my heartfelt apology and best wishes, with the warmest thanks of Volterra himself. Yours Tullio Levi-Civita

Illustre Presidente, Mi consenta di ricorrere al Suo benevolo, lungi operante appoggio nell'interesse del mio distinto ex assistente Prof. Ing. Enrico Volterra, figlio del nostro eminente collega Sen. Vito, il quale, naturalmente, unisce la sua preghiera alla mia. Poco dopo l'allontanamento di tutti gli ebrei dalle Università Italiane, Enrico Volterra aveva posto, per mio consiglio, la sua candidatura alla istituenda cattedra di scienza delle Costruzioni, coll'incarico di organizzare ex-novo un laboratorio di Elasticità, Stabilità, e Resistenza dei materiali, alla Universidad Nacional del Litoral in Rosario (Argentina). Egli era egregiamente preparato a tale compito per pregevoli e numerosi lavori teorici, e per speciale competenza sperimentale, acquisita nei più importanti laboratori di Europa. Effettivamente, in base ai suoi ottimi titoli, egli fu senz'altro prescelto dal Consiglio della Facoltà di Scienze di Rosario, e nominato all'unanimità il 14 Aprile 1939. Gli fu subito, dal Decano della Facoltà, Prof. Ing. Cortés Pla, comunicata la nomina per la durata di tre anni, accompagnata dalla bozza del contratto. Il punto essenziale è che il Decano, nel comunicargli quanto sopra, aggiungeva che egli avrebbe ricevuto quanto prima il permesso di sbarco in Argentina. Da allora, nonostante la nomina ufficiale e le multiformi sollecitazioni esercitate presso il Decano e presso altri, non fu possibile al Volterra di ottenere l'agognato permesso di entrare in Argentina. Recentemente egli si rivolse, e consigliò la stessa Facoltà di rivolgersi per appoggio a Monsignor Giuseppe Canovai, suo amico d'infanzia, Uditore di Nunziatura a Buenos Aires. Monsignor Canovai prese subito la cosa a cuore, come risulta dalla lettera inclusa, ricevuta ieri dalla Signora Virginia Volterra madre di Enrico. Apparisce da tale

---

[86]Both Giovanni Lampariello (1903–1964) and Enrico Volterra had been Levi-Civita's assistant. In 1939 Lampariello had won a competition for the chair of Rational Mechanics at the University of Messina. Giovanni Battista Montini (1897–1978), Substitute of the Secretariat of State from 1937, carried out intense activity in the Vatican Information Office, dealing with the exchange of information on both civilian and military prisoners of war and of the assistance that the Church provided to refugees and Jews.

lettera che, nonostante i buoni uffici del Canovai, la pratica non va avanti. Bisognerebbe potere esercitare una forte pressione in alto loco. Il dott. Giovanni Lampariello, ora professore all'Università di Messina, che già fu collega del Volterra, quando erano insieme assistenti a Roma, attirò un giorno, sull'angosciosa attesa del Volterra, l'attenzione di Monsignor Montini, Aggiunto alla Segreteria di Stato Vaticana: questi disse che volentieri se ne sarebbe ricordato all'occasione. Non si sa però se Egli abbia potuto occuparsene, né alcuno di noi potrebbe disturbarlo ulteriormente. Così stando le cose, oso far appello a Lei, pur sapendo quanto Ella sia assorbito da mille impegni, per chiederle se le è possibile, ricorrendo ai grandi mezzi, di cui Ella, anche internazionalmente dispone, di sistemare una situazione che dovrebbe esser liscia, ma pare invece seriamente compromessa dai recenti atteggiamenti del Governo Argentino in materia di immigrazione. Voglia comunque scusarmi e accogliere, coi più vivi ringraziamenti dei Volterra, i miei sentitissimi. Suo obbl. mo Tullio Levi-Civita

Volterra.12 A. Gemelli to T. Levi-Civita, [Rome] 3.6.1940

ANL-Levi-Civita, fol. 1r

Your Excellency, in response to your letter, in returning to you the letter of Archbishop Canovai, I also send you a letter to H.E. Monsignor Montini. Please go to him, speak to him, and explain to him how things are. Msgr. Montini, with the charitable character that distinguishes him and that you already know, will study together with you what should be done. It seems to me that this is the best solution. You could, in order to more easily contact Bishop Montini, make a phone call to Dr. Salviucci, who can put you in touch. I hope that this application will soon have a good solution, although I do not attempt to deny the difficulties. With devout regards Father Agostino Gemelli

Eccellenza, in riscontro alla sua lettera, mentre le rendo la lettera di Mons. Canovai, le compiego una lettera per S.E. Mons. Montini. Ella si rechi da Lui, gli parli, spiegandogli come stanno le cose. Mons. Montini, con quella carità che lo distingue e che Ella conosce, con Lei studierà che cosa conviene fare. Mi pare che questa sia la migliore soluzione. Ella potrebbe, per più facilmente arrivare a Mons. Montini, dare un colpo di telefono al Dott. Salviucci, perché si metta a sua disposizione. Io mi auguro che la pratica abbia presto una buona soluzione, per quanto non mi nascondo le difficoltà. Con devoti ossequi fr. Agostino Gemelli

Volterra.13 T. Levi-Civita to O. Veblen, Rome 17.6.1940

OVP, *EdV,* **fol. 1r**

Dear Friend, I dare to ask your benevolent help for a distinguished son of our beloved colleague Volterra. This son, Edoardo, has been, from 1930 till 1938 (dismissal of Jews from Italian universities) full professor of Roman Law in the University of Bologna, a chair celebrated since the Middle Ages. This Professor Volterra, Junior, is a nice man, 38 years old, enjoying perfect health. Besides his exceptional value as jurist, he is a brilliant speaker, teacher, and lecturer. His highest aspiration would obviously be to reach a new academic situation in U.S.A., but the moral discomfort urges him to emigrate as soon as possible, ready to undertake any work in a free country. He is waiting for an affidavit, having fulfilled all other conditions required to enter in the U.S.A. I (and with me obviously Professor Volterra, father and son) beg for your precious, preliminary support in the aim of abridging, as feasible, the local practices to get here an emigration visa for the U.S.A. The American officer, on whom the business depends, is your consul in Naples.[87] It would be very important if you or some other influential countryman could write to the consul, introducing to him, favourably, Professor Edoardo Volterra with the statement that, on account of his general and specific attitudes, he is undoubtedly useful for the U.S.A.

For eventual future opportunity of academic functions, I take the liberty to send here enclosed a curriculum with list of Volterra's publications and to add that Professor Salvatore Riccobono, the greatest scholar of Roman Law in our time (whose name has been given to the American Seminar of Roman Law in the Catholic University of Washington) has affirmed, in a testimonial concerning Edoardo Volterra, "that he considers him the best among Italian Romanists, having also a wide and profound preparation, which may be said, rather unique than rare."[88] I hope that, in spite of the heavy trouble I dare to procure to you, you will conserve to me your affectionate and benevolent cordiality. For this and the past proofs I express my warmest thanks. Under the constraint of a very reduced physical and intellectual way of living, I am well, and now about to leave for the Appennines with my wife. Also on her behalf I beg you to accept, with Mrs. Veblen, our best and grateful remembrances. Sincerely yours, Tullio Levi-Civita

---

[87] Thomas D. Bowman, American consul general in Naples from 1936.

[88] Salvatore Riccobono (1864–1958), professor of Roman Law at the University of Palermo (1897–1931) and Rome (1932–1935). In the academic year 1928–29, at the invitation of the Catholic University of America, he had held a course of Biblical Exegesis in Washington.

Volterra.14 B. Levi to T. Levi-Civita, Rosario 23.7.1940

ANL-*Volterra*, fol. 1r

Dear Professor, About ten days ago, I received your airmail of 26.6 with your recommendation regarding Volterra. I took care of it immediately, as indeed I had dealt with it just a short time before. The route that the Dean [Cortés Pla] advises me to follow here is a little long; namely, since this is a matter of relations with the Church, he advises putting it back in the hands of the assistant-Dean who seems to have particularly good relations with that body. This leads me to the fact that the answers come to me very slowly and unfortunately, so far, with results not dissimilar to those obtained by you; that is, declarations of goodwill and nothing else. So far, from the last attempt, I have no answer. Even from England I have had no news. I confirm that I do and will always do everything possible. It is only a great pity that V. has not found a way to 'jump the ditch' [i.e. overcome the difficulty] with his own means as I had repeatedly urged him to do. I affectionately send you my best wishes and ask you to pass on my greetings to Prof. Enriques and Volterra. Yours B. Levi

Caro Professore, Ho ricevuto una decina di giorni sono il Suo aereo del 26.6 colla Sua raccomandazione relativa al Volterra. Me ne sono occupato subito, come d'altronde me ne ero occupato anche già poco tempo prima. Il tramite che qui il Decano mi consiglia è un poco lungo; e cioè, trattandosi di relazioni colla Curia, di rimettere la cosa nelle mani del Vicedecano che pare avere con questa relazioni particolarmente buone. La cosa porta che le risposte mi giungono molto lentamente e disgraziatamente finora con risultati non dissimili da quelli ottenuti da Lei; cioè dichiarazioni di buona volontà e non altro. Finora all'ultimo tentativo non ho risposta. Anche dall'Inghilterra non ho più avute notizie. Le confermo che faccio e farò sempre tutto il possibile e solo è un gran peccato che V. non abbia trovato la strada per saltare il fosso coi proprii mezzi come replicatamente l'avevo sollecitato. La saluto affettuosamente e La prego di fare i miei saluti ai prof. Enriques e Volterra. Suo B. Levi

Volterra.15 O. Veblen to the American Consul in Naples, [Princeton] 27.7.1940

OVP, *EdV*, fol. 1r

Sir, I am informed that Professor Edoardo Volterra, formerly of the University of Bologna, has applied for an immigration visa for the United States. I know the father of Professor Edoardo Volterra, Professor Vito Volterra, a former Roman Senator, as a distinguished mathematician, and I should be glad to be of assistance to his son. I understand that the son, Professor Edoardo Volterra, also is very distinguished in his field, Roman Law, and that he is a man of splendid character.

I hope you will do everything possible to help him. Respectfully, Professor Oswald Veblen

Volterra.16 Father J. F. O'Hara to O. Veblen, New York 30.7.1940

OVP, *EdV*, fol. 1r

Dear Professor Veblen: I don't know how to apologize for neglecting so long to answer your letter of July 11th. The real reason is that I have been so hard pressed for additional chaplains that I have had no opportunity to give attention to anything else. The War Department asked for 250 priests in a month. Since we have no priests of our own and must beg them of other Bishops, the situation has been difficult. If I can help in any way to relieve the situation of Professor Edoardo Volterra, I shall be happy to do so. On your suggestion I am making inquiry of Professor Turley at Notre Dame who spent a year in Italy recently studying Roman Law.[89] I have seen too little of Notre Dame lately and I miss it. I may have a few hours there this week, and if I succeed in this plan, I shall extend your greetings to your good friends there. Sincerely yours, John F. O'Hara

Volterra.17 A. Rosenblatt to O. Veblen, Miraflores 16.8.1940

OVP, *EdV*, fol. 1r

Dear Sir, I have spoken with Mr. G. García on the affair of Mr. V. But he thinks that it is not possible to find here an University Situation for Mr. V. Indeed, he says that the Faculty of Law is of the opinion that only Peruvians can adequately lecture on law in Peru. The Dean of the Faculty of Law, now Minister of Public Instruction, is a nationalist.[90] Besides nobody knows here Latin sufficiently to understand Roman Law. It seems to day for the first time, as if a change in Europe was possible sooner or later. We would be extremely thankful to you, if you could send us a paper to be published in the *Revista de Ciencias* or in our Academy.[91] With highest regard, yours very respectful Alfred Rosenblatt

---

[89] John Patrick Turley, professor of Latin at the Notre Dame University (1943–1950).

[90] Pedro Máximo Oliveira Sayán (1882–1958) lawyer, jurist, professor, diplomat, and politician, served as Dean of the Faculty of Law of the University of San Marcos (1935–1941) and Minister of Public Instruction (1939–1943).

[91] Together with G. García, Rosenblatt had participated in creating the *Academia de Ciencias Exactas, Físicas y Naturales* in Lima, which published the journal *Actas de la Academia Nacional de Ciencias Exactas, Físicas y Naturales de Lima*. He was also a member of the editing board of the journal *Revista de Ciencias*, edited by the Universidad Nacional Mayor de San Marcos.

Volterra.18 The American Consul in Naples to O. Veblen, Naples 28.8.1940

OVP, *EdV*, fol. 1r
Sir: I have to acknowledge the receipt of your letter of July 22, 1940, regarding Professor Eduardo Volterra, who desires to apply for an immigration visa for the United States. The consulate general has not yet been able to consider Professor Volterra's formal application for an immigration visa, in the absence of sufficient documentary evidence of support. I am now awaiting such evidence from persons in the United States who are interested in Professor Volterra's case, and I shall be glad to consider, in addition, any financial guarantee which you yourself may desire to make in connection with this application. For your guidance, I enclose a direction sheet describing the nature of such documentary evidence of support as is usually required.[92] Very truly yours, Thomas D. Bowman, American consul general

Volterra.19 O. Veblen to Father J. F. O'Hara, [Princeton] 23.9.1940

OVP, *EdV*, fol. 1r
Dear Father O'Hara: May I take the liberty of disturbing you once more about the situation of Professor Edoardo Volterra? In your letter of July 30, you said that you had written to Professor Turley of Notre Dame to ask his opinion. Have you yet had any reply? Following a suggestion that Roman Law might be more appreciated in South America than here, I wrote to an acquaintance in Lima [Alfred Rosenblatt], Peru, who went into the question very carefully but reports that the Faculty of Law there "is of the opinion that only Peruvians can adequately lecture on law in Peru." He intimates also that the Minister of Public Instruction [Pedro Máximo Oliveira Sayán] is unsympathetic to the idea. So there appears to be no chance in that quarter. If you could make any other suggestion, I should be very much obliged. Yours sincerely, Oswald Veblen

Volterra.20 G. D. Birkhoff to V. Volterra, Cambridge Mass. 16.10.1940

ANL-*Volterra*, fol. 1r-v
Dear Professor Volterra: Your letter of September 22 has been received, and I am writing at once to get an answer to the question about the whereabouts of your

---

[92] OVP, *EV*: American consulate general Naples, Italy, *Suggestive Directions for the preparation of visa evidence*, fol. 1r.

son. As soon as I receive an answer, I will immediately write to you. I often think of the various pleasant contacts I have had with you and always with the utmost admiration. With highest compliments and kindest remembrances, sincerely yours, George D. Birkhoff

We are enjoying very much the presence at Harvard of Professor Zariski, who will remain with us all of the present academic year. Of course, you will know him very well.[93] His wife and children are here with him. During the present summer we have stayed here in Cambridge or in a little house almost 30 miles away which we recently had built. We recall so vividly our visit to your beautiful villa and the delightful hospitality with which you received us.[94] You have always been a great inspiration to American mathematicians generally and certainly to me in particular! Will you please extend our kindest remembrances to Madame Volterra as well. If by any chance I find any opportunity that might be of interest to your son, I will write him at once. Yours as always George D. Birkhoff

Volterra.21 V. Volterra to C. E. Inglis, 1940

ANL-*Volterra*, fol. 1r-v

Dear Sir, Some time ago I had the honour to write to you as an acknowledgment of my gratitude for all that you have done for my son Henry, who is actually in Cambridge. I received a very kind answer from you.[95] I wish to write to you again to send you, once more, my deepest thanks. Your proposal makes my son very proud as it enables him to prepare in order to obtain a degree at King's College.[96] This proposal is very flattering for my son and very promising for what concerns his future career. This most wished for degree, which you suggest, shows the consideration that you have for my son and the liking that you have taken to him. I feel very honoured both by this consideration and this great kindness of yours. The fact of undertaking studies and researches under such a learned guide as you are is a guarantee of success and gives the certainty of obtaining the best and most useful results for science as for teaching. Being enabled to frequent a laboratory, which for your merit enjoys such a high reputation in all the Institutions which

[93] Zariski had spent six years in Rome, from 1921 to 1927. Oscar had met his wife, Yole Cagli (1901–1997), while in Rome and they were exchanging lessons in Italian and Russian. They were married in 1924, in Kobryn. They had two children: Raphael (born in Rome in 1925) and Vera (born in 1932).

[94] Birkhoff refers to the Villino Volterra in Ariccia, near Rome.

[95] VOLTERRA.8.

[96] Enrico would earn a PhD in Engineering from King's College in 1941.

dedicate to the application of science, will set my son in the best conditions for acquiring that knowledge which may be a help and a benefit to put to execution the most difficult and greatest works. Believe me with renewed thanks. Yours sincerely [V. Volterra]

Chapter 6
Under Another Heaven

6.1 Upon Arrival: Placement, Relocation, and the Search of a New Identity for Uprooted Individuals

Ultimately, six Italian Jewish mathematicians managed to emigrate:

	Univ. of Departure	Destination	Date of Entry
Gino Fano	Turin	Switzerland	November 1938
Guido Fubini	Turin	United States	15.3.1939
Beniamino Segre	Bologna	United Kingdom	19.4.1939
Alessandro Terracini	Turin	Argentina	9.10.1939
Beppo Levi	Bologna	Argentina	6.11.1939
Bonaparte Colombo	Turin	Switzerland	24.11.1943

Their experiences were, of course, very different from one another, because their relocation and integration decisively depended on the country in which they were able to land, and the professional and cultural environment that they found there. The search for a new research identity depended on the local communities with which Italian scholars had to relate and interface. For some of them, the academic insertion was a shock. The sense of intellectual uprooting felt in the first months of the new university situation was equal to (and sometimes even stronger) than that felt in detaching oneself from homes and possessions. For others, adjustment was less traumatic; sometimes it was facilitated by previous contacts with a certain scholarly group, by the linguistic factor, by affinities between Italian society and that of the country in which they arrived, or by reconnecting with Italian colleagues and friends in a similar position. For still others, there was no insertion at all, and from a purely professional point of view, the immigration experience assumed the same traits as the study sojourns that university professors had always spent abroad.

The three destinations—English-speaking countries, South America, and Switzerland—were very different in terms of the integration of displaced Italian scholars. In the former, insertion was much more difficult. Language was the first serious obstacle. For people accustomed to expressing themselves with ease and a certain lexical elegance, it was desolating and humiliating to find themselves reduced to using a child's vocabulary.

Moreover, in the United Kingdom and the United States, the structure of the university system was very far removed from that of Italy: university professors had a different social status, and they interacted with each other and students in a very different way. The model of the School, with a Master and his protégés working on a common research project, could not be replicated in these new settings. Understanding an academic system which so differed from the only one they had known was extremely complicated for Fubini and Segre. How could they interpret, for example, the strong competition that existed also between senior scholars, and the mobility of researchers across a country? In their eyes, the first was inelegant and the second unjust since it appeared not as a choice but as a condemnation of the scholar and his family to a condition of eternal professional precariousness and erratic private life.

In the third place, some typically Italian forms of mathematical sociability (the Seminar, the library, the salon) had equivalents in name, but not in fact. The King's College London Geometry Seminar, for example, was something completely different from the Mathematical Seminar of Turin or Rome, in which old professors and young students, even undergraduates, met weekly to chat about their research, the books they had read, etc. In the American university libraries, one could not work in a team, the Master with his pupils sitting at the same table, as was customary in the mathematical libraries of Italian universities. On 13 October 1939, A. Flexner wrote to O. Veblen in this regard:

> Professor Fubini came to see me this morning to pay his respects, and in the course of our conversations it developed that he might want to bring his books to Fuld Hall and work here. Inasmuch as there are several vacant rooms, I told him that I was sure that there would be no objection to his working here rather than in Fine Hall, where the workers are more numerous. Should the space be required in the future for some other purpose there will be no difficulty about changing arrangement.[1]

For these reasons, integration in the English-speaking academia generally took place only to a partial extent. Neither in America, nor in England, did the Italians change their concept of the rights and duties of a university professor, and their Italian way of performing the role. Only Segre made a more assertive attempt to understand and make his own the dynamics of English academic life, but without arriving at a redefinition of his professional and research identity.

The situation of refugees in Latin America was very different. In addition to the more favourable linguistic conditions, the organisation of mathematics in South American universities was more like that of Italy. Terracini and Levi encountered no obstacles in recreating in Tucumán and Rosario the structures and infrastructures of scientific activity in which they had been trained and worked. Even the desire to rebuild a School within a non-Italian context was not viewed negatively: on the

---

[1] IAS, *GF*: A. Flexner to O. Veblen, Princeton 13.10.1939.

contrary! The decisions to entrust the direction of the Institute of Mathematics in Rosario to Levi and to call Terracini to Tucumán were determined in good measure by the guarantee that these two geometers would do talent scouting and exercise a mentoring function on the new generations of Argentine mathematicians. The integration of Italian displaced scholars was therefore generally complete and quick. Even in this area, however, Italians were and remained 'very Italian' in their way of placing themselves. As the mathematicians of the missions to São Paulo and Rio had done, Terracini and Levi imported to Tucumán and Rosario a certain style of research and teaching, and certain methods of organisation and communication of mathematics, typical of the institutes of Turin and Bologna. In this sense, their times in Argentina represent small-scale experiences that open an intriguing new chapter in the history of circulation, appropriation, and cross-contamination of mathematical know-how.

Finally, Switzerland: here a further distinction must be traced between Fano's case and those of the refugees of the last wave, after the Nazi occupation of Italy. Fano did not have to honour any special debt of gratitude to his Swiss colleagues and had no need to renegotiate his position with them. For him, it is improper to speak of insertion, since Fano did not aspire in the least to enter the mathematical environment of Lausanne or to restart his career there. He asked the governance of the *École d'Ingénieurs* for permission to access the university library, which was promptly granted, and nothing else. He continued his research activity as a freelancer, from his hotel room. With Swiss mathematicians he had cordial relations, he gladly met them at the *Cercle mathématique* of which he was a frequent visitor, he helped them where possible, agreeing for example to stand in for J. Marchand in the lessons of Descriptive Geometry at the *École d'Ingénieurs* (spring of 1945). However, we cannot speak of integration or self-relocation for Fano in Lausanne. For Colombo, Friberti, and Dedò, it also makes no sense to do so. Forced to escape after 8 September 1943, they sold or left everything in Italy. None of these scholars took with them books, reprints, or manuscripts. Life in internment camps prevented any kind of study. Only Colombo managed to return to teaching but in an entirely Italian context such as the IUC. None of these persons had interactions with Swiss mathematicians or with the structures of mathematics in Switzerland.

Above and beyond the divergence of personal experiences in the various venues where refugees found themselves living and working, some common traits are identifiable. First, the desire to affirm themselves and regain position and prestige in the global scientific positioning of host countries, a determination to show gratitude to the country that had welcomed them, as well as the moral imperative to do one's duty all acted as motivational stimuli and instilled new enthusiasm into the victims of persecution. Thus, regardless of age (some refugees were more than 60 years old), the forced sojourns abroad were productive. All Italian displaced mathematicians except Colombo, for the reasons we mentioned above, resumed publishing. All of them took up research activity, even those who like Fubini felt themselves 'already finished'.

Author	Publications in the years 1939–1945
Bonaparte Colombo	0
Gino Fano	11
Guido Fubini	15 (in 4 years, 1939–1943)
Beniamino Segre	18
Alessandro Terracini	22
Beppo Levi	38

Production, in quantitative terms, was impressive and sometimes even higher than that in the last years before emigration. Resumption, however, almost always took place in the wake of continuity and none of the expats was able to re-target comprehensively their activity, in step with the new scientific frameworks where they had relocated. From the qualitative point of view, the most consistent nucleus of publications consisted of translations, reprints, and reprises of works which had appeared in Italy before 1938, together with articles which were visibly in the wake of the studies conducted before emigration. The fact of having been deprived of their patrimonies had a weight in all of this.[2] Contrary to the stereotypical view, according to which mathematical research has no need of access to material assets, it was very difficult, if not impossible, to write a paper or even just to complete an article which was already drafted, without having their own papers, manuscripts and books available, or the opportunity to consult journals and collections.

More than in research, the contribution of Italians was intense and of high quality in the field of teaching, both basic and advanced, including problematic contexts such as the Italian University Camp in Lausanne. In other cases, it was not a question of classroom practice, but of contributing to the structuring of school systems and teaching programmes in the countries of adoption, in Argentina for example, and of the work done in the field of teacher training and continuing teacher training.

At the same time, there was an articulate and effective contribution to the organisation of mathematics, in South America especially. The commitment of Terracini and Levi in this field was intense; indeed, it can be said that the Argentine stay represented for them the stage of their work as organisers and 'sowers of ideas in a land that was virgin but eager to produce'.[3] Their spectrum of activity included the creation and publication of two mathematics journals, the development and management of the mathematical libraries of Rosario and Tucumán, the foundation of the mathematical Seminar of the Universidad Nacional del Litoral, the organisation of the *Jornadas matemáticas*, and the curatorship of a report on means for the

---

[2]Fubini left his books to a friend; Terracini would regain hold of his library in 1948, luckily safe; Segre had his manuscripts and books taken away at the time of internment on the Isle of Man. See for example in IAS, *GF*: G. Fubini to G. Blake, [Princeton] 21.11.1939: 'When I fled from my country, I did not know where I could go; and, also in order not to spend too much money, I could not bring with me my books, which I gave later to one of my friends. I have consequently only *one copy* of my papers; and now I want it'.

[3]Santaló (1962), p. XXVII: *la etapa de su labor como organizador y sembrador de ideas en terreno virgen, pero ávido de producir.*

advancement of mathematical research in Argentina, entrusted to Terracini by the *Sociedad Científica Argentina*.[4] Terracini and Levi expanded partnerships with their foreign colleagues at a time when international cooperation was being threatened. They tried to attract other talented students to Argentine laboratories and universities, and they identified potential mathematics incubators and research programmes. For these contributions, Terracini as President of the *Unión Matemática Argentina*, and Levi in his role as Director of the Institute of Mathematics in Rosario, received more recognition than they would ever have had in the Italian academic system.

Another form of transversal commitment, which united Italian displaced scholars regardless of their host country, was the communication activity. It gave rise to cycles of conferences to illustrate the Italian mathematical traditions, along with radio broadcasts and celebrations of the disappeared Masters (Volterra, Levi-Civita and Fubini) (Fig. 6.1).

The desire to feel like protagonists again, speaking to colleagues about their homeland's mathematical glories, was accompanied by the desire to keep high the name of Italy and the hope of maintaining cultural ties with the homeland and those who had remained there. The translation into Spanish of the works of important Italian mathematicians and the exchange of periodicals denote the value that refugees gave to this type of activity, particularly useful at a time of great fragility of international relations.

Finally, an important sphere of commitment was solidarity: all those who had managed to leave then strove to help those who had not yet been successful, circulating news of positions that opened abroad, sending money and (during the period of occupation) collecting and transmitting information on the victims of roundups and deportations.

6.2 Research

Fubini, Fano, Segre, Terracini, and Levi all belonged to a single School, and that was not without consequences. In fact, a peculiar feature of Jewish mathematical emigration from fascist Italy is precisely that it was not a set of individual emigration experiences but the diaspora of a large part of a research group with a well-defined identity. All the refugees were pure mathematicians, linked by strong cultural bonds with the University of Turin where they had been students and/or professors for decades consecutively. They were almost equally divided in algebraic geometers (Fano, Segre, Levi) and projective-differential geometers (Fubini and Terracini). Only Fubini and Levi had an output that was not entirely geometric, with papers in analysis and theory of functions, mathematical logic (Levi) and applied mathematics (Fubini). Their common membership of the Italian School, the sharing of the intentions and projects of this team, and the cultural and ideal affinities existing

---

[4]Santaló (1968); Santaló (2001).

Fig. 6.1 Volterra's celebration in Rosario, 1941. Top: Terracini is the first from the left and Levi the second from the right; middle: Terracini is the first from the left and Levi is lecturing; bottom (next page): the public attending Volterra's celebration, ANL-*Volterra*

Fig. 6.1 (continued)

among its members meant that their research activity led to intriguing episodes of cultural hybridism, but never to cases of real turn-off.

As far as projective-differential geometry is concerned, during the American period Fubini published only a couple of works on W congruences, in which two distinct lines of investigation converged: the Turin one initiated by Fubini and E. Čech in the twenties and the American one (developed by E.P. Lane, E. Wilczynski, and H. Jonas).[5] His contribution, not only in the geometric field but more generally, was so limited as to cause him to think that he did not deserve the stipend, however small, that the Institute for Advanced Study paid him.

Terracini reproposed some studies already published or started in Italy. This is the case, for example, of the essays *El invariante de Mehmke-Segre y los sistemas lineales* and *Sobre la existencia de superficies cuyas líneas principales son dadas*.[6] The former constituted a tribute to the mastership of Terracini's beloved mentor Corrado Segre, at the beginning of his new scientific life in Argentina.[7] The simple projective characterisations of this invariant had been given by Segre for two plane curves in 1897 and by Buzano for two surfaces in S_n ($n > 2$). In two papers dated 1936, Terracini had interpreted projectively this invariant by virtue of the conception of density of dualistic correspondences. In a 1940 reprint, he provided

---

[5]Fubini (1940a, b).

[6]Terracini A. (1940a, b).

[7]Terracini A. (1968), p. 130.

further applications of the preceding concepts to the theory of line congruences.[8] The second essay relied on the contents of two notes presented by Terracini in 1937 and in 1939, respectively.[9] Submitted for publication in English to the *Annals of Mathematics* in the winter of 1938–1939, the article finally appeared in Spanish as a monographic issue of the *Unión Matemática Argentina*. In this paper, chiefly dedicated to the links between the geometry of planar webs and the projective differential geometry of surfaces, Terracini responded to a question asked by W. Blaschke and G. Bol in their *Geometrie der Gewebe* (Berlin: Springer, 1938). Terracini obtained a characterization of Segre's 5-webs as solutions to a certain non-linear differential system. Under additional simplifying hypotheses, he succeeded in integrating explicitly the resulting system.

Terracini, however, went further and in Tucumán inaugurated also a new line of investigation in projective-differential geometry, which he would continue to develop even after his return to Italy in 1948. In 1941, appraising the works by E. Kasner and J. De Cicco, Terracini started studying a particular type of ordinary third-order differential equations and their integral line systems, which he called equations and systems (F) and (G), some of which had emerged in the study of the trajectories of positional forces fields in a plane.[10] According to Togliatti:

> we are here in the realm of the so-called geometry of differential equations. Moreover, they are evident and implicit - I would say that they are in the very nature of the contents - the links not only of these works, but among all of Terracini's projective-differential research, with the theory of differential equations, both ordinary and partial [. . .].[11]

Details of the appropriation of these works by the local audience would deserve in-depth analysis, but they can be credited for the creation of a working group in differential geometry which would be joined by many young Argentinian mathematicians, from L. Santaló to F. Herrera and M. Cotlar.[12]

In the case of classic algebraic geometry, Italian refugees embodied a singular phenomenon that deserves to be described in more detail: that of isolation or self-isolation both towards other emigrants and in front of resident geometers. In certain terms, Italians displaced scholars were exiles not in a political but in a mathematical sense. To properly understand this, one should remember that the Italian School of Algebraic Geometry had been at the top of the international scene until the early

---

[8]Terracini A. (1936a, b).

[9]Terracini A. (1937, 1939).

[10]Terracini A. (1941a).

[11]Togliatti (1969), p. 402: *Siamo qui nel campo della cosiddetta geometria delle equazioni differenziali; sono del resto evidenti ed impliciti, direi anzi che sono nella natura delle cose, i legami non solo di queste ma di tutte le ricerche proiettivo-differenziali di Terracini con la teoria delle equazioni differenziali, sia ordinarie che alle derivate parziali [. . .].*

[12]Santaló was a penetrating geometer, renowned for his contributions to integral geometry. Herrera was mainly an analyst, who dealt with function theory and Fourier series. A first-rate mathematician and a man of great humanity and righteousness, Cotlar worked in many areas of mathematics, including functional analysis, harmonic analysis, ergodic theory, and spectral theory.

twenties.[13] At that time, among the members of the team, there were algebraists such as G. Scorza, and number theorists like A. M. Bedarida. In 1921 Scorza had published the volume *Corpi numerici e algebre* (Messina, Principato), a text that together with those of L. Bianchi (*Lezioni sulla teoria dei numeri algebrici*, Pisa, Spoerri, 1923) and M. Cipolla (*Analisi Algebrica*, Palermo, Capozzi, 1921) placed Italy at the forefront in abstract algebra. In the thirties, however, the School had lost its momentum and had increasingly turned in on itself. Young students like Buzano, Bassi and Conforto who had gone abroad for a study sojourn now felt the delay accumulated by the Italian geometers compared to the foreign colleagues.[14] Van der Waerden's *Modern Algebra* (1930), which outlined the topics covered by E. Noether in the courses taught in academic years 1924–25 and 1927–28, had had a very limited circulation in Italy. 'Modern algebra does not exist in the Italian school system', Conforto wrote disconsolately to Bompiani in 1932.[15] In topology, the distance was just as great. Comessatti and Severi sensed the potential of topological tools to deal with old and new problems in algebraic geometry, but the desire to safeguard the Italian synthetic style prevailed over their farsightedness. The fundamental books of Lefschetz or Veblen had made only a modest impact on Italian geometers. Suffice it to say that Fano and Terracini had not bought them and that they had entered the mathematical libraries of Turin and Rome at a decidedly late date. In summary, as P. Salmon would caustically state: 'all Italian mathematicians were a bit ignorant, but especially the researchers of algebraic geometry'.[16]

The lack of interpenetration of algebra and topology with classic algebraic geometry developed according to the Italian style corresponds to the failure of integration of Italian displaced scholars with their new environments. This can be clearly appreciated in the cases of Fano and Segre.[17]

Despite his age, in Lausanne Fano continued his study of some special three-dimensional algebraic varieties, known today as Fano threefolds and Fano–Enriques threefolds, respectively.[18] Starting from analysis of the cubic hypersurfaces in the

---

[13] Conforto (1939).

[14] Bassi (1936, 1938).

[15] F. Conforto to E. Bompiani, Göttingen 24.1.1932, in https://media.accademiaxl.it/pubblicazioni/ Matematica/link/conforto.pdf: *Da noi invece l'Algebra moderna non è per niente entrata nell'ambito scolastico.*

[16] Salmon (1994–1995), p. 232: *Tutti i matematici italiani erano un po' ignoranti, ma soprattutto i cultori di geometria algebrica.*

[17] Also Levi reprised with variations some themes in algebraic geometry that he had dealt with many years before, at the time of university studies and the beginnings of his research career, without taking into consideration the new topological developments. During the Argentinian period most of Levi's energies, however, were directed to other tasks: teaching and activities that today fall under the umbrella of the so-called third mission and public engagement.

[18] Fano threefolds are smooth projective varieties V whose anticanonical system $|-KV|$ is ample or, equivalently, whose first Chern class is ample. Fano–Enriques threefolds are irreducible three-dimensional varieties V' whose general hyperplane sections are Enriques surfaces—i.e. algebraic surfaces S with irregularity $q = 0$, such that $K_S \neq 0$ and $2K_S = 0$ in $Pic(S)$, where K_S is the canonical divisor associated with S and $Pic(S)$ is its Picard group—and such that V' is not a cone over S. For a

complex four-dimensional projective space, Fano had opened this line of investigation in the early twentieth century. With the aim of extending the results of rationality and classification of algebraic surfaces achieved by Castelnuovo and Enriques, Fano had envisaged an articulated research project, which he had developed for many years mainly alone (with sporadic contributions by Severi). The centrality of the issues addressed by him within the agenda of the Italian School had been underlined by Castelnuovo in his plenary lecture at the 1928 ICM in Bologna:

> How can we decide whether an assigned equation with four unknowns represents a rational or a semi-rational variety? We know nothing about this, not even for the lowest values of the degree, higher than 2. Actually, research that Fano has been carrying out for several years, and which he will present during his communication, shows how complex the question is. He examines the varieties that have all the genera and plurigenera equal to zero and he distributes them into a finite number of families: the first one is composed of rational varieties, the second of semi-rational varieties, and the others of varieties that become more detached from rationality. An accurate classification of these types would shed light on a question that needs to be solved for the future development of algebraic geometry.[19]

In the same year, 1928, Fano had given a first classification of Fano threefolds based on the invariant Ω_2. In the thirties, he had continued working hard in this field. English geometers H. F. Baker, L. Roth, P. Du Val, D. Babbage, and J. Todd had in turn seized on the patrimony of concepts and tools pioneered by Fano to provide new ideas for advancing research on three-dimensional varieties.[20] Just before leaving Italy, in the memoir *Sulle varietà algebriche a tre dimensioni le cui sezioni iperpiane sono superficie di genere zero e bigenere uno*, Fano had proved that the general Fano–Enriques threefold is birational to a Fano threefold, providing a geometric classification of these varieties that was, however, based on a restrictive hypothesis on their singular points.[21]

In the first three years spent in Lausanne Fano continued to work on these themes, arriving at the proof of irrationality of V_3^3 (1943).[22] The originality of Fano's proof was widely recognised, all the more so as he tackled the problems 'almost emptyhanded because there was no foundation for higher dimensional algebraic

---

detailed analysis of Fano's contribution in this subject (Fano G. 1931b, 1938a and 1938b), see Luciano and Scalambro (2022); Scalambro (2023).

[19] Castelnuovo G. (1929), p. 200: *Come decidere se una equazione assegnata a quattro incognite rappresenti una varietà razionale o semirazionale? Nulla sappiamo in proposito, nemmeno per i più bassi valori del grado, superiori a 2. Anzi, ricerche che il Fano prosegue da vari anni, e di cui vi parlerà in una sua comunicazione, fanno vedere quanto la questione sia complessa. Egli prende in esame le varietà che hanno nulli tutti i generi e i plurigeneri e le distribuisce in un numero finito di famiglie, di cui la prima si compone di varietà razionali, la seconda di varietà semirazionali e le altre di varietà che si staccano sempre più dalla razionalità. Una classificazione accurata di questi tipi getterebbe molta luce sopra una questione che è necessario risolvere per lo sviluppo futuro della geometria algebrica.*

[20] Barrow-Green and Gray (2006).

[21] Fano G. (1938b).

[22] Fano G. (1945, 1947, 1950).

varieties'.[23] However, the paper *Nuove ricerche sulle varietà algebriche a tre dimensioni a curve sezioni canoniche* and *Osservazioni varie sulle superficie regolari di genere zero e bigenere uno* testify to the use of an arsenal of concepts and deductive tools which had encountered no significant changes.[24]

Although Fano had the opportunity to meet and compare ideas with various young 'topologising geometers', neither the approach to the problems nor the argumentative form of his results had varied. The sources mobilized were the same ones that Fano would have used before expatriation. The distinctive features of Fano's research were unchanged: in-depth culture in projective geometry, with the exploitation of its properties to enlighten the birational properties of algebraic varieties; geometrisation of algebraic language; consideration of hyperspatial projective geometry as a concrete model to dimensional linear algebra; transformational approach, i.e. widespread use of the so-called Cremona transformations (birational maps). The style of exposition was the same: a sort of heuristic itinerary, relying on intuition and projective methods and characterised by the praxis of extending by analogy the results progressively obtained to higher spaces and to general cases. Explicit citations, both in the footnotes and in the text, clearly accounted for a phenomenon of cultural endogamy. The fact that Fano never used, nor indeed mentioned, the new techniques in algebraic geometry developed abroad in any of his publications of the period 1938–1953, gives evidence of the Italian School's basic mistrust of the structural and topological methods and of the consequent difficulty of its members in addressing foreign colleagues, even those closest to the Italian approach (such as the Swiss).

It can be objected that Fano did not have the intellectual freshness and energy to behave differently but this objection does not apply to the other classical algebraic geometer displaced from racist Italy: Segre. Free from teaching assignments until 1942, Segre was the only Italian refugee to dedicate himself solely to research, for three years. A pupil of Severi and Enriques, in Rome Segre had been given the opportunity to gain a culture in abstract algebra and topology. He met Lefschetz and van der Waerden, worked with Zariski, followed Severi's courses in algebraic geometry and published the handouts.[25] His stay in Paris with Cartan had been useful for him to emancipate himself from Severi, to broaden the horizons of his research projects and to appropriate a new heritage of algebro-topologising techniques. In addition, Segre was a voracious reader who had reviewed for the *Bollettino dell'UMI* a large number of foreign texts in algebraic geometry. For these reasons, in the late twenties Segre was perhaps the least 'Italian' among the members of the Italian Geometric School, a characteristic also recognised by Severi, who wrote to him:

---

[23] Murre (1994), p. 224.

[24] Fano G. (1944c) and BSM-*Fano*: *Scritti.3*, fols. 1-6.

[25] F. Severi, *Conferenze di geometria algebrica raccolte da Beniamino Segre. Concetti fondamentali - Topologia - La teoria della base*, Roma, Stabilimento tipo-litografico del genio civile, 1927.

Lefschetz wrote to me two days ago and among other things he tells me: "I notice and follow from afar the work of your young assistant Segre. It seems to me that he is a very promising algebraic geometer. I hope he will topologise a bit." This evaluation of a scientist who undoubtedly has a lot of talent will certainly please you. Anyone who sees the issue with criteria a little more ... sensible than those of our 'friend' [Enriques], is able to judge better. But our friend's criteria will end up being increasingly discredited. There's no point worrying about it.[26]

In light of these facts, Segre appears to be the one displaced scholar who, more than any other, could have benefitted from a relocation. Segre made a serious effort to fit into the English-speaking mathematical reality in more than just a nominal way. Publishing eighteen works in less than seven years gives indication of his intense productivity. The panorama of the issues addressed was broad and some lines of study were new, namely arithmetic geometry. The attempt to acquire the style and practices of English publishing is evident. In addition to writing in English, Segre tried to 'think in English': to structure his papers with a subdivision of content, a type of prose and exposition different from those of his Italian works. His English friends (Todd, Roth, etc.), helped him in this exercise of mathematical re-nationalisation, i.e. in acquiring a style in algebraic geometry different from his own native one. However, the outcomes were variable. Segre achieved good results and praise, but also criticisms and failures.

A good example of this is provided by the two books on algebraic geometry for The Clarendon Press on which he worked in the years 1940–1946 (D6.2.1–D6.2.19). The first volume, *Geometry upon algebraic surfaces and varieties*, was to be written by Segre and Todd.[27] The collaboration between the two was suggested by W. V. D. Hodge; Todd and Segre promptly agreed.[28] The time limit for the completion of the manuscript was set in 18 months. If Segre managed to stay in England until he secured the Manchester lectureship, it was thanks to Clarendon which paid him two outright payments of £50 to relieve him of financial anxiety while he worked on the book. The editorial board, and especially K. Sisam, was sure that, independently of not being an English-trained mathematician, to have that work

---

[26]CA, *BSP*: F. Severi to B. Segre, [1929]: [*Mi ha*] *scritto in questi due giorni il Lefschetz e tra l'altro mi dice:* «*Je remarque et je suive de loin les travaux de votre jeune assistant Segre. Il me semble promettre beaucoup comme géomètre algébrisant. J'espère qu'il topologisera un peu.*» *Questo giudizio di uno scienziato che ha indubbiamente molto ingegno le farà certo piacere. Chi vede la cosa con criteri un po' più ... assennati di quelli del nostro amico* [Enriques], *giudica meglio. Ma i criteri del nostro amico finiranno coll'essere sempre più screditati. Non giova preoccuparsene. Mi dica se lo Snyder è tuttora costà. Gli scrissi: non si è fatto più vivo.*

[27]John Arthur Todd (1908–1994) was awarded an entrance scholarship to Trinity College in Cambridge in October 1925. After graduation Todd remained at Trinity to study for his doctorate in geometry under H. F. Baker's supervision. L. Mordell's assistant in Manchester (1931), Todd spent the session 1933–1934 at Princeton with a Rockefeller Scholarship. In 1937 he was appointed a lecturer at Cambridge, where he would remain until retirement.

[28]CA, *BSP*: K. Sisam [The Clarendon Press] to B. Segre, Oxford 20.3.1940; J.A. Todd to B. Segre, Shipley Yorks 27.3.1940; B. Segre to W. V. D. Hodge, [Cambridge] 29.3.1940; B. Segre to K. Sisam [The Clarendon Press], [Cambridge] 29.3.1940. Kenneth Sisam (1887–1971), a New Zealand academic and publisher, worked at The Clarendon Press from 1922 to 1942.

done would greatly advance Segre's prospects.[29] Moreover, he felt that it was the duty of a university press to help scholars in difficulty: 'we particularly like to help authors who are getting learned work done, and you are, I think, the first of the refugee scholars to finish a book for us, as well as make good progress with a second'.[30]

The publishing house relied heavily on the treatise, to be entitled *Geometry upon algebraic surfaces and varieties*. The English publishing market was lacking a seminal work on the subject, suitable also as a textbook for university courses. Segre was guaranteed to have the skills and expertise to write it within a reasonable timeframe. Moreover, he was an Italian, and Italians were considered excellent authors of university texts. He came from a School that had produced internationally appreciated classics, translated into other languages, such as the *Lezioni sulla teoria delle superficie algebriche* by L. Campedelli and Enriques (1932–1934), *Lezioni sulla teoria geometrica delle equazioni e delle funzioni algebriche* by Chisini and Enriques (1934), and *Trattato di geometria algebrica* by Severi (1926).

However, as the Segre–Todd correspondence shows, irreconcilable divergences in approach soon manifested. The differences, it should be noted, were of a purely mathematical nature; personal relations between Segre and Todd were excellent in 1939 and would remain so throughout the entire period spent by Segre in England. Yet Segre and Todd had a different view of algebraic geometry, its problems, and methods. Their clash was on substantive issues: contents, proportions of the chapters, their interconnections, and the style of exposition. Disagreements emerged immediately after the signing of the contract when Todd, frankly, stated to Segre:

> I have been reading Severi's book on series and systems of equivalence. His point of view is so very different from mine that I may be in danger, when I came to write out the stuff, of doing less than justice to its importance. What do you think ought to be stressed most? The problems of structure elaborated by Severi or the invariant systems and their applications?[31]

Segre responded:

> I think that the problems of structure elaborated by Severi are far less essential than your theory of invariant system. Yet, they ought to be enclosed in a book of didactical character, just on account of their comparative simplicity, as a first survey of some relevant notions and an introduction to other important topics of more constructive character already considered by Severi himself (as, for instance, the questions about the dimension of a complete series of equivalence); such topics, however, being not yet satisfactorily settled, could only be explained summarily, possibly in an Appendix.[32]

---

[29] CA, *BSP*: K. Sisam [The Clarendon Press] to B. Segre, Oxford 24.6.1941. See also K. Sisam [The Clarendon Press] to B. Segre, Oxford 24.7.1942, 31.7.1942, 28.8.1942, 1.1.1943, and 15.1.1943.

[30] CA, *BSP*: K. Sisam [The Clarendon Press] to B. Segre, Oxford 31.7.1942.

[31] CA, *BSP*: J. A. Todd to B. Segre, Shipley Yorks 27.3.1940. Todd referred to F. Severi, *Serie, sistemi di equivalenza e corrispondenze algebriche sulle varietà algebriche*, ed. by F. Conforto and E. Martinelli, Roma, Istituto Matematico della R. Università, 1938.

[32] CA, *BSP*: B. Segre to J. A. Todd, [Cambridge] 29.3.1940.

The theory of equivalence on algebraic varieties, the contents of a long introduction *à la* van der Waerden that Todd planned to write, the chapter on singularities were all points on which Segre and Todd failed to find agreement. Equally irreconcilable were their opinions on the general structure and proportions of the treatise: should it be modelled, as Segre implicitly hoped, on the *Lezioni* by Chisini and Enriques? Should it contain only classic results or also new ones, such as Du Val's proof of the resolution of singularities of a surface? Should new proofs of old results be included? And if so, where? Segre proposed to include appendices and additional sections dedicated to specific aspects of the theory of varieties. Todd was pushing to cut substantially settled questions, in order to try to inspire fresh trends of ideas and new researchers. Finally, there was the question of the degree of rigor. Todd, especially after he started writing his book on analytic and projective geometry (1944), began to denounce the lack of rigor in the usual treatment of projective geometry. Segre, hearing his Italian Masters under accusation, replied annoyed. It was the only moment in which relations between Segre and Todd were on the verge of breaking down.

In the meanwhile, postponements and lengthy interruptions occurred in the drafting of the manuscript, mainly due to Todd's very demanding teaching load.[33] Sisam and Hodge tried to mediate by proposing the solution of dividing the treatise into two volumes. At the end of September of 1945, after more than three years of delay with respect to the expected time limit for the completion of the manuscript, Sisam made a final attempt and suggested a personal meeting of the authors.[34] To help matters along, the Clarendon Press was willing to pay the expenses of a visit from Segre to Todd or vice versa. Todd—as Sisam explained to Segre—was worried because 'he was now rather rusty in the subject and there had been some major advances in America which he had not had time to study'. Thus, the invitation to Segre: 'You must try and persuade him that something good can be done, even if it doesn't cover all the ground he would wish to explore'.[35] Finally, Segre agreed with Todd a plan for producing the book in two volumes. Sisam was glad to know the decision and commented:

> I suppose that will mean that the first volume can be cleared fairly soon and will be able to stand as an independent contribution to the subject. If so, it will give you both more time for the problems of the second volume.[36]

---

[33] See for example CA, *BSP*: J. A. Todd to B. Segre, Shipley Yorks 28.12.1942 and K. Sisam [The Clarendon Press] to B. Segre, Oxford 1.1.1943: 'I heard from Professor Hodge recently, and I hope you and Dr. Todd are making good progress in difficult but more cheerful times. I am not pressing for the MS. at present because we are so much occupied with Government work, but I do want it to be ready for the Printer when the war ends, and his war work can give place to learning'.

[34] CA, *BSP*: K. Sisam [The Clarendon Press] to B. Segre, Oxford 25.9.1945.

[35] CA, *BSP*: K. Sisam [The Clarendon Press] to B. Segre, Oxford 28.9.1945.

[36] CA, *BSP*: K. Sisam [The Clarendon Press] to B. Segre, Oxford 8.10.1945.

Todd got back to work but immediately stopped, again when faced with the obstacles in writing the introductive chapter *à la* van der Waerden:

I got down to work at the book in the week before term started and found it very hard work; I find I don't follow the v.d.W. stuff as I used to do and was getting rather tangled in the elimination theory part. During term I have been too busy to look at it, but the rac. lies ahead and I hope to get on with it. But progress to date is, I fear, something reaches 90 pages.[37]

In the end, Todd abandoned the project. Of the parts he was supposed to write, there remain only the manuscript folios. Segre, on the other hand, exploited at least in part the materials he had produced for his *Lezioni di geometria moderna* (Bologna, Zanichelli, 1948), the text that sanctioned his return to the Italian mathematical world after his stay in England.

The second volume of which Segre (alone) was charged, entitled *The non-singular cubic surfaces. A new method of investigation with special reference to questions of reality*, had a decidedly easier history. Segre began writing it roughly at the same time as the *Geometry upon algebraic surfaces and varieties*, however in June 1941 he already submitted the definitive manuscript to Clarendon. In April 1942 the book was already in an advanced state of work so much so that the editorial staff cared about its advertising proposal: 2781 free copies to be sent on Segre's recommendation and copies with a review request for *The Times, Nature, The Mathematical Gazette, Science Progress* and *The Cambridge Review*.[38] Segre suggested that it would be convenient to extend the publicity also abroad, as far as the war circumstances would allow, and he proposed to add another five journals to the reviewer list (*The American Mathematical Monthly, Bulletin of the American Mathematical Society, Revista Matemática Hispano-Americana, Mathematical Student—Annamalai University South India*, and *Mathematical Reviews*), three institutions (Biblioteca de la Facultad de Ciencias Exactas, Físicas y Naturales de Buenos Aires, IAS, Société Mathématique de Moscou), and 15 colleagues: A. B. Coble and A. Emch (University of Illinois Urbana); T. R. Hollcroft (Wells College Aurora-on-Cayuga, New York); H. Lewy (University of California, Berkeley); A. Longhi (Liceo Cantonale in Lugano, Switzerland); Lefschetz, Veblen, Zariski, Snyder, Rosenblatt, Rey Pastor, Levi, Terracini, G. Albanese, and G. Wataghin (Instituto de Matemática da la Universidade S. Paolo Brasil).[39] He was satisfied.

As Baker stated, the treatise *The non-singular cubic surfaces* 'is a very remarkable monograph; is a direct product of war circumstances. [...] The book is one for which, for long years, the reader will be grateful to the author and to the Clarendon Press'.[40] In this book, the Italian synthetic tradition reached its apogee. By contrast, the new stimuli recognition that Segre may have achieved through his association with the English geometers can be barely perceived.

---

[37] CA, *BSP*: J.A. Todd to B. Segre, 10.3.1946.

[38] CA, *BSP*: Oxford University Press to B. Segre, London 21.4.1942.

[39] CA, *BSP*: B. Segre to Oxford University Press, [Cambridge] 25.4.1942.

[40] Baker (1943), pp. 39–40. See also CA, *BSP*: H. F. Baker to B. Segre, [Cambridge] 12.1.1943.

Documents

D6.2.1 Draft of a Letter by B. Segre to J. A. Todd, [Cambridge] 29.3.1940

CA, *BSP*, fol. 1r

Dear Todd, I think that the problems of structure elaborated by Severi are far less essential than your theory of invariant system. Yet, they ought to be emphasized in a book of didactical character, just on account of their comparative simplicity, as a first survey of some relevant notions and an introduction to other important topics of more constructive character already considered by Severi himself (as, for instance, the questions about the dimension of complete series of equivalence); such topics, however, being not yet satisfactorily settled, could only be explained summarily, possibly in an Appendix.

D6.2.2 K. Sisam [Clarendon] to B. Segre, [Oxford] 24.6.1941

CA, *BSP*, fols. 1r, 2r

Dear Professor Segre, I have had some correspondence with Professor Hodge about your affairs and your books, and have assured him that we shall be able to do something for you if you are not fortunate enough to secure the Southampton lectureship, for which I gather there is competition.[41] The only condition is that you should work on despite all the distractions of the times at the book on *Geometry upon Algebraic Surfaces and Varieties*, in which you are collaborating with Dr. J.A. Todd, for I am sure it is in the interests of the subject, as well as your own, that that book should be finished. I may add that, should you get the Southampton lectureship, and if the salary does not become due until the autumn term, we should like to make a grant of £50 anyhow, to bridge the gap and keep you cheerfully at work in geometry. I now come to the other work on *Cubic Surfaces*. I have already explained to Professor Hodge that we are in some difficulties with printing because we are so much preoccupied with Government work that requires the best-trained compositors, who do mathematics. But if you are content to accept that risk of delay, we should be prepared to accept the book in its present length as an independent book, and to pay you, at your choice, *either* the sum you would have got for a Cambridge Tract.,[42] *or* a royalty of 12 1/2 % of the United Kingdom published price on all copies sold in the United Kingdom, and 10% of the United Kingdom published price on all sales made in the U.S.A. or for Export, on which our return is less. I note that this last arrangement is

---

[41] Segre will fail at Southampton. The position will go to H. Hamburger. See Chapter 10, pp. 565–567.

[42] The series Cambridge Tracts in Mathematics and Mathematical Physics had been founded in 1912.

considerably better than a royalty depending on the return from sales in U.S.A., which is now the principal market. You would receive 10 presentation copies, and we should send out the necessary review copies in consultation with you. We should like to have the English presentation further revised for a substantive book, and Professor Hodge has suggested that Dr. D. Pedoe of Southampton might be willing to do it.[43] In these days, when labour is precious, and heavily corrected proofs are liable to stand untouched indefinitely, we should be glad to give you a little more time in order to get the MS. perfectly finished, so that if the compositor does his work well, there will be no question of alteration, and the difficult passage through the press may be shortened. We regularly set a correction limit on author's corrections in proof, apart from printer's errors, of 25s. per sheet of 16 pages, and find it is ample on good copy. We could draw here any necessary figures from rough sketches. I should be glad to know whether these proposals are acceptable to you, or to have any counter-suggestions. I very much hope you will be successful at Southampton, and so have an opportunity to carry on your work with an assured position. Yours very truly, Kenneth Sisam

D6.2.3 J.A. Todd to B. Segre, Cambridge 10.1.1943

CA, *BSP*, fol. 1r-v
Dear Segre, My last letter was not written in a spirit of reproach. Like you, I shall be glad to get the book finished, but I can't see that I shall have much time to get on with it before the summer, as it is a very busy year for all of us here. I suggest that you explain to Professor Sisam that I am making what progress I can, but that under war conditions the long vacation is the only time when I can devote much attention to it. Whether the book will be ready for the printer at the end of the war depends largely on factors beyond my control (such as the length of the war). [I *don't* advise you to put it that way to Sisam . . .] At any rate, I *did* get on with it last September, and reached (speaking from memory) somewhere about p. 75 of your MS., a good deal of which was plain sailing. My own MSS. reaches p. 120, not counting the long introductory chapter *à la* van der Waerden, and I left it (when last term started) in the middle of the stuff about proper singularities (and wondering whether all your detailed results were really essential).[44] Yours sincerely J.A. Todd

---

[43] Daniel Pedoe (1910–1998) had studied geometry at Cambridge University, under the guidance of H. F. Baker. After a sojourn at the Institute for Advanced Study, where he worked with S. Lefschetz, in 1936 was appointed as an assistant lecturer at the University College, Southampton.

[44] Todd alludes to a chapter to be structured in the wake of the volume by B. L. van der Waerden, *Einführung in die algebraische Geometrie*, Die Grundlehren d. math. Wiss. in Einzeldarstell. mit besonderer Berücksichtigung d. Anwendungsgebiete 51, Berlin, Springer, 1939.

D6.2.4 J.A. Todd to B. Segre, Cambridge 5.9.1943

CA, *BSP*, fol. 1r-v
Dear Segre, How is the world treating you these days? I haven't heard from you
since Xmas. I have begun to get down to your MSS. again after a long interval: it
has been a very strenuous year here and I didn't feel equal to the task until quite
recently. The more I look at the way things are going the less happy I feel about
the proportions of the book. At present I am working up some of your stuff on
resolutions, neighbourhoods, and analysis of singularities. There seems to be
about 80 pages of this, and I can't help feeling that this is far too big a slice of the
4-5 hundred pages which the O.U.P. [Oxford University Press] are contemplating, particularly as it doesn't include the main theorem on the removal of
singularities of surfaces. I am trying to cut it down a bit by leaving out inessen-
tials, but this is not easy in your closely-knit account. I sometimes wonder
whether you haven't got *Enriques-Chisini*[45] in mind as a model: you seem to
aim at an encyclopaedical thoroughness of treatment which I am not sure is what
is really wanted. One thing is clear, and that is that it will not be possible to fix a
date for the completion even of my first draft. The war is lasting long enough, but
I wish I could foresee the end of the assault on Surfaces and Varieties as clearly as
I can foresee the collapse of Fascism and Nazism in Europe. Talking of the latter,
what do you think of recent events in Italy? Do you suppose the Italians will be as
well disposed towards us as the Sicilians appear to have been, or are the Sicilians
(as some people have suggested) a sort of Italian "Irish problem"?[46] I hope you
are finding life in Manchester tolerable: it is not a beautiful city but there is good
country within reach if you can get out at weekends; and the people there are good
folk. Yours sincerely J.A. Todd

D6.2.5 Draft of a letter by B. Segre to J. A. Todd, Manchester 7.9.1943

CA, *BSP*, fol. 1r-v[47]
Dear Todd, I apologize for having been so many months without writing but I
have been very busy. Indeed I have written several papers and have just finished

---

[45] F. Enriques, O. Chisini, *Lezioni sulla teoria geometrica delle equazioni e delle funzioni algebriche*, Bologna, Zanichelli, 1934, 2 vols.

[46] Todd refers to the Allied invasion of Sicily, began on the night of 9–10 July 1943 and ended on
17 August. The military operation had a decisive influence in Italy: it favoured the dismissal of
Mussolini, the fall of the fascist regime and the subsequent armistice of Cassibile.

[47] The letter presents many deletions and corrections. In the first version, Segre had written: 'I
apologize for having let so many months to pass, without writing. One of the reasons for this reason
is that I have been very busy, and I have just finished the summer term lectures (since we have here
a summer term). But the chief reason is that I could scarcely have written to you without mentioning
the book, and I was not quite sure about how you would have like the argument. Now I am glad

my lectures of the summer term.[48] Now I see with pleasure that you have again time to deal with my MS. but am sorry about your feelings. When I wrote it, I had no models in mind; I only tried to put on sound bases the theory, without worrying about limitations of space. These were in fact put to us clearly only *after* I had completed most of my work. And so, besides other more important tasks, you have now the one, neither easy nor agreable, of reducing and cutting. I suggest that some of the collateral results, you consider interesting or useful enough, could be given without proof, or only with some hints about their demonstration. Additional sections (to be put at the end of each chapter and possibly to be printed in smaller types) could deal with this stuff, as well as with historical notes and bibliographical references. I know that it will be a long and not always amusing job. But I hope the book at the end will be a success, and I am prepared to contribute to it as much as I can. I also can tell you by experience that you will probably become more interested in the work, when you shall go into it more deeply. But perhaps you are too much unhappy in going on with the job, so I feel my duty to tell you that I should not mind if you wish to leave it. As much as I would regret it, I should be sorry if you continue only on account of the previous engagements. Recent events in Italy did not astonish me, since I have always thought that the situation of this country under Mussolini was rather similar to that of France under Laval.[49] The Sicilians are as good Italians as any others, but regional differences exist, comparable to those between Scotsmen and Englishmen. I am now quite accustomed to the life in Manchester, and do not dislike it. Especially I enjoy the trips into the country I am sometimes doing with my wife and children. I have proved that the square of the discriminant of any binary form (of degree >2) equals the resultant of the form and its Hessian. I feel this ought to be a well-known result, but I could not find any reference on the subject. What do you think about it?

---

to see that you have again time to deal with my MS., but not quite happy about your feelings. I have no doubts it will be a long and not always amusing job, but I am prepared to contribute to it as much as I can. I can also tell you by experience that after you would have gone deeper into the work, you will become the more interested in the work, the more deep you will go into it. The essential point is that you are quite made up your mind really willing, and not too much unhappy, in doing so. Otherwise, you had better to cease your collaboration'.

[48] B. Segre, *A note on arithmetical properties of cubic surfaces*, Journal of the London Mathematical Society, 18, 1943, pp. 24–31; *On a parametric solution of the equations $x^3 + y^3 + az^3 = b$, and on ternary forms representing every rational number*, Journal of the London Mathematical Society, 18, 1943, pp. 31–34; *On ternary non-homogeneous cubic equations with more than one rational solution*, Journal of the London Mathematical Society, 18, 1943, pp. 88–100; *A parametric solution of the indeterminate cubic equation $z^2 = f(x, y)$*, Journal of the London Mathematical Society, 18, 1943, pp. 226–233.

[49] Pierre Laval (1883–1945), president of the Council of Ministers of the French Republic from 18 April 1942 to 20 August 1944. After the end of the war he was tried and sentenced to death for his leading role in the Vichy Republic and for having been mainly responsible for the policy of collaboration with Germans.

D6.2.6 J. A. Todd to B. Segre, Cambridge 8.9.1943

CA, *BSP*, fol. 1r-v

Dear Segre, Thanks for your letter. I shouldn't *dream* of throwing my hand in after being at the job—off and on—for three years; I was only trying to hint to you that it will be a long time before I get through. I am getting ahead with the chapter on singularities, pruning it here and there; the analysis of double points is an interesting and valuable illustration of the theory of neighbourhood relations, and—within reasonable limits—it certainly should be described. I think the general principle which it might be useful to follow is that the book is to certain a logical treatment of the theory of surfaces (and varieties?) as it exists and am tempted to cut out obviously new material except insofar as it is necessary to make a rigorous treatment possible or where it provides *simple* and significant applications of the theory. After all, the place for essentially new results is in a paper and I have no desire to steal partial credit for work which is entirely your own. My complaint in my last letter—which you have taken a little too seriously—was simply a feeling that your account had a tendency to over-elaboration and generalisation. The above can't be taken *too* literally; for instance, if I were in possession of a criterion for the rationality of a V_3 (which of course I am not) I would not exclude it on the grounds of novelty. But a number of things of less importance could be shortened. I am trying—as an experiment—to make some slight rearrangements in your order of treatment in the chapter on singularities, but as I am only part of the way through, yet I cannot be certain whether they will be an improvement. Your remark about the Hessians of binary forms is new to me and interesting, but if I start playing about with it, I shall neglect more important matters. I will mention it to some of our invariant experts sometime. Yours very sincerely J.A. Todd

D6.2.7 Draft of a Letter by B. Segre to J. A. Todd, [Manchester] 10.9.1943

CA, *BSP*, fol. 1r-v

Dear Todd, I was very glad of receiving your clarifying letter. I understand and appreciate your delicacy in fearing to steal credit, but there is no need for it. Personally, I think that our collaboration could better succeed if no distinctions about our respective contributions were made, apart perhaps in a few lines of the Preface. On the other hand, if you think that the author of new results has to be mentioned, this can be done every time at its proper place, or in the additional sections I have suggested in my last letter. The important point is that no significant result has to be sacrified on account of such personal, even if praise-worthy reasons. On this respect, I should find rather difficult to give a criterion for defining what you call an "essentially new result" and for deciding *where* such results could exceptionally be included. For instance, I gave in my MS. many 'new proofs' of 'old results', whose known proofs were incomplete or, for other reasons, unsatisfactory. Some of these 'new proofs' could not be avoided without impairing the structure of the book, and, on the other hand, it seems to me that your reasons of delicacy would stand also for them. The question is different for new but unimportant results, and I agree on the need of cutting them altogether, or reducing them as suggested in my previous letter. But I warn you that some apparently trifling results may be of consequence for further developments in the M.S. For instance, some of my 'new results' on singularities are essential in a subsequent 'new proof' of [Max] Noether's result about rational quartic surfaces, as well as in the discussion (only hinted in the MS) of the rational quintic surfaces. Moreover, they could probably be of importance in other respects, as e.g., the study of the systems of equivalence. In conclusion I think that we should not merely give a rigorous account of already substantially settled questions, but also try to inspire fresh trends of ideas and new researches. This task is certainly difficult, especially in view of the restricted space allotted to us, but by no means impossible. Yours very sincerely

D6.2.8 J. A. Todd to B. Segre, Cambridge 3.10.1943

CA, *BSP*, (B). fol. 1r-v
Dear Segre, Just a note to report progress before term starts and puts an end to everything. I have completed what I think is an adequate account of singularities, including a careful exposition of Du Val's proof of the resolution of singularities of a surface.[50] I then started off on the chapter on postulation formulae (the last in my possession) and here I got badly side-tracked by trying to generalise one of your theorems. You define inductively the linear systems of forms satisfying certain conditions (B) by what amounts to specifying the nature of the general prime section and adding conditions imposed at isolated points; and you prove that in certain cases [. . .] it is enough to consider the behaviour of a general prime section belonging to a pencil not specially related to (B). It seems to me that this ought to be sufficient in all cases, and I endeavoured to construct a general proof, but without success, although I think I can get an algebraic (not algebro-geometrical) proof of the theorem for surfaces in [3]. I suppose there is no

---

[50] Patrick Du Val (1903–1987), PhD in Cambridge (1930), belonged to H. F. Baker's research group, together with Donald Coxeter, Leonard Edge, William Hodge, John Semple, and John Arthur Todd. Du Val visited Rome (1932), Princeton (1934) and in 1936 was appointed as an assistant lecturer at Manchester University. Du Val submitted an essay on the resolution of singularities of an algebraic surface for the Adams Prize at Cambridge in 1936 (American Journal of Mathematics, 58, 1936, pp. 285–289). Todd probably refers also to P. Du Val's papers, *The unloading problem for plane curves*, American Journal of Mathematics, 62, 1940, pp. 307–311 and *The Jacobian algorithm and the multiplicity sequence of an algebraic branch*, Revue de la Faculté des Sciences de l'Université d'Istanbul, ser. A, 7, 1942, pp. 107–112.

reasonable doubt that the theorem is true. I cannot imagine the sort of circumstance which would make it fail. Have you anything to say about it? Anyhow,
I spent quite a lot of time on it without getting anywhere. I have just seen your two
most interesting notes in the *Journal* and am looking forward eagerly to reading
your detailed account of the subject elsewhere. You have broken entirely new
ground here, and it all seems very interesting.[51] I hope you will not consider it an
impertinence if I congratulate you on the splendid clear way in which you have
announced your results; your mastery of English has improved unbelievably. I am
pretty sure that the style is your own: if any improvement has been affected in it,
I would assert with confidence from what textual critics call 'internal evidence',
that Mordell had no hand in it.[52] Remember me to your wife, and to Prof.
Mordell; also to Lord if he is still in the Department. Yours sincerely J.A. Todd

D6.2.9 Draft of a Letter by B. Segre to J. A. Todd, [Manchester] 8.10.1943

CA, *BSP*, fol. 1r-v

My dear Todd, I was very glad to receive your letter, and I thank you for your
friendly appreciation of my two notes.[53] These will be followed in the Journal by
other two.[54] In one of them, I prove the result quoted by Mordell in his note just
appeared (but presented *after* mine!), while in the remaining one I put the final
touch to theorem 1 of my note I, by solving parametrically the cubic Diophantine
equation $z^2 = f(x, y)$, where f is any rational cubic polynomial in x, y which cannot
be written as a polynomial in a single linear function of x and y. I have already
obtained several additional results on cubic surfaces.[55] One of them, by means of
which theorem VIII of note I follows at once from theorem VII of the same note,
is that "a non-singular cubic surface contains no homaloidal linear system of

---

[51] B. Segre, *A note on arithmetical properties of cubic surfaces*, Journal of the London Mathematical Society, 18, 1943, pp. 24–31; *A parametric solution of the indeterminate cubic equation* $z^2 = f$
(x, y), Journal of the London Mathematical Society, 18, 1943, pp. 226–233.

[52] Louis Joel Mordell (1888–1972), Fielden Chair of Pure Mathematics at the University of
Manchester (1923–1945), had introduced Segre in the geometry of numbers, of which he was
considered a leading specialist.

[53] B. Segre, *A note on arithmetical properties of cubic surfaces*, Journal of the London Mathematical Society, 18, 1943, pp. 24–31; *A parametric solution of the indeterminate cubic equation* $z^2 = f$
(x, y), Journal of the London Mathematical Society, 18, 1943, pp. 226–233.

[54] B. Segre, *On ternary non-homogeneous cubic equations with more than one rational solution,*
Journal of the London Mathematical Society, 18, 1943, pp. 88–100; *A parametric solution of the
indeterminate cubic equation* $z^2 = f(x, y)$, Journal of the London Mathematical Society, 18, 1943,
pp. 226–233.

[55] B. Segre, *On arithmetical properties of singular cubic surfaces,* Journal of the London Mathematical Society, 19, 1944, pp. 84–91.

complete intersections." An extension of this result to the non-singular V^3_3 in [4], would obviously prove its irrationality. I was told by Fano that this irrationality has been very recently proved by him, on considering the linear systems of surfaces of genera 1 lying on V^3_3, but I have not seen the proof.[56] I feel that one should be able to obtain the result also by my methods, but I have not yet had time of thinking seriously about this. I have instead accomplished an extensive arithmetical research on the quartic surfaces containing some lines (incidentally, I have proved that 64 is the *maximum* number of lines which may lie on a non-singular quartic).[57] Moreover, I have in mind a number of arithmetical questions upon algebraic surfaces of any orders. I should like to deal systematically with this subject in a book, for which, as I said, I already have much material. I knew through Mordell that Hardy would accept a Cambridge Tract; but, I am afraid, the stuff would be too crammed in it. Hence, I suggested to Professor Sisam to write another book for the Oxford Press.[58] My idea was, first, accepted in principle, then rejected on the ground that my work (which of course is now far from being finished) is a specialized original research to be published as a memoir and not as an ordinary book. So, I have just written to Veblen, in order to see whether I could be more successful in America. At present, I consequently do not know when and where you will be able to see a detailed account of my work. However, you could hear more of it, if you will come to a Lecture I shall probably give to the London Math. Soc. on *Arithmetic upon an algebraic surface*.[59] I should be glad to meet you on such an occasion, and we could then also have a talk about our book [*Geometry upon Algebraic Surfaces and Varieties*]. When writing the chapter on postulation formulae, I too have spent a lot of time in trying to establish (b) in [*n*]. But, for $n > 3$, I succeeded in it only by adding some further hypotheses, like (a). As far as I recollect now, I rather doubted then of the unconditional validity of (b) in [*n*]. This question is implicitly suggested by my MS., and many more questions arise similarly from the following chapters. Such questions could be especially stated in the additions, which (as I have previously suggested) we could put at the end of each chapter. My wife, Professor Mordell and Lord, all kindly remember you. Yours Segre

---

[56] The work by G. Fano, *Nuove ricerche sulle varietà algebriche a tre dimensioni a curve-sezioni canoniche* (Commentationes. Pontificia Academia Scientarum, 11, 1948, pp. 635–720), ready since 1942, had been presented by Severi to the Pontifical Academy on February 21, 1943, but was published only in 1948.

[57] B. Segre, *The maximum number of lines lying on a quartic surface*, Quarterly Journal of Mathematics, 14, 1943, pp. 86–96.

[58] Segre proposed to the Clarendon Press a plan of a book on *Arithmetical properties of Cubic Surfaces* in July 1943 but the Delegates' advisers refused because 'it was essentially a memoir on a specialized original research, and not an ordinary book such as a book publisher could undertake'. See CA, *BSP*: K. Sisam [The Clarendon Press] to B. Segre, Oxford 12.7.1943 and 28.7.1943.

[59] B. Segre, *Arithmetic upon an algebraic surface, a lecture given at the London Mathematical Society* on December 16, 1943, published in Bulletin of the American Mathematical Society, 51, 1945, pp. 152–161.

D6.2.10 J. A. Todd to B. Segre, Cambridge 25.7.1944

CA, *BSP*, fols. 1r-3r

Dear Segre, I was horrified to find that I have been two months without answering your last letter. My friends know I am a very poor correspondent, but I will try to make up for last time now. I have had the busiest year of my life and can't say it was all very pleasant. During the first two terms I was teaching something like 27 hours a week, most of it elementary work with scientists. The third term was a bit easier and now the vacation has come I have got down to work. I have been asked to write a textbook on projective geometry, suitable for Part II of the Tripos here, and have been busily engaged on the first draft of it for the last five or six weeks.[60] Scott is giving to have a look at it when I have finished (and it is going well), and then the real job will be putting it into decent shape.[61] In the course of writing this draft I have made several alarming discoveries, chief among which is that the objections raised by non-geometers as to the lack of rigour in the usual treatment of projective geometry are amply justified. (We discovered the same sort of thing a few years back, at a higher level, but I'd no idea that the rot extended down to quite simple matters). For instance, the usual argument which defines a twisted cubic curve as the product of three projective pencils of planes, and deduces the form $(\theta^3, \theta^2, \theta, 1)$ for the coordinates of the general point is full of assumptions that need careful consideration [as is clear from the fact that the current accounts seem to be silent about the *possibility* that the cubic is degenerate, let alone giving conditions which will ensure that it is not]. I am tempted (most reluctantly) to the conclusion that the 'elegance' and 'simplicity' of many of the familiar synthetic arguments used in projective geometry arise from the fact that inconvenient possibilities and special cases are ignored without any discussion at all. My account of the subject will therefore scandalise the 'pure' geometer by its excessive use of analytic methods. As these consist in part of a systematic use of matrices, it ought to be a useful sequel to a course of Algebra based (say) on Ferrar's book.[62] Incidentally, I propose to talk about invariant factors and elementary divisors. These usually come in works on algebra where no one ever reads them, and in my view, it is nonsense to discuss (say) pencils of quadric surfaces without them. My aim is a textbook, not too long, which contains the *essential* ideas in projective geometry in S_2 and S_3, without the elaborate discussion of (e.g.) special properties of conics which lead to obscure the really

---

[60] J. A. Todd, *Projective and analytical geometry*, New York, Pitman Publishing Corporation, 1946.

[61] David Bernard Scott (1915–1993), a pupil of W. V. D. Hodge, was a lecturer in mathematics at Queen Mary's College, London (1939–1946).

[62] W. L. Ferrar, *Higher Algebra*, Oxford, University Press, 1943.

fundamental theorems of the subject. It has been great fun writing it! I should like a copy of your *Quarterly Journal* paper on lines on a quartic surface, if you have one to spare.[63] I took the liberty of borrowing a result you once showed me (about the cross ratio (ABCD): (A'B'CD) where AA', BB', CC', DD' formed an involution on a conic) to set in the Tripos this year, and you will be pleased to hear that most of the candidates found it quite easy. With best wishes, Yours sincerely J.A. Todd

D6.2.11 Draft of a Letter by B. Segre to J. A. Todd, [Manchester] 28.7.1944

CA, *BSP*, fol. 1r
Dear Todd, I have read with interest your last letter, even if not all in it is quite clear and acceptable to me. I do not know, for instance, what you mean exactly by the lack of rigour in the "*usual treatment* of projective geometry". In fact, I know more than half a dozen perfectly satisfying accounts of the subject which of course does not imply that nothing valuable can yet be added to it. The authors of these accounts (including Castelnuovo, Enriques, Severi, Fano and Bertini with whose treatises you are apparently unacquainted) may sometimes use the expression 'in general', but I do not deem this to be a worse crime than to say, 'let *a* be an arbitrary real number'. In either case the exact significance, or better the significance for the *constructive* point of view, of such an expression may be questioned; but the mere fact of not bothering about this, cannot be called lack of rigour and still less 'rot'. I understand, however, that you may have slightly overstated your case, and I am eager to see your book before being 'scandalized'.[64] If at present you are not using my MS. on algebraic surfaces and varieties, would you please send it to me, as I have need of it myself?[65]

D6.2.12 J. A. Todd to B. Segre, Shipley Yorks 5.8.1944

CA, *BSP*, fols. 1-11
Dear Segre, Your letter has been forwarded to me here, where I am spending a short holiday (I return to Cambridge next Wednesday). You have evidently

---

[63] B. Segre, *The maximum number of lines lying on a quartic surface*, The Quarterly Journal of Mathematics, 14, 1943, pp. 86–96.

[64] J. A. Todd, *Projective and analytical geometry*, New York, Pitman Publishing Corporation, 1946.

[65] It is the manuscript of the volume by Segre and Todd, *Geometry upon Algebraic Surfaces and Varieties*.

misunderstood me on two points, and in both cases, it is my own fault. In the first place, the word 'rot' was used in an expression ('stop the rot' I believe) which is a colloquialism. I ought never to have inflicted on you, and was not intended to imply that the objectionable things I referred to were 'rubbish'. The more serious point is the '*usual treatment*' by which I meant the treatment in the current *English* textbooks, which are about the only ones largely read here except the lovely work of Reye (read only by specialists) and the elegant but rather limited little book by Duporcq.[66] I have often heard - and can well believe—that the Italian texts are excellent. Unfortunately, they are little known over here. I have never read them (though I should very much like to see one) and there are none in the Faculty Library at Cambridge. I have not looked in the University Library yet, but with your list of authors I shall certainly do so. If you consider this ignorance on my part reprehensible, I can only say that most of my serious geometrical work has been in the algebraic field, on which I possess (and have mostly read) all or almost all the Italian works I have heard of. Perhaps it might be to the point to make my objections a little more precise. I speak only of books written from a modern point of view, which deal fairly comprehensively with the subject, which rules out books like Filon (who seems to think projective geometry is a branch of descriptive geometry!) and O'Hara & Ward (which is too elementary).[67] This leaves, as far as I can see, only three works

(a) *Veblen & Young*
(b) *Baker*
(c) *Sommerville*.[68]

The latter is analytical and largely metrical, though in its way a very good book indeed (especially his *Analytical Conics*, which is the best account since Salmon in English and more up-to-date).[69] As to *Baker*, its defects *as a students text-book* are pretty obvious, and are due partly to the author's rather difficult style but mainly to the enormous mass of detail which makes it difficult to pick out the fundamental ideas from the less important things. To my mind, *Veblen & Young* is the best of the lot, and I can't understand why it isn't better known. But its very abstract method makes it 'difficult' for undergraduates. My aim is a book on projective and analytic geometry which covers the main topics adequately and

---

[66]T. Reye, *Lectures on the geometry of position. Part I. Translated and edited by Thomas F. Holgate*, London, Macmillan & Co., 1898; E. Duporcq, *Premiers Principes de géométrie moderne*, Paris, Gauthier-Villars, 3rd ed. 1938.

[67]L. N. G. Filon, *Introduction to projective geometry*, London, Arnold & Co., 1935; C. W. O'Hara, D. R. Ward, *An introduction to projective geometry*, Oxford, Clarendon Press, 1937.

[68]O. Veblen, J. W. Young, *Projective Geometry*, Boston and London, Ginn, 2 vols. 1919–20; H. F. Baker, *Principles of Geometry. Vol. 1 Foundations, vol. 2 Plane Geometry*, Cambridge, University Press, 1922; D. M. Y. Sommerville, *An introduction to the geometry of n dimensions*, London, Methuen & Co., 1929.

[69]G. Salmon, *A treatise on the analytic geometry of three dimensions*, sixth edit. revised by R. A. P. Rogers, London, Longmans, Green & Co., 1914.

accurately. The sort of difficulty that I alluded to may be illustrated by a concrete example: the deduction of the parametric form $(\theta^3, \theta^2, \theta, 1)$ of the variable point of a twisted cubic curve when this is defined as the locus of points of intersection of corresponding planes of 3 projective pencils. The argument in *Baker* runs in essence as follows. The variable point of the curve is given by $u_i - \theta v_i = 0$ $(i = 1, 2, 3)$ where u_i, v_i are six general linear forms (so that for general θ the three equations are linearly independent). From these we get $x{:}y{:}z{:}t = f(\theta){:}g(\theta){:}h(\theta){:}k(\theta)$ (1) where $f\ g\ h\ k$ are certain cubic polynomials in θ. If these are linearly independent, then we can solve for $(\theta^3, \theta^2, \theta, 1)$ and by a change of coordinate system obtain what we want. It remains to prove their independence, and here the 'snags' arise. Baker (who is rather more cautious than some others—e.g. myself in lectures on several occasions) proceeds—if I remember rightly—as follows. The equations (1) show that our curve possesses three points in every plane. [N.B. It is important for the argument that for a general plane these should be *distinct*, but this point is overlooked]. Hence if $f\ g\ h\ k$ are not independent the curve (or part of it) is a plane curve and hence (as it lies on two quadrics) degenerates into two or more parts, one of which (at least) must be a line. Baker then states *as an assumption* that if the projectivity is a general one, then no set of three corresponding planes ever meet in a line (and from this the result follows). I have quoted here from memory, but that is, I think, the gist of it. The point is that *there are two assumptions* which ought to be justified when the projectivity is general;

(a) that the polynomial $af(\theta) + bg(\theta) + ch(\theta) + dk(\theta)$ has no repeated factor for *every* value of a, b, c, d and
(b) essentially that the polynomials $f\ g\ h\ k$ have no common factor involving θ (and it is a simple matter to show that if (b) is proved (a) follows, e.g. by considering the discriminant of a pencil of binary cubics. I know all about Bertini's theorem, but that's too highbrow here!)

The point is that Baker never proves (b) except by the unsatisfactory a posteriori argument (that if you start off with $\theta^3\ \theta^2\ \theta\ \iota$ you can get your projective pencils). Something more direct is surely desirable. Incidentally it is a curious characteristic of Baker (for whom I have the highest personal regard) that he generally seems to recognize the difficulties which arise, and then deliberately evades them by the ingenuous remark 'we shall suppose'. Some of us feel that this sort of thing isn't quite good enough in 1944. Hence my attempt to tidy up things a bit. If you want another example, I suggest that you try to construct a correct statement and proof of the important converse of Pascal's theorem [the 6 points being distinct]—important because it is a criterion for 6 points to lie on a conic. The difficulty here is in the possibility of the conic being degenerate, and I found the complete discussion took rather longer than might have been expected, as several special cases have to be considered. I could go on like this for hours, but I fear I have already tired your patience: I hope at any rate I have allayed your fears. I will send your MSS book when I get to Cambridge next week. I will also send on what I have written. I've nearly forgotten all I ever knew about varieties: teaching elementary mathematics for everything up to 27 hours a week is not good for the memory where serious matters are concerned. I really think the war in Europe won't last very much longer—at any rate at the present rate of progress. At any rate I am more optimistic than I have been for a long time. I hope you are having pleasant vacation and will get a holiday sometime in it. My kind regards to your wife, Yours very sincerely J.A. Todd

D6.2.13 J. A. Todd to B. Segre, Cambridge 3.3.1945

CA, *BSP*, fols. 1r-2v

Dear Segre, Thank you for your letter. I was interested in your news of Enriques and Castelnuovo. It is good to hear that Enriques is restored to his chair, but surely he must be nearly due to retire: he cannot be a young man.[70] I must frankly say that I read the first part of your letter with mixed feelings. It made me realise very clearly how much geometry I have forgotten in the past five years. In my present state the idea of writing about algebraic surfaces and varieties horrifies me. This is all my fault, I know, but I have had little inclination for serious working these last few years. When the question of the book [*Geometry upon Algebraic Surfaces and Varieties*] was first mentioned you may remember that the suggestion of a collaboration was made, in the first instance, because you had had little experience of writing in English for an English audience. I am quite certain that these grounds are no longer valid, as for the past few years you have been proving out papers at a great rate! I feel, now, that I should only be acting as a brake on your efforts if I tried to get down to the job again; and I mistrust my judgement in dealing with the delicate points of the theory after such a long time during which I have been unable to think about it. Because I am anxious to see the book in print with the minimum of delay, I feel that it would probably be best if you undertook the task yourself. You would, of course, then be the sole author of the book. You are very welcome to make any use you please of the manuscript that I wrote out (or of any part of it). There is just one point here, and that is that Du Val's permission ought to be obtained before you include (if you decide to) his unpublished proof of the resolution of singularities of a surface.[71] I shall be very pleased to read your MSS when you have prepared it (if that is your wish) and to help with the proof sheets. But I don't honestly feel able to write myself. I hope you will not be offended at this suggestion. Had times been normal this would have been over and done with years ago. Your teaching duties have been less enacting than mine and fortunately have left you with enough mental vigour to produce some substantial mathematics. *My* mathematics seems to me, at the

---

[70] Enriques, Castelnuovo and their families had fortunately escaped deportation. After the liberation of Rome (4.6.1944) Enriques, 73 years old, had been reinstated in the chair of Higher Geometry.

[71] See D6.2.8.

moment, to have been a war casualty! So go right ahead, and the best of luck to you! I shall be staying in Cambridge this vacation, so shall not be able to see you in the North! But I expect to be in the North of England sometime during the summer and should be delighted to renew our acquaintance. I have been trying out part of my book on projective geometry this term in lectures.[72] It has been quite good fun. I hope to get *this* manuscript to the printers during the summer. It is of course a much less ambitious work than *Surfaces and Varieties*, but I believe it fills a gap in the literature. Yours very sincerely J.A. Todd

P.S. I don't quite know how much about 'varieties' you want to put into the book. The original idea (or one of them) was that I should try and give some account of the covariant systems of equivalence (as in my LMS papers).[73] I think I got rather discouraged about this when I started trying to make things 'rigorous'. But perhaps this was the left over till volume 2. Keep my MS. in case it's [of] any use to you. But you might let me have my Enriques (on surfaces of genus zero) book when you've finished with it, as it is rather inaccessible here.[74]

D6.2.14 Draft of a Letter by B. Segre to J. A. Todd, [Manchester] 8.3.1945

CA, *BSP*, fol. 1r-v
Dear Todd, Thank you for your friendly letter. No, I am not offended at your suggestion, but I should regret very much to have to do without your collaboration; the more so, since your present loss of interest in algebraic surfaces and varieties may be a temporary one. If this is so, and your proposal is chiefly due to your fear of delaying the book too much, I should suggest to wait still for a while before taking any decision, say until the summer, when you will have finished your book on projective geometry and we shall be able to meet. If, however, you really think you will not change your mind in a few months time, then I shall have to go on myself with the book. In this case, I think we should make the Clarendon Press acquainted with our change of programme; perhaps you could explain the situation to them and ask for their approval. I shall then gladly accept your very generous offer of availing myself of your MSS as well as of your help with the final MSS and the proof sheets; I thank you now most heartily for your kindness, which I shall of course acknowledge publicly in due time. I do not know yet how much about 'varieties' can be put in the book. The more ambitious plan of giving a fairly complete and rigorous account of the subject would probably entail new

---

[72] J. A. Todd, *Projective and analytical geometry*, New York, Pitman Publishing Corporation, 1946.

[73] J. A. Todd, *The geometrical invariants of algebraic loci*, Proceedings of the London Mathematical Society, s. 2, 45, 1939, pp. 410–424 and *Invariant and covariant systems on an algebraic variety*, Proceedings of the London Mathematical Society, s. 2, 46, 1940, pp. 199–230.

[74] F. Enriques, *Sulla classificazione delle superficie algebriche particolarmente di genere zero. Lezioni raccolte dal dott. Luigi Campedelli*, Roma, Tipografia del Senato, 1934.

extensive researches, and I doubt very much whether it can be carried out at the present stage. An important element for deciding about this may be given by Severi's book on series and systems of equivalence republished recently, of which, however, I have only been able to see the first few fascicles, appeared just before the war.[75] Be sure that I shall let you have your Enriques back as soon as I have finished with it (or at any moment you should require it).[76] Yours very sincerely [B. Segre]

D6.2.15 K. Sisam [The Clarendon Press] to B. Segre, Oxford 25.9.1945

CA, *BSP*, fol. 1r
Private.

Dear Professor Segre, Your letter of 17th July was brought to my attention when I came back from holiday, but my colleague here has been away for one reason or another ever since, and I must seem neglectful. Before writing to you, I thought I had better make enquiries in Cambridge about Todd. I hear that he has been extraordinarily busy, and has been diverted to the writing of a textbook for undergraduates, which was bound to deflect his mind and time from the stiffer task of an advanced book.[77] I believe the textbook has now been completed: Todd is still at 'The Hermitage', Silver Street: and it may be that he would now be more ready to come back to the work. I gather that one of his difficulties was that he was attempting to fill the defects in the literature of his part and that the task was too large. Professor Hodge tells me that he thinks the best chance of continuing the collaboration would be for you to have a conference with Todd, and discuss the plan, and possibly redistribute some of the work. It is almost impossible to collaborate by long-range correspondence, and I think we, in England, are least happy in that art. Anyhow, I am writing to Todd to ask him whether a personal consultation could not be arranged, and I have some hope that this may lead to a clarification. Professor Hodge, as always, is very anxious that the plan should be carried through, and ready to help with his advice. But this you must treat as private information. I have been diverted from the question of your future. We have all too few geometers here, but I think it likely that you will be urged to go to Italy, which must also be in great need of scholars, and that if you could defer the actual going for perhaps a year, until things have settled down there, you would go back under good conditions. At least, that seems to be the best opinion here. In this country there are always difficulties of naturalization and familiarity with

---

[75] F. Severi, *Serie, sistemi d'equivalenza e corrispondenze algebriche sulle varietà algebriche*, ed. by F. Conforto and E. Martinelli, Roma, Cremonese, 1938, republished in 1942.

[76] F. Enriques, *Sulla classificazione delle superficie algebriche particolarmente di genere zero. Lezioni raccolte dal dott. Luigi Campedelli*, Roma, Tipografia del Senato, 1934.

[77] J. A. Todd, *Projective and analytical geometry*, New York, Pitman Publishing Corporation, 1946.

local ways which make it hard to work at any but the most advanced levels; and at those levels there are comparatively few posts outside of Cambridge. In any case, I hope that the project you started with Todd will somehow end in a first-rate contribution to the subject. Yours sincerely, Kenneth Sisam

D6.2.16 J. A. Todd to B. Segre, Cambridge 26.9.1945

CA, *BSP*, fols. 1r-3v

Dear Segre, On arriving back here this evening from a day visit to London, where I delivered the manuscript of my textbook to the publishers,[78] I found a letter from Sisam awaiting me. He is apparently anxious that we should get together and have a talk about the book [*Geometry upon Algebraic Surfaces and Varieties*]. I think this is a good idea: we have rather lost contact recently, and if anyone is to blame it is probably myself. With my textbook off my hands and (as I hope) a more normal term in front of me, I hope to be able to start thinking seriously about algebraic geometry again. You will remember that when I returned your manuscript early in the summer, I sent you a long manuscript of my own, whose contents, I must admit, I am now somewhat fancy about. My impression was that it consisted (a) of an account of the fundamental ideas, *à la* van der Waerden, which looks short just about at the point where it looked like being useful because I got into difficulties over virtual intersections or something and (b) a second 'chapter' based on the early part of your manuscript with a good many glosses (both literary and mathematical) of my own.[79] I forgot how far this went, but I fancy it embodied Du Val's proof of the reduction of singularities of a surface.[80] What exactly have you been doing with your end of it; I think even before you left Cambridge there was a substantial part of your pencil copy that I hadn't read? You suggested in your letter, in reply to the one I enclosed with your MS. that you hoped I should reconsider the idea of collaboration. Sisam also urges this, and I am disposed to agree. But the difficulty from my point of view is (i) that I have forgotten a lot about the subject in the last few years and (ii) that we always seemed to be rather at loggerheads as to the scale and scope of the work. To which I should add that I can't help wondering whether Zariski's recent work (which I have not yet found time to read) doesn't render any other formulation of the fundamental principles out of date.[81] This is merely a doubt, but if anyone asks me to give an account of Zariski's papers I shall go on strike! Shall you be visiting Cambridge in the near future? If so, it would be a good thing to get

---

[78] J. A. Todd, *Projective and analytical geometry*, New York, Pitman Publishing Corporation, 1946.

[79] See D6.2.3.

[80] See D6.2.8.

[81] In the war years (1939–1945) Zariski had published 18 works of algebraic geometry. Due to the war, his reprints had had limited circulation outside the United States.

together and have a talk; I should be very glad to see you. Perhaps you could give me some idea as to how the joint manuscript stands, how much remains to be done, and how much of it is in a 'fair copy' state. We should then, at any rate, know where we were. There might be something to be said for concentrating on surfaces, and leaving the more advanced theory of varieties (canonical systems, which I believe I invented but have forgotten all about) to a second volume if the 'surfaces' part is within sight of completion. A nice manuscript would obviously be balm to Sisam's heart anyhow! Term starts here in just under a fortnight. If you could get up here next week for a day or two, I should be delighted to see you, and could probably find somewhere for you to sleep. Yours sincerely J.A. Todd

PS. By the way, my publishers seem anxious to get my textbook out as soon as possible. Hence *hope* to have it on sale in a year. I am interested in the reception it gets from geometers! I'm sending this to Burton Rd. Hodge told me he thought you'd moved, but didn't give me your new address.

D6.2.17 Draft of a Letter by B. Segre to J. A. Todd, [Manchester] 29.9.1945

CA, *BSP*, fol. 1r-v
Dear Todd, I have just received your letter, and am glad to know that, after all, you are prepared to reconsider the idea of collaboration. (I hope not only in order to please Sisam). In the incertitude I was about your decision, I have left the joint MS. as it stood. Hence, I myself have forgotten a lot about it, but you may be able to reorientate yourself from the index of my MS. I enclose). As far as I can remember, I had previously noticed that the two parts of your MS were not, somewhat, so closely related as one could wish; your 2nd part follows my 1st part, discarding however some results, as e.g. those on postulation formulae, and including what you call Du Val's proof of the reduction of singularities of a surface.[82] I had developed this subject since my part 1 deals with V_k and the 2nd gives the more advanced results on surfaces; I think my analysis of the question is far longer than yours which is probably necessary for the rigour (Du Val himself told me that he has been working recently on some such lines). My second part is completed excepting for the last two paragraphs, and the MS. (although in pencil) is quite legible. I agree with you about what you say on Zariski's work. If, as I have always thought to be preferable, the work can be distributed in two books, then book I could be ready for publication in a comparatively short time, and I

---

[82] See D6.2.8, D6.2.13, and D6.2.16.

hope you would undertake this job. At the moment I could give only rather limited help, since the preoccupation for my future gives me now little respite for serious work. I think, however, our respective roles could be reciprocated when preparing book II, thus putting our collaboration on fairer terms. I shall be delighted to see you again. I regret, however, that I cannot go to Cambridge next week, since term starts here on Thursday next, and I have still to prepare some lectures for pass students I shall give now for the first time. But I hope you will come to Manchester before starting your lectures, so that we can have a nice talk together and discuss the matter in detail. I shall certainly be interested in your textbook, and should like to go through the proofs, if you wish to have my impressions (and possible suggestions) on it before publication. Yours sincerely [B. Segre]

P.S. Will you please communicate my new address to Hodge.

D6.2.18 J.A. Todd to B. Segre, Cambridge 3.10.1945

CA, *BSP*, fols. 1r-2v
Dear Segre, Thanks for your letter. Our meeting will evidently have to be postponed, as I rather expected with term approaching. I took the liberty of showing your letter to Hodge, who, as you know, has always been interested in our project. He and I are both inclined strongly to your view (tacitly expressed in your letter) that the only apparent chance of getting anything moving in a finite and predictable time is to publish in two volumes. Question new arises, what are the contents of vol. I to be. Both on scientific and practical (i.e. from the point of view of available material) grounds it seems to me that this might usefully consist of general theory—i.e. your first part (substantially) together with anything suitable in my treatment on van der Waerden's lines.[83] This leaves Part II for the detailed theory of surfaces. Whether canonical systems come in Part I, Part II or a hypothetical Part III is a subject for discussion. If you agree (in principle) with this scheme it has the very definite advantage that for most of the stuff there is a fair copy to work on. Subject to your approval, I would then suggest the following programme. Would you read my two manuscripts carefully, comparing my recension of your Part I with yours and noticing any omissions you feel to be important (I had reasons of my own when I left anything out, which may or may not have been good: my version, based on yours, is only to be taken as a suggestion) and draw up suggestions as to how far the available work will serve as a draft for Part I. When we have got this settled, I will try and prepare a copy for press. I have the feeling that most of what I wrote out before will not

---

[83] See D6.2.3 and D6.2.16.

require much alteration. But you will have views of your own, and I shall expect to hear them. I should like to know what you think of this scheme as a working plan. Thank you for your offer about the proof sheets.[84] As a matter of fact, I have arranged with Maxwell that he should help me: he has the tactical advantage of living next door, which is useful when speed is a consideration (and everyone concerned wants to get this thing out as soon as possible).[85] So I hope you won't feel offended if I decline your kind offer, on the grounds that the extra time in the feat might cause delay. Besides, you might be busy on Part II yourself then! If time were not so pressing your impressions would be most valuable and I should have been very glad to have them. But Scott's verdict on my MS. contained so few points of criticism that I thought it safe to go ahead more or less simple handed.[86] I will send you a copy of the book when it comes out. Yours sincerely J.A. Todd

D6.2.19 J.A. Todd to B. Segre, Cambridge 7.12.1945

CA, *BSP*, fols. 1r-2r

Dear Segre, Thank you for the manuscript, which arrived safely the other morning. Parts of your covering letter are not very clear. For instance, which *Cambridge Tract* are you referring to? I know of none that is relevant. I shall have to get down to a serious studying of the work in the light of your comments, as soon as I have cleared off one or two old jobs of my own (including writing up a longish paper on the invariant theory of the pencil of conics, on which there is a lot to be said that hasn't been said before).[87] I expect to be coming north (to Shipley) just after Xmas to stay with relations there. This will be a holiday visit in the main, but I could easily get over to Manchester from there to see you for a day (it is only about an hour's run by train) and I think there will be a lot of points to discuss that we shall have to talk over together. I hope we shall be able to arrange a meeting. I shall be in Shipley probably 10 days or a fortnight (from about 28 Dec.). The manuscript you sent me is longer than I had expected, but as it's all in my own handwriting I suppose I really did write it once upon a time. But I'm not joking when I say that I've forgotten a great deal of this sort of geometry. I have been very busy again this term and didn't look like having much less teaching next term either. But I have been doing some work of my own—on

---

[84] Todd is referring to the proof sheets of his volume *Projective and analytical geometry*, New York, Pitman Publishing Corporation, 1946.

[85] Edwin Arthur Maxwell (1907–1987) a pupil of H. F. Baker, was a junior bursar at Queens' College, Cambridge (1933–1946).

[86] See D6.2.10.

[87] J. A. Todd, *The 'odd' number six*, Proceedings of the Cambridge Philosophical Society, 41, 1945, pp. 66–68; *Covariant line complexes of a pair of quadric surfaces*, Proceedings of the Cambridge Philosophical Society, 41, 1945, pp. 127–135.

invariants of multiple binary forms and so on, and have worked out complete systems for the double binary forms $a_x^3 a_y'$, $a_x^4 a_y'$ and for the form $A_x^2 a_y$ where x is a ternary variable and y a binary variable. This last I want to write out, and it will take a bit of time because the Algebra involved is intricate. I haven't had time to do more than glance at your last note in the Cambr. Phil. Soc.; but it looks very entertaining.[88] Yours sincerely J.A. Todd

6.3 'As a Missionary Would Explain the Gospel to the Cannibals': Teaching

Italian refugee mathematicians had extensive teaching experience, both in advanced and basic courses. Fubini and Fano had taught Infinitesimal Analysis and Descriptive Geometry, respectively, in the two-year programme of the University and Polytechnic of Turin for 30 years consecutively. Fubini and Terracini had held the two courses of Higher Analysis and Higher Geometry for the masterclasses from 1910 to 1938, and from 1925 to 1938, respectively. Levi had a still more varied teaching history, having worked both in preparatory courses, such as Algebraic Analysis, and in advanced classes. In Bologna, he had lectured in Higher Geometry. The most inexperienced from the point of view of didactic responsibilities was Segre, who had taught just six years (1932–1938). However, he too, in the years of assistantship to Fano and Severi, had carried out laboratory activities, tutored students, supervised graduate students and occasionally acted as a substitute teacher. The drafting of the handouts of Severi's lectures had been taken over almost completely by Segre and he had mentored many of Severi's pupils who spent a period of study in Rome.

Furthermore, by historical and social custom, since the Unification the role of university professor in Italy had been associated to research *and* teaching in equal proportion; for full professors (*ordinari*), the second duty even prevailed over the first. All the mathematicians who emigrated in 1938 were full professors, brought up with the example of C. Segre, who from the moment he was awarded the chair had dedicated himself entirely to teaching, even at the expense of his publications. For 36 consecutive years (1888–1924), Segre had spent the summer preparing his lectures in Higher Geometry. He was the man who, when Castelnuovo and Enriques amicably reproached him for his delay in submitting a work to the *Annali di Matematica pura ed applicata*, replied:

---

[88] B. Segre, *The biaxial surfaces, and the equivalence of binary forms*, Proceedings of the Cambridge Philosophical Society, 41, 1945, pp. 187–209.

[89] C. Segre to M. Pieri, Turin 20.11.1901, in Arrighi 1997, p. 115: *Tu trovi ancora il tempo di far ricerche: io no! Ad avere più giovani da far lavorare c'è l'inconveniente che non si ha più il tempo*

You still find time to do research: I don't! In having more young people to tutor there is the inconvenience that you no longer have the time to work for yourself! But we end up considering the work of our 'children' as our own work ...[89]

In light of these considerations, it is understandable that the expats tried as far as possible to be involved in education, proportionately more than the refugees from Central and Eastern Europe. Those who could not teach (like Fubini and Segre in the initial period) felt a strong diminution of their professional status. Testimonies agree that, even more than the research at the IAS, it was the course in Ballistics at the Columbia University that returned a smile to Fubini's face. Segre's enthusiasm when he returned to teaching and mentoring students in Manchester aroused the amused amazement of Todd, who considered this kind of duty a burden. Those who were able to immediately resume lecturing—Terracini and Levi—did so with deep joy and began to prepare the courses even before they had embarked.

Often, the courses proposed were revised versions of those held in Italy before emigration, but this does not mean that the refugees did not carry out some serious work of critical review and adaptation of contents to their new classes. Even in a highly problematic context such as that of IUC, Fano and Colombo did not allow themselves any sloppiness or superficiality in their preparation. Despite addressing students with very deep cultural gaps and veterans of tragic life experiences, they did not trivialise or simplify the content of their lectures, instead presenting the entire programmes they had outlined and examining candidates with the same toughness and no do-goodism.

The displaced scholar who stands out above the others as far as teaching is concerned is Terracini who held 17 courses in Argentina over nine years (1939–48). Terracini had a frenetic teaching commitment: three courses par year, one in Didactics within the *Profesorado*, one in Analytic Geometry at graduate level and one of Advanced Mathematics for the master's degree. In addition to presenting themes that were new in Latin America such as groups theory, Lie algebras, complex variable functions, algebraic functions, and calculus of variations, he contributed with his teaching to spreading the vision and didactical assumptions typical of the Italian School of Geometry.

Two of the courses held by Terracini in Tucumán are particularly significant (Fig. 6.2). The first, entitled *Metodología*, was a two-year foundational mathematics course, aimed at students enrolled on the teacher training programme.[90] Drafted in Turin in the summer of 1939 and refined on board the Augustus transatlantic which carried him and his family to Argentina, it was a lucid synthesis of the two Turin-oriented currents of thought which Terracini had had the opportunity to compare: those of Segre and Peano. In the first part, by adopting as a reference the Enriquesian series *Questioni riguardanti le Matematiche Elementari* (Bologna: Zanichelli,

---

per lavorare noi! Ma si finisce per considerare l'opera dei nostri figlioli come nostra propria opera
...

[90]BSM-*Terracini*: Terracini, notebook no. 19 *Metodología*.

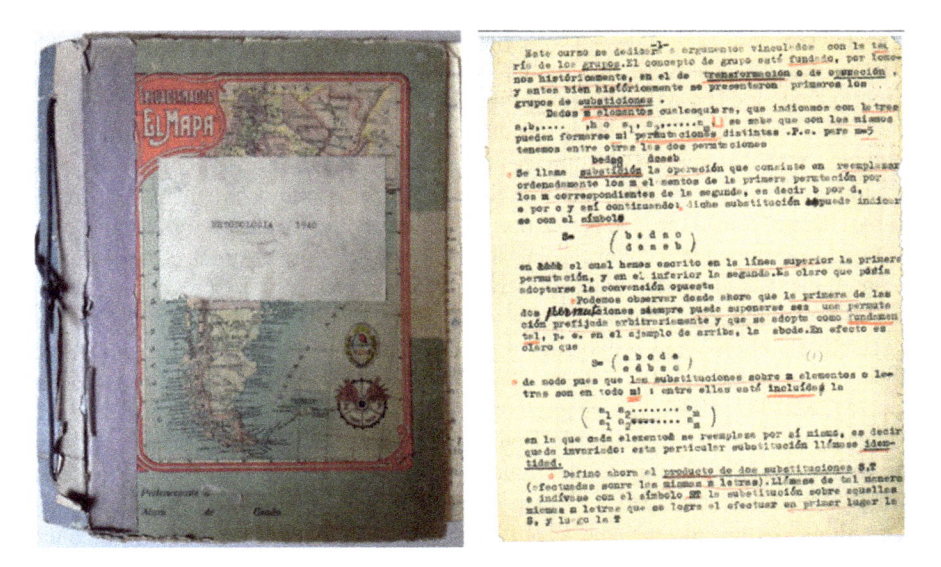

Fig. 6.2 *Metodología* and *Matématicas Superiores*, notebooks No.s 19 (left) and 28 (right), BSM-*Terracini*

1924–1927), the axioms of elementary geometry were dealt with in their technical, historical, and methodological aspects. In presenting the latter topics, Terracini reported (often *verbatim*) the contents of the lectures delivered by Segre at the Teacher Training School in Turin, which he had attended in 1910–11.[91] In his university years, however, Terracini had also been trained by Peano and, although distrusting the potential of a strict formal approach to mathematics, he had appraised the key assumptions of such procedure. Thus, in the second part of his lectures in Methodology, he presented the axioms of arithmetic using the *Formulaire de mathématiques* edited by the Peanian group (Torino: Bocca, 1894–1908), highlighted the most recent debates on meta-mathematical questions (consistence, independence, categoricity) and even taught Argentinian students to read the logico-ideographic language.

Likewise, the course of *Matemáticas Superiores* held in 1943 took up the subjects of Terracini's lectures in theory of groups and topology delivered in Turin in the year 1935–1936[92]: discontinuous groups, generatrices and their fundamental relations, the introduction to homology groups, and topological spaces, and the fundamental group had been, then, new topics for mathematics students in Turin.[93] Compared to

---

[91] Terracini's manuscripts on methodology are literal translations into Spanish of several passages of Segre's lectures delivered at the Teacher Training School from 1888 to 1921. Segre's lectures are edited in Conte et al. (2020).

[92] BSM-*Terracini*: notebook no. 28 *Matemáticas Superiores*; notebook no. 14 *Argomenti vari di geometria (topologia)*.

[93] BSM-*Terracini*: notebook no. 14 *Argomenti vari di geometria (topologia)*, pp. 1–25, 121–125, 237–281, 471–499.

the chapters dedicated to the surface characterization, homeomorphisms, the groups of transformations and Abelian groups, these concepts had represented the difficult and most advanced part of the course intended for students who wanted to do their thesis with Terracini and perhaps embark on a research career. In Tucumán, the audience was different. The sources that students could access were much more limited. Terracini took up the same subject but integrated the contents in a different frame. There was more focus on the introductory sections. A larger space was devoted to groups theory, developing the following points: abstract groups; generating elements and fundamental properties; Abelian groups with a finite number of generating elements; finite groups; Galois group and application of groups theory to algebraic equations; and polyhedral groups.[94] The most advanced chapters of the Turin lectures of 1935-1936 were reduced to the bare essentials while, from another course (a.y. 1931–1932, *Due geometri del secolo XIX Luigi Cremona e Sophus Lie*), Terracini drew some more classic chapters: group automorphisms; conjugate subgroups, invariant subgroups; substitution groups; Cremonian transformations; and plane contact transformations.[95] In proposing a first primer of Cremonian transformations, Steiner surface, and F^3 surfaces, he was convinced to offer his audience a fine portrayal of the Italian geometry Risorgimento. In turn, the lessons on Lie's theory of continuous transformation groups led him to celebrate the work of the Norwegian geometer in Argentina, on the occasion of the centenary of his birth, but simultaneously gave him the opportunity to illustrate the role of the Italian geometers (Segre and his followers) in the diffusion of this address of studies.

The success of Terracini's tuition was enormous and all students who had the chance to attend his lectures 'retained an indelible memory of the impression made by the high intellectual level, learning and clarity of presentation of this great Master'.[96] An entire generation of young Argentinian mathematicians knew algebraic and differential geometry, projective differential geometry in hyperspaces, and geometry on an algebraic curve from the synthetic point of view of Terracini's interpretations and narratives. As Herrera, one of the five students who took the 1939–40 course in Advanced Mathematics, would recall:

> A wonderful world opened up in my mind. [...] I think that none of the students present in that class had the slightest idea of what a group was, in its specific mathematical meaning, but ten minutes of the very clear explanation by Professor Terracini, provided with simple examples taken from Elementary Geometry, Algebra and Mathematics, were enough for all

---

[94]BSM-*Terracini*: notebook no. 28 *Matemáticas Superiores*, pp. 105, 131, 137, 149, 177, 185.

[95]Compare notebook no. 9 *Geometria superiore 1931–32—Due geometri del secolo XIX Luigi Cremona e Sophus Lie*, with no. 28 *Matemáticas Superiores*, pp. 79, 87, 163, 523, 553.

[96]*Evolución de las ciencias en la República Argentina* ..., 1979, p. 202: *conservan como un recuerdo imborrable la impresión que les produjo el alto nivel intelectual, la sapiencia y la lucidez de exposición de este gran Maestro.*

[97]Herrera (2000), pp. 105–106: *Un mundo maravilloso se abrió ante mi intelecto. [...]. Creo che ninguno de los presentes en aquella clase tenía la menor idea de lo que, en su específica connotación matemática, era un grupo, pero bastaron diez minutos de la clarísima exposición, ilustrada con sencillos ejemplos de la Geometría Elemental, del Álgebra y de la Matemática, para*

of us to grasp the substance of such an important concept without any intellectual stress. Over the years, I consolidated the idea that, in those unforgettable lectures, Prof. Terracini explained to us the foundations of the aforementioned theory, in much the same way as an experienced missionary would explain the gospel to the cannibals.[97]

The didactic talent of displaced Italian scholars was not only expressed in oral teaching but also in the production of textbooks and treatises for university courses. The publication of *La Matematica dell'ingegnere e le sue applicazioni* represented Fubini's greatest scientific commitment, to which he devoted himself until the day of his death. The book was published posthumously (1949, 1954) by his pupil G. Albenga in Italy, and not in the United States as Fubini had hoped. The lessons of Analytic, Descriptive, and Projective Geometry held by Fano at the Italian University Camp were cyclostyled and used not only by university students interned but also in Rome, in the immediate post-war period, by veteran students. Levi was the author of a handbook aimed at geometry training, *Leyendo a Euclides* (1947), which enjoyed great success in Argentina. Only Segre, for the reasons afore mentioned, failed to fully capitalise on the reputation of the Italians as treatise writers, and was forced to give up on the publication of the book *Geometry upon algebraic surfaces and varieties*.

A final mention goes to one teaching not in mathematics, but in history of science. Giorgio De Santillana, a graduate in Physics (1925), had studied philosophy in Paris, before returning to Italy as an assistant at the Physics Institute of the University of Milan. In 1929 he had been called to Rome by Enriques, as a collaborator and assistant in the organisation of a National Institute for the History of Science to be established at the University of Rome. The two had collaborated in publishing the volumes *Histoire de la pensée scientifique* (Paris, Hermann, 3 vols. 1936–1939) and *Compendio di storia del pensiero scientifico dall'antichità fino ai tempi moderni* (Bologna, Zanichelli, 1937). From Enriques De Santillana had bestowed a unique style of teaching on the history of science, permeated with a 360° perspective that inserted history of scientific thought into epistemology, philosophy, philology, history of technique, cultural and material history of knowledge.

De Santillana left Rome in 1935 and, after a stay in Paris, arrived in the United States in 1936. Among the very few Italians to be selected by the RKF rescue programme, he obtained a position in Harvard, then at MIT, thanks to his teaching talent, since he proposed courses in History of Science with such a new and original character that it aroused the interest of recruiters (D6.3.1 and D6.3.2).

---

que todos captáramos sin mayor esfuerzo mental la sustancia de tan importante concepto. A lo largo los años, afiancé la idea que, en aquella inolvidable clase, el prof. Terracini nos explicó lo fundamentos de la teoría aludida, como un misionero avezado hubiera explicado el evangelio a los caníbales.

Documents

D6.3.1 G. Sarton to M. Ascoli, Cambridge Mass. 6.8.1940

ECADFS Records, *GDS*, fol. 1r

Dear sir: Prof. Giorgio D. de Santillana has been teaching the history of science very successfully in Harvard University for the last two years, and has thus materially increased the reputation he had already obtained in Rome in cooperation with Prof. Federigo Enriques. I much hope not only for his sake and mine, but also for the sake of our studies, that means may be found to enable him to continue his work as a historian of science. Sincerely yours George Sarton[98]

D6.3.2 L. H. Seelye to S. Buchanan, 14.3.1941

ECADFS Records, *GDS*, fol. 1r

Dear Scott Buchanan:[99] Since I talked with you, not long ago, I have kept in mind your search for a man who could device the laboratory work for the study of the history of science. Yesterday afternoon there came to my office with letters of introduction from Max Ascoli and others, Dr. Giorgio de Santillana who has been teaching history of science at Harvard for the last two years. I had an enjoyable conversation with him and asked him whether he would be interested in the idea of doing something which was new in American college life, namely, the teaching of history of science through the construction by teachers and students of the changing apparatus for the instrumentation of science. He said this did interest him, especially as he had recently had a conversation with Dr. Conant of Harvard in which he had been extremely critical of current laboratory work in the science, while Dr. Conant had defended it.[100] Without burdening you with further detail I am hastening this letter off, for I feel sure you would like to meet this young Dr. de Santillana. [...] Cordially yours, Laurens H. Seelye[101]

---

[98] George Sarton (1884–1956), a Belgian chemist and historian of science emigrated in America in 1915, was professor of History of Science at Harvard University (1940–1951).

[99] Scott Buchanan (1895–1968) was a philosopher and an educator. He conceived an ambitious program for reviving American education and democracy through mass training in the traditional liberal arts by means of the Socratic method and the Great Books curriculum. As Dean of St. John's College in Annapolis, Maryland (1937–1947), Buchanan reorganized the school around the Great Books 'New Program'.

[100] James Conant (1893–1978), President of Harvard University (1933–1953).

[101] Laurens Hickok Seelye (1889–1960) former President of St. Lawrence University, Canton, New York (1935–1940), a member of the Executive Committee of ECADFS from 1940 to 1944.

6.4 'Bringing Over There the Voice of Italy': Lectures and Popular Events on Italian Mathematical Traditions

Holding conferences had always been a typical activity of university professors, performed through opening addresses, plenary lectures, talks, popular lectures, etc. One peculiarity of the Italians is that many of these had concerned the Italian Schools and traditions of research. The key-note lecture by Castelnuovo at the 1928 ICM had included the word School also in the title: *La geometria algebrica e la Scuola italiana*. Almost all the geometers had delivered lectures dedicated specifically to their School, in which they had retraced its evolution in relation to external influences and internal dynamics, and in relation to other geometry teams. In some cases, such a re-readings had aroused unpleasant quarrels among the members of the School itself, as happened to Segre and Severi in 1932.

In view of these elements, it is easy to understand why displaced Italian scholars spent a lot of time promoting the contributions of Italian research Schools, and of the Italian Geometry School in particular, holding conferences, seminars, and events destined to a broader audience, and even radio broadcasts in the host countries.[102]

Segre was the first to give a lecture entitled *Geometry in Italy from 1860 till now* at the London Mathematical Society on May 9, 1940, one month before Italy entered the war. In September 1941, Terracini held some meetings on *Origines de algunos conceptos geométricos* in Rosario, at Levi's invitation.[103] These conversations recapped the main contents of a series of lectures held in Turin in 1934 at the Mathematical Seminar of the University and the Polytechnic of Turin, then published in *Periodico di Matematiche* ((4), 15, 1935: 1–21). Shortly thereafter, Terracini returned the date, asking Levi to deliver in Tucumán a cycle of seminars on mathematical logic.[104] These talks, informed by Terracini and Levi's previous common background at Peano's classes, provided some stimulating insights of the studies in algebra of logic and propositional calculus carried out by Peanians between the late nineteenth and early twentieth centuries. Beyond their intrinsic value, diminished by some misunderstandings and imprecisions,[105] they largely contributed to the diffusion in Latin America of Italian logic insofar as they made its basic elements comprehensible for a public which, in most cases, was neither qualified to understand Peano's papers, nor in the material conditions for sourcing them. Levi's *Correría en la lógica* would be recollected by Terracini as:

---

[102]For example, Terracini Collection retains the typewritten text of the radio podcast devoted to *Geometría descriptiva y su valor formativo*, 1942, fols. 1–5.

[103]Terracini A. (1941b).

[104]Levi B. (1942a).

[105]For instance, Levi's idea of a proof is out of date; logic and metalogic levels are frequently confused.

some pleasant days, lived in Tucumán as a dip—at that time and in that place—into the Italian intellectual world. Beppo Levi had come to Tucumán to give those lectures, and at the same time the friend Leone Lattes, professor of Forensic Medicine at the University of Pavia, then living in Buenos Aires, who I had invited on behalf of the Argentine Scientific Society to speak about blood groups, came to Tucumán. Also in Tucumán were my brother Benvenuto and another friend of mine, Renato Treves, now a professor at the University of Milan. Really what we had constituted in those days was a Little Italy.[106]

Holding conferences on the recent history of geometry in Italy was Fano's main activity during his stay in Lausanne (see §9.5). Between 1942 and 1944, Fano gave five lectures at the *Cercle mathématique* in Lausanne.[107] The first entitled *Quelques aperçus sur le développement de la Géométrie algébrique en Italie pendant le dernier siècle* is particularly appropriate in the contest of an activity of School self-promotion. Fano had extensive communication and popularisation experience, beginning in Göttingen in 1893 when he had given two conferences *Über die neuesten Untersuchungen der italienischen Geometrie*, then continued in Aberystwyth (1923) and in Kazan (1929). In all these contexts, the common thread of his interventions had always been to promote to foreign colleagues the results of the School to which he fiercely proclaimed his membership.[108]

His first conversation at the *Cercle mathématique*, under the title *Quelques aperçus sur le développement de la Géométrie algébrique en Italie pendant le dernier siècle*, is along the same lines. It is not the most original from the point of view of contents, nor is it the most significant in terms of critical rereading of Fano's personal contributions, but it expresses nicely what the Italian School of Geometry was for its members, what it meant to be part of it, and how the birth, image and evolution of this research group was interpreted abroad. The notion of School, according to Fano, and the main characters of the historical fresco offered in the *Aperçus* can be summarised in the following points:

– Continuity and stability of the path taken by Italian geometry from Cremona to Severi

---

[106]Terracini A. (1963), p. 599: *vissute a Tucumán come in un tuffo - in quel momento e in quel luogo - nel mondo intellettuale italiano. Erano venuti a Tucumán Beppo Levi per tenere quelle conferenze, e contemporaneamente l'amico Leone Lattes, professore di Medicina Legale all'Università di Pavia, residente allora a Buenos Aires, che avevo invitato per conto della Società scientifica argentina a parlare dei gruppi sanguigni. Erano anche a Tucumán, come professori, mio fratello Benvenuto e l'altro amico Renato Treves, oggi professore all'Università di Milano. Era veramente una piccola Italia quella che avevamo costituita in quei giorni.*

[107]On the *Cercle mathématique* see Chatterji and Ojanguren (2013).

[108]Suffice to mention the conclusion of his first public lecture in Aberystwyth (BSM-*Fano*: [*Aberystwyth Lectures 1923*], Scritti.4, fol. 17bis): 'My first word to-day is to express to the Principal of the Council of this University my deepest gratitude for the great honour they have done to me with their kind invitation to deliver here some lectures on geometry. This honour is to Italy's contri[butions]. If Italy's contributions to modern geometry are very important, I am only a very modest representative of this geometrical School. I hope to succeed in suggesting you the interest of the matter'.

- Importance of the social dimension for the flourishing, competitiveness, and attractiveness of research Schools
- Emergence of a national identity in mathematical studies only after the Risorgimento
- A need for 'grafts' (*innesti*) of foreign trends of studies (primarily from Germany) onto the trunk of Italian geometric production
- Scientific and moral role of the Masters, both at the moment of creation of a School, and in the successive phases of development, expression, and affirmation of the working group beyond geo-political borders
- Added value of collaboration: the School is not only a team of scholars who share a research project or a workplace, but also a network and, in some ways, a family.

These are aspects which Fano had often thought about in the past and on which he would continue to reflect until the last months of his life when, preparing the commemoration of Castelnuovo, he lingered in thought on the Schools that in the last decades of the 19th century had determined the revival of geometric studies in Italy, and on those men (above all, Luigi Cremona) [...] who had made it possible.[109]

These were not entirely original topics. Some points of view unite the Lausanne conversation with other historical-informative interventions by members of the Italian geometric School, such as the plenary conferences of Severi and Castelnuovo at the ICMs in Toronto (1924) and Bologna (1928).[110] Instead of diminishing the interest of the Fano conference, however, these echoes, these convergences, constitute a clear testimony of the affinity of thought among the Italian geometers and their shared vision of the distinctive features of their School and salient stages of its history.

Presented at a terrible juncture in history, the lecture *Quelques aperçus sur le développement de la géométrie algébrique en Italie* is characterised by some national markers which at first reading might appear singular, almost bizarre. Yet they are definitely not so, in the measure that they similarly shape other series of seminars and conversations held by Italian refugees, for example the *Correrías en la lógica matemática* by Levi.[111] As Levi brought to Tucumán the voices of Peano and his protégés, so Fano intended to revive in Lausanne the studies and figures of those who had been his Masters and friends (Cremona, Segre, Castelnuovo, Klein), that golden season of Italian geometric research, which he had experienced first as an observer, and later as a player. The two contexts are obviously very different. Lausanne was not Tucumán, the audience was different, but the underlying spirit and basic aims were analogous: the claim, even during the immigrant experience, of

---

[109]Terracini A. (1953a), p. 702: *indugiato col pensiero sulle Scuole che negli ultimi decenni del secolo scorso avevano determinato il rifiorire degli studi geometrici in Italia, e sugli uomini, in primo luogo Luigi Cremona [...]. Uno dei molteplici indizi della continuità del pensiero geometrico italiano.*

[110]Castelnuovo G. (1929); Severi (1924).

[111]Levi B. (1942a). Analogous accents in Levi B. (1941c, 1943, 1946c).

their own affiliation to mathematical traditions that had led Italy to achieve an international leading position.

6.5   Publishing

Terracini and Levi systematically devoted themselves, for the first time in their lives, to publishing. In fact, having immediately 'felt the distance of the local scientific environment from Europe and its production'[112] and wanting to affirm their importance in the scientific field of the universities that had welcomed them, they founded and directed three journals: the *Publicaciones* of the *Instituto de Matemáticas de la Universidad Nacional del Litoral* (1939), the *Mathematicae Notae. Boletín del Instituto de Matemática* (September 1940) and the *Revista de Matemáticas y Física Teórica* of the *Universidad Nacional de Tucumán* (December 1940). They sanctioned the debut of the specialised mathematical press in Argentina and excellently filled a gap in South American mathematical publishing.

Co-directed by Terracini and the physicist F. Cernuschi, the first issue of *Revista* was published in December 1940, less than a year after Terracini's arrival in Tucumán. It published articles written in five languages (Spanish, Italian, French, English, and German) and from the first volume onward, it boasted a prestigious board of authors. Thanks to Fubini, the second volume of *Revista* included a paper by Einstein, *Demostración de la no existencia de campos gravitacionales sin singularidades de masa total no nula* (1941, pp. 5–11). Einstein personally asked Terracini and Fubini to translate his work, because he did not want to publish in the language of the Reich (D6.5.1 and Fig. 6.3).

To launch the *Revista* and lift it to high quality standards, avoiding the risk of creating a journal with local circulation only, Terracini and the co-director Cernuschi spared no effort.[113] Terracini himself enhanced the journal with a dozen papers and reviews. He solicited collaboration from Einstein, Veblen, Cartan, Courant, P. Erdös, L. Godeaux, etc., and asked for help from colleagues and other displaced scholars (D6.5.2). In fact, the *Revista* gave Levi-Civita, Enriques, Fubini, Fano, Segre, Ascoli, Loria, and other Italian mathematicians silenced by the racial laws, the opportunity to restart publishing.[114]

Awards were not long in coming. Papers were submitted to the *Revista*'s editorial board from all four corners of the world and, in turn, in his role as chief editor,

---

[112] A. Terracini to T. Levi-Civita, Tucumán 18.10.1939, in Nastasi and Tazzioli (2000), p. 403.

[113] Terracini's whole family was busy typing manuscripts, translating texts, proofreading, etc. See ATCET: G. Sacerdote Terracini to A. Sacerdote, Tucumán 1.2.1945.

[114] See for example D6.5.2 and OVP, *AT*: A. Terracini to O. Veblen, Tucumán 26.1.1943; O. Veblen to A. Terracini, [Princeton] 9.2.1943 and Archives de l'Académie des Sciences, Paris, Fonds Élie Cartan: A. Terracini to É. Cartan, Tucumán 6.1.1946.

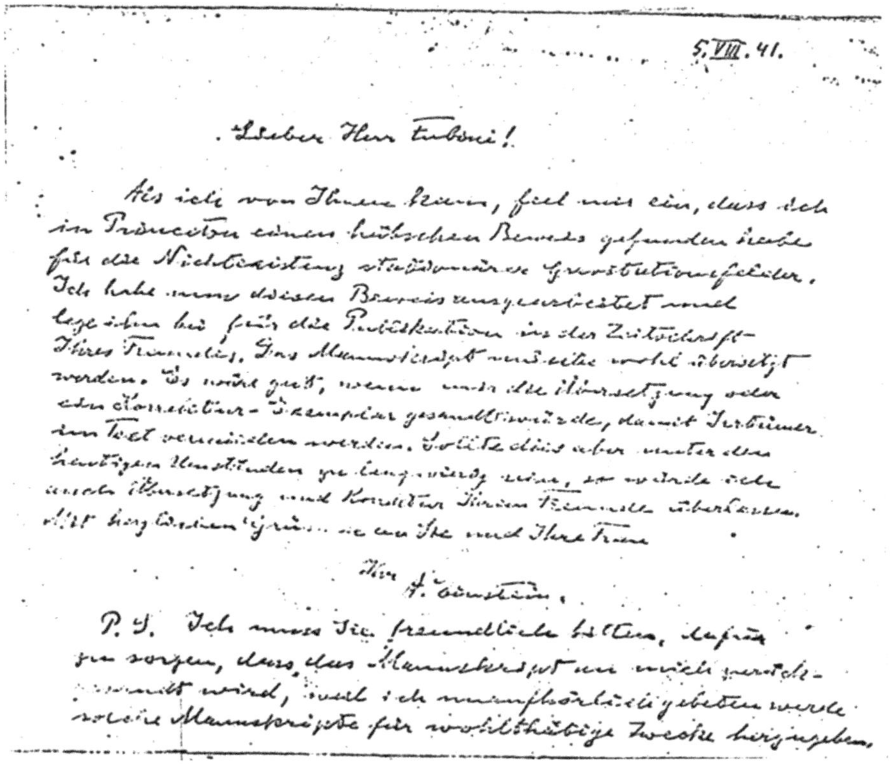
5.VIII.41.

Lieber Herr Fubini!

Als ich von Ihnen kam, fiel mir ein, dass ich in Princeton einen hübschen Beweis gefunden habe für die Nichtexistenz stationärer Gravitationsfelder. Ich habe nun diesen Beweis ausgearbeitet und lege ihn bei für die Publikation in der Zeitschrift Ihres Freundes. Das Manuskript muss aber wohl übersetzt werden. Es wäre gut, wenn mir die Übersetzung oder ein Korrektur-Exemplar geschickt würde, damit Irrtümer im Text vermieden werden. Sollte dies aber unter den heutigen Umständen zu kompliziert sein, so würde ich auch Übersetzung und Korrektur Ihrem Freunde überlassen.
Mit herzlichen Grüssen an Sie und Ihre Frau

Ihr A. Einstein.

P.S. Ich muss Sie freundlich bitten, dafür zu sorgen, dass das Manuskript an mich zurückgesandt wird, weil ich unaufhörlich gebeten werde solche Manuskripte für wohltätige Zwecke herzugeben.

Fig. 6.3 A. Einstein to G. Fubini, 5.8.1941, BSM

Terracini built up a network of international contacts much larger than that which he had enjoyed before emigration.

Similar considerations apply to the journals founded by Levi. The *Publicaciones*, which were a real novelty in the Argentine publishing market, were attached to the *Facultad de Ciencias Matemáticas* of the Universidad Nacional del Litoral and were addressed particularly to researchers attending the Mathematics Seminar in Rosario. They hosted the work of Argentine scholars[115] as well as essays by Italian and foreign authors such as Fubini, Terracini, Rosenblatt, F. Amodeo, and P. Montel. The *Publicaciones* came out in eight volumes between 1939 and 1946, two of which were entirely dedicated to the *Memorias ofrecidas por varios amigos, alumnos y admiradores en homenaje al Dr. Julio Rey Pastor*. Levi published nine long essays, almost monographic books.

Mathematicae Notae were a sort of Argentine hybrid between the *Giornale di Matematiche* by G. Battaglini and the *Bollettino dell'UMI*. Levi was editor of the

[115] M. Cotlar, L. A. A. Santalò, J. Babini, F. Gaspar, J. L. Massera, A. Komischke, R. Laguardia, J. C. Vignaux, J. V. Uspensky, F. I. Toranzos, E. A. Sagastume Berra.

journal from its foundation until the year of his death (1961). The journal published 'simple articles, without pretensions of research in high spheres'[116] generally submitted by students and mathematicians trained in Rosario, alongside research papers and articles concerning the connections among mathematics and other branches of the science. They also included methodological contributions, and obituaries of great mathematicians such as Levi-Civita and Volterra, who could not be commemorated publicly in Italy being Jewish.[117]

Like *Publicaciones*, the *Mathematicae Notae* had a distinctly international character: under Levi's supervision, they exchanged their issues with 146 journals from 31 different countries. Levi here published 42 articles. There were also numerous reviews of mathematics and physics texts, written mostly by Levi himself and his daughter Laura. The heart of *Mathematicae notae*—in accordance with the epistemological and didactic project carried out by Levi in Argentina—was however made up of three sections of questions–answers and exercises (*Ejercicios y problemas*, *Cuestiones* and *Flores y Hojas*), formulated and selected to create and consolidate a mixed readership, composed of professional mathematicians, amateurs and practitioners. The first section proposed exercises and problems inherent to the various sectors of pure and applied mathematics. Their solutions were collected and published by the editorial board, who often accompanied them with comments and annotations, including proofs of theorems used in the resolution, alternative solutions and generalisations inspired by these problems. The *Cuestiones* column was directed to a more specialised audience, to students eager to embark on the path of academic research; the questions proposed were of an advanced nature and required a strong mathematical background. The *Flores y Hojas* section, which appeared only between 1941 and 1943, proposed questions that were not excessively technical, but which required a solid logical-deductive rigour; for example, readers were asked to find faults, errors, and gaps within reasonings or arguments. At the end of each year, the board of directors awarded two prizes, one to students of the University of Rosario, who had provided the best solutions to the questions proposed, and a prize to the readers who had participated from Argentina or abroad. The prizes were books and offprints of mathematics and physics.[118]

The creation of the *Mathematicae Notae* and the *Revista de Matemáticas y Física Teórica* opened up unexpected scenarios of cultural appropriation by an entire generation of young Latin American mathematicians. The benefit was, however, reciprocal, because in their capacity as chief editors constantly on the lookout for articles for their journals, Levi and Terracini were able to build networks of relationships with colleagues from all over the world, on a much larger scale than would have been possible before emigration.

---

[116]Levi B. (1941a), p. 7: *artículos sencillos, sin pretensiones de investigación en altas esferas.* Interestingly, the journal was distributed free of charge to the students of the Faculty, so as to facilitate their involvement.

[117]Levi B. (1942b); *Homenaje a la memoria de V. Volterra y J.J. Thomson* 1941.

[118]*Antecedentes de la creación de las Mathematicæ Notæ* . . . 1941, p. 5.

Documents

D6.5.1 A. Terracini to A. Einstein, Tucumán 21.10.1941

AEA, AT, fol. 1r

Dear Professor, just a few days ago, returning from a lecture trip to Rosario and Buenos Aires, I found your kind letter of 29 September. Please excuse me for the fact that I am a little late in expressing my heartfelt thanks to you. At your request, instead of the German text of your essay, we will publish an English translation together with the Spanish one. Accordingly, we shall take care of the translation personally: if, as is to be expected, the interpretation presents no particular difficulty, we intend to quickly submit the corrections to our translation for your approval and revision. Hoping you agree with that, with respect and gratitude, your obedient A. Terracini

Sehr geehrter Herr Professor, Erst vor einigen Tagen, aus einer Vorträgen Reise nach Rosario und Buenos Aires zurückgekehrt, fand ich Ihren werten Brief vom 29 IX vor, möge dieser Umstand mich entschuldigen, venn ich mit etwas Verspätung meinen besten Dank dafür Ihnen ausspreche. Ihrem Wunsch nach, anstatt des deutschen Textes Ihres Aufsatzes, werden wir nebst der spanischen eine englische Übersetzung veröffentlichen. Dementsprechend werden wir die Übertragung persönlich besorgen: falls, wie voraussichtlich ist, die Interpretation keine besondere Schwierigkeit darbietet, gedenken wir erst die Korrekturen unserer Übersetzung Ihrer Genehmigung und Revision vorzulegen. Hoffentlich sind Sie damit einverstanden. Mit hochachtungsvollen und dankbaren Grüßen Ihr ergebener A. Terracini

D6.5.2 A. Terracini to O. Veblen, Tucumán 26.1.1943

OVP, AT, fol. 1r

Dear Professor Veblen, The first issue of vol. 3 of the *Revista de Matemáticas y Física Teórica* edited by this University has just been released, and I send it to you; the second issue of the same volume is now in press and will be followed shortly by the fourth volume. On this occasion, I express to you the strong desire of this University and my staff to continue receiving contributions from American mathematicians; we would be delighted if you should choose to promote our *Revista* by encouraging the submission of some work from your School and your colleagues. *Revista* is now published in fascicules, instead of in a single annual volume, which prevents delay in the publication of works. The review of the manuscripts by the North American censors makes their journey a little slower.

However, I have been able to verify that the delay is not significant and that the journal arrives regularly. Repeating my warmest thanks, I greet you cordially Alejandro Terracini

Estimado Profesor Veblen, Acaba de salir el fasc. 1 del vol. 3° de la Revista de Matemáticas y Física Teórica de esta Universidad, que le hice remitir; está imprimiéndose ahora el fasc. 2 del mismo volumen, al cual va a seguir dentro de poco el vol. 4°. Le expreso en esta ocasión el vivo deseo de esta Universidad y mío personal de seguir recibiendo colaboraciones de matemáticos estadounidenses; sería un gran placer si usted quisiera favorecer nuestra Revista fomentando el envío de algún trabajo de su Escuela y de sus colegas. La Revista se publica ahora en fascículos, en vez que en un solo volumen anual, lo que evita el retraso en la publicación de los trabajos. La revisión de los originales por parte de la censura norte-americana hace un poco más lento su viaje. No obstante, he podido comprobar que el retraso es de poca monta y que todo llega regularmente. Repitiéndole mis más vivos agradecimientos, le saludo cordialmente Alejandro Terracini

6.6   Mathematical Libraries

Italian refugees had significant experience in managing mathematical libraries. Fano was Director of the Special Mathematics Library in Turin from the death of Corrado Segre (1924) to 1938; Terracini collaborated with him in the Mathematical Seminar, attached to the library, from 1925 to 1938. Levi and B. Segre experienced the libraries of the Institutes of Mathematics in Bologna and Rome. Handling or collaborating in the management of a library had led them to acquire specific skills in the field of maintenance, conservation, and valorisation of mathematical collections of books, journals, models, and instruments. Deciding which books to buy, which subscriptions to subscribe to or suspend, how to organise a reading room, how to interact with booksellers and publishers, were all tasks carried out with passion and professionalism.

Compelled to abandon their personal libraries, which they would fortunately find intact on their return home, Italian refugees in South America immediately felt it essential to create mathematics libraries. The first step was to have their books sent to their new workplaces. Levi's collections arrived in Rosario in 1939. The personal libraries and miscellanies (i.e. collections of reprints) belonging to Levi, Mortara, A. Bassi, A. Mieli, and Rosenblatt still today represent a historic heritage of high value in the libraries of Rio, Buenos Aires and Lima.

The creation of mathematical libraries was important because it offered the material sources indispensable for development of research activity in two peripheral settings such as Rosario and Tucumán. Teaching also benefitted, especially in advanced courses. The enrichment was, however, bi-directional because when Terracini was nominated to rebuild the Special Mathematics Library of Turin (which had been damaged by the Allied bombings), he could count on the experience gathered in Argentina as library director, and on the contacts he had made with publishers, editorial boards of mathematical journals and foreign booksellers.

Chapter 7
Coming Back to Italy

7.1 The Bureaucracy of Reintegration into Academia

The racial laws were repealed by the Allied government in 1944. Decrees no. 9 of January 6, no. 25 of January 20, and no. 301 of October 19 readmitted professors to service. For economic purposes, the measure took effect from January 1, 1944 onwards, and the five years of forced absence (1938–1943) were considered as paid leave, equivalent to normal work as far as salary and pension contributions were concerned.

As at the initial time of discrimination, scholars received a standard letter signed by the Minister of Public Instruction, Vincenzo Arangio-Ruiz, in which he informed them of the reinstatement and 'formulated his best wishes for the further scientific and didactic activity that the professor himself, as for the past, will wish to offer the Italian University'.[1] To obtain reinstatement, a specific procedure had to be followed: communicating their residential details, stating whether they had emigrated abroad, producing a document proving the maintenance of Italian citizenship, and a whole score of documents to obtain the so-called career reconstruction and to request stipends and pensions in arrears.[2]

While university offices had been terribly efficient in 1938, they were now equally disorganised. Universities processed the reinstatements at different times and in different ways. The financial side of the issue alone was the source of abuses and cases of bad bureaucracy. Suffice it to mention Ascoli, who needed eighteen months and a dozen of reminders to the Ministry of Public Instruction to collect his

---

[1] See, for example, ACS-*Fano*: MPI [V. Arangio-Ruiz] to M. Allara [Rector of the University of Turin], Rome 29.10.1945: *formulare al professore in oggetto i miei voti migliori per l'ulteriore attività scientifica e didattica che, come per il passato, vorrà dare alla Università italiana.*

[2] The career reconstruction was the recognition of salary promotions to which one would have been entitled in the years 1938–43 if not for the racial laws. It should be kept in mind that pension payments to deposed Jewish scholars had been interrupted in the autumn of 1943 and, during the period of the RSI, the Jews (both those residents in Italy and abroad) had not received any money.

E. Luciano, *The Jewish Mathematical Diaspora from Fascist Italy*, Science Networks. Historical Studies 64, https://doi.org/10.1007/978-3-031-64896-0_7

salaries in arrears. This, briefly, is his story, which is unfortunately only one of many. In March 1946, the Ministry of Public Instruction asked the Ministry of Foreign Affairs to inform the Rector of the University of Milan that, before readmission, Ascoli had to make known in advance if he had kept his Italian citizenship, if he wished for readmission into service, and his current family status.[3] The Ministry made a mistake. Perhaps they confused the mathematician Guido Ascoli with the professor of History of French literature at the Sorbonne Guido Ascoli, who had emigrated to the United States in 1941, or with Alberto Ascoli, professor of General Pathology in Milan, who also emigrated to America. Ascoli immediately clarified the issue:

> I believe that in drafting this note there was a misunderstanding or rather an error in person, since I have never left Italy and since May I have made myself available to the University and resumed service in mid-June 1945. I believe instead that Prof. *Alberto* Ascoli, of the Faculty of Veterinary Medicine is in America, and has not yet reported his desire to resume teaching, and I deem that in fact the note concerns this colleague of mine. What concerns me, however, is the fact that from the document itself it appears that the Superior Ministry is unaware that I have been rehired, which also gives me the right to a review of my career and the consequent advancement of rank, for which I have been waiting for many months. I am sure that the University has transmitted to the Ministry the necessary information about me; in any case, I would be very grateful if you could take the opportunity to check and thus hasten the paperwork for the definitive settlement of my career.[4]

The Ministry acknowledged the error only at the beginning of June 1946. Since, according to the Prefecture, Ascoli had not resided in Milan from September 1942 onwards, they had accepted the unofficial news that Ascoli had taken refuge abroad, 'especially since it was commonly known that numerous former university professors had taken refuge in Switzerland during the period of the German occupation of Northern Italy'.[5] Despite the pressure of the Rector of Milan, it would still

---

[3] APICE-*Ascoli*: MPI to MFA, Rome 22.3.1946: *trovandosi egli all'Estero (risulta essere negli Stati Uniti d'America) era necessario facesse preventivamente conoscere se aveva conservato la cittadinanza italiana, se gradiva la riammissione in servizio ed il suo attuale stato di famiglia.*

[4] APICE-*Ascoli*: G. Ascoli to F. Perussia, Milan 4.4.1946: *Ritengo che nella redazione di detta nota sia occorso un equivoco o meglio un errore di persona, giacché io non ho mai abbandonato l'Italia e sino dal maggio decorso mi sono messo a disposizione dell'Università riprendendo poi effettivamente servizio alla metà di giugno 1945. Credo invece che si trovi in America e non abbia ancora segnalato il suo desiderio di riprendere l'insegnamento il prof. Alberto Ascoli, della Facoltà di Medicina Veterinaria, e che nel fatto la nota riguardi questo collega. Ciò che però mi riguarda è il fatto che dal documento stesso risulterebbe che presso il Superiore Ministero non sembra nota la mia avvenuta riassunzione in servizio, la quale mi dà anche diritto alla revisione della carriera e al conseguente avanzamento di grado, di cui sono da molti mesi in inutile attesa. Sono sicuro che l'Università avrà a suo tempo trasmesso al Ministero le dovute informazioni a mio riguardo; a ogni modo Le sarò molto grato se vorrà cogliere l'occasione per rinnovarle ed affrettare così le pratiche per la sistemazione definitiva della mia carriera.*

[5] APICE-*Ascoli*: MPI to F. Perussia, Rome 7.6.1946: *tanto più che si era saputo che numerosi professori ed ex professori universitari si erano rifugiati in Svizzera durante il periodo dell'occupazione tedesca dell'Italia settentrionale.*

take Ascoli six months and many reminders to be reinstated and apply for the back pay.[6]

Beyond the back wages, the financial aspect was notably delicate regarding return journeys. Professors dismissed in 1938 often moved away from the cities where they had resided. Especially in the period of occupation, those who remained in Italy left their resident cities, where the risk of falling into the raids was greater and moved elsewhere. Ascoli left Milan for Turin, Maroni moved from Pavia to Florence, Bemporad from Turin to Rome, etc. For displaced scholars in Italy, the Ministry of Public Instruction, in agreement with the Treasury, ruled that they should receive reimbursement of expenses to return to their place of work. Their journeys were considered equivalent to those of service of state employees. The subject was of course much more sensitive for scholars who had emigrated abroad. Except for Switzerland, where the return journey by train was affordable, the costs were prohibitive. The Ministry estimated that it took about 500,000 lire each to return from a European country, and one million for people returning from overseas.

University professors had generally emigrated with their families, who of course must be reunited, especially if there were young children. Given the expenses they had incurred to emigrate, and the fact that for over two years they had not received the pension, refugees did not have the economic means to return. Hence the requests for help to universities and to colleagues who had influence in ministerial and political chambers. In the case of Terracini, for example, Arangio-Ruiz signed the decree of readmission to service on 20 December 1944. The document specified that Terracini 'would reach his institute in Turin as soon as possible'.[7] In May the Italian embassy in Buenos Aires informed Terracini of the provision.[8] Terracini strongly desired to return but the costs of the trip were unsustainable, which is why Terracini turned to the MPI asking for help. At its turn the University of Turin intervened to ask the Ministries of the Treasury and of Foreign Affairs for economic support that allowed the Terracinis (six people in all) to buy tickets for the ocean liner that would bring them back from Argentina.[9] This request was not processed. At that point,

---

[6]See for example APICE-*Ascoli*: F. Perussia to G. Ascoli, Milan 20.8.1946, MPI to F. Perussia, Rome 15.11.1946, F. Perussia to Ufficio Provinciale del Tesoro, Milan 6.12.1946.

[7]ACS-*Terracini*: MPI decree, Rome 20.12.1944.

[8]ACS-*Terracini*: MPI to MFA, Rome 18.5.1945 and 29.8.1945.

[9]ACS-*Terracini*: E. Molé [MPI] to MFA, Rome 21.12.1945 and ASUT-*Terracini*: excerpt of the meeting of the full professors of the Faculty of Sciences, 14.11.1945: 'The Faculty, learned that financial difficulties hinder the prompt repatriation from Argentina of its colleague Prof. Alessandro Terracini, recently reinstated in the chair of Analytic and Projective Geometry of our University, a chair which Terracini had held with honor for over fifteen years; convinced that everything possible must be done so that the Italian University can as soon as possible reacquire a professor of such value, who is also an exemplary citizen, well-liked by all; expresses his vote that the government finds a way to speed up the professor's return as much as possible. To whom may concern: the return journey of this professor and his family is a journey for work reasons'. (*Dopo questo, la Facoltà venuta a conoscenza che difficoltà di ordine finanziario ostacolano il pronto rimpatrio dall'Argentina del collega Prof. Alessandro Terracini, recentemente reintegrato nella cattedra di Geometria Analitica e Proiettiva della nostra Università, cattedra che il Terracini aveva ricoperta*

Picone organised a petition. 86 mathematicians (practically all the Italian university professors of this discipline) signed a statement wherein they requested aid to support the return of Terracini, whose presence was considered essential for the reconstruction of the country (D7.1.2 and D7.1.3).[10] This plea, too, was ignored. Finally, Castelnuovo intervened and, as president of the Lincei Academy, submitted to the new Minister of Public Instruction, Guido Gonella, a note on the situation of Terracini, asking him for an intervention *ad personam* (D7.1.6, D7.1.7, and D7.1.8). In fact, Terracini managed to return only in February 1948, paying all the expenses himself.

For Segre and his family (four people), it was Castelnuovo, followed by Severi, who took action, first with Arangio-Ruiz, then with Gonella (D7.1.1, D7.1.4, and D7.1.5). Also in this case, a detailed report on Segre's condition was presented to the Minister, without effect.

Having taken note of the situation, and of the numerous requests that were arriving, in November 1946 the Ministry of Public Instruction finally asked the Treasury for an exceptional measure. According to the polls, it emerged that at that date, there were 29 university professors still residing abroad; professors who would have liked to return but who could not do so because they were not able to bear the travel expenses.[11] Taking into account the countries in which they had settled, and the size of the families involved, a sum of 24,400,000 lire was required. The measure invoked was rejected. Expats who returned would be forced to do so entirely at their own expense.

---

con onore per oltre tre lustri; convinta che debba essere fatto tutto il possibile acciocché l'Università Italiana possa al più presto riacquistare un docente di tanto valore, che inoltre è un cittadino esemplare, da tutti benvoluto; esprime il voto affinché il governo trovi il modo di accelerare il più possibile il ritorno del prof. Terracini, facendo presente a chi di ragione che il viaggio di ritorno di questo professore e della sua famiglia, è un viaggio per ragioni di servizio.)

[10] Picone and Terracini were good friends: they had shared the war experience and in 1917 they had worked together to calculate the firing tables for the use of heavy artillery in the mountains.

[11] It is not clear how the data were collected. For England, the names were most likely provided by Edoardo Ruffini (1901–1983). Together with his father, the Liberal Senator Francesco, Edoardo Ruffini had been among the very few Italian professors who refused to take the oath of loyalty to the fascist regime in 1931, and for this reason he had been dismissed from the chair of History of Law at the University of Perugia. From 1945 to 1947, Ruffini served as cultural attaché with the rank of counselor at the Italian embassy in London. In this capacity, in September 1946 he prepared for the SPSL a list of Italian persecuted scholars. Most of the files of Italian scholars in the SPSL archive contain the note: 'from Ruffini's list' or something similar. The list clearly presented errors and omissions: for example, Castelnuovo was not dismissed under the racial laws (because he was already retired), nor did he emigrate, nor was he reinstated.

Documents

D7.1.1 G. Castelnuovo to B. Segre, Rome 8.9.1945

CA, *BSP*, fol. 1r-v
Dear Professor, Your letter of August 8, which reached me now only clarifies a mystery that I had not been able to explain. Indeed, a dozen days ago I received a package from Boston, but the name of the sender was unknown to me. [. . .] I only regret that you bothered for me, since I have to recognize that the very little I have done for you is too small compared to your gift. I hope to have better luck later in my action at the Ministry. I can tell you that Ugo Amaldi also dealt with the matter, without success. Perhaps now that the war is over and merchant ships will start sailing again for civilian travellers, it will be easier for you to obtain a travel permission. I can't believe that the difficulties come from Rome and I don't see the reason. As I already wrote to you, the Ministry of Public Instruction would move with greater energy if it could count on your resuming teaching as early as next November. Of the professors who left Italy for political reasons, very few returned, either sent by allied governments, or requested by our government for their specific competences (Colonnetti, Marchesi, Einaudi, . . .).[12] Gino Fano wrote to me from Lausanne at the beginning of August. He hoped to be able to return to Italy by the end of that month, and I replied to him at the address in Mantua, but I still don't know if he is there. I know nothing about Terracini, nor about Beppo Levi; they certainly haven't returned yet. I don't understand what my previous letter could contain that attracted the attention of the censure, perhaps some misinterpreted symbol or mathematical term. Many cordial greetings and once again many thanks from your most affectionate G. Castelnuovo

Caro Professore, La Sua lettera dell'8 agosto giuntami ora soltanto mi chiarisce un mistero che non avevo saputo spiegare. Effettivamente una dozzina di giorni fa ricevetti un pacco da Boston ma il nome dello speditore mi riusciva ignoto. [. . .] Solo mi rincresce che Ella si sia disturbata per me, giacché devo riconoscere che quel pochissimo che ho fatto per Lei è troppo piccola cosa di fronte al dono. Spero di essere più fortunato in seguito nelle mie pratiche al Ministero per il Suo ritorno. Posso dirle che se n'è occupato anche Ugo Amaldi, egli pure senza successo. Forse ora che la guerra è terminata e le navi mercantili riprenderanno a circolare per i viaggiatori civili, sarà più facile a Lei di ottenere il permesso. Non posso credere che le difficoltà provengano da Roma e non ne vedrei la ragione. Come già Le scrivevo il Ministero della P.I. si muoverebbe con maggiore energia se potesse contare che Ella riprendesse i corsi fin dal prossimo novembre. Di professori usciti dall'Italia per ragioni politiche pochissimi son rientrati, o mandati dai Governi alleati, o richiesti dal nostro per competenze

---

[12]Gustavo Colonnetti (1886–1968), Concetto Marchesi (1878–1957), and Luigi Einaudi (1874–1961). See §3.24.

specifiche (Colonnetti, Marchesi, Einaudi, ...). Anche Gino Fano mi ha scritto da Losanna al principio di agosto. Sperava di poter rientrare in Italia per la fine di quel mese, ed io gli ho risposto all'indirizzo di Mantova, ma non so ancora se egli si trovi colà. Di Terracini non so nulla, né di Beppo Levi; certo non son ritornati finora. Non capisco che cosa potesse contenere la mia lettera precedente da attirare l'attenzione della censura, forse qualche simbolo o termine matematico male interpretato. Molti saluti cordiali e di nuovo mille ringraziamenti dal suo aff.^{mo} G. Castelnuovo

D7.1.2 M. Picone, A. Signorini and E. Bompiani to the Italian Mathematicians, Rome 15.1.1946

Archivio Storico INAC, fol. 1r

Dear Colleague, we have recently had good news about Alessandro Terracini from his brother-in-law, engineer Aldo Sacerdoti [sic!],[13] and we learned, with the greatest pleasure, that he wishes to return to his homeland as soon as possible. The fulfilment of his wish is opposed by the prohibitive cost of the return journey for himself and his large family. Since, on the other hand, it is obviously in the precise interest of the Italian University to reacquire Terracini's work, we have decided to undertake action with the higher authorities aimed at obtaining those measures that can make Terracinis' return journey possible. We would like this action to be supported by all mathematicians in Italy and for this purpose we take the liberty of sending you the enclosed copy of the memorandum to be presented to the Minister of Public Instruction, proposing that you sign it and present it to the mathematicians of your University so that they too may sign it.[14] While waiting for your kind reply, we send you our warmest regards.

P.S. Please send the signed copy of the memorial to Prof. Mauro Picone—National Institute for the Applications of the Calculus, Piazzale delle Scienze 7—Rome

Caro Collega, abbiamo avuto recenti buone notizie di Alessandro Terracini da suo cognato, ing. Aldo Sacerdoti [sic!], ed abbiamo appreso, col più vivo piacere, che egli desidera tornare in Patria al più presto. All'esaudimento di questo suo desiderio si oppone il costo proibitivo del viaggio di ritorno per sé e per la sua numerosa famiglia. Siccome d'altra parte è, ovviamente, nel preciso interesse dell'Università italiana, riacquistare l'opera del Terracini, abbiamo deciso di intraprendere un'azione presso le superiori autorità intesa ad ottenere quei provvedimenti che possano rendere possibile il viaggio di ritorno di Terracini con la sua famiglia. Vorremmo che a tale azione fosse data l'adesione di tutti i

---

[13] Aldo Sacerdote (1897–1979), brother of Giulia Sacerdote Terracini.

[14] The document is missing.

matematici d'Italia e a tale scopo ci permettiamo di inviarle l'acclusa copia del memoriale da presentare al Ministro della Pubblica Istruzione proponendole di apporvi la propria firma e di presentarla ai matematici di codesta Università affinché anch'essi possano eventualmente firmarla. In attesa di un Suo cortese sollecito riscontro Le inviamo i più cordiali saluti.

P.S. Si prega inviare la copia del memoriale firmata al Prof. Mauro Picone— Istituto Nazionale per le Applicazioni del Calcolo, Piazzale delle Scienze 7— Roma

D7.1.3 A. Terracini to M. Picone, Tucumán 15.7.1946

Terracini Coll., fol. 1r-v

Dearest Mauro, Your dear letter of 21 June reached me very quickly at the end of the same month. I have postponed my answer in order to be able to tell you about my conversation with the Italian embassy in Buenos Aires. I was very sorry for the news of Enriques' death.[15] As soon as I received the sad news from you, I wrote to Castelnuovo to send him my condolences and ask him to pass them on to the whole family and Faculty. Not knowing for sure if, after all the hardships of the past few years, the address of Castelnuovo is still that of via Boncompagni 16, I added the University to the envelope. If by chance, despite these precautions, my letter did not arrive, please tell him. In any case, I also ask you to convey to your Faculty the expression of my deep sorrow for the very serious loss suffered. I left at the beginning of the month for Buenos Aires, to chair the meeting of the Argentine Mathematical Union. On that occasion, I announced the deaths of Comessatti, Tonelli, De Franchis and Enriques and I also held the commemoration of the latter, after which Rey Pastor took the floor to remember in a special way his visit to Buenos Aires. The Argentine Mathematical Union has also decided to send its condolences to the family via Castelnuovo: I suppose that the relevant letter will come later.[16] I will now tell you about the conversation I had at the embassy about our return to Italy. I tell you right away that, through an apparent 'courtesy', I had the impression that they do not intend to deal with it thoroughly unless pushed, and strongly pushed, by Rome. The embassy received a telegram from the Ministry of Foreign Affairs (which, according to what my brother-in-law Aldo Sacerdote wrote to me, must date back to the end of March) inviting them to "facilitate" our return to Italy; the embassy replied

---

[15] Enriques had passed away on June 14, 1946.

[16] As president of the *Unión Matemática Argentina*, Terracini announced the deaths of Comessatti, Tonelli, De Franchis and commemorated Enriques at the meeting held on July 8, 1946. The condolence letter sent to Castelnuovo and the eulogy pronounced by Terracini were published in *Federigo Enriques. Homenaje de la Unión Matemática Argentina. Palabras de los doctores A. Terracini y J. Rey Pastor*, Revista de la Unión Matemática Argentina, XII, 1, 1946, pp. 3–9.

telegraphically; they did not want to show us the text of the telegram, hiding instead behind the excuse of official confidentiality; the date of the answer was all that could be known, and it would be early June. From what we were told, in the answer:

(a) the embassy asked Rome to give precise instructions, accompanied by the general nature of the verb "facilitate". And in this it seems to me that it is right because the Ministry of Foreign Affairs should give concrete instructions so that our trip is considered as a service trip, that is, at the expense of the State.

(b) the embassy, as far as I understand, does not consider my case (or that of Benvenuto and his family[17]) as an isolated case, but asked for general instructions; I do not know if these are for all Italian citizens or for all state employees and families, who want to repatriate.

Now it seems to me that my case (so that of Benvenuto) must be considered absolutely separate and "detached" from any similar cases (such, as I think, is also your opinion, when, in the letter of June 21, you speak of "exceptional measures"). In the first place there is the statement signed by 86 Italian mathematicians, who, thanks to your never sufficiently-appreciated initiative, presents my case as that of a person, whose presence in Italy—deservedly or not—is considered as particularly useful (D7.1.2). For Benvenuto there is a vote of the Faculty of Milan, just as for me there is a similar one from my Faculty of Turin, dated last November.

To this is added another circumstance in our regard. Law no. 587 of May 23, 1940, granted university professors suspended for the well-known reasons in 1938, a four-year allowance, equal to the difference between salary and pension from December 1938 to December 1942. This allowance was granted to us by subsequent judgment of the Council of State of 24 September 1941, no. 302, and liquidated by the Ministry on 19 October 1942. Having ascertained that the check was payable abroad, at the official exchange rate of the time, the relative roles for payment arrived in Argentina in May 1943, but shortly afterwards the pension payments were suspended here; we were denied payment of the four-year check accrued long before. We Italian university professors residing in Argentina believed and believe that this non-payment was not justified, for the twofold reason that it was not a pension, and that the payment should have taken place much earlier. But our repeated requests in this regard have always remained in vain, even when we noted that the scarcity of means available to the embassy did not, and does not, prevent the payment of salaries to its own staff. This four-year cheque was to the value of about 15 thousand pesos for me, as for Benvenuto, at the exchange rate of that time; and we are still owed this sum, which would have been amply sufficient for the expenses of the return journey.

---

[17]Benvenuto Terracini (1886–1968) was the brother of Alessandro. Benvenuto, his daughter Eva (1914–1986) and the grandmother Eugenia Levi (1862–1952) had joined Alessandro and his family in Tucumán in July 1941. They would return to Italy in April 1947.

So, it should be at least decided that the travel expenses be borne by the Italian State, in compensation for non-payment of the aforementioned 15 thousand pesos (approximately). Therefore, in my opinion, all these reasons should be presented by you to the competent Ministries to obtain that the matter is resolved favourably, and that, on the part of the same Ministries, precise and concrete instructions are sent to the embassy in Buenos Aires regarding our return. The generic word "facilitations" is not sufficient for this purpose, although other facilitations are in fact desirable, e.g., the free issue of passports and so on. According to the embassy, the telegram directed to the Ministry of Foreign Affairs apparently in early June has not had any response so far. I add some private information I had outside the embassy, i.e., some journeys have already taken place, without payment by the person concerned. We could not check the truthfulness of this information: but certainly, at the embassy they told Benvenuto that three daughters of an Italian actually had their trip paid. I should now respond to many other points in your letter … [Alessandro Terracini]

Carissimo Mauro, La tua carissima lettera del 21 giugno mi è giunta con grande rapidità alla fine dello stesso mese. Ho rimandato la risposta per poterti riferire in merito al mio colloquio con l'ambasciata italiana in Buenos Aires. Mi ha fatta molta pena la notizia della morte di Enriques. Appena ricevuta da te la triste notizia scrissi a Castelnuovo per mandargli le mie condoglianze e pregarlo di trasmetterle a tutta la famiglia e alla Facoltà. Non sapendo con sicurezza se, dopo tutte le traversie degli scorsi anni, l'indirizzo di Castelnuovo è sempre quello di via Boncompagni 16, ho aggiunto sulla busta anche l'indirizzo dell'Università. Se per caso nonostante queste precauzioni la mia lettera non fosse arrivata, ti prego di volerglielo dire. In ogni caso prego anche te di trasmettere alla vostra Facoltà l'espressione del mio vivo dolore per la gravissima perdita subita. Sono partito ai primi del mese per Buenos Aires, per presiedere la riunione dell'Unione Matematica Argentina. Nella medesima ho dato comunicazione delle morti di Comessatti, Tonelli, De Franchis e Enriques e tenni pure una commemorazione di quest'ultimo, dopo la quale prese la parola Rey Pastor per ricordare in modo particolare la Sua visita a Buenos Aires. La Unione Matematica Argentina ha anche deciso di mandare le sue condoglianze alla famiglia nella persona di Castelnuovo: suppongo che la relativa lettera giungerà più avanti. Vengo ora a parlarti in merito al colloquio da me avuto all'ambasciata sopra il nostro rientro in Italia. Ti dico subito che, attraverso un'apparente cortesia, ho avuto l'impressione che non abbiano l'intenzione di occuparsene a fondo se non sono spinti, e fortemente spinti, da Roma. L'ambasciata ha ricevuto un telegramma del Ministero degli Esteri (che secondo mi scrisse a suo tempo mio cognato Aldo Sacerdote, dovrebbe essere di fine marzo) che la invitava a "facilitare" il nostro ritorno in Italia; a questo telegramma l'ambasciata rispose telegraficamente: il testo del telegramma non ce lo vollero mostrare trincerandosi dietro un "segreto di ufficio": la data della risposta è tutto quello che si poté sapere, e sarebbe solo dei primi di giugno. Da quanto ci fu detto, nella risposta

(a) *l'ambasciata chiede a Roma che dia istruzioni precise, allegando la genericità del verbo "facilitare". E in questo mi pare abbia ragione perché il Ministero degli Esteri dovrebbe dare concretamente istruzioni affinché il nostro viaggio sia considerato come di servizio, e cioè avvenga a spese dello Stato.*

(b) *l'ambasciata, da quanto capisco, non considera il mio caso (e così quello di Benvenuto e sua famiglia) come un caso a sé, bensì domandò istruzioni di carattere generale, non so se per tutti i cittadini italiani oppure tutti gli impiegati dello Stato e famiglie, che vogliano rimpatriare.*

Ora a me pare che il caso mio (così quello di Benvenuto) debba essere considerato assolutamente a parte e "sganciato" da eventuali casi analoghi (tale, del resto, mi pare poter dedurre sia anche la tua opinione, quando, nella lettera del 21 giugno, parli di "provvedimenti eccezionali"). In primo luogo, c'è l'esposto degli 86 matematici italiani, che, grazie alla tua mai abbastanza apprezzata iniziativa, presenta il mio caso come quello di una persona, la cui presenza in Italia – meritatamente o no – viene considerata come particolarmente utile. Per Benvenuto c'è un voto in proposito della Facoltà di Milano, così come per me ce n'è anche uno analogo dello scorso novembre della mia Facoltà di Torino. A ciò si aggiunge nei nostri riguardi un'altra circostanza. La legge 23 maggio 1940 n° 587 concesse a noi professori universitari esonerati per le note ragioni nel 1938 un assegno quadriennale pari alla differenza fra stipendio e pensione dal dicembre 1938 al dicembre 1942. Tale assegno ci fu riconosciuto da successiva sentenza del Consiglio di Stato del 24 settembre 1941, n° 302, e liquidato dal Ministero in data 19 ottobre 1942. Ottenutosi così che l'assegno fosse pagabile all'estero al cambio ufficiale di allora, i relativi ruoli per il pagamento giunsero in Argentina nel maggio 43 ma poco dopo furono qui sospesi i pagamenti della pensione e, adducendo tale ragione, ci fu negato il pagamento del predetto assegno quadriennale maturato molto tempo prima. Noi professori universitari italiani residenti in Argentina ritenemmo e riteniamo che tale mancato pagamento non fosse giustificato, per la duplice ragione che non si trattava di pensione, e che al pagamento avrebbe dovuto farsi luogo assai prima. Ma le nostre ripetute richieste in merito sono sempre rimaste vane, anche quando abbiamo allegato che la scarsità di mezzi di cui disponeva l'ambasciata non impediva e non impedisce i pagamenti degli stipendi al personale alle sue dipendenze. Detto assegno quadriennale importava circa 15 mila pesos per me, e altrettanti per Benvenuto, al cambio di allora; e siamo sempre ancora creditori di tale somma, che sarebbe stata ampiamente sufficiente per le spese del viaggio di ritorno. Quindi almeno si dovrebbe ottenere che le spese di viaggio fossero sostenute dallo Stato italiano, in compenso del non avvenuto pagamento dei predetti 15 mila pesos (circa). Quindi, secondo me, tutte queste ragioni dovrebbero essere da te esposte ai competenti Ministeri per ottenere che si risolva favorevolmente la questione, e che, da parte dei medesimi si mandino all'Ambasciata di Buenos Aires istruzioni precise e concrete in merito al nostro ritorno. La parola generica "facilitazioni" non è sufficiente a tal uopo, per quanto altre "facilitazioni" siano desiderabili, p. e. quella del rilascio gratuito dei passaporti e simili. Al dire dell'Ambasciata, il telegramma diretto dalla medesima al Ministero degli Esteri secondo quanto pare ai primi di giugno non ebbe sinora risposta alcuna. Ti aggiungo ancora che informazioni private da me avute fuori dell'ambasciata direbbero che qualche viaggio ha già avuto luogo, senza pagamento da parte dell'interessato. Non abbiamo potuto controllare la veridicità di tali informazioni: però certo all'Ambasciata hanno detto a Benvenuto che tre figlie di un italiano hanno avuto effettivamente il viaggio pagato. A moltissimi altri punti della tua lettera dovrei ora rispondere . . . [A. Terracini]

D7.1.4 F. Severi to B. Segre, Rome 3.9.1946

CA, *BSP*, fol. 1r
Dearest Segre, When I came to Rome yesterday, I immediately tried to make contact with Minister Gonella and Colonnetti.[18] Unfortunately, neither one nor the other is in Rome. Gonella will return perhaps on Friday 6, and I will try to see him immediately; Colonnetti is in Switzerland, and I have been informed that he will make a quick trip to Rome around the 18th for the Council of Christian Democrats, before returning immediately to the north and again to Rome at the end of September. In the meantime, it is likely that Colonnetti will come to Piedmont. You know that he has a villa in Pollone in the province of Vercelli. Once you are sure he is there, you should pay him a visit, regarding our project which he fully approved. Meanwhile, looking for information at the Research Council, I learned the following from the official who deals with the matter: the Council of the C.N.R. has approved the President's plan regarding the removal from teaching of some highly qualified university professors, so that abroad or in their institutions, or in another Italian university, they deal with scientific research only, with full economic tranquillity for a two-year period at least. Minister Gonella has already unofficially approved this plan. Practices have been initiated with the Ministry of the Treasury for the expenditure necessary to pay the substitutes of the seconded professors. There should be no serious obstacles to this because it is a minimal expense. I will talk to Minister Gonella about this and then I will rewrite to you. In the meantime, I wanted to write to you immediately so that you would not miss the opportunity to talk to Colonnetti if by chance in these days you come to Piedmont. The volumes *Serie di equivalenza* and *Funzioni*

---

[18]Guido Gonella (1905–1982), Minister of Public Education from 1946 to 1951. Gustavo Colonnetti, after the return to Italy, actively involved in politics with the Christian Democrats, as a member of the Consulta (1945–1946) and of the Constituent Assembly (1946–1948). In 1946 he was president of the CNR and member of the Higher Council of Public Education.

abeliane (3 copies for each volume) have been sent to Roth.[19] The price is L. 500 per copy for the *Serie di equivalenza* and L. 200 per copy for *Funzioni abeliane*. At your convenience, the sum can be sent by you directly to the National Institute of Advanced Mathematics, University City—Rome, for example, by bank transfer made out to the President of the Institute of Advanced Mathematics. Again, our cordial greetings also to your wife. Yours sincerely F. Severi

Carissimo Segre, Venuto ieri a Roma ho subito cercato di prendere contatto col Ministro Gonella e con Colonnetti. Disgraziatamente né l'uno né l'altro sono a Roma. Gonella vi tornerà forse venerdì 6 e cercherò di vederlo subito; Colonnetti è invece in Svizzera e mi hanno informato che farà una rapida scappata a Roma verso il 18 del mese per il Consiglio della Democrazia Cristiana, per tornare subito al nord e ritornare a Roma verso la fine di settembre. Nel frattempo, è probabile che Colonnetti venga in Piemonte. Ella sa che egli ha una villa a Pollone in provincia di Vercelli. Una volta assicuratosi che fosse là, sarebbe bene che Ella lo andasse a trovare, riferendosi al nostro progetto che egli approvò pienamente. Frattanto, assunte informazioni al Consiglio della Ricerche, dal funzionario che si occupa della questione ho saputo quanto segue: il Consiglio della Presidenza del C.N.R. ha approvato il piano del Presidente circa il distacco dall'insegnamento di alcuni professori universitari, particolarmente qualificati, perché all'estero o nella loro sede, o in altra sede universitaria italiana, si occupino con piena tranquillità economica, per un biennio almeno, di sole ricerche scientifiche. Il Ministro Gonella ha già approvato in via ufficiosa questo piano. Sono state iniziate pratiche col Ministero del Tesoro per la spesa occorrente a pagare i supplenti dei professori distaccati. Non ci dovrebbero essere ostacoli gravi per questo, perché trattasi di una spesa minima. Parlerò della cosa al Ministro Gonella e dopo Le riscriverò. Frattanto ho voluto scriverle subito affinché Lei non perdesse l'occasione di parlare con Colonnetti se per caso in questi giorni venisse in Piemonte. I volumi delle Serie di equivalenza *e delle* Funzioni abeliane *(3 copie per ciascun volume) furono spediti a Roth. Il prezzo è di L. 500 per copia per le* Serie di equivalenza *e di L. 200 per copia per le* Funzioni abeliane. *Quando crede, la somma può essere da Lei spedita direttamente all'Istituto Nazionale di Alta Matematica, Città Universitaria – Roma, per es. a mezzo di assegno bancario sbarrato intestato al Presidente dell'Istituto di Alta Matematica. Di nuovo cordialissimi saluti nostri anche alla Signora. Suo aff.^mo F. Severi*

---

[19]F. Severi, *Serie, sistemi di equivalenza e corrispondenze algebriche sulle varietà algebriche*, ed. by F. Conforto and E. Martinelli, Roma, Istituto Matematico della R. Università, 1938, reprinted in 1942; F. Conforto, *Funzioni abeliane e matrici di Riemann. 1*, Roma, Libreria dell'Università, 1942. Leonard Roth, lecturer in Mathematics at the Imperial College in London, had worked in Rome in 1930–31, as a RKF fellow, and had entered in close contact with Segre, Castelnuovo, Enriques, and Severi.

D7.1.5 F. Severi to B. Segre, Rome 10.9.1946

CA, *BSP*, fol. 1r-v

Dearest Segre, I received yours of the 5th (with the check) and today yours of the 7th. In the meantime, you will have received mine of the 7th with the long letter prepared for Hodge on which I await your opinion. Yesterday I had a long conversation with the Minister [Gonella] about the Institute [of Advanced Mathematics], and I left him a draft law with the new constitution and funding. He is very much in favour and will try to overcome the difficulties. I also told him about you and left him the Memorandum of which I enclose a copy. He will assist; but what you wrote to me after the interview with Colonnetti, who is mainly responsible for the action, is not very encouraging. I will speak myself with Colonnetti as soon as he comes to Rome. [. . .] Meanwhile, I ask you if you would be willing to accept a transfer for Higher Geometry, in which case I would immediately resign from the assignment and temporarily stand in at the Faculty on condition you were called on. Answer me immediately, perhaps by cable, so that I have your answer before going to Arezzo and can talk to the Dean in advance. What I can do, to the utmost of my ability to improve your treatment, will be done! For example, if they give me the funds to carry on the Institute while the law is being made, I will propose to my Council that we start again to allocate courses in addition to those held by the professors at the Institute. I would like to pay for these assignments a little more respectfully than was done in Rome for tenured professors. In short, the sooner you come to Rome, the better it will be for the work that we must carry out together. After completing the call in Cagliari, I should go for a couple of months to Spain (Switzerland also insist), but I'm still not sure if I will actually go. We will talk about Franchetta at the first opportunity.[20] The course that you would do at the Institute and the book that you would like to prepare are both excellent.[21] You will certainly have seen the lectures that Deuring gave us in 1941 (Rend. of Rome, vol. 2, p. 361).[22] Remind us to your wife and please accept my cordial greetings. Yours most affectionate F. Severi

---

[20] Alfredo Franchetta (1916–2011), a pupil of Enriques, was an assistant and lecturer at the University of Rome. Segre had found some errors in one of his works and Severi had suggested: 'I don't see why you can't put the matter right, with earnest citation of his errors. He took that problem from me, but he approached the research completely captivated by Enriques. Now the widows of Enriques and Castelnuovo want him to work with me'. (*Non vedo perché Ella non possa rimettere a posto la questione, con sobria citazione degli errori di lui. Egli ha tratto da me quel problema, ma si avvicinava al lavoro tutto preso da Enriques. Ora le vedove di Enriques e Castelnuovo desiderano ch'egli lavori con me.*) See CA, *BSP*: F. Severi to B. Segre, Soriano 29.8.1946.

[21] Segre had proposed to give at INDAM a course in abstract algebra and to publish a text on modern geometry. The book *Lezioni di geometria moderna*, vol. 1, Zanichelli, Bologna, would appear in 1948.

[22] M. Deuring, *La teoria aritmetica delle funzioni algebriche di una variabile*, 25.6.1941, Rendiconti del Seminario Matematico dell'Università di Roma, s. 2, V, 1941, pp. 361–412.

Memorandum *re.* **Prof. Segre**

CA, *BSP***, fols. 1r, 2r**

Prof. Beniamino Segre, one of the most talented of my students and my former assistant in Rome, was affected in 1938 by racial measures, while he was a full professor at the University of Bologna. He emigrated to England with his family (his wife and three children) and there, after many painful events (he lost a small daughter in London, because of the aerial bombings), he was able to settle as a teacher at the University of Manchester, where he remained until the current school year.[23] Returning to Italy two or three months ago to resume contacts with the homeland and on speaking with me (who kept in touch with him as long as it was possible and reconnected after liberation), he decided to return definitively. On arrival in Italy, the thought of leaving his homeland again resurrected strong nostalgia in him and he therefore preferred to face greater sacrifices here rather than return to live abroad. He resigned from the University of Manchester and refused a post he had been offered at the University of Aberdeen with a salary of £700 a year. This gesture deserves consideration no less than that in which his remarkable scientific merits must be taken. Since I trust that the National Institute of Advanced Mathematics will be put in a position to actively resume its activity, I can express the opinion that among young people there is no one more worthy than Beniamino Segre to add to the list of current teachers in the Institute. The transfer to Rome would be possible for him even immediately if the Faculty called him (as I do not doubt it will do) to the chair of Higher Geometry, which remained vacant after the death of Federigo Enriques. But the economic situation in which B. Segre would find himself would not be relieved by an income that a much more consistent course can offer (through propines and lithographs) and he could not afford the large expenses that the transfer to the Capital would involve, especially since the Ministry of the Treasury has not granted any contribution to the costs of his return to the homeland. I therefore believe that it is more convenient to try to have Segre in Rome immediately by taking advantage of one of the grants for scientific research only, which the Research Council proposes (with the consent of the Minister of the P.I.) to offer particularly deserving professors, seconded at the C.N.R. At the same time, Segre could also hold courses at the Institute. Maintaining this situation for two years, at the end of the C.N.R. check, you could find for him a definitive placement, which is also economically satisfactory. I strongly recommend this solution to the authority of H.E. the Minister, as I am eager that Segre can begin to contact the Institute in

---

[23] Segre had emigrated with his wife Fernanda Coen (1908–1976) and his children Sergio (born on 22 May 1933), Silvana (14 May 1934), and Ornella (15 June 1937). Ornella died in London in 1940 while her father was interned.

preparation for the approaching moment in which my own university life will
end. Francesco Severi Rome, 9 September 1946

*Carissimo Segre, Ho ricevuto la Sua del 5 (coll'assegno) e oggi la Sua del 7. Nel
frattempo Ella avrà ricevuto la mia del 7 con la lunga lettera preparata per
Hodge sulla quale attendo il Suo avviso. Parlai ieri a lungo col Ministro
dell'Istituto e gli lasciai uno schema di legge con la nuova costituzione e pel
finanziamento. Egli è favorevolissimo e cercherà di superare le difficoltà. Gli
parlai anche di Lei e gli lasciai il promemoria di cui Le accludo copia. Egli
agevolerà, ma ciò che mi scrive dopo il colloquio con Colonnetti, al quale spetta
in linea principale l'azione, è poco incoraggiante. Parlerò io stesso col
Colonnetti, appena verrà a Roma. [...] Intanto le chiedo se sarebbe disposto
ad accettare il trasferimento per la Geometria Superiore, nel qual caso io
rinuncerei subito all'incarico e agirei in Facoltà purché si facesse la sua
chiamata. Mi risponda subito, magari per espresso, onde io abbia la risposta
prima di andare ad Arezzo e ne possa preventivamente parlare al Preside. Quello
che potrà farsi nei limiti di ogni mia possibilità per migliorare il trattamento,
sarà fatto! P. es. se mi danno i fondi per tirare avanti l'Istituto intanto che si fa la
legge, proporrò al mio Consiglio che si ricominci a dare incarichi per corsi oltre
quelli dei professori dell'Istituto. Vorrei pagare questi incarichi un po' meno
indegnamente di quanto non si faceva a Roma pei professori di ruolo. Insomma,
più presto Ella verrà a Roma e meglio sarà per l'opera che insieme dobbiamo
svolgere. Io, dopo 'espletato' il concorso per Cagliari, dovrei andare per un paio
di mesi in Spagna (insistono anche dalla Svizzera), ma non son ancora sicuro se
ci andrò. Parleremo di Franchetta alla prima occasione. Ottimi il corso ch'Ella
farebbe all'Istituto e il libro che vorrebbe poi preparare. Avrà visto certo le
conferenze che ci fece Deuring nel 1941 (Rend. di Roma, vol. 2, p. 361). Ci
ricordi alla Signora e si abbia i miei saluti cordiali. Suo aff. F. Severi*

Promemoria riguardante il prof. Segre

*Il prof. Beniamino Segre, uno dei più valenti fra i miei discepoli e già mio
assistente a Roma, fu colpito nel 1938 dai provvedimenti razziali, mentre era
professore ordinario all'Università di Bologna. Emigrò in Inghilterra con la
famiglia (la moglie e tre bambini) e là dopo molte vicende dolorose (perdette a
Londra una piccola figlia, in conseguenza dei bombardamenti aerei), poté
sistemarsi come insegnante all'Università di Manchester, dove è restato fino a
tutto il corrente anno scolastico. Tornato in Italia due o tre mesi or sono per
riprender contatto con la Patria e consigliarsi con me (che avevo tenuto le
comunicazioni con lui finché fu possibile e le avevo riallacciate a Liberazione
avvenuta), decise di rientrare definitivamente. Rimesso il piede in Italia, il
pensiero di abbandonare di nuovo la Patria risuscitava in lui sofferenze
nostalgiche ed egli ha perciò preferito di affrontare qui maggiori sacrifici
piuttosto che tornare a vivere all'Estero. Si è così dimesso dall'Università di
Manchester ed ha rifiutato un posto che gli era stato offerto all'Università di*

Aberdeen con uno stipendio di 700 sterline annue. Questo suo gesto merita considerazione non inferiore a quella in cui debbon tenersi i suoi notevolissimi meriti scientifici. Poiché confido che l'Istituto Nazionale di Alta Matematica sarà posto in condizione di riprendere attivamente la propria vita, posso esprimer l'avviso che fra i giovani nessuno meglio di Beniamino Segre potrebbe aggiungersi a coloro che già insegnano nell'Istituto. Il trasferimento a Roma sarebbe per lui possibile anche subito se la Facoltà lo chiamasse (come non dubito farebbe) alla cattedra di Geometria Superiore, restata vacante per la morte di Federigo Enriques. Ma la situazione economica in cui B. Segre verrebbe a trovarsi non sarebbe alleviata dai proventi che può offrire (attraverso le propine e le litografie) un corso molto più numeroso ed egli non potrebbe sobbarcarsi alle spese ingenti che importerebbe il trasferimento nella Capitale, tanto più che finora il Ministero del Tesoro non ha concesso alcun contributo alle spese di rientro del Segre in Patria. Credo pertanto che sia più conveniente cercar di aver subito il Segre a Roma profittando di uno degli assegni che il Consiglio della Ricerche si propone (coll'assenso del Ministro della P.I.) di destinare a professori, particolarmente meritevoli, distaccati fuori sede per sole ricerche scientifiche. Nel fatto poi il Segre potrebbe tener corsi anche all'Istituto. Protraendo questa situazione per un biennio, si potrebbe nel frattempo trovare per lui una sistemazione definitiva conveniente anche economicamente al cessare dell'assegno del C.N.R. Raccomando vivamente questa soluzione all'autorità di S.E. il Ministro, desideroso come sono che il Segre possa cominciare a prender contatto coll'Istituto in preparazione del momento ormai non lontano in cui si chiuderà la mia vita universitaria. Francesco Severi Roma, 9 settembre 1946

D7.1.6 G. Castelnuovo to MPI, [November 1946]

ACS-*Terracini*, fol. 1r-v

To the Ministry of Public Instruction: Alessandro Terracini, professor of Analytic Geometry at the University of Turin, was forced in 1939 by the racial laws to leave his post and emigrate to the Argentine Republic, where he was appointed professor of Mathematics at the University of Tucumán. There, with his teaching and writings, he kept the prestige of the Italian School high and founded a Journal of Mathematics, which he currently directs and where Italian and Argentine works have been published. Now the reasons that forced him to expatriate have ceased to exist, and Terracini wishes to return to his place at the University of Turin, but his return to his homeland is impeded by the very significant travel expenses, given that the family that would return to Italy is composed not only of himself, but also his mother, wife and three children.[24] A provision in favour of Terracini seems necessary, especially since he wishes to return to his country despite a state of financial conditions lower than those he enjoys in Argentina at a time when many Italian professors are emigrating, attracted by more lavish earnings. This comes after the recent losses of many distinguished mathematicians, such as Tonelli, Enriques, Comessatti, and De Franchis;[25] the Italian School needs an injection with scientists of the value of Terracini, if it does not want to lose the position of privilege that foreign nations envy us. All the Italian mathematicians have already pronounced themselves in favour of Terracini's return, as shown in the memorial dated March 6, 1946, a copy of which is sent (Annex A). The Italian embassy in Buenos Aires has also approved the measure to facilitate his return, as shown in the letter from the embassy to Prof. Terracini, of which a copy is attached (Annex B).[26] It should be added that the financial effort requested on the part of the Italian Government could be considerably lightened if, with the consent of the interested scholar, the sums that this Ministry owes to Terracini were drawn from his salary from 1 January 1944, from the difference between salary and pension during the four years 1940-43 and from pension arrears. Such sums could not so far be transmitted to the Argentine Republic, but the Government could use them to pay, at least in part, the costs of the sea voyage. We are confident that, for the above reasons, this Ministry, in agreement with the Ministry of Foreign Affairs, will implement the measure requested in favour of Prof. Alessandro Terracini and his family. Prof. Guido Castelnuovo—President of the National Academy of Lincei.

I would add that the Faculty of Sciences of Turin voted for the recall of A. Terracini, asking that his return and that of his family be considered as a service trip (November 1945).[27]

Al Ministro della P. Istruzione: Alessandro Terracini, professore di Geometria Analitica all'Università di Torino, fu costretto nel 1939, in seguito alle leggi razziali, a lasciare il suo posto ed emigrare alla Repubblica Argentina, dove fu nominato professore di Matematica all'Università di Tucumán. Colà con l'insegnamento e con gli scritti tenne alto il prestigio della Scuola Italiana e fondò un Giornale di Matematica, che attualmente dirige, ove vennero pubblicati lavori italiani e argentini. Cessate ora le ragioni che lo hanno costretto ad espatriare, il Terracini desidera rioccupare il suo posto all'Università di Torino; al suo ritorno in patria fanno ostacolo le spese di viaggio molto rilevanti, dato che la famiglia che rientrerebbe in Italia si compone, oltre che di lui, della

---

[24]Eugenia Levi Terracini (1862–1952), Giulia Sacerdote Terracini (1899–1974), Lore (1925–1995), Cesare (1927–1972), and Benedetto (1931).

[25]Leonida Tonelli (1885–1946), Federigo Enriques (1871–1946), Annibale Comessatti (1886–1945), Michele De Franchis (1875–1946).

[26]Annexes A and B are missing.

[27]See footnote 9, p. 419.

madre, della moglie e di tre figlioli. Un provvedimento a favore del Terracini che desidera ritornare nel proprio paese, a condizioni finanziarie inferiori a quelle di cui gode in Argentina, mentre tanti professori italiani emigrano attratti da più lauti guadagni, sembra imporsi, tanto più che con le perdite recenti di molti insigni matematici, quali il Tonelli, l'Enriques, il Comessatti, il De Franchis, la Scuola Italiana ha bisogno di essere rinsanguata con scienziati del valore del Terracini, se non vuole perdere una posizione di privilegio che nazioni straniere ci invidiano. A favore del ritorno del Terracini si son già pronunziati tutti i matematici italiani, come risulta dal memoriale in data 6 marzo 1946, di cui si trasmette copia (allegato A). Al provvedimento per facilitare il ritorno si è pure dimostrata favorevole l'Ambasciata d'Italia a Buenos Aires, come risulta dalla lettera dell'Ambasciata al Prof. Terracini, di cui è allegata una copia (allegato B). Si aggiunga che l'aggravio finanziario richiesto dal Governo italiano potrebbe essere notevolmente alleggerito se col consenso dell'interessato si disponesse delle somme che codesto Ministero deve al Terracini per stipendio dal 1° gennaio 1944, per differenza tra stipendio e pensione durante il quadriennio 1940-43 e per pensione arretrata, somme che non poterono essere trasmesse finora alla Repubblica Argentina, ma di cui il Governo potrebbe far uso per pagare, almeno in parte, le spese del viaggio marittimo. Si confida che codesto Ministero, d'accordo col Ministero degli Esteri, voglia per le ragioni su esposte attuare il provvedimento richiesto a favore del Prof. Alessandro Terracini e della sua famiglia. Prof. Guido Castelnuovo – Presidente dell'Accademia Nazionale dei Lincei

Aggiungo che la Facoltà di Scienze di Torino ha con suo voto richiesto il richiamo di A. Terracini domandando che il ritorno di lui e della famiglia sia considerato come viaggio di servizio (novembre 1945).

D7.1.7 MPI to the Ministry of the Treasury, cc. MFA, Rome 6.11.1946

ACS-*Terracini*, fol. 1r-3v
This Ministry, with letter no. 12570 of 2 July 1946, addressed to the Presidency of the Council of Ministers, cc. to this Dicastery, envisaged the opportunity of granting special facilities to university professors readmitted to service, to reach their headquarters, especially with regard to those who were abroad at the time of reinstatement. Attention was drawn to the case of Prof. Alessandro Terracini, readmitted to service as Professor of Analytic Geometry with elements of Projective and Descriptive Geometry with drawing at the University of Turin, for whom the mathematics professors in the Italian Universities intervened with a statement asking to provide him with the means to return to Italy, in consideration of his special merits (D7.1.2). A letter has now been received from the President of the Academy of Lincei, here attached, with which a measure is invoked to facilitate the return to homeland of the aforementioned Prof. Alessandro Terracini (D7.1.6); it states that his return would be of great interest to the Italian School, as

the recent loss of many distinguished mathematicians such as Tonelli, De Franchis, Comessatti and Enriques means that it needs an injection of new blood with scientists of the level of Terracini, if it does not want to lose that position of privilege that foreign nations envy us. Therefore, this Ministry is asked to examine with all due attention the possibility of issuing a decree that allows the reimbursement of travel expenses to university professors who were deposed for political or racial reasons, and who now return here following readmission to service. According to our calculations, there are 29 university professors who at the time of reinstatement were resident abroad:

(1) Ascarelli Tullio, Professor of Commercial Law at the University of Bologna, resident in Brazil
(2) Ascoli Alberto, Professor of General Pathology at the University of Milan, resident in the United States of America
(3) Bachi Roberto, Professor of Statistics at the University of Genoa, resident in Palestine
(4) Bachi Riccardo, Professor of Political Economy at the University of Rome, resident in Palestine
(5) Cassuto Umberto, Professor of Hebrew and Semitic Languages at the University of Rome, resident in Palestine
(6) Foà Bruno, Professor of Political Economy at the University of Bari, resident in the United States of America
(7) Foà Carlo, Professor of Human Physiology at the University of Milan, resident in Brazil
(8) Franco Salomone Enrico, Professor of Anatomy and Pathological Histology at the University of Pisa, resident in Palestine
(9) Herlitzka Amedeo, Professor of Human Physiology at the University of Turin, resident in Argentina
(10) Levi Beppo, Professor of Mathematical Analysis at the University of Bologna, resident in Argentina
(11) Lattes Leone, Professor of Forensic Medicine and Insurance at the University of Pavia, resident in Argentina
(12) Levi Della Vida Giorgio, Professor of Hebrew and Semitic Languages at the University of Rome, resident in the United States
(13) Levi Giorgio Renato, Professor of General Chemistry at the University of Pavia, resident in Brazil
(14) Levi Teodoro, Professor of Archaeology and History of Ancient Art at the University of Cagliari, resident in the United States
(15) Liebman Enrico, Professor of Civil Procedure Law at the University of Parma, resident in Brazil
(16) Momigliano Arnaldo, extraordinary Professor of Roman History at the University of Turin, resident in England
(17) Mondolfo Rodolfo, Professor of History of Philosophy at the University of Bologna, resident in Argentina

(18) Ottolenghi Michelangelo, extraordinary Professor of Anatomy of Domestic Animals at the University of Sassari, resident in Ecuador

(19) Ravà Renzo, extraordinary Professor of Labour Law at the University of Florence, resident in the United States of America

(20) Rossi Bruno, Professor of Experimental Physics at the University of Padua, resident in the United States of America

(21) Segré Beniamino, Professor of Analytic Geometry at the University of Bologna, resident in England

(22) Segré Angelo, Professor of Economic History at the University of Trieste, resident in the United States of America

(23) Segré Emilio, Professor of Higher Physics at the University of Palermo, resident in the United States of America

(24) Sereni Angelo, extraordinary Professor of International Law at the University of Ferrara, resident in the United States of America

(25) Terracini Aron, Professor of Glottology at the University of Milan, resident in Argentina

(26) Terracini Alessandro, Professor of Analytic Geometry at the University of Turin, resident in Argentina

(27) Tedeschi Guido, extraordinary Professor of Civil Law at the University of Siena, resident in Palestine

(28) Viterbo Camillo, extraordinary Professor of Commercial Law at the University of Cagliari, resident in Argentina

(29) Zamorani Vittorio, Professor of Paediatric Clinic at the University of Pavia, resident in Venezuela.

Only four of these have returned to Italy: Professors Bachi Riccardo, Levi Della Vida Giorgio, Levi Teodoro, and Momigliano Arnaldo. Travel expenses for residents in America (including family members) average L. 1,000,000; the cost of travel for those who reside in Palestine or in other countries of Europe rises to L. 400,000.

Considering that in the various countries of the Americas there are 22 professors and 7 live in Palestine or other countries of Europe, the need for reimbursement of travel expenses to these professors can be calculated as L. 24,400,000 (twenty-four million four hundred thousand lire). We are waiting to know the opinion of this Dicastery regarding the situation, pointing out the urgency of the action requested. Minister Gonella

Questo Ministero con lettera n. 12570 del 2 luglio 1946, diretta alla Presidenza del Consiglio dei Ministri trasmessa per conoscenza a codesto Dicastero, prospettava la opportunità di concedere particolari agevolazioni ai professori universitari riammessi in servizio per raggiungere la loro sede, avuto riguardo specialmente a coloro che all'atto della riammissione si trovavano all'estero. In particolare, si richiamava l'attenzione sul caso del prof. Alessandro Terracini riammesso in servizio quale ordinario di Geometria Analitica con elementi di Proiettiva e Geometria Descrittiva con disegno nell'Università di Torino, per il quale i professori titolari di insegnamenti matematici nelle Università Italiane facevano voto perché gli fossero forniti i mezzi per il ritorno in Italia tenuto conto delle speciali benemerenze del professore. È ora pervenuta una lettera del Presidente dell'Accademia dei Lincei, che si unisce in copia, con la quale viene caldeggiato un provvedimento per facilitare il ritorno in patria del predetto Prof. Alessandro Terracini, che sarebbe di grande interesse per la Scuola italiana, la quale con la recente perdita di molti insigni matematici quali il Tonelli, il De Franchis, il Comessatti e l'Enriques, ha bisogno di essere rinsanguata con scienziati del valore di Terracini, se non vuol perdere una posizione di privilegio che le nazioni straniere ci invidiano. Pertanto, si prega codesto Ministero di voler esaminare con ogni attenzione la possibilità di emanare un provvedimento che consenta il rimborso delle spese di viaggio ai professori universitari che allontanatisi dalla loro sede in seguito a dispensa per motivi politici o razziali, vi ritornino in seguito alla riammissione in servizio. I professori universitari che all'atto della riammissione trovavansi all'estero sono, in seguito a definitivi accertamenti, i seguenti in numero di 29. [. . .] Di essi solo quattro sono rientrati in Italia, i Proff. Bachi Riccardo, Levi Della Vida Giorgio, Levi Teodoro, Momigliano Arnaldo. La spesa del viaggio per coloro che si trovano in America (compresi i famigliari) si aggira in media a L. 1.000.000; la spesa del viaggio per coloro che risiedono in Palestina od in altri paesi d'Europa si può fare ascendere a L. 400.000. Considerato che nei vari paesi delle Americhe risiedono 22 professori e 7 risiedono in Palestina od altri paesi d'Europa, il fabbisogno per il rimborso delle spese di viaggio ai detti professori può calcolarsi in L. 24.400.000 (ventiquattro milioni quattrocentomila lire). Si rimane in attesa di conoscere l'avviso di codesto Dicastero in merito a quanto esposto, segnalando l'urgenza del provvedimento invocato. Il Ministro Gonella

D7.1.8 Memorandum *re.* Prof. Terracini, Naples 2.12.1946

ACS-*Terracini*, fol. 1r

Alessandro Terracini, full professor of Analytic Geometry at the University of Turin, was exempted from service in 1938 on racial grounds. He was re-admitted to service in 1944. Professor Terracini moved with his family in 1939 to Tucumán (Argentina) where he teaches at the University. Reintegrated into service, he now wishes to return to Italy with his family to resume teaching at the University of Turin and his scientific activity in Italy. Eighty-six mathematicians, professors in various Italian Universities, and the Academy of Lincei have pointed out to the Minister of Public Instruction that Prof. Terracini's return to Italy is of undisputed benefit to Italian school and science (D7.1.2 and D7.1.6). This return is not possible if Prof. Terracini is not given the means to support the expenses of his transfer and those of the people making up his family (wife and three children) amounting to approximately 4,000 dollars. The Ministry of P.I., convinced of the need to favour in every way the prompt repatriation of Prof. Terracini, on 6 November 1946, presented the matter to the Ministry of the Treasury, putting forward every useful element to reach a favourable decision (D7.1.7). In February 1946, the Ministry of the Treasury had pointed out to the Ministry of Public Instruction that, according to current regulations, a request from the latter Ministry could not be accepted for all displaced professors and their families to be repatriated at the expense of the State; but the same Ministry of the Treasury did not rule out the possibility of granting some financial aid, in the form of subsidies, after evaluating individual cases and the costs incurred in reaching the places of employment. It is asked that Prof. Terracini be granted a subsidy of 4,000 dollars or the equivalent in Argentine pesos, which should be made available to him in Argentina to allow him to pay for the travel expenses of himself and his family. It should be noted that without this grant, Prof. Terracini finds himself in the absolute impossibility of repatriating, with sure damage to Italian school and science, and that the sum in dollars that the State should pay out to allow his repatriation is lower or equal to that which the State pays for trips abroad of figures from the fields of science, literature, industry or for other reasons (sporting, artistic events, etc.), which are certainly no more useful than the repatriation of Prof. Terracini.

Il prof. Alessandro Terracini, ordinario di Geometria Analitica nell'Università di Torino, è stato, per ragioni razziali, esonerato dal servizio nel 1938. È stato riammesso in servizio nel 1944. Il prof. Terracini si è trasferito nel 1939, con la famiglia, a Tucumán (Argentina) dove insegna in quella Università. Riammesso in servizio, egli desidera rientrare in Italia con la Sua famiglia per riprendere l'insegnamento all'Università di Torino e la sua attività scientifica in Italia. Ottantasei matematici, docenti nelle varie Università italiane, e l'Accademia dei Lincei hanno fatto presente al Ministro della Pubblica Istruzione che il ritorno in Italia del Prof. Terracini è di indiscussa utilità della scuola e della scienza italiana. Detto ritorno non è possibile se al Prof. Terracini non vengono dati i mezzi per sostenere le spese del trasferimento suo e delle persone costituenti la sua famiglia (moglie e tre figli) ammontanti a circa 4000 dollari. Il Ministero della P.I., convinto della necessità di favorire in ogni modo il sollecito rimpatrio del prof. Terracini ha, in data 6 nov. c.a. prospettata la questione al Ministero del Tesoro, prospettando ogni utile elemento per addivenire ad una favorevole decisione. Il Ministero del Tesoro, nel febbraio 1946, aveva fatto presente a quello della P.I. che, in base alle norme vigenti, non poteva essere accolta una richiesta di quest'ultimo Ministero perché tutti i professori espatriati e le loro famiglie fossero rimpatriati a spese dello Stato; ma lo stesso Ministero del Tesoro non escludeva che si potesse concedere qualche aiuto finanziario, sotto forma di sussidio, previa valutazione dei singoli casi e degli oneri da sostenere per raggiungere le sedi di servizio. Si chiede che al Prof. Terracini venga concesso un sussidio di 4000 dollari o l'equivalente in pesos argentini, che dovrebbero essere messi a sua disposizione in Argentina per poter sostenere le spese per il rimpatrio suo e della sua famiglia. Si fa presente che senza questo sussidio il Prof. Terracini si trova nell'assoluta impossibilità di rimpatriare con comprovato danno della scuola e della scienza italiana, e che la somma in dollari che lo Stato dovrebbe sborsare per permettere questo rimpatrio è inferiore o pari a quella che lo Stato sborsa per viaggi all'estero di personalità del campo delle scienze, delle lettere, dell'industria o per altri motivi (manifestazioni sportive, artistiche, ecc.) non certo più utili di quanto è per le scienze italiane il rimpatrio del prof. Terracini.

7.2 To Stay or Return?

Return was associated with three basic questions closely linked: whether, where and how to return? Expats who intended to come back were faced with reinstatement legislation that had significant flaws. Reinstatement was provided for in the institution from which they had been dismissed in 1938 and with the same position in the hierarchical system (full professor, extraordinary professor, assistant, lecturer, etc.). However, the authorities had not considered the changes in the meantime: universities had replaced teachers who had been dismissed as a result of the racial laws. Their chairs were now occupied by other scholars, some of whom had been appointed through regular competitions. The *libere docenze*, revoked in 1938, were not considered in law. It was therefore unclear whether and how lecturers should be reinstated. Their precarious positions had sometimes been suppressed. If they still existed, they were occupied by others.

The seven years which had passed since the purge had also brought about changes in the lives of deposed scholars. Some, like Fubini, had disappeared. Others were now close to retirement and reluctant to resume teaching for a few months. The retirements of Fano and Enriques, expected in 1946, meant that the Faculties of Turin and Rome had to decide how to fill their places (D7.2.9). Fano wanted to pass on his chair to Segre. On the contrary, Severi hoped that Segre would return to Rome, to the chair left vacant by Enriques. Both proposals were welcomed by the people concerned, who simply wished to return. Both, however, caused problems for the University of Bologna, where Segre should be reinstated. Segre himself did not want to disrespect his colleagues by returning to Bologna only for a short time before moving on to Turin or Rome.

The academic staff, meanwhile, was not always ready and willing to welcome returning emigrants. The situations varied widely from place to place. In Turin, at the time of departure of Fubini and Terracini, Tricomi had managed to obtain precise promises: Buzano would return the chair of Higher Geometry to Terracini, and Tricomi himself would give up the chair of Higher Analysis to Fubini, as soon as they could return. Reinstatements thus took place serenely. The colleagues were ready to reunite with the displaced scholars, and they did so with a certain warmth. In this case, however, much depended on the fact that former expats met up again with true friends, such as Tricomi, Persico, and Buzano, with whom they had stayed in contact throughout the period spent abroad. Elsewhere, however, this was not the case. In Bologna, for example, Levi perceived a resistance to his return. In Rome,

Bompiani and Picone were not at all happy with Severi's proposal to call Segre to the chair that had been Enriques'. Their opposition was tenacious, to the point that it would take months of battles in the Faculty before Segre was hired.

Further complicating the picture was the purge procedure, which had opened in 1945 against faculty members charged with having carried out activities in support of fascism and of having collaborated with the Republican fascist government (RSI). Both the local and central commissions, which encompassed formerly persecuted scholars such as Guido Ascoli, Giuseppe Levi, and Edoardo Volterra, tried to take serious action but without success. The purge was substantially a farce.[28] None of the university professors, not even those most colluded with the regime, was sanctioned. Severi was accused of having made repeated displays of apologies of fascism and having collaborated with the RSI by participating in the meeting of the Academy of Italy in Florence (March 1944). After a first deliberation that resulted in Severi's dismissal, he presented an appeal in the form of a lengthy, detailed document in his own defence (D7.2.2). After various vicissitudes and following testimony in his favour, the central commission accepted the appeal and reduced the accusations, arriving at the following conclusion:

> Severi did not receive from fascism anything more than what he merited; he did however consent that his reputed name, his moral rectitude, his high scientific merits, and his past as an anti-fascist intellectual be used, not only for the good of Italy, but also for the political aims of the regime. A much more serious fault is that this adhesion, by a personage such as Severi, effectively constituted for the regime a noteworthy reinforcement. [...] Ultimately, since the activity carried out by Severi was of a predominantly scientific nature, it was not deemed sufficiently serious to lead to a declaration of his being "unworthy of serving the State"; the commission sentenced him to a lesser penalty, that is, a mere censure.[29]

Similarly, Picone and Bompiani, who were initially convicted, presented a defence appeal corroborated with a mass of 'whitewash papers', and were finally also acquitted with a mere censure.

Ultimately, most universities would adopt the solution of the so-called supranumerary professorships; in other words, the Jews were reinstated alongside those who had replaced them in 1938. The supranumerary chairs were then to be suppressed upon the retirement of reinstated scholars. Terracini, for example, returned in 1948 as a supranumerary professor of Analytic Geometry; this post was abolished on his retirement in 1964.

---

[28] Guerraggio and Nastasi (2018a).

[29] ACS, MPI, Direzione Generale Istruzione Superiore, *Fascicoli professori epurati*, b. 31, f. *Severi Francesco*, 9.5.1945: *il prof. Severi non ha ricevuto dal fascismo nulla di più di quello che egli meritasse; ha, però, consentito che il suo nome insigne, la sua dirittura morale, le sue alte benemerenze scientifiche, ed il suo passato di intellettuale antifascista, oltre che al bene dell'Italia, servissero anche alle finalità politiche del regime. Responsabilità tanto più grave, in quanto detta adesione, riferibile ad una personalità come quella del Severi, costituì effettivamente, per il regime, un notevole potenziamento [...]. Trattasi, però, di una attività che, nel suo complesso, e per il suo aspetto scientifico, prevalente, non può assolutamente portare ad una dichiarazione di indegnità a servire lo Stato, ma solo ad una sanzione minore.*

In addition to the supranumeracy, there were two other possibilities. The National Research Council (CNR), of which Colonnetti was chairman from 1944 to 1956, and the Ministry of Foreign Affairs granted some secondments. These were temporary assignments to carry out research activities, on lines of studies considered strategic for the cultural reconstruction of Italy, in Italian or foreign institutions. Severi proposed that one of these assignments went to Segre, but this did not happen. Segre, understandably, was not enthusiastic about the prospect of leaving Manchester (where he had a temporary position but with acceptable prospects of becoming permanent) for a secondment (which not only was a temporary position but could not exceed two years). Levi asked to be *distaccato*, i.e. given paid leave by the Ministry of Foreign Affairs, at the Institute of Mathematics in Rosario (D7.2.10 and D7.2.11). Finally, there was the INDAM, which from 1939 invited Italian and foreign scholars of authority to hold advanced courses aimed at fellows of the Institute, providing them with a modest salary. Severi invited Segre to teach a course in abstract algebra in 1946, so as to facilitate his return. Funds, however, were lacking and the project was cancelled. Due to the very difficult economic conditions, the contribution of INDAM in facilitating the return of Jewish displaced scholars was non-existent.

Just as it had been a difficult choice to leave or stay in 1938, so it was in 1945. Beyond the financial capabilities that conditioned expats in the decision, there were other factors to consider. From the family point of view, some refugees had families that had now been integrated into their second homelands. The sons of Fano and Fubini did not plan to return to Italy; nor did Eva Terracini and Laura Levi. The sons of Fubini and Fano held positions of indisputable importance and prestige. They had settled down in their adopted countries; some of them had gotten engaged and/or married, and sometimes their children had been born there.

From a professional point of view, the decision was equally difficult. Grateful for the country that had hosted them with affection and generosity, the desire not to abandon for the second time what they had built (the Institutes, the journals, the pupils etc.), had a significant weight on individual decisions.

Even more relevant was the uncertainty of what awaited them in Italy. News of the reinstatements arrived at the beginning of 1944, through newspapers or diplomatic representatives. At that time, communicating with Italy was very difficult. For several months, displaced scholars did not receive any news of relatives who had remained in Italy. When communications resumed, the news was terrible: deportations of people of all ages and conditions, including old people and children, to unidentified camps and colonies in Central Europe; trains loaded with individuals who were seen leaving in the night; massive destructions of cities by bombings; the impossibility of finding accommodation, except at unaffordable prices (D7.2.1, D7.2.5, D7.2.6, D7.2.7). The material conditions of the country were described as frightening, and they truly were so (D7.2.3, D7.2.4, D7.2.8).

Two other factors also contributed to stoking the doubts and ambivalence of feelings: fear of antisemitism, and disappointment at the failure of the purge of fascists (D7.2.8). As to the first point, returning scholars received general assurance of their safety. Castelnuovo for example reassured Segre that persecution had been something artificially created through political action, which was not matched by any real anti-Semitic sentiment on the part of the Italian population. The incredible Togliatti amnesty, which 'put back in circulation common [. . .] and political criminals, including Jew-rakers and torturers, except those whose tortures had been particularly heinous (sic)',[30] and the failure to purge the fascists from academia saw Italian and former refugees united in condemnation (D7.2.4).

The decision of whether to stay abroad or return was not made alone. Displaced mathematicians communicated with each other and, if in 1938 it had been Levi-Civita who had helped them to leave, now it was Castelnuovo to help them return. The first letters exchanged by Fano, Segre, Terracini and Castelnuovo are profoundly moving:

> Our science has also been hit by serious losses in recent times. [. . .] We must count on you and the few of your generation who work seriously so that the high traditions of the Italian mathematical School do not die out.[31]

It was a School that found itself again, but also a group of colleagues and friends who had been separated by the persecution and who, at that moment were called on to decide whether to have their lives and careers in upheaval for the second time. Severi and Picone backed Castelnuovo, aware that in order to revive Italy and restore its international prestige, it was imperative to bring back the expat colleagues. Their action was effective, more than that of the physicists, who failed in convincing many of the former Panisperna boys to come back.

For different reasons, and with different timing, the Italian Jewish mathematicians who had emigrated almost all chose to return. The first to do so were the refugees in Switzerland: Colombo (May 1945), Fano (September 1945), and Friberti. Segre returned at the end of 1946; Terracini was the last, in February 1948. In this, they differed sharply from other professional categories, in which the percentage of those who remained abroad is much larger. The hope of what Lore Terracini called an 'affective and ideological recovery', the conviction that 'between Italy and fascism it was possible to distinguish; between Germany and nazism no'[32] and the will to contribute to the reconstruction of the country prevailed over other considerations. Levi was the only one to make a contrary decision for family, professional and ideological reasons (the fear of a third world war, this time with the use of atomic bombs). Even Levi, however, fully reconnected with Bologna and with Italian mathematics.

Conversely, all the returning emigrants maintained contacts with the countries in which they had taken refuge. Terracini, for example, continued to edit the *Revista* remotely for some ten years, by suggesting articles, mentoring authors, etc.

---

[30] D7.2.8.

[31] CA, *BSP*: G. Castelnuovo to B. Segre, Rome 15.3.1946: *La nostra scienza è stata anche colpita da gravi perdite in questi ultimi tempi. [. . .] Su Lei e sui pochi della Sua generazione che lavorano seriamente dobbiamo contare perché non si spengano le alte tradizioni della Scuola matematica italiana.*

[32] Terracini L. (1989), p. 363: *recuperaciones afectivas e ideológicas. Entre Italia y fascismo era posible distinguir; entre Alemania y nazismo, no.*

Furthermore, his correspondence with F. Herrera, F. Cernuschi and many others clearly reveal the tangible legacy he left behind after his sojourn in Argentina.[33] Fano spent the last years of his life as a commuter, spending six months in Italy and six in the United States with his children. As a matter of fact, he now had more regular and assiduous contacts with American mathematicians than he had enjoyed in all his life. Segre too did not neglect connections with Great Britain. His correspondence with English colleagues lasted for the rest of their lives and he continued to follow the activities of the SPSL, also sending donations, until 1960. Segre would also come back to the United Kingdom regularly, for example, in the spring of 1950 for a series of lectures on *Arithmetical questions on Algebraic Varieties* (King's College London, March 11 and Oxford, April 12–14) which was described in these terms in *Nature*:

> In a brilliant performance, worthy of the tradition of the great Italian School of Geometry, he demonstrated the solution of a problem which was proposed to him only a fortnight before by Professor J.G. Semple, of King's College, London, namely, to characterise the inflexional curve of an algebraic surface in four dimensions.[34]

Documents

D7.2.1 G. and A. Terracini to A. Sacerdote, Tucumán 2.9.1944

ATCET, fols. 1r

Dearest Aldo, your letters from March and April continue to arrive, but unfortunately still nothing after June.[35] We still don't know anything about your meeting with Alma and Micky, and I can't help but be quite uneasy about you.[36] More and more it seems impossible to me that you have not yet been able to let me know your news after the liberation of Rome. Your most recent letter is dated May 17th. Through Eugenio Fubini, we got a note from Mario Segre from Rome in which there are only the names of Rita, Aldo, Alma, Adolfo, Rina, Giulia and Dario, all

---

[33] See for example, in BSM-*Terracini*: F. Herrera to A. Terracini, 5.2.1948, 14.3.1948, 18.3.1948, 31.3.1948, 20.8.1948, 2.11.1948, 1.12.1948, 20.4.1948, 18.5.1948, 6.6.1948, 1.7.1948, 20.4.1952; F. Cernuschi to A. Terracini, 3.1.1948; R. Mondolfo to A. Terracini, 16.4.1948; L. Romaña to A. Terracini, 19.4.1948; J. Würschmidt to A. Terracini, 25.10.1948, etc.

[34] K.H. Hirsch, *British Mathematical Colloquium at Oxford*, Nature, vol. 165, n. 4204, 27.5.1950, p. 845.

[35] See also TERRACINI.10. Aldo Sacerdote was Alessandro's brother-in-law.

[36] Alma Belloni Sacerdote (1905–1986), wife of Aldo Sacerdote and Alessandro's sister-in-law. Micaela Sacerdote (born in 1939), daughter of Aldo and Alma. Alma and Micaela were in Abruzzo. For several months, from September 1943 to June 1944, they had no news of Aldo, who had remained in Naples.

well in July.[37] We can't wait to know more details about everyone!! And I really hope that by now your isolation from your family has come to an end. But when will we know?! The course of the war makes us happier and more confident in an imminent and glorious end. Unfortunately, the loved ones of the north are not yet freed, and I am always anxious for them, but maybe it will be a matter of a short time and we hope to receive some good news soon!! I am very pleased to hear of your various and varied activities, which, of course, have served to distract you during the anguished months of solitude. But I don't want you to overwork yourself and damage your health. We were very upset to hear of your weight loss, and I am very worried about your health. [. . .] We imagined that food prices were so high, but I hope you won't lack anything. Terrible news continues to arrive here of what happened before the liberation by the Allies and I tremble for the fate of so many loved ones!! Nothing new from us! We've had the usual warm winter and we're all very tired of it, especially considering that we're heading towards summer, and we haven't had any cool weather this year. Our hope is that this will be our last summer here and we are ready to. . . sweat for a few more months, without complaining too much. Once again, I express to you our desire to return as soon as possible, and I beg you (I believe that there will be no need for us to insist so much) to do everything possible so that Sandro has his position restored. I don't think we should be too confident that we will stay here for much longer, unless things change in the near future. The boys are well, always busy, and always very interested in the events of the war. [. . .] Sandro has also written to Castelnuovo and Amaldi.[38] [. . .] We'll see you again soon??? I hope so infinitely. Lots of kisses to everyone. We had telegraphed to Lt. Giorgio Lattes (brother of Fiamma Treves) who fights with the French army, to send us news of you, but it seems that he did not look for you, because he did not send us any news of you.[39] I hope that when this letter arrives, we will already be calm about the fate of Rita and everyone. In any case, we don't think it's prudent for you to look into it. Giulia and Sandro

---

[37] Eugenio Fubini (1913–1997), son of Guido Fubini. Mario Segre, cousin of Giulia Sacerdote Terracini. Rita Sacerdote Artom (1895–1978), sister of Giulia, was Alessandro's sister-in-law. Aldo, brother of Giulia, and his wife Alma Belloni. Adolfo and Rina Sacerdote were first cousins of Cesare Sacerdote, father of Giulia Sacerdote Terracini. Giulia Vitale Dalmonte and Dario Dalmonte. Giulia Vitale's mother was first cousin of Giulia Sacerdote Terracini.

[38] Ugo Amaldi (1875–1957), mathematician and father of the physicist Edoardo Amaldi, was professor of Algebra and Analytic Geometry at the University of Rome (1924–1944).

[39] Giorgio Lattes (born in 1913) was the son of Leone Lattes, former professor of Forensic Medicine at the University of Pavia. Leone Lattes, his wife Virginia Rabbeno, and children Giorgio, Camilla, and Lisa Fiamma had emigrated to Argentina in 1938. Fiamma had married Renato Treves, former lecturer of Philosophy of Law at the University of Urbino, emigrated to Tucumán in 1939. In 1939 Giorgio had returned to Europe in order to enroll in the French Foreign Legion and fought in an Allied military unit.

Carissimo Aldo, continuano ad arrivare le tue lettere di marzo e aprile, ma purtroppo ancora nulla posteriore al giugno. Non sappiamo ancora nulla del tuo incontro con Alma e Micky, e non posso fare a meno di essere abbastanza intranquilla sul vostro conto. Sempre più mi pare impossibile che tu non abbia ancora potuto farmi sapere vostre notizie dopo la liberazione di Roma. La tua più recente lettera è quella del 17 maggio. Per mezzo di Eugenio Fubini, abbiamo avuto un biglietto di Mario Segre da Roma in cui soltanto vi sono i nomi di Rita, Aldo, Alma, Adolfo, Rina, Giulia e Dario, tutti bene in luglio. Siamo ansiosissimi di sapere altri particolari su tutti!! E spero proprio che ormai il tuo isolamento dalla tua famiglia abbia avuto una fine. Ma quando lo sapremo noi?! L'andamento della guerra ci rende sempre più contenti e fiduciosi in una fine prossima e gloriosa. Purtroppo i cari del nord non sono ancora liberati e io sono sempre in ansia per loro, ma forse sarà questione di poco tempo ancora e speriamo di ricevere presto qualche buona notizia!! Mi fa molto piacere sapere delle tue diverse e varie attività, che certo hanno servito a distrarti nei mesi angosciosi di solitudine. Non vorrei però che tu ti sovraccaricassi di lavoro e ne risentisse la tua salute. Ci ha assai impressionati sapere del tuo peso così ridotto e sono assai in pensiero per la tua salute. [. . .] Immaginavamo che i prezzi dei viveri erano così elevati, ma io spero che a te non mancherà nulla. Continuano a giungere qui notizie terribili di quanto è avvenuto costì prima della liberazione da parte degli alleati e io tremo per la sorte di tante persone care!! Di noi nulla di nuovo! Abbiamo avuto un inverno solitamente caldo e ne siamo tutti assai stanchi, specialmente pensando che ci avviamo verso l'estate e che di fresco quest'anno non ne abbiamo proprio avuto. La nostra speranza è che questa sia la nostra ultima estate qui e siamo pronti a . . . sudare ancora per qualche mese, senza troppo lagnarci. Ancora una volta ti manifesto il nostro desiderio di ritornare appena sarà possibile, e ti prego (credo del resto che non ci sarà bisogno che insistiamo tanto) di fare tutto il possibile perché Sandro riabbia il suo posto costì. Non credo che dobbiamo confidare troppo di rimanere ancora tanto tempo qui, a meno che le cose cambino in un prossimo futuro. I ragazzi stanno bene, sempre occupati, e sempre interessatissimi alle vicende della guerra. [. . .] Sandro ha scritto anche a Castelnuovo e Amaldi. [. . .] Ci rivedremo presto??? Lo spero infinitamente. Moltissimi baci a tutti. Avevamo telegrafato al ten. Giorgio Lattes (fratello di Fiamma Treves) che combatte costì con l'esercito francese, perché ci mandasse tue notizie, ma pare che non ti ha cercato, perché non ci ha mandato nessuna notizia tua. Spero che al giungere la presente si sia già tranquilli sulla sorte di Rita e tutti. In ogni caso non ci sembra prudente che tu ne faccia ricerca. Giulia e Sandro

D7.2.2 Memorandum by F. Severi, *Azioni nell'interesse di persone invise al fascismo,* **[November-December 1944]**

ACS, MPI, *Professori epurati,* **b. 31** *Francesco Severi,* **N. 372, fols. 2r, 3r, 4r, 5r**
Excerpt from a letter (produced in original) from B. Segre to Severi, 20.12.1938, immediately after the racial laws. Segre is preparing the documents to emigrate to England or America, and I am helping him.

B. Segre to F. Severi, Bologna, 20.12.1938. Dear and illustrious Master, I am very grateful to you for your constant affectionate interest. Unfortunately, so far, I have not decided, nor have I been able to decide, anything about my future, even the most immediate. Here's my situation in a nutshell... Allow me, also on behalf of my wife, to send you and your kind wife my best wishes and best regards. Yours faithfully, Beniamino Segre

Copy of letter (produced in original) from B. Segre to Severi, 16.4.1939. Severi had assisted him in obtaining a pension that could be cashed abroad and in obtaining the authorisation to exchange a sum into foreign currency.

B. Segre to F. Severi, Turin, 16.4.1939. Dear and illustrious Professor, I thank you sincerely for the affectionate interest you have always shown me and for the last two letters, the last of which was very useful to me in preparing the operation of those 15,000 lire, which I was able to carry out yesterday without difficulty. I will only stay here for 2 or 3 days; and I will give you my new address immediately, as soon as I have a relatively stable one. In the event you need to communicate with me in the meantime, you can send letters to my father [Samuele Segre] in Turin, via dei Mille 4. The deep sadness of having to leave the people and things dearest to us is mitigated by the thought of the continuity of the most intimate feelings and spiritual bonds, many of which are indeed strengthened by this trial. On behalf of my wife too, I express my most fervent wishes and offer my best regards to you and to your kind wife.[40] Your faithful and affectionate Beniamino Segre

Extract from a letter 7.6.1939 (produced in original) from B. Segre to Severi, after Segre had arrived in England. B. Segre to F. Severi, Cockfosters, 7.6.1939. Dear and illustrious Professor, I am very grateful to you for your kind thought, all the more so because I know how precious your time is. My physical and psychological conditions, which have returned almost to normal, have allowed me to resume my scientific work: which was also facilitated by this serene and welcoming environment... Your faithful and affectionate Beniamino Segre.

Copy of letter 5.1.1941 (produced in original) from Arturo Maroni (Jew) to Severi, after Maroni had to leave his university chair in Florence following the

---

[40]Rosanna Orlandini (1878–1952).

racial laws.[41] I had offered to help him in continuing his studies and to send him my works.

A. Maroni to F. Severi, Florence, 5.1.1941. Your Excellency, I am very grateful to you for your letter. Of course, I am always very interested in mathematical studies; my biggest regret now is that I don't have time to devote myself to research. In fact, I am forced to give private lessons from morning to night in order to keep the wolf from the door with my wife and three sisters.[42] I look forward to receiving your latest work, and I thank you in advance. My address is always the same: Via dei Conti 7, Florence. With renewed thanks and devoted greetings, yours, Arturo Maroni

Copy of letter 23.11.1930 (produced in original) from Guido Ascoli (Jew) to Severi. Ascoli was not my disciple. However, appreciating his work, I advised him to compete at the University to escape middle school and thus be able to produce better academic work. I was President of the Commission that judged him in the competition he won.[43]

G. Ascoli to F. Severi, Turin, 23.11.1930. Your Excellency, I was impressed and moved by your exquisite and spontaneous courtesy, by your flattering words, and I am infinitely grateful. The Commission, and you in particular, have shown great benevolence towards me. I hope I shall not show myself unworthy of it in the future, especially if placed in far better working conditions than at present. I have begun taking steps to know the precise situation of the vacant chairs especially in Padua and Ferrara, and the intentions of the colleague ranked second, so as not to lose the possibility, which I believe is very scarce, of an immediate position. This will help me, if necessary, to set out a course of action for next year. I am, of course, very little known, and I do not believe that the faculties will spontaneously ask for my humble self. Would it be too much, if necessary, for me to ask you for a word—one single word will suffice—of introduction to any eventual post available to me? Excuse me for my boldness and, once again, please accept my warmest thanks. Yours faithfully, Guido Ascoli

Copy of a letter dated 19.12.1942 (produced in original) by Prof. Gino Fano (Jew), former full professor of the Royal University of Turin, from which it appears that I helped him for the publication of his scientific works.

G. Fano to F. Severi, Lausanne, 19.12.1942. Dear Severi, I receive your letter dated December 11. By recorded delivery I am sending you the Note on F^4 containing a network of curves of genus 2 and I thank you for presenting it at the

---

[41] See §4.2.

[42] Two of his sisters, Rita (1882–1943) and Dora (1882–1943), were arrested in Florence. Deported to Auschwitz, they did not survive.

[43] See §4.1.

Pontifical Academy.[44] In footnote 14) there is a slight correction to be made to the number of modules on which a F^4 with involutory homography depends: if you wish, you can also modify or delete it. I had seen a mention of your travel to Zurich in the *Squilla Italica*—newspaper of the Italians of Switzerland.[45] Swiss French newspapers—at least the ones I usually see—didn't talk about it. Some days ago, I got some more details from Mr. Eckmann, who brought me your greetings.[46] The fire of the 28th fortunately spared my studio, but it damaged some rooms both in our flat and on the floor below.[47] And more serious injuries to the building were caused by the explosion of a bomb not far away, in a subsequent raid. So, the house is no longer habitable, and we are transporting the furniture to Colognola or Mantua.[48] With our best wishes to you and your wife for the holidays and New Year's Eve, and our most cordial greetings. Gino Fano

Without reproducing the copies, I will limit myself to mentioning two letters (one from Prof. [Ettore] Bortolotti and one from Prof. Guido Castelnuovo, a Jew) which show how I took care (and successfully) succeeded in preventing Castelnuovo's book on the origins of Infinitesimal Calculus from being taken out of circulation after the racial laws.[49] I will produce the letters to the Commission. Similarly, without reproducing a copy, I mention a letter dated 30.11.1943 proving that I passed some money to a Jew (Erich Lorant) who had taken refuge in Casentino.[50] The money was sent to him, through me, by another Jew from Rome. I will produce the letter to the Commission.

Estratto di lettera 20.12.1938 di B. Segre al Severi, subito dopo la legge razziale *(prodotto in originale). Il Segre fa pratiche per emigrare in Inghilterra o in America e io lo aiuto. B. Segre a F. Severi, Bologna, 20.12.1938. Caro ed illustre Maestro, Vi sono molto grato del Vostro costante affettuoso interessamento. Non ho purtroppo deciso né potuto decidere nulla finora, circa il mio avvenire anche il più immediato. Ecco in breve la mia situazione... Permettete che, anche a nome di mia moglie, mandi a Voi ed alla gentile Signora fervidi auguri e distinti saluti. Vostro aff.mo Beniamino Segre*

---

[44] G. Fano, *Superficie del 4° ordine contenenti una rete di curve di genere 2*, Commentationes. Pontificia Academia Scientarum, 7, 1943, pp. 185–205. Severi had presented the paper in the session of 21 February 1943.

[45] *La squilla italica. Giornale degli italiani nella Svizzera* was a fascist weekly journal published in Lugano since 1923.

[46] Beno Eckmann (1917–2008), a topologist, PhD from the Eidgenössische Technische Hochschule in Zürich (1941), was a lecturer at the University of Lausanne.

[47] On 28 November 1942 Turin was bombed by the English Air Force (RAF). The bombing, which for the first time used 'blockbuster bombs', was particularly dramatic for Turin.

[48] Fano refers to Villa Fano in Colognola ai Colli (Verona) and to his birthplace home in Mantua.

[49] G. Castelnuovo, *Le origini del calcolo infinitesimale nell'era moderna*, Bologna, Zanichelli, 1938.

[50] Casentino is a mountainous area in Tuscany.

Copia di lettera 16.4.1939 (prodotta in originale) di B. Segre al Severi, *che lo aveva assistito affinché gli fosse assegnata una pensione riscuotibile all'estero e gli fosse consentito di convertire una certa somma in valuta estera. B. Segre a F. Severi, Torino, 16.4.1939. Caro ed illustre Professore, Vi ringrazio sentitamente per l'affettuoso interessamento che sempre mi dimostrate e per le ultime due lettere, l'ultima delle quali mi è stata utilissima per predisporre l'operazione di quelle 15.000 lire, che ieri ho potuto compiere senza difficoltà. Mi tratterrò qui soltanto 2 o 3 giorni; e Vi darò subito il mio nuovo recapito, appena ne avrò uno relativamente stabile. Nel caso che nel frattempo abbiate a comunicare con me, potrete indirizzare a Torino, via dei Mille 4, presso mio padre. La profonda tristezza di dover lasciare le persone e le cose più care è mitigata dal pensiero della continuità dei più intimi sentimenti e dei legami spirituali, molti dei quali risultano anzi rafforzati da questa prova. Anche a nome di mia moglie, formulo i più fervidi auguri e porgo distinti saluti a Voi ed alla gentile Signora. Vostro dev.mo e aff.mo Beniamino Segre*

Estratto di lettera 7.6.1939 (prodotta in originale) di B. Segre al Severi, *dopo che Segre era giunto in Inghilterra. B. Segre a F. Severi, Cockfosters, 7.6.1939. Caro ed illustre Professore, Le sono assai grato del buon ricordo; e ciò tanto più in quanto so come sia prezioso il suo tempo. Le mie condizioni fisiche e psicologiche, ritornate quasi normali, mi hanno consentito la ripresa del lavoro scientifico: il che mi è stato anche agevolato da questo ambiente sereno ed accogliente... Suo dev.mo e aff.mo Beniamino Segre*

Copia di lettera 5.1.1941 (prodotta in originale) di Arturo Maroni (ebreo) al Severi, *dopo che il Maroni aveva dovuto lasciare la cattedra universitaria a Firenze in seguito alla legge razziale. Io mi ero offerto di aiutarlo nella continuazione dei suoi studi e di mandargli i miei lavori. A. Maroni a F. Severi, Firenze, 5.1.1941. Eccellenza, Vi sono molto grato per la Vostra lettera. Certo gli studi di matematica mi interessano sempre molto; ed il mio più grande rincrescimento è ora quello di non aver tempo per dedicarsi alla ricerca. Infatti, sono costretto a fare lezioni dalla mattina alla sera per poter sbarcare il lunario insieme con mia moglie e le mie tre sorelle. Sarò molto lieto di ricevere i Vostri ultimi lavori, e ve ne ringrazio in anticipo. Il mio indirizzo è sempre il solito: Via dei Conti 7, Firenze. Con rinnovati ringraziamenti e devoti saluti, aff.mo Arturo Maroni*

Copia di lettera 23.11.1930 (prodotta in originale) di Guido Ascoli (ebreo) al Severi. *L'Ascoli non era mio discepolo. Tuttavia, apprezzandone i lavori, lo consigliai a concorrere all'Università per togliersi dalle scuole medie e poter così produrre meglio per la scienza. Fui Presidente della Commissione che lo giudicò, nel concorso da lui vinto. G. Ascoli a F. Severi, Torino, 23.11.1930. Eccellenza, sono rimasto confuso e commosso dalla Sua squisita e spontanea cortesia, dalle Sue parole così lusinghiere, e gliene sono infinitamente grato. La Commissione, e Lei in particolare, hanno mostrato verso di me molta benevolenza; spero di non mostrarmene indegno nel futuro, specie se posto in* *condizioni di lavoro di gran lunga migliori delle attuali. Ho iniziato le pratiche per conoscere la situazione precisa delle cattedre vacanti in modo speciale a Padova e a Ferrara, e le intenzioni del secondo in graduatoria, per non perdere la possibilità, che però credo scarsissima, di una sistemazione immediata. Ciò mi servirà, in caso, a segnarmi per l'anno prossimo una linea di condotta. Io sono, naturalmente, ben poco noto, e non credo che le Facoltà si daranno spontaneamente da fare per la mia modesta persona. Abuserò troppo della sua bontà, se occorrendo Le domanderò una parola – una basterà – di presentazione per la sede che mi restasse disponibile? Mi scusi l'ardire e si abbia, di nuovo, i miei più vivi ringraziamenti. Dell'E.V. dev.mo Guido Ascoli*

Copia di lettera 19.12.1942 (prodotta in originale) del prof. Gino Fano (ebreo) *già ordinario della R. Università di Torino, dalla quale risulta com'io lo aiutassi per la pubblicazione dei suoi lavori scientifici. G. Fano a F. Severi, Lausanne, 19.12.1942. Caro Severi, Ricevo la tua lettera 11 corr. Con piego a parte raccom. ti spedisco la Nota sulle F^4 contenenti una rete di curve di genere 2, e ti ringrazio della prossima presentazione all'Accademia Pontificia. Nella nota 14) è indicata la lieve rettifica al numero dei moduli da cui dipende una F^4 con omografia involutoria: se credi, puoi anche modificarla o sopprimerla. Della tua venuta a Zurigo avevo visto cenno sulla* Squilla Italica *– giornale degli Italiani della Svizzera. I giornali svizzeri in lingua francese – almeno quelli che vedo abitualmente – non ne hanno parlato. Qualche maggiore dettaglio ho avuto giorni sono dal sig. Eckmann, che mi ha portati i tuoi saluti. L'incendio del 28 u.s. ha fortunatamente risparmiato quanto era nel mio studio, ma ha danneggiato alcuni ambienti sia del nostro alloggio cha del piano sottostante. E più gravi lesioni allo stabile ha prodotto lo scoppio, non lontano, di una bomba, in un'incursione successiva. Sicché la casa non è più abitabile, e stiamo provvedendo a trasportare il mobilio a Colognola o a Mantova. Alla Signora e a te i nostri migliori auguri per le prossime feste e capo d'anno, e i nostri più cordiali saluti. Gino Fano*

Mi limito, senza riprodurre la copia, ad accennare a due lettere (una del prof. Bortolotti ed una del prof. Guido Castelnuovo, ebreo) le quali dimostrano com'io mi sia occupato (e con successo) d'impedire che il libro sulle origini del Calcolo infinitesimale del Castelnuovo fosse tolto di circolazione, dopo la legge razziale. Produrrò le lettere alla Commissione. Similmente accenno, senza riprodurre copia, ad una lettera 30.11.1943 comprovante che ho fatto pervenire ad un ebreo (Erich Lorant) rifugiato in Casentino un soccorso in denaro che gli era stato inviato a mio mezzo da un ebreo di Roma. Produrrò la lettera alla Commissione.

D7.2.3 G. Castelnuovo to B. Segre, Rome 7.1.1945

CA, *BSP*, fol. 1r-v

Dear Professor, My friend Roth gave me news of you and your address, so I
immediately take the opportunity to write to you.[51] I am pleased that you have
found a position, even temporary, as a professor at that University [Manchester]
and that you have spent these years of war with fewer worries than we did. Last
winter we had to stay hidden with false names, but fortunately neither we, nor the
three children who were with us in Rome, suffered any harm, and since last June
we have returned to our apartment.[52] Unfortunately, it's been six months since we
had any news of one of our sons, who lived in Venice and then in the province of
Trento; we only know that he was in a concentration camp in the province of
Modena, and we hope, if he did not manage to escape, that he is still there.[53] Roth
writes to me that you are worried about the future. Now I can assure you that as
soon as you have returned to Italy, you will automatically find yourself reinstated
in the chair with the salary that starts from 1 January 1944. Of course, you cannot
resume your place in Bologna until this city is liberated;[54] but you can reside in
any place in liberated Italy without any obligation, and you can do some courses,
if you like, in Rome, or Naples or Florence. Colleagues here would certainly be
happy to entrust you with some teaching. I will take steps at the Ministry of Public
Instruction to facilitate, through the British Government, the possible granting of
a permit to return to Italy. However, I must advise you to think carefully before
leaving your place in Manchester. Here, life is very much more expensive; basic
food costs at least 10 times what it costed when you left Italy, while salaries have
increased by only 70 or 80%. If you could write to me, tell me what you have
done in this period and tell me about scientific life in England; here we are in the
dark about everything. Prof. Enriques has resumed the teaching of Higher
Geometry. I was Commissioner of the CNR for four months, and I had there
the valid help of Tricomi, refugee in Rome since September 1943, who sends you
many greetings.[55] Today, the Council, with a technical orientation, is in the hands
of Colonnetti. Please accept cordial greetings and best wishes from all of us to
you and yours; most affectionate G. Castelnuovo

---

[51] Leonard Roth (1904–1968) had spent the academic year 1930–31 in Italy, working in algebraic
geometry under the mentorship of Castelnuovo, Enriques, and Severi.

[52] Guido Castelnuovo and his wife Elbina Enriques (1870–1960) had remained in Rome with their
children Maria (1899–1991), Gino (1903–1995), and Emma (1913–2014).

[53] Mario Castelnuovo (born 1897) escaped deportation to Auschwitz by jumping from the moving
train before stopping at the Bolzano station.

[54] Bologna would be liberated on April 21, 1945.

[55] Castelnuovo was Special Commissioner of the CNR from June 1944 until December 1944. He
was replaced by G. Colonnetti. Tricomi spent many months in Rome (September 1943–April 1945),
engaging in the anti-fascist partisan struggle, as a member of the Action Party.

Caro Professore, L'amico Roth mi comunica le Sue notizie e il Suo indirizzo di cui approfitto subito per scriverle. Ho piacere che abbia trovato un posto, almeno provvisorio, di professore in codesta Università e che abbia passato questi anni di guerra con minori preoccupazioni di noi. L'inverno passato siamo dovuti star nascosti, con falsi nomi, ma fortunatamente né noi, né i tre figli che erano con noi a Roma, non abbiamo subito danni, e dal giugno scorso siamo rientrati nel nostro appartamento. Purtroppo, di un figlio, che abitava a Venezia e poi in Provincia di Trento, non abbiamo notizie da sei mesi; sappiamo solo che era in un campo di concentramento in Provincia di Modena, e speriamo, se non ha potuto fuggire, che sia ancora lì. Il Roth mi scrive che Ella è preoccupato per l'avvenire. Ora posso assicurarla che appena Ella sia ritornato in Italia, si trova automaticamente reintegrato nella cattedra con lo stipendio che decorre dal 1° gennaio 1944. Naturalmente non può riprendere il Suo posto a Bologna, finché questa città non sia liberata; può risiedere in qualsiasi posto dell'Italia liberata senza nessun obbligo, e può far qualche corso, se preferisce, a Roma, o Napoli o Firenze. I colleghi di qua sarebbero certo lieti di affidarle qualche insegnamento. Farò dei passi al Ministero della P. Istruzione perché facilitino, presso il Governo inglese, la concessione di un eventuale permesso per ritornare in Italia. Devo però consigliarle di riflettere bene prima di lasciare il posto che ha a Manchester. Qui la vita è molto rincarata; i generi di prima necessità costano per lo meno 10 volte quel che costavano quando Ella ha lasciato l'Italia, mentre gli stipendi sono aumentati solo del 70 od 80%. Se ha occasione di scrivermi mi racconti ciò che ha fatto in questo tempo e mi parli della vita scientifica in Inghilterra; qui siamo all'oscuro di tutto. Il Prof. Enriques ha ripreso l'insegnamento di Geometria Superiore. Io sono stato per quattro mesi Commissario al Consiglio Naz. delle Ricerche ed ho avuto colà il valido aiuto di Tricomi, profugo a Roma dal settembre 1943, il quale Le manda molti saluti. Oggi il Consiglio, con indirizzo tecnico, è in mano di Colonnetti. Accolga cordiali saluti ed auguri da tutti noi per Lei e per i Suoi; aff.^{mo} G. Castelnuovo

D7.2.4 G. Castelnuovo to B. Segre, Rome 25.3.1945

CA, *BSP*, fols. 1r-2r

Dear Professor, Thank you very much for your letter of 17 February, and for the interesting news you give me. I am sorry for the vicissitudes you have gone through and above all I regret to hear of the loss of your child, a misfortune that I was not aware of, and for which I send, very late, my heartfelt condolences.[56] I will gladly read your works when they arrive; I see that you have worked a lot in

---

[56]Ornella Segre (1937–1940) had contracted measles. The incessant bombing of London (the so-called Battle of Britain) prevented her from being transferred to hospital. She died while her father was interned on the Isle of Man.

these troubled years, and I hope that you will soon be able to complete the two volumes which you tell me of. Are there any new results from Hodge or his School?[57] I have been looking for a geometric proof of Hodge's theorem on the non-existence of double integrals of 1st species with zero periods, with Severi's approach (semi-exact integrals), but I find it so difficult that I sometimes doubt whether the theorem is true. I now answer your most urgent request regarding the conditions in which you would find Italy in coming here. As I already wrote to you, the most serious difficulties are economic. An official statistic published in these days shows that the price of food is now 17 times what it was in 1940. What is certain is that today even a modest family cannot spend less than 110 or 120 L. per day per person on food, and that a place in a 3rd or 4th class trattoria costs no less than 120-130 L. Salaries, on the contrary, even considering the latest increases, have merely doubled. A full professor at the highest grade of career will be able to collect a maximum of L.10,000 per month, i.e., 25 pounds at the official exchange rate, less than 20 at the real exchange rate. On the other hand, you have no need to worry in the slightest about the anti-Semitic issue, which has never existed in our country outside the official spheres. It was Mussolini's anti-Semitic policy that drove many honest fascists away from the party. The Nazi persecution then produced a very strong reaction. I cannot describe to you what displays of affection and sympathy the Jews of Rome received from the local people, even the humblest, even from the Police itself, during the eight months of German occupation. It is thanks to the protection of many Catholic families, the immense majority of concierges, religious institutes and the Vatican that 9000 of the 12,000 Jews of Rome have been saved; all or almost all those who were not found in their homes in the early days of the raids. I therefore believe that I can assure you that for many years no anti-Semitic campaign will rise again among us. It is difficult to give you advice on what you should do as soon as the war is over. I think, however, it would not be convenient for you to burn your bridges about a possible return. You should, if this is allowed, ask for leave for a few months, and study the situation here before deciding. If you have the opportunity to see Prof. Baker or to write to him, please give him my best wishes; I always remember him with great affection.[58] Many cordial greetings to you from your most affectionate G. Castelnuovo

---

[57] W. V. D. Hodge had obtained an international reputation solving a problem about integrals on a surface which had been posed by Severi using Lefschetz's topological methods (1930). In March 1936 Hodge had been appointed as Lowndean professor of Astronomy and Geometry, replacing H. F. Baker. During the war, he had a very demanding teaching load but continued his research activity and began a collaboration with Daniel Pedoe, which led to the book *Methods of Algebraic Geometry* (Cambridge, University Press, 3 vols. 1947–1954).

[58] H. F. Baker, Lowndean Professor of Astronomy and Geometry at St. John's College, Cambridge, had interwoven a network of relationships with the Italian School of Geometry since the end of nineteenth century.

Caro Professore, La ringrazio molto della Sua lettera del 17 febbraio e delle interessanti notizie che mi dà. Mi dispiace delle peripezie che hanno attraversato e soprattutto mi duole della disgrazia della loro bambina, disgrazia che ignoravo e per la quale mando, molto in ritardo, le mie vive condoglianze. Leggerò con piacere i Suoi lavori quando arriveranno; vedo che ha lavorato molto in questi anni così agitati, e spero che potrà presto condurre a termine i due volumi di cui mi scrive. C'è nessun risultato nuovo di Hodge o della sua Scuola? Sto da un pezzo cercando una dimostrazione geometrica del teorema di H. sull'inesistenza di integrali doppi di 1ᵃ specie a periodi nulli, nell'indirizzo di Severi (integrali semiesatti), ma trovo tali difficoltà da farmi dubitare qualche volta se il teorema sia vero. Rispondo ora alla Sua richiesta più urgente relativa alle condizioni in cui troverebbe l'Italia venendo qui. Come già le scrissi le condizioni più gravi sono le economiche. Da una statistica ufficiale pubblicata in questi giorni risulta che il numero indice dei prezzi dei generi alimentari è oggi 17 volte quel che era nel 1940. Quel che è certo è che una famiglia anche modesta non può spendere oggi di vitto meno di 110 o 120 L. al giorno per persona, e che un posto in una trattoria di 3° o 4° ordine non costa meno di 120-130 L. Gli stipendi invece, anche tenendo conto degli ultimi aumenti, sono appena raddoppiati. Un professore ordinario al grado più alto potrà riscuotere al massimo 10000 L. al mese, cioè 25 sterline al cambio ufficiale, meno di 20 al cambio reale. Ella non ha invece da preoccuparsi minimamente della questione antisemita, che in realtà non è mai esistita nel nostro paese fuori dalle sfere ufficiali. Fu la politica antisemita di Mussolini ad allontanare dal partito molti dei fascisti in buona fede. La persecuzione nazista ha poi prodotto una fortissima reazione. Non so dirle quali prove di affetto e di simpatia abbiano avuto gli ebrei di Roma da parte della popolazione, anche la più umile, da parte della stessa Questura, durante gli otto mesi di occupazione tedesca. È grazie alla protezione di moltissime famiglie cattoliche, della immensa maggioranza dei portieri, degli istituti religiosi e del Vaticano, che si son salvati 9000 dei 12000 ebrei di Roma; tutti o quasi tutti quelli che non erano stati trovati nelle loro abitazioni i primi giorni delle razzie. Credo quindi poterle assicurare che per molti e molti anni non risorgerà tra noi una campagna antisemita. È difficile darle un consiglio su ciò che Le converrà fare appena finita la guerra. Credo però non le converrebbe tagliarsi i ponti nei riguardi di un possibile ritorno costì. Dovrebbe, se la cosa è ammessa, farsi concedere un permesso di qualche mese, e studiare qui la situazione prima di decidere. Se ha occasione di vedere il Prof. Baker o di scrivergli, lo saluti molto da parte mia; lo ricordo sempre con molta simpatia. A Lei molti saluti cordiali dal suo aff.mo G. Castelnuovo

D7.2.5 B. Segre to J. B. Skemp [SPSL], Manchester 18.5.1945

SPSL, *BS*, fol. 208

Dear Miss Skemp, I enclose the questionnaires you have sent me to fill up, and take with pleasure this opportunity for giving additional information about my activity in this Country during the last six years.[59] I have delivered several occasional lectures in Cambridge and London, and lectured continuously during the last three years in Manchester University. I have published a book[60] and a score of mathematical papers, and also prepared abundant material for further publications. Among several acknowledgements of my research work in this period, I may quote the one by Professor Virgil Snyder (one of the editors of the Bulletin of the American Mathematical Society), who writes me about it: 'While I want to congratulate you on your splendid showing, I am thankful that the world can still produce first-class results'.[61] I shall never forget the support of the Society P.S.L., without which it would have been impossible to me to accomplish anything I have done during the last six years, and I beg you to convey my feelings of gratitude to all the kind persons who have helped me through the Society. I have not at present any definite post-war plans.[62] I am glad to tell you that the Italian Government has invited me to go back to my chair in Italy. I have not yet taken a decision, since I am reluctant to leave for good this Country, to which I feel much attached; moreover, I think I should encounter serious economical difficulties if I would go back now to Italy with my family. Very probably, I shall therefore stay here at least for the whole academical year 1945–1946; but I wish I could be able to go to Italy for a few weeks during the next summer, in order to discuss my position on the spot, and also to see again my parents and other relatives. My indebtedness to the Society would still increase if you could try and obtain me the necessary permits from the British authorities. Yours sincerely, B. Segre

D7.2.6 L. Godeaux to B. Segre, Liège 13.8.1945

CA, *BSP*, fol. 1r

My dear Colleague, Chance put me in possession of your address, and I took the opportunity to find out about you. I hope you and yours have not suffered too much from the war. I can immediately reassure you about my fate and that of my loved ones. We came out alive, but the last five years were not happy, especially

---

[59]The questionnaire is kept in SPSL, *BS*, fol. 207. Joseph Bright Skemp (1910–1992) was the Secretary of the SPSL from 1944 to 1946.

[60]B. Segre, *The non-singular cubic surfaces*, Oxford, The Clarendon Press, 1942.

[61]CA, *BSP*: V. Snyder to B. Segre, Ithaca 30.3.1945.

[62]See Chap. 10.

for those, like me, who had taken part in the previous war. I only have a little news from Italy: a letter from Enriques. He is in good health, as is his family, and has returned to his chair of Higher Geometry at the University of Rome. One of his nieces, a daughter of the zoologist Paolo Enriques, was shot by the Germans in Florence.[63] Do you plan to return to Italy and take up your chair in Bologna? I would allow myself to wish this, just as I wish Mr. Terracini to return to Turin. During the war, I had some contact with Mr. Fano, a refugee in Lausanne; he succeeded in demonstrating the irrationality of the cubic variety of four-dimensional space, but I do not yet know his proof.[64] I have a whole bunch of offprints to send you, but I'll wait for you to tell me where I can send them. Please accept, my dear colleague, the expression of my best devotion. Lucien Godeaux[65]

Mon cher Collègue, Le hasard me met en possession de votre adresse et j'en profite pour prendre de vos nouvelles. J'espère que vous et les vôtres n'avez pas trop souffert de la guerre. Je vous rassure tout de suite sur mon sort et celui de mes proches. Nous en sommes sortis vivants, mais ces cinq dernières années ne furent pas gaies, surtout pour ceux qui, comme moi, avaient pris part à la guerre précédente. Je n'ai que peu de nouvelles d'Italie : une lettre d'Enriques. Il est en bonne santé, ainsi que sa famille et a repris sa chaire de Géométrie supérieure à l'Université de Rome. Une de ses nièces, une fille du Zoologiste Paolo Enriques, a été fusillée par les Allemands à Florence. Comptez-vous rentrer en Italie et reprendre vous aussi votre chaire de Bologne ? Je me permettrai de le souhaiter, de même que je souhaite la rentrée à Turin de M. Terracini. Pendant la guerre, j'ai eu quelques relations avec M. Fano, réfugié à Lausanne ; il a réussi à démontrer l'irrationalité de la variété cubique de l'espace à quatre dimensions, mais je ne connais pas encore sa démonstration. J'ai tout un paquet de tirés-à-part à vous envoyer, mais j'attendrai que vous me disiez où je peux les envoyer. Veuillez agréer, mon cher Collègue, l'expression de mes meilleurs sentiments. Lucien Godeaux

---

[63] Anna Maria Enriques Agnoletti (1907–1944), daughter of Paolo Enriques, was an archivist at the State Archives of Florence (1932–1938), then a paleographer at the Vatican Library. After German occupation, she joined the anti-fascist groups operating in Tuscany and participated in an espionage organization (Radio CORA) intended to collect information to transmit, via radio, to the Allies. She was arrested on 12.5.1944, tortured and shot on 12.6.1944.

[64] The work *Nuove ricerche sulle varietà algebriche a tre dimensioni a curve-sezioni canoniche* (Commentationes. Pontificia Academia Scientarum, 11, 1948, pp. 635–720), ready since 1942, had been presented by Severi to the Pontifical Academy on February 21, 1943, but was published only in 1948.

[65] Lucien Godeaux (1887–1975), professor of Geometry at the University of Liège (1925–1958), had spent a study sojourn in Bologna in 1912 in order to work under the guidance of Enriques.

D7.2.7 Draft of a Letter by B. Segre to F. Severi, [Manchester] 9.5.1946

CA, *BSP*, fols. 1r-2r

Dear Professor, I now receive your long letter of the 2nd of this month (D7.3.3), and I am grateful to you, especially for the expressions so full of understanding and affection in my regard. I am sorry to hear that unfortunately your vicissitudes are not over yet, and that other trials and annoyances await you. I fully understand your bitterness, and I regret that your activity is thus partly hampered; but I am pleased that, in your equanimity, you can judge certain excesses with a serene spirit, and that—strengthened by your work—you are waiting calmly and confidently for the outcome of this further trial. All that remains on my part is to wish you a speedy solution, certain that, in a short time, this will be nothing more for you than a pale memory of this tempestuous era. [. . .] It is my intention to go to Italy as soon as possible, because I feel a deep nostalgia for my people and my country, and I plan to leave at the end of June or early July. Once there, I will be able to judge whether to return permanently with my wife and children by the summer, or to stay another year in Manchester, where they would be happy to have me again. I would much prefer the first alternative, also to escape the instability and uncertainty that has weighed on my life for over seven years, and I am therefore willing to face the many difficulties of the moment: but I would not like these difficulties to prove insurmountable, or such as to block my scientific activity. I am mostly worried by the problem of housing, taught by the example of my parents [Samuele Segre and Leonilda Segré], who for over a year lived at the Mauriziano hospital in Turin for lack of better accommodation. But I do not lack adaptability, after having to change accommodation in England for 10 times, and not always in happy conditions. Do you think it is really difficult to find, in Bologna or Rome, modest accommodation for me and my family, which now—after the loss in painful circumstances of the youngest in London in 1940—is reduced to my wife, a 13-years-old boy and a 12-years-old girl?[66] Perhaps the State or the I.N.C.I.S.[67] could be of some help? Another issue, of lesser import but not indifferent for me, is that of the expenses for our final return trip and for the shipment of our books and personal belongings. The embassy in London, which I consulted, asked Rome whether these expenses could be reimbursed in whole or partially; we are awaiting response. However, as I told you, I will soon be there in person; and I hope that in this way I will have the opportunity to see you again. When do you think that the question of the succession of Enriques will be discussed by the Faculty.?[68] With heartfelt and

---

[66] Fernanda Coen Segre (1908–1976), Sergio (born on 22 May 1933), and Silvana (14 May 1934).

[67] The National Institute for State Employees' Homes (*Istituto Nazionale per le Case degli Impiegati Statali*), founded in 1924, was an Italian public body established to build homes and manage their assignment to public employees, at subsidized rent.

[68] In January 1946, Federigo Enriques had been retired due to age limits.

affectionate wishes, I send you and your wife our warmest regards. Your most devoted and affectionate [B. Segre]

Carissimo Professore, mi giunge ora la sua lunga lettera del 2 c.m., e gliene sono gratissimo, specie per le espressioni così piene di comprensione ed affetto a mio riguardo. Mi spiace di sentire che le sue traversie non siano purtroppo ancora finite, e che altre prove e seccature non lievi l'attendano. Comprendo bene la sua amarezza, e rammarico che la sua attività ne sia di conseguenza in parte intralciata; ma mi compiaccio che, nella sua equanimità, Ella riesca a giudicare certi eccessi con spirito sereno, e che – forte del suo operato – Ella attenda con calma sicura l'esito di questa prova ulteriore. A me non resta che augurarle un rapido scioglimento, certo che, fra qualche tempo, ciò non sarà più per Lei che un pallido ricordo di quest'epoca procellosa. [. . .] È mia intenzione di andare in Italia appena possibile, poiché sento profondamente la nostalgia della mia gente e del mio paese e conto di partire di qui io solo verso la fine di giugno od ai primi di luglio. Sul posto potrò giudicare se ritornare definitivamente con moglie e figli entro l'estate, oppure restare un altro anno a Manchester, dove sarebbero lieti di avermi ancora. Io preferirei di gran lunga la prima alternativa, anche per sottrarmi all'instabilità ed incertezza che da oltre sette anni pesa sulla mia vita, e sono disposto ad affrontare perciò le molteplici difficoltà del momento: ma non vorrei che tali difficoltà si dimostrassero poi insuperabili, o tali da inceppare la mia attività scientifica. Specialmente mi spaventa il problema dell'alloggio, ammaestrato dall'esempio dei miei genitori, che per oltre un anno hanno vissuto all'ospedale Mauriziano di Torino per mancanza di una sistemazione migliore. Ma l'adattabilità non mi manca, dopo aver dovuto cambiare alloggio in Inghilterra per ben 10 volte, e non sempre in condizioni felici. Crede Lei veramente difficile di trovare a Bologna o Roma un alloggio modesto per me e la mia famiglia, che ora – dopo la scomparsa in penose circostanze della più piccola a Londra nel '40 – è ridotta alla moglie, un ragazzo di 13 anni ed una ragazza di 12? O che forse lo Stato o l'I.N.C.I.S mi potrebbero essere di qualche aiuto? Un'altra questione, di minor momento, ma per me non indifferente, è quella delle spese per il nostro viaggio di ritorno definitivo e per la spedizione dei nostri libri ed effetti personali. L'Ambasciata a Londra, da me interpellata, ha chiesto a Roma se tali spese mi potranno venir rimborsate tutte o in parte; e sono in attesa di risposta. Comunque, come Le dissi, mi recherò fra non molto in avanscoperta; e spero che così avrò modo di rivederLa. Quando crede che ad un dipresso la questione della successione di Enriques sarà dibattuta dalla Facoltà? Con auguri vivi ed affettuosi, invio a Lei ed alla gentile Signora i nostri più distinti saluti. Suo dev.^{mo} ed aff.^{mo} [B. Segre]

D7.2.8 Draft of a Letter by E. Persico to A. Terracini, Gressoney-La-Trinité 7.8.1946

APR, fols. 1r-2v

Dearest Sandro, I inaugurate the Alpine holidays by responding to your letter of July 3rd, after having agreed, with Tricomi (who left Turin a few days ago for Beatenberg), that he would answer re. the economic and financial aspect.[69] I return with pleasure the Republican greetings, glad to see that your vote would have been in agreement with mine. I was, of course, pleased with the outcome of the referendum. I do not conceal from you, however, that the pleasure for the outcome of the referendum would have been greater if things had been carried out in full regularity, even from a formal point of view, until the very end, without giving the king [Umberto II] the opportunity to leave the floor to the well-known declaration and the royalists crying out for abuse.[70] All the more so since, as far as we know, there was no need for it. Unfortunately, even afterwards, the new government has not spared us disappointments (even for those who are not inveterately grumpy as I realise I have become): for example, the astonishing amnesty, which has put back in circulation common criminals (whose absence, in fact, was starting to be felt, given that the assaults and robberies had begun to thin out) and political criminals, including Jew-rakers and torturers, except those whose tortures had been particularly heinous (sic).[71] It is precisely the regret of not having been, at the time, a moderately brutal torturer. And the alleged aim of "pacifying the country" has not been achieved at all, because a great many people are indignant. Sometimes some of the amnestied are summarily executed, while on the other hand the fascists, considered the act as a dutiful if incomplete repentance, reoccupy the old posts and demand back wages. Maybe it's because you can read articles signed by the most fanatical defenders of the so-called Republic of Salò, who as soon as they got out of prison openly resumed their propaganda activity, sneering in their hearts about the advantages of freedom of the press and the vote that among n litigants the $(n+1)^{th}$ enjoys. As a tribute to that precept that your left hand does not know what the right hand is doing, on the same page of the newspaper in which new severe measures against the black market, the concealment of goods etc. were threatened, the extension of the

---

[69] Enrico Persico (1900–1969), Terracini and F. G. Tricomi were close friends. They had been colleagues at the University of Turin from 1930 to 1938.

[70] The Italian Republic was born following the results of the institutional referendum, called for 2 June 1946. For the first time women also participated in a national political consultation. 12,717,923 citizens voted in favour of the republic and 10,719,284 in favour of the monarchy. The former King Umberto II voluntarily left the country on 13 June 1946, without waiting for the results of the referendum. The appeals presented by the monarchist party were rejected by the Court of Cassation on 18 June 1946.

[71] The so-called Togliatti amnesty was a provision of law (Presidential Decree 22 June 1946, n. 4) for the extinction of sentences. It was proposed by the Minister of Grace and Justice Palmiro Togliatti and approved by the De Gasperi first government. The provision was introduced in a country torn by conflict and the civil war that broke out after the armistice (8.9.1943).

amnesty to food crimes was announced, with results that you can imagine. The fact that a law as fundamental as the one on the referendum has been formulated with the same clarity and precision of language with which our worst students give the definition of the specific heat (hence the not inconsistent controversy about the way in which the majority is evaluated) seemed to me to be a very bad omen for our future legislation. But this is the story of a month ago: today very strict new measures are announced, etc. And as an immediate response, the same newspapers as well as the top exponents of the black market, in a meeting held in Piazza Navona, have allocated large sums for the corruption of officials. But let's leave aside these trifles, to which perhaps I give excessive symptomatic "value", and let's come to firmer stuff: what will the lira do? How much is a matter of sincere opinion and how much of manoeuvring? As a manoeuvre, it has undoubtedly been very effective so far, having managed to give a certain confidence that has stabilised exchange rates and prices for a while, but now, as you know, these have started to rise again and the current opinion, even among economists, is that another sharp devaluation is inevitable. Here, however, I realise that I am invading the field of Tricomi, who will have illustrated the situation more fully. The prospects of the peace treaty are certainly not very cheerful, but I will tell you that I do not share the general pessimism, partly because I hope that some improvement can still be obtained, and partly because, by temperament and principle, I am not very sensitive to territorial questions (people, I do not know why, are much more concerned with such questions than with economic ones). Of course, we will not be able to avoid application of the mitigated principle 'who breaks-pays', but this was to be expected. [...] As I write this letter (a few lines per day, so as not to tire myself too much!) I receive (11/8) yours of July 26th. I gladly authorise the translation of my inaugural lecture.[72] In fact, speaking of translations, I have to ask you a favour. I am in negotiations with Eng. Levialdi who wants to translate into Spanish my *Meccanica Atomica* for a publishing house in Buenos Aires. I know from Castelnuovo that you have taken an interest in the analogous translation of his *Geometria Analitica*, and I would like to ask you first of all to tell me, confidentially, if the translator and the publisher seem recommendable, and to suggest to me some of your friends who may be able to watch my interests a little, since by the time the matter is agreed you will probably have already returned.[73] I had thought about Beppo Levi, but I don't know if he

---

[72] In December 1945 Persico had held in Turin the inaugural speech of the academic year 1945–46 entitled *Il nuovo fuoco*. The text was published in the journal *Scientia*, 79, 1946, pp. 83–92.

[73] Andrea Levialdi (1911–1968) earned his PhD from the University of Rome (1937) and emigrated to Argentina in October 1941. Levialdi had proposed to translate into Spanish the book by E. Persico, *Fondamenti della meccanica atomica* (Bologna, Zanichelli, 1936) for the publishing house Mundo Científico, which had published the G. Castelnuovo's *Lecciones de Geometría analítica. Geometría analítica del plano y del espacio, conceptos fundamentales de geometría proyectiva, curvas y superficies de segundo orden* (La Plata-Buenos Aires, Mundo Científico, 1943, transl. by Andrea Levialdi and Manuel Sadosky). Terracini advised against asking the publishing house Mundo Científico. See APR: A. Terracini to E. Persico, Tucumán 31.10.1946 and 26.12.1946.

will come back too or if he is the right person. I sincerely hope that the laborious paperwork for your return will be successfully completed, and that consequently you will soon arrive in port as well. In these days, looking at Monte Rosa, I, too, often thought of the beautiful days in Macugnaga, which we hope to repeat next summer, there or elsewhere. In the meantime, please receive with all your family my affectionate greetings. [E. Persico]

Carissimo Sandro, inauguro gli ozii alpini rispondendo alla tua del 3 luglio, dopo aver convenuto con Tricomi (partito giorni fa da Torino per Beatenberg), che alla parte economica e finanziaria avrebbe risposto lui. Ricambio con piacere il saluto repubblicano, lieto di vedere che il tuo voto sarebbe stato concorde al mio. Mi ha fatto naturalmente piacere l'esito del referendum. Non ti nascondo però il fatto che il piacere dell'esito del referendum sarebbe stato per me maggiore se le cose si fossero svolte in piena regolarità anche formale fino all'ultimo, senza dare occasione al re di lasciarsi dietro il noto proclama e ai monarchici di gridare al sopruso, tanto più che, per quanto ne sappiamo, non ce ne era affatto bisogno. E purtroppo anche in seguito, il nuovo governo non ci ha risparmiato le delusioni (anche per chi non è brontolone inveterato come mi accorgo di esser diventato io): per esempio la sbalorditiva amnistia, che ha rimesso in circolazione delinquenti comuni (di cui veramente da un po' di tempo si sentiva la mancanza, essendo che le aggressioni e le rapine cominciavano a diradarsi) e politici, sia rastrellatori e torturatori, eccetto quelli le cui torture erano particolarmente efferate (sic). Viene proprio il rimpianto di non aver fatto, a suo tempo, il torturatore moderatamente efferato. E l'asserito scopo di "pacificare il paese" non è stato affatto raggiunto perché moltissima gente è indignata, e ogni tanto qualcuno degli amnistiati viene giustiziato sommariamente, mentre d'altra parte i fascisti hanno preso la cosa come una doverosa benché incompleta resipiscenza, rioccupano gli antichi posti e pretendono gli stipendi arretrati. Sarà che si possono leggere articoli firmati dai più fanatici difensori della cosiddetta Repubblica di Salò, che appena usciti di prigione hanno ripreso scopertamente la loro attività propagandistica, sghignazzando in cuor loro sui vantaggi della libertà di stampa e sul voto che tra gli n litiganti l'(n+1)-esimo gode. In omaggio poi al precetto che la mano sinistra non sappia quel che fa la destra, nella stessa pagina di giornale in cui erano minacciati nuovi severi provvedimenti contro il mercato nero, l'occultamento di merci, etc., era annunciata l'estensione dell'amnistia ai reati annonari, col risultato che puoi immaginarti. Il fatto poi che una legge così fondamentale come quella sul referendum, sia stata formulata con la stessa chiarezza e precisione di linguaggio con cui i nostri peggiori allievi definiscono il calore specifico (donde la controversia non priva di fondamento sul modo di valutare la maggioranza) mi è sembrato di pessimo augurio per la nostra futura legislazione. Ma questa è storia di un mese fa: oggi sono annunciati nuovi severissimi provvedimenti, etc. E come immediata risposta, gli stessi giornali come i massimi esponenti del mercato nero, in una riunione tenuta a Piazza

Navona, hanno stanziate forti somme per la corruzione dei funzionari. Ma lasciamo da parte queste piccolezze, a cui io do forse eccessivo "valore" sintomatico, e veniamo a cose più sode: che farà la lira? Quanto c'è di opinione sincera e quanto c'è di manovra? Come manovra, finora è stata senza dubbio assai efficace, essendo riuscita a dare una certa fiducia che ha per un po' stabilizzato cambi e prezzi, ma ora questi, come sai, hanno ripreso a salire e l'opinione corrente, anche fra gli economisti, è che un'altra forte svalutazione è inevitabile. Qui però mi accorgo di invadere il campo di Tricomi, che ti avrà illustrato più ampiamente la situazione. Le prospettive del trattato di pace sono certamente poco allegre, ma ti dirò che in questo io non condivido il generale pessimismo, un po' perché spero che qualche miglioramento si possa ancora ottenere, un po' perché, per temperamento e per principio, sono poco sensibile alle questioni territoriali (di cui la gente, non so perché, si preoccupa assai più di quelle economiche). Certamente, non potremo evitare l'applicazione, sia pure attenuata, del «chi rompe paga», ma questo bisognava ben aspettarselo. [...] Mentre scrivo questa lettera (a poche righe per giorno, per non affaticarmi troppo!) mi arriva (11/8) la tua del 26 luglio. Ben volentieri autorizzo la traduzione della mia conferenza inaugurale. Anzi, a proposito di traduzioni devo chiederti un favore. Sono in trattativa con l'ing. Levialdi che vuol tradurre in spagnolo la mia Meccanica Atomica *per una casa di Buenos Aires. So da Castelnuovo che tu ti sei interessato della analoga traduzione della sua* Geometria Analitica *e vorrei pregarti anzitutto di dirmi, in via riservata, se il traduttore e l'editore sembrano raccomandabili; di suggerirmi qualcuno poi dei tuoi amici di costì che possano sorvegliare un po' i miei interessi visto che, quando la cosa sarà concordata tu sarai probabilmente già tornato. Avevo pensato a Beppo Levi, ma non so se torni anche lui e se è la persona adatta. Spero vivamente che le laboriose pratiche per il vostro ritorno arrivino finalmente in porto, e che di conseguenza arriviate presto in porto anche voi. Anch'io in questi giorni, guardando il Monte Rosa, ho ripensato spesso ai bei giorni di Macugnaga, che speriamo di rinnovare l'estate prossima, lì o altrove. Gradisci intanto, con tutti i tuoi, i miei più affettuosi saluti.* [E. Persico]

D7.2.9 F. Tricomi to B. Segre, Turin 26.10.1946

CA, *BSP*, fol. 1r-v

Dear Segre, For a long time now, I have needed to send an answer to your letter of 30 September but—for practical reasons—first I wanted to speak of your situation in the Faculty; but after several postponements, the Faculty met for the first time after the holidays only last night. Regarding what you wrote to me, the Faculty, on my proposal, has decided to postpone the matter until next spring, seeing as it was not possible to elect a successor to Prof. Fano in the current academic year. That is, it was not believed possible to proceed now with a generic call for Geometry because if you should regrettably retract your previous decision to come here to Turin, most likely we would instead take different measures, perhaps offering a second chair of Analysis. In the light of this, and on behalf of all colleagues, please let us know as soon as possible (no later than May 1947) of your final decisions so that we can act promptly in arranging a call for the next school year 1947-48. It's a long time since I had any direct news of Terracini, who was recently left low by strong tonsillitis. The indirect news (through the Artoms)[74] always leaves me hopeful of seeing him return here soon. Perhaps his brother Benvenuto will arrive first, as I have recently heard he should embark in early January. Kind regards from your F. Tricomi

Caro Segre, Da lungo tempo sono in debito di una risposta alla tua del 30.IX ma – per potere essere concreto – volevo parlar prima della tua cosa in Facoltà, e questa, dopo vari rinvii, si è riunita per la prima volta dopo le vacanze soltanto ieri sera. Considerato quanto mi hai scritto, la Facoltà, su mia conforme proposta, ha deciso di rimandare la questione alla prossima primavera, visto che, comunque, non era possibile dare un successore al Prof. Fano già nell'incipiente anno accademico. Non si è cioè creduto di procedere già ora ad un primo passo chiedendo genericamente di poter provvedere ad una chiamata per la Geometria, perché se tu, con nostro rincrescimento, rinunzierai definitivamente al precedente desiderio di venire qui a Torino, molto probabilmente si provvederà invece per altra materia, forse per una seconda cattedra di Analisi. In tali condizioni, anche a nome dei colleghi tutti, ti prego di farci conoscere al più presto possibile (comunque, non oltre il prossimo maggio 1947) le tue definitive decisioni onde poter provvedere in tempo utile ad una chiamata pel prossimo anno scolastico 1947-48. Da parecchio tempo non ho notizie dirette di Terracini che in questi ultimi tempi è stato incomodato da una forte tonsillite. Le notizie indirette (attraverso gli Artom) mi lasciano però sempre sperare di vederlo presto qui ritornare. Forse arriverà prima il fratello Benvenuto che, a quanto ho ultimamente udito, dovrebbe imbarcarsi già ai primi di gennaio. Cordialissimi saluti dal tuo F. Tricomi

D7.2.10 B. Levi to the Rector of the University of Bologna, Rosario 19.12.1947

ACS-Levi, fol. 1r

To the Rector of the University of Bologna, Only now am I able to give a proper answer to the communication you sent to me on 18 August, signed by Prof. Barbieri[75] on behalf of the Rector [Guido Guerrini], concerning my placement at

---

[74] Terracini's brother-in-law Alberto Artom and his wife Rita Sacerdote.

[75] Giuseppe Antonio Barbieri (1880–1956), chemist, Dean of the Faculty of Agricultural Sciences of the University of Bologna and pro-Rector from 1945 to 1947.

the disposal of the Ministry of Foreign Affairs. I must begin by justifying myself for not giving an answer earlier, even if inconclusive; the reason is that this letter reached me at the end of September, nor did it explicitly ask for a response, while identical communication sent to me more than a month before by the embassy asked for an answer, and I replied. Therefore, the Ministry was already informed of the situation. The reason for the delay lies in this fact: while the Faculty (to which the Institute of Mathematics that I am directing is attached) expressed to me the desire for my stay, for various reasons, I did not consider it possible at that time to bind myself beyond the period established by the existing contract which expired on 31 March 1948, as the Ministry knew from previous communications and as I confirmed in the reply given through the embassy. Only a few days ago (precisely on the current 15th) a new contract was signed to replace the former, which, satisfying my demands, confirmed me in my role virtually as long as my life will allow. I must state that I always feel strongly bound to my university and I would like to hope that the regular exchange of publications between my current Institute and the other (whose classrooms are constantly before my eyes), will be worth keeping this bond alive also for my colleagues. I am deeply grateful for the words with which, on behalf of the Rector, Prof. Barbieri sent the communication of 18 August, which certainly would have resulted in me taking another decision if I did not think that in the meantime other valiant colleagues could, and are, fulfilling the teaching functions that would have been assigned to me. For these reasons, I ask you to intercede with the Ministry to settle my position however you deem best, taking into account that, *re.* the question contained in the ministerial note, my position here seems to be assured for a period of several years as the Government wishes (I believe that you would not like to issue a decree for a period that is too limited). Personally, since the new contract allows me to ask for a permit for a trip to Italy when I deem appropriate, I hope that I will not be too long in seeing again the University and the colleagues left in Bologna, to whom I ask you to convey my affectionate feelings. Please, Sir, receive my cordial greetings and the expression of my utmost observance. [Beppo Levi]

Magnifico Sig. Rettore dell'Università di Bologna, Solo ora mi trovo in condizione di rispondere in modo sensato alla comunicazione che mi fu inviata in data 18 agosto a firma, per il Rettore, del Prof. Barbieri, relativa al mio collocamento a disposizione del Ministero degli Affari Esteri. Debbo tuttavia incominciare col giustificarmi di non aver dato prima d'ora una risposta, anche inconcludente se tale era il caso; e la ragione si è che detta lettera mi giunse a fine settembre, né per la sua redazione chiedeva esplicitamente risposta, mentre identica comunicazione mi era stata rimessa più di un mese prima dall'Ambasciata, alla quale avevo risposto e quindi il Ministero già era stato posto al corrente della situazione. La ragione del ritardo sta nel fatto che, mentre la Facoltà alla quale è annesso l'Istituto di Matematica che io sto dirigendo mi manifestava il desiderio della mia permanenza, per ragioni varie io non ritenevo possibile in quel momento di vincolarmi al di là del periodo stabilito dal contratto in corso la cui scadenza – come era a conoscenza del Ministero per comunicazioni anteriori e come ho confermato nella risposta data per mezzo dell'Ambasciata – era fissata per il 31 marzo 1948. Solo da pochi giorni (precisamente il 15 corrente) è stato firmato un nuovo contratto in sostituzione dell'anteriore, il quale [in] soddisfacimento a talune mie esigenze, mi conferma nella direzione virtualmente per tutto il tempo che le mie possibilità lo consentano. Debbo ora dichiararle che mi sento sempre legato alla mia Università dal più vivo affetto e vorrei sperare che l'invio regolare delle pubblicazioni del mio Istituto attuale all'altro le cui aule mi stanno continuamente davanti agli occhi, valga a mantenere vivo questo legame anche da parte dei colleghi. Sono vivamente grato per le parole con cui, a nome del Rettore, il prof. Barbieri ha voluto accompagnare la comunicazione del 18 Agosto, le quali certamente mi avrebbero determinato per altra decisione, se non pensassi che nel frattempo altri valorosi colleghi hanno disimpegnato e stanno disimpegnando costì le funzioni didattiche che potrebbero essermi assegnate. Per queste ragioni mi rivolgo a Lei pregandola di intercedere presso il superiore Ministero nel senso che la mia posizione sia regolarizzata nel modo che si consideri migliore tenendo in conto che, per ciò che riguarda la domanda contenuta nella comunicazione ministeriale e che io credo di dover interpretare nel senso che non si vorrebbe emettere un decreto per un periodo di tempo troppo limitato, la mia posizione qui pare assicurata per un periodo di più anni come il Governo lo voglia consentire. Per parte mia, poiché il nuovo contratto mi consente di chiedere, nel momento che lo ritenga opportuno, una licenza sufficiente per un viaggio in Italia, spero che non tarderò eccessivamente a rivedere l'Università e i colleghi lasciati a Bologna, ai quali la prego di far giungere i miei sentimenti affettuosi. Gradisca, Magnifico Rettore, i miei saluti cordiali e l'espressione della mia massima osservanza. [Beppo Levi]

D7.2.11 MFA to MPI, Rome 12.3.1949

ACS-Levi, fols. 1r, 2r
Pursuant to the cable of this Ministry no. 12594/1656 of December 20 *re.* Beppo Levi, professor of Mathematical Analysis at the University of Bologna, the following report (dated February 5) of the consul general of Italy in Rosario is transcribed, for the provisions of competence and begging benevolent consideration. While I was about to answer your cable, I received a statement from Prof. Levi from Bologna (where he recently arrived to spend a short holiday) in which he represented his wonder and apprehension for the provision referred to. I believe that the Prof.'s apprehensions are more than justified in relation to the fruits of his work, as Director of this Institute of Mathematics and as a diffuser of Italian culture, in general, in this centre. Only recently, as can be seen from the report attached to the cable of this consul general no. 1897/138, January 20, has the activity of this celebrated mathematician begun to profile itself in its entirety with the assumption of organic teaching courses that will have to highlight the

degree of development of our scientific culture. To achieve this goal, Prof. Levi has had to overcome immense difficulties and it would be a shame if just when the fruits of his effort should be reaped, he was to be removed, interrupting a work of great honour for Italy and its scientific sector. The esteem he enjoys, both among the students and the teaching staff of the Universidad Nacional del Litoral, makes Prof. Levi a cultural example that we are envied, and many foreigners would gladly see entrusted to their own men the task with which he is entrusted here. Let us add that by reaching his 75th birthday in 1950, i.e., the age of retirement in Italy, Prof. Levi would be precluded from continuing his work both in Italy and here where instead his presence is of great use for Italian national purposes. Furthermore, Prof. Levi, after many years of absence from Italy for the well-known racial measures, recently repatriated with limited personal baggage such as that allowed by air transport, to spend a short period of holidays in Italy. Therefore, he would still have to return to Argentina to close his home and transport his furniture to Italy, etc., which would constitute a considerable and, perhaps, impossible financial burden. For all these reasons, but mainly for that relating to the dissemination and affirmation of Italian culture abroad, I believe that the reinstatement of Prof. Levi should be postponed, and that the period of availability to this Ministry for cultural purposes abroad should be extended at least by a duration equal to that of the new contract that the Professor has signed or will sign with the University of Santa Fe. This General Directorate, for its part, can only associate itself with the considerations set out by the consul general of Italy in Rosario on the advisability of further extending the post abroad of Prof. Levi, and waits to know the decisions that this Ministry will adopt in this regard.

A seguito del telespresso di questo Ministero n. 12594/1656 del 20 dicembre u.s. relativo al Prof. Beppo Levi, ordinario di Analisi Matematica nell'Università di Bologna si trascrive, per i provvedimenti di competenza e con preghiera di benevolo esame, il seguente rapporto in data 5 febbraio u.s. del Consolato Generale d'Italia a Rosario. Mentre mi accingevo a riscontrare il telespresso in riferimento, mi giunge dal Prof. Levi da Bologna, dove è giunto da poco per trascorrere un breve periodo di vacanza, un esposto in cui l'interessato mi significa la sua meraviglia ed apprensione per il provvedimento di cui al precitato foglio. Ritengo che le apprensioni del Prof. siano più che giustificate in relazione ai frutti della sua opera, come Direttore di questo Istituto di Matematica e come diffusore della cultura italiana, in genere, in questo centro. Soltanto da poco tempo, come si rileva dalla relazione allegata al telespresso di questo Consolato Generale n. 1897/138 del 20 gennaio u.s., l'attività di questo illustre matematico comincia a profilarsi nella sua interezza con l'assunzione di corsi di insegnamento organici che dovranno mettere in evidenza il grado di sviluppo della nostra cultura scientifica. Per giungere a questo il Prof. Levi ha dovuto superare immense difficoltà e sarebbe veramente un peccato se quando si dovrebbero raccogliere i frutti di questo sforzo egli venisse allontanato interrompendo un'opera di grande onore per l'Italia e per le sue scienze. La stima di cui gode, sia tra gli studenti che nel corpo insegnante dell'Università del Litoral, fa del Prof. Levi un'illustrazione culturale che ci è invidiata e molti stranieri vedrebbero volentieri affidato ad elementi della loro nazionalità, l'incarico di cui è qui investito. E a ciò si aggiunge che compiendo il Prof. Levi nel 1950 il suo 75° anno di età, epoca nella quale egli sarebbe collocato a riposo, si verrebbe a precludergli la possibilità di continuare la sua opera sia in Italia che qui dove invece la sua presenza riesce di grande utilità ai fini anche nazionali italiani. Inoltre il Prof. Levi, dopo molti anni di sua assenza dall'Italia per le note misure razziali, è recentemente rimpatriato con un ristretto bagaglio personale quale quello consentito dai trasporti aerei, per trascorrere in Italia un breve periodo di riposo. Dovrebbe quindi sempre rientrare in Argentina per liquidare la sua casa e trasportare in Italia i propri mobili ecc. il che costituirebbe un aggravio finanziario non indifferente e, forse, non sopportabile. Per tutti questi motivi, ma principalmente per quello relativo alla diffusione ed all'affermazione della cultura italiana all'estero, ritengo che si dovrebbe soprassedere al provvedimento di rientro del Prof. Levi, il cui periodo di messa a disposizione di codesto Ministero per i suoi fini culturali all'estero dovrebbe essere prolungato per lo meno di una durata pari a quella del nuovo contratto che il predetto Professore ha fatto o farà con la Università di Santa Fé. Questa Direzione Generale non può da parte sua che associarsi alle considerazioni esposte dal Consolato Generale d'Italia a Rosario sulla opportunità che sia ulteriormente prorogato il comando all'estero del Prof. Levi, e rimane in attesa di conoscere le decisioni che codesto Ministero avrà ritenuto di adottare al riguardo.

7.3   Remigration and Post-War Cultural Reconstruction of Italy

Those who returned did so with the aim of contributing personally to the cultural reconstruction of the country. This would unite the expats for racial reasons (Terracini and Segre *in primis*), but also those who had remained in Italy, such as Ascoli, Persico, and Tricomi (D7.3.4, D7.3.6, and D7.3.7). Enthusiasm was great, not only on the part of young people but also of the older Masters, such as Fano, Colonnetti, and Castelnuovo (D7.3.10). Colonnetti assumed the presidency of the CNR, Castelnuovo that of Lincei, coordinating its reconstitution (the Lincei had been annexed to the Academy of Italy in 1939) and resumption of activities.

The disappointing experience of the Constituent Assembly cooled the enthusiasm with which some of them had approached politics but did not dampen the energy with which they engaged in the cultural reconstruction of the country. This fact justifies at least in part an element that would otherwise be inexplicable, i.e. the speed with which formerly persecuted scholars 'forgave' their colleagues and resumed collaborating with them. Obviously, there was not complete awareness of what had occurred in 1938. Terracini, for example, knew that Severi and Bompiani were strong fascists, but he was not fully aware of the anti-Jewish campaign carried out by Bompiani within the Italian Mathematical Union. In other cases, however, events unfolded differently. Segre, for example, was aware that it had been Severi who had expelled him from *Annali di Matematica pura ed applicata* and saw the statement with which UMI affirmed that Italian mathematics was the result of scholars of the Aryan race only. Despite this, relations were re-established promptly. Only Terracini refused at first to be reinstated in the UMI but, after Segre's intervention, and in order not to cause President Berzolari any displeasure, he soon changed his mind and accepted the reinstatement (D7.3.8 and D7.3.9).

The speed in re-establishing contacts with scholars who had compromised themselves with the regime derived largely from some convictions shared among the former expats: priority was the cultural reconstruction of the country; restoration of Italy's reputation in the eyes of the world passed through its cultural rehabilitation; as it had been a responsibility and a duty for expats to contribute to positive evaluation of the work accomplished by Italians abroad, so now it was the responsibility and duty for returning emigrants to contribute to restoring Italy to the top of the international mathematical scene. In other words, the right to compensation for what one had suffered individually must be postponed in favour of fulfilling the duty of citizens and men of science towards the homeland.

For the world of mathematics, reconstruction basically passed through the following four points: strengthening of relations with the Anglo-Saxon countries, especially with America; de-provincialisation of national research; expansion towards new fields of study both in pure mathematics (abstract algebra, topology), and in applied mathematics (statistics, information theory, etc.). In particular, everybody agreed on the fact that it was necessary to turn towards the English-speaking world. It is thus not surprising that, among the displaced scholars, it was Segre's experience that was to be given priority as an asset. Severi, on behalf of the Pontifical Academy of Sciences, asked Segre to draw up the survey on mathematical research in English-speaking countries during the war years. Segre agreed to geometry and suggested Hodge for analysis (real field and complex field), number theory, group theory and topology (D7.3.2). Segre was also asked to publish a report on the new type of differential analyser created in Cambridge in 1942, with the support of the RKF Foundation, and one on the school system in England, and to review (for the *Bollettino dell'UMI*) various texts that appeared in the United States and the United Kingdom in the war years, and in the immediate post-war period.[76]

Loss and Gain: Jewish mathematical emigration from the Third Reich has often been described in these terms which, for Italy, are not appropriate. The question to be asked, in the Italian case, appears to be another: of those who returned, did they return changed? Fano, certainly not: he resumed his studies on varieties and devoted his last years to the project of publishing a volume summarising his results on threefolds. From a professional point of view, the experience in Argentina changed Terracini, although not regarding research activity. The sojourn in Tucumán left him

---

[76] Segre B. 1947 and Segre B. 1948.

with an interest in history of mathematics and methodology that he would continue to develop in the last twenty years of his life (1948–1968). Above all, when there, Terracini gained experience as an organiser and entrepreneur of mathematics, which he was then able to exploit as Director of the Special Mathematics Library in Turin (1948–1964), editor of the *Rendiconti del Seminario Matematico dell'Università e del Politecnico di Torino*, vice president (1952–1958), then president of UMI (1958–1963). Segre was the only one to bring back from his stay in England (and to import into Italy) some trends in algebraic geometry which he had approached: arithmetic geometry, above all. The move to Rome, in an environment where Severi and Enriques remained anchored to the typical Italian research style, and in which there were few interlocutors (A. Franchetta, G. Zappa, F. Conforto) caused his initial momentum to be lost (D7.3.1, D7.3.3, and D7.3.5). With the publication of the book *Lezioni di Geometria Moderna* (1948), any echo of his research deriving from the English period can be considered substantially extinguished.

However, just as expatriation is not merely a matter of place but a condition of the soul, so it is with the return. It would be unusual for scholars who had never stopped living in Italy with their minds and hearts, to return there physically with different minds and hearts compared with 1938.

Documents

D7.3.1 F. Enriques to Segre, Rome 11.9.1945

CA, *BSP*, fol. 1r-v
Dear Segre, I thank you for your kind letter of 2 August (received yesterday) and I congratulate you on your scientific activity. Even the arithmetic questions you are talking about interest me; since I read the paper by Poincaré on cubic curves and surfaces, I proposed the problem (existence of a dense set of rational points) to Beppo Levi, who happily solved it.[77] But personally, I am particularly interested in the question of the continuous system over irregular surfaces, in view of my lecture notes on surfaces, which were written in 1942, and are now being published.[78] This is an extremely sensitive issue. I wasn't able, at the time, to reconstruct the proof you showed, based on the indications you yourself gave me, before my trip to Paris. Severi, who had had more extensive indications from you, believed that he had succeeded in this goal. His demonstration seemed to me

---

[77] Enriques refers to the following papers: B. Segre, *On limits of algebraic varieties, in particular of their intersections and tangential forms*, Proceedings of the London Mathematical Society, s. 2, 47, 1942, pp. 351–403; H. Poincaré, *Sur les propriétés arithmétiques des courbes algébriques*, Journal de Mathématiques Pures et Appliquées, s. 5, 7, 1901, pp. 161–234 and B. Levi, *Saggio per una teoria aritmetica delle forme cubiche ternarie*, Nota I, XLI, 1905–06, pp. 739–764; Nota II, Nota III, Nota IV, XLIII, 1907–08, pp. 99–120, 413–434, 672–681.

[78] F. Enriques, *Le superficie algebriche*, Bologna, Zanichelli, 1949, ed. by G. Castelnuovo.

obscure and therefore doubtful; I believed, however, (in the paper in *Commentarii Helvetici*) that I had overcome the difficulty: in reality, my demonstration was wrong; but from the observation also derived the error of Severi's demonstration.[79] I was not allowed then to add anything to the paper of the *Commentarii Helvetici*; S[everi] instead was given the possibility of writing an apostille in which he said he had obtained a more general theorem (he dealt with algebraic entities instead of clarifying the thing in the simplest case).[80] But shortly afterwards S[everi] himself, who was exposing these theories in the lectures at the Institute of Advanced Mathematics, realised that the proposed proof was vitiated by a radical error. I would like very much this question to be accommodated. Otherwise, it is better to return to my first proof based on the infinitely close curves of the various orders, which I have re-examined, and I believe is substantially right, even if not strictly complete. I do not hide that I am a little worried about it, because now I have other works in the pipeline and therefore, I do not know what difficulty the examination of your proof may present me, assuming that I can see it before the publication of my book. But you might help me by sending me an exposition in Italian of the proof, preferably limited to a particular case, for example, to the case of the surfaces with gender $p_a = -1$ and $p_g = 1$ (system $| C |$ of genus π and degree $2\pi-2$). I add that a reassuring element would be for me that you yourself examine the demonstration contained in the paper by Severi, published by the Academy of Italy, and identify the error that it contains; an error which, on the other hand, has also been recognised by me and by Castelnuovo in conversations on the subject, and which, as I told you, is recognised by the author himself.[81] The delicate point lies here: when a curve of $| 2C |$ with $2\pi-3$ double points tends to the limit, so that it breaks, one does not see how to determine the limit series that becomes indeterminate: infinitesimals of the various orders can come into play, which gives rise to difficulties. Kind regards, yours truly F. Enriques

Caro Segre, La ringrazio della Sua buona lettera del 2 Agosto (pervenutami l'altro giorno) e mi congratulo della Sua attività scientifica. Anche le questioni aritmetiche di cui mi discorre mi interessano, fin da quando lessi la memoria sulle cubiche di Poincaré, e proposi a Beppo Levi il problema (sull'esistenza di un insieme denso di punti razionali), che egli ha felicemente risolto. Ma personalmente sono interessato in ispecie alla questione del sistema continuo sopra le superficie irregolari, e ciò in vista del libro delle mie lezioni sulle superficie, che è stato redatto nel 1942, e che ora è in pubblicazione. La

---

[79]F. Enriques, *Sui sistemi continui di curve appartenenti ad una superficie algebrica*, Commentarii Mathematici Helvetici, XV, 1943, pp. 227–237.

[80]F. Severi, *Intorno ai sistemi continui di curve sopra una superficie algebrica*, Commentarii Mathematici Helvetici, XV, 1943, pp. 238–248.

[81]F. Severi, *La teoria generale dei sistemi continui di curve sopra una superficie algebrica*, Memorie della Reale Accademia d'Italia, Classe di Scienze Fisiche, Matematiche e Naturali, s. 7, XII, 1941, pp. 337–430.

questione è estremamente delicata. Io non riuscii, a suo tempo, a ricostruire la dimostrazione da Lei indicata, sulla base delle indicazioni da Lei stesso fornitemi, prima della mia andata a Parigi. Severi, che aveva avuto da Lei più ampie indicazioni, credette essere riuscito allo scopo. La Sua esposizione sembrò a me oscura e quindi dubbia; credetti però (nella memoria dei Commentarii Helvetici*) di aver superato la difficoltà: in realtà la mia dimostrazione era sbagliata; ma dal constatarlo derivava anche l'errore della dimostrazione del Severi. A me non si permise allora di aggiungere nulla alla memoria dei* Commentarii Helvetici *e al S. fu dato invece di scrivere una postilla in cui diceva di avere ottenuto un teorema più generale (egli si attaccava al caso di enti algebroidi invece di chiarire la cosa nel caso più semplice). Ma poco dopo il S. stesso, che stava esponendo coteste teorie nelle lezioni dell'Istituto di Alta Matematica, ebbe ad accorgersi che la dimostrazione proposta era viziata da un errore radicale. Io desidero vivamente che la cosa si possa accomodare. Altrimenti conviene ritornare alla mia prima dimostrazione basata sulle curve infinitamente vicine dei varii ordini, che ho riesaminato e credo sostanzialmente giusta, anche se non rigorosamente completa. Non le nascondo che la cosa mi preoccupa un poco, perché ho ora a mano altri lavori e quindi non so quale difficoltà possa offrirmi l'esame della Sua esposizione, ammesso che io possa vederla prima della pubblicazione del mio libro. Ma Lei potrebbe aiutarmi, inviandomi un esposto in lingua italiana della dimostrazione, preferibilmente limitata ad un caso particolare, per esempio, al caso delle sup. di genere $p_a = $ -1 e $p_g = $ 1. (sist.[ema] | C | di gen.[ere] π e grado 2π - 2). Aggiungo che un elemento tranquillizzante sarebbe per me che Lei stesso esamini la dimostrazione contenuta nella memoria di Severi dell'Accademia d'Italia e si renda conto dell'errore che essa contiene, che d'altra parte è stato riconosciuto anche da me e dal Castelnuovo in conversazioni sull'argomento, e che, come Le ho detto, è riconosciuto dall'autore stesso. Il punto delicato sta in ciò che, quando si fa tendere al limite una curva di | 2C | con 2π - 3 punti doppi, in guisa che si spezzi, non si vede come determinare la serie limite che diventa indeterminata: possono entrare in gioco infinitesimi dei varii ordini, che dan luogo a difficoltà. Cordiali saluti suo aff. mo F. Enriques*

D7.3.2 Draft of a Letter by B. Segre to V.D. Hodge, 19.2.1946

CA, *BSP*, fol. 1r

Dear Hodge, The *Pontificia Academia Scientiarum* has the intention of publishing, as soon as possible, a number of Reports on the scientific developments during the war period 1939-45 in the Allied and Axis Countries.[82] I know Professor Whittaker has been asked to prepare the Report on Analysis and kindred topics in the Anglo-Saxon Countries, but he has refused.[83] Do you think someone in Cambridge would undertake to write this report, or, alternatively, the task could be divided among a few people dealing with different subjects (functions of real and complex variables, theory of numbers, theory of groups, topology, differential geometry, etc.)? If you could make any suggestion, I shall be glad to forward it to the *Pontificia Academia Scientiarum*. The Reports should include an accurate Bibliography and give a short account of the trend of ideas and main results (I myself have already written the one on algebraic geometry). Severi has told me he would be prepared to go to Cambridge for a month or two, even if no payment is possible, but an invitation from Cambridge University should be required in order to obtain the necessary permits.[84] Do you think accommodation for him could be found at a college? Also, could you please indicate the approximate expense for living in a college or hotel? With kindest regards Your [B. Segre]

D7.3.3 F. Severi to B. Segre, Rome 2.5.1946

CA, *BSP*, fols. 1r, 2r, 3r
Dearest Segre, Almost two months have passed since your last kind letter and I answer only today, although I have always had it in the back of my mind to answer you immediately. The delay was determined by a desire to accompany my letter with a package of my most recent works, reciprocating your papers, the last of which (always interesting) on projective-differential geometry and algebraic-arithmetic geometry I have received in these days. And I could not prepare the packet before today. [. . .] The packet of my works is a bit large and heavy, and I fear that by ordinary mail it might get lost, so I send it through the English embassy at the Vatican, both to you and to Hodge. The two packages will leave in these days. They also include my first volume published so far on *Serie di*

---

[82] In the perspective of a prompt and broad restoration of international scientific relations, in 1945 the Pontifical Academy of Sciences took the initiative to publish a *Consuntivo del lavoro di ricerca scientifica compiutosi nel mondo civile dal 1939 a oggi*. As far as Mathematics is concerned, the following *Relationes de Auctis Scientiis Tempore Belli a. 1939–1945* were published: N. 8, F. Conforto, G. Zappa, *La geometria algebrica in Italia dal 1939 a tutto il 1945*, pp. 3–43; N. 11. B. Segre, *Geometria algebrica nei Paesi anglo-sassoni*, pp. 3–51; N. 12. P. Buzano, *La Geometria differenziale in Italia*, pp. 3–27; N. 15. A. Signorini, *La meccanica razionale e la fisica matematica nell'Italia centrale e meridionale dal 1939 a oggi*, pp. 3–17; N. 22. S. Cinquini, L. Amerio, A. Ghizzetti, *Analisi matematica in Italia nel campo reale*, pp. 3–85. There were no *Relationes* written by foreign mathematicians.
[83] SEGRE.30.
[84] SEGRE.25.

equivalenza etc. (over 400 pages) and the volumes of the Volta 1939 Mathematical Conference (not held) and the 1942 Mathematical Conference, held in Rome on the initiative of the Institute of Advanced Mathematics.[85] I sent all the reprints of my works that I have available, based on my notes of the last that I had sent to you and Hodge. Some reprints are missing, but they are not essential, because the war has produced losses and disorder even in my library both here and in Arezzo and I had tried to put aside what I wanted most. I could not add the second volume of my *Lezioni di Analisi* (published together with G. Scorza Dragoni: now the 3rd is being prepared), because I no longer have copies.[86] It will be for later. From my works you will see the state of the question of continuous systems of curves on a surface and the reasons that drop the theorem for certain continuous systems. If Zappa did not send you the counterexamples he found, he will do so.[87] My essay on continuous systems (published with the other forty-year *Memoria* on enumerative geometry, successively transformed into a book of the Springer series, which was lost before being printed), also needs to be reviewed in many points and I approached the revision work in my lectures at the Institute of Advanced Mathematics, which will be published in vol. II of my *Lezioni sulle serie di equivalenza* (forthcoming).[88] Please see how objectively I quoted Castelnuovo and Enriques back in earlier times. However, they have behaved and continue to behave badly with me (especially Enriques, whom you know very well). But what can you do... [...] Enriques, according to the law that rightly readmitted him and those who were in his condition, will have to retire at the end of the current school year, because he reaches 75 years of age in January. I repeat to you (the thing should remain confidential for now) my very strong desire that you come here for the Higher Geometry (the other geometry chairs are covered) and I propose to launch the issue as soon as possible. Even if the appointment were made next year and you did not feel comfortable about coming immediately (given the difficult present conditions), I could easily obtain a delay. But it is useless to talk about it

---

[85] F. Severi, *Serie, sistemi di equivalenza e corrispondenze algebriche sulle varietà algebriche*, ed. by F. Conforto and E. Martinelli, Roma, Istituto Matematico della R. Università, 1942; *Atti del Convegno Matematico tenuto in Roma dall'8 al 12 novembre 1942*, Roma, INDAM, 1945; Reale Accademia d'Italia, Fondazione Alessandro Volta, *Convegno di Scienze Fisiche Matematiche e Naturali 1939. Matematica Contemporanea e sue applicazioni*, Atti dei Convegni 9, Roma, Reale Accademia d'Italia, 1943.

[86] F. Severi, G. Scorza Dragoni, *Lezioni di Analisi*, Bologna, Zanichelli, vol. 2, part I, 1942.

[87] G. Zappa, *Sull'esistenza di curve algebricamente non isolate, a serie caratteristica non completa, sopra una rigata algebrica*, Acta. Pontificia Academia Scientarum, VII, 2, 1943, pp. 1–5 and *Sull'esistenza, sopra le superficie algebriche, di sistemi continui infiniti, la cui curva generica è a serie caratteristica incompleta*, Acta. Pontificia Academia Scientarum, IX, 9, 1945, pp. 91–93.

[88] F. Severi, *La teoria generale dei sistemi continui di curve sopra una superficie algebrica*, Memorie della Reale Accademia d'Italia, Classe di Scienze Fisiche, Matematiche e Naturali, s. 7, 12, 1941, pp. 337–430; *I fondamenti della geometria numerativa*, Annali di Matematica pura ed applicata, s. 4, 19, 1940, pp. 153–242; *Lezioni sulle Serie, sistemi d'equivalenza e corrispondenze algebriche sulle varietà algebriche (raccolte da F. Conforto ed E. Martinelli)*, Roma, Cremonese, t. 1, 1942.

first. My dream, before dying, is to spend some time with you (hopefully, a few years) and deliver you into the Institute of Advanced Mathematics together with the traditions to be preserved and strengthened of Italian algebraic geometry. You may know that poor Comessatti passed away a few months ago (from cancer!) and that De Franchis too has recently died.[89] It is perhaps twenty days since Tonelli passed on prematurely! How many losses! Returning to my work, I will tell you that the provisional paper on periodic functions of several variables that you will read contains only a small part of my results achieved in a *Memoria* of almost 400 pages (in progress) on the quasi-abelian functions.[90] The theorems of structure and existence are missing (I did not have time to include them) because they wanted the summary note, even incomplete, to insert in the book in honour of Speiser.[91] I have two other *Memorie* in the pipeline and on other topics; I'll tell you about them another time. [. . .] All things considered, it is impossible for me to move now, because former academics of Italy have been recently included among alleged profiteers of the regime. It's a bad story, because of course, also former academics may be liable for prosecution, but they cannot presume to stain the reputation so indiscriminately of people of undisputed fame who were admitted to the Academy of Italy as a scientific distinction.[92] The suffering to which Italian people were subjected during the war, however, must mean that certain allowances be made for such excesses. Of course, I will come out of the painful test immaculately clean as I have always been, but I cannot be absent just when the law begins to operate (it was issued on 23 April) and I must clarify issues, document them, etc. I know your affection for me, and therefore I add, for your peace of mind, that this does not alter in any way my university position, which was perfectly defined in the purge. Everything will be resolved but it will take two or three months. [. . .] Yours truly F. Severi

Carissimo Segre, Son passati quasi due mesi dalla ultima Sua gentile ed io rispondo soltanto oggi, pur essendo sempre restato colla spina di risponderle subito. Il ritardo è stato determinato dal desiderio di accompagnare la mia lettera con un pacco de' miei più recenti lavori, ricambiando i Suoi, gli ultimi

---

[89] Annibale Comessatti and Michele De Franchis, both members of the Italian Geometric School, died on 13 September 1945 and 19 February 1946, respectively.

[90] F. Severi, *Funzioni quasi-abeliane*, Pontificia Academia Scientarum. Scripta Varia, 4, Roma, Pontificia Academia Scientarum, 1947, p. 327.

[91] F. Severi, *Le funzioni periodiche di più variabili*, Commentarii Mathematici Helvetici, 18, 1945, pp. 16–29. As the Editorial Board of Commentarii specified: 'Die vorliegende Arbeit war für die Festschrift Andreas Speiser, Orell Füssli, 1945, berechnet. Wegen verspätetem Eintreffen konnte sie leider nicht mehr in diese aufgenommen werden'.

[92] After the liberation of Rome (June 4, 1944) the Academy of Italy had been suppressed, at the suggestion of B. Croce. Scholars who, like Severi, had participated in the sittings held after the advent of the Social Republic were subjected to the purge procedure. Severi had attended the meeting of 20 March 1944 (in Florence). See ACS: MPI, Direzione Generale Istruzione Superiore, *Fascicoli professori epurati*, b. 31, *Severi Francesco*.

dei quali (sempre interessanti) di geometria proiettivo-differenziale e di geometria algebro-aritmetica ho ricevuto in questi giorni. E non ho potuto preparare il pacco prima di oggi. [...] Il pacco de' miei lavori siccome è un po' grande e pesante e temo che per posta ordinaria si smarrirebbe, lo mando pel tramite della Ambasciata inglese presso il Vaticano, tanto a Lei che a Hodge. I due pacchi partiranno in questi giorni. Son in essi compresi anche il mio I volume finora pubblicato sulle Serie di equivalenza *ecc. (di oltre 400 pagine) e i volumi del Convegno Matematico Volta 1939 (non tenuto) e del Convegno Matematico 1942, tenutosi in Roma per iniziativa dell'Istituto di Alta Matematica. De' miei lavori ho mandato tutti quelli che avevo, riattaccandomi a ciò che prima avevo annotato vicino al Suo nome e a quello di Hodge. Ne mancano taluni, ma non essenziali, perché la guerra ha prodotto perdite e disordine anche nella mia biblioteca sia qui che in Arezzo ed io avevo cercato di mettere in disparte quel che più mi premeva. Non ho potuto aggiungere il II volume delle mie* Lezioni di Analisi *(pubblicato insieme a G. Scorza Dragoni: ora è in preparazione il 3°), perché non ne ho più copie. Sarà per più tardi. Dai miei lavori vedrà lo stato della questione dei sistemi continui di curve sopra una superficie e le ragioni che fanno cadere il teorema per certi sistemi continui. Se Zappa non gli ha mandato gli esempi che ha trovato, lo farà. La mia memoria quarantennale pei sistemi continui (pubblicata coll'altra* Memoria quarantennale sulla geometria numerativa, *trasformata in un libro della collezione Springer, che è andato perduto prima di esser stampato), ha bisogno perciò anch'essa di esser riveduta in molti punti e al lavoro di revisione mi son accinto nelle mie lezioni all'Istituto di Alta Matematica, che vedranno la luce nel II vol. in preparazione delle* Lezioni sulle serie di equivalenza. *La prego di vedere con quale obiettività ho citato Castelnuovo ed Enriques in tempi non sospetti. Tuttavia, loro si son portati e si portano male con me (specialmente Enriques, che Lei ben conosce). Che farci? [...] Enriques, secondo la legge che ha giustamente riammesso lui e quelli che si trovavan nella sua condizione, dovrà andare in pensione al termine del corrente anno scolastico, perché ha finito 75 anni in gennaio. Le ripeto (e resti per ora la cosa fra noi) il mio vivissimo desiderio che Lei venga qui per la Geometria Superiore (le altre cattedre geometriche son coperte) e mi propongo di lanciare la cosa non appena possibile. Anche se la nomina sarà fatta col venturo anno e Lei non si troverà in comodo di venir subito (date le difficili condizioni del momento) si otterrà facilmente una dilazione. Ma è inutile parlarne prima. Il mio sogno, prima di morire, è di vivere qualche tempo con Lei (speriamo qualche anno) e di consegnarle l'Istituto di Alta Matematica insieme alle tradizioni da conservare e da rafforzare della geometria algebrica italiana. Lei forse saprà che il povero Comessatti è scomparso da qualche mese (per un cancro!) e che è morto da poco anche il De Franchis. Sono forse venti giorni che è scomparso immaturamente il Tonelli! Quanti vuoti! Tornando ai miei lavori Le dirò che la nota preventiva sulle funzioni periodiche di più variabili ch'Ella leggerà non contiene riassunti che una piccola parte de' miei risultati conseguiti in una* Memoria *di quasi 400 pagine di stampa (in corso) sulle funzioni quasi abeliane. Mancano i teoremi di struttura e di esistenza, che non feci a tempo a inserire,*

*perché volevano la nota riassuntiva, anche incompleta, per onorare Speiser. Ho
in corso altre due* Memorie *e su altri argomenti; gliene parlerò un'altra volta.
[...] Tutto considerato, mi è impossibile per ora di muovermi di qui, perché gli ex
accademici d'Italia sono stati inclusi di recente, come presunti profittatori del
regime, nella legge pei profitti di regime. Cosa enorme, perché si potevano
perseguire le persone anche se ex accademici, ma non macchiare con una tal
presunzione persone di fama indiscussa che avevano avuto una distinzione
scientifica. La sofferenza cui il popolo italiano è stato soggetto durante la guerra
deve tuttavia fare giudicare con spirito sereno anche certi eccessi. Naturalmente
io escirò dalla prova penosa immacolato come sono sempre stato, ma non mi
posso assentare proprio nel momento in cui la legge comincia a funzionare (è
andata in vigore il 23 aprile) e io devo chiarire, documentare, ecc. So che Lei mi
vuol bene ed aggiungo perciò per Sua tranquillità che ciò non altera in nulla la
mia posizione universitaria perfettamente definita in sede epurativa. Tutto si
risolverà bene, ma ci vorrà due o tre mesi. [...] Suo aff. F. Severi*

D7.3.4 A. Terracini to F. Tricomi and E. Persico, Tucumán 3.7.1946

APR, fols. 1r, 2r, 3r, 4r

Dearest friends Tricomi and Persico, it is not out of meanness that I am writing
you a joint letter, and to prove it I am sending you a copy each. The reason is that I
leave tomorrow for Buenos Aires (for a week); I do not want to delay further
writing to you and I am reduced to the last moment, as when (according to
Franco, and also according to the truth of the facts) I went to buy cigarettes just as
the locomotive was about to whistle, and the train left with my friends and my
skis inside. I have been planning to write to you for a long time. Your letters dated
17 and 21 March, respectively—despite the forecasts—took some time to cross
the Atlantic, and even longer to get out of the trunk of the gentleman who brought
them to Tucumán. Then I learned from Picone (whose letter was still much slower
than yours) of your adhesion to the memorial that he had presented to the Minister
of the P.I.; adhesion for which I thank you both, and I thank you with fraternal
affection; I beg you to offer my heartfelt thanks also to my Turin mathematician
colleagues (to some of the non-Turin ones I wrote directly; of course, I could not
thank all 86! I thanked a mathematician for each Faculty).[93] [...] During my trip
to Buenos Aires, I will naturally go to the embassy to hear a little about my return.
Picone tells me, in a very recent letter, that he was counting on reactivating the
procedures with the new ministry. But before talking about the new ministry, let's
talk for a moment about the new Italian regime, and let's send each other the first
republican greeting.[94] From here we have followed the elections to the Constituent and the subsequent developments with great trepidation. It seems to me a

---

[93] See D7.1.2 and D7.1.3.
[94] See D7.2.8.

good sign: (a) that the elections have had the effect they have had; (b) that subsequent attempts to cloud the situation have been aborted, and that the difficulties have been overcome with a lot of common sense, and with a political maturity on the part of Italians, which one could hardly expect after so many years of fascism. This leads me to be very optimistic (albeit in a relative sense; in an absolute sense, I believe that optimism is no longer possible for our generations) about the future of Italy. The news of the last few days on the treatment given to Italy is very unflattering; and, even from a strictly personal point of view, the idea that, if you are ever able to ski again at the Piccolo San Bernardo, you will have to show your passport, does not fail to dishearten me. But the last word has not yet been said; in any case I am convinced that Italy will still be able to take its place and lead its life. Perhaps you will find my judgment a little influenced by the usual American optimism; but you know that it is corrected by my personal predisposition, which generally does not lean towards optimism. The trouble is also, about optimism or pessimism in personal evaluations of events, that the vision of world events that one can have on this side of the ocean is generally distorted by the correspondents of news agencies, and by newspapers in general, which are often arranged to see and evaluate things with a culture and spirit different from ours. For example, it is certain that there are probably not many things to say about De Nicola; but it made a certain impression on me to read, as one of the most important facts about him, that he has three grandchildren in Buenos Aires.[95] And, although this will probably seem stupid to you, this is precisely one of the things that make the desire to return more acute in me: the desire to review Italian things with Italian eyes. Reviewing your letters to answer you, I notice that Franco too begins by admitting his delay in writing to me: I am therefore in good company. Neither dare he to tell me that preparing his courses takes him much time: what should I say, with four courses that afflict me (and it particularly afflicts me that I held 4 courses, and they pay me only for three;[96] but I do not want to complain because they are good people, and for example now they are paying for my trip to Buenos Aires). The world can go to rack and ruin, as Persico said, but none of us will ever adapt to presenting botched courses. I read with great pleasure the various news that you communicate to me, and I appreciated your answers all the more, because not always are people able to get such precise information. Speaking of libraries [...], I wrote to the United States to find out some information about a central agency which, from what I have vaguely heard, would take care of sending books to war-damaged libraries; I also talked about the works not acquired after 1939 and now very difficult to acquire (mentioning in particular the book by Szegö that is so desired by Tricomi); and as soon as I know

---

[95] Enrico De Nicola (1877–1959) was elected Provisional Head of State by the Constitutional Assembly on 28 June 1946 and held this position from 1 July of the same year to 31 December 1947.

[96] *Matemáticas Superiores (Introducción a la geometría superior), Metodología* and *Geometría analítica.*

something I will let you know.[97] Two or three years ago I had a chat here in Tucumán with Stone, then President of Am. Math. Soc.; and at least at that moment it seemed to me that there were good intentions, albeit vague, for the future.[98] We will now see if these intentions can somehow materialize, or if they are destined to the fate of the Atlantic chart. I am very happy with the idea that the new rooms of our library are larger than the old ones: I hope it will be possible to study in the library, since, according to what they write to me, my house will be very [bad] reduced. And in the library, is there still the expensive carpet that we had my wife buy? The news of the deaths is very bad: I sent my condolences directly to the relatives of some of the late colleagues, including Einaudi, but I ask you, on the occasion, to renew them to everyone.[99] And just today I wrote other condolences to Castelnuovo, for the death of Enriques announced to me by Picone.[100] It's a terrible death toll and cause of standstill, which I beg you to stop immediately. Thank you for information about salaries: Tricomi's 17,500 lire net per month cheers me up by comparing the data with another less favourable figure I received from another source. Could you write to me exactly about my net salary with wife and three children? I suppose that on my return I will have the post; but please if you confirm it. [...] I have not heard anything about Fano, to whom I wrote weeks ago in Turin (at Rabbeno): how did you find him?[101] Special thanks to Franco for what he did with the Principato publishing house: I hope that his denial of the *Algebra* book is not due to the shame of having an ugly child, but only to his friendship for me![102] Here the copy I have is circulating considerably among the students of my course in Methodology who must take trial lessons, on middle school topics. [...] I wish APPU a long life, although I must confess that it is not APPU that makes me especially want to return (excuse me, Tricomi).[103] For the seven years that I have been here, I have heard (or not heard) so many discussions on university problems

---

[97] Tricomi had taken the direction of the Special Mathematics Library of the University of Turin in 1938, at the dismissal of Gino Fano. The library had suffered significant damages in the bombings of 1942. Upon its reopening in 1945 Tricomi asked Terracini and Fano for help to rebuild the library's heritage.

[98] Terracini had taken advantage of the meetings with G. D. Birkhoff and M. H. Stone, who visited Tucumán in the autumn of 1942. He had highlighted the difficulties of those scholars who, living far from research institutions, had to depend on the help of colleagues and friends. Hence the proposal to establish a central agency for bibliographical information under the aegis of the American Mathematical Society. See *Propuesta de institución de un comité central para informaciones bibliográficas matemáticas*, Revista de Matemáticas y Física Teórica 3, 1942, pp. 369–379.

[99] Renato Einaudi had tragically lost his wife Maria Luisa, who died of botulism.

[100] See D7.1.3.

[101] Adele Errera Rabbeno was a close friend of Gino Fano's sisters. Adele's daughter Virginia had married Leone Lattes, who was Gino Fano's father-in-law.

[102] Terracini had published under a false name (i.e. signed by F. G. Tricomi) the book *Algebra elementare ad uso dei Licei*, Messina, Principato, 1940.

[103] Tricomi was president of the Piedmonts' section of the Association of University Professors (*Associazione dei Professori Universitari di ruolo*) from 1945 to 1948.

(discussions generally only abstract and made in view of ideal universities that have nothing to do with those that really exist), that now my desire is to operate in a good university rather than arguing about a better one. But if it pleases Tricomi, I will join APPU (although I don't like the name too much). Persico's inaugural speech was successful not only in Europe, but also in America, and right now our graduate still has its reprint which I loaned him (he is a Trieste-born, named Battig,[104] who lives in Tucumán, but works with Guido Beck, now in Cordoba). [...] The local situation here is unchanged, with respect to what Tricomi mentioned in his letter: the university repercussions so far are perceived to a lesser extent than in past years; however, in some Universities, yes, they are felt, and also in some Faculties of our University. Technically, in our case, we are fortunate that there is a person at the helm with scientific interests, so that on that side we do not have to fear; however, the scientific aspect cannot be isolated from the others. Returning to the first issue, Picone has great confidence in his insistence with the new Minister regarding our return. I very much hope for everything that can come out of this. Keep me up to date with all the news on this issue as well as the rest. If coming back we will have to tighten our belts, so much the better to facilitate digestion! Write to me soon. To both of you, an affectionate hug from Sandro.

My wife doesn't hug you, but she says hello too, and all my guys do the same. And my mom too. Benvenuto and Eva are in Buenos Aires: the former permanently, the latter on a visit. I also greet you on their behalf Sandro

Carissimi amici Tricomi e Persico, non è per grettezza che vi scrivo una lettera in due, e per dimostrarvelo ve ne mando una copia per ciascuno. Il fatto è, invece, che parto domani per Buenos Aires (per una settimana); non voglio ritardare ulteriormente a scrivervi e mi sono ridotto all'ultimo momento, come quando (secondo Franco, e anche secondo la verità dei fatti) andavo a comperare le sigarette proprio nel momento in cui stava per fischiare la locomotiva, e stava per partire il treno con dentro i miei amici e i miei ski. E sì che da parecchio ho in mente di scrivervi. Mi sono giunte a suo tempo le vostre lettere del 17 e 21 marzo rispettivamente, che tuttavia – nonostante le previsioni – hanno impiegato un certo tempo a traversare l'Atlantico, e anche più a uscire dal baule del signore che le ha portate, per essere avviate a Tucumán. E poi ho saputo da Picone (la cui lettera è ancora stata molto più lenta della vostra) della vostra adesione al memoriale che ha fatto presentare al Ministro della P.I.; adesione per cui vi ringrazio entrambi, e vi ringrazio proprio con affetto fraterno; mentre vi prego di essere interpreti dei miei vivi ringraziamenti anche presso i colleghi matematici

---

[104] In December 1945 Persico had held in Turin the inaugural speech of the academic year 1945–1946 entitled *Il nuovo fuoco*. The text had been published in the journal Scientia, 79, 1946, pp. 83–92. Augusto Battig, physicist, would write his PhD thesis (*Aportes teóricos al estudio del efecto Cherenkov*, Universidad Nacional de Tucumán, Instituto de Física, 1951) under the guidance of Guido Beck (1903–1988), a Czech Jewish-born physicist, emigrated to Argentina in 1943.

torinesi ([ad] alcuni dei non torinesi ho scritto direttamente per ringraziare; naturalmente non potei farlo con tutti gli 86! Ho ringraziato un matematico per Facoltà). [...] Nel mio viaggio a Buenos Aires andrò naturalmente all'ambasciata per sentire un po' qualche cosa sul mio ritorno. Mi dice Picone, in una sua recentissima, che contava riattivare le pratiche col nuovo ministero. Ma prima di parlare del nuovo ministero, parliamo un momento del nuovo regime italiano, e mandiamoci il primo saluto repubblicano. Abbiamo seguito da qua con molta trepidazione le elezioni alla Costituente, e gli sviluppi successivi. Mi pare un buon segno: a) che le elezioni abbiano avuto l'effetto che hanno avuto; b) che i successivi tentativi di intorbidare la situazione non si siano risolti in niente, e che le difficoltà si siano superate con molto buon senso, e con una maturità politica da parte degli Italiani, che si poteva anche sospettare non ci sarebbe stata dopo tanti anni di fascismo. Ciò mi induce a molto ottimismo (sia pure in senso relativo, ma in senso assoluto l'ottimismo credo non sia più possibile per le nostre generazioni) sull'avvenire dell'Italia. Veramente le notizie degli ultimissimi giorni sul trattamento fatto all'Italia sono assai poco lusinghiere; e, anche da un punto di vista strettamente personale, l'idea che, se mai si potrà ancora andare in ski al Piccolo San Bernardo, ci sarà da esibire il passaporto, non manca di avvilirmi. Ma non è ancora detta l'ultima parola, e comunque sono convinto che l'Italia potrà ancora prendere il suo posto e fare la sua vita. Forse voi troverete il mio giudizio un poco influenzato dal solito ottimismo americano, ma sapete che esso è corretto dalla mia personale predisposizione, la quale non propende generalmente all'ottimismo. Il guaio è anche, a proposito di ottimismo o pessimismo, e delle valutazioni personali degli avvenimenti, che la visione degli avvenimenti mondiali che si può avere di qua è generalmente deformata dai corrispondenti delle agenzie giornalistiche, e dai giornali in genere, che sono portati spesso a vedere e valutare le cose con preparazione e con animo diversi dai nostri. Per esempio, è certo che su De Nicola probabilmente non ci sono molte cose da dire; ma mi fece una certa impressione leggere, come uno dei dati più importanti su di lui, che ha tre nipoti a Buenos Aires. E, per quanto questo probabilmente vi parrà stupido, è proprio questa una delle cose che rendono più acuto in me il desiderio di tornare costì: desiderio di rivedere cose italiane con occhi italiani. Rivedendo le vostre lettere per rispondervi, osservo che anche Franco comincia la sua esprimendo il suo ritardo a scrivermi: sono dunque in buona compagnia. Né mi dica il medesimo che impiega tanto tempo nella preparazione dei suoi corsi: che cosa dovrei dire io dei ben quattro che mi affliggono (e particolarmente mi affligge il fatto che di corsi ne faccio 4, e mi pagano solo per tre; ma non mi voglio lamentare perché sono brava gente, e per esempio ora mi pagano il viaggio a Buenos Aires). Può andare il mondo alla carlona, come dice Persico, ma lui e tutti noi non ci adatteremo mai a fare dei corsi raffazzonati. Con molto piacere ho lette le varie notizie che mi comunicate, e tanto più ho apprezzate le vostre risposte, perché non sempre da tutti si hanno ragguagli così precisi. A proposito di biblioteche [...] ho scritto negli Stati Uniti per sapere qualche cosa su una centrale che, a quanto ho udito vagamente, si occuperebbe di favorire l'invio di libri a biblioteche danneggiate dalla guerra; ho parlato anche delle opere non acquisite dopo il 1939 e ora molto difficilmente acquisibili (facendo in particolare il nome del libro di Szegö che sta tanto a cuore a Tricomi); e appena saprò qualche cosa ve lo farò sapere. Due o tre anni fa conversai qua in Tucumán con Stone, allora Presidente della Am. Math. Soc.; e almeno in quel momento mi parve che vi fossero delle buone intenzioni, sebbene vaghe, per l'avvenire. Vedremo ora se queste intenzioni in qualche modo potranno concretarsi, o se sono invece destinate alla sorte della carta dell'Atlantico. Mi rallegra moltissimo l'idea che, come scrive Franco, i nuovi locali della nostra biblioteca siano più ampi dei vecchi: spero possa così essere possibile di studiare in biblioteca, dato che, per quanto mi scrivono, la mia casa sarà molto [mal] ridotta. E in biblioteca c'è ancora il ricco tappeto che avevamo fatto comperare da mia moglie? Molto male le notizie mortuarie: ai parenti di alcuni fra gli scomparsi, tra cui Einaudi, ho mandate direttamente le mie condoglianze, ma vi prego, all'occasione, di rinnovarle a tutti. E proprio oggi ho scritto altre condoglianze a Castelnuovo, per la morte di Enriques annunciatami da Picone. È una vera moria, alla quale vi prego di mettere termine immediatamente. Ringrazio quanto mi scrivete degli stipendi: le 17,500 lire mensili di Tricomi, nette, mi rallegrano confrontando il dato con altro meno favorevole in mio potere. Potreste scrivermi esattissimo, com'è il mio stipendio netto con moglie e tre figli? Suppongo che al mio ritorno avrò l'incarico, ma mi farete il piacere di confermarmelo. [...] Non ho più saputo nulla di Fano, al quale ho scritto settimane fa a Torino (presso Rabbeno): come lo avete trovato? Un ringraziamento speciale a Franco per quanto ha fatto con l'editore Principato: spero che il rinnegamento del libro di Algebra da parte sua non sia dovuto alla vergogna di avere un figlio brutto, ma solo alla sua amicizia per me! Qui la copia che ho circola molto fra gli studenti del corso di Metodologia che devono fare delle lezioni di prova, con temi di scuole medie. [...] Auguro lunga vita alla APPU, sebbene devo confessare che non è essa (mi scusi Tricomi) quella che mi fa specialmente desiderare il mio ritorno. Da sette anni che sono qua, ho sentito (o non sentito) tante discussioni sui problemi universitari (discussioni generalmente solo astratte, e fatte in vista di università ideali che non hanno nulla a che fare con quelle realmente esistenti), che ormai il mio desiderio è quello di operare in una università buona piuttosto che di discutere su una migliore. Ma se ciò fa piacere a Tricomi mi assocerò alla APPU (il nome non mi piace poi troppo). Il discorso inaugurale di Persico ha avuto successo non solo in Europa, ma anche in America, e proprio ora ce lo ha ancora un nostro laureato al quale l'ho imprestato (un triestino, di nome Battig, che sta a Tucumán, ma lavora con Guido Beck, ora a Cordoba). [...] La situazione locale qua è come la sapete, e già me ne accennava Tricomi nella sua lettera: le ripercussioni universitarie sinora si sentono in misura più leggera che in anni passati; però in alcune Università sì, e anche in alcune Facoltà della nostra. Tecnicamente, nella nostra, abbiamo la fortuna che c'è a capo una persona di interessi scientifici, cosicché da quel lato non abbiamo da temere; però non si può isolare l'aspetto scientifico dagli altri. Tornando al primo detto, Picone ha una grande fiducia nella sua insistenza col nuovo Ministro a proposito del nostro ritorno. Spero molto vivamente in tutto quanto potrà derivarne. Mantenetemi al corrente di tutte le notizie in merito nonché di tutte le altre. Se al ritorno dovremo stringere la cintola, tanto meglio per facilitare le digestioni! Scrivetemi presto. A entrambi un affettuoso abbraccio da Sandro.

Non vi abbraccia, ma vi saluta anche mia moglie, e tutti i miei ragazzi fanno lo stesso. Idem mia mamma. Benvenuto e Eva sono a Buenos Aires: la prima permanentemente, il secondo in visita. Vi saluto anche a nome loro, Sandro

D7.3.5 G. Castelnuovo to B. Segre, Rome 19.12.1946

CA, *BSP*, fol. 1r-v

Dear Segre, First, thank you for your good wishes, which I cordially reciprocate to you and yours. The first wish is that you find suitable accommodation for your family, which will allow you to devote yourself quietly to your studies. Prof. Fano was ill in Boston, but now he is fine, he wrote me a long letter, and I have frequent news of him from my daughter Gina.[105] His address is 3510 Rodman str., Washington D.C. N.W. The *Memoria* on the cubic variety to be published by the Pontifical Academy has not yet come out; you will have seen the short extract published in the *Rendiconti dei Lincei*.[106] In these days, a young man from here, very clever, has communicated to me a very simple and brief demonstration of the irrationality of the cubic variety based on topological considerations. But I need to think about it again. I am pleased with what you tell me about the honours paid to my dear friend Pincherle.[107] If you inform the President of the Academy of Lincei of the date of the ceremony, we will immediately arrange to have the Academy represented by some member of Bologna, and to send a telegram of tribute. Many cordial greetings from your most affectionate G. Castelnuovo

---

[105] After the war the Fanos visited their children in the United States. They were to spend a little time in Boston with Robert and then come to live with Ugo and his family in Washington. Unfortunately, while in Boston Gino had a heart attack and for several weeks he was not allowed to travel. By early fall, they finally joined Ugo, 'he an invalid and she totally disoriented by a lifestyle she had never seen before' (Fano U. 2000, p. 193).

[106] G. Fano, *Nuove ricerche sulle varietà algebriche a tre dimensioni a curve-sezioni canoniche*, Commentationes. Pontificia Academia Scientarum, 11, 1948, pp. 635–720. Castelnuovo was wrong. The extract had been published in Acta. Pontificia Academia Scientarum, 9, 1945, pp. 163–167.

[107] On 11 March 1947, on the occasion of the unveiling of a bust of Salvatore Pincherle in the Mathematical Institute of the University of Bologna, Segre held a *Discorso commemorativo dell'insigne matematico Salvatore Pincherle*, published in the Annuario dell'Università di Bologna 1946–48, pp. 126–133, and in Rivista di Matematica della Università di Parma, 4, 1953, pp. 3–10.

Caro Segre, Grazie anzitutto dei Suoi auguri che ricambio cordialmente a Lei e ai Suoi. Il primo augurio è di trovare un alloggio conveniente per la Sua famiglia, che Le permetta di dedicarsi tranquillamente ai Suoi studi. Il Prof. Fano è stato malato a Boston, ma ora sta benino, mi ha scritto egli stesso una lunga lettera, e di lui ho frequenti notizie dalla mia figlia Gina. Il suo indirizzo è 3510 Rodman str., Washington D.C. N.W. La Memoria *sulla varietà cubica che deve esser pubblicata dall'Ac. Pontificia non è ancora uscita; avrà visto il breve estratto pubblicato nei* Rendiconti dei Lincei. *In questi giorni un giovane di qua, molto intelligente, mi ha comunicato una dimostrazione molto semplice e breve della irrazionalità della varietà cubica fondata su considerazioni topologiche. Ma ho bisogno di pensare ancora alla cosa. Mi fa piacere quanto mi scrive sulle onoranze al mio caro amico Pincherle. Se Ella fa scrivere al Presidente dell'Acc. dei L. una lettera con cui si comunichi la data della cerimonia, provvederemo subito a far rappresentare l'Accademia da qualche socio di Bologna, e a mandare un telegramma di omaggio. Molti cordiali saluti dal suo aff.mo G. Castelnuovo*

D7.3.6 B. Segre to I. J. Ursell [SPSL], Bologna 2.11.1947

SPSL, *BS*, fol. 221
Dear Miss Ursell,[108] I apologize for not having written before. I have pleasant recollections of the seven years I have passed in England and always remember the good friends I have left there. When I came to Italy, during the summer of the last year, I intended to pay a visit to this Country and examine at the same time the possibility of returning here for good with my family. I decided then to do so for several reasons. The principal was that I had been offered to be restored to my ancient position, and so I thought I could help in the reconstruction of this unfortunate Country and do my work better in such a leading capacity. This was also in agreement with the suggestions I had from Professor L. J. Mordell; on the other hand, Professor M.H.A. Newmann told me before I left England that I could have kept my job in Manchester only for at most another year.[109] I found the situation in Italy very dark, worsened by great difficulties both in the material and the moral field. The people were irregularly fed and disconcerted, the housing problem was terrific, and the communications almost disrupted. However, I was glad to see everywhere a hard will to work and forget the past madness, and to start a new life. Now in fact things are already getting better in many ways; the trains are running a little slowly but almost regularly, and many public buildings have already been reconstructed (including hundreds of bridges and railway

---

[108] Ilse J. Ursell, the post-war secretary of the SPSL.

[109] Louis Mordell (1888–1972) had passed from Manchester to Cambridge in 1945. Max Newmann (1897–1984) had replaced Mordell as Fielden Professor of Mathematics at Manchester.

stations, and most of Bologna University). Academical life is also gradually improving. The number of university students is now increased enormously (there are about 20,000 of them only in Bologna!), and the means at our disposal are very scanty; but we are doing our best, and not without success, to counteract the difficulties. I have already written the first volume of a book on Geometry (*Lezioni di Geometria Moderna*) which ought to be out shortly,[110] and many papers on different subjects. The latter have already appeared or will soon appear in the *Annali di Matematica*, the *Rendiconti Accademia dei Lincei* and the *Bollettino dell'Unione Matematica Italiana*, which have resumed publication. I should be very glad to have the opportunity of presenting some mathematical papers by English Scholars for publication in any of these periodicals, and, more generally, of cooperating in any scheme for restoring and increasing the understanding and intercourse between English and Italian mathematicians. For this purpose, I have written a report dealing with the development of Geometry in England and America during the war (already published by the *Pontificia Academia*) and delivered a lecture on the Schools in England (shortly to be issued as a pamphlet, of which I shall send you a copy).[111] With grateful remembrance of the Society, Yours sincerely, Beniamino Segre

D7.3.7 I.J. Ursell [SPSL] to B. Segre, Cambridge 7.11.1947

SPSL, *BS*, fol. 222

Dear Professor Segre, I was very glad to have your letter of November 2nd with the full report concerning your final decision to return to Italy, and your work since your return. I expect Professor Mordell would have mentioned your return to me if he had known that I might have been interested in the information. As you know he is now in Cambridge and lives in fact two houses away from me.[112] I have during the last few months had various reports about Italy, and three weeks ago listened to a lecture given by Dr. Pettoello at the International Club House here.[113] It is good to know that such determined efforts are being made towards reconstruction, so that in due course life will return to normal or at least as near normal as is possible in our disturbed world. I am very glad to know that you are

---

[110] B. Segre, *Lezioni di Geometria Moderna Vol. 1*, Bologna, Zanichelli, 1948.

[111] B. Segre, *Geometria algebrica nei Paesi anglo-sassoni*, Relationes de auctis scientiis tempore belli, Pontificia Academia Scientiarum, 1946, pp. 3–51; *La Scuola in Inghilterra*, Il Filomate, 1, 1947, pp. 53–58.

[112] Louis Mordell had passed from Manchester to Cambridge in 1945. Segre had informed Mordell about his decision in August 1946. See CA, *BSP*: L. Mordell to B. Segre, Cambridge 16.8.1946.

[113] Decio Egberto Saadi Pettoello (1886–1984), graduated in Philosophy at the University of Turin in 1919. In 1922 he had emigrated to Cambridge where he taught at the University. He was a member of the anti-fascist exile association Free Italy Committee.

doing something to acquaint the Italian University world with wartime develop-ments in England. This is one of the best ways to further the understanding between two countries, and I wish you the best of success. I am looking forward to having a copy of your pamphlet and will make an attempt to brush up my Italian by reading it. I must admit however that my knowledge of it is extremely slight. Yours sincerely, Ilse J. Ursell - Secretary

D7.3.8 B. Segre to A. Terracini, Bologna 16.3.1948

BSM-*Terracini*, fol. 1r-v
Dear Sandro, I thank you for your kind letter and for your decision not to justify the rejection with the report of that meeting. I read it with amazement and regret; but I was told that it was drafted along these lines by Ettore Bortolotti, without it fully reflecting what had been said, and that someone (Amaldi for ex.) already protested at the time. However, I was assured that the presidency of the U.M.I. will take the opportunity of the Congress in Pisa to make an act of resipiscence.[114] I know that your refusal is here the subject of regret, and I think Berzolari will write to you shortly, trying to convince you to rethink your decision.[115] It is my opinion that it would be appropriate for you to do so, also considering what I have written to you before. I saw you again with great pleasure, and I hope that now we can meet more frequently. It is difficult for me to return to Turin for Easter because in April I have to go to Rome and Trieste for conferences, then again in May to Rome for the Geometry competition (Catania chair); all this keeps me very busy. Also, on behalf of my wife, I send you all the best Easter wishes. To you, an affectionate hug, yours, Beniamino

Carissimo Sandro, ti ringrazio della lettera gentile e della tua decisione di non motivare il rifiuto colla relazione di quella seduta dell'U.M.I. Ho letto

---

[114]Terracini had been invited to hold one of the plenary lectures at the UMI Congress in Pisa (23–26 September 1948). He initially refused. Terracini had in fact read in the *Bolletino dell'UMI* the report of the meeting of 10 December 1938 in which the Scientific Commission not only had not raised any doubts or perplexities about the racial laws but had reiterated that 'The Italian mathe-matical School, which has acquired widespread fame throughout the scientific world, is almost entirely the creation of scientists of the Italic (Aryan) race'. As secretary, Ettore Bortolotti had drawn up the minutes of the meeting. Ugo Amaldi did not attend it because he did not sit in the Scientific Commission. No documents have been found in the UMI Archive that demonstrate protests against this declaration. From the official speeches there emerges no 'act of resipiscence' as had been promised. L. Berzolari did not participate in the Congress for health reasons, but sent a message, read by E. Bompiani, where he only spoke of the resumption of international relations. Bompiani, in his talk as vice president, quickly and generically mentioned the 'sad period passed' (*triste periodo passato*) and the 'shipwreck overcome' (*naufragio superato*). See Giacardi and Tazzioli 2021.

[115]D7.3.9.

quest'ultima con stupore e dispiacere; ma mi si è detto ch'essa fu compilata così a suo tempo da Ettore Bortolotti, senza ch'essa rispecchiasse appieno ciò ch'era stato detto in quella seduta, e che qualcuno (come Amaldi) già allora aveva protestato. Comunque, mi si è assicurato che la presidenza dell'U.M.I. coglierà l'occasione del Congresso di Pisa per fare atto di resipiscenza. So che qui il tuo rifiuto ha fatto dispiacere, e ritengo che Berzolari ti scriverà fra breve cercando di farti ritornare sulla tua decisione. È mia opinione che sarebbe opportuno se tu volessi farlo, tenuto anche conto di quanto dianzi ti ho scritto. Ti ho riveduto con vivissimo piacere, e spero che ormai ci si potrà trovare con una certa frequenza. È difficile che torni a Torino per Pasqua poiché in aprile dovrò andare a Roma ed a Trieste per conferenze, e poi di nuovo in maggio a Roma per il concorso di Geometria per Catania, e tutto ciò mi tiene assai occupato. Anche a nome di mia moglie anticipo a voi tutti i migliori auguri pasquali. A te un abbraccio affettuoso dal tuo Beniamino

D7.3.9 L. Berzolari to A. Terracini, Pavia 18.3.1948

BSM-*Terracini*, fol. 1r
Dear Professor, Prof. Villa, the secretary of U.M.I., informs me that you have declined the invitation to give a plenary lecture at the third Congress of U.M.I. in Pisa.[116] This pains me. At the inauguration of the Congress, the colleagues who return to Italy after a long, cruel and unjust exile will be explicitly welcomed, thus restoring the validity of their work in favour of Italian mathematics; the invitation addressed to you by the Scientific Commission also implied such recognition; for this reason, allow me to tell you that your refusal would give me great regret. I therefore renew to you my prayer that you will give the plenary lecture; of course, you have ample freedom to deal with the subject that you most like. With cordial greetings, I am your most devoted and affectionate

Caro Professore, Il prof. Villa segretario dell'U.M.I., mi comunica che Ella ha declinato l'incarico di tenere un discorso generale al terzo Congresso dell'U.M. I. a Pisa, e ciò mi addolora. All'inaugurazione del Congresso verrà dato esplicitamente il benvenuto ai colleghi rientrati in Italia dopo un lungo, crudele e ingiusto esilio, ridonando così la loro valida opera a favore della matematica

---

[116]Mario Villa (1907–1973) professor of Analytic and Projective Geometry at the University of Bologna from 1940, was the secretary of the UMI from 1945 to 1961. In the end Terracini accepted and decided to hold a plenary lecture of projective geometry, focused on Guido Fubini, as an homage to a victim of the racial laws and a friend who passed away before he could see the end of the Second World War: A. Terracini, *Guido Fubini e la geometria proiettiva differenziale*, in *Atti del terzo Congresso dell'Unione Matematica Italiana, tenuto in Pisa nei giorni 23-26 settembre 1948*, Roma, Cremonese, 1951, pp. 41–44 and in Rendiconti del Seminario Matematico. Università e Politecnico di Torino, 9, 1949–50, pp. 97–123.

italiana, e l'invito rivoltole dalla Commissione Scientifica aveva anche significato di un tale riconoscimento; per questa ragione mi consenta di dirLe che la Sua non accettazione mi darebbe gran dispiacere. Le rinnovo pertanto la preghiera che Ella si sobbarchi a tenere il discorso, per il quale, beninteso, Ella ha ampia libertà di trattare l'argomento che più Le sarà gradito. Con cordiali saluti, mi abbia suo dev.^{mo} e aff.^{mo} L. Berzolari

D7.3.10 G. Fano to B. Segre, New York 21.3.1949

CA, *BSP*, fol. 1r

Dear Prof. Segre, thank you for your kind letter of 15 February, which arrived a few days ago. I am surprised to hear that you wrote to me in Mantua without success; I am always informed of the letters that arrive there in my absence, and when necessary, they are forwarded to me. To tell the truth, I still have to thank you for your volume of *Geometria moderna* that I saw with interest; it came to me while I was still in Italy, in October I think, but on the verge of returning here; and this explains my silence at that time.[117] Perhaps the letter to which you refer was addressed to me at the same time? Your Note in the U.M.I. Bulletin on the maximum number of nodes of a S^n hypersurface, did not have any follow-up by Severi?[118] You and Terracini have made yourselves truly commendable representatives of the country, resuming your university chairs. Castelnuovo, in a letter I had this morning, complains about the continuous departure from Italy of good young people, and the consequent lowering of the level of Italian Universities. And among the less young, in Turin, Tricomi and Persico are missing; in Rome, Bompiani is always away; Bernardini (now there … for a few months); Severi celebrates his 70th birthday; Sansone I don't know if he will arrive at this age.[119] I plan to be back in Italy at the end of May; in Mantua in June, except for the days of the Lincei meetings; then in Colognola ai Colli (Verona). So, I hope to see you. In theory, I also intend to attend next year's International Congress.[120] With the most cordial greetings, yours aff. Gino Fano

---

[117] B. Segre, *Lezioni di Geometria Moderna Vol. 1*, Bologna, Zanichelli, 1948.

[118] B. Segre, *Sul massimo numero di nodi delle superficie di dato ordine*, Bollettino dell'UMI, s. 3, 2, 1947, pp. 12–16.

[119] From 1948 to 1951 Francesco Tricomi worked at the CALTECH in Pasadena, collaborating with A. Erdélyi, W. Magnus, and F. Oberhettinger on the Bateman Manuscript Project, which aimed to publish a text on special functions and classical functional transformations. Discouraged by the post-war conditions, Enrico Persico had decided, in 1947, to move to the Laval Physics Institute in Quebec. Gilberto Bernardini (1871–1933), professor of Spectroscopy and Experimental Physics at the University of Rome, was invited as visiting professor at Columbia University in 1949. Severi was born in 1879; Giovanni Sansone in 1888.

[120] The International Congress of Mathematicians will be held in Cambridge, Massachusetts, from 30 August to 6 September 1950.

Carissimo Prof. Segre, La ringrazio della sua gentile lettera del 15 febbraio, giuntami qualche giorno fa. Mi sorprendo che lei mi abbia scritto a Mantova senza successo; perché delle lettere che giungono là in mia assenza io sono sempre informato, e occorrendo mi vengono inoltrate. A dir vero, devo ancora ringraziarla del suo volume di Geometria moderna *che ho veduto con interesse; mi è giunto mentre ero ancora in Italia, in ottobre mi pare, ma sul punto di tornare qui; e questo le spieghi il mio silenzio di allora. Forse la lettera cui lei accenna mi era stata diretta contemporaneamente? La sua Nota sul Bollettino dell'UMI sul massimo numero di nodi di una ipersuperficie di uno spazio qualsiasi, non ha avuto alcun seguito da parte di Severi? Lei e Terracini si sono resi veramente benemeriti del Paese, riprendendo le loro cattedre universitarie. Castelnuovo, in una lettera che ho avuta stamane, lamenta la continua partenza dall'Italia di buoni giovani, e il conseguente abbassamento di livello delle Università Italiane. E dei meno giovani, a Torino, mancano Tricomi e Persico; a Roma, Bompiani, via tutti i momenti; Bernardini (ora lì ... per qualche mese); Severi compie i 70; Sansone non so ancora se vi andrà* [sic!]. *Io conto di essere di nuovo in Italia a fine maggio; a Mantova per giugno, salvo i giorni delle sedute Lincee; poi a Colognola ai Colli (Verona). Spero quindi vederla. In massima, ho anche intenzione di prender parte al Congresso internazionale dell'anno prossimo. Coi più cordiali saluti, suo aff. Gino Fano*

Part II
Individuals

October 1939 - Arrival of the Terracini family in Tucumán, Argentina, Terracini Coll.

Chapter 8
'An Illustrious Migrant': Guido Fubini in Princeton

In 1968, Laura Capon (wife of Nobel Prize winner for Physics Enrico Fermi) published the book *Illustrious migrants*, dedicated to intellectuals who had shared the experience of emigration for political and racial reasons with herself and her husband. The only mathematician mentioned in the volume was Guido Fubini.

Born in Venice on 19 January 1879, and trained at the *Scuola Normale Superiore* in Pisa, Fubini was truly one of the greatest Italian mathematicians of his generation. Analyst, projective-differential geometer, and applied mathematician, Fubini had a multifaceted production for which he obtained the Royal Prize for Mathematics in 1919.[1] Among other things, inspired by his teacher Bianchi, he was among the very few Italian specialists in algebra and number theory. Assigned the teaching post of Higher Analysis in Catania in 1903, he won the chair of Mathematical Analysis in Genoa in 1906, and two years later arrived in Turin. From 1908 to 1938, he was a full professor of Infinitesimal Analysis at the Polytechnic and a tenured professor (*incaricato*) of Higher Analysis at the University. In the same year as his move to Turin, Fubini married Annetta Fubini-Ghiron (1892–1973), who gave him two sons: Gino (1911–1965) and Eugenio (1913–1997).

Refused due to his height, he did not take part in the Great War, but his strong feelings of Italianness and patriotism were unquestionable. During the 20 years of fascism, Fubini maintained a low profile: he took the oath of allegiance to the party in 1931, he joined the Turin fascist session with his children in 1933, he made donations to some fascist organisations, and he accepted—with ill-concealed annoyance—the impositions of the regime: wearing the pin and black shirt at the faculty meetings and on official occasions, inserting the fascist year reference at the bottom of letters, documents, and publications, Italian-ising the names, calling Lagrange 'Lagrangia' in one of his conferences in Turin. Friends were to remember his impatience with these meaningless rituals:

---

[1] On Fubini mathematical work, see Terracini A. (1944), Picone (1946), and Segre (1954). See also Nastasi (1993).

E. Luciano, *The Jewish Mathematical Diaspora from Fascist Italy*, Science Networks. Historical Studies 64, https://doi.org/10.1007/978-3-031-64896-0_8

he could not feel at ease under fascism, whose intolerance and empty prosopopoeia affected him, and whose catastrophe he clearly anticipated. He thus closed more and more in himself, dividing his life between family, study and teaching.[2]

As for anti-Semitism, Fubini had a vision clearer than that of several of his colleagues. Perhaps because of his contacts with German mathematicians and the relationship of scientific collaboration and friendship with the Czechoslovak mathematician Eduard Čech, from the mid 20s, Fubini understood that Italian high culture was not immune from the academic anti-Semitism that was breathing in Central and Eastern European institutions: his election as a fellow of the Academy of Italy was boycotted while his applications to become a member of Lincei and the Academy of Sciences of Turin encountered various difficulties.[3] In early 1938, the CNR refused to publish a monograph on the Laplace transform, written by Fubini and his assistant A. Ghizzetti 'because one of the authors was Jewish'.[4] Fubini, however, did not pay much attention to these episodes, declassifying them as acts of intra-academia rivalry. At the start of the summer holidays of 1938, he did not mention any anxieties or worries for the future; rather, he spoke with Colonnetti and Levi-Civita about the book he was writing with G. Albenga, and the strenuous autumn reopening that awaited him.

At the time of the Italian racist turn, Fubini was (with Fano, Levi-Civita, and Enriques) among the most internationally renowned Italian mathematicians for his results in analysis, projective and differential geometry, mathematical engineering, and elasticity theory. He represented the excellence of the Italian mathematical tradition, 'the most penetrating and ingenious living Italian analyst after Volterra',[5] as Levi-Civita wrote to Veblen. He had to his credit hundreds of publications, in all the leading European journals, and had travelled a lot, for work and leisure, so much so that in the family they jokingly nicknamed him 'the wandering Jew'.[6] However, he had never been away for long periods and had limited contacts with the Anglo-Saxon world; in particular, he had no direct contact with American mathematicians,

---

[2] Segre (1954), p. 287: *non poteva conseguentemente trovarsi bene sotto il fascismo, di cui risentiva l'intolleranza e la vuota prosopopea, e di cui antivedeva con chiarezza la catastrofe. Si rinchiuse quindi sempre più in sé, dividendo la Sua vita tra la famiglia, lo studio e l'insegnamento.*

[3] See, for example, G. Fubini to T. Levi-Civita, [Turin] 4.6.1932 in Nastasi and Tazzioli (2003), p. 125: 'Thank you for the affectionate care with which you wanted to participate in the Lincei vote, which naturally pleased me very much. As for Turin, as I have been told, it was only anti-Semitism, which easily triumphed because 4/5 of the votes are required for the appointment'. (*Grazie della affettuosa premura con cui hai voluto partecriparmi la votazione lincea, che mi ha naturalmente fatto piacere assai. Quanto a Torino, a quanto mi è stato riferito, si trattava solo di antisemitismo; il quale facilmente trionfava perché per la nomina si richiedono i 4/5 dei voti*).

[4] Ghizzetti (1982), p. 19: *perché uno degli Autori era ebreo.*

[5] FUBINI.1.

[6] G. Fubini to T. Levi-Civita, Turin 8.1.1935, in Nastasi and Tazzioli (2003), p. 127: *l'ebreo errante.*

with whom he had only exchanged offprints. He had no command of English; indeed, he envied Levi-Civita who was able to write fluently in this language.[7]

8.1   The Complicated Pre-departure Preparation Activities

It is not clear whether, at the enactment of the racial laws, Fubini had already returned from his holidays in Turin, but certainly on 23 October, he was in Paris, with his wife and children, and he had settled at the Hotel d'Albany at 202 Rue de Rivoli. He would never see Turin again.

When he received the letter of dismissal from service, Fubini had in fact already decided to leave Italy, to the point that he applied for counter-discrimination for scientific merits (he would obtain it in March 1940) but had so little confidence in this practice as to scribble barely a few lines.[8] The effect of racial discrimination, however, was a shame that he could not accept. Fubini was therefore one of the earliest refugees, the first to leave the country.

Persecution took away all his energy and prevented him from concentrating on research. At 59, Fubini felt himself to be a finished man (Figs. 8.1 and 8.2).

This is the reason why all the attempts to seek accommodation in the States did not concern himself personally, but rather his children. It was only for them that, in two letters written five days apart, Fubini asked Levi-Civita for help:

> I am here provisionally to try, with the help of friends and relatives residing here, to find a way for my children to work and lay down a new path in their life. Some people advise me to think about the United States. Unfortunately, this is not very simple. It would be, I am told, greatly simpler, if I received any call for employment whatsoever over there; in that case, I was told that I could easily obtain a residence permit for myself and my family. I thought I'd

---

[7]G. Fubini a T. Levi Civita, Paris 1.11.1938, in Nastasi and Tazzioli (2003), p. 144. Seven months after settling in Princeton, Fubini would still ask the Institute's secretary: 'Will you be so kind as to verify whether I have written without mistakes? Am I a "fellow" of the Institute or not? Must I change this word "fellow"?' (IAS, *GF*: G. Fubini to G. Blake, [Princeton] 27.10.1939).

[8]See ASPoliTo-*Fubini*: G. Vallauri [Director of the R. Polytechnic of Turin] to G. Bottai [MEN], Turin 26.10.1938: 'Your Excellency, in connection with the cable dated 18 this month, which I immediately communicated to teachers of the Jewish race, the enclosed letter from Prof. Guido Fubini-Ghiron arrived, which I dutifully transmit to this Superior Dicastery (D2.6.2). I must report that the scientific value of Prof. Fubini has so far made him one of the illustrious living mathematicians at an international level; in fulfilling his teaching duties, he has always shown great commitment, perfect discipline, and full understanding of the special needs of mathematics education for students in engineering, thus achieving particular didactic effectiveness'. (*Eccellenza, in relazione con la circolare telegrafica in data 18 corrente mese, che ho immediatamente comunicata ai docenti di razza ebrea, mi perviene l'acclusa lettera del prof. Guido Fubini-Ghiron, che trasmetto doverosamente a codesto Superiore Dicastero. Debbo riferire che il valore scientifico del prof. Fubini lo ha fatto finora considerare in sede internazionale come uno fra gl'illustri matematici viventi e che nell'adempimento dei doveri scolastici egli ha sempre dato prova di grande impegno, di perfetta disciplina e di piena comprensione delle speciali esigenze dell'insegnamento delle matematiche per gli allievi ingegneri, raggiungendo così particolare efficacia didattica.) See also D2.6.2 and D2.6.4.

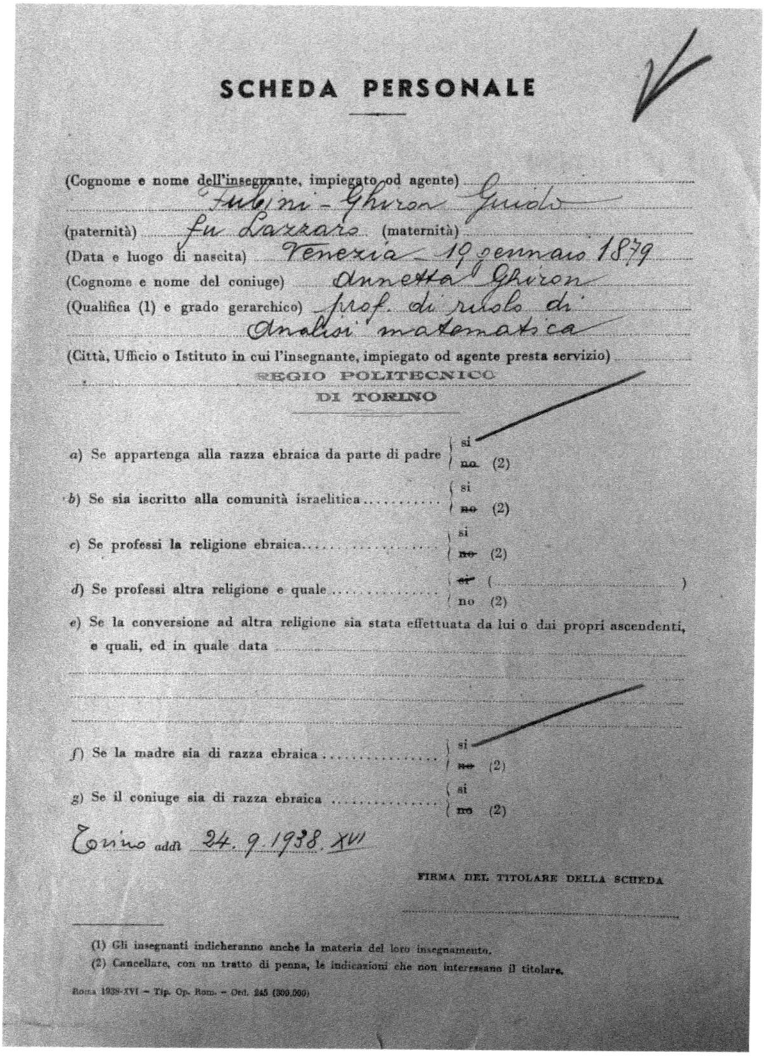

SCHEDA PERSONALE

(Cognome e nome dell'insegnante, impiegato od agente) Fubini - Ghiron Guido

(paternità) fu Lazzaro (maternità)

(Data e luogo di nascita) Venezia - 19 gennaio 1879

(Cognome e nome del coniuge) Annetta Ghiron

(Qualifica (1) e grado gerarchico) prof. di ruolo di Analisi matematica

(Città, Ufficio o Istituto in cui l'insegnante, impiegato od agente presta servizio)

REGIO POLITECNICO DI TORINO

a) Se appartenga alla razza ebraica da parte di padre { si / ~~no~~ (2)

b) Se sia iscritto alla comunità israelitica .... { si / ~~no~~ (2)

c) Se professi la religione ebraica .... { si / ~~no~~ (2)

d) Se professi altra religione e quale .... { ~~si~~ (..........) / no (2)

e) Se la conversione ad altra religione sia stata effettuata da lui o dai propri ascendenti, e quali, ed in quale data

f) Se la madre sia di razza ebraica .... { si / ~~no~~ (2)

g) Se il coniuge sia di razza ebraica .... { si / ~~no~~ (2)

Torino addì 24.9.1938.XVI

FIRMA DEL TITOLARE DELLA SCHEDA

(1) Gli insegnanti indicheranno anche la materia del loro insegnamento.
(2) Cancellare, con un tratto di penna, le indicazioni che non interessano il titolare.

Roma 1938-XVI - Tip. Op. Rom. - Ord. 245 (300.000)

Fig. 8.1 Racial census form compiled by G. Fubini, ACS-*Razzismo* 1938

try; since, as I am told, Birkhoff is omnipotent there, I would like to ask you to write to him and to interest him in my case. I am sure that you will forgive me for the hassle, because a father, for his children, must leave no stone unturned.[9]

[9]G. Fubini to T. Levi-Civita, Paris 23.10.1938, in Nastasi and Tazzioli (2003), p. 142: *Sono qui provvisoriamente per cercare, con l'aiuto di amici e di congiunti qui residenti, di trovare una via affinché i miei figlioli possano lavorare ed aprirsi una nuova strada nella vita. Alcuni mi consigliano di pensare agli Stati Uniti. La cosa non è purtroppo molto semplice. Essa sarebbe,*

Fig. 8.2 Guido Fubini in 1939, Fubini Coll.

Personally, I consider myself out of the question: my life is now in its sunset. What I did, I did. I lack the serenity necessary for new studies, even if only to continue the research started on the science of construction. But I have a duty to think of my children, to try to make it easier for them to start a new life. I would be glad to go to the United States only if I were allowed to take my wife and children there: I would not go there if I had to go alone, or if I had a stay limited to six months or so. With Veblen I had only an exchange of papers, with Birkhoff I never had any relations. Since I do not know personally either one or the other, I dare not write to them. You know the environment better than I do; if you can write to someone (whomever you think is more appropriate) to facilitate the achievement of my purpose, I would be very grateful. I don't know what to decide. If I could find other ways (which for now are completely nebulous) to help my children, then I would give up everything, for myself. Could you give me some advice?[10]

---

mi si dice, di gran lunga semplificata, se io ottenessi una qualsiasi chiamata laggiù; mi si dice che in tal caso potrei facilmente ottenere il permesso di soggiorno per me e per i miei. Ho pensato di tentare; e, siccome, a quanto mi si racconta, là Birkhoff è onnipotente, ti vorrei pregare di scriverne a lui e di interessarlo al mio caso. Sono certo che tu mi perdonerai la seccatura, perché un papà, per i suoi figlioli, non deve lasciare nulla di intentato.

[10] G. Fubini to T. Levi-Civita, Paris 28.10.1938, in Nastasi and Tazzioli (2003), p. 143: *Personalmente io mi considero fuori questione: la mia vita volge al tramonto. Quello che ho fatto, ho fatto. Mi manca la serenità necessaria per nuovi studii, anche soltanto per continuare le ricerche iniziate sulla scienza delle costruzioni. Io ho il dovere di pensare ai miei figli, di cercare di facilitare loro l'apertura di una nuova strada. Andrei volentieri negli Stati Uniti, soltanto se mi fosse concesso di portarvi mia moglie e i miei figli: non ci andrei se dovessi andar solo, oppure se avessi un soggiorno limitato a sei mesi o poco più. Col Veblen ho avuto soltanto scambio di memorie, col Birkhoff non ho mai avuto rapporti. Non conoscendo personalmente né l'uno né l'altro, non oso scrivere loro. Tu conosci l'ambiente meglio di me; se puoi scrivere a qualcuno*

The communication with Levi-Civita was the first and only step that Fubini took: there was no curriculum, no request to the SPSL or ECADFS, and no contact with foreign mathematicians. His personal file was compiled by Veblen's secretary, based on information received from Levi-Civita. Only upon his arrival at Princeton did Fubini produce the following document:

> Curriculum vitae of the Prof. Guido Fubini
> > Prof. in the Universities of *Catania* and *Genova* from 1903
> > Prof. in the Polytechnykum [sic!] and at the University of Turin from 1908
> > Member of the *R. Accademia dei Lincei*, of the Society of XL, of the *Accademia di Torino*, di *Bologna*, of the *Instituto Lombardo*, of the mathem. Societies of *Czech-Slovaquia* [sic!], of *Russian*, of *Madrid* etc
> > > Royal prize of mathematics
> > > Golden medal for mathematics.
> > > Fubini has studied:
> > > 1) the geometrical applications of the theory of the continuous groups of Lie,
> > > 2) the theory of automorphic functions and discontinuous groups
> > > 3) calculus of variations and principle of *minimum*
> > > 4) geometry project-differential [sic!]
> > > 5) differential and integral equations
> > > 6) researches on the problems of technical sciences.[11]

Fubini was spared the process of making requests for help to foreign mathematicians and applications to rescue agencies because Levi-Civita moved with extraordinary speed and incisiveness. Two days after receiving Fubini's plea for help, he wrote Veblen a fervent recommendation letter, to bring the case to his attention.[12] Fubini's was the first testimonial that Levi-Civita produced for an Italian scholar. Levi-Civita was aware that, in the United States, few posts remained for European mathematicians, but he was confident that, for men with Fubini's international reputation, there would be no shortage of opportunities. In addition to his talent, Fubini was in fact a 'man of profound intelligence and ready wit, who was celebrated among Italian scholars as a model of a clear and brilliant expositor'. Veblen immediately took action and asked A. Flexner, Director of the Institute for Advanced Study, to offer Fubini a fellowship at the Institute.[13] On December 1, Flexner informed Levi-Civita that the School of Mathematics had unanimously invited Fubini as a temporary member with a stipend for the second semester of the 1938–1939 academic year.[14] Veblen informed Fubini of the decision and his file was closed.[15]

(a chi ti sembrerà più opportuno) per facilitare il raggiungimento del mio scopo, io te ne sarò molto grato. Io non so che cosa risolvere. Se trovassi altre vie (che per ora sono del tutto nebulose) per aiutare i miei figli, allora io rinuncerei, per me, ad ogni cosa. Potresti tu darmi qualche consiglio?

[11] IAS, *GF*: Guido Fubini's typewritten curriculum vitae, attached to the letter from T. Levi-Civita to O. Veblen, Rome 30.10.1938.

[12] FUBINI.1. See also D3.11.1 and D3.11.2.

[13] FUBINI.2.

[14] IAS, *GF*: A. Flexner to T. Levi-Civita, [Princeton] 1.12.1938.

[15] FUBINI.3.

From that moment on, the spasmodic search began at the American consulate for the documents necessary to obtain the extension of passports and entry visas for the United States.[16] The bureaucracy was slow and cumbersome, and the regime had not defined clear rules for the issuance of these documents. No one, either in the administrative offices of the universities or in the consulates, knew what papers were to be presented. Could the invitation from Princeton, for example, function as an *affidavit*? To enter the non-quota immigrants' category, Fubini had to present a non-legalised copy and a legalised copy of a document proving that he had been a professor in an Italian university for the past two years. In addition, he also had to prepare documents regarding the pension. Here, too, the set of documents to be retrieved included hard-to-find papers such as the birth certificate, the military reform certificate, and the marriage certificate. Fubini had no intention of returning to Turin, not even for the short time necessary to complete these requirements, which is why he turned to colleagues and friends such as G. Vallauri, G. Albenga, and E. Perucca, as well as to employees of the Polytechnic secretariat, ready to pay 'not only legal expenses, but also travel (and ancillary) expenses' for anyone who helped him.[17]

The inefficiency, inflexibility, and dishonesty of certain pen-pushers which the Fubinis had to tolerate were countless. Suffice it to mention the story of the 'state of service' of Guido and Eugenio. The certificates attesting to the service provided at the Polytechnic of Turin were immediately prepared, but the Dean was in Rome and could not sign them. When they were finally signed, Secretary N. Vigna found that only one copy had been legalised, as requested by the interested party. In the meantime, though, rules had changed, and the legalisation of both copies was now necessary. Vigna trusted that the certificates would arrive on time, but Fubini paced up and down in anxiety as he needed to arrive in Princeton in mid-January.[18] Legalisation was a ministerial act, which is why it took another week for the state of service document to be sent and returned from Rome in a legalised version. Ten days later, rules for emigration had changed once again and Vigna relayed the following information:

> The Ministry, to which the certificate requested by you for legalisation had been sent, asserted that it is not up to the School to issue this document, and replaced it with another – which I attach to you – issued by the Ministry itself and therefore not legalised. If such legalisation is necessary, I think that the only competent institution be the Ministry of Foreign Affairs through the embassy or consulate. In any case, I send you a copy of the last annual report of the Polytechnic from which it appears that you were a full professor and representative of the professors in the Board of Trustees. As for the birth certificate, the conscription certificate, and the copy of the marriage certificate for the purposes of

---

[16] See D3.4.3, Fubini.4 and Fubini.5.

[17] ASPoliTo-*Fubini*: G. Fubini to N. Vigna, Paris [Dicember 1938]: *non solo le spese legali, ma anche le spese di viaggio (e accessorie).*

[18] ASPoliTo-*Fubini*: G. Fubini to N. Vigna, Paris 16.12.1938.

retirement, they were already sent to the Ministry in 1935 and I believe that no more are needed.[19]

At the beginning of January 1939, Fubini was exhausted by the inertia that proved rampant in the offices and, since he understood that it was not a very easy matter to obtain at short notice from the Italian offices all the documents required without resorting to some personal high-ranking contact, he turned to Flexner and Veblen:

> I ask you if you will kindly intervene at the American consulate in Paris in order to facilitate the granting of the visa. Prof. Lefschetz, who is now in Paris and who will go tomorrow with me to the consulate, has advised me to write to you. I hope you will forgive me for troubling you.[20]

To help him, Flexner wrote to H.R. Marraro, Director of the Italian Interuniversity Bureau and through him arrived at the Italian consul general in New York, who asked the Italian authorities in Paris to expedite the situation of the Fubini family.[21] The odyssey, however, had not come to an end. By mistake, Flexner's letter was mailed to the Italian consul G. Vecchiotti, instead of Marraro.[22] Without receiving an answer, Flexner solicited Marraro to speed up the processing of Fubini's passport.[23] Other precious time got lost.

In mid-January, the passport validation procedure was still not complete and Fubini was still going crazy to collect the documents he needed:

> Among those which (neither I nor the consulate can understand the mystery) do not arrive is my marriage certificate. To obtain it sooner, I ask you to lend me the certificate that is in my file in the office: I will make the authenticated copies here and send it back to Turin. You would be doing me a great favour.[24]

---

[19] ASPoliTo-*Fubini*: N. Vigna to G. Fubini, Turin 29.12.1938: *Il Ministero, al quale era stato inviato il certificato da Lei richiestoci per la legalizzazione, asserisce che non spetta alla Scuola rilasciare tale documento e lo sostituì con altro - che accluso Le rimetto - del Ministero stesso e quindi non legalizzato. Qualora proprio tale legalizzazione Le fosse necessaria, penso che unico organo competente dovrebbe essere il Ministero degli Esteri pel tramite dell'Ambasciata o del Consolato. Ad ogni modo Le invio a parte copia dell'annuario ultimo del Politecnico dal quale risulta che Ella era professore di ruolo e rappresentante dei professori nel Consiglio di Amministrazione. Quanto alla fede di nascita, al certificato di leva ed alla copia dell'atto di matrimonio ai fini della pensione, già furono inviati al Ministero fin dal 1935 e ritengo non ne occorrano altri.*

[20] IAS, *GF*: G. Fubini to A. Flexner and O. Veblen, Paris 2.1.1939.

[21] IAS, *GF*: A. Flexner to H.R. Marraro, [Princeton] 3.1.1939 and H.R. Marraro to A. Flexner, New York 5.1.1939.

[22] IAS, *GF*: M.C. Eichelser [Secretary of A. Flexner] to H.R. Marraro, Princeton 7.1.1939; H.R. Marraro to A. Flexner, New York 12.1.1939; A. Flexner to H.R. Marraro, [Princeton] 13.1.1939.

[23] IAS, *GF*: A. Flexner to H.R. Marraro, [Princeton] 18.1.1939.

[24] ASPoliTo-*Fubini*: G. Fubini to N. Vigna, Paris 13.1.1939: *sto impazzendo per raccogliere i documenti che mi sono necessarii. Tra quelli che (né io né il Consolato riusciamo a capire il mistero) non arrivano c'è il mio certificato di matrimonio. Per far più presto la pregherei di prestarmi il certificato che è nel mio incarto in Segreteria: ne farò fare qui le copie autentiche e lo reinvierò a Torino. Ella mi farebbe proprio un grande favore.*

While the delay in the issue of pension was annoying, but not dramatic as Fubini had enough money to support his family and buy tickets to New York, the difficulties he encountered in receiving visas seriously undermined his health.

Meanwhile, Fubini's son Eugenio, a physicist and a pupil of Fermi, who since 1936 had worked at the Polytechnic of Turin as a researcher and lecturer for the master's degree course in Electrotechnics, was encountering other difficulties. His state of service document was not approved because it was written in Italian. The English translation took a month and a half, from December 14, 1938 to January 25, 1939, despite the fact that it was only about ten lines long![25]

In mid-February, the Italian consulate had still not extended the validity of the Fubinis' passports. Marraro could no longer help because the Italian Interuniversity Bureau had been closed and passports and visas were now handled in the office of the Royal Italian consul general:

> Since my last letter to you I have resigned as Director of the Italian Interuniversity Bureau. I am, therefore, no longer in a position to intervene on your behalf. However, I am sure that your name is sufficiently well known to carry considerable weight with the authorities. I would suggest you writing directly to Comm. Gaetano Vecchiotti, Royal Italian consul general, 626 - 5th Avenue, New York City. [. . .] I am very sorry that in this and other similar matters I can no longer be of any service to you. But if I can be of any assistance to you in non-official affairs, you may be sure that I shall always be happy to place myself at your disposal.[26]

Flexner and Veblen were extremely sorry because they could not imagine 'anyone who could have been more effective in the position'.[27] However, as Marraro suggested, Flexner appealed to the Italian consul general, pleading for help in the granting of the Fubinis' passports.

On 6 March 1939, Guido and Anna appeared with witnesses in front of a notary in Paris for an act (*atto di notorietà*) rectifying Anna's passport, where only the name Anna appeared instead of the two names Annetta Estella. It was the last step. Two days later, the consul general informed Flexner that no reply had been received from the Ministry of Foreign Affairs, but that there was reason to be optimistic.[28] That same week, Fubini and his family sailed from Le Havre. They landed in New York on March 15, and on March 23, they arrived at Princeton. Guido and Anna settled in

---

[25] IAS, *GF*: G. Vallauri [Director of the R. Polytechnic of Turin], Turin 14.12.1938: *Il Dott. Eugenio Fubini-Ghiron è stato alle dipendenze di questo Istituto dal 1°.5.1936 al 4.12.1938-XVII in qualità di 'Ricercatore'. Dal 1°.2.1937 egli fu incaricato del compito di capo della Sezione Radiofisica del Reparto Comunicazioni dell'Istituto. Durante la sua permanenza nell'Istituto egli svolse attività organizzativa, collaborando nel periodo di attrezzatura dell'Istituto, sperimentale, sia per ricerche di carattere originale, sia per prove a servizio di altri enti, didattica, come incaricato di insegnamenti monografici al Corso di Perfezionamento in Elettrotecnica, Sezione Comunicazioni Elettriche. Nello svolgimento di queste mansioni egli ha sempre dimostrato ottima preparazione tecnica e scientifica, molta operosità e larga iniziativa. Lascia l'Istituto in dipendenza delle disposizioni contenute nel R.D.L 17.11.38-XVII n.1728.*

[26] IAS, *GF*: H.R. Marraro to A. Flexner, New York 14.2.1939.

[27] IAS, *GF*: A. Flexner to H.R. Marraro, [Princeton] 16.2.1939.

[28] IAS, *GF*: Cafere [consul general of Italy] to A. Flexner, New York 8.3.1939.

a rented apartment, at 94 Bayard Lane, whereas their children went to live in New York.[29]

8.2   A Welcome Displaced Scholar Who Felt as If He Was Stealing His Salary

The call to the United States, a world without fascism and persecution, offered a new lease of life to Fubini. He asked for (and obtained) naturalisation very quickly. His first request for the U.S. naturalisation document, released in Trenton, is dated 14 September 1939.

Asked to outline the reasons for his emigration, in 1941, in the field 'voluntary statement of any other relevant facts which may be of possible use to the United States Government, including plans for citizenship', he declared: 'I came into U.S., because my family could no longer live under the 'fascist' government'.[30] This last consideration is interesting insofar as it proves how Fubini reacted to expatriation. Until 1938, he had not declared any opinion on politics and could not be labelled as anti-fascist. The purge shocked him and led him to critically review his conduct as a scholar, neutral with respect to the events of life and society. He soon considered the United States his new homeland and appreciated the desire (indeed, the urgency) of his sons to take an active part in national defence projects. It was a source of great pride for him that Gino and Eugenio put their scientific skills at the service of the war effort. Colleagues, from Veblen to Aydelotte, witnessed this evolution and on at least two occasions underlined Fubinis' loyalty to America and their anti-fascist sentiments. The first opportunity was provided by Gino's request to visit his father and therefore to move freely between Princeton and New York; the second was linked to the request for references on Eugenio's moral and political conduct when he had to rent an apartment. At both times, Aydelotte and Veblen were clear:

> Our friend, Professor Guido Fubini-Ghiron, has been with us here for several years in Princeton. We know him and his family very well, and they are most enthusiastic and loyal friends of this country, who hope to become citizens as soon as the necessary formalities have been completed. Professor Fubini is not in very good health, and it is a great comfort to him to be visited from time to time by his son, because he himself is not well enough to make the necessary journeys. May we suggest that the most liberal policy possible be followed in permitting *Gino Fubini-Ghiron* to come to New York or to Princeton to visit his father and mother. It may be of interest to add that the young Fubini is engaged in important work connected with the national defence.[31]

> Mr. Fubini-Ghiron is the son of Professor Guido Fubini-Ghiron, who has been a member of the Institute for Advanced Study (of which I am Director) since 1938. Professor

---

[29] IAS, *GF*: G. Fubini to A. Flexner, 13.3.1939 and E.S. Bailey [Secretary] to whom it may concern, Princeton 24.4.1939.

[30] IAS, *GF*: information sheet filled in by Guido Fubini, 13.12.1941.

[31] IAS, *GF*: O. Veblen to United States District Attorney Boston Massachusetts, [Princeton] 16.1.1942.

Fubini-Ghiron is an eminent Italian mathematician. He and his family were driven into exile for political and racial reasons. He is a member of the Jewish race and is also a liberal, totally opposed to the whole philosophy of fascism. He is a man of high character and a cultivated gentleman. His two sons are notably successful in American business life.[32]

At Princeton, Fubini resumed his scientific activity, and in October 1939 he enrolled in the American Mathematical Society.[33] On the recommendation of the professors of the School of Mathematics, his contract was renewed for three academic years, although with a trivial salary of $1000, which Flexner judged completely inadequate to his abilities and experience.[34] His health deteriorated rapidly, however, forcing him to leave Princeton in 1942. Upon refusing (with sincere regret) the assignment for the academic year 1942–1943, Fubini felt compelled to admit that he had not earned his salary:

> I am deeply thankful to you and to the Faculty for your kind offer, and I regret to be obliged to decline it. In the years that I spent in Princeton I learned to love this town and to enjoy the privilege of living in the Institute, but I realized too that I don't earn my stipend, however small. I have found myself in such a position that I could do nothing of use for the Institute; and I regretted this very much.[35]

Fubini was very hard on himself, so much so that F. Aydelotte (the new Director of IAS) immediately replied:

> I hope very much that you will talk the whole matter over with our mathematicians before coming to a final decision. I know they will not agree with you as to the value of your contribution to the scientific work of the Institute. Over and above that, I should like to say what a great satisfaction it gives me to have you and Mrs. Fubini here as individuals. I understand the personal and family reasons which may make you feel that, in spite of everything, you ought to return to New York, and if that is your decision, I shall respect it. I want you to know at the same time, however, how highly your presence in Princeton is valued by all of us.[36]

Fubini's opinion was, in fact, quite unwarranted. In the four years spent in Princeton, he had been engaged on several fronts, including teaching, and solidarity. His contribution was essential in evaluating the profiles of mathematicians willing to leave Italy. Fubini was consulted by Veblen and Weyl regarding the requests of Segre, Ascoli, and Terracini.[37] Although his voice was not always heard, his views were widely appreciated. Moreover, Fubini was behind many gestures made 'by anonymous benefactors' in favour of displaced Italian scholars.

He contributed a substantial donation, equal only to Veblen's, to the scholarship that allowed Gina Castelnuovo to attend a Summer School in Wolfeboro and to work at the Mt. Desert Island Biological Laboratory (Fig. 8.3). It was Fubini who

---

[32] IAS, *GF*: F. Aydelotte to BING&BING, [Princeton] 12.2.1942.

[33] IAS, *GF*: G. Blake [Secretary, School of Mathematics] to G. Fubini, Princeton 1.11.1939.

[34] IAS, *GF*: A. Flexner to G. Fubini, [Princeton] 6.5.1939.

[35] Fubini.8.

[36] IAS, *GF*: F. Aydelotte to G. Fubini, [Princeton] 3.12.1941.

[37] Fubini.6.

Fig. 8.3 Contributions for
Gina Castelnuovo's grant,
OVP, *GC*

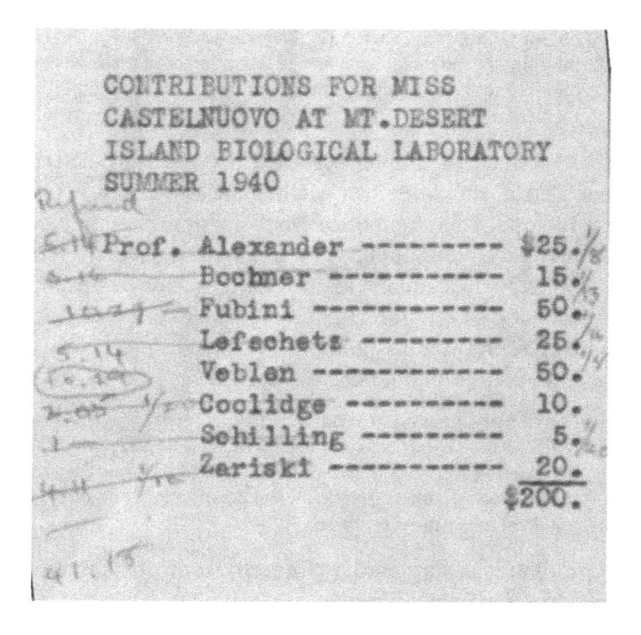

supported Bonaparte Colombo and his family so that they could escape from Turin
and take refuge in Switzerland (see Chap. 13). Terracini and Tricomi, who remained
in correspondence with Fubini until his death and who evaluated his manuscripts of
the American period for possible publication, testified that despite health problems
Fubini resumed his studies in Princeton. He published some works of analysis, one
in projective-differential geometry, and, above all, he devoted himself to completing
his book of mathematics for engineering on which he had worked in Italy until
1938.[38]

He also followed, even if without taking an active part, the new trends of research
cultivated at the Institute. He gladly talked about relativity with Einstein, who at
Fubini's request agreed to contribute a paper to the *Revista* published by Terracini.[39]

At Princeton, Fubini gave various lectures and, as suggested by the correspondence
of Severi and Bompiani, in all probability, he would have been invited as a plenary
lecturer at the 1940 ICM in Cambridge Mass. To his great regret, he did not return to
teaching, except for one course on Ballistics taught at Columbia in 1942–1943.

Despite the restrictions placed on age and health, American colleagues cared
about his presence and his contribution. When the time came for planning the yearly
budget, each time Flexner and Aydelotte had Fubini in their minds and arranged for
the continuation of his stipend, because: 'everyone likes Professor Fubini'.[40] Einstein in particular had very friendly relationships with him till his death (Fig. 8.4).

---

[38]Fubini.10.

[39]Fubini.11.

[40]IAS, *GF*: F. Aydelotte to A. Flexner, [Princeton] 4.12.1940.

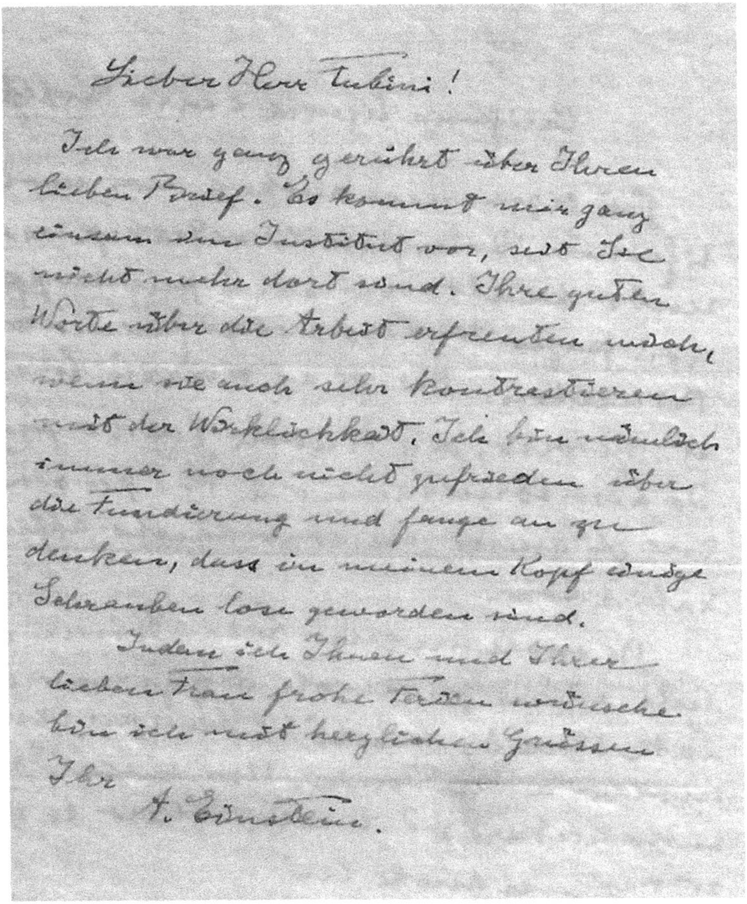

Lieber Herr Fubini!

Ich war ganz gerührt über Ihren lieben Brief. Es kommt mir ganz einsam im Institut vor, seit Sie nicht mehr dort sind. Ihre guten Worte über die Arbeit erfreuten mich, wenn sie auch sehr kontrastieren mit der Wirklichkeit. Ich bin nämlich immer noch nicht zufrieden über die Fundierung und fange an zu denken, dass in meinem Kopf einige Schrauben lose geworden sind.

Indem ich Ihnen und Ihrer lieben Frau frohe Ferien wünsche bin ich mit herzlichen Grüssen

Ihr

A. Einstein.

Fig. 8.4 A. Einstein to G. Fubini, [4.6.1943], Fubini Coll.

Flexner and Aydelotte were both aware that it seemed unlikely that Fubini could find a post since he was over 60, but on a suggestion from R. Courant, an association with New York University was considered.[41] The suggestion was timely. Fubini would be invited to lecture at Columbia University in 1940, while his appointment at the Institute for Advanced Study would be continuously renewed. In recognition of his work, the Institute invited Fubini to be a member *in absentia* for the years 1942–1943 and 1943–1944.[42] These gestures were greatly appreciated by Fubini

[41] IAS, *GF*: F. Aydelotte to A. Flexner, [Princeton] 4.12.1940 and FUBINI.7.

[42] IAS, *GF*: F. Aydelotte to G. Fubini, [Princeton] 23.5.1942; FUBINI.9 and F. Aydelotte to G. Fubini, [Princeton] 17.5.1943.

who at each renewal never failed to emphasise how happy he was at Princeton and how proud he was to be one of its fellows:

> I thank you very much for your letter and for the kind offer of the renewal of my appointment at the Institute for next academic year. I am very grateful to you and to the Faculty of Mathematics, and I accept your offer with a great pleasure. I am so well in Princeton; and a man who wishes to work and to study cannot find a better place than Princeton and its Institute.[43]

> It will be an honour, for me, to belong always to 'the Institute', where I lived three happy years. Even my physician told me yesterday that in this time my heart turned young again and that my conditions are much better.[44]

> I thank very much you and the School of Mathematics for your kind letter and invitation. I am proud of it. Last winter and spring my health went from bad to worse. But now, from some symptoms, the physician deduces I will improve very soon. If it is so, as I hope, the first thing to do will be for me to pay a visit to the Institute, its Director, and so many friends I have there. And I can perhaps begin again to study. I would be so happy![45]

8.3   Dead in Exile

Fubini did not see the end of the war. His health continued to deteriorate until his death in New York on June 6, 1943. The mourning was intense, both in America and in Italy. Among the first to express his condolences to Anna and their sons was Einstein, followed by Aydelotte and by many Italian expats, including Ugo Fano and Gina Castelnuovo, the Lattes and the Terracinis (FUBINI. 12–17 and Fig. 8.5).

An obituary of Fubini appeared in the *New York Times* and a notice was published in *Science*. Colleagues asked and obtained that he be commemorated at the Polytechnic of Turin on 27 July 1943:

> The Dean recalled the high scientific value, combined with exceptional clarity and evidence of presentation that made the late colleague a sure Master of thought and mathematical algorithm; certain of expressing the feelings of all colleagues, he addressed a respectful greeting to the deceased and expressed to the family the deep condolences of the Faculty.[46]

---

[43] IAS, *GF*: G. Fubini to F. Aydelotte, Princeton 12.3.1941.

[44] IAS, *GF*: G. Fubini to F. Aydelotte, New York 28.5.1942. Also the son Eugenio underlined how much Fubini was happy at the IAS: 'The peace of mind offered to him by the Institute and the friendship of some of its members made happier the last years of my father's life' (E. Fubini to F. Aydelotte, New York 11.5.1944).

[45] IAS, *GF*: G. Fubini to F. Aydelotte, New York 18.5.1943.

[46] ASPoliTo-*Fubini*: A. Bibolini to P. Calliano, Turin 27.7.1943: *Il Preside, ricordato l'alto valore scientifico, congiunto ad una eccezionale chiarezza ed evidenza di esposizione che fecero del compianto collega un Maestro sicuro del pensiero e dell'algoritmo matematico, certo di rendersi interprete dei sentimenti di tutti i colleghi, ha rivolto allo Scomparso un saluto deferente ed ha espresso alla Famiglia le vive condoglianze della Facoltà.*

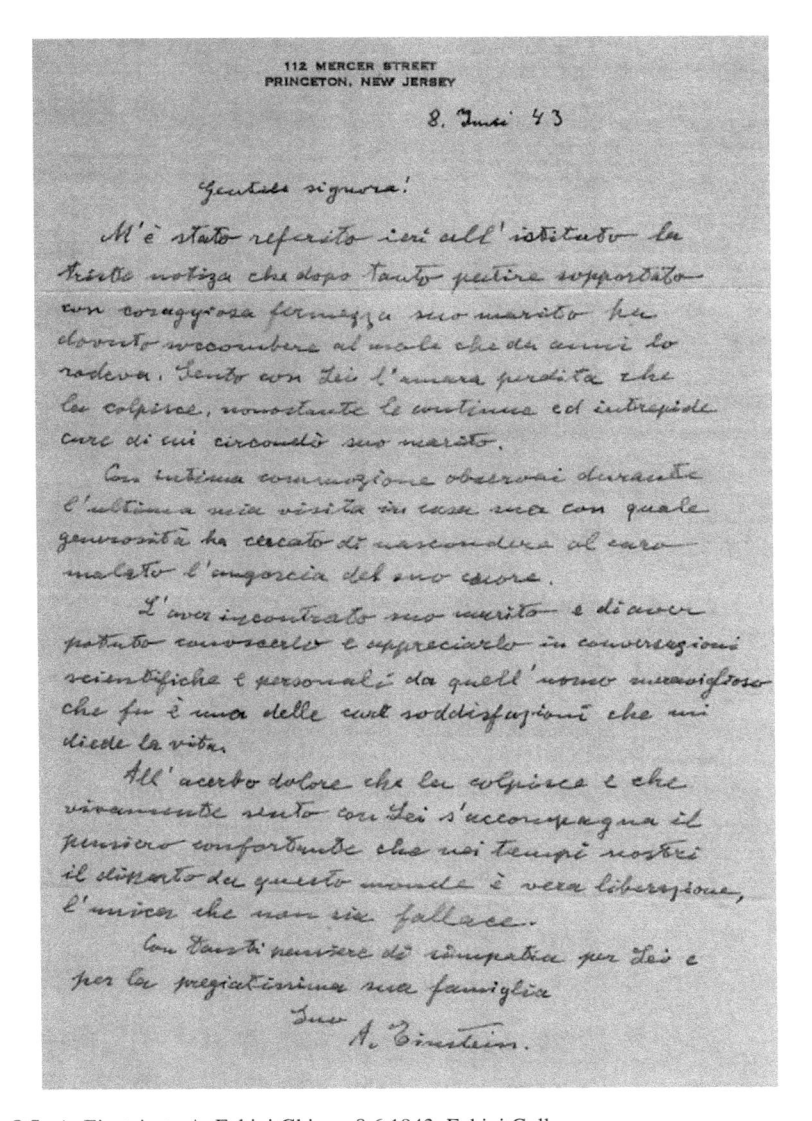

Fig. 8.5 A. Einstein to A. Fubini Ghiron, 8.6.1943, Fubini Coll.

Perhaps the most moving tribute, however, was the one Fubini received from Terracini (FUBINI.15) and which was published in the *Revista de la Unión Matemática Argentina*:

Firstly, the figure of Fubini, full of enthusiasm and overflowing with vivacity, can hardly be detached from his works, characterised by the brilliance of their style, the adamantine clarity in the way of posing problems and placing them within the framework of already-known questions, the way of undertaking problems without fear of difficulties. His figure always in motion, the rapid accent of his strong voice full of expressive inflections always surface from his pages. Fubini gave us a confirmation that mathematics is not a frigid and impersonal thing. On the other hand, his thinking was largely unitary, even though he seemed to take an interest in different topics. [...] Fubini had didactic qualities of the first order: I think that the talent of researcher and speaker are rarely found together in the same person as in Fubini. His students at the Turin Engineering School idolised him and, with good reason, since several thousand engineers who came out of that institution owed him their solid mathematical training.[47]

Like Volterra and Levi-Civita, Fubini would be officially commemorated in Italy only after the Second World War, at the Lincei and UMI by Castelnuovo and Picone, respectively. Speaking at the UMI, Picone would make a harsh attack on the 'foolish, infamous racial measures, eternal shame for Italy' that had caused 'the death in the pain of exile of a great mathematician who had served the homeland, with the noblest conception of his duties as a citizen and teacher and who highly honoured it with distinguished works, and loved it in a manner not possible for most'.[48]

Documents

Fubini.1 T. Levi-Civita to O. Veblen, Rome 30.10.1938

IAS, *GF*, fols. 1r, 2r
Dear Professor Veblen, As you probably know, all Italian teachers, racially Jew, from elementary schools to universities, have been dismissed; furthermore, some other antisemitic rules have been established, or foreseen in this country. Among the Italian mathematicians thus stroken, some one's [sic!], as Enriques and I, have attained the retiring age;[49] but there are some distinguished colleagues still young, or hardly mature, who had expected a better future for them and their families. This being premised as a matter of fact, I remember that once you have

---

[47]Terracini (1944), pp. 28–29: *En primer término la persona de Fubini, llena de entusiasmos y rebosante de vivacidad, mal puede separarse de sus trabajos con el brillo de su estilo, con su nítida claridad en la forma del plantear los problemas y colocarlos en el marco de problemas ya conocidos, con su manera de acometerlos sin temor a las dificultades. Casi parece que su figura siempre en movimiento, que el rápido acento de su voz fuerte y llena de inflexiones expresivas surjan de sus páginas. Nos da Fubini una confirmación de que la matemática no es cosa frígida e impersonal. Por otro lado, su pensamiento en gran parte es unitario, aun cuando parece tomar interés en tópicos distintos. [...] Fubini ha tenido calidades didácticas de primer orden: creo que raramente se encuentran reunidas en una misma persona condiciones de investigador y de expositor como en Fubini. Sus estudiantes de la Escuela de Ingeniería de Turín lo idolatraban y, con mucha razón, pues varios millares de ingenieros que han salido de esa casa de estudio a él le debieron su sólida formación matemática.*

[48]Picone (1946), p. 56: *dopo aver [...] servito la Patria, con la più nobile concezione dei propri doveri di cittadino e di docente e altamente onorata con opere insigni, fu costretto a staccarsene per gli stolti, infami provvedimenti razziali del 1939, eterna vergogna per questa Sua Italia che egli amò come di più non è possibile.*

[49]Enriques and Levi-Civita were born in 1871 and 1873, respectively.

said to me that unfortunately there is no place more in America (better in the U.S.A.) for European mathematicians, except perhaps for men of international renown. Such is undoubtedly my dear friend Professor Guido Fubini, till now full professor of Calculus in the Polytechnic-school and lecturer for Higher Analysis at the University of Turin. I scarcely need to remind his genial work in differential geometry, where he has open new pathways with a great deal of papers and a book on projective differential geometry;[50] his researches on movements and on conformal and geodesic groups being equally fundamental. After Volterra, he is considered the most penetrating and ingenious living Italian analyst, having supplied essential contributions in many vitals fields: for instance, automorphic functions of several variables, also in connection to Hermitian forms; old and new problems in the calculus of variations; integral equations with non-symmetric polar kernels (said also equations of the third kind); partial differential equations (extensions of the methods of Riemann and Picard to equations of higher order; asymptotic behavior of certain differential equations encountered in the radiotechnique; modern critics of foundations (multiple integration, mean-theorem, minimizing [minimum] principle). In the last time his chief interest turned himself toward the mathematical theory of elasticity and other questions of engineering science; he has found the way of attacking and resolving the difficult problem of the flexion of beams with curvilinear axis, and has otherwise in print (with professor Albenga) a big volume of engineering mathematics and its applications.[51] Professor Fubini is (as long as he will not be expelled as a Jew) a National Member of the Academies dei Lincei, of Turin, of The XL, and several others. He is a vivid man of profound intelligence and ready wit, who was celebrated among Italian scholars as a model of clear and brilliant expositor. Of course, if it is possible to offer him a satisfactory, stable situation, nothing better; but it would be in any case an objectively desirable policy for The Institute of Advanced Study to invite him as a temporary fellow. Professor Fubini, who is now in Paris (Albany Hotel, 202, Rue de Rivoli) would accept with great pleasure and gratitude such an invitation, not only as a conspicuous honour, but also as a precious occasion given to him of transferring to America his family (wife and two sons). With my highest compliments to Mrs. Veblen, also on behalf of my wife, I beg you to receive my best thanks, greetings and excuses if I dare to call upon the scientific and human mission of the Institute through your illuminate patronage. Very faithfully yours, Tullio Levi-Civita

---

[50] G. Fubini, E. Čech, *Geometria proiettiva differenziale*, Bologna, Zanichelli, 1926–1927.

[51] G. Fubini, G. Albenga, *La matematica dell'ingegnere e le sue applicazioni*, Bologna, Zanichelli, 2 vols., 1949, 1954.

Fubini.2 O. Veblen to A. Flexner, [Princeton] 22.11.1938

IAS, *GF*, fols. 1r, 2r

Dear Doctor Flexner: There was a meeting of the mathematical professors of the Institute this morning from which come the following recommendations: 1. We have received from Professor Levi-Civita a letter suggesting that Professor Guido Fubini, formerly of Turin (whose address now is Albany Hotel, 202 Rue de Rivoli, Paris) be invited to the Institute as a temporary member for the remainder of the present academic year. Professor Fubini is characterized by Levi-Civita as the leading analyst of Italy after Volterra (Volterra is now an extremely old man.[52]) It was generally agreed that Fubini is a mathematician of very high distinction, who would be a real addition to our group. It was decided to request you to issue this invitation if any funds can be found which are available for the purpose. We think that such an amount as $500 or perhaps $750 might possibly serve. Professor Fubini's wife and two sons are with him in Paris. [...] I shall be very glad to explain any of these matters further if you so desire. Yours sincerely, Oswald Veblen

Fubini.3 O. Veblen to G. Fubini, [Princeton] 1.12.1938

IAS, *GF*, fol. 1r

Dear Professor Fubini: You will very soon receive a letter from Dr. [Abraham] Flexner, the Director of our Institute, inviting you to become a temporary member for the second term of the present academic year. This term begins on January 17 and ends on the 1st of May. The stipend which is being offered is a very small one due to the fact that our funds had practically all been allocated before I heard from Professor Levi-Civita that you might like to come to this country. We hope that in any case the invitation will be of use to you in getting a visa for your American trip. To me personally it will be a source of great satisfaction to have the opportunity to become better acquainted with you and your scientific work. Yours sincerely, Oswald Veblen

Copy: December 1, 1938

My dear Professor Fubini:

Upon the recommendation of the professors of the School of Mathematics of the Institute for Advanced Study I have pleasure in inviting you to be a member of the Institute in the second term of the academic year, 1938-1939. The second term opens January 17 and closes May 1, 1939. You will receive a stipend of $ 500.00. Under separate cover I am sending you the latest Bulletin of the Institute. Sincerely yours, Abraham Flexner

---

[52]Volterra was 78 years old.

Fubini.4 G. Fubini to O. Veblen, Paris 12.12.1938

IAS, *GF*, fol. 1r-v
Dear Professor Veblen, After an absence of two days I have received your kind
letter, for which I thank you very much. Your invitation allows me to hope that I
may recreate my life and that of my family; and this is naturally for me a very
important consideration. But I shall be specially happy to know you and your
eminent colleagues, whose works I have admired for a long time. Please accept
my thanks and my real gratitude. Unfortunately there are many documents
required by the American consul. The most difficult one to obtain is the visa of
the Italian government, but I hope that this will be possible to have. Should it be
necessary, I hope that I shall be able to obtain a recommendation from Princeton
for the American consul in Paris. I shall leave as soon as possible. Yours faithfully
Guido Fubini

Fubini.5 G. Fubini to A. Flexner, Paris 12.12.1938

IAS, *GF*, fol. 1r-v
Dear Professor Flexner, I must thank you very much for your invitation which, I
hope, will give me the opportunity to find a new home for my family and to create
a new life, and enable me to continue my studies and research work. It will be a
great pleasure to work among so many eminent men of science. The American
consul has already indicated the documents which are necessary, and I have taken
steps to obtain them. The most difficult one to obtain is the visa of the Italian
government; but I hope this will not be an unsurmountable difficulty. I shall leave
on receiving these documents. Yours faithfully Guido Fubini

Fubini.6 B. Drury [ECADFS] to E. Fubini, New York, 14.7.1939

ECADFS Records, *AT*, fol. 1r
Dear Dr. Fubini-Ghiron: In reply to your letter of July 11 I am exceedingly sorry
to say that there is very little that the Emergency Committee in Aid of Displaced
Foreign Scholars can do to assist Professor Alessandro Terracini to find a position
in the United States. As it happens the Emergency Committee cannot because of
its regulations bring the availability of displaced scholars to the notice of colleges
and universities unless asked to do so by such institutions. Upon receiving such a
request from a college or university the Committee will suggest candidates
suitable for the position to be filled. I am enclosing herewith a copy of our last
annual report to give you a little fuller idea of our program. If Professor Terracini
is to find a place in this country it will have to be through the efforts of his friends

and colleagues here. It is, as you know, exceedingly difficult to find positions for displaced scholars as the number of exiles is becoming very great. The field of mathematics is practically saturated. We already have on file several sets of Professor Terracini's papers, one of which we received directly from him. If you wish I shall be very glad to return to you the curriculum you sent us. Sincerely yours, Betty Drury – Assistant Secretary

Fubini.7 Extract of a Letter from O. Veblen to R. Courant, [Princeton] 14.12.1940

IAS, *GF*, fol. 1r

With regard to Fubini, I think that the best solution would be the indefinite continuation of his present very modest stipend, which would enable him to live quietly in Princeton and continue his scientific work. He reports that his health is better here than it has been anywhere else for many years, and he is gradually becoming a most useful fixture in this environment. I wish that you could get Flexner to see that if he could bring such an arrangement about, he would be doing his Institute a real service. Yours sincerely, Oswald Veblen

Fubini.8 G. Fubini to F. Aydelotte, Princeton 30.11.1941

IAS, *GF*, fol. 1r

My dear Director,[53] Thank you very much for your friendly letter. I am deeply thankful to you and to the Faculty for your kind offer, and I regret to be obliged to decline it. In the years that I spent in Princeton I learned to love this town and to enjoy the privilege of living in the Institute, but I realized too that I don't earn my stipend, however small. I have found myself in such a position that I could do nothing of use for the Institute; and I regretted this very much. On the other hand, I realized that I could no longer overlook the fact that, with my family divided in two or three parts, the stipend I received was far from counterbalancing the increased expenses.[54] It is indeed very difficult and very sad for me to leave Princeton. This Institute can offer and offered me the most magnificent place to

---

[53] Franklin Ridgetway Aydelotte (1880–1956), Director of the Institute for Advanced Study (1940–1947).

[54] Gino (1911–1965) lived in Cambridge and studied structural engineering at the Massachusetts Institute of Technology. Eugenio (1913–1997), a technician at CBS, lived in New York in a room at the International House, a non-profit organization that provided affordable housing for students and young researchers. Veblen echoed Fubini's worries (OVP, *TYT*: O. Veblen to T.Y. Thomas, [Princeton] 5.5.1939): 'Fubini came here this spring with his wife and two sons. In his own country he was a very rich man, but he has been able to bring away with him only a small amount of money'.

study and live. In it I have had the privilege to find so many famous scientists and good friends; leaving it means ultimately to abandon the studies to which I have devoted more than forty-five years of life. I will always be grateful to you and to the Institute for all I have received during the past years and formulate my best wishes for the future of this great and perhaps unique Institution. With my best thanks and regards, Yours truly, Guido Fubini

Fubini.9 G. Fubini to F. Aydelotte, New York 5.10.1942

IAS, *GF*, fol. 1r

My dear Director, Because of your kindness, I have always the honour of being a member of the Institute, even *in absentia*. I hoped I was allowed to *vivere eius vitam*, to come very often to Princeton, to study in your library, to meet the Director and so many dear friends. But my health grew worse after I left Princeton. The physician gave only the following prescription: I must remain many hours on the sofa, I must rest, I must not travel. And therefore, I am obliged to remain all the day in my room with my papers; but I always hope my health will recover, and I shall be able to come to Princeton and to meet you. With my best regards, sincerely yours Guido Fubini

F.10 G. Fubini to G. Blake, 22.1.1943

IAS, *GF*, fols. 1r–2r

Dear Miss Blake,[55] My physician does not allow me to come to Princeton: the stairs at Princeton junction, the cold temperature at the station of Princeton, the danger of a chill, etc. etc. are great difficulties, according to his opinion. My wife, of course, joins the physician; and in this manner I am obliged to remain here. But perhaps the physician is right; I know that a chill is for me a three or four month's illness, because of my chronic bronchitis. And therefore, I shall send you my manuscript (some pages are already printed; other pages are written by hand, and have *many* corrections).[56] Of course, I shall pay as soon as I know my debt (I can also pay in advance). My manuscript will arrive to you *insured*. It is divided into four parts. [...] So far, there are $405 + 522 + 18 + 46$ pages $= 991$ pages. When everything comes back to New York, it is better than the microfilms and the manuscript don't travel with the same mail. If, by chance, both were lost, it would be a ruin for me, because I have not another copy. And I worked *many* and *many*

---

[55] Gwen Blake was the Secretary of the School of Mathematics at the Institute for Advanced Study.

[56] The book by G. Fubini and G. Albenga, *La matematica dell'ingegnere e le sue applicazioni* would be published in Bologna, by Zanichelli, in 1949–1954.

years in order to write this book. I thank you very much for your kindness. I am sure you will excuse me. But everything is for me so difficult. I cannot move, and I remain all the day at my table in my room. My best thanks and regards G. Fubini

P.S. Can you, please, write me when my package has arrived to you? I hope it will leave N.Y. next Monday or next Tuesday.

Fubini.11 A. Einstein to G. Fubini, 1943

Fubini Coll., fol. 1r

Dear Mr. Fubini! I was very touched by your lovely letter. We feel all alone at the Institute since you're no longer there. Your good words about my work made me happy, even though they contrasted with reality. I'm still not satisfied with the foundations and I'm starting to think that some screws in my head have become loose. With best regards, I wish you and your dear wife happy days, yours A. Einstein

Lieber Herr Fubini! Ich war ganz gerührt über Ihren lieben Brief. Es kommt mir ganz einsam im Institut vor, seit Sie nicht mehr dort sind. Ihre guten Worte über die Arbeit erfreuten mich, wenn sie auch sehr kontrastieren mit der Wirklichkeit. Ich bin nämlich immer noch nicht zufrieden über die Fundierung und fange an zu denken, dass in meinem Kopf einige Schrauben lose geworden sind. Indem ich Ihnen und Ihrer lieben Frau frohe Ferien wünsche, bin ich mit herzlichen Grüßen. Ihr A. Einstein

Fubini.12 F. Aydelotte to E. Fubini, [Princeton] 7.6.1943

IAS, *GF*, fol. 1r

Dear Mr. Fubini: I was deeply grieved to receive your telegram this morning, informing me of your father's death, and all the members of the School of Mathematics have asked me to convey to you and to the members of your family the deep sense of loss which we feel and our sympathy with you in your bereavement. I think your father was happy in his work at the Institute and I constantly heard echoes of the value which his colleagues, young and old, attached to their association with him. We were proud to continue him as a member *in absentia* after he found it necessary to return to New York and we always enjoyed his occasional visits to Princeton. I should appreciate it if you would give your mother affectionate messages of sympathy from Mrs. Aydelotte [Marie Jeanette Osgood] and myself. Yours sincerely, Frank Aydelotte, Director

Fubini.13 A. Einstein to A. Ghiron Fubini, Princeton 8.6.1943

Fubini Coll., fol. 1r
Dear Madam! I was told yesterday at the Institute the sad news that after so much suffering endured with courageous firmness, your husband had to succumb to the evil that had been gnawing at him for years. I share with you the bitter loss that strikes you, despite the continuous and dauntless care which you gave your husband. Deeply moved, I observed during my last visit to your house with what generosity you tried to hide from the dear sick man the anguish of your heart. I met your husband, and I was able to know and appreciate him in scientific and personal conversations; he was a wonderful man and the fact of having encountered him is one of the great satisfactions of my life. The bitter pain that afflicts you, and that I share in, is accompanied by the comforting thought that, in our times, departure from this world is true liberation, the only one that is not hollow. With many thoughts of sympathy for you and for your precious family. Yours A. Einstein

Gentile Signora! Mi è stata riferita ieri all'Istituto la triste notizia che dopo tanto patire sopportato con coraggiosa fermezza suo marito ha dovuto soccombere al male che da anni lo rodeva. Sento con Lei l'amara perdita che la colpisce, nonostante le continue ed intrepide cure di cui circondò suo marito. Con intima commozione osservai durante l'ultima mia visita in casa sua con quale generosità ha cercato di nascondere al caro malato l'angoscia del suo cuore. L'aver incontrato suo marito e aver potuto conoscerlo e apprezzarlo in conversazioni scientifiche e personali da quell'uomo meraviglioso che fu è una delle care soddisfazioni che mi diede la vita. All'acerbo dolore che la colpisce e che vivamente sento con Lei s'accompagna il pensiero confortante che nei tempi nostri il diparto da questo mondo è vera liberazione, l'unica che non sia fallace. Con tanti pensieri di simpatia per Lei e per la pregiatissima sua famiglia. Suo A. Einstein

Fubini.14 U. Fano to A. Ghiron Fubini, Cold Spring Harbor 8.6.1943

Fubini Coll., fol. 1r
Dear Madam, dear guys, the loss of the Professor hit me too, very hard. My thoughts are closely with you in these days. And bitterness grows at the doubt that this tragedy is due to those circumstances that had so deeply wounded him. I also think how important are, now more than ever, the bonds in your family, which are perhaps more intimate and stronger than in any other family I know. While such bonds may result in the deepest suffering, being so extraordinarily united will certainly help you. I would like in some way to try to convey to you also the feelings of my parents [Gino Fano and Rosetta Cassin Fano], which I can well

imagine but which perhaps will not reach you directly. The long and shared life, at work and on holiday, and the resulting friendship, have left such a trace that I dare not think of how they will suffer in receiving the news. For me personally, friendship is identified in the many memories that unite me to all of you, but also and especially in the essential part that the Professor had in my education.[57] His ever-generous help, at school and outside of school, has left an impression not only in terms of the things I have learnt but (especially) in the way I have been taught. We would be grateful if you could let us know if and when any of you will enjoy seeing us. We will certainly find a way to overcome the small difficulties that keep us away from the city. Affectionately, Ugo

Cara Signora, cari ragazzi, La Perdita del Professore ha colpito anche me molto duramente. Il mio pensiero vi segue da vicino in questi giorni. E l'amarezza cresce nel dubbio che questa tragedia sia dovuta a quelle circostanze che l'avevano così profondamente ferito. Penso pure quanto importanti siano, ora più che mai, i legami nella vostra famiglia, che sono forse più intimi e forti che in qualsiasi altra famiglia io conosca. Se questi legami faranno subire il colpo più duro a voi, l'essere così straordinariamente uniti vi sarà certo di aiuto. Vorrei in qualche modo cercare di trasmettervi anche i sentimenti dei miei genitori, che ben posso immaginare ma che forse non potranno giungervi direttamente. La tanto lunga comunanza di vita, in lavoro e in vacanza, e la amicizia che ne è risultata, hanno lasciato tale traccia che non oso pensare all'impressione che subiranno nel ricevere la notizia. Per me personalmente, l'amicizia si identifica nei tanti ricordi che mi uniscono a voi tutti, ma anche e specialmente nella parte essenziale che il Professore ha avuto nella mia educazione. Il suo sempre generoso aiuto, a scuola e fuori di scuola, ha lasciato traccia non solo nelle cose insegnate ma specialmente per il modo con cui mi sono state insegnate. Vi saremo grati se ci farete sapere se e quando chiunque di voi avrà piacere di vederci. Troveremo certo modo di superare le piccole difficoltà che ci tengon lontani dalla città. Affettuosamente Ugo

Fubini.15 A. Terracini, G. Sacerdote Terracini, E. Levi Terracini to E. Fubini, Tucumán 10.6.1943

Fubini Coll., fols. 1r, 2r
Dear Eugene, yesterday morning I received your telegram [and] I assure you that I have not yet recovered from the shock. It is not banal to tell you that my thoughts were continually directed to your poor father. When I last saw him almost four

---

[57] The Fubini and Fano families were linked by a thirty-year friendship. Ugo Fano (1912–2001) had been Fubini's student in the courses of Higher Analysis at the University of Turin (1932–1934) and a fellow of Eugenio Fubini in Turin and Rome.

years ago, he seemed so brimming with health, although saddened by events, that throughout his illness it was difficult for me to imagine as a sick person the energetic and full-of-life Fubini that I knew. What can I tell you, dear Eugene? With your father's passing, I lose a man whom I knew for about twenty years;[58] perhaps the bonds of friendship had become even more intense in these last four years when many kilometres separated us, without, however, detaching us from each other. I always felt him to be a friend, I can tell you frankly, my best friend. I met your father in a year that seems so distant, when for the first time he taught the course in Higher Analysis; I was his student, attending the fourth year of university at that time. I was also his colleague for some 15 or 20 years, during which we exchanged ideas on an incredible number of mathematical and non-mathematical matters. If we were not away, at least three times a week we talked at the University, at the Polytechnic, or walking along Via S. Francesco da Paola. How many conversations this street witnessed, conversations in which, as time went by, political considerations and his catastrophic predictions (how many were right!) about the chaos of Europe, and the sad fate of so many homeless people, and the misfortunes to which the crazy policy of its leaders were going to drag poor Italy, became more and more mixed up. I entered the Polytechnic, and the name of Fubini was heard everywhere; for the students, he was without a doubt the most beloved teacher.[59] Giacomino spoke only of Fubini, and all his colleagues had an extraordinary love and esteem for him. His kindness was renowned and admired by everyone who knew how many people your father had helped. Not to mention his mathematical works, in particular those of differential projective geometry that I can almost say I saw born in his scientific production, and to which he was still dedicating some of his latest works. I will never forget the help that your father gave me to strengthen our *Revista de Matemáticas y Física Teórica*, either with his own contributions from the very beginnings, or with those from other individuals that he procured from the United States, up to that of Einstein.[60] I will never forget a generous offer he made to me at a time when it seemed that certain difficulties might prevent my brother's departure!![61] I don't want to think that I will never ever meet him again: you know that, on certain spiritual matters, your father and I had very different points of view (although this often did not prevent us from reaching similar practical results). Whatever it was, I couldn't help but think of him when I read in a ritual

---

[58] Terracini considered Fubini one of his Masters. They had been colleagues from 1925 to 1938.

[59] Terracini had taught Analytic Geometry at the Polytechnic from 1923 to 1925.

[60] G. Fubini, *Equazioni differenziali per i periodi di un integrale iperellittico*, Revista. Serie A. Matemáticas y Física Teórica, I, 1940, p. 73–79; *On hyperautomorphic functions*, Revista. Serie A. Matemáticas y Física Teórica, I, 1940, p. 87–94. Fubini had procured for the Revista a paper by A. Einstein, *Demostración de la no existencia de campos gravitacionales sin singularidades de masa total no nula, Demonstration of the non-existence of gravitational fields with a non-vanishing total mass free of singularities*, Revista. Serie A. Matemáticas y Física Teórica, II, 1941, p. 5–15. See D6.5.1.

[61] Benvenuto Terracini (1886–1968) was able to leave Italy for Tucumán only in July 1941.

book «We will then see again those friends, who have already passed beyond this life, and to whom our soul is still so closely linked; and then we will be able to rest from the fatigues, free of the dangers we encounter in the stormy sea of life.» All my relatives participate intimately in your mourning. I beg you to pass on to your mother and your brother my deep feelings, and I hug you affectionately. Alejandro

P.S. I would like to publish in my mathematical journal an extensive biographical portrait of your father.[62] At your earliest convenience, you should send me the most detailed biographical data, a photograph of your father (which, after publication, I would be glad to keep personally), and all the reprints of his publications that you can find (unfortunately I have here only those of the last three years). Now is not the time to write to you on other matters: however, I would ask you the favour of investigating what has happened to the second half of the manuscript of his book, which has never reached me.[63] This is exactly what the last letter from him, arrived on the afternoon of the 9th (yesterday), referred to. If it were necessary to send another copy, you should remember that I still lack the title of the book, and the preface (from our correspondence, I deduce that he intended to clarify here some circumstances related to the genesis of the work). If by bad luck this has been lost, and there is no copy of the preface, you should write another one yourself. Excuse me if I speak to you now of such matters: we live so far away, that we must take into account the time it takes for the correspondence to arrive. Colleagues at this University, and particularly Dr. Cernuschi, entrusted me with expressing their deep condolences to you as well.[64]

Dear Mrs., at this painful moment my thoughts are affectionately directed to you and your sons to express how much we regret the loss of your Husband. The friendship that linked us to him was so pleasant for all of us!!! My husband is very distressed, and it is difficult for us to ease his pain. I greet you and your sons with all my affection, Giulia.

Eugene Querido, ayer por la mañana he recibido tu telegrama [y te] aseguro que aún no me he recobrado de la impresión que me he recibido. No es un lugar común lo de decirte que mi pensamiento estaba continuamente dirigido a tu pobre Padre. Cuando lo vi por última vez casi son cuatro años parecía tan rebosante de salud, aunque entristecido por los acontecimientos, que durante el largo de su enfermedad me resultaba difícil imaginar al Fubini siempre en movimiento y lleno de vida que yo conocía como a una persona enferma. ¿Qué

---

[62] Fubini's obituary would be published in the Revista de la Unión Matemática Argentina: A. Terracini, *Guido Fubini (1879-1943)*, s. 1, 10, 1944, p. 27–30.

[63] The manuscript by G. Fubini and G. Albenga, *La matematica dell'ingegnere e le sue applicazioni* will be published in Italy, by Zanichelli, 2 vols., 1949–1954.

[64] Félix Cernuschi (1907–1999), professor of Theoretical Physics and Applied Mathematics at the University of Tucumán from 1939, was co-editor of the *Revista de Matemáticas y Física Teórica*.

puedo decirte, querido Eugene? Desaparece para mí con tu Padre un hombre cerca del cual he vivido puede decirse unos veinte [años], tal vez los vínculos de la amistad aún más [intensos] en estos últimos cuatro años, en los cuales muchos kilómetros nos separaban, y sin embargo nuca nos habían separados. Siempre sentí en él un amigo, pero puedo decirte que ahora [mi mejor] amigo. Conocí a tu Padre en ese año que parece tan alejado, en el cual por primera vez él dictó el curso de análisis superior y yo he sido su alumno frecuentando en ese entonces el cuarto año de la universidad. He sido su colega unos 15 o 20 años, durante los cuales nos hemos cambiado ideas sobre una cantidad increíble de asuntos matemáticos o no matemáticos. Si no nos encontrábamos en otras partes per lo menos tres veces por semana conversábamos en la universidad, en la escuela politécnica, o recorríamos junto esa calle de S. Francesco de Pola, que ha asistido a un sinnúmero de conversaciones matemáticas a las cuales con el pasar del tiempo siempre más se mezclaban sus consideraciones políticas y sus previsiones catastróficas (y cuanto sido acertadas!) sobre el caos de Europa, y el triste destino de tantos desamparados, y las desgracias a las cuales la política loca de sus dirigentes iba a arrastrar a esa pobre Italia. Yo entraba a la escuela politécnica, y el nombre de Fubini se oía por todas partes; para los estudiantes era sin duda el profesor más querido. Giacomino hablaba que de Fubini, y todos los colegas le tenían un cariño y una estimación extraordinarias. Su bondad era conocida y admirada por todo el mundo que sabía a cuántas personas tu Padre ha ayudado. Ni hablo de los recuerdos matemáticos, en particular de los de esa geometría proyectiva diferencial que puedo decir casi he visto nacer en su producción científica, y a la cual todavía está dedicado alguno de sus últimos trabajos. Nunca olvidaré la ayuda que tu Padre me ha prestado para afianzar nuestra Revista de Matemáticas y Física Teórica, ya sea con sus propias colaboraciones desde el comienzo, como con las otras que él supo proporcionarme de Estados Unidos, hasta la de Einstein. ¡¡Nunca olvidaré una oferta generosa que él mi hizo en un momento en el cual parecía que ciertas dificultades pudieran impedir la partida de mi hermano!! No quiero pensar que nunca jamás Le encontraré: tú sabes que, en ciertos asuntos místicos, tu Padre y yo teníamos puntos de vista muy distintos (aunque a menudo esto no impedía que llegáramos a consecuencias prácticas parecidas). Sea lo que fuere, no pude no pensar en él al leer en un libro ritual «Reveremos entonces aquellos Amigos, que ya han pasado más allá de esta vida, y a los cuales nuestra alma sigue tan vinculada; y entonces podremos descansar de las fatigas libres de los peligros que encontramos en el mar tormentoso de la vida.» Todos mis familiares participan íntimamente a vuestro luto. Yo te ruego que expreses a tu madre y tu hermano mi profundo sentimiento, y te abrazo afectuosamente. Alejandro

P.D. Quisiera publicar en mi revista de matemáticas una amplia noticia biográfica sobre tu Padre. Apenas te resulte posible pensar en estos asuntos, deberían enviarme los más detallados datos biográficos, una fotografía de tu Padre (que, después de la publicación, seríamos grato tener entre mis recuerdos), y todo lo que puedas de copias de sus publicaciones (de las cuales por mala suerte tengo aquí tan solo las de estos últimos tres años). No es ahora el momento para escribirte de otros asuntos: sin embargo, me harías un favor si puedes investigar lo que he ocurrido con la segunda mitad del original de su libro, el que nunca me ha llegado. Justamente a esto se refería la última carta de Él, que me ha llegado en la tarde del día 9 (ayer). Si fuera preciso enviar otra copia, deberías recordarte que todavía me faltan el título del libro, y el prefacio (el en cual de nuestra correspondencia deduzco que El aclararía algunas circunstancias vinculadas sobre la génesis de la obra). Si por mala suerte esto se hubiera perdido, y no existiera copia del prefacio, deberías escribir otro [tu] mismo. Discúlpame si te hablo ahora de tales asuntos: sin embargo, vivimos tan lejos, que hay que tener en cuenta el tiempo que se necesita para que llegue la correspondencia. Los colegas de esta Universidad, y particularmente el dr. Cernuschi me han encargado de que os exprese también sus profundos pésames.

Querida Señora, en este momento tan doloroso mi pensamiento se dirige con afecto a Ud. y a sus hijos para expresarle cuanto lamentamos la perdida de su Esposo. ¡La amistad que nos vinculaba a él era tan grata para todos nosotros!!! Mi esposo está afligidísimo y resulta difícil para nosotros aliviar su dolor. Le saludo con todo cariño a Ud. y a sus hijos, Giulia

Fubini.16 Leone and Laura Lattes to A. Ghiron Fubini, Buenos Aires 13.6.1943

Fubini Coll., fol. 1r

Dear Mrs. Anna, From Renato we learned the sad news, which fills us with pain.[65] A great mind has been extinguished, a great heart has ceased to beat, all over the world your torment finds deep echoes. Those who knew your Guido well realise what you have lost. For us, who were so close to each other with such cordial friendship, the happy memories of Pacol, Vercelli, of your hospitable house in Via Pietro Micca, and those less cheerful, but nevertheless sweet, memories of the Parisian chats, crowd the mind and give us the anguish of the irreparable. We experienced the same tragedy, and even though geography separated us, our souls remained united, and from afar we followed the bitterness first of his transplant, his compromised health later, and his sad withdrawal from the valuable work that had been the reason and substance of his noble life. I am sure that his last days, of which I would like to know something, were illuminated by the glow of the coming dawn, and that he had the consolation of knowing that better days are being prepared for our, for your beloved children above all. So be it soon, dear Mrs. Anna; we will then be able to think of him and of the infinite

---

[65] Renato Treves, professor of General Theory of Law, Philosophy of Law and Sociology at the University of Tucumán since 1939. Treves was Lattes' son in law.

victims of the scourge, with a calm and, perhaps, serene soul. Meanwhile, for a long time we have not heard from our Giorgio, and you can imagine our anxiety waiting for news![66] Our sacrifice is hard, even though we are sustained by an unshakable hope. Please express to Gino and Eugenio all our sympathy, and believe me, now and always, your affectionate friend. L. Lattes[67]

Dear Anna, with deep infinite anguish I think of you, and I share your pain. I had such admiration, such sympathy, such a lively friendship for your husband that I understand how immense and unbridgeable is the void that his death leaves. At least he had the comfort of seeing his children serenely on their way to a new life. For a long time, we have been without news of Giorgio, our resistance (not only moral but also physical) has reached the limit, we can no longer bear it. Laura

Cara Signora Anna, Da Renato abbiamo saputo la tristissima notizia, che ci ha riempito di dolore. Una gran mente si spense, un gran cuore cessò di battere, in tutto il mondo il suo strazio trova echi profondi. Chi ha conosciuto il Suo Guido ben si rende conto di ciò che ha perduto. Per noi poi, che c'eravamo stretti di tanta cordiale amicizia, i ricordi felici del Pacol, di Vercelli, della casa ospitale di via Pietro Micca, e quelli meno allegri, ma tuttavia dolci, delle chiacchiere parigine, si affollano alla mente e ci danno l'angoscia dell'irreparabile. Abbiamo vissuto la stessa tragedia, e se anche la geografia ci ha separati, gli animi erano rimasti uniti, e da lontano abbiamo seguito l'amarezza del suo trapianto prima, della sua salute compromessa in seguito, e del suo triste ritiro dall'alto lavoro che era stato ragione e sostanza della sua nobile vita. Son certo che i suoi ultimi giorni, di cui gradirei sapere qualcosa, sono stati illuminati dai bagliori della prossima aurora, e che ha avuto la consolazione di sapere che giorni migliori si preparano per i nostri, per i suoi diletti figli. Così sia presto, cara Signora Anna; e potremo allora pensare a Lui ed alle infinite vittime del flagello, con animo placato e, forse, sereno. Frattanto, da lungo tempo non abbiamo notizie del nostro Giorgio, e può pensare con quale animo ansioso le aspettiamo! È duro il nostro sacrificio, malgrado ci sorregga una incrollabile speranza. Voglia esprimere a Gino ed Eugenio tutta la nostra simpatia, e mi creda, ora e sempre, suo affezionato amico. L. Lattes

Cara Anna, con profonda infinita angoscia penso a voi e condivido il vostro dolore. Avevo tanta ammirazione, tanta simpatia, così viva amicizia per tuo marito che comprendo quanto immenso e incolmabile sia il vuoto che lascia la sua scomparsa. Almeno egli ebbe il conforto di vedere i suoi figlioli serenamente avviati verso una nuova vita. Da molto tempo siamo senza notizie di Giorgio, la

---

[66] Giorgio Lattes (born in 1913) was a close friend of Ugo Fano. He had emigrated in Tucumán, but, in 1939, he had returned to Europe in order to enroll in the French Foreign Legion and fought in Allied military units.

[67] Leone Lattes, his wife Virginia Rabbeno and children Giorgio, Camilla, and Lisa Fiamma had emigrated in Argentina in 1938, after a sojourn in Paris.

nostra resistenza non solo morale ma anche fisica è giunta all'estremo, non se ne può più. Laura

Fubini.17 Gina Castelnuovo to A. Ghiron Fubini, Philadelphia 22.6.1943

Fubini Coll., fol. 1r-v

Dearest Madam, only today, on opening the latest issue of *Science,* did I learn of the death of the poor Professor.[68] You cannot imagine how strong my shock was, and even now that I am writing to you, I cannot believe it. I never thought that the last time I was in NY would be the last time I would see the Professor. I always hoped that the ailments that afflicted him would be temporary. You cannot know what the Professor represented for me at this time; it was just like coming back to my family when I came to visit him. I know the affection that my father had for him and how much he esteemed and loved him, and I also know that this was reciprocated by the Professor. All this made me feel, now that I am far from my family, even closer to him and it was always with great pleasure that I came to visit him. Not only his fame of a great Scientist but also (and above all) his great honesty, integrity, and generosity led me to esteem and appreciate him. Both he and my father [Guido Castelnuovo] are the type of people who are admired even unintentionally and who, unfortunately, are not frequently found in this world. Fortunately, the Professor had the comfort of having his whole family close to him and of being at peace by having them in this country. I know how much pain it would be for my father to learn of his death; unable to let him know, I send to you and the boys not only all my deepest condolences but also those of my family who I know would join you in your great pain. Accept, with your children, all my most affectionate thoughts and my dear memories. Affectionately yours, Gina Castelnuovo

Carissima Signora, solamente oggi aprendo l'ultimo numero di Science *apprendo della morte del povero Professore. Non si può immaginare quanto forte sia stata la mia impressione e ancora adesso che Le scrivo non me ne so capacitare. Non pensavo mai che l'ultima volta che ero a NY sarebbe stata l'ultima che avrei visto il Professore. Speravo sempre che i disturbi di cui soffriva sarebbero stati temporanei. Non sa quanto il Professore rappresentasse per me in questo momento; era per me proprio come tornare in famiglia quando venivo a trovarlo. So l'affetto che mio padre aveva per Lui e quanto Lo stimasse e Gli volesse bene, e so anche che ciò era ricambiato dal Professore. Tutto ciò mi aveva fatto sentire, adesso che ero lontana dai miei, ancora più vicina a Lui ed era sempre con un gran piacere che venivo a trovarlo. Non solo era la fama di grande Scienziato che aveva, ma soprattutto la grande onestà, integrità e*

---

[68] *Recent deaths*, Science, vol. 97, issue 2529, 18 June 1943, p. 547.

generosità che me lo facevano stimare e apprezzare. Sia Lui come mio padre sono persone che si ammirano anche senza volerlo e che purtroppo non si trovano frequentemente in questo mondo. Fortunatamente il Professore ha avuto il conforto di avere vicino a sé tutta la sua famiglia e di essere tranquillo per averli in questo paese. So quanto dolore sarebbe per mio padre il sapere della sua morte e non potendo farglielo sapere mando a Lei e ai ragazzi non solo tutte le mie più vive condoglianze ma anche quelle della mia famiglia che so si unirebbe a Lei nel Suo grande dolore. Riceva insieme ai figliuoli tutti i miei più affettuosi pensieri e il mio caro ricordo. Affettuosamente sua Gina Castelnuovo

Chapter 9
'Never Go to a Country Likely to Be at War with Italy': Gino Fano in Switzerland

The eldest among the mathematicians uprooted from racist Italy, and one of the major protagonists of the Italian School of Algebraic Geometry, Gino Fano left Turin for Switzerland in the winter of 1938 at the age of 70 years (Fig. 9.1).

He settled in Lausanne, where he would remain until 1945, engaging in three areas: solidarity, helping Jewish aid associations to 'track down people who had unfortunately disappeared forever';[1] geometry teaching, in the courses organised for Italian university students interned in the Lausanne and Huttwil camps; and dissemination of knowledge, through a series of conferences that he held at the *Cercle mathématique*. The seven years that he spent in Switzerland are generally dismissed as the unpleasant and unjust epilogue of a highly successful scientific life, but this is definitely not the case, as they actually yield a broader narrative: that of Jewish scholars, trained in the Belle Époque of scientific internationalism, who saw the principles of the rule of law denied by race theories, and who witnessed the perversion of collective consciences under totalitarian regimes.

A nineteenth-century gentleman, a nationalist, and far removed from Judaism, Fano came to the decision to leave the country only as a result of pressure from his children. Emigration was not, for him, a choice; the stay in Lausanne was neither a new beginning nor an escape. His feelings of Italianness and sense of patriotism were strong, to the point of making him experience emigration more as an exile than as an elected move. Discrimination did not bring him to an awareness or rediscovery of his identity roots. In the face of persecution, loyalty to his country prevailed over feelings of anger or revenge. Thus, instead of disavowing his Italianness, Fano agreed to work in a 'highly patriotic' initiative and, in front of his colleagues at the *Cercle mathématique,* celebrated the leading position achieved by the Italian geometric tradition in the golden age of mathematical cosmopolitanism (1860–1914). In doing so, he wrote his own chapter in the history of communication

---

[1] Terracini A. (1952a), p. 487: *rintracciare persone purtroppo scomparse per sempre.* See also Terracini A. (1952–53) and Terracini A. (1953b).

strategies and international promotion of the tradition of the Italian School of
Algebraic Geometry, giving us a revealing insight into the sense of belonging to
this research group.

9.1   Nationalist, Atheist but Not Freemason

Born in Mantua into a family where the patriotic tradition was alive and which had
instilled in him 'high feelings of Italianness,[2] Fano was frankly nationalist. In addition,
he had had a military training, with four years at the College of Milan and a very short
transition to the Military Academy of Turin, before undertaking university studies in
Mathematics. His own belief, however, was not merely the Garibaldian sentiment
shared by many Jews of the first post-Risorgimento generations.

In the Great War, as an interventionist of the first hour, he switched his civilian
clothes for uniform and committed himself personally to managing the Industrial
Mobilisation Regional Committee in Piedmont and making his contribution to the
'spiritual assistance of the nation [. . .] that is, to maintain the public spirit in full and
continuous agreement with the supreme directives of the government'.[3] An excellent
connoisseur of recent and contemporary political history, he gave various

---

[2] Segre B. (1952), p. 262: *alti sentimenti di italianità*. His father Ugo had been a Garibaldian
volunteer and took part in the campaign of 1866. Fano always maintained his residence in Mantua
and remained a member of the Jewish community of Mantua.

[3] Fano G. (1915), p. 1: *Assistenza spirituale della Nazione; a dar opera cioè a mantenere lo spirito
pubblico in accordo pieno e continuo colle supreme direttive del governo*. See also Fano G. (1919).

propaganda speeches, including a conference at the Society of Culture in Turin entitled *Il confine del Trentino e le trattative dello scorso aprile con la monarchia austro-ungarica* (June 1915). A small but personal effort, 'inspired by love for the country',[4] as he railed against the Versailles camarilla led by those who 'administrated European politics, and in particular Italy, as it was hardly conceivable that the Austria of the Holy Alliance and the prince of Metternich could do'.[5]

The pathway that links Risorgimento patriotism, interventionism, and nationalism is fully embodied in Fano. Disgusted by the barbarisation of the political debate during the red 2-year period, in the face of rising fascism, he reacted in the manner of an old liberal. His son Ugo recalled one episode. In October 1922, Fano's wife, returning from a train journey in which she had met a group of squads, commented in front of her father-in-law and her husband 'the coup is imminent'. The reactions of the two men were very different:

> My grandfather was distraught. I do not know whether he saw further than we did, whether his sense of fair play and devotion to the rule of law made it impossible for him to accept the sudden departure from constitutional practice, or whether he was mainly shocked by the bad manners and gross behavior of the fascists. Anyway, until he died, he was vociferously opposed though too old to do anything about it. I think my father [Gino Fano] was also displeased but much calmer. He was involved in his scientific work, which the fascists did not disturb. He was certainly strongly nationalistic, Italy was 'my country for right or wrong', and my impression is that he considered Mussolini and his cohorts like a childhood disease of a very young nation, a terrible nuisance but a stage that would pass.[6]

Two years later, in March 1924, when Fano held the conference *Intenti, carattere e valore formativo della matematica* at the Turin War School, he had now fully assimilated some cornerstones of fascist rhetoric: the army as an 'eminent factor of life and greatness of a nation';[7] the importance of scientific education that had made it possible to 'temper the intellectual, physical, and moral qualities of the leaders who had led Italy to victory and had opened new, higher destinies to the country';[8] the link between political and cultural Risorgimento; the eulogy of L. Cremona, who 'at the first whispers of the unified nation'[9] had invited his fellow academics to glorify the homeland with their studies; and the worthy response given to his appeal by Italian mathematicians, who had arrived in a few years at a leading position 'of which Italians should be legitimately proud and which should be of incitement and hope for the future'.[10] Therefore, it is not surprising that Fano, who was not

---

[4]Fano G. (1915), p. 1: *ispirati dall'amore al Paese.*

[5]Fano G. (1915), p. 10: *trattano la politica europea e in particolare l'Italia come appena appena si può comprendere che le trattasse [...] l'Austria della Santa Alleanza e del Principe di Metternich.*

[6]Fano U. (2000), pp. 183–184.

[7]Fano G. (1924), p. 9: *fattore così eminente della vita e della grandezza della Nazione.*

[8]Fano G. (1924), p. 9: *temprarono le doti intellettuali, fisiche, morali, dei condottieri che l'Esercito guidarono alla vittoria, e al Paese aprirono nuovi, più alti destini.*

[9]Fano G. (1924), p. 31: *quando l'Italia, come Nazione riunita, era ancora ai primi vagiti.*

[10]Fano G. (1924), p. 32: *Con orgoglio legittimo di Italiani, prendiamo atto, traendone incitamento ed auspici per l'avvenire del nostro Paese!*

politically neutral or unaware like many colleagues, did not back down vis-à-vis the two *vulnera* inflicted by the regime on the rule of law: the oath of allegiance to the regime (1931) and registration with the PNF (1933).

As far as relationship with Judaism is concerned, Fano was a typical 'Yom Kippur Jew': previous to his generation, all members of his family had been enrolled in the Community of Mantua, and their professional achievements had been promptly published in the pages of *Vessillo Israelitico,* but in the 30s, they were a typical integrated and assimilated family working, living, studying, and socialising in a town like Turin, where less than 1% of the population was made up of Jews:

> Some were more religious than others, some not religious at all but respectful of family traditions. [...] The Jews were full citizens like everybody else, were very few in number, and had little idea of the past cycles of liberation and repression accompanying the ebb and flow of archaic autocratic governments.[11]

Perhaps convinced that there was nothing to fear as long as one 'kept one's nose clean of politics', Fano maintained this feeling until he failed to be elected as a member of the Royal Academy of Italy. In fact, Mussolini himself opposed the candidacy which had been proposed by F. Severi, alleging Fano's affiliation with the Judeo-Pluto-Masonic conspiracy as the motive. For Fano, this brought mortification, which, to some extent, prepared him for the onslaught on arrival:

> Father was never a Mason; he despised any kind of secret society; everything in his life had to be clear and above-board. [...] While there was not yet any overt antagonism, a pervasive kind of anti-Semitism was tacitly acknowledged. Father was shattered by the rejection. I think possibly his deep disappointment prepared him for the storm to come.[12]

Of Jewish descent from both paternal and maternal lineages, Fano was dismissed from his posts at the University and the Polytechnic on 29 November and 7 October 1938, respectively; he was removed from the direction of the Special Mathematics Library, and expelled from all the academic and scientific societies to which he belonged (Lincei, UMI, *Accademia Virgiliana,* Academy of Sciences of Turin, etc.). For Fano, this represented the collapse of the 'three pillars of his life: family, the homeland and the profession'.[13]

The shock of professional demotion, social exclusion, and complete marginalisation from academia was traumatic, although Fano did not reveal his dismay to anyone but family and friends, even responding to the letter of forced resignation from the Polytechnic. As a matter of fact, his situation was simpler than that of other colleagues. First, in 1938, Fano had only three years left until retirement. This meant that he was paid an indemnity that was almost the maximum amount. His family, moreover, held a large fortune and extensive assets and estates,

---

[11] Fano U. (2000), pp. 179, 184.

[12] Fano U. (2000), p. 184.

[13] Fano R. (2004), p. 3.

which allowed them to live with dignity even without paid income.[14] In third place, he enjoyed such good intellectual and social standings as to trust in the success of the reverse discrimination procedure. Fano's sister Alina Regina (born in 1874) was the widow of Leone Wollemborg, Senator of the Historical Left and ex-Minister of Economy and Finance of the Zanardelli government; another sister, Maria Fano Ettlinger (1871–1966) was an accredited translator of novels and other texts from English; his wife, Rosa, was a relative of the economist Roberto Michels. Fano's application for counter-discrimination read as follows:

> Compliant with the letter of the 19th of this month from the Rector of the Royal University of Turin, I am sending here a Memorial (with a short summary, separately) and various documents (as per the list), in order to obtain, pursuant to declaration 6th of this month from the Grand Council of Fascism, that my family (documents 1-2) is considered to belong to the 7th category (families with exceptional merits). I am not, strictly speaking, a war volunteer of the World War 1 (category 2°); however, my military service, in peace and in war, was entirely voluntary, consequent to the nomination as an officer that I requested, without which I would never have been called to war. Conscious and proud of civil, patriotic traditions, benefice of my family, I also carried out activities in various fields and for several decades, with the faith and passion of an Italian and a fascist, with complete and absolute adherence to the events, the ideals, the spirit of the Nation, trying to do good to others and to the Country wherever possible without ever even remotely considering, in my attitudes or in my work, the state of belonging by birth to a religion different from that of almost all Italians. My wife, children, sisters always fully shared such conduct. I trust you will not want to exclude us from national life, and that, above all, you would want to exclude my children, educated and brought up in the religion of the homeland and in the atmosphere of fascism.[15]

The Fanos' applications for counter-discrimination would indeed all be accepted, with that of Gino being approved in 1940, but news of the positive outcome of the procedure was to arrive when the family had already dispersed.

---

[14] Filling in the questionnaire to be submitted to the SPSL (SPSL, *GF*, fol. 297), in the section 'Sources of Income before dismissal', Fano stated: 'About Italian L. 3000 monthly from Univ. position, further, a good family position'. Their properties were confiscated by a special public agency, the *Ente di Gestione e Liquidazione Immobiliare* (Agency for Real Estate Management and Liquidation) in 1943.

[15] ACS-*Contro-discriminazione*: G. Fano to MEN, Turin 31.10.1938: *Conforme a lettera 19 corr. del Sig. Rettore della R. Università di Torino, invio qui unito un Memoriale (con breve sommario a parte) e documenti vari come da elenco, per ottener, a sensi della dichiarazione 6 corr. del Gran Consiglio del Fascismo, che la mia famiglia (docum. 1-2) sia considerata appartenente alla categoria 7° (famiglie aventi eccezionali benemerenze). Non sono, a stretto rigore, un "volontario di Guerra" della Guerra mondiale (categoria 2°); ma il mio servizio militare, in pace e in guerra, fu tutto volontario, conseguente alla nomina ad Ufficiale da me chiesta, senza di che non sarei mai stato chiamato alle armi. Conscio e orgoglioso delle tradizioni civili, patriottiche, benefiche della mia famiglia, ho io pure esplicato attività in vari campi e per parecchi decenni, con fede e passione di Italiano e di Fascista, con completa ed assoluta adesione alle vicende, agli ideali, allo spirito della Nazione, cercando di fare del bene ad altri ed al Paese dovunque ho potuto, senza mai nemmeno lontanamente ricordarmi nelle mie vedute, nei miei atteggiamenti, nell'opera mia, di appartenere per nascita ad una religione diversa da quella della quasi totalità degli Italiani. A ciò mia moglie, i miei figli, le mie sorelle si sono sempre pienamente associati. Confido non si vorrà escluderci dalla vita Nazionale, e che meno che mai si vorrà escluderne i miei figli, educati e cresciuti nella religione della Patria e nell'atmosfera del Fascismo.*

The situation of the young Fanos was decidedly more serious. Fano had two sons: Ugo (1912–2001) and Roberto (1917–2016). The latter was to start his fourth year at the Turin Polytechnic expecting to graduate in 1939, which was guaranteed by law. The elder son, Ugo, who graduated in Mathematics and Physics in 1934, was a pupil of E. Persico and E. Fermi, and a promising atomic physicist but was at the beginning of his career. Ugo—the 'Colossal Fanaccio' (*Fanaccio colossale*) as he was affectionately nicknamed—and Roberto would be the first to make up their minds about emigration, encouraged by their cousin Giulio Racah and Ugo's girlfriend, Camilla Lattes:

> There was no more doubt that we must get out. Where and how? Encouraged by Lilla, I took responsibility for moving my family to safe harbor. There followed months of planning, deciding where we might settle [. . .]. Father was very much against the idea, but he let us do it because he understood that the survival of the family was at stake.[16]

9.2   A Golden Migrant Experience

In one of his last interviews, Robert Fano recalled the family debate around the choice—staying or leaving—which the racial laws forced them to make:

> FANO: And at that point my family, with the exception of my father, decided it was time to go.
> INTERVIEWER: And your father?
> FANO: Well, we persuaded my father to move too.
> INTERVIEWER: So, you all got out?
> FANO: We all got out. Yes.
> INTERVIEWER: You were lucky.
> FANO: [. . .] Basically there was a family emergency reunion, on my birthday as a matter of fact - November 17 - in our country home near Verona. And basically, we decided that we had to scramble, because that war was coming and God only knows what's going to happen.[17]

To flee from Italy, three prerequisites had to be fulfilled: adequate financial assets, a network of international relationships, and a certain mentality, too. The Fano family had the resources because, through highly dangerous smuggling activity, Roberto had succeeded in transferring the family's patrimony to Switzerland. This included the money necessary for his parents to settle in Lausanne, and for himself and his brother to obtain visas for America without asking for the aid of international rescue committees. The inner strength to rebuild one's existence and career as a stranger in a foreign country is not for everyone. Fano did not have it; nor, unlike some colleagues such as G. Fubini, was he willing to accept any solution to keep the family together.

---

[16]Fano U. (2000), p. 187.

[17]Transcript of the Interview MIT 150 | Robert M. Fano '41, ScD '47, in https://infinite.mit.edu/video/robert-m-fano-%E2%80%9941-scd-%E2%80%9947

Disagreement emerged between himself, his wife, and his children, mainly pivoting around 'the American route'. Fano could turn to many personal contacts in the United States, collected throughout his long professional trajectory, and had sojourned in America a few times: the first in St. Louis in 1904 and the last in Los Angeles in 1932.[18] However, he belonged to the generation which identified Paris, Berlin, and Göttingen as traditional destinations of academic mobility and did not understand the young researchers who grew up considering English to be the *lingua franca* and looking to the United States as a new mecca for scientific studies.[19] The desire to appropriate the so-called 'American way of life' was inconceivable for him. Finally, there was also an ideological reason: Fano refused to take into consideration the idea of emigrating to any country likely to be at war with Italy.[20]

Faced with his determination, his wife and children could do nothing but capitulate and be content with persuading him to take refuge in Switzerland. This choice of destination was largely counter-trend. Indeed, Switzerland was only to become the frontier of hope for victims of persecution after the armistice of 8 September 1943. In 1938, when the Fano family relocated there, very few considered it as a destination and generally those who chose it did so for political, not racial, reasons. The Swiss Confederation represented at most a free transit for those who intended to move on towards other destinations. For Fano, by contrast, it was not a temporary solution, but the only one that he was willing to accept.

Once the decision had been made, all Fano's family left and dispersed. The first to flee Italy were Gino and Rosa, who entered Switzerland in November 1938, settling in Lausanne at the Hotel Élite, a quite familiar environment where in the past they had often stayed for business trips or on holiday.[21] Ugo fled to Paris in February 1939. After an unsuccessful attempt to obtain a post in Norway, he managed to get a visa to Argentina. Embarking in Bordeaux, he reached Buenos Aires in July 1939, thence went to New York, and finally landed in Washington, where he was offered a position at the Carnegie Institution's Department of Terrestrial Magnetism.[22] Roberto postponed his departure for a few months to finish his exams, and the delay was almost fatal. Due to the outbreak of the war, he was no longer able to reach Bordeaux, one of the main French ports of embarkation to the United States. In Lausanne, where he went to greet his parents, however, he met his cousin Leo

[18] ASUT-*Fano*: G. Fano to S. Pivano (Rector of the University of Turin), Turin 26.6.1932. In its itinerary, the Fano family would pass through New York (July 25), Washington (July 28–29), S. Francisco (August 6–9), Los Angeles (August 10–14), Boston, and New York (August 20).

[19] In the questionnaire form submitted to the SPSL (SPSL, *GF*, fols. 296, 298), as far as language proficiency is concerned, Fano states that he is fluent in French and German while he can read, write, and speak English 'rather well'.

[20] Fano R. (2004), p. 3: 'my father, as he told me before my departure, would never go to a country likely to be at war with Italy'.

[21] From an interview with R. Fano, it would seem that his parents had reached Switzerland already in November 1938, perhaps directly from Colognola ai colli. Fano had a renewed passport for 1938, so he did not encounter any problems at the frontier.

[22] FANO.1, FANO.2, and FANO.9.

Wollemborg, a writer and journalist who (thanks to the assistance of a high-ranking priest at the French embassy in Zurich) managed to obtain two sets of papers to allow passage through France. Roberto thus reached the United States in October 1939, an immigrant who mitigated the suffering of separation from relatives, friends, and places with the enthusiasm of a 20-year-old, the determination to keep his family's reputation strong, and the certainty of reuniting with his brother. Fano's grandson, Giulio Racah, a former professor of Theoretical Physics at the University of Pisa, settled in Palestine in September 1939, after an intermediate stage in London. The last to leave, in 1940, was Fano's sister-in-law, Ines Cassin Treves (1844–1966), who left for the United States with her sons Enrico, Gino Roberto, and Pietro Giorgio.[23]

In the meantime, from the beginning of 1939, difficulties had multiplied also for those who already lived abroad. In fact, the Fanos could not access the assets they had moved to Switzerland and deposited at the *Union de Banques Suisses*. Thus, at the suggestion of Racah, who paid a visit to the SPSL offices, Ugo wrote to the Society proposing a sort of agreement: an anonymous Swiss person (his father himself) would issue a bank transfer to the organisation, which the SPSL would then pay to him in the form of grants in order to finance his research.[24] The SPSL accepted and sent Gino Fano the questionnaire to be completed so that his case could fall under the administrative laws concerning asylum seekers.[25] Filling in these documents, although it was a mere formality, was unpleasant, even humiliating, for a scholar who had prepared his last curriculum vitae for a job vacancy 32 years earlier.[26] On 6 March 1939, the *Union de Banques Suisses* informed the SPSL that one of its clients (who did not want to be named) had allocated 15,000 francs to Gino Fano to enable him to continue his scientific activity in Lausanne.[27] Therefore, a practice began of monthly scholarships which, albeit with various obstacles due to Foreign Exchange Control, would allow the Fanos to survive the war years with relative peace of mind.[28] In this sense, Fano enjoyed (at least to some extent) a golden experience of emigration.

---

[23] Her husband, Elia Emanuele Treves (1877–1943), would instead be arrested during the German occupation of Turin and deported to Germany. See G. Fano to C. Somigliana, New York 10.12.1947, in D'Agostino (2007), p. 93. His last known whereabouts are on the convoy that departed from Milan on 6.12.1943, destination Auschwitz.

[24] Fano.3, Fano.4, and Fano.5.

[25] Fano.7 and SPSL, *GF*: G. Fano to SPSL, Lausanne 14.3.1939: 'I received your letter of the 8th March. Please to accept my best thanks for your kindness, for the interest you showed to me, and for your friendly help in my present situation'.

[26] SPSL, *GF*: General & Confidential Information, fols. 296–298 and Fano.6.

[27] Fano.8.

[28] Fano's dossier kept in the SPSL Archive includes extensive correspondence between the Foreign Exchange Control, the SPSL, and the *Union de Banques Suisses* concerning the grant payment, which had not gone unnoticed during checks on foreign deposits in 1940 and 1941.

9.3 The Latest Studies on Varieties

Although it was a fictitious grant, the SPSL financed Fano to continue his geometric research and he was required to report annually on his outputs. As a matter of fact, despite his age, Fano was productive: between 1939 and 1945, he published 12 works in *Commentarii Mathematici Helvetici, Revista de Matemáticas y Física Teórica,* and *Acta* of the *Pontificia Academia Scientiarum.*[29] The themes are those typical of Fano's scientific portfolio: algebraic curves, non-rationality and birational geometry in dimension 3, cubic threefolds, Fano threefolds, Fano-Enriques three-folds, Enriques surfaces, and their automorphisms.

Production was renewed in the wake of continuity, as was normal for senior mathematicians displaced from Italy, such as Fubini and B. Levi, and in some ways even mandatory since these scholars had been forced to leave behind their libraries, collections, and manuscripts. From this point of view, Fano benefitted from the fact that, until the summer of 1939, his son Roberto forwarded to him the books and offprints received in Turin.[30] After his departure for the United States, this practice ceased, and he had to 'make do' with what he could find in Swiss libraries.

Although there is no trace of Fano's insertion into the Swiss mathematical scene, which counted leading names such as G. De Rham, P. Finsler, A. Speiser, M. Plancherel, R. Fueter, and so on, those of the Swiss period are not merely remakes or translations of publications that appeared before Fano's emigration. For example, the note *Sulle curve ovunque tangenti a una quintica piana generale* (submitted to the *Commentarii Mathematici Helvetici* on 10 October 1939, and here published in vol. 12, 1940, pp. 172–190) gives evidence of a web of relationships between Fano and the Cambridge geometric School that followed his departure from Turin.[31] But it was mainly in the field of threefolds that Fano achieved his most significant successes, arriving in Lausanne at some important results on classification and rationality problems for cubic varieties. The paper *Nuove ricerche sulle varietà algebriche a tre dimensioni a curve-sezioni canoniche*, ready since 1942, was presented by Severi to the Pontifical Academy in February 1943 but would be published only in 1947.[32] Its contents, however, were partially known both at national and international levels. For example, Segre, J.A. Todd, L. Godeaux, and G. Castelnuovo discussed them in their letters:

> I have already obtained several additional results on cubic surfaces. One of them [...] is that 'a non-singular cubic surface contains no homaloid linear system of complete intersections'. An extension of this result to the non-regular V^3_3 would obviously prove its irrationality. I was told by Fano that this irrationality has been very recently proved by him, on considering the linear system of surfaces of genera 1 lying on V^3_3, but I have not seen the proof. I feel

---

[29] References are listed in Collino, Conte, and Verra (2014), p. 57.

[30] ANL-*Levi-Civita*: G. Fano to T. Levi-Civita, Lausanne 9.2.1939.

[31] Fano G. (1940).

[32] Fano G. (1945, 1947).

that one should be able to obtain the result also by my methods, but I have not yet had time to think seriously about this.[33]

During the war, I had some communication with M. Fano, a refugee in Lausanne; he managed to demonstrate the irrationality of the cubic variety of four-dimensional space, but I don't know his proof yet.[34]

Prof. Fano was ill in Boston, but now is well; he wrote me a long letter, and I have frequent news of him from my daughter Gina. His address is 3510 Rodman str., Washington D.C. N.W. The Memoir on the cubic variety which must be published by the Pontifical Academy has not yet come out; you will have seen the short summary published in the Lincei *Rendiconti*. In these days, a very intelligent young man working here communicated to me a very simple and brief demonstration of the irrationality of the cubic variety based on topological considerations. But I need to think again about the matter.[35]

For Fano, the work on V^3_3 would substantially be his last original article, since he was to stop research in the middle of the war because he felt himself inefficient, as 'too often he had to read papers for a second time'.[36] However, after returning to Italy and until his death, he continued to think about drafting 'a global exposition of these last works, in which, hopefully, some hypotheses of work (it is not known whether and to what extent restrictive ones) made by him should have been removed'.[37]

9.4   The Return to Teaching in the Italian University Camp in Lausanne

Fano was among the first to be recruited by G. Colonnetti for teaching in the Italian University Camps. The two had been colleagues at the Polytechnic of Turin for almost 20 years, from 1919 to 1938. Fano immediately agreed to resume his profession with such energy and genuine enthusiasm as to deserve a personal letter of thanks from Colonnetti at the end of the first semester. In addition to lecturing in Lausanne, Fano also signed up to teach in the University *Studia* of Mürren and Huttwil where, thanks to himself and Modesto Dedò, 'valid disciples were attracted to geometric studies'.[38] In Huttwil, he was both teacher and President of the

---

[33] D6.2.9.

[34] D7.2.6.

[35] D7.3.5.

[36] Fano R. (2004), p. 3.

[37] Segre B. (1952), p. 263: *un'esposizione d'assieme di questi ultimi lavori, nella quale, possibilmente, avrebbero dovuto essere rimosse certe ipotesi di lavoro (non si sa se ed in qual misura limitative) da Lui ammesse [. . .]. Non ci è però noto fino a qual punto Egli abbia potuto dare corso a tale progetto; e sarebbe indubbiamente di grandissimo interesse se, fra le Sue carte, si potesse ritrovare qualcosa di conclusivo al riguardo.* The manuscript plan of the work is kept in BSM-*Fano: Scritti vari*, fols. 45–52, 128–133.

[38] Terracini A. (1953a), p. 709: *discepoli valorosi di questi corsi sono stati attirati verso gli studi geometrici.* See §3.24, D3.24.1, and in ACT: E. Carletto to G. Fano, Mürren 15.11.1943.

examination boards for all mathematics courses and in just one semester he set and held 166 oral tests, assisted by two colleagues only: Alessandro Levi (philosopher of law) and Paolo D'Ancona (historian of art).

Of the geometry teaching imparted by Fano, there remains an evocative textual trace in two volumes of handouts: the *Lezioni di Geometria descrittiva* written by Roberto Ballarati and Franco Brindisi and the lecture notes in analytic and projective geometry compiled by an anonymous student.[39] Fano's expertise in this field was enormous. Suffice it to say that in Turin he had been the only full professor of Projective and Descriptive Geometry from 1901 until the two courses were merged. At the Polytechnic of Turin, he had taught Projective Geometry from the foundation of the School of Engineering in 1908, right up to the time of the racial laws. Equally impressive was his output as a textbook author, which included many treatises on descriptive, projective, and analytic geometry, two of which were co-authored by A. Terracini.[40] The second edition of *Lezioni di geometria analitica e proiettiva* was their last joint project before the departure of Fano for Lausanne and of Terracini for Tucumán.[41] The handouts of Fano's lessons reflect his expertise in this sector. In those of analytic and projective geometry, a sub-discipline to which he had always attributed special value for its insight into the development of geometric studies in Italy, one can grasp the clarity of his didactic style, his way of proceeding 'from a few premises to the construction of large theories, [a way] that left the students full of admiration and almost amazed, especially after the more complicated apparatus of elementary geometry, learned in secondary schools'.[42] Equally evident is the legacy of his classical texts in descriptive geometry, especially the third edition of the *Lezioni di geometria descrittiva date nel R. Politecnico di Torino*, on which he based both theory chapters and the applications (instruments for construction machinery and equipment, photogrammetry, etc.).

Alongside teaching, both Colonnetti and Fano provided personal solidarity. With an Italy divided in two by the Gothic Line, dozens of families turned to them in desperation, looking for news of their children, parents, and relatives. Fano carried out painstaking and commendable work from this point of view, both personally and

---

[39] Fano G. (1944a, 1944b).

[40] G. Fano, *Lezioni di geometria descrittiva*, Torino, litogr. 1903; *Lezioni di geometria descrittiva date nel R. Politecnico di Torino*, Torino, Paravia, 1909, 2nd ed. 1914, 3rd ed. 1926; *Geometria Proiettiva. Lezioni raccolte da D. Pastore e E. Ponzano*, Torino, litogr. 1907; G. Fano, A. Terracini, *Lezioni di geometria analitica e proiettiva*, Torino, litogr. 1926, then Torino, Paravia, 1930, 2nd ed. 1940, 3rd ed. 1948.

[41] Fano returned to Turin a few times, up to 1940, to make arrangements with Paravia in view of the second edition of *Lezioni*. See BSM-*Terracini*: G. Fano to G. Sacerdote Terracini, New York 16.12.1947 and G. Fano to A. Terracini, New York 6.2.1948, Mantua 7.4.1948, Mantua 9.5.1948, Mantua 20.5.1948.

[42] (Terracini A. 1952–53), p. 325: *da poche premesse alla costruzione di teorie di larga portata [...] che lasciava gli studenti ammirati e quasi meravigliati, soprattutto dopo l'apparato più macchinoso della geometria elementare, appresa nelle scuole secondarie.*

in collaboration with spontaneous rescue groups.[43] He was among the first Italian refugee mathematicians to know of the horror of the trains destined for the extermination camps, and he was among those who fought, with the weapons of academic prestige and diplomacy, to stop the horror of deportations.

9.5   Keynote Lectures at the *Cercle mathématique*

Fano is credited as one of the main promoters of the Italian geometric culture abroad, a kind of mathematics communication exercise which he began at the *Mathematische Gesellschaft* in Göttingen in 1893, continued in Aberystwyth 1923, and ended in Lausanne.[44] Here, from May 1942 to February 1944, at the *Cercle mathématique,* Fano held five lectures dedicated to Italian algebraic geometry, from a historical and 'School' perspective, which met with success and gave rise to lively discussions with his new Swiss friends and colleagues G. De Rham, G. Dumas, G. Juvet, and J. Marchand.[45] The history of the Italian geometric Risorgimento, Castelnuovo and Enriques' classification of algebraic surfaces, contributions on threefolds, and on birational transformations allowed Fano to celebrate the results of a tradition to which he strongly felt he belonged. The desire to exhibit, to display in front of foreign colleagues, the best achievements of his own School is not simply the nostalgic recollection of an old geometer at the end of his career but, on the contrary, is a sort of mission that permeates the experiences of dissemination of the Italian refugee mathematicians.[46]

In the first conversation, *Quelques aperçus sur le développement de la géométrie algébrique en Italie pendant le dernier siècle* (evening of 4 May 1942), Fano traced the recent and contemporary history of mathematics in Italy (Fig. 9.2).

After having outlined the difficult beginnings of Italian mathematics, which for a long time remained on the sidelines of the international research scene due to the unfavourable historical-political conditions in which the country found itself, Fano dwells on the geometric Risorgimento which, for him, is parallel to its political counterpart. Recapping the famous plenary lecture pronounced by V. Volterra at the 1900 ICM, Fano underlines the role played by F. Brioschi, E. Betti, and F. Casorati in the development of research and its internationalisation. However, it was Cremona, whom Fano had met in Rome in his youth, who set the foundations for the birth of the Italian School of Geometry, with his works on algebraic curves and

---

[43] FANO.10 and Sarfatti (1979).

[44] See, for example, Fano G. (1923).

[45] Minutes of the sittings of the *Cercle* (Fonds De Rham, [*Séances au Cercle mathématique*] 4.5.1942, 11.5.1942, 2.2.1943, 13.5.1943, 10.2.1944), register the presence of at least twenty participants in each conversation, both *Cercle* associates and invited guests. The manuscript drafts of the five lectures are held in BSM-*Fano* and are digitized in the website edited by L. Giacardi (http://www.corradosegre.unito.it/fondo_fano_s.php).

[46] On this aspect, see Sect. 6.4.

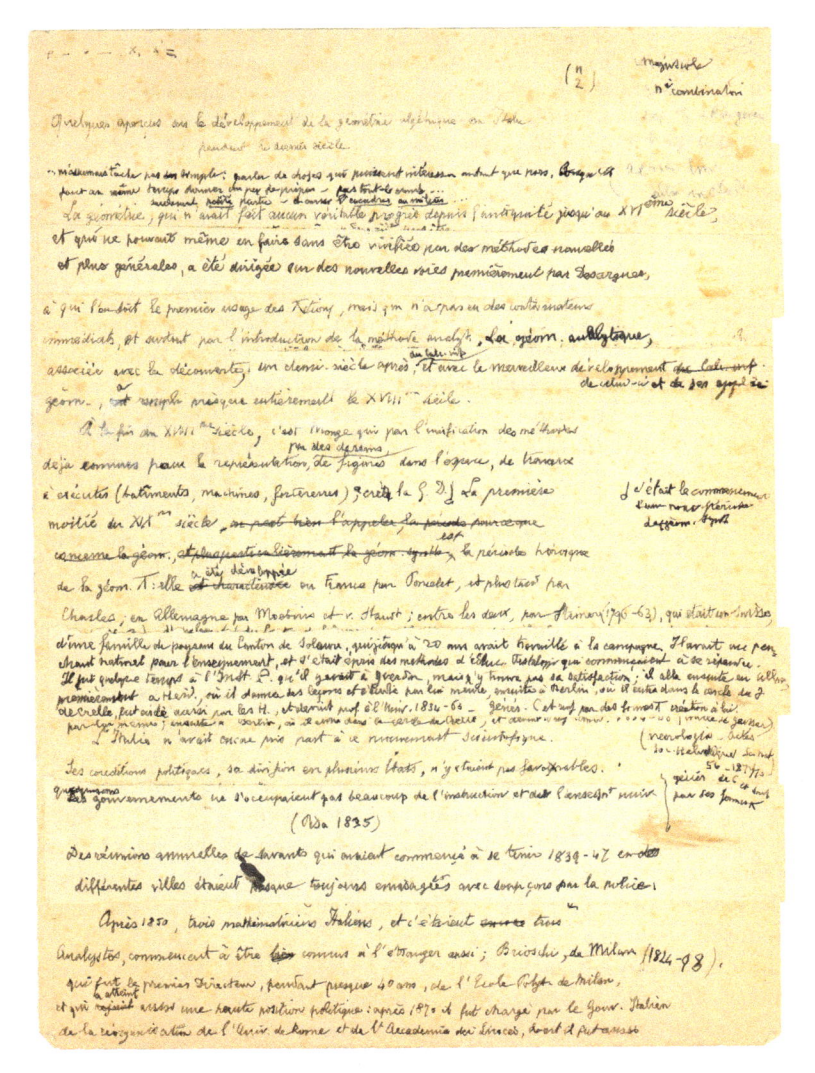

Fig. 9.2 Manuscrit draft of the lecture *Quelques aperçus sur le développement de la géométrie algébrique en Italie pendant le dernier siècle*, fol. 53r, BSM-*Fano*

surfaces, and on birational transformations. This School, characterised by a synthetic and intuitive approach, and strongly influenced by F. Klein, would reach universally acknowledged positions of excellence under the guidance of great Masters such as Veronese, Segre, Bertini, Castelnuovo, Enriques, and Severi.

The second lecture, *Géométrie sur les surfaces algébriques* (11 May 1942) focuses on one of the masterpieces of the Italian School: the theory of algebraic surfaces by Castelnuovo and Enriques. De Rham thus summarised the contents of this talk that was 'full of life' (*plein de vie*):

Les premiers essais d'extension aux surfaces de la géométrie algébrique des courbes rencontrent tout de suite des difficultés : on se contente alors de traiter des cas très particuliers, pour essayer d'y voir plus clair. La notion de genre s'introduit par la considération des séries linéaires. Mais la considération des éléments multiples de la surface introduit une autre définition de genre, ne coïncidant pas avec la première, du moins pour des surfaces 'exceptionnelles', les conditions d'existence d'intégrales simples de 1ère espèce introduisent aussi des phénomènes peu compréhensibles au débout : ils seront élucidés par une étude topologique. M. Fano parle encore, pour terminer, de la rationalité des surfaces, et des courbes exceptionnelles dans les correspondances birationnelles.[47]

The third lecture at the *Cercle*, *Aperçu général sur les surfaces du 3ème ordre* (2 February 1943) deals with a subject particularly dear to Fano: K3 surfaces, Enriques surfaces, and their automorphisms. Various aspects are dealt with: number of straight lines contained in a cubic surface, ruled surfaces of the third order, general surfaces, definition of fundamental points and questions of reality, canonical and Jacobian series, geometric and arithmetic genre of a curve, and a first glimpse in the theory of Abelian integrals according to Picard. Emphasis is given to the problem of the 27 straight lines on the general cubic surface. Two main reference sources are used by Fano: the *Enzyklopädie der mathematischen Wissenschaften*—and this is not surprising—but also an educational text, the lithograph of his course *Complementi di Geometria* edited by A. Bassi, Fano's assistant in Turin in the academic year 1934–1935. In that course, Fano had offered some basic chapters of geometry that did not find adequate space in his 2-year courses and which, nevertheless, he considered essential elements 'of general geometric culture, as is commonly conceived'.[48] Among these, there were, in fact, the surfaces of the third order. With obvious distinctions, due to the vast difference in the preparation of the audience, there are also various similarities between the third conference at *Cercle mathématique* and chapters IV (pp. 123–141) and VI (pp. 177–242) of the lecture notes in *Complementi di Geometria*. Although it is a 'remarkable speech'[49] from the point of view of framing the theme within the state of the art, this is the conference in which the passage of time and the distance accumulated by Italian geometers from other groups of research are more evident.

The fourth lecture, *Les surfaces du 4ème ordre* (13 May 1943), was published posthumously by Aldo Andreotti who, after taking shelter in Lausanne in October 1942, to avoid deportation to labour camps in Germany, attended the courses of De Rham and B. Eckmann at the University and those of Fano in the IUC. Returning to Italy at the end of the war, in 1951, Andreotti took over from Fano in the chair of Analytic and Projective Geometry. The lecture *Les surfaces du 4ème ordre* represents the natural continuation of the previous one, focusing on quartic surfaces. Recalling the studies of M. Noether and W.F. Meyer, Fano gives the general form of the fourth-order surfaces, derives the equation of the tangent plane, and examines the

---

[47]Fonds De Rham, [*Séances au Cercle mathématique*], 11.5.1942, fol. 85.

[48]Fano G. (1935), p. IX: *di quella cultura generale geometrica, che viene comunemente presupposta.*

[49]Fonds De Rham, [*Séances au Cercle mathématique*], 2.2.1943, fol. 89: *remarquable exposé.*

singular points, paying special attention to the surfaces of Kummer, Veronese, and Steiner. Again, the literature cited is extensive but very classic.[50]

In the last one, *Transformations de contact birationnelles dans le plan* (10 February 1944), Fano outlines his contributions concerning the birational geometry in dimension 3.[51] On this occasion, he had the opportunity to recapitulate some of his past studies presented at the International Congress of Mathematicians in Bologna in 1928 and to highlight two important issues, on which he had resumed work in recent times: systems of curves that correspond, under a birational contact transformation, to points of the plane or to straight lines, and the determination of the simplest operations with which to obtain, as products, the totality of birational transformations.[52]

9.6   A Commuter Between Two Continents

Fano was both among the first to leave Italy and among the first to come back. The time in Switzerland was a parenthesis for him, to be closed as soon as possible in view of the future. As he wrote to his friend Carlo Somigliana, if the second half of his life 'took place in a less quiet period, I could say more agitated than the previous [...] we have happily overcome it, and we trust well for the future'.[53] This future was envisaged by Fano in Italy, where he returned immediately after the liberation. What Terracini defined 'el dilema de la vuelta' did not affect him.

However, the case of his sons was quite different. Italian Jews who arrived in America on the eve of the war can be somewhat schematically divided into two groups: those who had come as immigrants and then tried to integrate into the new world, and those who had come as refugees, and were always ready to return to their homeland.[54] The young Fanos fall into the former category. Both were now settled, married, and with children; they had obtained American citizenship in 1946 and had completed the path from de-nationalisation to re-nationalisation. Ugo was in Washington, Head of the Nuclear Physics section at the Bureau of Standards.

---

[50] From this point of view, it must indeed be said that the antiquated horizon of Fano's cultural references could have aroused some perplexity in Andreotti who, in fact, took care to note (Fano G. (Andreotti) 1953–54, p. 301) that the text of Fano gave 'with great elegance, a very complete and panoramic idea of the theory of fourth-order surfaces, but it was neither conceived nor drafted for printing'. (*dà con molta eleganza un'idea assai completa e panoramica degli argomenti trattati. È da tenere presente che il testo [...] non era redatto per la stampa.*)

[51] Fano G. (1931a, 1932).

[52] BSM-*Fano: Transformations de contact birationnelles dans le plan*, Scritti.3, fol. 2r: *negli ultimi mesi.*

[53] G. Fano to C. Somigliana, New York 18.10.1950, in D'Agostino (2007), p. 96: *si è svolta in un periodo meno tranquillo, potrei dire più agitato del precedente; ma l'abbiamo felicemente superato, e confidiamo bene per l'avvenire.*

[54] Pontecorboli (2013), p. 144.

Roberto had received his doctorate in 1941 and was now a professor of Electrical Communications and Computer Science at the Massachusetts Institute of Technology. The desire to establish themselves in their new homeland and contribute to determining its technological-scientific primacy prevailed over the sense of dispossession that they had experienced after racial discrimination.[55]

Faced with their determination to stay in America, this time it was the father who capitulated, agreeing to spend the last few years of his life partly in Italy and partly in the United States, where he arrived for the first time after the war in August 1946. Thus, although reinstated in service on 29 October 1945, Fano only nominally resumed teaching from May to November 1946, when he retired.[56] Aware that his return to the chair was fictitious, he immediately told his colleagues that the time had come to think about his succession, a succession that—as he confided to his friends Tricomi and Castelnuovo—he hoped would be ensured by Segre or Terracini, who 'were meritorious of the country, having resumed their university positions'.[57]

On the contrary, Fano continued to maintain his contacts with the Italian and American milieu and made a decisive contribution to the reconstruction of the heritage of the Special Mathematics Library of Turin, which had been partially destroyed in the bombings of 1942–43, by donating his entire collection of over 5,000 offprints. Declared an emeritus professor in the summer of 1948,[58] he dedicated the last years of his life to writing a monograph concerning the demonstration of the irrationality of the V_3^3 of P^4. The idea, which Fano 'does not know if he feels like it to carry on',[59] was to draw up an overview of his works, simplifying some points and trying to remove some restrictive hypotheses, relating to the basic curves of linear systems. Among his manuscripts kept in Turin, however, no significant traces of this synthesis have been found. His last talk, held at the Mathematical Seminar of the University and Polytechnic of Turin in February 1950 and published in his *Festschrift* volume, focused on this subject.[60]

---

[55] See Campanile (2018), p. 363.

[56] Fano.11, Fano.12 and ASUT-*Fano*: M. Allara (Rector of the University of Turin) to G. Fano, Turin 5.11.1945; G. Fano to M. Allara, Lausanne 8.12.1945.

[57] CA, *BSP*: G. Fano to B. Segre, New York 21.3.1945 (D7.3.10): *Lei e Terracini si sono resi veramente benemeriti del Paese, riprendendo le loro cattedre universitarie.* See also Segre.32 and D7.2.9.

[58] ASUT-*Fano*: M. Allara (Rector of the University of Turin) to MPI, Turin 18.6.1948; MPI to M. Allara, Rome 19.7.1948; M. Allara to G. Fano, Turin 24.7.1948; G. Fano to M. Allara, Colognola ai Colli 1.8.1948. The proposal to declare Fano emeritus came from Terracini, who also drew up the report concerning 'his merits as teacher and scientist' (*meriti come maestro e come scienziato*). The report, dated 7 June 1948, is kept in ASUT-*Fano*.

[59] BSM-*Terracini*: G. Fano to A. Terracini, Colognola ai colli 12.7.1951: *non so se me la sento!* See also BSM-*Terracini*: G. Fano to A. Terracini, 25.7.1951, 21.11.1951.

[60] ASUT, Correspondence of BSM: A. Terracini to G. Fano, Turin 15.2.1950, G. Fano to A. Terracini, New York 5.1.1951 and New York 23.1.1951, A. Terracini to G. Fano, Turin 29.1.1951.

Documents

Fano.1 U. Fano to T. Levi-Civita, Paris 1.2.1939

ANL-*Levi-Civita*, p.c.
Distinguished Professor, A friend of mine in Trondheim is trying to get me a place there and warns me that it would be useful to have a few written lines of presentation sent to the mathematician Prof. Viggo Brun of the local *Norges Teknisk Høgskole*.[61] Could I ask you to take an interest in this? I am here now, but I would not like to remain; yesterday in Lausanne I saw Dad, who is fine. With the warmest thanks, and greetings to you, your wife and to Sig.na Cornelia, Yours, Ugo Fano

Chiar.mo Professore, Un amico mio di Trondheim sta tentando di farmi avere un posto colà e mi avverte che mi sarebbe utile una riga di presentazione inviata al matematico Prof. Viggo Brun della locale Norges Teknisk Høgskole. *Potrei pregarla di interessarsi di questa cosa? Io sono ora qui, ma non vorrei rimanerci; ieri a Losanna ho visto papà, che sta benino. Coi più vivi anticipati ringraziamenti, ed ossequi a Lei, alla Signora e alla Sig.na Cornelia, Suo Ugo Fano*

Fano.2 U. Fano to Levi-Civita, Paris 15.2.1939

ANL-*Levi-Civita*, fol. 1r
Distinguished Professor, Many, many thanks for taking care of me so kindly, and again for your encouragement, which is very welcome. I know from Trondheim that your letter has benefited me greatly, although it could not avoid a decision unfavourable to me. However, the letter itself has been passed on to other Norwegian professors who are now attempting to obtain at least my residence permit. With renewed thanks, I offer you my respects. Yours, Ugo Fano

Chiar.mo Professore, Mille, mille grazie di essersi occupato così gentilmente di me, e ancora per i suoi rallegramenti, graditissimi. So da Trondheim che la sua lettera mi ha giovato molto, benché non abbia potuto evitare una decisione sfavorevole a me. Tuttavia, la lettera stessa è stata passata a altri professori norvegesi che ora stanno occupandosi di procurarmi almeno il permesso di soggiorno. Con rinnovati ringraziamenti, Le porgo i miei ossequi. Suo Ugo Fano

---

[61] Fano, who had met the physicists Johan Holtsmark (1894–1975) at the informal Copenhagen conference of 1936, tried to join the Trondheim accelerator laboratory. Holtsmark rejected Fano's application and excused the rejection in a letter to E. Fermi on the ground that Norway did not fund foreign scientists. Viggo Brun (1885–1978), mathematician and number theorist, was a professor at the Technical University in Trondheim (1923–1946).

Fano.3 U. Fano to SPSL, Paris 15.2.1939

SPSL, *GF*, fol. 303

To the Society for Protection of Science and Learning – London. During the last November I had some contact with you and then I sent you my curriculum vitae with questionnaire and publications; on that time, I lived in Italy at Turin. The present letter doesn't regard myself, but my father, Prof. Gino Fano, 68 years old, former professor of Projective Geometry at the University of Turin, who has been dismissed "as a Jew" in the last autumn and had still no contact with you. He doesn't think to get another position abroad for his age, but he lives now at Lausanne (Switzerland) and works there at the mathematical institution of the University as a private scientist. He has fortunately own money to support himself and his family, but he meets now the following question. He has to give to the Italian consulate an evidence regarding his possibility of support in Switzerland, without showing his real economical resources, because it is forbidden to have money abroad. (On the same ground the present communication to you has a very reserved character). This difficulty is a merely formal one, but we couldn't still overcome it. Now I hope that you can help me, perhaps agreeing with an arrangement of the following sort. A Swiss person, Mr. N.N. (if necessary, a real name could be given), would write you and offer a sum (for instance Frs. 15,000, i.e. 750 Lg.), showing the desire, that you employ it to give a corresponding grant for an year to Prof. G.F. "for scientific work at the University of Lausanne"? Is it possible for you to agree with such an arrangement? Recently my cousin, Prof. Racah, has payed a visit to you in London and spoken about such questions; then you told him, that such arrangements aren't impossible. I hope to hear directly your opinion about it. With my best thanks and regards, Yours sincerely Ugo Fano

Fano.4 D.C. Thomson [SPSL] to U. Fano, London 17.2.1939

SPSL, *GF*, fol. 304

Dear Dr. Fano, In reply to your letter of the 15th February concerning your father Professor Fano, I write to enquire whether the procedure you outline in your last paragraph would involve the money actually being sent by a Swiss person to our funds and reforwarded to Professor Fano. In event of its being an actual transaction of which the description by us would be fictional we are prepared to collaborate. We feel however that we could not write about a transaction which had not taken place at all. We are willing to receive and forward the money describing it as you suggest. I am, Yours very truly, David Cleghorn Thomson – General Secretary

Fano.5 U. Fano to D.C. Thomson [SPSL], Paris 19.2.1939

SPSL, *GF*, fol. 305
Dear Sir, in reply to your letter of the 17th February, I am heartily grateful to you that you are prepared to help us. Of course, we are prepared to make a real transaction and to send you actually the money. Concerning the practical application, we could for instance follow this procedure: I beg you now to send directly to my father (Prof. G. Fano, Hotel Élite, Lausanne) a copy of your "Questionnaire", in order that he would be regularly inscribed in your lists of displaced students, and he would officially beg you of interesting yourselves with him. As an answer you may then outline you have some hope of really being in condition to give him a grant. On that time, you will receive a letter from a Swiss person, who will be disposed to send you the money for the purpose of giving a grant to Prof. Fano. In short it seems to me that the procedure should be as like as possible to the ordinary cases you are dealing with. In connection with this question, I should like to know whether the order of magnitude of the sum I have proposed you (that is about 60 Lg. monthly) seems you to be the right one or perhaps too big. In conclusion, I thank you once more very much. With best regards, I am, Yours sincerely, Ugo Fano

Fano.6 Gino Fano Curriculum Vitae

SPSL, *GF*, fol. 302
1888-1892 Studied at the University of Torino under Professors C. Segre, G. Castelnuovo (assistant), E. D'Ovidio, G. Peano

1892 Doctor's degree in Torino, with a thesis on algebraic curves

1892-93 Assistant to Prof. D'Ovidio in Turin

1893-94 Scholarship at the Mathem. Institute of Göttingen (Germany) under Prof. F. Klein

1894-99 Assistant to Prof. G. Castelnuovo in Rome

1895 *Privat-Dozent* for Analytical and Projective Geometry, University of Rome

1899-901 Extraordinary (adjoint) Professor for Algebra and Analytical Geometry, University of Messina.

1899 I received an invitation as extraord. Professor at Göttingen but did not accept it.

since 1901 Professor

1905-1938 Ordinary Professor of Geometry in Turin

Since 1926 also Director of the Mathem. Institute and Library

1923 Accepting an invitation of the University of Aberystwyth (Wales), I delivered there a series of lectures on Italian Geometry, during the last 50 years.

It is impossible to me to send from here a complete list of my publications, the number of which is *certainly more than one hundred* (1892-1938). The greatest number concern *algebraic Geometry* (Postulates of Geometry – Geometry of the line, particularly congruences of the 2nd and 3rd order – Surfaces and M₃ with infinite projective or birational transformations – Cremonian Transformations in the plane, in the space, their groups, their infinitesimal transformations, applications to projective Geometry & differential equations – Birational contact-transformations. Many questions on algebraic Curves, surfaces, M_3, ... particularly, during the last years, on M_3^{2p-2} of S_p, which have all genera zero, and are not rational. These researches are still in course).

Volumes published: *Geometria descrittiva* (460 pg., 400 fig.) – 1910 – 3rd edit. (1925)

Geometria analitica e proiettiva (640 pg.) – 1930 – 2nd edit. in course

Geometria non euclidea (250 pg.) – 1935

Many autographed courses of lectures. The last one:

Complementi di geometria (246 pg.) – 1935.[62]

Fano.7 D.C. Thomson [SPSL] to U. Fano, London 20.2.1939

SPSL, *GF*, fol. 306

Dear Dr. Fano, I have to-day sent a copy of our questionnaire to your father at Lausanne. With regard to your enquiry about the size of the grant, our grants are never bigger than £250 a year unless they are given from an outside source and earmarked, but it could be perfectly simple in this case to give a grant of whatever amount is mentioned, because we should simply say it was an earmarked grant. I am, Yours sincerely, David Cleghorn Thomson - General Secretary

---

[62] G. Fano, *Lezioni di geometria descrittiva date nel R. Politecnico di Torino*, Torino, G.B. Paravia, 1910; G. Fano, A. Terracini, *Lezioni di geometria analitica e proiettiva*, Torino, G.B. Paravia, 1930; G. Fano, *Geometria non euclidea. Introduzione geometrica alla teoria della relatività*, Bologna, Zanichelli, 1935; G. Fano, *Complementi di Geometria*, Torino, G.U.F., 1935.

Fano.8 Union Bank of Switzerland to SPSL, Zurich 6.3.1939

SPSL, *GF*, fol. 308
Dear Sirs, One of our clients who does not wish to be named has been informed that you should like to find a method of financing in order to enable *Prof. Gino FANO, formerly Prof. at the University of Torino, at present residing at Lausanne*, to continue his researches. Our client – being desirous of facilitating the matter of his financial assistance – asked us to inform you that he had decided to put at your disposal an amount of S. Fr. 10,000 at the condition that you pay to Prof. Gino Fano a monthly scholarship of Lg. 40.-/- for the period of one year, from April 1st 1939 until March 31st 1940, "for his scientific researches at the Mathematics Institution of the University of Lausanne." The above scholarship could be sent by you to Prof. Fano directly, in the course of the year, in one amount or in several instalments, or through the medium of our Lausanne Branch. Any communications of you in the matter should be addressed directly to Prof. Gino Fano, whose address is Hotel Élite, Lausanne, however without mentioning in any way by whom the money has been deposited. As soon as we shall have received your agreement as to above system, we shall remit you the amount of Fr. 10,000. According to the authorisation of our customer we confirm to you that a small difference between the amount of Fr. 10,000 and your remittances of Lg. 480. -/- would be destined for the funds of your Society. Yours faithfully, Union Bank of Switzerland

Fano.9 Agenda, 21.11.1939, Applications for New People

ECADFS Records, *UF*, fol. 1r, 2r, 3r
United States Public Health Service – Ugo Fano physicist

$1,000 for six-month period. Permanency not assured

Age: 27

Status: Professore Incaricato (Assistant Professor?) at the Architectural Faculty, University of Rome

Specialty: theoretical physics

Now in the United States

"I am writing in behalf of Dr. Ugo Fano who has been appointed Research Fellow with the Washington Biophysical Institute for a six-months' period without compensation. While we have as yet no real appraisal of Dr. Fano's ability, because simply of the shortness of duration of our acquaintance, we have been well impressed with his personality, excellence of training and previous associations.

He proposes to investigate from the standpoint of theoretical physics, the bio-physical data which we are obtaining on the effectiveness of both visible and

X-radiation in producing lethal and genetic changes. In this work he will be associated not only with our group but also with those of the National Institute of Health, the National Cancer Institute, and Dr. Tuve's group in the Terrestrial Magnetism Laboratory of the Carnegie Institution.[63]

We hope to be able to offer Dr. Fano a real opportunity to demonstrate his ability in a field of research which specially commanded his interest. The six-months' period which I have mentioned may be extended if Dr. Fano's work shows promise as, of course, the time is too short for extensive investigation.

While Dr. Fano has offered to work without compensation I believe his opportunity for productive research would be greatly improved if reasonable compensation could be furnished him. Since we have no funds available for this, I wish to request in his behalf that a grant-in-aid be made to him". – F.S. Brackett, Principal Physicist, Division of Industrial Hygiene, National Institute of Health, United States Public Health Service, and Director of the Washington Biophysical Institute, October 30, 1939.[64]

On November 16 Dr. Brackett wrote again: "Dr. Fano has made application to the Guggenheim Foundation for appointment as Fellow. While I do not feel at all sure that there will be favorable action, you may want to await their decision before proceeding with consideration of this request.

As to the amount, a man of Dr. Fano's training would normally receive a salary in the range from $2,000 to $2.400. No doubt, however, Dr. Fano would appreciate any smaller substantial grant that would ease his financial situation. Though not in actual distress at this time, obviously his meager reserves could not last over a protracted period.

I would suggest therefore that for a six-months period, a grant of $1,000 might be considered, subject, of course, to the action of the Guggenheim Foundation."

Professor Enrico Fermi, Nobel Prize Winner in Physics, who is now at Columbia University, wrote on November 4 concerning Dr. Fano: "Dr. Ugo Fano graduated in Physics in Turin and afterwards worked with me in Rome. He is a young man of serious training and of good will, whose scientifical knowledge is based on solid ground. His present interest is particularly directed towards the application of physics in the field of biology, and he is planning to pursue this kind of researches with the fellowship that he might obtain from you."

Professor Heisenberg of the University of Leipzig wrote on March 22, 1939, as follows: "Dr. Ugo Fano has worked scientifically at the Institute for Theoretical Physics of the University of Leipzig in the year 1937. From his work and from

---

[63] Merle Tuve (1901–1982) was the Director of Terrestrial Magnetism Research at the Carnegie Institution for Science (1946–1966).

[64] Frederick Sumner Brackett (1896–1988), physicist and spectroscopist, Director of Biophysics Research at the National Institute of Health from 1936.

many discussions with him I have come to the conclusion that Dr. Fano is completely master of the methods of modern theoretical physics and that he has a very good understanding of experimental atomic physics. Dr. Fano attacks scientific problems with great energy and vigor and his activity has contributed essentially to the life of scientific management of the Institute."[65]

Dr. Fano states that in 1936-37 he won a fellowship to study abroad and went to Leipzig to work under Professor Heisenberg. In 1938 he won a competition for a position of Assistant of Experimental Physics in Italian universities, but resigned in order to accept an appointment as Professore Incaricato at the University of Rome. In the fall of 1938, he was dismissed from this position as a consequence of the racial laws.

Supplement to Agenda. November 30, 1939. The Committee will be interested in the following letter of November 20 from Dr. M.A. Tuve of the Department of Terrestrial Magnetism of the Carnegie Institution of Washington. "I am writing to advise you that I think highly enough of the qualifications of Dr. Ugo Fano to have made special arrangements at considerable trouble for him to work in Washington during this winter. I think that he is a scholarly type of man which we should be glad to help in every way in this country. His interests and training are unusual, in that he is one of the three or four men I know in physics who are making serious efforts to apply the ideas of physics to the analysis of biological problems. He has worked for several months here in our laboratory, and I have helped to arrange for him to be a guest investigator at the National Cancer Institute under the auspices of the Washington Biophysical Institute this winter.

Any help that you may be able to give Dr. Fano will be to the advantage of research progress."

Fano.10 G. Fano to G. Colonnetti, Lausanne 3.12.1944

ACS-*Fano*, fol. 1r
Dear Professor, I am again here to disturb you, to beg you to add this postcard to the letter for my sister, to inform her of the following. Her postcard (that is, of my sister) of 17/10 reached me yesterday afternoon, while of her previous letters, which she confirms having sent, nothing has so far arrived. I phoned with the news of the Montalcinis to Aldo Levi: more precisely, I spoke with his wife, from whom I learned that they had recently had a telegram "Gino Bianca Montalcini well." I also communicate the news to Elda M. and Carolina. I also see that nobody yet knows nothing, unfortunately, about Pia S. and Clara, with nothing

---

[65] Werner Heisenberg (1901–1976) was the Head of the Department of Physics at the University of Leipzig (1927–1943).

being discovered here so far about the latter. Thank you again, with best regards and best wishes for a good trip. Gino Fano

Egregio Prof.^e^, Sono di nuovo a disturbarla, per pregarla unire alla lettera per mia sorella anche la presente cartolina, per soggiungerle quanto segue. La sua cartolina (cioè di mia sorella) 17/10 mi è giunta ieri pomeriggio, mentre delle sue precedenti missive, che mi conferma, nulla mi è finora pervenuto. Ho telefonato iersera stessa le notizie dei Montalcini a Aldo Levi: più esattamente, ho parlato colla moglie, dalla quale ho saputo che avevano avuto recentemente un telegramma "Gino Bianca Montalcini bene." Comunico pure le notizie a Elda M. e a Carolina. Vedo anche nulla aveva ancora saputo, purtroppo, della Pia S. e Clara, della quale ultima qui finora null'altro si è potuto sapere. A Lei grazie di nuovo, coi migliori saluti e auguri di buon viaggio. Gino Fano

Fano. 11 Memorandum for the Ministry of Higher Instruction, General Direction, Lausanne 30.3.1946

ACT, fol. 1r-v

In the year 1943-44, according to agreements made with Prof. Gustavo Colonnetti, I held 2 courses in Analytic and Projective Geometry (4 hours per week, given the lack of preparation of the students) and in Descriptive Geometry (3 hours) to the Italian military students interned in Lausanne (about 150 lessons in total). In 1944-45, the students of the first two years of the engineering programme, who were interned in Switzerland, were gathered in the camp of Huttwil (a village in the Canton of Bern), where - not being able to move there - I went twice a week, with 3 hours of travel both on the way out and on the way back, presenting 2 lessons of Analytic and Projective Geometry each time, with a short interval, and 1 of Descriptive Geometry. This must have been communicated to the Ministry through the documents sent by the Swiss camps to the Royal Legation of Bern. For these teachings I received only a small allowance. In October 1945 I had to have a cataract surgery on my left eye. The operation was very successful, but due to some other slight ailment that occurred earlier (perhaps due, it was said, to the overwork of the previous months), I was advised not to resume teaching for some time. In November, while I was still recovering from the operation, I was informed by the Rector of the University of Turin [Mario Allara] of my return to service, with a decree in progress, effective for economic aspects from January 1, 1944. I replied by thanking him for the communication, without being able to specify when it would be possible for me to be in Turin. Meanwhile, I was about to reach, (and have now done so) my 75th birthday, on 5 January 1946. While last year I submitted without hesitation to the fatigue of the trip and the lessons in Huttwil (because I was convinced that, given the circumstances, it was my duty to do so), this year I could not [help] but wonder what advantage the students of Turin would have if I resumed my course for a few months, replacing those who had taught it in previous years and who must take it up again next year. Therefore, I ask the Ministry to examine whether, considered the teachings I imparted here in the years 1943-44 and 1944-45 and the other issues set out above, even though I have not now resumed teaching in Turin, my resumption of service can equally be considered effective, in compliance with the aforementioned decree and with effect from 1 January 1944, up to that time that the Ministry itself will establish, before retiring me again. As far as I know, my pension was paid to me through the bank account until January 1945, and then it was suspended. Prof. Gino Fano

Pro-Memoria pel Direttore Generale dell'Istruzione Superiore

Nell'anno 1943-44, in seguito a accordi col Prof. Gustavo Colonnetti, ho tenuto agli studenti italiani militari internati a Losanna i 2 corsi di Geometria Analitica e Proiettiva (4 ore settimanali, data la poca preparazione dei giovani) e di Geometria Descrittiva (3 ore, con complessivamente circa 150 lezioni). Nel 1944-45 gli studenti del 1° biennio di ingegneria internati in Svizzera furono raccolti nel campo di Huttwil (villaggio nel Cantone di Berna), dove – non potendomi colà trasferire – io mi recavo 2 volte per settimana, con 3 ore di viaggio sia all'andata che al ritorno, tenendovi ogni volta 2 lezioni di Geometria Analitica e Proiettiva, con breve intervallo, e 1 di Geometria Descrittiva. Ciò deve risultare al Ministero dai documenti inviati dai campi Svizzeri alla R. Legazione di Berna. Per questi insegnamenti ho percepito soltanto una piccola indennità. Nell'ottobre 1945 ho dovuto farmi operare di cataratta all'occhio sinistro. L'operazione è riuscita ottimamente, ma per qualche altro lieve disturbo manifestatosi prima (dovuto forse, fu detto, al lavoro dei mesi precedenti), mi fu consigliato di non riprendere per qualche tempo l'insegnamento. Nel novembre, mentre ero ancora convalescente dell'operazione, mi fu comunicata dal Rettore dell'Università di Torino la mia riassunzione in servizio, con Decreto in corso, e decorrenza agli effetti economici dal 1° gennaio 1944. Risposi ringraziando della comunicazione, senza poter precisare quando mi sarebbe stato possibile essere a Torino. D'altra parte, io stavo per compiere e ho oramai compiuto, il 5 gennaio 1946, il 75° anno di età. E mentre l'anno scorso mi sono sottoposto senza esitare alla fatica del viaggio e delle lezioni a Huttwil, perché convinto che, date le circostanze, era mio dovere farlo, quest'anno non ho potuto [fare] a meno di domandarmi di che vantaggio sarebbe stato per gli studenti di Torino un mio corso di pochi mesi, sostituendomi io a chi l'aveva tenuto negli anni precedenti e l'avrebbe ripreso l'anno prossimo. Prego pertanto il Ministero di esaminare se, dati gli insegnamenti da me qui impartiti negli anni 1943-44 e 1944-45 e quanto altro sopra esposto, possa egualmente tenersi ferma – pur non avendo ora ripreso l'insegnamento a Torino – la mia riassunzione in servizio, conforme al Decreto citato e con decorrenza agli effetti economici dal 1° gennaio 1944, fino a quell'epoca che il Ministero stesso crederà stabilire, collocandomi poi di nuovo a riposo. Per quanto mi consta, la pensione di cui già godevo mi fu versata in c.c. postale fino a gennaio 1945, e poi sospesa. Prof. Gino Fano

Fano.12 G. Castelnuovo to Sangiorgio [MPI], Rome 6.4.1946

ACS-*Fano*, fol. 1r-2r

Dear Comm. Sangiorgio, Gino Fano, professor of Analytic Geometry at the University of Turin, asks me to give you this memorandum. Rather than wasting your time with my visit, I prefer to send you the memorandum and beg you to give it your kind attention. Fano, a refugee in Lausanne during the German domination, was unable to return to Turin to take his place after the liberation of Piedmont, having to undergo a serious operation in one eye. Only today will he return to Italy but having celebrated his 75th birthday in January he is about to retire.[66] Is it worth it that in these conditions he resumes teaching at the University of Turin for three or four weeks, finishing a course already started by others? The useful courses he gave in Switzerland to young Italians who had taken refuge there could be regarded as equivalent to the teaching in Italy required for his readmission to service? Prof. Colonnetti, who organised them, spoke to me very warmly of the courses given by Fano. That there is *moral* equivalence between the courses given in Italy and those given to Italian citizens sheltered in Switzerland is undisputable. It remains to see how an equivalence can be justified from a *formal* point of view, to place Fano in the same conditions in which he would find himself if he had resumed teaching in Italy. Such is the question that Fano and I submit to your judgment. I recommend that you examine the matter carefully and communicate the result of your exam to Fano or to myself. If you would like to meet me and talk about it, please call me at 43259. I thank you very much for the care with which you will study the case, and I take this opportunity to send you my best regards and cordial greetings, yours, much obliged, G. Castelnuovo

Egregio Comm. Sangiorgio, Gino Fano, professore di Geometria analitica all'Università di Torino, mi prega di consegnarle l'unito pro-memoria. Piuttosto che farle perder tempo con una mia visita, preferisco mandarle il memoriale e pregarla di dedicarvi la sua benevola attenzione. Il Fano, profugo a Losanna durante la dominazione tedesca, non poté far ritorno a Torino per riprendere il suo posto dopo la liberazione del Piemonte dovendo subire una grave operazione ad un occhio. Soltanto ora rientrerà in Italia, ma avendo compiuto in gennaio il 75° anno sta per esser collocato a riposo. Val la pena che in queste condizioni egli riprenda per tre o quattro settimane l'insegnamento all'Università di Torino, per terminare un corso già iniziato da altri? Non potrebbero esser riguardati come equivalenti all'insegnamento in Italia che sarebbe richiesto per la sua riammissione in servizio i corsi veramente utili che egli ha dato in Svizzera ai giovani italiani che si erano rifugiati colà? Di questi corsi impartiti dal Fano mi parlava tempo fa con molto calore il Prof. Colonnetti che li aveva organizzati. Che vi sia equivalenza morale tra i corsi dati in Italia e quelli dati ai cittadini italiani riparati in Svizzera, direi di sì. Resta a vedere come dal punto di vista

---

[66]Fano was born on January 5, 1871.

formale *si possa giustificare una equivalenza, in guisa da porre il Fano nelle stesse condizioni in cui si troverebbe se avesse ripreso l'insegnamento in Italia. Tale è la questione che il Fano ed io sottoponiamo al Suo giudizio. Le raccomando di esaminare attentamente la cosa e di comunicare al Fano o a me il risultato del Suo esame. Se desidera vedermi e parlare con me della cosa, non ha che da telefonarmi al n° 43259. La ringrazio in ogni modo della cura con cui studierà il caso, e colgo l'occasione per inviarle i miei distinti e cordiali saluti, obbl. G. Castelnuovo*

Chapter 10
Bringing to England 'the Foremost of the Younger School of Italian Geometers': Beniamino Segre

10.1 The Darling Pupil of Francesco Severi

Beniamino Segre was the youngest among the mathematicians who left Italy after the introduction of the racial laws (Fig. 10.1).

Born in Turin in 1903, at the age of 16, he enrolled at the University of Turin with a scholarship. Here, his teachers were C. Segre (cousin of his mother), G. Fano, and G. Fubini. He graduated with a thesis on algebraic geometry entitled *Genere della curva doppia per la varietà di S_4 che annulla un determinante simmetrico*, with supervisor Segre. In the same year, 1923, he published his first work, *Sul moto sferico vorticoso di un fluido incompressibile* (Annali di Matematica pura ed applicata, (4), 1, 1923–24, p. 31–56), revealing an ability to deal with very different topics both in pure and applied mathematics. In 1926, thanks to a Rockefeller scholarship, he went to Paris for advanced studies under the guidance of É. Cartan.

Back in Italy, he became assistant to Fano in Turin and later to Severi in Rome. The relationship with Severi was uncommon.[1] In the 30s, Severi was the 'man alone in command' of Italian mathematics and had immense academic power. He was a charismatic figure and generous Master, in some ways, who put Segre in contact with important foreign scholars and presented many of his works to the Lincei and the Academy of Italy. To Segre's request for research topics, Severi always responded willingly, sending him his preprints 'so that he could draw from them a topic for his own research, developing issues that were indicated here and there'.[2] Severi, however, was a demanding head, who did not hesitate to make use of Segre's 'intelligent and always zealous cooperation'.[3] Thus, Segre taught Severi's students

---

[1] See Babbit and Goodstein (2009) and Babbit and Goodstein (2012).

[2] CA, *BSP*: F. Severi to B. Segre, Rome 16.2.1932: *perché ne abbia la presunzione e possa trarne argomento per ricerche sue, intorno ad alcune questioni che ho indicato qua e là.*

[3] CA, *BSP*: F. Severi to B. Segre, Rome 16.2.1932: *cooperazione intelligente e zelante come sempre.*

E. Luciano, *The Jewish Mathematical Diaspora from Fascist Italy*, Science Networks. Historical Studies 64, https://doi.org/10.1007/978-3-031-64896-0_10

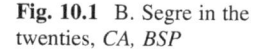

Fig. 10.1 B. Segre in the twenties, *CA, BSP*

in his place, took care of the promotion of his textbooks, and performed many other tasks, more like a personal secretary than an assistant. Severi found it normal, for example, to ask him to correct the drafts of his articles 'to see how he coped with the necessary footnotes'.[4]

Severi had a project in algebraic geometry and intended to develop it with his protégés. He had very definite evaluation criteria, both regarding his colleagues and the various approaches. This fact, combined with a difficult character, led him to exercise his role as Master more like a Duce than a leader. This was at the expense of his assistants, such as Segre, who were pulled in different directions by the two opposing Roman branches of the School: on the one hand Severi and on the other F. Enriques and G. Castelnuovo. In such a mood of tension, rivalry, and distrust, it was not easy for a young man like Segre who daily received letters of this tenor from his mentor:

> As for the completeness of the characteristic series is concerned, in these mornings [...] I have demonstrated it brilliantly and irreproachably. I send today (or tomorrow if I do not have time today) a *Nota* to the Lincei. I will show you the demonstration in Rome. I do not show you now because I do not want that when meeting that gentleman [Enriques] you inadvertently let slip a few leading phrases. We will see the Enriquesian demonstration, if there be one; but we want to see it without new ... embezzlements. On the contrary, I entrust myself to your seriousness that you do not tell anyone – much less that gentleman – that I

---

[4]CA, *BSP*: F. Severi to B. Segre, Rome 16.2.1932: *vedere come se la cava coi necessari complementi a piè di pagina.*

have resolved the matter. I will tell you why, which is understandable. Do we understand each other? Answer me, with confirmation.[5]

The relationship between Segre and Severi was very close and remained so even after Segre, as the winner of the chair of Geometry, left Rome for Bologna, together with his wife Fernanda Coen (1908–1976) and his first son, Sergio, born in 1933.[6] In Bologna, his other two children were born: Silvana in 1934 and Ornella in 1937. In the new headquarters, Segre continued his research according to the Italian tradition but, even without completely emancipating himself from Severi's tutelage, broadened his horizons to other themes: algebra and topology, theory of complex functions and differential geometry, number theory, and history of mathematics.[7] Compared with his fellow men, he thus acquired a good algebraic culture, which was recognised by his colleagues.[8]

In the four years preceding discrimination, Segre took over management of the Mathematical Institute of Bologna, climbed to the top of the UMI governance, and took over from S. Pincherle as co-director of the *Annali di Matematica pura e applicata*. That his reputation was now firmly established is confirmed by the invitation to be one of the members of the mathematical mission in Brazil. Segre refused, for the sake of his university, and his gesture was applauded by the academic authorities of Bologna. This rapid rise, however, aroused the jealousies of some colleagues, in particular of Ettore Bortolotti, and within the UMI Scientific Commission, there began to circulate a trickle of gossip against the Judeo-Bolognese binomial constituted by Segre and Levi (Figs. 10.2 and 10.3).

---

[5]CA, *BSP*: F. Severi to B. Segre, Arezzo 2.10.1932: *Quanto alla completezza della serie caratteristica, in queste mattine [. . .] ho dimostrato in modo brillante e ineccepibile la cosa. Spedisco oggi stesso (o domani se non faccio in tempo oggi) una* Nota *ai Lincei. Le dirò a Roma la dimostrazione. Non gliela dico ora perché non voglio che rivedendo quel signore* [Enriques] *inavvertitamente le scappi di bocca qualche frase orientatrice. Vedremo la dimostrazione enriquesiana, se ci sarà; ma la vogliamo vedere senza nuove . . . appropriazioni indebite. Mi affido anzi alla sua serietà perché Ella non dica per ora a nessuno – e tanto meno a quel signore – che ho risoluto la questione. A voce le dirò il perché, che del resto è intuibile. Ci siamo intesi? Mi risponda, dandomi atto.*

[6]Severi considered Segre like a son and with a play on words he called him 'his' Benjamin. See CA, *BSP*: F. Severi to S. Segre [father of B. Segre], Rome 5.12.1932.

[7]Martinelli (1978–79), pp. 6–7; Tallini (1983).

[8]For example, when B. de Finetti submitted to G. Castelnuovo a problem inherent in an algebraic structure of the calculus of probabilities, Castelnuovo redirected it to Segre 'who certainly knows better than anyone else the theories and literature on similar topics'. See CA, *BSP*: B. de Finetti to B. Segre, Trieste 18.7.1934; draft of a letter from B. Segre to B. de Finetti, [Rome July 1934] and B. de Finetti to B. Segre, Trieste 18.10.1934.

Fig. 10.2 B. Segre in the thirties, *CA, BSP*

10.2 A Mathematical Italian Identity Which Struggles With Resettlement

Dispensed from service in October, Segre was purged from the editorial management of the *Annali di Matematica pura ed applicata* and the board of UMI. The expulsion from the *Annali* was particularly painful since Segre learned that it was Severi who had plunged the knife. The very evening of the promulgation of the racial laws, Segre and his wife were the protagonists of an outrageous event. This is the report sent by Segre to the Rector of the University of Bologna:

> Hon. Rector, I think it appropriate for me to inform Your Excellency of an unfortunate event that occurred to me yesterday evening. I will present it to you without neglecting any detail and hopefully with photographic fidelity, so to speak. Yesterday evening, returning home around 10 1/2 p.m. with my wife, I thought of presenting myself to the Becocci District Group – on which I depend – to justify my forced absence from the meeting called for the morning of the 5th of this month, since – at the same time – I was supposed to attend the inaugural ceremony of the S.I.P.S. Congress to be held on that day at the Royal University. In fact, I explained this justification to Mr. Brighenti, Vice-Trustee of the above-mentioned group, to whom I was introduced. The latter, after listening to me, looked at me, and abruptly addressed me «Beniamino Segre» «I am he,» I replied. «Are you a member of the Party?»

Fig. 10.3 Beniamino Segre and Beppo Levi, May 1938, *CA, BSP*

«Yes, sir.» «So why aren't you wearing the pin?» «I left it at home.» «Bad, very bad! The first duty of the fascist is to wear the pin always, in any circumstance.» In the unkind tone in which these words were uttered, it seemed to me to bear a sardonic vein and an allusion to recent measures, well known to all. Whereupon I replied, «I am a Jew, and as such ... » However, I could not go on. Mr. Brighenti came up to me, with clenched fists and a contemptuous air, interrupting me: «What a tongue you have! How dare you speak like that, here, in the Casa del Fascio?» I then tried to finish my explanation and said: «After H.E. the Head of Government declared that a Jew is not an Italian citizen, I considered it implicit that I could no longer continue to wear the pin of the P.N.F.» «That might have been interpreted as an insincere display,» I would have liked to add. But Mr. Brighenti didn't give me time: «Do you have the card?» he interrupted me again. «Yes, at home.» «Go and get it and bring it to me here immediately. I'll give you fifteen minutes.» Meanwhile my wife, who had hitherto been waiting on the road, was surprised by my delay, and had peeped into the room where I was. I was already preparing to obey, going promptly home with her, when Mr. Brighenti called me back, saying that he would immediately arrange for me to be arrested, and that it was useless for me to resist. I replied that I knew that the power was on his side, and I allowed myself to be escorted to another room under the supervision of 3 or 4 of the many young people present. I was called back a little later and was asked to repeat some of my previous answers. I then went out with my wife, who in the meantime had been waiting for me in the waiting room, declaring that I would only take her home and that I would return immediately, at the disposal of the Group, but I was violently prevented from doing so. My wife was told to return home without me, which she did. After another forced and supervised wait, during which I found a way to let my supervisors know that – as Full

Professor of our University and Director of its Mathematical Institute – I was not the last to arrive on the block, I was escorted back into the presence of Mr. the Vice-Trustee, who said to me: «You must thank me, Mr. Professor, that I did not follow the first impulse to punch you, as I should have done as a fascist and a soldier; the fact that I didn't is only out of respect for this Casa del Fascio. Now there remain two issues pending between us. Firstly, between you and me in my role as the Representative of the P.N.F., I will report you to the political authority for having had the audacity to criticise the Head of Government. Secondly, of a personal nature: the first time I meet you on the street, I will smash your face.» I objected that, far from criticising, I had not even had the intention of judging the unquestionable work of the Head of Government, but the Vice-Trustee replied that I could make my case elsewhere and that I could now leave. When I went back outside again, it was already 11.30 p.m. With the most distinguished greetings, Yours, Beniamino Segre

P.S. – I would add that, when I presented myself last night at the Becocci Group to justify my absence, I had not remembered that I was not wearing the pin, so that I could not even imagine the incident that I now regret; otherwise I would certainly have fulfilled my duty as a fascist by letter or by sending another person in my place.[9]

---

[9] ACS-*contro-discriminazione*: B. Segre to A. Ghigi [Rector of the University of Bologna], Bologna 3.9.1938: *On. Rettore, ritengo opportuno che io informi la M.V. Ill.ma di un fatto increscioso occorsomi ieri sera. Ve lo esporrò senza trascurare alcun particolare, con fedeltà che cercherò di rendere – per così dire – fotografica. Ieri sera, rincasando verso le 22 ½ con mia moglie, pensai di presentarmi al Gruppo Rionale Becocci – da cui dipendo – per giustificare la mia forzata assenza all'adunata indetta per la mattina del 5 corrente mese, dato che – in pari tempo – avrei dovuto presenziare alla cerimonia inaugurale del Congresso della S.I.P.S. che dovrà tenersi in tale giorno presso la Regia Università. Esposi infatti questa giustificazione al Sig. Brighenti, Vice-Fiduciario del gruppo suddetto, presso il quale fui introdotto; questi, dopo avermi ascoltato mi squadrò, e «Segre Beniamino» mi interpellò bruscamente. «Sono io» risposi. «Sei iscritto al Partito?» «Sissignore». «Com'è allora che non hai il distintivo?» «L'ho lasciato a casa». «Male, malissimo! Primo dovere del Fascista è di portare il distintivo sempre, in qualunque circostanza». Nel tono poco benevolo con cui queste parole furono profferite, a me parve riscontrare una venatura sardonica ed un'allusione a recenti provvedimenti, a tutti ben noti; onde di primo impeto replicai: «Sono Ebreo, e come tale...» Non potei però proseguire; il sig. Brighenti mi si fece sotto, coi pugni chiusi e con aria sprezzante, così interrompendomi: «Ma che lingua hai! Come osi parlare in tal modo, qui, nella Casa del Fascio?» Cercai allora di completare il mio pensiero e dissi: «Dopo che S.E. il Capo del Governo ha dichiarato che un Ebreo non è italiano, ho ritenuto implicito di non poter più continuare a portare il distintivo del P.N.F.» ... «ciò che si sarebbe potuto interpretare come un'ostentazione poco sincera», avrei voluto aggiungere; ma il sig. Brighenti non me ne lasciò il tempo: «Hai la tessera?» ancora mi interruppe. «Sì, a casa» «Valla a prendere e portamela qui subito, ti do tempo un quarto d'ora.» Intanto mia moglie – che era fin'allora rimasta in attesa sulla strada – impressionata dal mio ritardo, aveva fatto capolino nella stanza dove mi trovavo; e già mi accingevo ad obbedire, andando prontamente a casa con lei, quando il sig. Brighenti mi richiamò dicendo che avrebbe provveduto subito a farmi arrestare, e che era inutile io opponessi resistenza. Replicai che sapevo che la forza era dalla sua e mi lasciai accompagnare in un'altra stanza, sotto la sorveglianza di 3 o 4 dei molti giovani presenti. Nei fui ancora richiamato poco appresso, e mi furono fatte ripetere alcune delle mie precedenti risposte. Feci allora per uscire con mia moglie, che nel frattempo mi attendeva in anticamera, dichiarando che l'avrei soltanto riaccompagnata a casa e che sarei tosto ritornato, a disposizione del Gruppo, ma ciò mi fu impedito con violenza; e mia moglie fu invitata a rientrare a casa senza di me, come infatti fece. Dopo un'altra attesa forzata e vigilata, durante la quale trovai modo di far sapere ai miei sorveglianti che – come professore ordinario della nostra Università e Direttore del relativo Istituto Matematico – non ero poi l'ultimo venuto, fui riaccompagnato alla presenza del sig. Vice-Fiduciario, che così mi disse: «Dovete ringraziare, sig. professore, che non ho seguito il primo impulso di prendervi a cazzotti, come avrei*

Segre immediately applied for counter-discrimination for distinct scientific merits (D2.11.2) but perhaps did not have great hopes that his request would be accepted and so immediately began to look for another haven in which to rebuild his family and professional life. The fact of having been dismissed with a minimum pension, after just few years of full professorship, put him in serious economic difficulties. A wife and three small children (the youngest was just one year old) depended on him. His older colleagues, from Castelnuovo to T. Levi-Civita, would be ready to support him, but Segre acted differently from other emigrants. First of all, he aimed exclusively at the English-speaking world. Secondly, he turned to Severi only. Perhaps he trusted in his help, perhaps he did not dare to address Castelnuovo, Enriques, and Levi-Civita. Severi, however, did very little for him. He only helped Segre to convert a sum of money into British pounds at a more advantageous exchange rate than the official one (D7.2.2).

At an international level, the network that Segre could activate in 1938 consisted of foreigners who had or had had an interaction with Severi and with the Rome geometric environment. The first to take action was O. Zariski, who in November 1938 brought Segre's situation to O. Veblen's attention.[10] As for Terracini, Veblen was hopeful about the possibility of bringing Segre to Princeton but S. Lefschetz responded negatively.[11] Veblen then wrote to various colleagues: T.Y. Thomas (California), E.P. Lane (Chicago), G.Y. Rainich (Michigan), and so on. Thomas, in particular, was interested in bringing men who were active scientifically into his environment, but Veblen did not know whether he would be sympathetic in the case or, indeed, sufficiently influential to accomplish anything. At the same time, V. Snyder talked with A. Coble (Illinois) and with the ECADFS.[12] On the ECADFS front, hopes for help appeared immediately very limited:

> I wish there were something encouraging that I could say as to the prospect of assistance for Dr. Segre. As you can well imagine, his is one of a great number of personal problems

---

dovuto fare quale squadrista e combattente; ma se non l'ho fatto è solo per rispetto di questa Casa del Fascio. Ora restano due cose da regolare fra noi. Per la prima fra me Rappresentante del P.N.F. e voi, provvederò denunciandovi all'Autorità politica per avere avuto l'ardire di criticare il Capo del Governo. Per la seconda, di carattere personale, provvederò fuori, rompendovi il muso la prima volta che vi incontrerò per la strada.» Alla mia obiezione che, lungi dal voler criticare, io neppure avevo avuto l'intenzione di giudicare l'opera insindacabile del Capo del Governo, il sig. Vice-Fiduciario replicò che avrei potuto far valere le mie ragioni in altra sede e che ora potevo andarmene. Quando tornai a riveder le stelle erano scoccate da poco le 23 ½. Colla più distinta osservanza Vostro dev.mo Beniamino Segre.

P.S. – Aggiungo che, presentandomi ieri sera al Gruppo Becocci per giustificare la mia assenza, non mi ero rammentato di essere senza distintivo, sì che ero ben lontano dal prevedere l'incidente che ora rammarico; altrimenti avrei certamente adempiuto il mio dovere di Fascista per lettera od inviando in mia vece altra persona.

[10] SEGRE.1.

[11] SEGRE.2. Veblen asked Zariski to write to T.Y. Thomas 'much the same sort of letter as the one you wrote to me, with the addition, however, of a paragraph or two about his [Segre's] personal qualities, based on your acquaintance with him.' (OVP: O. Veblen to O. Zariski, [Princeton] 8.11.1938).

[12] SEGRE.3.

created by the situation in Central Europe and now in Italy. [...]. The most that I can do is to file your letter with the office of our committee and if an inquiry from an American Institution comes in for a man of his training, we would be glad to call attention to Dr. Segre's availability. The picture that is presented by the existing situation in Europe is pathetic in the extreme and it makes one very blue as to the future of the intellectual life in those countries.[13]

In order not to work at cross purposes an amplified record for Segre, to be substituted for the file previously prepared in December 1938, was sent to the American Friends Service Committee.[14] Meanwhile, Snyder and Veblen compared ideas. At that moment, a temporary position in Michigan had opened up, but both were doubtful about the possibility of winning the post. Veblen had suggested Rainich Segre's name along with those of R. Frucht and K. Löwner, but he feared that the other names would appeal to him more.[15] At his turn, Snyder would be surprised if Segre were chosen because: 'at least part of the duties of the man to be appointed at Michigan would be routine undergraduate teaching. If so, it seems to me that it would be unfair to the students and to Segre to turn him loose on that job without some preparation'.[16]

These perplexities led Veblen and Snyder to cooperate in drawing up an alternative scheme for taking care of Segre. Since Coble intended to ask for a sabbatical leave for the first term of 1939–1940, he could use Segre to look after candidates writing their theses while he himself was away. If this worked out, Segre could come to Cornell for the second term. Finally, Veblen would have the possibility of a place in the Institute for the next year 1940–1941. With this project in hand, Snyder returned to the issue with ECADFS and presented the financial coverage of the outlined plan: for 1939–40, the University of Illinois would provide Segre with a salary of $1200, Cornell would arrange $450 and Veblen would propose Segre for an Institute stipend for the following year, provided they had sufficient funds available when the time comes. ECADFS was asked for a contribution to this rescue plan, particularly deserving because Segre was 'generally rated as the ablest young man in his field of algebraic geometry to be found anywhere'.[17] ECADFS was glad to hear of the encouraging progress Snyder and Veblen had made with regard to a possible position for Segre, and Farrand was truly optimistic: if they felt that the outlook for a permanent position was reasonably well assured, it could be taken as read that the appeal would be greeted sympathetically by the Committee, provided the amount called for was fairly small.[18]

Like other would-be emigrants, Segre had meanwhile turned to the SPSL. In October 1938, the Society had become aware of his case, presented by G.B. Jeffrey,

---

[13]ECADFS Records, *BS*: L. Farrand to V. Snyder, New York 14.2.1939.

[14]OVP, *RefGen*: G. Blake to H. Kraus [American Friends Service Committee], Princeton 10.3.1939.

[15]OVP, *BS*: O. Veblen to V. Snyder, [Princeton] 7.3.1939.

[16]OVP, *BS*: V. Snyder to O. Veblen, Ithaca 10.3.1939.

[17]ECADFS Records, *BS*: V. Snyder to L. Farrand [ECADFS], Ithaca 25.3.1939.

[18]ECADFS Records, *BS*: L. Farrand [ECADFS] to V. Snyder, New York 29.3.1939.

President of the London Mathematical Society. In the United Kingdom, Segre was not only welcome, but wanted. On 20 February 1939, J.G. Semple, professor at King's College, London, and one of the founders of the London Geometry Seminar, informed the SPSL of the intention to set up a fund to help the Italian colleague.[19] The SPSL offered itself as a trustee. In just a few weeks, Semple and Hodge managed to raise 128 pounds; the SPSL added 122. This amount was enough for Segre to leave Italy with his family and to guarantee him a year's stay in England. On 11 March, SPSL informed Segre that the Committee had decided to give him a grant for a year and that they would reconsider the question of giving a further grant for the next year.[20] SPSL also took active steps in helping Segre to make all preparations to come to the country as soon as possible. It asked the Home Office to release a permit to enter England and assisted the Segres in looking for lodgings either in London or in Cambridge.[21] In mid-April, the Segre family left Bologna and, after a stop-over in Turin to greet Beniamino's parents, arrived in London on 19 April 1939. At Semple's suggestion, they temporarily settled in Cockfosters, near London. On 1 August 1939, they moved to Cambridge.[22]

In Segre's plans, however, England was only a transitory destination on the path to the United States. The project arranged by Snyder and Veblen had not seen further developments but everything seemed to be proceeding for the best. That was the case until May 1939 when Zariski revealed to W.V.D. Hodge his intention to come to Cambridge around Easter time, to meet geometers in the country and have some informal discussions on matters of common interest. Hodge was excited about the opportunity to hear Zariski's views on algebra and topology. He wanted to invite him to lecture at university and at his seminar. Here, however, a problem arose. In Cambridge, there was a vigorous economy campaign, and Hodge wished for an analogous invitation to go to Segre since the fee payable would help him considerably. Hodge did not imagine that he would succeed in having both requests granted. Therefore, in view of Segre's greater need, he proposed to Zariski that he accept a purely nominal fee, or even, forgo the payment in lieu of a promise that Hodge would deliver some lectures for free when he visited Baltimore in 1941. In that context, Hodge talked to Zariski for the first time about Segre's complicated situation:

This brings me to your question about Segre's position. It is rather complicated. He first applied to our central committee for refugees, but they had no funds with which to help him. Then about ten of us geometers raised a private subscription for him, hoping to be able to provide for him for two years. We did not raise quite enough for this purpose, but we came to an arrangement with the committee which enabled us to give him the assurance of a small income (250 per annum) for two years. We handed over our subscriptions to the Society, who were able to make up the difference. Thus, although most of his income is provided by private subscriptions, the provision made for him is subject to the regulations of the Society, and I do not think the grant would be continued if he was in receipt of an income from

---

[19] SEGRE.4, SEGRE.6.

[20] SPSL, *BS*: D.C. Thomson [SPSL] to B. Segre, London 11.3.1939.

[21] D3.4.9 and SPSL, *BS*: J.G. Semple to D.C. Thomson [SPSL], Eastbourne 31.3.1939,

[22] SPSL, *BS*: B. Segre to J.B. Skemp [SPSL], Manchester 17.11.1945.

Illinois or Cornell. The maximum income from all sources which the Society would allow him would be 300. If I may suggest it, I believe that Coble could be of most assistance to Segre if he would postpone his plan for helping him for a year. If Segre has not found a permanent post by 1941, he may find himself in serious difficulty, and Coble's plan might than save the situation.[23]

These were the same days that Levi-Civita was contacted by Argentina to propose names of persecuted Italian scholars to be sent to Rosario. Levi-Civita immediately thought of Segre but the latter declined the invitation:

> I am very grateful to you for suggesting my name and for so kindly informing me. In the current conditions, I do not urgently need to find a workplace, but this is certainly a problem that worries me right now: and I would gladly be a pioneer if I am not given the opportunity to find something better. Here I am 1/2 hour on the "Tube" from the centre of London, discreetly settled with my wife and three children, and I am very happy. I was welcomed very cordially by the mathematicians here, and now I can return to my studies with a certain tranquillity of mind.[24]

On the same days that Hodge wrote to Zariski, Veblen told Coble that he had just heard from Fubini that Segre was now in England and was living on a small stipend collected by British mathematicians. However, Veblen 'imagined it could be arranged that he would not lose this when he comes to the U.S.'[25]

From here on, the story is unclear. Snyder informed Coble that Segre was displaced in England but he still aimed to come to the United States and concluded:

> I wrote him that you and I both feel that he will make more progress there than here for the present and suggested that he spend the next academic year in England, but next November he should communicate with Veblen concerning a possible position at the Institute for the following year. I also stated that the whatever plans you and I had in mind for the year 1939-40 would be abandoned. [. . .] But during the year at the Institute Segre will at least have time to look around. At any rate, I am thankful for your cooperation and in one sense am not sorry it has turned out as it has.[26]

Probably, there was a misunderstanding between Snyder and Coble because the latter summed up the situation to Veblen as follows: Cornell authorities were no longer prepared to support Snyder's proposal and therefore it would be inadvisable for Segre to come to the United States. Under these circumstances, he thought that it might be better to take no further steps in the matter.[27] For the first time since the

---

[23] OVP, *BS*: W.V.D. Hodge to O. Zariski, Cambridge, 8.5.1939.

[24] ANL-*Levi-Civita*: B. Segre to T. Levi-Civita, Cockfosters 8.5.1939: *Le sono molto grato di aver fatto il mio nome e di avermene così gentilmente avvertito. Nelle attuali condizioni non ho urgente bisogno di trovare un posto, ma è questo certamente un problema che fin d'ora mi preoccupa: e farei assai di buon grado il pioniere, se non mi sarà dato di trovare qualcosa di meglio. Qui sono a ½ ora di "tube" dal centro di Londra, discretamente sistemato colla moglie ed i tre bimbi, e ne sono lietissimo. Sono stato accolto molto cordialmente dai matematici di qui, ed ora posso ritornare ai miei studi con una certa tranquillità d'animo.*

[25] OVP, *BS*: O. Veblen to A.B. Coble, [Princeton] 10.5.1939. See also SEGRE.5.

[26] OVP, *BS*: V. Snyder to A.B. Coble, Ithaca 13.5.1939. See also CA, *BSP*: V. Snyder to B. Segre, Ithaca 13.5.1939.

[27] OVP, *BS*: A.B. Coble to O. Veblen, Urbana 17.5.1939.

beginning of the correspondence, Coble also mentioned the question of linguistic skills and asked Veblen and Zariski whether Segre would be able to lecture in intelligible English. Veblen was baffled: he was aware that the University of Illinois' offer depended on Coble's sabbatical leave during the first term of 1939–1940. He trusted that something could be arranged with the English group who were supporting Segre, even if he had to go back to England. Doubts about whether an American university might be able to give him a stipend applied to all refugees, since: 'no one knows what funds might be available at that time or what other demands there will be'. Above all, Veblen had a strong conviction:

> It always seems to me best to make use of opportunities when they exist. The opportunity to give Segre a trial at Illinois seems like a very real one, especially when combined with the opportunity to return to England for the second term. Since Segre has already been in England for some weeks, I feel that we can count confidently on his being able to use English reasonably well by next autumn.[28]

However, Coble was uncompromising. He confirmed that the only cogent reason which made Segre's appointment in Illinois particularly timely for the first semester of 1939–1940 alone was that A. Emch would be retired and that he would be on leave of absence, 'so that the usual course in algebraic geometry could not be given unless they imported some person'.[29] At this point, however, he was convinced that Segre would be taking an unnecessary chance in coming to America for the somewhat limited appointment.[30] The tone of the letter to Veblen was very dry:

> I must say rather definitely and categorically that there is no possibility whatever of placing Segre here. I know too well the feeling of the administrative officers, who would have to approve such an appointment, to even make an effort in that direction. On the other hand, I rather infer from two sentences in your letter that you are hopeful that he might remain here if he were given a trial. I am sure that the only arguments which enabled me to get the conditional approval which I secured for the one semester appointment were our immediate need for the coming semester in that field and the assurance that Professor Segre would be provided for elsewhere after that period. Since this assurance is no longer to be counted on, I am myself reluctant to take any further steps in the matter.[31]

Veblen made a last attempt, trying to involve Lefschetz and Zariski, and asked Coble to think about the advantages that Segre could derive from a stay in the States:

> he would have a chance to make himself known personally to people in this country. The fact that there would be no permanent opening for him in Illinois detracts from this advantage only so far as this one university is concerned. I don't feel that I ought to urge the matter, but I cannot resist making the logical point which is involved.[32]

Coble remained adamant in his position and Veblen capitulated, not without expressing his bitterness to Zariski and Lefschetz:

---

[28] OVP, *BS*: O. Veblen to A.B. Coble, [Princeton] 22.5.1939.

[29] OVP, *BS*: A.B. Coble to O. Zariski, Urbana 22.5.1939.

[30] OVP, *BS*: A.B. Coble to O. Zariski, Urbana 22.5.1939.

[31] OVP, *BS*: A.B. Coble to O. Veblen, Urbana 27.5.1939.

[32] OVP, *BS*: O. Veblen to A.B. Coble, [Princeton] 29.5.1939.

> It seems to me a terrible pity that this decision should be taken, for it seems to kill Segre's one good chance. Since he has a stipend in England it would have been extremely easy for him to come over here for the term at Illinois and then to return to England. This would have involved a minimum of responsibility on the part of people in this country and given him a chance to make valuable acquaintanceships.[33]

The failure of the entire programme with respect to Segre's coming to the United States, communicated by Coble himself to Segre at the beginning of June with a good dose of hypocritical regret, amazed British mathematicians.[34] Hodge asked Zariski about the matter, who in turn justified himself with him and Veblen: 'I am afraid that it would not work to Segre's advantage. One must also not forget that Segre has three children and that a trip to America for a short period means either five round trip tickets (if his family accompanies him) or two separate items of living expenses (if his family remains in England)'.[35]

United Kingdom's entry into the war (September 3, 1939) marked a drastic change in Segre's position. On the very same day, Segre made himself available to Great Britain: 'after the events of today I am willing to repeat my offer to serve in any capacity in which I can be of use to the Nation',[36] but he clearly understood that his position, already precarious, was destined to become even more so. The officers of the Society felt strongly that the refugee scholars must, as far as possible, be encouraged to find employment in work of national importance rather than to continue academic research when a British person would not be in a position to do this. A special meeting of the Allocation Committee was scheduled to consider the general policy to be adopted in providing research grants and to consider individually all grants made from earmarked donations.[37] Segre's position was discussed in November 1939.[38] Hodge and Semple were unable to provide concrete employment prospects.[39] The new set of circumstances also forced some contributors to the Segre Fund to review their financial responsibilities and modify their contributions. Segre then resumed interviews with the Americans. He recalled his case to Snyder and Veblen and their half-promise of the possibility of having him go to Princeton for the year 1940–1941, a welcome opportunity which constituted one of his old and steadfast aspirations.[40] He pointed out that until the war broke out, his prospects for a definite position at Cambridge were good and reassured them that the use of English was not so much of a task. Snyder invited him not to be discouraged and

---

[33] OVP, *BS*: O. Veblen to O. Zariski, cc. S. Lefschetz [Princeton] 29.5.1939.

[34] CA, *BSP*: A.B. Coble to B. Segre, Urbana 7.6.1939.

[35] OVP, *BS*: O. Zariski to O. Veblen, Baltimore 21.6.1939.

[36] SPSL, *BS*: B. Segre to SPSL, Cambridge 3.9.1939. See also Segre.7.

[37] SPSL, *BS*: N. Searle [SPSL] to W.V.D. Hodge, London 17.10.1939.

[38] SPSL, *BS*: N. Searle [SPSL] to B. Segre, London 23.11.1939.

[39] SPSL, *BS*: W.V.D. Hodge to N. Searle [SPSL], Cambridge 14.10.1939.

[40] OVP, *BS*: B. Segre to O. Veblen, Cambridge 10.11.1939.

expressed the wish 'to be able to greet him in the United States next summer'.[41] This time it was Fubini who advised against it:

> I have spoken to Prof. Veblen; he is *completely* of my opinion, and absolutely advises against coming here. At the Institute, you would have no guarantee of stability, and entering the University is *a difficult* thing; and you should start your career again from the *first step*. I came here *not for myself*, but *for my children*. And in two years at most I will reunite with them and *abandon science*. In the opinion of Veblen and myself, you have to remain in London until after the final victory; and then, if that's the case, you can think about America. You know well that, in any case, I am always ready to do what I can for my dear Segre.[42]

This time, Fubini's opinion was evidently heard by Veblen who shortly afterwards replied to Segre:

> the most desirable course of action would be for you to stay in England as long as you can so as to acquire as complete command of the English language as possible before your initial experiment with an American university. This would enable you to make your first steps with a minimum of handicap on account of difficulty with the language. I do not need to tell you that so many Europeans have recently been absorbed into the mathematical community of this country that it has become extremely difficult to find openings.[43]

The road to South America, meanwhile, had also closed, as the two vacant positions for mathematics (Rosario and Tucumán) had gone to Levi and Terracini. At the beginning of December 1939, SPSL informed Segre that the Society had been reluctantly compelled to take a drastic step: all grants would be considerably reduced in size, and in most cases, it would not be possible to continue them at all. No decisions had been made in any individual cases, but the Society urged refugees to try to find some other means of subsistence before the expiry of their grant.[44]

In February 1940, Veblen heard incidentally through F. Kottler that there were two lectureships in mathematics available at the University of Melbourne. Segre could be a very desirable candidate for one of these posts, so he suggested that he apply.[45] Veblen knew very little about these openings, but he wrote to T. MacFarland Cherry to endorse Segre:

> He is a very good geometer, and I think has a very adaptable personality so that he would find relatively little difficulty in fitting himself into a new environment. I have also mentioned the opening to Professor Weyl, who has written to his friend Professor Hans

---

[41] CA, *BSP*: V. Snyder to B. Segre, Ithaca 28.11.1939.

[42] CA, *BSP*: G. Fubini to B. Segre, Princeton 1.12.1939: *Ho parlato col prof. Veblen; egli è completamente del mio parere, e ti sconsiglia assolutamente di venir qua. All'Institute non avresti alcuna garanzia di stabilità, ed entrare all'Università è cosa difficile, e dovresti cominciar la carriera dal primo gradino. Io sono venuto qui non per me, ma per i miei figlioli. E tra due anni al massimo mi riunisco ad essi e abbandono la scienza. Secondo il parere di Veblen e mio, resta a Londra fin dopo la vittoria finale, e poi, se sarà il caso, penserai all'America. Sai bene del resto che, in ogni caso, io sono sempre pronto a fare quanto posso per il mio carissimo Segre.* See also SEGRE.7.

[43] OVP, *BS*: O. Veblen to B. Segre, Princeton 9.12.1939.

[44] SPSL, *BS*: E. Simpson [SPSL] to B. Segre, London 6.12.1939.

[45] OVP, *BS*: O. Veblen to B. Segre, [Princeton] 28.2.1940.

Hamburger. [...] I hope you will not feel that I have been officious in passing the information on to these two other candidates. We have absorbed a great many of the German and Italian refugees in the United States. The results have been uniformly good scientifically, and there have been very few cases in which there has been any real difficulty in the way of personal adjustments.[46]

Segre was very obliged to Veblen and immediately applied through the London Universities Bureau of the British Empire. At the same time, he became acquainted with Cherry himself by cable.[47] As chance would have it, Hodge was charged with interviewing Segre and he prepared him:

I am just writing this note to warn to be prepared to see me at the interview. [...]. Do not be surprised if I ask you a number of questions to which I already know the answer. I shall do so in order to bring out points which I think other members of the committee ought to know.[48]

Evidently, the interview went wrong because Segre notified his American colleagues of the failure of his application. His situation was thus becoming more and more difficult, especially on account of several restrictions imposed upon every alien, and he feared being interned should Italy enter the war.[49] His command of the English language was steadily increasing, to the point that he had already been able to give lectures to the London Mathematical Society, participate in Hodge's and Hardy's Seminars, and write some long papers in English. Veblen was sincerely sorry to hear that he did not win the position at Melbourne and promised to continue to look out for any opening which may arise. His thoughts, however, were turned elsewhere: 'at present I feel, and doubtless you agree with me, that nothing else is as important as to bring the help of this country to England'.[50] However, Veblen recommended Segre to A. Johnson, for eventual inclusion in the RKF's rescue plan for 1940–1941.[51] Segre was ranked in a low position and did not receive the grant.

In March 1940, the time had come for the SPSL to arrange for Segre's livelihood for the second year. Segre had done massive research work during the period spent in England, and he had also registered through the special emergency list of scientists with the Ministry of Labour for any work which he might be specially qualified to do in the war effort. The Society hoped very much that despite his nationality he would become absorbed in some defence activity when the demand for scientists increased. However, the Society did not know how to deal with his case. At the beginning of the war, there had been a special meeting of the Executive Committee to discuss the question of grants during war time. The pre-war maximum grant of £250 p.a. for married men had been reduced to £ 200 p.a. and the Allocation Committee would not

---

[46] OVP, *BS*: O. Veblen to T. MacFarland Cherry, [Princeton] 28.2.1940.

[47] OVP, *BS*: B. Segre to O. Veblen, Cambridge 13.3.1940.

[48] CA, *BSP*: W.V.D. Hodge to B. Segre, Cambridge 28.3.1940.

[49] OVP, *BS*: B. Segre to O. Veblen, cc. A.B. Coble, S. Lefschetz, V. Snyder and O. Zariski, Cambridge 1.6.1940.

[50] CA, *BSP*: O. Veblen to B. Segre, Princeton 26.6.1940.

[51] OVP, *BS*: A. Johnson to O. Veblen, New York 17.9.1940.

agree to supplementary income received through other sources to a higher level than this. It had also been agreed that grants should be made for 3-monthly periods, partly because the financial position made it difficult to plan, and partly because it was thought that refugee scientists should be persuaded to accept any alternative which might arise. However, Segre's case was quite different because his grant was derived from private donations. All contributors committed to continue their contributions and the SPSL agreed to supplement the donated funds for a further 12 months.[52] Enclosing his cheque, one of the contributors wrote: 'I hope that he will find a permanent place among us. We are fortunate in having him as a guest'.[53]

After settling down, Segre had started working at King's College London with Semple.[54] Despite the concerns arising from linguistic issues and those of the insertion into an academic environment organised in a very different way from the Italian context, he had resumed his research activity and worked with intensity. At the end of the first year in the United Kingdom, he had already almost finished writing a major paper *On limits of algebraic varieties, and particularly of their intersections and envelopes* (Proceedings of the London Mathematical Society, (2), 47, 1942, pp. 35–403). He had given three lectures at the Hodge's Seminar, one at the Hardy's Seminar, and another at the London Mathematical Society. He had attended some courses given by Semple in London and by Hodge and J.A. Todd in Cambridge and he was negotiating with the Clarendon Press about writing a book on *Geometry upon algebraic surfaces and varieties.*[55]

Segre's path of integration into the English mathematical community came to an abrupt halt in June 1940. On June 10, Mussolini announced Italy's entry into the war against France and England; Italians residing in Great Britain were declared 'enemy aliens' and interned. The SPSL went into a frenzy: its statutes did not permit the continued payment of research grants to scholars who, by becoming interned, were no longer able to continue their scientific work, and precluded it from paying relief indefinitely to the families. When the first internments took place, the Society assumed that the government would soon make adequate provision for remaining family members. Since this was not the case, for the interim period it renewed grants at a relief rate for a further month. Among the subscribers to the Segre Fund, there was unanimous agreement that the money should be used for the benefit of Mrs. Segre and the children.[56] At the end of June, however, the Committee requested that the Jewish Refugees Committee, Italian Department, administer a special grant,

---

[52] SPSL, *BS*: N. Searle [SPSL] to J.G. Semple, London 18.3.1940; N. Searle [SPSL] to W.V.D. Hodge, London 28.3.1940; N. Searle [SPSL] to J.A. Todd, H.S. Ruse, F.P. White, P. Fraser, V.C. Morton, J.L. Wren and L. Roth, London 29.3.1940.

[53] SPSL, *BS*: H.S. Ruse to SPSL, Southampton 3.4.1940.

[54] Two months after his arrival Segre had been elected Honorary Member of the Senior Common Room of King's College (see CA, *BSP*: S. Davis [Secretary] to B. Segre, London 13.6.1939).

[55] SPSL, *BS*: B. Segre to E. Simpson [SPSL], Cambridge 1.3.1940.

[56] SPSL, *BS*: W.V.D. Hodge to E. Simpson [SPSL], Cambridge 1.7.1940.

destined to the family.[57] Obtaining this was very complicated for Segre's wife, Fernanda, who had to deal with endless paperwork in a situation of great distress.[58]

At the beginning of July Segre entered the Bury Internment Camp on the Isle of Man, where he again met Vito Volterra's son, Enrico. The SPSL immediately moved to request his release and very quickly put together a dossier encompassing a memorandum from Segre and two heartfelt letters from Hodge and Todd.[59] Segre's release was finally ordered on 16 August 1940.[60] Meanwhile, Fernanda and the children had moved from Cambridge to London, as guests in L. Roth's house.[61] These were the days of the Battle of Britain, and the youngest Segre daughter, Ornella, died of measles when the bombings prevented her from being hospitalised. For Segre, it was an immense tragedy, barely relieved by the affection of their English friends:

> I have meant, ever since I heard of it, to write to tell you how very very grieved both I and my wife were to know of your sad misfortune. Will you tell Mrs. Segre how much we sympathise. Time will blur, but cannot erase, the loss. I wish you and yours all the happiness that this season is supposed to bring to all of us: we shall work through this time; we shall hardly forgive those who have caused our unhappiness.[62]

The period between August 1940 and August 1942 was, for Segre, the hardest part of his stay in England. In February 1941, the SPSL informed Hodge that the Segre Fund was definitively exhausted and asked if there was any possibility of the fund being continued or, failing this, whether there was any likelihood of a post for this Italian scholar. The Allocation Committee had been compelled not only to reduce the number of grants but also to discontinue some existing grants owing to lack of funds, and it was doubtful whether they could maintain Segre for more than three months. The only possibility was to raise a supplementary sum of money among his friends.[63] Hodge responded with pessimism:

> While I do not want to say definitely that there is no hope of the fund raised on his behalf being continued, I think that there is very little hope of this being done. I shall write to Professor Semple, who was, with me, a joint organizer of the fund, and find out his views, but when we last talked about the matter, we agreed that there was little prospect of success for a further appeal. A number of those who subscribed are now facing increased expenditure on account of the war, and I know that others feel that enough has been done for an individual and prefer to give their money to organizations. Nor can I say that I feel confident that Segre will obtain a post in the near future. I am trying to help him to obtain one of a few appointments which I have heard about, but there is nothing to indicate that his chances of success are outstanding, and after so many disappointments I am inclined to be pessimistic. [. . .] I think it would be useful if you were to write to Segre to impress on him the desirability

---

[57] SPSL, *BS*: E. Simpson [SPSL] to The Jewish Refugees Committee, Italian Department, London 25.6.1940.

[58] See e.g. SPSL, *BS*: E. Simpson [SPSL] to F. Coen Segre, London 26.6.1940.

[59] SEGRE.8, SEGRE.9 and SEGRE.10.

[60] SEGRE.11 and SEGRE.12.

[61] Segre B. (1976).

[62] CA, *BSP*: H.F. Baker to B. Segre, Cambridge 20.12.1940.

[63] SPSL, *BS*: B. Johnson [SPSL] to W.V.D. Hodge, London 20.2.1941.

of obtaining a post, whatever its kind, in the near future, which would tide him over present difficulties. I am of the opinion that he considers that an academic post is the only thing for him, and that he must simply wait until one turns up for him. It would, of course be a great pity if a mathematician of his worth were forced to go out of academic life, but I am afraid that he may have to face this, temporally at least. Is it within your powers to find some kind of a job for him?[64]

The SPSL was aware of the difficulties. There were quite a few openings for teachers of mathematics in British schools, a good knowledge of the English language was essential and some openings, for example, one at Liverpool University, were in a protected area where the authorities insisted on having friendly aliens and ruled out Germans, Austrians, and Italians. The SPSL suggested that Segre contact the International Labour Branch since this new Department of the Ministry of Labour was trying very hard to absorb skilled aliens with employment.[65] Segre replied, disheartened: 'I have already tried several times to get some teaching job, but till now without success. I shall go on trying'.[66] New attempts were followed by new failures. The first was in Southampton. It was H.S. Ruse who signalled the opening of a position, although the conditions were not very advantageous, with the salary he was allowed to offer being £300 for the year:

> I must confess that I feel a certain discomfort in having to write letters such as this to a colleague of your distinction. I am sure you will understand that I hate the circumstances that have led to it and that I should consider it a privilege to have you as a member of the staff.[67]

The place went to another candidate. Overseas, Snyder proposed Segre to R.P. Agnew and W.B. Carver, respectively, chairman and acting chairman of the Mathematics Department at Cornell, but Carver felt it would be an insult to offer an instructorship to a person who had held a full professorship at the University of Bologna. Moreover, both Agnew and Carver were rather dubious about the capacity of Segre to teach routine geometry and calculus to freshmen, and Snyder shared the same feeling:

> Some adjustments might be made, but the other teachers of the same rank would have to be considered. And of course, it is outside of my job to suggest any changes, and I have no voice in any selection. [...] I now feel that we have done everything that we could do, but the circumstances are such that we are simply helpless.[68]

Just when Segre was losing hope, two anchors of salvation arrived: the Relief Committee for Refugees from Italy provided some funds for a further grant to Segre and the Clarendon Press offered him a contract to publish a text on cubic surfaces. This book (*The non-singular cubic surfaces. A new method of investigation with special reference to questions of reality*), the preparation, for Clarendon again, of another volume on the geometry of varieties, in collaboration with Todd, and many

---

[64] SPSL, *BS*: W.V.D. Hodge to SPSL, Cambridge 21.2.1941.

[65] SPSL, *BS*: E. Simpson [SPSL] to W.V.D. Hodge, London 24.2.1941.

[66] SPSL, *BS*: B. Segre to E. Simpson [SPSL], Cambridge 1.4.1941.

[67] SPSL, *BS*: H.S. Ruse to B. Segre, Southampton 29.4.1941.

[68] OVP, *BS*: V. Snyder to O. Veblen, Ithaca 7.6.1941.

work in arithmetic geometry kept Segre busy throughout the 2-year period 1942–1944.[69]

In the first half of 1942, Segre faced the last three rejections at Birmingham, Sheffield, and Swansea. In Birmingham, there was a temporary lectureship for which the governance sought a man who had already had a fair amount of experience of teaching Mathematics to Engineering students in England. Segre was not judged the right candidate for a post 'for which not a Mathematician but a teacher of 'utilitarian' Mathematics was required'.[70] For the other two positions, he was considered (on paper) to be among the strongest candidates, especially since he presented excellent reference letters from Todd, Semple, Hodge, and Baker, but he failed.[71]

In July 1942, Segre was finally appointed assistant lecturer at the University of Manchester. For him, this would be the only success achieved in the United Kingdom.[72] On 29 September, the family settled there. Having obtained a semi-stable position restored to Segre the serenity to devote himself to his studies. Despite the discouraging war news, and although devoid of news of parents and relatives in Italy,[73] his productivity was extraordinary: a book and 23 papers in 7 years on a multiplicity of geometric themes, both old and new. Among the new subjects is to be mentioned arithmetic geometry, which he first tackled thanks to L. Mordell and K. Mahler.[74] In October 1943, Segre informed Veblen that he had just finished writing a paper *On arithmetical properties of quartic surfaces* which was:

> a self-contained research, 48 pages long, about certain quartic Diophantine equations of a rather general equation. In it new methods, chiefly geometric in character, are applied, and several new results are obtained. I have already used similar methods with success in a systematic research on arithmetical properties of cubic surfaces. Some of my results on this subject will be announced in four *Notes* presented to the London Math. Soc.[75]; but I shall develop the matter in an extensive work, which in the opinion of Professor Mordell, could be published as a Cambridge Tract. I have, however, a still more ambitious plan. If, as I hope, I

[69] SPSL, *BS*: G. Peiser [SPSL] to B. Pritchard [The Relief Committee for Refugees from Italy], London 21.5.1941 and Segre.18–21.

[70] CA, *BSP*: G.N. Watson to B. Segre, Birmingham 19.6.1942.

[71] Segre.13–16.

[72] SPSL, *BS*: B. Segre to E. Simpson [SPSL], Cambridge 24.7.1942. See also E. Simpson [SPSL] to B. Segre, London 25.7.1942: 'I am sure you will enjoy your work in Manchester, in spite of the climate and ugliness of the town, as it has such a lively intellectual life, and you will meet very interesting and stimulating people. I am very glad of your success.'

[73] Segre had no news of parents and relatives in Italy for many months. See, for example, CA, *BSP*: B. Segre to W.V.D. Hodge, [Manchester] 3.9.1944 and H.F. Baker to B. Segre, Cambridge 24.1.1945: 'For your family anxieties I feel, we both feel, the most profound sympathy. I hope the events may not be so bad as you naturally fear. Perhaps you will write if you get good news'.

[74] See Sect. 6.2. On Segre's mathematical output, see Vesentini (2009).

[75] B. Segre, *A note on arithmetical properties of cubic surfaces*, Journal of the London Mathematical Society, 18, 1943, p. 24–31; *On a parametric solution of the equations $x^3 + y^3 + az^3 = b$, and on ternary forms representing every rational number*, Journal of the London Mathematical Society, 18, 1943, p. 31–34; *A parametric solution of the indeterminate cubic equation $z^2 = f(x, y)$*, Journal of the London Mathematical Society, 18, 1943, p. 226–33; *On arithmetical properties of singular cubic surfaces*, Journal of the London Mathematical Society, 19, 1944, p. 84–91.

shall succeed in dealing with other significant problems, I has not had time yet to investigate, I should like to publish instead a small book on the subject, of about 200 pages.[76]

Veblen was positively impressed since he was 'so much taken up with other matters at present that he hardly found any time for mathematical studies'.[77]

In the war years, contacts with the English-speaking mathematical milieu continued to proceed along a double track. The British openly appreciated Segre's results. Semple, for example, defined those on double-forms and related quintuplets 'new, [...] very interesting indeed, perhaps charming is the word for this kind of result which is so simple and so striking'.[78] The only episode of contrariness was Clarendon's refusal to publish the manuscript by Segre on *Arithmetical Properties of Cubic Surfaces* as a Cambridge Tract; this refusal was dictated, however, by economic rather than scientific reasons.[79] In America, meanwhile, the reception of Segre's work was more problematic. The paper *Arithmetic upon an algebraic surface,* submitted to the *Duke Mathematical Journal,* was rejected because it did not fall within the scope of the journal.[80] The editor-in-chief suggested that Segre consider the possibility of resubmitting it to the *Bulletin of the American Mathematical Society* and mailed the manuscript to Veblen, for a possible publication in *Annals of Mathematics.* The *Annals* rejected the paper, which in the end appeared in the *Bulletin of the American Mathematical Society* (51(2), 1945, pp. 152–161). Another manuscript by Segre, *The algebraic equations of degree 5, 9, 157, ..., and the arithmetic upon an algebraic variety,* was forwarded by Zariski to A.A. Albert, for possible publication in the *Transactions of the American Mathematical Society* since it was 'essentially algebraic in nature, and Albert takes care of the algebra papers'.[81] It was rejected but eventually appeared in the *Annals of Mathematics* (s. 2, 46, 1945, pp. 287–301). However, some senior scholars, such as J. Coolidge and Snyder, followed Segre's work closely and viewed his achievements very positively:

> I am glad to see that you are still advancing that sort of mathematics that seems to me to have lasting interest and importance.[82]
>
> From time to time, I had an opportunity to see results of your sustained activity, but now the post-man has brought a whole library of interesting and valuable mathematical memoirs. While I want to congratulate you on this splendid showing, I am thankful that the world still can produce first-class results.[83]
>
> The postman has just brought me a whole library of reprints, all from your pen. They include:
>
> *On arithmetical properties of singular cubic surfaces* (Journal of the London Mathematical Society)

---

[76] CA, *BSP*: B. Segre to O. Veblen, Manchester 1.10.1943. See also D6.2.9.

[77] CA, *BSP*: O. Veblen to B. Segre, Princeton 1.11.1943.

[78] CA, *BSP*: J.G. Semple to B. Segre, London 12.3.1944.

[79] See CA, BSP: K. Sisam [Clarendon Press] to B. Segre, 12.7.1943 and 28.7.1943.

[80] CA, *BSP*: J.M. Thomas to B. Segre, Durham 6.6.1944.

[81] CA, *BSP*: O. Zariski to B. Segre, Baltimore 2.6.1944.

[82] CA, *BSP*: J.L. Coolidge to B. Segre, Cambridge Mass. 14.3.1945.

[83] CA, *BSP*: V. Snyder to B. Segre, Ithaca 30.3.1945.

The Algebraic Equations of Degrees 5, 9, 157 ... and their Arithmetic upon an Algebraic Variety (Annals of Mathematics, April 1945)

Arithmetic upon an Algebraic Surface (Bulletin American Mathematical Society, Feb. 1945)

A Four-dimensional analogue of Pascal's Theorem for Conics (American Mathematical Monthly, March 1945)

Lattice Points in Infinite Domains and Asymmetric Diophantine Approximations (Duke Mathematical Journal, June 1945).

This is a formidable list of subjects, and each is treated in a skillful manner. I congratulate the mathematical public on its good fortune in having such a rich grist. Here's hoping the University of Manchester may long continue to reap the benefits of your labors, and that you and your family may also enjoy life there. My own activity is in the past tense, but I still get a decided thrill in sharing the advances of others.[84]

By way of contrast, 'Lefschetz-trained' geometers identified elsewhere the pivotal horizons of research in algebraic geometry. So for example, thanking for the same reprints on arithmetic geometry mentioned above, Zariski informed Segre:

There is more interest here for algebraic geometry now than there ever was in the past. Outstanding algebraists like Chevalley and André Weil are now working mainly in algebraic geometry. They have developed a quite general intersection theory on abstract algebraic varieties. I am quite sure that when the war is over, we will have many capable young students working in this field.[85]

10.3   The Dilemma of Returning: Cambridge or Rome?

On 21 April 1945, Bologna was liberated. The war seemed to be coming to an end. Baker correctly imagined Segre was giving a good deal of thought to his future and gave him some important news: at least two university lectureships in Cambridge would be opening up in the near future; if he was interested, he should watch the papers for particulars. Neither Baker nor Hodge—with whom Baker had talked— had any part in the selection. They did not know Segre's plans for the future and were not sure he would enjoy having to lecture for the Tripos examination, 'a very different thing from being your own master, as you would be if you were a professor, in England, or Italy'.[86] In the past, the same Hodge had appeared a little discouraging when the question of a lectureship for Segre in Cambridge came up for discussion; this was because he felt that the conditions of appointments were not very suitable for a case like Segre's, and he was 'unwilling to advise him to take a step which he might regret'.[87] Conditions—said Hodge—had now improved considerably, so that, if it was Segre's wish to remain in England, a position in Cambridge might suit him. The competition was not easy, because as Hodge explained:

---

[84] CA, *BSP*: V. Snyder to B. Segre, Ithaca 20.9.1945.

[85] CA, *BSP*: O. Zariski to B. Segre, Baltimore 2.6.1944.

[86] CA, *BSP*: H.F. Baker to B. Segre, Cambridge 12.4.1945. The appointments were advertised in *The Times* of Wednesday May 9, 1945.

[87] CA, *BSP*: W.V.D. Hodge to B. Segre, Cambridge 27.4.1945.

appointments are made by an Appointments Committee of the Faculty, of which I do not happen to be a member this year. No doubt the committee will consult me on your qualifications as a geometer, (and you need have no fear of what I shall say!), but other questions affecting general policy in the Faculty and University will rest entirely with the committee and I shall have no say in these. I suppose the inevitable question of appointing a foreigner will come up, and this will rest with the committee; I have no reason to suppose that they will take any particular view on this question, but it is better that you should realise that there may be questions other than those of mathematical qualifications which will be taken into account. I myself would like to see you here very much. [...] If you decide to go back to Italy, and I could not blame you if that is your decision, I hope you will not in any circumstances leave without seeing me again. I have been half out of the geometrical world throughout the war, and I have felt I have got very much out of touch with you as a result, but I am laying all my plans to return to the full-time study of geometry by next October, and I am hoping to resume all my old contacts. I occasionally hear news of you through Todd, but like me he has felt the results of overwork badly these last years.[88]

In those days, Segre had just been approached by the representative of the Italian Government, E. Ruffini, and invited to go back to his post in Italy, but he had not yet decided, nor did he have definite post-war plans.[89] He was torn at the idea of leaving the country that had welcomed him but he wanted to return to Italy, at least for a short time, so as to reunite with family. The possibility of spending again some time in Cambridge, however, so strongly appealed to him that he decided to apply for one of these lectureships.[90] Hodge was glad to learn that the prospect of a post in Cambridge had some attractions for Segre and suggested that he be open with the appointments committee and give some indication of his intentions, writing for instance in the application:

when conditions become more stable, I should like to consider the possibility of returning to my professorship in Bologna, but I expect to be in England for at least two more years. If the University were to appoint me to a lectureship, I should do my best to remain in its service for at least two years, and I should give as long notice as possible of any decision to leave.[91]

At the same time, a position also opened at the University College of North Wales (Bangor) and, while congratulating him on the liberation of Northern Italy, Baker informed Segre about this further opportunity:

I do not know whether, after a while, you may wish to return to Italy. For my part I hope you will be content to remain in England where your work will be of great value for English Mathematics. I have recently been reading again the volume on the Cubic Surface which you wrote. My admiration for it is greatly increased as I know it better. Probably you have seen the enclosed announcement from the *Times* of yesterday. Bangor is some distance from other places; but it might be agreeable to Signorina Segre.[92]

[88] CA, *BSP*: W.V.D. Hodge to B. Segre, Cambridge 27.4.1945.

[89] SEGRE.22.

[90] CA, *BSP*: B. Segre to H.F. Baker, Manchester 9.5.1945.

[91] CA, *BSP*: W.V.D. Hodge to B. Segre, Cambridge 4.5.1945.

[92] CA, *BSP*: H.F. Baker to B. Segre, Cambridge 5.5.1945.

Segre however, would not apply for the Bangor chair because 'Professor Mordell told me that the Wales people would even object to having an Englishman, and so I suppose they would not be keen on having a foreigner'.[93]

In the summer, Segre presented his application for a chair in Cambridge along the lines that Hodge had dictated to him.[94] He trusted in the outcome of the selection, and so did his colleagues, Semple notably.[95] On 12 July, the damper arrived: Hodge communicated to Segre that the Appointments Committee had made their decisions and that he was not one of those appointed:

> I don't think you need fear they did not appreciate your eminence; a number of them had asked me about you, and I was satisfied when I left them that they knew your mathematical studies. But, of course, there are many other things to be taken into account and I cannot say what reasons they had. But I can say that no geometer was appointed.[96]

The most disappointed was Semple who called into question some twist of local university politics to understand why they had not appointed him.[97] For Segre, rejection was the straw that broke the camel's back:

> You may be sure I do not doubt my work in this country has been appreciated, even if it has had so far no recognition of any kind. I could not yet make up my mind about my going back to my country, owing to the present state of confusion; in fact, up to now, I have received no letters from any of the many friends and relations I had in Bologna, Turin and Milan, while the letters from Castelnuovo were rather discouraging. However, now I quite understand it is perhaps better for me to think and try to go as soon as possible to Italy or, alternatively, to some other country.[98]

Meanwhile, in May 1945, Segre had communicated to the SPSL that the Italian Government had invited him to return to his chair in Bologna and asked permission from the British authorities to visit Italy during the summer.[99] The Society had helped him in securing the expatriation document.[100] After the failure at Cambridge, Segre returned to the point and reminded the SPSL that he hoped to be able to travel to Italy, or alternatively, to go to the United States. At the same time, he was hoping to avail himself of a statement made in the House of Commons on 13 November, which outlined certain categories of distressed persons on the continent who had relatives in England able and willing to look after them, and to whom the British Government were prepared to grant visas. He thus asked for help from the Society for his mother-in-law, Gilda Coen, who would be very glad if she could come and

---

[93] B. Segre to H.F. Baker, Manchester 9.5.1945.

[94] SEGRE.23.

[95] CA, *BSP*: W.V.D. Hodge to B. Segre, Cambridge 5.7.1945 and J.G. Semple to B. Segre, London 7.7.1945.

[96] CA, *BSP*: W.V.D. Hodge to B. Segre, Cambridge 12.7.1945. The Committee appointed H.G. Booker, M.H.L. Pryce and A.J. Ward.

[97] CA, *BSP*: J.G. Semple to B. Segre, Redhill Surrey 10.8.1945.

[98] CA, *BSP*: B. Segre to W.V.D. Hodge, Manchester 17.7.1945.

[99] D7.2.5.

[100] SPSL, *BS*: J.B. Skemp [SPSL] to B. Segre, London, 22.5.1945.

live with them for a while.[101] Understandably, the Society was a little doubtful. If Segre had really made his mind up to accept the invitation to Bologna, his mother-in-law would return with him in a year's time. Consequently, the advice was given to ask only for a permit for a recuperative holiday for her in England.[102]

The reasons for Segre's uncertainty depended on the fact that he negotiated the return with various interlocutors (Castelnuovo, Severi, Tricomi, and Colonnetti) and took into consideration various options: return to Bologna to his chair (this was the proposal that had been made to him by the government); accept the reinstatement in Bologna, but ask for a secondment abroad, in the United Kingdom or United States, for a couple of years; go to Turin to take over Fano's chair, since Fano would have just a few months of work before retirement; go to Rome to take over Enriques' chair, since Enriques too was close to retirement; and return as a seconded professor in service at INDAM or at the CNR.[103] Italian colleagues, of course, had conflicting opinions about what was best for him. Castelnuovo, the first with whom Segre reconnected, recommended Bologna or Turin.[104] Severi, on the other hand, wanted Segre in Rome.[105] Making amends for his behaviour in 1938, rehabilitating himself in the eyes of colleagues, even personal gain were all concomitant causes for Severi engaging in a battle to bring Segre back to Rome. For his favourite pupil, Severi had great plans: to reinstate him to the board of the *Annali di Matematica pura ed applicata*; to involve him in the relaunch of INDAM, before bequeathing it to him; to relaunch (with Segre's help) the Italian Geometric School, which had been decimated by deaths and emigrations. For this last point, it was essential to capitalise on Segre's experience in England, hence the task of drafting the reports on algebraic geometry in the Anglo-Saxon countries during the war years, the invitation to hold a course in abstract algebra at INDAM and to publish, possibly with the contribution of the CNR, a treatise on modern geometry.[106] There were also personal reasons why Severi wanted to quickly reconnect with Segre. Severi was in fact facing the purge process and hoped that Segre would procure him an invitation to Cambridge, so as to get away from Italy for a while.[107] It is quite disconcerting that Segre immediately submitted to all Severi's demands, intervening to clarify, in front of Baker and Hodge, Severi's position towards fascism and his acquittal by the purge commission, with a simple censure.[108]

In May 1946, Segre planned to return to Italy for a brief visit during the summer. As he wrote to the SPSL, he wished to go and see on the spot what the situation was

---

[101] SPSL, *BS*: B. Segre to J.B. Skemp [SPSL], Manchester 17.11.1945.

[102] SPSL, *BS*: J.B. Skemp [SPSL] to B. Segre, London 21.11.1945.

[103] See D7.1.4, D7.1.5, D7.2.3 and D7.2.4.

[104] SEGRE.24 and SEGRE.27.

[105] SEGRE.25 and SEGRE.28.

[106] SEGRE.29 and SEGRE.30.

[107] D7.1.5 and SEGRE.25.

[108] SEGRE.26 and SEGRE.28.

like.[109] The family remained in Manchester.[110] It was a journey with multiple objectives: to see family members, to reconnect with colleagues, to understand the conditions of the country, and to definitively evaluate if and where to return. Both Baker and Hodge were convinced that Segre had not yet definite post-war plans.[111] Mordell, by contrast, clearly suspected that it was a definitive return:

> I am glad that you find conditions in Italy not too bad and that you will be resuming your proper position as a distinguished professor of maths in your own right. I was very sorry that my departure from Manchester prevented me from doing more for you. I was very glad to have you there and profited both mathematically and linguistically from your presence.[112]

Upon return, Segre knew that both his family, and the family of his wife, had escaped the deportation to the extermination camps. Colleagues welcomed him very warmly.[113] Even L. Berzolari (who, in 1938, had not defended Segre within the UMI Scientific Commission) praised him for the 'intense research activity carried out in England'.[114] A. Maroni wished for his definitive return to Italy 'a return that is surely of great benefit to Italian mathematics'.[115] L. Campedelli 'was glad to see re-starting the periodical and pleasing arrival of his works, substantial and always interesting'.[116] On the other hand, the political situation in Italy appeared still very uncertain

---

[109] SPSL, *BS*: B. Segre to SPSL, Manchester 28.5.1946.

[110] Segre asked SPSL to sponsor him with the Home Office, so he may obtain the re-entry permit, either to return to fetch his family immediately or to stay in UK himself with his family for another period before returning. See I.J. Ursell [SPSL] to The Under-Secretary of State, Home Office Aliens Department, Cambridge 30.5.1946.

[111] CA, *BSP*: H.F. Baker to B. Segre, Cambridge 11.5.1946 and W.V.D. Hodge to B. Segre, Cambridge 17.5.1946.

[112] CA, *BSP*: L.J. Mordell to B. Segre, Cambridge 16.8.1946.

[113] See, for example, CA, *BSP*: G. Batta Bonino to B. Segre, Pavia 6.9.1946: 'The colleague and common friend prof. Villa has communicated to me (with your current address) the very welcome news of your return to Italy and your upcoming resumption of activity in our Faculty. It is a source of deep and sincere satisfaction for me to be able to offer a warm greeting to my dear friend who returns to the Bolognese Chair to restore to our university the luster that comes from his distinguished scientific personality. In offering yourself, I express to you my best wishes (which are a certainty) that the greatest satisfactions may gladden your fruitful work and possibly may cushion the memory of a painful parenthesis'. (*Il Collega e comune Amico prof. Villa mi ha comunicato (con il tuo indirizzo attuale) la graditissima notizia del tuo ritorno in Italia e della tua prossima ripresa dell'attività nella nostra Facoltà. È per me motivo di viva e sincera soddisfazione poter porgere un caldo saluto all'Amico carissimo il quale ritorna alla Cattedra Bolognese per ridonare alla nostra Università il Lustro che proviene dalla sua Insigne Personalità scientifica. Nel porgerti questo saluto profondamente sentito, formulo a Te l'Augurio (che è certezza) affinché le maggiori soddisfazioni abbiano ad allietare la tua opera feconda e possibilmente possano attutire il ricordo di una parentesi dolorosa.*)

[114] CA, *BSP*: L. Berzolari to B. Segre, Pavia 28.11.1946: *feconda attività scientifica da Lei svolta in Inghilterra.*

[115] CA, *BSP*: A. Maroni to B. Segre, Florence 1.12.1946: *ritorno che è senza dubbio un ottimo beneficio per la Matematica italiana.*

[116] CA, *BSP*: L. Campedelli to B. Segre, Florence 2.12.1946: *vedo riprendere il periodico e gradito arrivo dei tuoi lavori, copiosi e sempre interessanti.*

and the living conditions of the country were difficult. In the cities affected by bombs, the housing problem was particularly serious. At this stage, Severi was decisive in convincing Segre to take the definitive step: remaining in Italy.[117]

After a period in Bologna, Segre would receive the transfer to Rome on the chair of Higher Geometry which had been occupied by Enriques (disappeared that same year on June 14). He would remain in Rome for the rest of his life, a major protagonist of Italian mathematics in the post-war period right up to the 60s.

Documents

Segre.1 O. Zariski to O. Veblen, Baltimore 3.11.1938

OVP, *BS*, fol. 1r
Dear Professor Veblen: I am enclosing a letter from B. Segre (please return) in which he informs me about his present predicament. It appears that he has lost his position of *Professore ordinario* at the University of Bologna, that he has been expelled from all Italian mathematical organizations, that he has been dismissed from his duties as co-editor of the *Annali di Matematica*, and finally, that he has lost all these rights without acquiring the right to compensation or to pension. He expresses a desire to come to the U.S., if he could get placed here on a relatively stable basis. As a friend of Segre, I should like to help him.[118] As a co-worker in the field of algebraic geometry, I have a great respect for his mathematical abilities, and I feel that it would be a pity if he was lost to mathematics. I realize that his poor knowledge of English would handicap him here in the beginning. Perhaps the Institute would be willing to make some arrangements by which he could spend a semester or two in Princeton. I am also writing to Lefschetz, and I sincerely hope that some way will be found to help Segre in his present predicament.[119] Sincerely yours, O. Zariski

Segre.2 O. Veblen to O. Zariski, Princeton 8.11.1938

OVP, *BS*, fol. 1r
Dear Zariski: I talked over the question about B. Segre with Lefschetz at some length this morning. He thinks that we ought not to bring him here to Princeton even for only a semester or two. His points are, in the first place, that the situation

---

[117] See D7.1.5 and SEGRE.31-35. His family joined him shortly after in October 1946.

[118] Zariski and Segre were close friends. In the years 1926–1927, they had met regularly at the Caffè Greco in Rome in order to gossip and play chess.

[119] Lefschetz had met Segre in Rome in 1931.

in this country is becoming so difficult for the Jews that we ought not to aggravate it by bringing anymore; and in the second place, that there are a number of mathematicians who either are, or are soon likely to be, as much in difficulties as Segre, and who are more valuable scientifically. I am loath to agree with him in either of these contentions; but I doubt whether it is advisable to bring an algebraic geometer to Princeton as long as Lefschetz takes this point of view.[120] With regard to Lefschetz's first point, I am inclined to think that we shall not substantially aggravate the situation by finding positions for Jews of high caliber, for there will be a great many others of less value who will come in spite of anything that anyone can do. And I think that it is likely to be, on the whole, the less valuable ones who will be crowded out. To put it in more general terms, I am not clear enough that the danger of provoking opposition by a liberal policy is as great as the danger from the moral weakness which would result from giving it up. I don't think that I succeeded in saying this very clearly to Lefschetz, and I am therefore giving him a copy of this letter. With regard to Lefschetz's second point, I have to admit that I do not know of any work of Segre's on differential geometry which I have been obliged to take account of, and I don't feel competent to contest what he or you have to say on the subject of his standing in algebraic geometry. [O. Veblen] This page *not* mailed, at Prof. Lefschetz's request.

Segre.3 V. Snyder to L. Farrand [ECADFS], Ithaca 9.2.1939

ECADFS Records, *BS*, fol. 1r
Dear Dr. Farrand: At the suggestion of Dean Richtmyer I am writing you about another Jew, realizing that your time and strength have already been pretty thoroughly taxed with this problem.[121] And the embarrassing part is that I have no definite plan, and at present wish simply to lay the matter before the Emergency Committee in Aid of Displaced Foreign Scholars, asking for proper procedure on my part. Under another cover I am sending you the outline of the biography of Beniamino Segre, in his own English, and a complete list of his published papers.[122] Ten years ago, while I was on a sabbatic leave in Rome, Italy, I met Dr. Segre, who at that time was an assistant to Professor Severi, with whom I was working. In the spring of 1929 Severi was away from Rome for months; he wrote me to confer with Segre from time to time on my work. The association lasted about four months. He was a polite young man, particularly congenial to talk with, and seemed interested in what I was doing. He is strikingly

---

[120] See also D3.13.1 and D3.13.2.

[121] Floyd Karker Richtmyer (1881–1939), professor of Physics at Cornell University (1918–1939), was Dean of the graduate school (1931–1939).

[122] D3.18.1.

fair, has blue eyes, and no trace of appearance or manner that would suggest a Jew. I have not seen him since; he has always sent me his reprints, and we exchange greetings from time to time, but that is all. He is generally recognized as one of the strongest young mathematicians in Italy. While his main interest is in algebraic geometry, he has also written on algebra and in differential equations. Any suggestions as to help for this young man would be deeply appreciated. Mrs. Snyder joins me in sending best greetings to you and Mrs. Farrand. Sincerely yours, Virgil Snyder

Segre.4 J.G. Semple to D.C. Thomson [SPSL], London 20.2.1939

SPSL, *BS*, fols. 26–27
Dear Sir, This is just to confirm the arrangements we made today in our interview, today, [sic!] *re B. Segre.* I informed you that friends of Beniamino Segre in this country had individually guaranteed sums of money amounting to £ 128 for two years, besides some few extra promises for single sums, to help to enable him to leave Italy & come to this country. And you undertook to recommend the Society for the Protection of Science & Learning to augment these private efforts, so as to guarantee Segre £250 for the first year, his case to be reviewed at the end of the year. You were also of the opinion that it might be possible to regard the extra single sums promised as additional to the £ 250, on account of Segre's family & the special efforts his friends have made on his behalf. You were also kind enough to offer (on behalf of the S.P.S.L.) to act as trustees for the "Segre Fund" as initiated by Professor Hodge & myself, so that subscriptions would be payable to the S.P.S.L.; and this would mean, I take it, that you would write to the persons who have promised contributions if & when such contributions are required. In this connection I can furnish you, when you require it, with a complete list of the full names & addresses of the persons concerned, as you have now only a rough note of names & amounts. I think, if it seems suitable to you, that the subscribers might prefer the "Segre Fund" idea under your auspices than a general merging of their subscriptions in the S.P.S.L. funds, though the distinction would be in fact nominal. But if this presents any difficulty, I person-ally would not object to any other arrangement; & I am sure the others would not. Finally, you agreed, if Segre was accepted by your Council, to provide the necessary guarantees etc. to facilitate his admittance to this country and to put him in touch with the various committees looking after the accommodation of refugees once here.[123] I need hardly say how much we appreciate the indispens-able cooperation of the S.P.S.L. in this matter and your own invaluable help & advice. I hope our united efforts will succeed in rescuing Segre, in the time being

---

[123]See D3.4.9.

at any rate, from his present impossible situation. Please correct me if I have in any way misunderstood you. Yours sincerely, J.G. Semple

Segre.5 O. Veblen to H. Shapley, [Princeton] 14.3.1939

OVP, *RefGen*, fols. 1r
Dear Shapley: This is to be a reply both to your letter to me of March 2 and to your letter to Weyl dated March 11.[124] I agree with you that we cannot do the necessary work to get together a selected group of scientists until I get something more definite to go on, and that I have not yet succeeded in doing. I don't know how long it will take, if ever accomplished. Russell handed me the statement about Jacchia, and I will put it in our file of refugees, ready to use in case anything materializes.[125] I also have heard from Bullitt about the University of Louisville and have talked with Weyl and von Neumann about possible candidates. Bullitt insists that what they want is a man who would be very active and productive, and I think we ought to come as near as possible to giving him what he wants. This will probably exclude some of the older men, for whom we are most sympathetic, and among the mathematicians the two names which seem most promising are those of Löwner (aged forty-five) and Hamburger (aged forty-nine).[126] But this will be a matter for further discussion when either or both of us have seen Bullitt. I believe you have the records of both of these men in your files. Beniamino Segre would also be a first-class candidate for this position. He is very prolific, and I have no doubt that the Louisville people would be well satisfied with him. There is, however, a good chance that he will be taken care of otherwise, because Snyder (emeritus, of Cornell) and Coble (Head of the Department at Illinois) are both interested in him and trying to work out a cooperative scheme to take care of him. His work is closely connected with theirs. This is perhaps as much as I can say on the subject until I hear further from you or from Bullitt. Yours sincerely, Oswald Veblen

---

[124]D3.13.9.

[125]Henry Norris Russell (1877–1957), Director of the Princeton University Observatory from 1912 to 1947. Luigi Jacchia (1910–1996) was an Italian astronomer. On his case, see D3.8.3.

[126]On this position, see Sect. 3.13 and D3.13.9. William Marshall Bullitt (1873–1957), an influential lawyer who taught at Harvard University, was a member of the Louisville Bar Association. The position in Louisville would be offered to Karl Löwner (1893–1968), a Czech mathematician dismissed from the Charles University of Prague.

Segre.6 J.G. Semple to D.C. Thomson [SPSL], London 20.4.1939

SPSL, *BS*, fol. 66

Dear Sir, With reference to your letter of April 1st, I have now to report that Professor Segre, with his wife and three children,[127] arrived in this country yesterday, and they are now living (temporary) in apartments at the following address: c/o Mrs. C.N. Woolnough, Fasibbi, Ashurst Road, Cockfosters. He has received the letter from you, addressed to him, c/o myself; and I dare say he will call on you in a few days time. I enclose, as you requested, a complete list of the promised subscriptions to the Segre Fund. You will notice that in addition to the promises for two years, there are promises amounting to £ 35 – 10 for single sums. I should imagine these, which were to be regarded as additional, would be particularly useful to Segre in the immediate future to enable him to move about and find a permanent settling place. But you will probably be able to advise him. His present lodgings, you will remember, are only a *pied à terre* till he decides on a permanent place of residence. Yours sincerely, J.G. Semple

List of Promised Subscriptions to Segre Fund
Professor W.V.D.H. Hodge, F.R.S., 28 Barrow Rd., Cambridge

£ 25 per annum for 2 years

Professor J.G. Semple, 4 Ruskin Close N.W. 11

£ 25 per annum for 2 years

P. Fraser, Esq. [sic!], M.A., The University Bristol

£ 20 per annum for 2 years

J.A. Todd, M.A. Ph.D., The Hermitage, Silver Street Cambridge

£ 15 per annum for 2 years

L. Roth, Esq. [sic!], M.A., 21 Bryndale Crescent, N. 14

£ 13 per annum for 2 years

T.L. Wren, Esq. [sic!], M.A., 32 Paddenswick Rd., W.6.

£ 10 per annum for 2 years

Professor V.C. Morton, M.A., Eryl Mor. Brynymor Road Aberystwyth

£ 10 per annum for 2 years

F. P. White, Esq. [sic!], M.A., St. John's College, Cambridge

---

[127]Fernanda Coen Segre (1908–1976), Sergio (born on 22 May 1933), Silvana (14 May 1934), and Ornella (15 June 1937).

£ 10 per annum for 2 years

Professor H.S. Ruse, M.A. D. Sc., University College, Southampton £ 5 per annum for 2 years

W.L. Edge, Esq. [sic!], M.A., The Mathematical Institute, 16 Ghennbiss St. Edinburgh

£ 10 – 10 – 0

Professor A.L. Dixon, Magdalen College Oxford

£ 10 – 0 – 0

D.W. Babbage, Ph.D., Magdalene College Cambridge

£ 10 – 0 – 0

Professor W.P. Milne, The University, Leeds

£ 5 – 0 – 0. (These last four are promises of *single sums*)[128]

Segre.7 B. Segre to T. Levi-Civita, Cambridge 11.12.1939

ANL-*Levi-Civita*, fol. 1r-v
Dear and illustrious Professor, I would like to send you my warmest and most heartfelt good wishes for the year that is about to begin, which – it is to be hoped – will repair so many injustices. We often talk about you with Volterra, from whom I have frequent news of you.[129] Even now I do not have any concrete prospect in sight, and my situation here has become more uncertain since the outbreak of war; for example, for this reason, I was not given a lectureship that otherwise – in all probability – I would have been entrusted with.[130] However, I

---

[128] Peter Fraser (1880–1959), professor at the University of Bristol (1906–1946); John Arthur Todd (1908–1994), lecturer at Cambridge (1937–1973); Leonard Roth (1904–1968), lecturer at Imperial College in London (1931–1965); Thomas Lancaster Wren (1889–1972), lecturer at University College London (1919–1954); Vernon Charles Morton (1896–1978), professor and Head of Pure Mathematics Department at the University College of Wales, Aberystwyth (1933–1961); Francis Puryer White (1893–1969), lecturer at St. John's College in Cambridge (1922–1961); Harold Stanley Ruse (1905–1974), professor of Mathematics at the University College, Southampton (1937–1946); William Leonard Edge (1904–1997), lecturer at University of Edinburgh (1932–1975); Arthur Lee Dixon (1867–1955) Savilian professor in Oxford (1922–1945); Dennis William Babbage (1909–1991), research fellow at Magdalene College in Oxford (1933–1939); and William Proctor Milne (1881–1967), professor at Leeds University (1919–1946).

[129] Enrico Volterra (1905–1973), son of Vito Volterra, and Levi-Civita's assistant, in February 1939, had left Italy for Cambridge. See Sect. 5.2.

[130] Segre had been asked to give a series of lectures at Cambridge, for £ 50 (3 lectures per week for a period of 8 weeks in autumn of 1939). They were cancelled on outbreak of war. See CA, *BSP*: SPSL to B. Segre, London 17.8.1939.

am satisfied with this stay, which allows me fruitful contacts with this interesting mathematical environment and gives me the opportunity to continue my studies without too many material concerns; I have already given some lectures and I plan to do others later. My wife and I have readily become accustomed to this change of lifestyle, and I am particularly pleased that the children have been able to settle in very well. With warm greetings, to you, your kind wife and your sister-in-law, and best regards from your most devoted and affectionate B. Segre

P.S. Hamburger, who was also able to come to England and stayed in Cambridge for a few days, asks me to pass on his greetings to you.[131]

Caro ed illustre Professore, Desidero le giungano i miei più vivi e sentiti auguri per l'anno che sta per iniziarsi, il quale – è da sperare – porrà riparo a tante ingiustizie. Parliamo sovente di Lei col Volterra, da cui ho frequenti sue notizie. Neppure ora ho nulla di concreto in vista, e la mia situazione qui è diventata più incerta dopo lo scoppio della guerra; ad esempio, per questo motivo, non mi è stato dato un incarico che altrimenti – con ogni probabilità – mi sarebbe stato affidato. Sono tuttavia soddisfatto di questo soggiorno, che mi permette proficui contatti con questo interessante ambiente matematico, e mi lascia il modo di continuare i miei studi senza troppe preoccupazioni materiali; ho già tenute alcune conferenze ed altre conto di tenere in seguito. I miei ed io ci siamo prontamente abituati al cambiamento di vita, e sono particolarmente lieto che i bimbi abbiano potuto ambientarsi benissimo. Ancora tante cose belle, anche per la Sua gentile Signora e per Sua cognata, e distinti saluti dal Suo dev.ᵐᵒ ed aff.ᵐᵒ B. Segre

P.S. L'Hamburger, che pure ha potuto venire in Inghilterra e rimase a Cambridge per alcuni giorni, mi lasciò di salutarla.

Segre.8 F. Coen Segre to N. Searle [SPSL], London 7.8.1940

SPSL, *BS*, fol. 164–165
Dear Miss Searle, I send you a copy of the letter my husband, professor B. Segre, at present interned in the Isle of Man, sent to the Home Office a month ago. I beg you to use your valuable influence in its support, and apologise for any inconvenience, that this may cause you. Thanking you in anticipation, Yours sincerely Fernanda Segre

---

[131] Hans Hamburger (1889–1956), former professor at the University of Cologne, had emigrated in the United Kingdom in 1939. From 1941 to 1946, he was lecturer at the University of Southampton.

Sir, I beg to apply for my release from internment, and I wish to call your kind attention on the following points, on which my request is based.

i) Till 1938 I was full time Professor and Director of the Mathematical Institute of the University of Bologna (Italy). Although I never had any political activity, on the 2nd of September 1938 – being a Jew – the Italian Government dismissed me, depriving me of my salary and of any possibility of scientific work.

ii) I came to England as a Refugee with my wife and three children on the 19th of April 1939, with the support of the Society for the Protection of Science and Learning and of the British Mathematicians who, for that purpose, had raisen a fund.

iii) As soon as I arrived in this country, I declared to the Home Office (through the Society for the Protection of Science and Learning) my willingness to join any civil or military service, should Great Britain enter a war.

Iv) At the outbreak of the war, on the 3rd of September 1939, I wrote a letter to the Society for the Protection of Science and Learning, expressing my deep wish to help the British Nation in any capacity.[132] The same wish I still have today.

v) During the past months I gave a lecture at the London Mathematical Society and many lectures at Cambridge University. I have moreover an important mathematical paper in print in the *Proceedings of the London Mathematical Society*, and I was about to write a book on higher Mathematics for the Oxford University Press, in collaboration with Dr. J.A. Todd (fellow of Trinity College, Cambridge).[133]

I am sure you will be able to verify quite easily these statements and also that Professor W.V. Hodge of Cambridge University and Professor J.G. Semple of London University will give any further information which you may require, so that all doubts as to my feelings with regard to the present struggle could be entirely and promptly removed. Yours faithfully Beniamino Segre

N. 58848, House N. 13 – Palace Internment Camp – Isle of Man

---

[132] SPSL, *BS*: B. Segre to SPSL, Cambridge 3.9.1939: 'Dear Sir, After the events of to-day I am willing to repeat my offer to serve in any capacity in which I can be of use to the Nation. Yours sincerely Beniamino Segre'

[133] B. Segre, *On limits of algebraic varieties, in particular of their intersections and tangential forms*, Proceedings of the London Mathematical Society, (2), 47, 1942, p. 351–403 and B. Segre, J.A. Todd, *Geometry upon Algebraic Surfaces and Varieties*, Oxford, Clarendon Press, expected for 1942.

Segre.9 W.V.D. Hodge to E. Simpson [SPSL], Cambridge 8.8.1940

SPSL, *BS*, fols. 167–168

Dear Miss Simpson, In reply to your enquiries about Professors H. Hamburger and B. Segre, I can state without reservation that in both cases their work represents a definite contribution to learning and science. I have a more detailed knowledge of Segre's work, which is in the top class, and includes some of the most important contributions to geometry made in recent years, but I also know enough of Hamburger's work to appreciate the contribution to mathematical knowledge which his researches have made. Perhaps I ought to deal with the two separately. [...] Segre is the foremost of the younger school of Italian geometers, and that means that he is one of the best of the younger generation of geometers in the world. He has made a considerable number of contributions to geometrical knowledge which are of first-class importance. I need not specify them in detail, since an adequate description of them would involve me in great technical details which could only be understood by an expert. The simple truth is that every geometer finds it well worth his while following Segre's work closely. His recent work has been concerned with problems concerning "limiting loci" and he has recently completed an important paper on this subject which is being published in the *Proceedings of the London Mathematical Society*.[134] He has also begun to write a book on Higher Geometry in collaboration with Dr. J.A. Todd, of Trinity College, and when completed this book will be a valuable contribution to geometrical literature, and will be in much demand in this country, U.S.A., and any other countries to which it can be exported.[135] It is to be published by the Oxford University Press. In addition to these activities, Segre has proved himself an invaluable member of my geometrical seminar, and has been a great source of help and encouragement to young geometers here. The answers to the three questions which you ask explicitly are:

(a) In view of the collaboration with Dr. Todd which I have mentioned, Segre would find residence in Cambridge most convenient, but residence elsewhere would not mean that the collaboration was impossible, only more inconvenient. I am not sure which areas are protected, but I consider that after Cambridge the most convenient places of residence, in my opinion, would be London, Bristol, Manchester, in that order.

(b) As far as I know, Segre's only source of income is his grant from the S.P.S.L.

(c) To the best of my belief, Segre's sympathies in the war are entirely on our side. He has offered his services for work of national importance, and very shortly before his internment I heard him express the wish that an Italian legion would be formed in this country. He is not interested in politics, and his attitude is similar to people like myself who are not normally very interested

---

[134]B. Segre, *On limits of algebraic varieties, in particular of their intersections and tangential forms*, Proceedings of the London Mathematical Society, (2), 47, 1942, p. 351–403.

[135]B. Segre, J.A. Todd, *Geometry upon Algebraic Surfaces and Varieties*, Oxford, Clarendon Press, expected for 1942.

in political matters, but who feel that at the present time national requirements must take precedence over personal interests. [. . .]

Yours sincerely, W.V.D. Hodge

Segre.10 J.A. Todd to E. Simpson [SPSL], Cambridge 9.8.1940

SPSL, *BS*, fol. 170

Dear Miss Simpson, It gives me great pleasure to support an application from the S.P.S.L. for the release of Prof. B. Segre from internment. Professor Segre has been engaged in mathematical teaching and research for some fifteen years, during which time he has, I believe, published well over a hundred original papers on Geometry. His work is often of striking originality, and I should have no hesitation in saying that, at the time when he was dismissed from his post at Bologna for reasons of race, he was the most prominent among the algebraic geometers of the younger generation in Italy. He has, as you know, been in Cambridge during the present year, and I personally have been impressed by his deep devotion to his science, and by the vigour of the interest with which he has continued to pursue it. Before he was interned, he and I had agreed to collaborate in the preparation of a book on *Algebraic Surfaces and Varieties* for the Oxford University Press.[136] I understand that the Press are also prepared to cooperate in urging his release. If he were released, he would be able to continue his studies in Cambridge or London without any difficulty. As regards his financial resources, the S.P.S.L. are probably better acquainted with the situation than I am, but I believe I am correct in saying that the Segre fund has still the best part of a year to run. I should have thought that there would be no great difficulty, even in these times, of ensuring his continued support, should he be unable to find employment of some kind. Personally, I have no doubts whatever as to his personal integrity and loyalty to this country; I believe in fact that he desires nothing better than to help us in any way within his capacity. I have no hesitation in recommending, as strongly as I can, that this distinguished scholar should be released at the earliest possible date. Yours sincerely, J.A. Todd

Segre.11 B. Segre to E. Simpson [SPSL], London 15.8.1940

SPSL, *BS*, fol. 173

Dear Miss Simpson, I am glad to be able to announce to the S.P.S.L. that I have been released from internment, and I wish to express my deepest gratitude for the

---

[136] B. Segre, J.A. Todd, *Geometry upon Algebraic Surfaces and Varieties*, Oxford, Clarendon Press, expected for 1942.

helpful support you so kindly gave to my wife and myself during my internment. I would willingly accept any work of National utility and would be thankful to you if you could help me in finding it. Should that not be possible, then I would continue my research work here in London, since Cambridge is now a protected area. I have not yet had back the scientific books, pamphlets and manuscripts which were taken away from me at Bury Internment Camp, before going to the Isle of Man; but I hope they will be soon returned to me, if possible through your kind concern. My address, from Monday 19th, is: B. Segre, c/o Miss Summers

2, Callard Avenue – London N. 13. Yours sincerely Beniamino Segre

Segre.12 E. Simpson [SPSL] to B. Segre, London 16.8.1940

SPSL, *BS*, fol. 174

Dear Professor Segre, It is not often that we now start the day with a piece of good news. I was very happy to have your letter, and Professor Hodge also called in to tell me that you were now free. We are very glad about this, and hope that you have not suffered too much by your unpleasant experience. I do not know whether a letter from us will carry any weight, but I am writing to the Commandant of the Bury Internment Camp to try to secure the return of the books and manuscripts that were taken away from you. It would be helpful if you could let me have a list of these, but I am writing without waiting for your reply, in case you do not remember exactly what was left at Bury. Yours sincerely, Esther Simpson - Secretary

Segre.13 J.A. Todd to Whom It May Concern, Cambridge 15.4.1942

CA, *BSP*, fol. 1r

I understand that Dr. B. Segre is a candidate for the lectureship in Mathematics at University College Swansea. I have known Dr. Segre personally for the last two and a half years, and by reputation for a much longer period. In my opinion he is the most outstanding of the younger generation of Italian geometers. His output of original work is enormous, and this work covers a remarkably wide field; I have no doubt that his appointment would be a source of considerable strength to any mathematical Faculty. I would strongly support his application. J.A. Todd. Lecturer in Mathematics in the University of Cambridge

Segre.14 J.G. Semple to Whom It May Concern, Bristol 16.4.1942

CA, *BSP*, fol. 1r-v

I understand Dr. Segre is applying for the vacant mathematics lectureship at Swansea, and it gives me great pleasure to say from my knowledge how well he is fitted for such a post. Of Dr. Segre's great eminence as a mathematician, it is quite unnecessary for me to speak: his name and the great volume of first class work he has produced and is producing, extending over geometry, differential equations, to topics in applied mathematics, speaks for itself. He is a mathematician of real eminence and unusual versatility, and without doubt would be a great acquisition for any University department of mathematics. From my personal acquaintance with Dr. Segre, I may also say without hesitation that he would be one of the best and most friendly and helpful of colleagues. Since coming to this country as a refugee, he has done his utmost to learn the methods of our universities and to equip himself to be most useful under new conditions. And I am sure that both students and staff at Swansea would find him most anxious and well able to play his full part in the work of the department. I would recommend Dr. Segre most warmly for the vacant post. J.G. Semple (Professor of Mathematics, University of London, King's College)

Segre.15 W.V.D. Hodge to The Registrar of the University College of Swansea, Cambridge 16.4.1942

CA, *BSP*, fol. 1r[137]

Dear Sir, Professor B. Segre is a geometer of international repute, and indeed is the most distinguished of the younger members of the Italian School of geometry. His researches have produced over 100 papers, on topics covering a wide range of subjects, including the theory of surfaces, the general theory of algebraic varieties, and projective differential geometry. It is not easy to select researches for special mention, since his contributions in various fields have been of great importance, but I feel that I must mention particularly his work on the theory of equivalence on algebraic varieties, of which he must be regarded as the leading exponent. Professor Segre held a chair of Mathematics at Bologna until he was removed from it under the racial laws introduced in Italy in 1938. He came to England in 1939. I had hoped to benefit from his presence in Cambridge by arranging for him to give courses of lectures to my advanced students, but the war has interfered with these plans.[138] Nevertheless, it has been possible to have him address my seminar in geometry on numerous occasions, and he has acquired

---

[137] One similar letter was sent by W.V.D. Hodge to The Registrar of the University of Sheffield, on June 11, 1942.

[138] See footnote 130.

some experience in lecturing in English. While he has still a marked foreign accent, he is easy to follow, and he impresses me with his knowledge of how to lecture. His lectures have, of necessity, been confined to topics of geometrical interest, but in Italy he has lectured on all the main branches of mathematics, pure or applied, and I have no doubt that he would be able to perform all the duties required of him if he were appointed to Swansea. I have much pleasure in supporting his application for a lectureship in Mathematics at the University College of Swansea. I am, Sir, Yours faithfully, W.V.D. Hodge Lowndean Professor of Astronomy and Geometry, University of Cambridge

Segre.16 H.F. Baker to Whom It May Concern, Cambridge 16.4.1942

CA, *BSP*, fol. 1r-v
Professor Beniamino Segre tells me that he is a candidate for the post of lecturer in Mathematics at the University of Swansea; I consider it a high privilege to bear witness to his eminence and fitness for the post. While I cannot claim to know at first hand his skill as a teacher, I know that he worked in close collaboration with Professor Severi in Rome, and was chosen to be Professor in Bologna, where he was also a successful Editor of the Italian Journal, the *Annali di Matematica*. Since he has been in England he has continued to work assiduously at his subject; and the Oxford University Press is printing two books at the present time which he has written in England.[139] His presence in a Mathematical School would be a constant source of inspiration to students and colleagues alike – by whom, I am sure, he would be much liked – and would add a great deal to the reputation of the School. Moreover his output would be a valuable addition to the British output, and be of permanent value. H.F. Baker lately Professor of Astronomy and Geometry in Cambridge; F.R.S.

Segre.17 J.G. Semple to B. Segre, London 27.3.1944

CA, *BSP*, fols. 1r-3v
Dear Segre, Very many thanks for your long and interesting letter. I am sorry if I sounded in any way vague about your paper: it was quite unintentional. I had in mind, probably, that you had not finished with it yet. Now you mention it, I think there might well be risk of censorship delay at the present time if you sent it to R.I.A. [Proceedings of the Royal Society of London. Series A], especially as it would have to cross and recross to be refereed. Why not send to the Edinburgh

---

[139]B. Segre, *The non-singular cubic surfaces*, Oxford, Clarendon Press, 1942 and B. Segre, J.A. Todd, *Geometry upon Algebraic Surfaces and Varieties*, Clarendon Press, expected for 1942.

Math. Soc., where Edge will take charge of it for you on the spot?[140] About your future, I may say quite frankly that I would do anything in my power to help you to obtain a permanent post in this country; indeed, I feel that our subject would be strongly stimulated and encouraged by your presence amongst us. If I mention difficulties, it is only in order to estimate what has to be overcome. In the first place I must admit that as regards positions and facilities and freedom, you would be much better off if you got yourself reinstated in Italy: for, as regards *chairs* in this country, most Universities and Colleges would be too nervous to put a department in charge of yourself who, however brilliant in research, might not understand the partial drudgery and routine attaching to a Professor's office here as opposed to the continental system. Your best chance, I imagine, would be to become lecturer or reader: there will be so vast an upheaval and demand as soon as the war ends, that I think you should not have too much difficulty in setting into that grade. I know of many departments who will want to fill gaps that have been left vacant because of the war, our own at King's College for example. But can you really be satisfied with that kind of position? Your present position – except as a war-time emergency – is so unworthy of you as to be absurd; at least I regard it so. If you get put on the lecturer grade, as a permanent appointment, you will still have a salary much below what is worthy of you and what you have a right to expect: also, you will have more routine work, examinations, corrections and so on than you have now by far, and probably less time for research. Is your objection to return to Italy so strong that you can put up with all that? For a time at any rate? The best alternative, and I should have thought it was the right thing, would be for you to be taken on at Cambridge (or Oxford), where the system is more elastic – at least in Cambridge. But there the trouble is that you are in a School opposed to Hodge's: Hodge belongs to the Zariski School of extreme rigour and the severest economy of the printed word, and, on the whole, for the elimination of anything that could be regarded as 'pretty' or 'charming' from Geometry. So, from what I have sensed, Hodge is not likely to press for you to be taken on at Cambridge, though naturally he has a great respect for you. It is, of course, possible he might change his mind if you stuck to your guns and took a post outside Cambridge! I wonder if Mordell could use his weighty influence to affect the issue.[141] I am very much looking forward to the end of the war when I can get back to research again and get rid of this incubus of overwork and shortage of staff and air-raids and Home-Guard and all the rest of it which makes life so generally wearisome. Perhaps it won't be so very long now. Very kind regards from my wife and myself to Mrs. Segre and the children. She often asks for Mrs. Segre and wonders how she manages in all the extreme difficulties and trials of these days. Yours sincerely, J.G. Semple

---

[140]William Leonard Edge (1904–1997), lecturer at University of Edinburgh (1932–1975) was a member of the Edinburgh Mathematical Society; he served it as President and for many years as Librarian.

[141]Louis Joel Mordell (1888–1972), professor at the University of Manchester (1923–1945), was the President of the London Mathematical Society (1943–1945).

Segre.18 K. Sisam [The Clarendon Press] to B. Segre, Oxford 5.4.1944

CA, *BSP*, fol. 1r

Dear Professor Segre, Had I not been so busy, I should have written earlier to thank you for Christmas greetings and to enquire how you were getting on. But as the War goes on business becomes more and more crowded. I am glad to hear of the review of your book, and perhaps before long Todd will write and enable you to report progress on *Geometry upon Algebraic Surfaces and Varieties.*[142] I think it would be very important for you both to get this major work finished. I am sorry I cannot manage book form for your arithmetical researches, but our resources seem to be occupied as far ahead as I can foresee: there is going to be a great shortage of school and university books, which may well hamper education after the War.[143] Offprints are nowadays very slow, and tend to be very untidy-looking. I am told that it may be some two months before our Printer clears your offprints in J.O.M. [Quarterly Journal of Mathematics, Oxford Series],[144] because he is so taken up with Government work. I can understand your anxiety about your relations in Northern Italy: I think they will be all right if they are not in one of the great industrial towns. There the tension and distress must be very serious, though my impression is that the Germans have now too much on their hands to spend time on hunting out ordinary people in the interests of their racial theories. Our best chance is that they may not be slowly driven back on all their lines, so that there will be the maximum devastation; but that they should collapse under many strains, as I believe they will before this year is out. In that case, they will leave occupied areas disorganized, but not beyond the reach of military organization: that must come first, because, as in Berlin itself after the bombing, civilian life is too disorganized to recover quickly by its own resources. For your own case I think you must still wait patiently until you know what is best to be done. I think it likely that scholars like yourself will be needed once Italy is liberated, and I hope you have some means of keeping in touch with the authorities who will have to deal with Italy. That is outside our field. But should Italy fail, there ought to be room in university teaching here, when it has to meet post-war needs, for a modern geometer: they are scarce in England except at Cambridge. I shall do anything I can if an opportunity should arise. With all good wishes. Yours sincerely, Kenneth Sisam

---

[142] B. Segre, J.A. Todd, *Geometry upon Algebraic Surfaces and Varieties*, Oxford, Clarendon Press, expected for 1942.

[143] See D6.2.9.

[144] The offprints of B. Segre's papers *The maximum number of lines lying on a quartic surface*, Quarterly Journal of Mathematics, Oxford Series, 14, 1943, p. 86–96 and *A remark on unicursal curves lying on the general quartic surface*, Quarterly Journal of Mathematics, Oxford Series, 15, 1944, p. 24–25.

Segre.19 L. Roth to B. Segre, London 20.9.1944

CA, *BSP*, fol. 1r-v

My dear Beniamino, During the last few days I have been thinking over the project of writing an account of algebraic threefolds, with the necessary critical remarks and exposition of the unsolved questions in the theory – it would thus be an amplification of the little work I wrote in French a good while back, and which has never been published.[145] I must confess that, over and above the labour involved in the new project, the difficulty of getting it published does not add to the attractions. I do not think it would find a publisher here; do you suppose it would fare better in the U.S.? Or better still, it could be rewritten in Italian and published in Italy! The account would include (a) projective theory (b) geometry on a V_3 (c) auto-transformations and questions of rationality (d) correspondence between V_3 – excluding Cremona transformations of S_3, which is a big subject in itself. With regard to (e) topology on the V_3 I would like your opinion. The only paper I know on the subject is by Todd, in the Proc. Edin. Math. Soc.[146] Are there any works of importance? Then, finally, (f) a brief account of varieties V_d by way of appendix, in the way that Rosenblatt has done it (*Atti Congresso Intern.* 1928).[147] Fortunately, I happen to possess a good number of the papers covering these fields, so the difficulty of trying to find them in blitzed libraries will be avoided. But the state of the general theory in (b) is serious, so much so that I do not see how to write critically of it without giving offence. For instance, I have been looking at Todd's paper on algebraic invariants of a V_d, in which he states and uses the fact that Severi has shown that the genus p_d is a numerical character of V_d.[148] But this is not true: Severi says that it would follow from a certain assumption – an assumption that has, I believe, not even been justified for curves. Then again, I seem to recollect that Severi has published papers (Mem. Acc. Ital.) in which he claims to have settled the questions concerning composite varieties of any dimension and the theory of the base for any V_d. Is this so, and is it not true that the papers are worthless? I do not know what could be done about such works; there is also the trouble about his early paper (Rend. Palermo 1909) which is full of dubious matter.[149] One cannot avoid quoting it, and one cannot rely on

---

[145] The book will be published in 1955: L. Roth, *Algebraic threefolds with special regard to problems of rationality*, Ergebnisse der Mathematik und ihrer Grenzgebiete, Reihe Algebraische Geometrie, Heft 6, Berlin, Springer, 1955.

[146] J.A. Todd, *On the topology of certain algebraic threefold loci*, Proceedings of the Edinburgh Mathematical Society, s. 4, II, 1935, p. 175–184.

[147] A. Rosenblatt, *Sopra le varietà algebriche a tre dimensioni fra i cui caratteri intercedono certe disuguaglianze*, in *Atti del Congresso Internazionale dei Matematici (Bologna, 3–10 Settembre 1928)*, Bologna, Zanichelli, vol. 4, 1931, p. 123–128.

[148] J.A. Todd, *Invariant and covariant systems on an algebraic variety*, Proceedings of the London Mathematical Society, s. 2, 46, 1940, p. 199–230.

[149] F. Severi, *Fondamenti per la geometria sulle varietà algebriche*, Rendiconti del Circolo Matematico di Palermo, 28, 1909, p. 33–87.

many of the results. If a critical account is to be of any value at all to the student, it must try to make clear which of the results seem dependable and which are merely suggestive. It appears to me that a great deal of the existing theory is still unproved – everyone takes it for granted. It may be divided into (1) results which might be rigorously established without too much difficulty, and (2) results whose validity is not doubted but which it would be exceedingly difficult – perhaps at present, impossible – to demonstrate. The projective results, many of which are straightforward, are in a better condition. But on the algebraic side, if we except your two large papers on the V_3, I do not think there is any reliable body of results to be found. Let me know what you think about these questions, when you can find the time. We were very sorry to hear that you have been so long deprived of news from the family. I do trust that you will soon hear good news both from the Argentine and from Italy. With all good wishes from us all and affectionate greetings from Yours Leonard

Segre.20 L. Roth to B. Segre, London 26.9.1944

CA, *BSP*, fol. 1r-v
My dear Beniamino, we were relieved to learn that the parcel arrived, although very late. In the haste of writing the note to enclose I forgot to thank you for the remarks about Haldane's problem which you sent in your last letter but one. I do so now, with apologies. I am very glad to know that you approve the general plan of my proposed report on threefolds.[150] I am sending a rough first scheme for your further approval. When you have the time, would you see whether I have omitted any important particulars or whether you can suggest improvements. I want to explain some other matters in connection with the scheme. First, as to the style, I think it would follow Zariski's report, save that there will be no 'proofs'.[151] The idea is to give a connected account which anyone who understands the main points of the theory of algebraic surfaces could follow. There will of course be a complete bibliography in the Encyclopaedia manner.[152] All the principal theorems will be enunciated (those that have really been proved) and others will be commented upon. Gaps in the theory will be described. It is the nature of the work which leads me to doubt whether a publisher could easily be found for it. It is not an original contribution, so no English mathematical society would take it; and it is not a treatise – in the present state of the theory, with all its

---

[150] See SEGRE.19.

[151] O. Zariski, *Algebraic surfaces*, Ergebnisse der Mathematik und ihrer Grenzgebiete, 3, Nr. 5. Berlin, Springer, 1935.

[152] Roth refers to the essays of the *Encyklopädie der mathematischen Wissenschaften*, published by Teubner from 1898 to 1935.

serious lacunae, that is impossible – so it could not appear as a mathematical book. I can only write a report, for I do not see myself succeeding where many of the great names in geometry have failed partially or totally to fill the gaps. That is the difficulty; however, I shall carry on with it and hope for a solution when the work is completed. I cannot tell how long the work will be, but it can hardly be of great length, in view of what I have in mind. Now, as regards the detailed scheme. The first section is much shorter than appears, for a lot of the topics can be dismissed with a mere reference to the relevant papers. With regard to the second section, has any work been done on the singularities of a threefold? I cannot think of any. Again, the existence of the superficial irregularity has still not been established by algebro-geometric methods. And, although I have not seen Severi's later paper on the theory of the base for a V_d, I suppose it is connected with the case d = 2, for which the groundwork is still transcendental. Now, in the fourth section, I will include the Cremona transformations, as you suggest, but I do not think an exhaustive account is practicable; besides, there is H.P. Hudson's book and Berzolari's article.[153] I think the best plan would be to give the outstanding results, and *full* references to work that has appeared since Berzolari, sending the reader to Berzolari for the rest. The same remark applies to the account of space involutions of which the literature is very considerable. I repeat that I will give you any help you may need with your book. I would even go further and, though very reluctantly, become a nominal joint 'author' if it would lighten your task. But in the first place, you have still to get rid of Todd, who does not appear willing to resign. And in the second place, think what my position would be vis-à-vis Todd when he knew that I had taken over from him. It is not a pleasant situation to contemplate. With all good wishes and affectionate regards from Yours Leonard [. . .][154]

Segre.21 L. Roth to B. Segre, Southgate 6.10.1944

CA, *BSP*, fol. 1r-v

My dear Beniamino, I was very pleased to know that you approve, with some reserves, of the detailed scheme I sent you.[155] I hope now, by explaining a few points more fully, to get your approval for the whole. The only thing that puzzles me is your surprise at seeing the plan. What else could it have contained? I want to write an account of the algebraic V_3, so I should imagine the subjects I have set

---

[153] H.P. Hudson, *Cremona transformations in plane and space*, Cambridge, University Press, 1927; L. Berzolari, *Algebraische Transformationen und Korrespondenzen*, in W.F. Meyer, H. Mohrmann (eds.), *Encyklopädie der mathematischen Wissenschaften*, Bd. 3-2-2b, Leipzig, Teubner, 1934, p. 1781–2218.

[154] The post-scriptum by Marcella Baldesi Roth to Fernanda Coen Segre has not been transcribed.

[155] See SEGRE.19 and SEGRE.20.

down were more or less obligatory. I am so glad that the work will not overlap yours – which would have been a pity. Now, as to your comments, I fully agree with you that a Report which sketches the main proofs is more valuable than the other sort. But I was thinking, in the first instance, of a work resembling a section of the *Encyclopaedia*, and I feared that if outline proofs were included the thing would be too lengthy. However, I incline to accept your advice and include sketches of the proofs, whenever they are of the requisite algebro-geometric character. I do not want to embark on proofs (or theory) of a transcendental or topological nature. When, as in the case of the superficial irregularity, the only known basis is transcendental, I want to explain the circumstance, note the gap in the theory, and pass on, without going into much detail. Although it is clear that a certain amount of critical, or rather adverse, comment would seem unavoidable, I must say I would rather it could be omitted. I would prefer to criticize the state of the existing theory than make unpleasant remarks about its authors – remarks that would necessarily be unwelcome however they were disguised. This is the aspect of the job that appeals to me very little. I thought you had seen reference to Castelnuovo's unpublished researches on systems of rational surfaces; it is in the *Memorie Scelte*, referred to in one of the added notes.[156] The problem is that of finding the maximum dimension of such a system in S_3. Apparently, the author's results are inconclusive. With regard to Zariski, I fancy that about 4 years ago he published a demonstration of the fact that the V_d whose singularities are multiple V_{d-1} can be transformed birationally into one without singularities.[157] That is all I remember about it. Two nights ago I had a phone call from Cantalamessa. He said he had heard from you (and was most gratified by this attention), and also, he said you are satisfied that he is on a promising line of research. I hope this is so, for he comes along to me occasionally, obviously looking for help, which I cannot give him in that particular field, as I know nothing whatever about it. So, I am relieved that he is on to something at last. With many thanks for your advice, and all good wishes from yours affectionately Leonard [. . .][158]

---

[156] G. Castelnuovo, *Memorie scelte. In occasione del suo giubileo scientifico*, Bologna, Zanichelli, 1937, p. 332–334.

[157] O. Zariski, *Pencils on an algebraic variety and a new proof of a theorem of Bertini*, Transactions of the American Mathematical Society, 50, 1941, p. 48–70.

[158] The post-scriptum by Marcella Baldesi Roth to Fernanda Coen Segre and the index of the projected book have not been transcribed.

Segre.22 Draft of a Letter by B. Segre to W.V.D. Hodge, Manchester 3.5.1945

CA, *BSP*, fol. 1r

Dear Hodge, I have been very glad to receive your letter for whom I thank you most heartily. I too should like if we could remain in touch after the war and I hope there will be ample scope and opportunity for it. My plans for the near future are not yet made up. The Italian Government, through its representative in London, has invited me to return to my post in Italy, but until now I have undertaken no steps for going back at once. My decision will in a large extent depend on the news I hope to receive soon from my parents in Turin.[159] However, even if I should agree to go back as soon as possible, I do not know how long it would take before I could be able to leave England. I should like very much to spend some time in Cambridge, and actually participate to the University life; but for the reason explained above, I do not know now how long I could stay there. If you think that, in the circumstances, this uncertainty will not annoy the Faculty, I shall apply for one of the lectureships you so kindly have pointed out to me. I quite understand the difficulties which can arise for appointing a foreigner, and so I more appreciate your support in face of them. The correspondence with Italy, even with Rome, is still very slow and irregular, and I have not heard from any mathematicians in Italy since receiving from Castelnuovo the news I have told Todd about.[160] I have recent news from Fano (Switzerland) and Terracini (Argentine); they are well, and working hard in many capacities; neither of them has yet decided whether and when to return to Italy. With kindest regards to Mrs. Hodge and yourself, Yours sincerely [B. Segre]

Segre.23 Draft of a Letter by B. Segre to R. Stoneley, Secretary of the Appointments Committee Pembroke College Cambridge, Manchester 25.5.1945

CA, *BSP*, fol. 1r

Dear Dr. Stoneley, I wish to apply for a Lectureship in Mathematics in the University of Cambridge. I came to this Country as a refugee from Italy, just over six years ago. After a period of research work in London and Cambridge, where I delivered a few occasional lectures, I was appointed Assistant Lecturer in Mathematics at the Victoria University, Manchester, as from September 29th, 1942, and am there now. I have done all the normal work of an ordinary member of the staff and have taught and examined both honours and pass students. During the last six years, I have written a book [*The non-singular cubic surfaces*],

---

[159] Samuele Segre and Leonilda Segré.
[160] D7.2.3, D7.2.4 and D6.2.13.

published by the Clarendon Press, Oxford, in 1942, and many mathematical papers in English. Up to now, twenty of these have been published or accepted for publication in England or the United States. Some of these have arisen from my association with Prof. Mordell and Dr. Mahler.[161] Professors Mordell and Hodge can speak of my scientific work, and my teaching, and are willing to give further information. Further particulars, if needed, should be have from Professor Mordell as soon as possible, since he will leave for Canada in the early part of June. I must inform the Committee that I have been asked by the Italian Government to go back to the chair in Mathematics I had in Bologna before 1939. I am sorry I cannot, in the circumstances, be definite about my plans for the future; I should, however, consider it as a privilege and honour to become now a Lecturer at Cambridge. If the University were to appoint me, I should do my best to remain in its service for at least two years, and I should give as long notice as possible of any decision to leave. Yours sincerely [B. Segre]

Segre.24 G. Castelnuovo to B. Segre, Rome 8.7.1945

CA, *BSP*, fols. 1r-2r
Dear Professor, I have received your letter of 17 May, and the reprints you have sent me, for which I am very grateful. I have reflected on what you would like from the Ministry, but the only practical solution seems to me the one I now describe to you. It is clear indeed that the chair of Bologna remains at your disposal until the day when you can come to Italy. Even if the Ministry might hasten your return, it would do so only in the face of an assurance on your part to occupy, as soon as you arrived, that chair. If you do not want to make commitments in this regard now, you should wait for the embassy in London to give you a travel permit, take advantage of it to come to Italy to see your parents, and then talk to the Ministry of Public Instruction. I do not know if it will be able to make a commitment to keep that or another chair free for one or two years, and even if the commitment were given, what guarantee would there be to see it preserved in the light of possible changes of Ministers and directives? You could, however, resume the chair of Bologna, and ask for a year of leave to meet the commitments made in England. All this is possible when you can come to Italy. I would like to add that in Bologna today there is the Rector Edoardo Volterra, my very good

---

[161] Segre refers in particular to the papers on arithmetic geometry: B. Segre, *A note on arithmetical properties of cubic surfaces*, Journal of the London Mathematical Society, 18, 1943, p. 24–31; *On a parametric solution of the equations $x^3 + y^3 + az^3 = b$, and on ternary forms representing every rational number*, Journal of the London Mathematical Society, 18, 1943, 31–34; *A parametric solution of the indeterminate cubic equation $z^2 = f(x, y)$*, Journal of the London Mathematical Society, 18, 1943, 226–233; *On arithmetical properties of singular cubic surfaces*, Journal of the London Mathematical Society, 19, 1944, p. 84–91 and the joint paper by B. Segre, K. Mahler, *On the densest packing of circles*, American Mathematical Monthly, 51, 1944, p. 261–270.

friend, and he could help you in requests for assignments or some other means that would allow you to round up your meagre salary.[162] He would gladly do it to bleed the anaemic Mathematical School of Bologna, and I would gladly encourage him in this sense. But anyway, I repeat that, in leaving England as soon as possible for a visit to Italy, you should not burn your bridges, but study here the possibilities that our universities offer you for a return, a return which would be desired by all Italian mathematicians. I have read with great interest your works on arithmetic geometry.[163] Neither from Terracini, nor from Fano have I received letters for a long time, and therefore I know nothing about their return. In Turin (or rather in Torre Pellice, via Amedeo Bert 47), there is now Tricomi.[164] Many cordial greetings from your most affectionate G. Castelnuovo

Caro Professore, Ho ricevuto la Sua del 17 maggio, e gli opuscoli che mi ha mandato, di cui Le sono molto grato. Ho riflettuto su ciò che Ella vorrebbe dal Ministero, ma la sola soluzione pratica mi sembra sia quella che Le espongo. È chiaro, intanto, che la cattedra di Bologna resta a Sua disposizione fino al giorno in cui Ella avrà la possibilità di venire in Italia. Ammesso pure che il Ministero avesse il modo di affrettare il Suo ritorno, esso lo farebbe soltanto di fronte ad un'assicurazione da parte Sua di occupare, appena arrivato, quella cattedra. Se Ella non vuol prendere impegni in tal senso fin d'ora, Le conviene aspettare che l'Ambasciata di Londra Le procuri il permesso di viaggio, profittarne per venire in Italia a vedere i Suoi, e poi parlare con la Dir. dell'Istruzione Sup. Non so se questa potrà prendere impegno di mantenere libera quella od un'altra cattedra per uno o due anni, ed anche se l'impegno venisse dato, quale garanzia ci sarebbe di vederlo conservato di fronte a possibili cambiamenti di Ministri e di indirizzi? Ella potrebbe tuttavia riprendere la cattedra di Bologna, e chiedere un anno di aspettativa per soddisfare gli impegni presi in Inghilterra. Tutto ciò quando Ella potrà venire in Italia. Aggiungo che a Bologna è oggi Rettore Edoardo Volterra, mio ottimo amico, e che egli potrebbe aiutarla in richieste di incarichi o di qualche altro mezzo che Le permettesse di arrotondare il magro stipendio. Lo farebbe volentieri per rinsanguare l'anemica Scuola Matematica di Bologna, ed io ben volentieri lo incoraggerei in questo senso. Ma comunque, Le ripeto che Ella, lasciando appena possibile l'Inghilterra per una visita in Italia, non dovrebbe rompere i ponti, ma dovrebbe studiar qui le possibilità che le

---

[162] Edoardo Volterra (1904–1984), son of the mathematician Vito Volterra, full professor of Roman Law up to the racial laws. After having tried in vain to emigrate, and after a period of wandering between Egypt and France, Volterra had returned to Italy in May 1940. He was arrested in March 1943 for anti-fascist activities (he was one of the founders of the Action Party). He participated in the war of liberation with the rank of partisans Commander and was awarded the silver medal for military valor and the war cross. In 1944, he was appointed Provincial Deputy of Rome, and in 1945–1946, he was elected member of the National Council. He was Rector of the University of Bologna from 1945 to 1947.

[163] See footnote 161.

[164] Francesco Tricomi returned to Turin in April 1945, after 18 months of partisan struggle in Rome.

nostre Università le offrono per un ritorno, che sarebbe desiderato da tutti i matematici italiani. Ho letto con molto interesse i Suoi lavori di aritmetica geometrica. Né da Terracini, né da Fano ho ricevuto lettere da molto tempo, e quindi nulla so sul loro ritorno. A Torino (o meglio a Torre Pellice, via Amedeo Bert 47) ora vi è il Tricomi. Molti cordiali saluti dal suo aff.ᵐᵒ G. Castelnuovo

Segre.25 F. Severi to B. Segre, Rome 15.10.1945

CA, *BSP*, fols. 1r, 2r, 3r, 4r

Dearest Segre, Finally I can write to you. Your letter of 2 August, sent to Arezzo (which I left at the end of September last year, without returning), arrived there on 12 September (!), and was redirected with great delay. I would add that, before answering you, I wanted to speak with the Minister of Education, Arangio-Ruiz, and that was not possible for me until yesterday.[165] I have asked the Minister that you be invited here to Rome, at the expense of the Italian Government and as soon as possible, to give some lectures on the progress of mathematics and especially of geometry and related theories during the war, with particular regard to the Anglo-Saxon countries. The Minister promised me that he would do it immediately. I myself would have invited you through the Institute of Advanced Mathematics (of whose intense life, which took place largely in the early years of the war, and which was about to begin when you had to leave Italy, you will perhaps have some fragmentary news[166]), if I had had the means and if I were not just now making attempts to solve the serious financial situation in which the Institute finds itself. Before moving on to something else, I would like to ask you, with regard to the Anglo-American mathematical contribution during the war, to accept the assignment of writing *a report covering from 1939 onwards, perhaps embracing all mathematics or at least algebraic geometry and neighbouring fields*, limited to the Anglo-Saxon world or to those countries that have been in scientific contact with it.[167] I address this prayer to you on behalf of the President of the Pontifical Academy of Sciences, which took the initiative of a scientific survey for all sciences and everywhere, precisely in the period 1939-1945. The reports must constitute *basic guidelines*, not encyclopaedic articles, but must report only on the fundamental results, without too much concern for completeness both as regards the results and the bibliography. The work that the Academy

---

[165] Vincenzo Arangio-Ruiz (1884–1964), Minister of Public Instruction (12.12.1944–8.12.1945).

[166] Segre had left Italy in April 1939. INDAM had been established in July and would be inaugurated in October 1939.

[167] B. Segre, *Geometria algebrica nei Paesi anglo-sassoni*, Relationes de auctis scientiis tempore belli, Pontificia Ac. Scientiarum, 1946, p. 3–51. See D7.3.2 and D7.3.6.

organises must be carried out as quickly as possible, because it will be of great use even if not complete, as long as it is rapid. The bibliography should be the most extensive (that is, also including minor works), but as far as possible without wasting time. Your arrival here, which I await with deep affection, will give you the means to make initial contact with the Italian environment and to prepare your return for 1946-47. At the end of the academic year, which is about to begin, Amaldi and Enriques will retire (the last temporarily recalled until the age of 75, in partial compensation for racial measures); and I (*between you and me*) would like to see you here for Higher Geometry as a successor to Enriques.[168] Speaking of vacant chairs: the chair of Geometry in Padua has now become vacant, after the premature passing on 13 September of my dear Comessatti, who died of colon cancer.[169] It is a serious loss for Italian algebraic geometry and a deep pain for me, as I am plunged more and more into solitude every day. I do not hide from you, dear Segre, that I was also very saddened by the fact that you had not yet written to me, while you had sent your news to others (e.g., to Martinelli, Castelnuovo, etc.). I wondered: did I offend him in some way? And I did not feel that I could in conscience reproach myself for anything; far from it! But I was a little consoled by the thoughtful question asked to Martinelli about my physical and moral state. I ended up thinking that there were reasons other than a slowing down of the bonds towards me, which had led you not to write to me and I had the consoling confirmation of this from your letter. About the exchange of correspondence with Martinelli, I must say that I learned only after Martinelli (very close to me) had taken the initiative, of the request he made for some English mathematicians (Baker, Hodge, etc.) to intervene in order to alleviate the situation in which I found myself.[170] If I had known it before, I would have dissuaded him, as this was obviously inappropriate. Please feel free, as soon as you can, to make my point of view known to those English colleagues to whom you had turned for this eventual reason. My present situation will be sufficiently clarified by the passages that I reproduce, contained in judgment of me by the purge Commission: The 1st degree Commission stated that "the moral rectitude and merits of Prof. Severi as a person and as a scientist are beyond question and on the other hand no one would dare to question him."[171] The liquidator Commissioner of the

---

[168] Ugo Amaldi and Federigo Enriques were born on 18 April 1875 and 5 January 1871, respectively.

[169] Annibale Comessatti (1886–1945), a pupil of Severi and his assistant for 12 years (1908–1920), taught Descriptive Geometry and its Applications at the University of Padua from 1922 to 1945.

[170] Enzo Martinelli (1911–1999), a pupil of Severi and his assistant until 1939, had asked some English mathematicians to testify in favor of his Master in the purge trial. Among the whitewash papers produced by Severi, there are no documents from English mathematicians. A number of rumors about Severi's activities as a collaborator reached Baker and Hodge. See CA, *BSP*: W.V.D. Hodge to B. Segre, Cambridge 17.12.1945.

[171] ACS: MPI, Direzione Generale Istruzione Superiore, *Fascicoli professori epurati*, b. 31, *Severi Francesco*: Conclusioni adottate dalla Commissione per l'epurazione del personale universitario, 23.12.1944, p. 4.

Accademia d'Italia [Vincenzo Rivera] declared to the 1st degree Commission that "personally he found in the activity of Prof. Severi a marked independence and courageous conduct to make the interests of science prevail even against the dominant policy."[172] And finally here is a passage of the judgment of the Central Commission: "to judge the personality of Prof. Severi as a whole, we must also take into account the fact that he, a member of the Socialist Party as a young man, opposed fascism in the most dangerous period: that is, in the first decade; he resigned in 1925 as Rector of the University of Rome in political protest after the Matteotti crime; signed the Manifesto Croce . . .; in the many competitions which he oversaw, he always defended the most deserving candidate, even if non-fascist or a Jew. . . . Finally, it should not be forgotten that after 8 September 1943, he clearly refused to give his support to republican fascism; he refused all the positions that with every enticement were offered to him; he carried out remarkable and commendable activity in favour of the partisans of his native land and in particular he managed to save from the Nazi-Fascist requisitions (despite serious threats) the funds of the Bank of Arezzo, as well as those of the victims of the city itself. All things considered, the Central Commission came to the conclusion that Prof. Severi did not receive from fascism anything more than he deserved. . .".[173] We have all suffered atrociously. I imagine your terrible odyssey and your sorrow for the loss of little Ornella. But I too, and even my wife [Rosanna Orlandini], who has shared with me every anxiety and every danger, have gone through dramatic periods, aggravated by my incoercible tendency to never escape responsibilities and duties, at the cost of any more perilous consequence. In 1943-44, I especially risked my life daily for my city under the bombings and under the action of the artillery during the battle: a very rare example in the face of general bewilderment. The importance of Arezzo as a road and railway junction has naturally produced serious damage and demolitions, especially since the real battle has lingered about 3 months in our province. Just think, I left my villa in Staggiano (which you know) in a safe position (and, in fact, I had limited damage from requisitions), to go and live voluntarily in another house of mine in the furiously affected area of the city, where I had created a duplicate of the Bank's offices, in anticipation that the headquarters could be hit by bombing, as in fact happened on 22 January 1944. And there I stayed day and night to rush out only when I had to help the poor displaced people: I stayed there because, encouraged by my example, the employees of the Bank would stay there for a few hours a day, to guarantee the minimum essential level of city and provincial economy that allowed the life of the population, and the partisans would survive. And I cannot tell you what threats and harassment from the German side I was constantly

---

[172] ACS: MPI, Direzione Generale Istruzione Superiore, *Fascicoli professori epurati*, b. 31, *Severi Francesco*: V. Rivera al Presidente della Commissione d'epurazione del personale universitario, 7.11.1944, fol. 49.

[173] ACS: MPI, Direzione Generale Istruzione Superiore, *Fascicoli professori epurati*, b. 31, *Severi Francesco*: Decisione della Commissione Centrale per l'Epurazione sul ricorso proposto da F. Severi, 23.5.1945, p. 3–4.

subjected to. I resisted in such a way as to be able to save everything, everything, from the removals: meanwhile, in my villa, 3 families of victims lived for free. Two are still there. Despite everything, I did as you did. The fortitude and serenity of spirit to work have never abandoned me and I have continued to work even amid the bombings and most serious concerns. Since the last work I gave you in 1939 there are about 40 other publications that have come from my pen, among them the II volume of the *Lezioni di Analisi* published in 1942 and the first volume of 415 pages of the *Lezioni sulle Serie, sistemi d'equivalenza e corrispondenze algebriche sulle varietà algebriche*, published already in 1942, in which the theory received a structure I believe to be almost definitive, with essential complements. (Now I already have about 200 typed folders of the 2nd volume, which will also include the transcendent and topological section).[174] That year, I then produced a *Memoria*, which in terms of size (over 350 typed pages) I think is practically unprecedented, on the general theory of periodic functions of several variables, of which Abelian functions are special cases.[175] You will also see in due time, if you have not already heard something of it through two short Notes, one by Enriques and the other mine, in the *Commentarii Helvetici*, the state of the art concerning the completeness theorem, for which all previous attempts have failed (even yours, which I myself had followed and extended, does not solve the question).[176] I realised this through a criticism that tortured me for a long time. As things stand, the theorem is true for arithmetically effective systems and for the curves that I call *emiregolari* (those on which the canonical system cuts a complete series). I and Zappa have given examples in which the theorem falls. But in the demonstration, there remains a transcendent element and I am of the opinion that it cannot now be eliminated if not by trying to demonstrate by topological means its principle of breaking ($p_g + 1$ minimum number of knots necessary for the breaking). I am very interested in your research in arithmetic geometry. It is an important and difficult field, which must be developed, especially since Italy has not given considerable contributions to number theory for some time, while in algebraic geometry there are the means to address these issues as well. Your contacts with Mordell and with the English School have been very useful to you on this subject.[177] About the English School, I have tried (so far, in vain) to obtain Hodge's book on functional

[174] F. Severi, G. Scorza Dragoni, *Lezioni di Analisi*, Bologna, Zanichelli, vol. 2, part I, 1942; F. Severi, *Lezioni sulle Serie, sistemi d'equivalenza e corrispondenze algebriche sulle varietà algebriche* (raccolte da F. Conforto ed E. Martinelli), Roma, Cremonese, t. 1, 1942.

[175] F. Severi, *Funzioni quasi-abeliane*, Pontificia Academia Scientarum. Scripta Varia, 4, Roma, Pontificia Academia Scientarum, 1947, 327 p.

[176] F. Enriques, *Sui sistemi continui di curve appartenenti ad una superficie algebrica*, Commentarii Mathematici Helvetici, XV, 1943, p. 227–237; F. Severi, *Intorno ai sistemi continui di curve sopra una superficie algebrica*, Commentarii Mathematici Helvetici, XV, 1943, p. 238–248.

[177] See SEGRE.23.

harmonics.[178] How can I have it? I await with real interest the book that you tell me you are preparing for the Clarendon Press.[179] How many things we need to discuss. But I would even like it if we could speak not only in Italy, but also over there. I would like to change air for a while and have some contact with the English mathematical world, since from our cooperation with that of young people like Hodge, Todd, etc. and experienced Masters like Baker (by the way, how is he?), much could be done. You see that I still have the energy and purpose of a young man. See if it is appropriate to express this desire of mine in Cambridge or Manchester. I would like to be invited for a stay of a few months. Write to me soon about everything, and also about this. I also remember your wife with great affection, and my wife and I send you our warmest wishes, to you and your children. An affectionate handshake, yours, F. Severi

Carissimo Segre, Finalmente posso scriverle. La Sua del 2 agosto, inviata ad Arezzo (che io ho lasciato alla fine di settembre dell'anno scorso, senza più tornarvi), arrivata là il 12 settembre (!), mi è stata respinta con grande ritardo. Aggiungasi che, innanzi di risponderle, volevo parlare col Ministro dell'Istruzione Arangio-Ruiz e ciò non mi è stato possibile che ieri. Ho chiesto al Ministro ch'Ella sia invitato qui a Roma, a spese del Governo italiano e al più presto possibile, per tenere qualche conferenza sopra i progressi delle matematiche e specialmente della geometria e teorie collegate, nel periodo della guerra, con particolare riguardo ai paesi anglo-sassoni. Il Ministro mi ha promesso che farebbe subito. Avrei io stesso provveduto ad invitarla per mezzo dell'Istituto di Alta Matematica (della cui intensa vita, svoltasi in gran parte nei primi anni della guerra, e che stava per iniziare quando Lei dovette lasciare l'Italia, Ella avrà forse qualche frammentaria notizia), se ne avessi avuto i mezzi e non stessi appunto ora facendo i tentativi per risolvere la grave situazione finanziaria in cui l'Istituto si trova. Prima di passare ad altro vorrei pregarla, appunto a proposito del contributo matematico anglo-americano durante la guerra, di volersi assumere la compilazione di un rapporto dal 1939 in poi, abbracciante magari tutta la matematica o per lo meno la geometria algebrica e campi vicini, *limitatamente al mondo anglosassone o a quei paesi che con esso sieno stati in contatto scientifico. Le rivolgo questa preghiera a nome del Presidente della Pontificia Accademia delle Scienze, la quale ha preso l'iniziativa di un consuntivo scientifico per tutte le scienze e dovunque, appunto nel periodo 1939-1945. I rapporti devon costituire primi orientamenti, non già articoli enciclopedici, epperò devon riferire soltanto sui risultati fondamentali, senza troppa preoccupazione della completezza sia per ciò che concerne i*

---

[178] W.V.D. Hodge, *The theory and applications of harmonic integrals*, Cambridge, Univ. Press. and New York, Macmillan, 1941.

[179] B. Segre, J.A. Todd, *Geometry upon Algebraic Surfaces and Varieties*, Oxford, Clarendon Press, expected for 1942.

risultati sia la bibliografia. Bisogna che il lavoro che l'Accademia organizza sia compiuto nel modo più rapido, perché sarà di grande utilità anche non completo, purché sia rapido. La bibliografia dovrebbe esser la più ampia (comprendendo cioè anche lavori minori), ma nei limiti del possibile, ossia senza perdere tempo. La Sua venuta qui, che io aspetto con viva affettuosità, Le darà mezzo di prendere un primo contatto coll'ambiente italiano e di preparare il Suo ritorno pel 1946-47. Colla fine dell'anno accademico, che sta per iniziarsi, andranno in pensione Amaldi ed Enriques (l'ultimo richiamato temporaneamente fino a 75 anni, a parziale risarcimento dei provvedimenti razziali); ed io (sia detto per ora fra noi) aspirerei a vederla qui per la Geometria superiore come successore di Enriques. A proposito di vacanza di cattedre: è restata ora vacante anche la cattedra di Geometria a Padova, essendo morto immaturamente il 13 settembre u.s. il mio caro Comessatti, che si è spento distrutto da un cancro al colon. È una grave perdita per la geometria algebrica italiana e un acuto dolore per me, che resto ogni giorno di più in solitudine. Non le nascondo, caro Segre, che mi aveva addolorato anche moltissimo il fatto ch'Ella non mi avesse ancora scritto, mentre aveva inviato sue nuove ad altri (p. es. a Martinelli, a Castelnuovo, ecc.). Mi domandavo: ho forse errato verso di lui? E non sentivo di potermi in coscienza rimproverar nulla; tutt'altro! Ma mi consolava un po' la premurosa domanda fatta al Martinelli del mio stato fisico e morale. Ho finito col pensare che vi fossero ragioni diverse da un rallentamento dei vincoli verso di me, che la avessero indotta a non scrivermi e ne ho avuto la consolante conferma dalla Sua lettera. A proposito dello scambio di corrispondenza col Martinelli Le debbo dire di avere appreso soltanto dopo che il Martinelli (molto legato a me) ne aveva preso l'iniziativa, della richiesta da lui fatta perché da parte di qualche matematico inglese (Baker, Hodge, ecc.) si compissero atti a me favorevoli onde districare la situazione in cui mi trovavo. Se lo avessi saputo prima, ne lo avrei dissuaso, essendo la cosa evidentemente poco opportuna. Voglia compiacersi, appena Le è possibile, di far conoscere questo mio punto di vista a quei colleghi inglesi a cui Ella si fosse allora per questo eventualmente rivolto. La mia situazione presente Le sarà a sufficienza chiarita dai passi che le riproduco, contenuti in giudizi che mi riguardano di organi di epurazione: La Commissione di 1° grado ha sentenziato che "la dirittura morale e le benemerenze del Prof. Severi come persona e come scienziato sono fuori discussione e d'altronde nessuno oserebbe metterlo in dubbio." Il Commissario liquidatore dell'Accademia d'Italia ha dichiarato alla Commissione di 1° grado che "personalmente gli risultava nell'attività del Prof. Severi una spiccata indipendenza e condotta coraggiosa per far prevalere gli interessi della scienza anche contro la politica dominante." Ed ecco infine un passo del giudizio della Commissione Centrale: "Per giudicare poi la personalità del Prof. Severi nel suo complesso, bisogna anche tener conto che egli, iscritto da giovane al partito socialista, ha avversato il fascismo nel periodo più pericoloso: e cioè nel primo decennio; si dimise nel 1925 da Rettore dell'Università di Roma per protesta politica dopo il delitto Matteotti; firmò il manifesto Croce ...; nei molti concorsi ai quali ha partecipato come giudice, ha sempre difeso il candidato più meritevole, anche se non fascista o se ebreo. ... Non va infine taciuto che dopo l'8 settembre 1943 si è rifiutato nettamente di dare la sua adesione al fascismo repubblicano; ha rifiutato tutte le cariche che con ogni allettamento gli sono state offerte; ha esplicato una notevole e commendevole attività a favore dei partigiani della sua terra nativa e in particolar modo è riuscito a salvare dalle requisizioni nazifasciste, nonostante gravi minacce, i fondi della Banca di Arezzo, nonché quelli dei sinistrati della città stessa. Tutto quindi considerato la Commissione Centrale è pervenuta nella conclusione che il Prof. Severi non ha ricevuto dal fascismo nulla di più di quanto egli meritasse...." Abbiamo tutti atrocemente sofferto. M'immagino la vostra odissea terribile e il vostro dolore per la perdita della piccola Ornella. Ma anch'io, anche mia moglie, che ha diviso con me ogni ansia ed ogni pericolo, abbiamo traversato periodi drammatici, aggravati dalla mia incoercibile tendenza a non sfuggire mai le responsabilità e i doveri, a costo di qualunque più perigliosa conseguenza. Nel 1943-44 specialmente ho esposto ogni giorno la vita per la mia città sotto i bombardamenti e sotto l'azione dell'artiglieria durante la battaglia: esempio rarissimo di fronte al generale smarrimento. L'importanza di Arezzo come nodo stradale e ferroviario ha naturalmente prodotto danni e demolizioni gravi, tanto più che la battaglia vera e propria si è attardata circa 3 mesi nella nostra provincia. Pensi che ho rinunciato alla mia villa di Staggiano (che Lei conosce) in posizione sicura (e difatti ha avuto danni limitati alle requisizioni), per andare ad abitare volontariamente in un'altra mia casa nella zona furiosamente colpita della città, dove avevo creato un duplicato degli uffici della Banca, in previsione che la sede potesse esser colpita da bombardamenti, come infatti avvenne il 22 gennaio 1944. E là rimanevo giorno e notte per accorrere soltanto dove dovevo onde aiutare la povera gente sbandata: vi rimanevo perché incitati dall'esempio vi restassero per poche ore al giorno gl'impiegati della Banca, che doveva necessariamente funzionare affinché sopravvivesse quel minimo di economia cittadina e provinciale che consentiva la vita alla popolazione e ai partigiani. E non Le so dire a quali minacce e vessazioni da parte tedesca ero di continuo sottoposto. Ho resistito in tal modo da poter salvare tutto, tutto dalle asportazioni: intanto nella mia villa vivevano gratuitamente 3 famiglie di sinistrati. Due ci sono ancora. Nonostante tutto ho fatto come Lei. La forza d'animo e la serenità di spirito per lavorare non mi hanno mai abbandonato e ho continuato a lavorare anche sotto i bombardamenti e fra le preoccupazioni più gravi. Dall'ultimo lavoro che Le diedi nel 1939 sono circa altre 40 pubblicazioni che sono uscite dalla mia penna e tra queste il II volume delle Lezioni di Analisi *uscite nel 1942 e il I volume di 415 pagine delle* Lezioni sulle Serie, sistemi d'equivalenza e corrispondenze algebriche sulle varietà algebriche, *uscito pure nel 1942, in cui la teoria ha ricevuto un assetto, che credo quasi definitivo, con complementi essenziali. (Ora ho già pronte circa 200 cartelle dattiloscritte del 2° volume, che conterrà anche la parte trascendente e topologica). In questo anno ho poi costruito in una* Memoria, *che per la sua mole (oltre 350 pagine dattiloscritte) credo non abbia molti precedenti, la teoria generale delle funzioni periodiche di più variabili, di cui le funzioni abeliane sono casi particolari. Ella vedrà anche a suo tempo, se già non ne ha avuto nozione approssimativa attraverso due brevi Note, una di Enriques e l'altra mia, nei* Commentarii Helvetici, *lo stato della questione concernente il teorema della completezza, pel quale tutti i tentativi precedenti hanno finito col fallire (anche il suo, ch'io stesso avevo seguito ed esteso non risolve la questione). Del che mi sono accorto attraverso una critica che mi ha torturato per tanto tempo. Allo stato delle cose il teorema è vero per sistemi aritmeticamente effettivi e per le curve ch'io chiamo emiregolari (quelle su cui il sistema canonico sega una serie completa). Da me e da Zappa sono stati addotti esempi in cui il teorema cade. Però nella dimostrazione resta un elemento trascendente ed io son d'avviso che non si possa ormai eliminarlo se non tentando di dimostrare per via topologica il suo principio di spezzamento (p_g + 1 minimo numero di nodi necessari per lo spezzamento).*

M'interessano molto le Sue ricerche aritmetiche di geometria algebrica. È un campo importante e difficile, che bisogna sviluppare, tanto più che l'Italia non ha dato da tempo notevoli apporti alla teoria dei numeri, mentre nella geometria algebrica ci sono i mezzi per affrontare anche tali questioni. I contatti con Mordell e colla Scuola inglese Le sono stati in materia molto utili. A proposito della Scuola inglese. Ho cercato finora invano di avere il libro di Hodge sui funzionali armonici. Come si può avere? Attendo con vero interesse il trattato ch'Ella mi dice di aver in preparazione presso la Clarendon Press. Di quante cose dobbiamo parlare. Ma io desidererei addirittura che non solo in Italia, ma anche costà potessimo parlare. Vorrei cambiare per un poco ambiente e mettermi a contatto col mondo matematico inglese, giacché dalla cooperazione nostra con quella di giovani come Hodge, Todd, ecc. e di Maestri sperimentati come Baker (a proposito, come sta?), molto si potrebbe cavare. Veda che ho ancora l'energia e i propositi di un giovanotto. Guardi se è il caso di esprimere questo mio desiderio a Cambridge o anche costà. Vorrei esser invitato per una permanenza di qualche mese. Mi riscriva presto su tutto e anche su questo. Anch'io ricordo la Sua Signora con molta simpatia ed io e mia moglie inviamo a Loro e ai figli i nostri più caldi auguri. Un'affettuosa stretta di mano, suo aff. F. Severi

Segre.26 W.V.D. Hodge to B. Segre, Cambridge 17.12.1945

CA, *BSP*, fols. 1r-v

Dear Segre, I was very glad to get your news of Severi. It happens that a number of rumors about Severi's activities as a collaborator had reached me, and it is a great relief to learn that they have been officially dispersed. I should be over-joyed if arrangements could be made for Severi to visit Cambridge. Had I had your letter a month ago I should have suggested to the Faculty Board that he be asked to be our Rouse Ball Lecturer for the coming year. (This involves giving one public lecture, and carries a fee of, I think, £50, and in addition the lecturer is asked to give a few informal lectures for an additional payment of £25). However, I thought Severi was ruled out, and the secretary has been asked to negotiate with certain other mathematicians from abroad. Except for this arrangement, the Faculty does not have any funds with which to bring distinguished mathematicians to Cambridge. So the question arises, would Severi be prepared to come at his own expense? Perhaps you could give me some idea of his present financial position. I could, too, ask the British Council if they could bring him over, either with or without payment, but I do not at the moment know whether they extend their activities to Italy, and I may add, confidentially, that I have always found them a very troublesome body to deal with. I think that the best thing I can do is to wait until the Faculty Board meets in mid-January. If it turns out that the secretary has failed to get anyone as Rouse Ball lecturer, I could then suggest Severi. But in view of the rumours which have been circulating, it would strengthen my case if you could tell me where I can obtain an official statement that he has been cleared by the Purge Committee. If, on the other hand, we have found someone to be Rouse Ball lecturer, I could suggest action by the Faculty, either through the British Council or otherwise, to get him here, but we should not then be able to pay him and the finance might raise a difficulty. But again I would like to have official proof of his innocence of collaboration charges. I may say that other distinguished foreign mathematicians have said they would like to visit Cambridge for about a couple of months, but the difficulties of finance have usually meant we could do nothing. I would very much like to obtain the mean of inviting them here, but I have not much hope of obtaining it internally. The University has all its work cut out to get money for the necessary expansion of its own activities. But I shall try to see whether anything can be done. With best wishes for Christmas and New Year to Mrs. Segre and you, Yours very sincerely W.V.D. Hodge

Segre.27 G. Castelnuovo to B. Segre, Rome 18.12.1945

CA, *BSP*, fols. 1r-v
Dear Professor, I received your letter dated 8th current, and I thank you, also on behalf of my family, for the good wishes which we cordially reciprocate. I also received a package of your works a few days ago and I congratulate you on your industriousness. I was particularly interested in the search for the maximum number of lines of a surface of the 4th order (is it the only Schur surface that has the maximum number of lines?) and the works of an arithmetic nature. I will then read the others. We discussed the paper *On limits of algebraic varieties* . . . with Enriques at length, and I know that he wrote to you about it.[180] Given the

---

[180]B. Segre, *On limits of algebraic varieties, in particular of their intersections and tangential forms*, Proceedings of the London Mathematical Society, (2), 47, 1942, p. 351–403.

delicacy of the issue, it would seem appropriate to carry out a work that aims exclusively to settle the issue involving the characteristic series of a continuous system of curves. Under what conditions can the completeness theorem of the characteristic series of a complete continuous system be used? The theorem has been used in fundamental research on surfaces and a fully satisfactory geometric proof would be needed. I am pleased to see that you have decided to resume your teaching position in Bologna for the next academic year. I have no doubt that before the summer you will be able to easily obtain the travel permit, and, if necessary, I will have the Ministry renew your invitation. I haven't known anything about Terracini and B. Levi for a while; the first wishes to return and I foresee that, as soon as the trip is possible, he will come to resume his place in Turin. I have frequent and recent direct news of Gino Fano. However, perhaps for health reasons, he does not talk to me about an immediate return, and I know that the sons are putting pressure on their parents to go and settle in the United States. On January 9th the Accademia dei Lincei reopens, under the provisional presidency of the archaeologist Prof. Rizzo and myself; and the publication of *Rendiconti* will resume on January 1st. If you have short works to send (5 or 6 pages of print) they will be appreciated. Cordial greetings and best wishes; Yours most affectionate G. Castelnuovo

Caro Professore, Ricevo la Sua dell'8 corr. e La ringrazio, anche a nome dei miei, degli auguri che ricambiamo cordialmente. Ho anche ricevuto giorni fa un pacco di Suoi lavori e mi rallegro con Lei della Sua operosità. Mi hanno interessato in particolare la ricerca del massimo numero delle rette di una sup. del 4° ordine (è la sola sup. di Schur a possedere il massimo numero di rette?) ed i lavori di carattere aritmetico. Leggerò poi gli altri. Con Enriques abbiamo discusso a lungo la Memoria On limits of algebraic varieties . . ., *e so che egli Le ha scritto in proposito. Data la delicatezza della questione, apparirebbe opportuno un lavoro che si proponesse esclusivamente di metter a posto la questione interessante la serie caratteristica di un sistema continuo di curve. Sotto quali condizioni si può adoperare il teorema della completezza della serie caratteristica di un sistema continuo completo? Il teorema è stato adoperato in fondamentali ricerche sulle superficie ed occorrerebbe una dimostrazione geometrica pienamente soddisfacente. Vedo con piacere che Ella si è deciso a riprendere la cattedra di Bologna per l'anno scolastico futuro. Non dubito che per l'estate Le riuscirà facile ottenere il permesso di viaggio, e, se sarà necessario, Le farò rinnovare l'invito dal Ministero. Di Terracini e B. Levi non so nulla da un pezzo; il primo desidera ritornare e prevedo che, appena sarà possibile il viaggio, verrà per riprendere il suo posto di Torino. Di Gino Fano ho frequenti e recenti notizie dirette; egli però, forse per ragioni di salute, non mi parla di un prossimo ritorno, e so che i figli esercitano pressioni perché i genitori vadano a stabilirsi agli Stati Uniti. Il 9 gennaio si riapre l'Accademia dei Lincei, sotto la presidenza provvisoria dell'archeologo Prof. Rizzo e mia; e col 1° gennaio riprenderà la pubblicazione dei* Rendiconti. *Se ha brevi lavori da mandare (5 o 6 pag. di stampa) saranno graditi. Di nuovo saluti ed auguri cordiali; aff.^{mo} G. Castelnuovo*

Segre.28 Draft of a Letter by B. Segre to F. Severi, Manchester 20.12.1945

CA, *BSP*, fols. 1r-2r

Dearest Professor, Your letter of October 15th reached me exactly two months later. [. . .] I see that you and your figure have courageously gone through terrible moments and periods, and I hope that now all that remains of all this is an unpleasant memory. I understand very well your desire to change your environment for some time, and I will be happy if I can help you in this attempt, also because in this way I will be able to have the welcome opportunity to see you again soon and to talk with you about many things that are close to my heart. I immediately began an exchange of correspondence on this matter with Hodge (who is today the most influential English Geometer), and I enclose a copy of the three letters we have exchanged so far, from which you will have a clear idea of how things turned out.[181] I will also write to Ruffini so that, if requested by Hodge, he is ready to provide the desired documentation.[182] In the meantime, you can tell me your impressions, and especially tell me if you would be willing to come here upon invitation, but at your expense. I believe it is very likely that, once in Cambridge, you may also be invited by other universities, including Manchester, although in England the interest in mathematical research is much less lively outside of Cambridge. I gladly accept the task of drafting the summary report requested by the President of *Pontif. Acc. delle Scienze*.[183] However, I will have to confine myself to algebraic geometry and nearby fields, in view of the enormous delay with which this request has reached me and its obvious urgency. I can add that the time I have is rather scarce, because of the many commitments I have at the University. In addition to lectures, seminars, etc., I must also think about frequently giving and correcting homework, preparing and correcting quarterly and final (written) exams, etc., so that part of my activity here is in

---

[181]CA, *BSP*: W.V.D. Hodge to B. Segre, Cambridge 17.12.1945, Segre.26 and B. Segre to W.V.D. Hodge, Manchester 19.12.1945.

[182]CA, *BSP*: Ministero della Pubblica Istruzione. Direzione Generale Dell'istruzione Superiore, Dichiarazione. Roma 10.1.1946: *Si dichiara che al Prof. Francesco Severi, Ordinario di Alta Geometria nell'Istituto Nazionale di Alta Matematica di Roma, non è stata inflitta alcuna sanzione ai sensi delle vigenti disposizioni per l'epurazione delle pubbliche amministrazioni. Si rilascia in carta libera su richiesta dell'interessato per gli usi consentiti dalla legge.* See also E. Ruffini, For the Italian Ambassador, London 31.1.1946: To Whom It May Concern: This is to certify that, according to a certificate issued by the Ministry of Education in Rome on January 10th, 1946, Professor Francesco Severi, of the Rome University, has not been subjected to any sanction provided for by current legislation on the epuration of the Italian Civil Service for fascist activities.

[183]B. Segre, *Geometria algebrica nei Paesi anglo-sassoni*, Relationes de auctis scientiis tempore belli, Pontificia Ac. Scientiarum, 1946, p. 3–51.

some ways similar to that of a middle school teacher in Italy, especially this year, after the coming of Newmann (a topologist who cares very little about geometry) to replace Mordell (called to Cambridge in place of Hardy, who retired).[184] I hope, however, that in about a month I will be able to draw up that report and send it to you, unless you tell me otherwise. [...] What you write to me about your scientific activity in the last 6 years has interested me very much and I look forward to being able to see what has been done by you and others in Europe during this period. I have not well understood what you write to me on the completeness theorem: have you seen, in this regard, my *Memoria* published in '42, which I sent you some time ago along with other works?[185] Hodge, Todd and Zariski, when they read it, found nothing to complain about. The untimely death of Comessatti, of which I already knew, has saddened me a lot, also because it further reduces the ranks of Italian geometers at a crucial moment, when the activity of its best exponents is desperately necessary for reconstruction.[186] However, I do not despair that things in Italy will soon take a better turn, and I would be pleased if, within the limits of my strength, I could contribute to this improvement. In this regard, I am extremely grateful to you for what you write to me about the chair in Rome, to which I would feel proud to be able to ascend; moreover, I would be even happier for the possibility that I would like to have constant personal contacts with you, with the certainty of deducing inspiration and stimulus in my work, and of making our emotional bonds more and more intimate. An affectionate handshake from your ever-mindful disciple.

P.S. I had written to Baker about your case to induce him to take some steps in your favour. Now I have not written about it again, since I had then presented that suggestion as my own. I add that Baker responded to me in a rather evasive manner, and I now understand better his answer after what Hodge wrote to me about the rumours that were circulating about you. I believe that these were due to an item appeared in some English newspaper concerning a Severi Minister of the so-called Republican Government; news that I now know to be false, i.e. referring to some homonym of you.[187]

Carissimo Professore, La sua lettera del 15 ottobre mi è giunta esattamente a distanza di due mesi. [...] Vedo che Ella e la sua figura hanno attraversato coraggiosamente momenti e periodi terribili, e spero che ora di tutto ciò non rimanga loro che il ricordo poco lieto. Comprendo benissimo il suo desiderio di cambiare ambiente per qualche tempo, e sarò lieto se potrò aiutarla in questo intento, anche perché così potrò avere la gradita opportunità di rivederla presto

---

[184] In 1945 M. Newman replaced L. Mordell in Manchester, when Mordell moved to Cambridge in order to replace G.H. Hardy, retired in 1942.

[185] B. Segre, *On limits of algebraic varieties, in particular of their intersections and tangential forms*, Proceedings of the London Mathematical Society, (2), 47, 1942, p. 351–403.

[186] D7.3.3.

[187] CA, *BSP*: W.V.D. Hodge to B. Segre, Cambridge 17.12.1945. Leonardo Severi (1882–1958) was the Minister of National Education from 25.7.1943 to 11.2.1944, so not during the RSI period.

e di intrattenermi con Lei su tante cose che mi stanno a cuore. Ho subito iniziato al riguardo uno scambio di corrispondenza con Hodge (che oggi è il Geometra inglese più influente), e le accludo copia delle tre lettere che ci siamo scritti finora, dalle quali avrà una chiara idea del come si sono messe le cose. Scriverò anche al Ruffini perché, se richiesto da Hodge, sia pronto a fornire la documentazione voluta. Intanto Ella potrà comunicarmi le sue impressioni in proposito, e specialmente dirmi se eventualmente sarebbe disposta a venir qui dietro invito, ma a sue spese. Ritengo assai probabile che, una volta a Cambridge, Ella potrà anche venir invitata da altre Università, Manchester compresa, per quanto in Inghilterra l'interesse per la ricerca matematica sia assai meno vivo fuori di Cambridge. Accetto volentieri l'incarico di stendere il rapporto sommario richiesto dal Presidente della Pontif. Acc. delle Scienze. *Dovrò però restringermi alla geometria algebrica e campi vicini, in vista dell'enorme ritardo con cui tale richiesta mi è pervenuta e dell'ovvia urgenza di darvi corso. Posso aggiungerle che il tempo di cui dispongo è piuttosto scarso, causa i molteplici impegni che ho all'Università. Oltre a lezioni, seminari, ecc., debbo infatti anche pensare a dare frequentemente e correggere compiti a casa, a preparare e correggere esami (scritti) trimestrali e finali, ecc., sicché qui parte della mia attività ha qualche analogia con quella di un professore di scuola media in Italia, specie quest'anno, dopo la venuta del Newmann (un topologo che tiene in assai poco conto la geometria) a sostituzione del Mordell, (chiamato a Cambridge al posto di Hardy, andato a riposo). Spero tuttavia che in circa un mese potrò redigere quel rapporto ed inviarglielo, salvo che Ella mi dica altrimenti. [...] Quanto Ella mi scrive della sua attività scientifica negli ultimi 6 anni mi ha interessato moltissimo ed attendo con viva impazienza di potermi mettere al corrente con quanto è stato fatto da Lei e altri in Europa durante tale periodo. Non ho ben compreso quanto Ella mi scrive sul teorema della completezza: ha visto al riguardo la mia* Memoria *uscita nel '42, che Le inviai tempo addietro assieme ad altri lavori? Hodge, Todd e Zariski, quando la lessero, non vi trovarono nulla da ridire. La prematura scomparsa del Comessatti, di cui già sapevo, mi ha molto addolorato, anche perché viene ad assottigliare le fila dei geometri italiani in un momento cruciale, quando l'attività dei migliori è disperatamente necessaria per la ricostruzione. Non dispero tuttavia che le cose in Italia prendano presto una piega migliore, e sarei lieto se, nei limiti delle mie forze, potessi contribuire a tale miglioramento. A questo proposito Le sono estremamente grato per quanto mi scrive circa la cattedra di Roma, alla quale mi sentirei fiero di poter salire; di ciò, inoltre, sarei anche più felice per la possibilità che me ne verrebbe di avere contatti personali costanti con Lei, colla certezza di dedurne ispirazione e stimolo nel mio lavoro, e di rendere vieppiù intimi i nostri legami affettivi. [...]. Un'affettuosa stretta di mano dal Suo sempre memore discepolo.*

P.S. Avevo scritto a suo tempo al Baker per indurlo a fare qualche passo in Suo favore. Ora non ho riscritto in proposito, poiché allora avevo presentato quel suggerimento come cosa mia. Le aggiungo che il Baker mi rispose in forma piuttosto evasiva, che ora meglio comprendo dopo quanto Hodge mi ha scritto circa le voci che circolavano sul suo conto. Credo che queste fossero dovute ad una notizia apparsa su qualche giornale inglese concernente un Severi Ministro del cosiddetto Governo Repubblicano; notizia che ora so essere falsa, oppure riferentesi ad un suo omonimo.

Segre.29 F. Severi to B. Segre, Rome 31.1.1946

CA, *BSP*, fol. 1r

Dearest Segre, I am once again indebted to you. Please attribute the delay in replying to your dear letter of December 20 to the desire I had to give you a complete reply to everything. Your interest, so affectionate, so full, so solicitous, was a source of moving and intense satisfaction for me, in seeing myself so well remembered by my beloved disciple. Above all, I wanted to have completed the action for your coming here during the Easter holidays, an action started with the previous Minister Arangio-Ruiz and then interrupted by the recent ministerial crisis.[188] I have not yet been able to obtain the decision of the new Minister Molé, above all because of the financial difficulties which are almost insurmountable in Italy today; but I don't abandon the field.[189] I would have liked to give you good news for you and very good for me. That's why I was late. However, when I received your report for the Pontifical Academy this morning,[190] I stopped delaying, even though I could not yet communicate to you what I would like. I thank you for the report, which offers a clear, complete, and luminous picture as you know to do. I'll pass it on to the press immediately and will take care of sending the proofs. This letter is, so to speak, interlocutory. I hope to have it followed soon by another. But I didn't want to delay in thanking you for what you have done with Hodge (to whom I will write directly as soon as possible: in the meantime, please thank him for me) and for the report.[191] I also want to send you my works and the volume I mentioned to you as soon as possible. I'm looking for the safest and fastest way. But I have a lot to do to put a lot of things back together! Sending via the Vatican, as I experienced, is not prompt. I don't know if regular mail will tolerate the weight and if it is safe. I'll look for it anyway. I had your works! Bravo! Thank you for this too. I cannot tell you how pleased I will in coming there, and I am so grateful, I repeat, to you and to Hodge. A fortnight ago I had the Foreign Ministry send to our embassy in London (with a letter from me

---

[188] Vincenzo Arangio-Ruiz (1884–1964), Minister of Public Instruction (12.12.1944–8.12.1945).

[189] Enrico Molé (1889–1963), Minister of Public Instruction in the first De Gasperi government (10.12.1945–1.7.1946).

[190] B. Segre, *Geometria algebrica nei Paesi anglo-sassoni*, Relationes de auctis scientiis tempore belli, Pontificia Ac. Scientiarum, 1946, p. 3–51.

[191] See SEGRE.25 and SEGRE.26.

to Prof. Ruffini) the Ministry of Education's declaration that no sanction was imposed on me during the purge trial, so that the embassy could respond with official document upon request from Cambridge.[192] A final act was carried out on the 20th of this month, according to power that the Government had until January 21 to put the officials of the first higher ranks to rest, even if they were acquitted in the purge. My name did not even surface in the Council of Ministers: my position of absolute acquittal is very clear and now definitive. I hope that Hodge will have requested the information in time before the Cambridge Faculty's further decision. In principle I accept to come even on onerous conditions. I would just like to have an approximate idea of the order of the expenses for living in a hotel in Cambridge or in other English cities, and of the travel expenses. Next time, I will talk to you about mathematics and in particular about continuous systems. [. . .] An affectionate handshake for you. Yours sincerely F. Severi

Carissimo Segre, Io sono nuovamente in debito con Lei. Voglia ascrivere il ritardo a rispondere alla Sua cara lettera del 20 dicembre al desiderio che avevo di scriverle replicando in modo completo a tutto; ché il Suo interessamento così affettuoso, così pieno, così sollecito è stato per me motivo di commossa ed intensa soddisfazione, nel vedermi tanto ben ricordato dal mio prediletto discepolo. Soprattutto volevo aver condotto a termine la pratica per la Sua venuta qui nelle vacanze pasquali, pratica avviata col precedente Ministro Arangio-Ruiz e interrotta poi dalla recente crisi ministeriale. Non ancora ho potuto ottenere la decisione del nuovo Ministro Molé, soprattutto per le difficoltà finanziarie oggi quasi insuperabili in Italia; ma non abbandono il campo. Avrei voluto scriverle per darle la notizia buona per Lei e buonissima per me. Perciò ho tardato. Ma avendo stamani ricevuto la Sua relazione per la Pontificia Accademia, rompo gli indugi, nonostante non possa ancora comunicarle quel che desiderei. La ringrazio della relazione, che offre un quadro chiaro, completo, luminoso come Lei sa fare. La passo subito alla stampa e curerò l'invio delle bozze. Questa mia lettera è per così dire interlocutoria. Presto spero di farla seguire da altra. Ma non volevo tardare a ringraziarla di quanto Ella ha fatto con Hodge (al quale scriverò direttamente al più presto: intanto voglia ringraziarlo da parte mia) e della relazione. Voglio anche spedirle quanto prima i miei lavori e il volume cui Le accennai. Cerco il mezzo più sicuro e più rapido. Ma ho tanto, tanto da fare per rimettere a sesto una quantità di cose! L'invio a mezzo Vaticano come ho sperimentato non è sollecito. La posta ordinaria non so se tollererà il peso e se è sicura. Cercherò insomma. Ebbi i suoi lavori! Bravissimo! La ringrazio anche di quelli. Non so dirle con quanto piacere verrò costà e son tanto grato, Le ripeto, a Lei e a Hodge. Da una quindicina di giorni ho fatto spedire dal Ministero Esteri alla nostra Ambasciata a Londra (con una mia lettera al prof. Ruffini) la dichiarazione del Ministero Istruzione che non mi è stata inflitta alcuna sanzione in sede epurativa, affinché

---

[192]See footnote 182, p. 607.

l'Ambasciata potesse rispondere con documento ufficiale a una richiesta da Cambridge. Un ultimo atto epurativo è stato compiuto il 20 corr. coll'esercizio della facoltà politica che il Governo aveva fino al 21 corr. di porre a riposo i funzionari dei primi 5 gradi, anche se prosciolti in sede epurativa. Il mio nome non è neppure affiorato in Consiglio dei Ministri: la mia posizione di assoluto proscioglimento è chiarissima e ormai definitiva. Spero che Hodge avrà richiesto in tempo l'informazione prima dell'ulteriore decisione della Facoltà di Cambridge. In massima io accetto di venire anche a condizioni per me onerose. Soltanto desidererei prima avere un'indicazione approssimativa dell'ordine di grandezza della spesa per la vita in Albergo a Cambridge o in altre città inglesi e della spesa per il viaggio. Quando Le scriverò Le parlerò di matematica e in particolare dei sistemi continui. A Lei un'affettuosa stretta di mano. Suo aff.^{mo}
F. Severi

Segre.30 F. Severi to B. Segre, Rome 7.2.1946

CA, *BSP*, fol. 1r-v
Dearest Segre, I am following up on my air mail dated January 31, which I hope reached you regularly. The day before yesterday, we had a meeting at the Pontifical Academy of the Commission which I chair for the publication of the scientific survey for the years 1939-45. The President of the Academy, Father Gemelli, was also present. Everyone was very satisfied with your report.[193] To accentuate the international, or rather, the super-national character of a publication of the Pontifical Academy as an Institution of the Holy See, it is advisable that the various reports be published in many different languages (naturally in the language in which they were submitted). Now, it is fine that your report, written by an Italian, be submitted in Italian, but it would seem appropriate to publish it in English, since it concerns the scientific movement in the Anglo-Saxon world. I think it won't be difficult for you to do or have the translation done quickly: I say quickly because these works have to be published as soon as possible. The Academy places at your disposal the modest sum of 5 pounds for the trouble, or for paying the translation expense. We would be very grateful to you if you could satisfy our request. I confirm that I am always insisting to obtain the means that allow you to come and visit us in Italy. In case, I will telegraph. I await news from you about Cambridge, and information of the order of magnitude of the expenses. In Cambridge perhaps I could sojourn in one of the colleges. I am sending to you and to Hodge, separately, through the kind intermediary of the English embassy at the Vatican, my recent works and some papers by Zappa, Martinelli and Conforto. I also attach, for each of them, a copy of the Proceedings

---

[193] B. Segre, *Geometria algebrica nei Paesi anglo-sassoni*, Relationes de auctis scientiis tempore belli, Pontificia Ac. Scientiarum, 1946, p. 3–51.

of the Mathematical Conference of 1942 and of my first volume on equivalence series, as well as for you, a copy of the report that Conforto and Zappa made on algebraic geometry in Italy from 1939 to 1945.[194] Finally, another prayer: we would need to find an English colleague to be charged with a report similar to yours for Analysis (both real and complex analysis, number theory, group theory, topology, etc.) in the Anglo-Saxon world from 1939 to 1945. If it were not possible to have such a complete report and it was necessary to divide the content between several authors, it would still be fine. I await some suggestions from you on this matter. We had turned to the Pontifical Academician Whittaker, but perhaps he is too old and has declined the invitation, nor has he given us any suggestions.[195] Our cordial greetings to you and your wife. Your most affectionate F. Severi

Carissimo Segre, Faccio seguito al mio aereo del 31/1, che spero Le sarà regolarmente pervenuto. Ieri l'altro abbiamo avuto all'Accademia Pontificia una riunione della Commissione che io presiedo per la pubblicazione del consuntivo scientifico 1939-45. Era presente anche il Presidente dell'Accademia, Padre Gemelli. Tutti sono stati molto contenti della Sua relazione. Per accentuare il carattere internazionale, o per meglio dire, super-nazionale dell'Accademia Pontificia come Istituzione della Santa Sede, conviene che le varie relazioni siano pubblicate nelle più diverse lingue: naturalmente ciascuna nella lingua in cui viene presentata. Ora, sta bene che la Sua, scritta da un italiano, sia stata presentata in lingua italiana, ma sembrerebbe opportuno, dal momento che si tratta di una relazione relativa al movimento scientifico nel mondo anglo-sassone, che fosse pubblicata in inglese. Penso che a Lei non sarà difficile fare o far fare rapidamente la traduzione: dico rapidamente perché questi lavori è meglio che siano pubblicati al più presto. L'Accademia mette a Sua disposizione per il disturbo o la spesa della traduzione, la modesta somma di 5 sterline. Mi farebbe cosa molto gradita se potesse accondiscendere alla nostra preghiera. Le confermo che sto continuando le mie insistenze per ottenere il mezzo affinché Lei possa venire a farci una visita in Italia. Se mai Le darò comunicazione telegrafica. Aspetto da parte Sua notizie circa Cambridge, che mi rendano edotto anche dell'ordine di grandezza della spesa. A Cambridge forse potrei andare pure in uno dei College. Mando a Lei, e a Hodge, in piego a parte, per il gentile tramite dell'Ambasciata inglese presso il Vaticano, lavori miei degli ultimi tempi e taluni di Zappa, Martinelli e Conforto. Unisco altresì, per ciascuno di Loro, copia degli Atti del Convegno Matematico del 1942 e del mio I Volume sulle Serie di equivalenza, nonché per Lei, una copia del rapporto che Conforto e

---

[194]F. Severi, *Atti del Convegno Matematico tenuto in Roma dall'8 al 12 novembre 1942*, Roma, INDAM, 1945; *Serie, sistemi di equivalenza e corrispondenze algebriche sulle varietà algebriche*, ed. by F. Conforto and E. Martinelli, Roma, Istituto Matematico della R. Università, 1942; F. Conforto, G. Zappa, *La geometria algebrica in Italia dal 1939 a tutto il 1945*, Relationes de auctis scientiis tempore belli, Pontificia Ac. Scientiarum, N. 8, 1946, p. 3-43.

[195]See D7.3.2.

Zappa hanno fatto sulla Geometria algebrica in Italia dal 1939 al 1945. In ultimo un'altra preghiera: avremmo bisogno di trovare presso le Università inglesi un collega che si occupasse di un rapporto analogo al Suo per l'Analisi (campo reale e campo complesso, teoria dei numeri, teoria dei gruppi, topologia, ecc.) nel mondo anglo-sassone dal 1939 al 1945. Se non fosse possibile avere il rapporto così complesso e occorresse dividerlo fra vari, la cosa andrebbe bene lo stesso. Aspetto qualche Suo suggerimento in proposito. Ci eravamo rivolti all'Accademico Pontificio Whittaker, ma egli è forse troppo vecchio e ha declinato l'incarico, né ci ha dato per ora nessun suggerimento. Di nuovo cordialissimi saluti nostri a Lei e alla Signora. Suo aff.mo F. Severi

Segre.31 F. Severi to B. Segre, Soriano Cimino (Viterbo) 7.8.1946

CA, *BSP*, fol. 1r-v

Dearest Segre, I hope that you will have received my previous letter, in which I communicated to you the information I got in Rome during my trip there on July 25, *re*. the possibility of your settling in Rome in the first half of September. It is always, as I told you, the arrangement proposed with Colonnetti, which would allow us to resolve, within the next two years, the question of your definitive settlement in Rome under tolerable economic conditions. I will give all my effort for this because I most earnestly desire that you can come to Rome immediately. We must work for the reorganization of the Institute [of Advanced Mathematics] and fully resume its activity and scientific research, since today our poor Italy has little possibility of success outside the realm of science and art. I was in Rome again yesterday (August 5) and I can only confirm the previous news. I am sending you the letter from the high official of the Foreign Ministry, to whom I addressed in your presence, regarding the contribution for your travel expenses for returning to Italy. As you can see, Foreign Affairs are washing their hands of it and at the competent General Directorate of the Ministry of Public Instruction have once again assured me that they have harassed the Ministry of the Treasury. But same old, same old: lack of money! They also told me that for Terracini they interested the UNRRA for arranging a travel with its steamers; but the organization responded in spades.[196] I will be commuting until August 31, having postponed my return to Rome until that day (at least). There is a bakery in Rome. At the beginning of September I will make personal contact with the new Minister (who responded with great cordiality to one of my letters) and I will once again raise the question of repatriations, the question of your coming to Rome and the

---

[196] The United Nations Relief and Rehabilitation Administration was an international relief agency founded in November 1943. Its purpose was to plan, coordinate, administer, or arrange for the administration of measures for the relief of victims of war in any area under the control of any of the United Nations through the provision of food, fuel, clothing, shelter and other basic necessities, medical and other essential services.

agreements to adapt the statute of the Institute of Advanced Mathematics to the present situation.[197] If the Institute wants to be preserved (and I hope so) the State will have to assign us an annual budget, which I have never requested until now, having provided it on my own initiative and by me. I received from Hodge the book on harmonic integrals and various papers, which however do not contain much that is original.[198] It seems that for the problem of the basis on Grassmannians he does not know, or he has forgotten my works. Yet he met Aprile, who could not ignore my papers because I laboriously advised him in some work on this topic while he followed the activities of the Institute. Please let me know when your wife and children will return. Kind regards also from my wife. Your most affectionate F. Severi

P.S. On August 5, I gave definitive instructions for the shipment to Roth of 3 copies of my book and of that of Conforto (but not of my works which I have not had time to collect).[199] It took me a while (it seems impossible!) to ascertain the price of the volumes (£200 for Conforto and £500 for mine). But please wait to send the money to the Institute, so that I acknowledge the shipment. The handwriting is not too clear, because I wrote sitting in the woods. But you are used to reading my letters.

Carissimo Segre, Spero ch'Ella avrà ricevuto la mia precedente di qui, nella quale Le comunicavo le informazioni da me assunte a Roma nel mio viaggio là il 25 luglio, circa la probabilità di una Sua sistemazione a Roma e della conoscenza di tale possibilità nella prima metà di settembre. Si tratta sempre, come Le dicevo, della sistemazione prospettata col Colonnetti, la quale ci consentirebbe di risolvere, nel biennio di respiro, la questione della Sua definitiva sistemazione a Roma a condizioni economiche tollerabili. Io darò per questo tutta l'opera mia, perché desidero vivissimamente ch'Ella possa venir subito a Roma. Dobbiamo lavorare per la nuova organizzazione dell'Istituto e riprenderne in pieno l'attività, anche con la ricerca scientifica, giacché poche possibilità di affermazione restano oggi alla nostra povera Italia, fuori del terreno della scienza e dell'arte. Sono stato a Roma di nuovo anche ieri l'altro 5 e non posso che confermare le notizie precedenti. Le compiego mia lettera dell'alto funzionario del Ministero Esteri, al quale mi rivolsi, in Sua presenza, circa il contributo per le Sue spese di rientro in Italia. Come Lei vede gli Esteri se ne lavano le mani e alla Direzione generale competente del Ministero Istruzione mi hanno di nuovo assicurato di aver tempestato il Ministero del Tesoro. Ma

[197] Guido Gonella, Minister of Public Instruction (14.7.1946–26.7.1951).

[198] W.V.D. Hodge, *The theory and applications of harmonic integrals*, Cambridge, Univ. Press. and New York, Macmillan, 1941.

[199] F. Severi, *Serie, sistemi di equivalenza e corrispondenze algebriche sulle varietà algebriche*, ed. by F. Conforto and E. Martinelli, Roma, Istituto Matematico della R. Università, 1942; F. Conforto, *Funzioni abeliane e matrici di Riemann. 1*, Roma, Libreria dell'Università, 1942. See D7.1.4.

siamo ormai lì: mancanza di baiocchi! Mi hanno anche detto che per Terracini avevano interessato l'UNRRA pel ritorno con piroscafi di questa; ma l'organizzazione ha risposto picche. Io tornerò a Roma periodicamente fino al 31 agosto, avendo rinviato fino a quel giorno (almeno) il nostro rientro là. C'è un forno a Roma. Ai primi di settembre prenderò contatto personale col nuovo Ministro (che ha risposto con grande cordialità ad una mia lettera) e prospetterò nuovamente la questione dei rimpatri, la questione della sua venuta a Roma e gli accordi per adeguare la costituzione dell'Istituto di Alta Matematica alla situazione presente. Se si vuole conservare l'Istituto (e lo spero) lo Stato dovrà assegnargli un contributo annuo, che io non avevo mai finora richiesto, avendo provveduto di mia iniziativa e da me. Da Hodge ho ricevuto il libro sugl'integrali armonici e varie Memorie, le quali però non contengono granché di originale. Sembra che pel problema della base sulle grassmanniane egli non conosca o abbia dimenticato i lavori miei. Egli si è poi incontrato (e questo non poteva saperlo) con Aprile al quale feci fare faticosamente un lavoro su questo argomento mentre seguiva l'attività dell'Istituto. Mi dia notizia quando la Signora e i figli rientreranno. Cordiali saluti anche da mia moglie. Suo aff.mo F. Severi

P.S. *Ho dato disposizione definitiva il 5 per la spedizione a Roth delle 3 copie del mio libro e di quello di Conforto (ma non de' miei lavori che non ho avuto tempo di riunire). Mi ci è voluto un po' (pare impossibile!) ad appurare il prezzo dei volumi (£200 quello di Conforto e £500 il mio). Ma Lei aspetti a mandare i denari all'Istituto, che io Le dia atto dell'avvenuta spedizione. La calligrafia non è troppo chiara, perché ho scritto seduto nel bosco. Ma Lei è abituato a leggere le mie lettere.*

Segre.32 G. Castelnuovo to B. Segre, Rome 3.10.1946

CA, *BSP*, fol. 1r

Dear Segre, I see from your letter that your whole family is now in Italy, and I rejoice in it. I see and it does not surprise me that you cannot find anywhere to settle with your family in Bologna. The same difficulties are encountered everywhere and especially in the towns destroyed by bombings. But is it appropriate to think about permanent accommodation? Do you have no hope of being called to Turin, where Fano's post is vacant?[200] My *Geometria Analitica* has always been published and sold, even in the epoch of greatest persecutions. A few months ago, a new reprint was released.[201] The publisher (Soc. editr. Dante Alighieri) will

---

[200] See D7.2.9.

[201] G. Castelnuovo, *Lezioni di geometria analitica*, Roma, Albrighi, Segati e C., 1935; Roma, Società Anonima Dante Alighieri, 1946.

send you a copy today. Thus, the extracts of your Lincei *Note* will be sent to you immediately.[202] Kind regards from your most affectionate G. Castelnuovo

Caro Segre, Vedo dalla Sua lettera che la famiglia è ormai tutta in Italia, del che mi rallegro. Vedo e non mi sorprende che Ella non riesce a trovare ove sistemarsi con i Suoi a Bologna. Le stesse difficoltà si incontrano dovunque e soprattutto nelle città sinistrate. Ma è il caso di pensare a una sistemazione stabile? Non ha speranza di esser chiamato a Torino, ove è rimasto vacante il posto di Fano? La mia Geometria analitica *è sempre stata pubblicata e venduta, anche nell'epoca delle maggiori persecuzioni. Pochi mesi fa è uscita una nuova ristampa. L'editore (Soc. editr. Dante Alighieri) Le manderà oggi stesso una copia. Così Le saranno subito spediti gli estratti delle Sue* Note *Lincee. Cordiali saluti dal suo aff.^{mo} G. Castelnuovo*

Segre.33 F. Severi to B. Segre, Rome 9.10.1946

CA, *BSP*, fols. 1r, 2r, 3r, 4r
Confidential.

Dearest Segre, I'm sorry, but we are still faced with setbacks, and moreover rather pessimistic horizons. Yesterday and the day before yesterday I conferred for a long time with Picone (who returned a few days ago), Bompiani and Conforto about your transfer.[203] Apart from Conforto, who is *toto corde* with me, the other two suggest difficulties not for yourself as a person about whom (especially Picone) they make statements of very deep and wide esteem, but for other considerations, which they consider substantial. You cannot fully understand the matter! Bompiani confirms his concern that Rome "lords it over" too much with other Universities and clarifies it by saying that in his recent contacts with professors of various Universities (especially in the North) he has found that they are up in arms against Rome, which they say has exercised for too long the "dictatorship of mathematics"; the arrival of a valiant scholar like Segre would aggravate a situation that he recognises is essentially due to jealousies, which have been the cause of subtle struggles fought in recent times against him, but above all against me. Personally, I don't worry about anything. For three years my spirit has been ready for anything, and I consider with serenity every

---

[202] B. Segre, *Un'estensione delle varietà di Veronese, ed un principio di dualità per forme algebriche*, Atti della Accademia Nazionale dei Lincei. Rendiconti. Classe di Scienze Fisiche, Matematiche e Naturali, (8), 1, 1946, p. 313–318 and 559–563.

[203] Mauro Picone (1885–1977), professor of Infinitesimal Analysis in Rome (1932–1960); Enrico Bompiani (1889–1975), professor of Geometry (1927–1959) and Head of the Mathematical Institute from 1939 to 1959; and Fabio Conforto (1909–1954), professor of Analytic and Projective Geometry (1939–1953).

eventuality, every malevolence, every lack of gratitude on the part of those who have been benefited by me and who do not possess that noble virtue of gratitude, which is the quality of men of great moral stature and . . . of dogs. I just need to do what I consider my duty, whatever might happen. In the endless discussions, I pointed out to Bompiani the following points: objectively, the anti-Roman insurgence is not justified; the University of the Capital is right in aspiring to a situation different from that of the provincial Universities (compare Paris with the provincial Universities of France); little by little other Universities too will emerge with young forces (competitions are now unfolding and others will soon follow, this is what the Minister assured me); Italy cannot, at least for now, afford the luxury of having many mathematical Schools that can be really effective in their activity; it is therefore natural to concentrate more forces in Rome; here there is the Institute of Advanced Mathematics (terrible flea in the ear for the envious and the very first cause of the struggles against me; time only, if I live, will prove me right); the Institute has a national function, accentuated by the last reform plan, which I presented to Gonella;[204] in Italy, especially after Scorza, modern algebra has so far been almost neglected;[205] Segre can give considerable impulses of proselytism in this field; his action in this field would take place with greater effectiveness and prestige for Italian science in Rome than in Bologna or elsewhere; I believe that I have a certain right to prepare my succession at the Institute (of which no one dares to disown the great usefulness and success in the action carried out until now, when it found itself in mortifying material conditions); the primacy of Italian algebraic geometry requires that the Chair held formerly by Cremona be filled not by temporary teachers, but by a full professor; etc., etc. Picone then sincerely expressed his concern about the prioritising of geometric prevalence over the analytic with the appointment of a full professor of Higher Geometry like Segre (and then, I exclaimed, let's transfer some other . . . and I jokingly inserted a name). Because, he says, in the Faculty there are already three geometers, against two analysts, of whom only one (Picone) has a School. I pointed out to him that in reality there are only two geometers in the Faculty (Bompiani and Conforto) as two are analysts, because I am in the Faculty as President of the Institute and not as a professor, so much so that until now I did not even have any teaching. On the other hand, if in Rome there is an extra geometer (me), there is also an extra analyst (Fantappié), and an extra professor of mechanics (Krall).[206] They replied with the compliment (of little substantial value) that, given my personality, everyone considers me as an integral part of the Faculty and as a catalysing figure for protégés who desert other disciplines

---

[204] Guido Gonella (1905–1982), Minister of Public Instruction from 1946 to 1951.

[205] Gaetano Scorza (1876–1939), professor of Projective and Descriptive Geometry in Rome (1934–1939), was among the few Italian algebraists and greatly contributed to the diffusion of the theory of algebras in Italy with the treatise *Corpi numerici e algebre* (Messina, Principato, 1921).

[206] In 1939, both Luigi Fantappié (1901–1956) and Giulio Krall (1901–1971) had been appointed as professors of High Analysis and Applied Mathematics, respectively, at INDAM.

(poor me!). I pointed out that here it is a question of covering a chair hitherto held by a full professor (Enriques); if there is no tenured position available for a chair of Higher Analysis, this is because, at the time, chairs of mathematics were taken away by naturalists and physicists; if you wanted to have another analyst (but they do not want it!) you should fight to get a chair back. Finally, the argument that posits everyone clearly in front of their responsibilities. The Dean, [Guido] Bargellini (who desires to help me in the battle), had already privately emphasised the fact (which I promptly communicated to my colleagues) that a temporary measure for Higher Geometry, as the teaching assignment entrusted to me is, would probably resurrect the old aspiration of the naturalists of Rome to have a chair in Genetics (it was recommended for D'Ancona,[207] son-in-law of Volterra) as well as one in Plant Physiology. And so, a post of mathematics may have torn us apart. I have warned of this danger since 1939, when I passed from Higher Geometry to Advanced Geometry. On that occasion I declared that I accepted the position *pro tempore* until the moment when it would be possible to permanently provide a tenured professor, in the view that it was necessary that in Rome the chair of Higher Geometry be covered by a tenured professor. And for this reason, I agreed with the Faculty to keep the post; without such assurance I would not have accepted the assignment. And I had formal and verbal guarantees. But after, there was the flood! Colleagues should therefore think about the responsibility they are taking on for themselves. For my part, without threatening any retaliation, I pointed out that I would personally conform to that course of action that would protect in the best way scientific interests, which I saw in danger; but probably, if we mathematicians had not agreed on the measure I proposed, I would have put the problem of whether or not to immediately provide for the chair before the Faculty, making available the teaching assigned to me. In fact, I added, today that it is a question of giving a successor to Enriques, it is easier for the Faculty to feel the moral imperative to give way to mathematics; by contrast, after I have held the position for the 3 years preceding my retirement, or even less, it will be very likely that the Faculty will end up giving the floor to other disciplines. And I wanted to discharge myself of that responsibility. In light of all this, my colleagues asked me to postpone any decision for 5 or 6 days, especially since it was necessary to inform Signorini, who only returned in the evening.[208] I shall therefore inform you of my response in a few days. But, as you can grasp, all these skirmishes do not cheer me. About the C.N.R., the news are no better.[209] Colonnetti is almost never in Rome and is not here now, and I have therefore not been able to chat again with him. On the other hand, the implementation of his plan (made more difficult by a considerable Frond wind that, for

---

[207] Umberto D'Ancona (1896–1964), biologist and naturalist, professor of Zoology in Padua (1936–1964). On 22 July 1926, he had married Luisa Volterra. Luisa and Umberto collaborated in biological research for many years.

[208] Antonio Signorini (1888–1963), professor of Rational Mechanics in Rome (1938–1958).

[209] D7.1.4, D7.1.5.

various reasons, is rising against him, even in Piedmont, one man tells me) is subject to the approval of a bill for the Committees of the Research Council, which has not yet come out and which will be laboriously applied. What Colonnetti assured me this summer, i.e., that in October he would be able to adopt the measure under which you could immediately come to Rome to the Institute, was pure illusion! However, even on this issue the last word has not been said. Nor is it said that I will not be able to obtain for you a secondment at the Institute. (Would you accept that?) Needless to say, measures of any kind, as required by university regulations, should be adopted by December; therefore, there is no danger that you could be transferred in the middle of the year. If a measure could be taken later, it would, however, become effective in the year 1947-48. Unfortunately, I do not know what to advise for your housing accommodation. See what you can find! Try not to cut ties: this is what I sincerely desire, anticipating the pleasure of having you working at my side, the pleasure of following the course, so thrilling, that you are preparing and to be able to appreciate more closely the very interesting results that you announce to me. I'm living a troubled life like never before. [. . .] I forgot to answer you on an important point. The Treasury, *re*. the contribution for professors returning to Italy, has invited the Ministry of Public Instruction to specify the amount of money and number of those concerned. It is imminent that a bill will be sent to the Treasury that specifies the number of those concerned as about 15 and asks for a sum of 15 million.[210] The Ministry of Public Instruction is not pessimistic: 15 million, today, is next to nothing. But in the Ministries, there is a huge sluggishness, and everything moves slowly. In contrast, the country is pervaded by a sense of lively industriousness, which we are envied abroad e.g., in France. Please, remind us to your wife when you write to her; please reply to me soon and receive my cordial greetings. Your most affectionate F. Severi

Riservata

Carissimo Segre, Sono spiacente, ma siamo ancora di fronte a battute d'aspetto e per giunta piuttosto pessimiste. Ho ieri l'altro e ieri conferito a lungo con Picone (tornato da pochi giorni), Bompiani, Conforto, circa il Suo trasferimento. A prescindere da Conforto che è toto corde *con me, gli altri due avanzano difficoltà non per la persona, di fronte alla quale (specialmente Picone) si profondono in dichiarazioni di amplissima stima, ma per altre considerazioni, ch'essi considerano obiettive. Capire fino in fondo non si può in tutto! Bompiani conferma la sua preoccupazione che Roma "pompi" troppo dalle altre Università e precisa la sua preoccupazione dicendo che nelle sue recenti prese di contatto con professori di varie Università (specie del Nord) ha constatato che si è in armi contro Roma, perché, dicono, ha esercitato troppo a lungo la "dittatura delle matematiche" e che la venuta di un valoroso come Segre*

---

[210]D7.1.7.

aggraverebbe da questo punto di vista la situazione ch'egli riconosce del resto essenzialmente dovuta a gelosie, che sono state non ultima causa delle lotte subdole combattute in questi ultimi tempi contro di lui, ma soprattutto contro di me. Per mio conto non mi preoccupo di nulla. Sono tre anni che il mio spirito è pronto a tutto e considero con serenità ogni evenienza, ogni malevolenza, ogni mancanza di riconoscenza da parte di chi è stato da me beneficato e che non possiede quella nobile virtù della gratitudine, che è qualità degli uomini di grande levatura morale e ... dei cani. Mi basta soltanto di compiere quello che reputo il mio dovere, accada quel che vuole accadere. Nelle interminabili discussioni di ore ho fatto osservare a Bompiani che obiettivamente la fronda antiromana non è giustificata e che d'altronde l'Università della Capitale può bene aspirare a una situazione diversa da quella delle Università di Provincia (si confronti Parigi colle Università provinciali della Francia); che a poco a poco anche le altre Università si risolleveranno con giovani forze (si fanno ora concorsi ai quali altri ne seguiranno presto; così mi ha assicurato il Ministro); che l'Italia non può, almeno per ora, permettersi il lusso di tante Scuole matematiche, che possano efficacemente agire; che perciò è naturale concentrare più forze a Roma; che qui c'è l'Istituto di Alta Matematica (terribile pulce nell'orecchio degl'invidi e primissima causa delle lotte contro di me, di cui il tempo, se io vivo, mi darà ragione); che l'Istituto ha una funzione Nazionale, accentuata coll'ultimo disegno di riforma, da me presentato a Gonella; che in Italia, specialmente dopo Scorza, si è finora quasi trascurata l'algebra moderna nella quale Segre può dare impulsi notevoli di proselitismo e che l'azione di lui in questo campo si svolgerebbe con maggior efficacia e maggior prestigio per la scienza italiana a Roma piuttosto che a Bologna o altrove; che credo anch'io di poter avere un certo tal quale diritto a preparare la mia successione all'Istituto (di cui nessuno osa disconoscere la grande utilità e il grande successo nell'azione svolta fino a quando non si trovò nelle mortificanti condizioni materiali presenti, che devon d'altronde cessare); che la situazione di primato della geometria algebrica italiana esige che la Cattedra che fu di Cremona sia coperta non con provvedimenti temporanei, ma con un professore ordinario; ecc. Picone poi ha manifestato sinceramente la sua preoccupazione di un accentuarsi della prevalenza geometrica sull'analitica colla nomina d'un professore ordinario di Geometria Superiore come Segre (e allora, ho esclamato io, trasferiamo qualche altro ... di cui ho umoristicamente fatto il nome). Perché, dice lui, in Facoltà siamo già tre geometri, contro due analisti, di cui uno solo (lui) ha una Scuola. Gli ho fatto osservare che in realtà i geometri della Facoltà sono soltanto due (Bompiani e Conforto) come gli analisti, perché io son in Facoltà come Presidente dell'Istituto e non come professore, tanto è vero che finora non vi avevo neppure alcun insegnamento. D'altronde se a Roma c'è con me un geometra in più, c'è anche un analista in più (Fantappié), come un cultore di meccanica (Krall). Mi hanno risposto col complimento (di scarso valore sostanziale) che, data la mia personalità, tutti mi considerano come parte integrante della Facoltà e come un richiamo di scolari che disertano altre discipline (povero me!). Ho fatto osservare che qui si tratta di coprire una cattedra finora tenuta da un ordinario (Enriques) e che se non c'è ora disponibile un posto di ruolo per una cattedra superiore di Analisi, ciò dipende dalla circostanza che a suo tempo cattedre di matematica furon portate via dai naturalisti e dai fisici e che se si volesse avere un altro analista (ma non lo vogliono!) si potrebbe cercare di riavere una cattedra. In ultimo l'argomento che pone tutti nettamente di fronte alle proprie responsabilità. Il Preside Bargellini (che ha desiderio di aiutarmi nella battaglia) pose già meco privatamente in rilievo il fatto, che mi affrettai a comunicare ai colleghi, che un provvedimento temporaneo per la Geometria Superiore, com'è l'incarico a me attribuito, farebbe con ogni probabilità risorgere la vecchia aspirazione dei naturalisti di Roma di avere una cattedra di Genetica (per cui era preconizzato D'Ancona, genero di Volterra) oltre ad una di Fisiologia vegetale. E così il posto di ruolo matematico potrebbe esserci strappato. Io avvisai il pericolo fin da quando passando nel 1939 dalla Geometria Superiore all'Alta Geometria dichiarai di conservare l'incarico pro tempore *fino al momento in cui si sarebbe potuto stabilmente provvedere con un professore di ruolo, ritenendo necessario che a Roma la cattedra di Geometria Superiore fosse coperta da un professore di ruolo. E per questo impegnavo la Facoltà a conservare il posto, senza di che non avrei accettato l'incarico. E allora ebbi affidamenti formali e verbale. Ma dopo c'è stato il diluvio! Pensino dunque i Colleghi alla responsabilità che si assumono. Per parte mia, senza minacciare alcuna rappresaglia, facevo osservare che mi sarei uniformato personalmente a quella linea d'azione che tutelasse nel miglior modo interessi scientifici, che vedevo in pericolo; epperò che probabilmente, se non ci si fosse accordati noi matematici sul provvedimento da me proposto, avrei messo il problema astratto della opportunità o meno di ricoprir subito la cattedra dinanzi alla Facoltà, ponendo a disposizione l'incarico attribuitomi. Infatti, soggiungevo, oggi che si tratta di dare un successore ad Enriques, è più facile che la Facoltà senta il dovere di lasciare il posto alla matematica; mentre dopo che io avessi tenuto l'incarico pei 3 anni che mi restano od anche per meno, è molto probabile che si finirebbe per attribuire il posto ad altre discipline. Ed io volevo scaricarmi da tale responsabilità. Di fronte a tutto ciò i colleghi domandarono di soprassedere ad ogni decisione per 5 o 6 giorni, tanto più che occorreva informare il Signorini, che soltanto ier sera è tornato. Le comunicherò dunque fra qualche giorno il responso. Ma, come può pensare, tutte queste schermaglie non mi allietano. Circa il C.N.R. le notizie non sono migliori. Colonnetti non sta quasi mai a Roma né in particolare vi è ora e non ho potuto pertanto prendere con lui nuovi contatti. D'altronde l'attuazione del suo piano (reso più difficile da un notevole vento di fronda che, per ragioni varie, si leva contro di lui, mi dicono anche in Piemonte) è subordinata all'approvazione di un disegno di legge pei Comitati del Consiglio Ricerche, che ancora non è venuto fuori e che sarà di laboriosa applicazione. Quel che Colonnetti mi assicurò quest'estate, che cioè a ottobre egli sperava di poter adottare il provvedimento, con cui Ella avrebbe potuto venir subito a Roma all'Istituto, era pura illusione! Comunque, anche da questo lato non è detta l'ultima parola. E neppure è detto che io non riesca ad ottenere per Lei un*

Comando presso l'Istituto. (Lo accetterebbe?). Va da sé che provvedimenti di qualsiasi genere, come vuole la legge universitaria, dovrebbero essere adottati entro dicembre e non ci sarebbe perciò pericolo ch'Ella potesse esser trasferito a metà anno. Se un provvedimento potesse esser preso dopo, la sua adozione avverrebbe sempre coll'anno 1947-48. Non so dunque purtroppo che cosa consigliarle per la Sua sistemazione casalinga. Veda Lei. Cerchi di non tagliare i ponti: ecco quel che io vivamente desidero, pregustando il piacere di averla al mio fianco a lavorare e il piacere anche di seguire il corso tanto ghiotto ch'Ella sta preparando e di poter apprezzare più davvicino gl'interessantissimi risultati che mi preannuncia. Io faccio vita tribolata come non mai. [. . .] Dimenticavo di risponderle sopra un punto importante. Il Tesoro circa il contributo pei professori rientrati in Italia ha invitato l'Istruzione a precisare l'entità della spesa e il numero degl'interessati. È imminente l'invio al Tesoro di un disegno di legge che precisa in circa 15 il numero degl'interessati e chiede una somma di 15 milioni. All'Istruzione non sono pessimisti: 15 milioni oggi vanno e vengono. Ma gli è che nei Ministeri c'è un'enorme fiacca e tutto va a rilento. E dire invece che il paese è pervaso da un senso di laboriosità, che ci è invidiato p. es. in Francia. Ci ricordi alla Signora quando Le scrive; mi risponda presto e si abbia i miei cordiali saluti. Suo aff.^{mo} F. Severi

Segre.34 F. Severi to Segre, Rome 12.10.1946

CA, *BSP*, fol. 1r

Dearest Segre, one old and one recent item of news. I hasten to communicate the latest news immediately, after I was regretfully obliged to give you not very good information with my previous letter. This morning, I talked with Colonnetti, who told me that yesterday, as soon as he returned to Rome, he went to the Ministry to request the general measure, which should go to the agenda of the next Councils of Ministers, and which will allow your secondment at the Institute of Advanced Mathematics. He also told me that, since the law concerning the new Committees of the Research Council cannot reasonably be approved before next year, he will personally assume responsibility for the choice of some seconded professors, including you, considering my proposal sufficient to discharge any responsibility. I am sure that once you arrived here, other issues would also be easily resolved, and doubts would be removed (the origin of which is not very clear indeed). I will soon have to propose to the Editorial Board of the *Annali* the replacements for Tonelli and De Franchis. I will propose you in place of De Franchis, and I believe that I will be seconded immediately in this matter, especially since I am the director of the *Annali*.[211] But keep it under your hat until it is approved. I have to

---

[211] Leonida Tonelli and Michele De Franchis had disappeared on 12 March 1946 and 19 February 1946 respectively. The proposal is disconcerting, if we take into consideration that it had been

talk about it with the co-director Sansone, who already agrees with your nomination. It is a matter of forwarding, together with your nomination, the name of an analyst, to Bompiani and Signorini. Cordial regards. Yours truly F. Severi

P.S. Colleagues still have not resolved their reservations after the interview with Signorini, of which I do not know the outcome. But for now, we must not disturb the hornet's nest.

Carissimo Segre, una fredda e una calda. Mi affretto a comunicarle subito la calda, dopo che avevo con dispiacere dovuto darle informazioni non ottime con la mia precedente. Stamani ho parlato con Colonnetti, il quale mi ha detto che ieri, appena tornato a Roma, è andato al Ministero per sollecitare il provvedimento generale, che dovrebbe andare a uno dei prossimi Consigli dei Ministri e che consentirà il Suo comando presso l'Istituto di Alta Matematica. Egli mi ha inoltre dichiarato che, non potendo avere la legge di costituzione dei nuovi Comitati del Consiglio delle Ricerche prima dell'anno venturo, si assumerà personalmente la responsabilità della scelta di alcuni professori comandati, tra i quali Lei, ritenendo sufficiente la mia proposta a scarico della sua responsabilità. Sono sicuro che quando Lei fosse qui per questa via, si risolverebbero facilmente anche altre questioni e si rimuoverebbero dubbi la cui origine non è chiarissima. Dovrò proporre prossimamente al Comitato di redazione degli Annali *la sostituzione di Tonelli e di De Franchis. Proporrò al posto di De Franchis Lei, e credo che in questo sarò seguito subito, tanto più che io sono il Direttore degli* Annali. *Ma tenga riservata la cosa finché non sia avvenuta. Ne dobbiamo parlare col Condirettore Sansone, il quale è del resto già d'accordo sul Suo nome. Si tratta di fare insieme con lui anche il nome di un analista agli altri, cioè a Bompiani e Signorini. Cordiali saluti. Suo aff.ᵐᵒ F. Severi*

P.S. Ancora i Colleghi non hanno sciolto le riserve dopo il colloquio con Signorini, di cui non conosco l'esito. Ma bisogna per ora non pungolarli.

Segre.35 F. Severi to B. Segre, Rome 22.10.1946

CA, *BSP*, fols. 1r, 2r

Dearest Segre, On the evening of 18th, we had a meeting with Bompiani, Picone and Signorini.[212] Conforto was missing, in South Tyrol for high school exams, but his presence was not necessary because he was in complete agreement with me. The meeting was convened after a step that had been taken in the morning,

---

Severi who had expelled Segre and Levi-Civita from the Editorial Board of the *Annali di Matematica pura ed applicata* in the autumn of 1938.

[212] See SEGRE.33.

and that had greatly disturbed me. Prof. Krall of our Institute (a man of great value whom I respect very much) had come to me, sent by Signorini, to ask me if I would have anything against his transfer to the post of Mathematical Physics. In what tenured position? In the post remaining vacant due to the death of Enriques, he told me. To which I opposed my firmest objection. This was obviously a way to circumvent my proposal or to postpone it to another time, and perhaps, why not? to take away a place from Geometry. The faculty meeting began on Friday night by reporting what Krall had told me; but immediately it was added that Krall, as instructed by me, had communicated that I could not adhere to such a proposal. The skirmishes with courteous weapons lasted a couple of hours, and I silenced my contradictors, so much so that Picone, adding a compliment to a realistic situation, said that my opinion was fascinating and that at first, nobody saw anything to object to but, on further consideration, came up with some objections. In any case, the conclusion to which I had to adapt in order not to bring to the Faculty the expression of a disagreement among mathematicians, which the three colleagues had skilfully tried to transfer from people to things, given that you rightly did not wish to come here in a position of struggle and that you had allowed me to understand that you did not mind remaining for this year in Bologna, was the following: I agreed to accept the teaching assignment provided that it was accompanied with a declaration about the temporary nature of that measure, so that the Faculty would not commit the post to another discipline. The faculty session was to take place, and did so, in the afternoon of the next day, Saturday 19. In the morning, however, the three colleagues tried to change things by saying that basically there was no need for any declaration, given that the danger that the vacancy would be available otherwise, after they withdrew the proposal to assign it to Mathematical Physics, did not exist, because the naturalists did not make any proposal for Genetics. I resisted this change in scenario and induced my colleagues to accept in silence the following declaration that I intended to make in the Faculty: «I accept the assignment of Higher Geometry; but I make my acceptance conditional to that same declaration that I made when the Faculty benevolently charged me with said role, following my transfer from tenured professorship of Higher Geometry to the professorship at the Institute of Advanced Mathematics, namely that: I consider the measure temporary, and I deem it necessary, for various specific reasons that I do not repeat today, that in Rome the chair of Higher Geometry be held by a tenured professor.» The Faculty took note of my declaration in the minutes of the meeting, with bated breath. With this act, I wanted to reserve the right to postpone to a more opportune moment the proposal for your transfer: that is, at a time when the competitions will have re-injected the provincial Universities with new elements, so that they cannot criticise the appointment of a new professor of mathematics in Rome. On the other hand, Signorini confided that their hidden motive in acting as they did was that I help them to win another post for Mathematical Physics ... So be it! I am sorry that you decided not to come to Rome even if there was the possibility (which seems almost certain) to implement Colonnetti's purpose: a secondment would have kept you here for two years and in the meantime the question of the transfer would have been resolved.[213] But I certainly do not dare to insist because I see the family difficulties and I appreciate the respect that you wish to award your colleagues in Bologna, who have welcomed you well. It means that we will arrange, if possible, for you to come this year to the Institute of Advanced Mathematics at least for a short course of conferences. Let me know your precise wishes, so that I can also talk about it with Colonnetti, with whom it is very difficult to conclude anything, because he is almost always away from Rome. [. . .] With regard to your book, there is no doubt that I will look for all the possible ways to facilitate its publication.[214] Tell me precisely how many pages would be in the usual format of publications made by Zanichelli under the auspices of the National Research Council. The Institute of Advanced Mathematics does not currently have availability for publications. I am now working to have a provisional allowance for this year, while the reform scheme of the constitutive law of the Institute, to which the Ministry has already given its maximum consent, is thoroughly underway. Our greetings to your family and a cordial handshake to you, with the deep regret that my most immediate desire is vanishing from the horizon! Yours truly F. Severi

P.S. Hodge has not yet answered to my letter. Did he write to you?

Carissimo Segre, La sera del 18 avemmo una riunione con Bompiani, Picone e Signorini. Mancava Conforto che trovasi in Alto Adige per esami di maturità, ma la cui presenza, del resto, non era necessaria perché egli era in tutto d'accordo con me. La riunione succedeva a un passo che era stato fatto la mattina presso di me, e che mi aveva notevolmente inquietato. Era venuto, mandato da Signorini, il Prof. Krall del nostro Istituto (uomo di grande valore che io stimo moltissimo) per chiedermi se avrei avuto nulla in contrario ch'egli accettasse il trasferimento in Facoltà alla Fisica Matematica. A quale posto di ruolo? Mi fu risposto: al posto di ruolo restato vacante per la morte di Enriques. Al che io opposi il mio più fermo diniego. Si trattava evidentemente di un modo per eludere la mia proposta o rinviarla ad altro momento, e magari anche, se possibile, per togliere un posto alla Geometria. Fu colla ripetizione di quello che mi aveva detto Krall che cominciò la riunione di venerdì sera; ma subito fu anche detto che Krall aveva, come ne era stato da me incaricato, comunicato che io non potevo aderire a una proposta del genere. Le schermaglie ad armi cortesi durarono un paio d'ore, ed io ridussi al silenzio i miei contradditori, tanto che Picone, aggiungendo un complimento a una situazione realistica, volle affermare che la mia parola era affascinante e che lì per lì non si trovava nulla da obiettare, ma che poi ripensandoci le obiezioni nascevano. Ad ogni modo la conclusione alla quale mi dovetti adattare per non portare in Facoltà l'espressione di un dissidio fra matematici, che i tre colleghi avevano cercato abilmente di trasferire dalle

---

[213] See SEGRE.33.

[214] B. Segre, *Lezioni di geometria moderna*, volume 1, Bologna, Zanichelli, 1948.

persone alle cose, tenuto conto che giustamente Lei non avrebbe desiderato venire qui in una posizione di lotta e che d'altronde mi aveva lasciato capire che in fondo non Le dispiaceva di rimanere ormai per quest'anno a Bologna, fu che io avrei finito col dichiarare di accettare l'incarico accompagnandolo con una dichiarazione che ritiene il provvedimento temporaneo, affinché la Facoltà non impegnasse con altra disciplina il posto di ruolo. La seduta di Facoltà doveva aver luogo, ed ebbe luogo, nel pomeriggio del giorno successivo, sabato 19. La mattina però si cercò da parte dei tre colleghi di mutare le cose dicendo che in fondo non c'era bisogno di nessuna mia dichiarazione, atteso che il pericolo che si disponesse altrimenti del posto di ruolo vacante, dopo che loro ritiravano la proposta di assegnarlo alla Fisica Matematica, non esisteva, perché i naturalisti non facevano per ora nessuna proposta per la Genetica. Io resistetti a questo mutamento di scena e indussi i colleghi ad accettare in silenzio la dichiarazione che dovevo fare in Facoltà e che fu a un dipresso la seguente: «Accetto l'incarico di Geometria Superiore; ma subordino la mia accettazione a quella medesima dichiarazione che feci quando la Facoltà volle analogamente attribuirmi l'incarico della stessa disciplina in seguito al mio trasferimento da professore di ruolo della Geometria Superiore a professore dell'Istituto di Alta Matematica, e cioè che considero il provvedimento temporaneo, ritenendo necessario che a Roma, per varie ragioni specifiche che oggi non ripeto, la cattedra di Geometria Superiore sia tenuta da un professore di ruolo.» La Facoltà prese atto della mia dichiarazione con iscrizione a verbale della provvisorietà del provvedimento senza che nessuno fiatasse. Ho voluto con ciò riservarmi di rinviare a momento più opportuno la proposta per il Suo trasferimento per la Geometria Superiore: al momento cioè in cui i concorsi avranno rinsanguato con nuovi elementi le Università di provincia, sicché non si possa criticare la chiamata di un nuovo professore di matematica a Roma. D'altronde Signorini mi confidò a nome di tutti che il loro secondo pensiero nell'agire come hanno agito, è che io li aiuti ad ottenere un altro posto di ruolo per la Fisica matematica ... Ammettiamolo pure. Mi spiace che Ella abbia deciso di rinunziare a venire a Roma anche se ci fosse la possibilità (che sembra certissima) di attuare il proposito di Colonnetti: il che l'avrebbe tenuto qui per due anni e nel frattempo la questione del trasferimento sarebbe stata risolta. Ma non oso certo insistere perché mi rendo ragione delle difficoltà familiari e dei riguardi che Ella vuole usare ai colleghi di Bologna, che l'hanno bene accolto. Vuol dire che vedremo se sarà possibile farla venire quest'anno all'Istituto di Alta Matematica almeno per un breve corso di conferenze. Voglia Lei ora specificarmi in modo preciso i Suoi desideri, in guisa che io ne possa parlare anche con Colonnetti, col quale è difficilissimo concludere qualcosa, perché è quasi sempre via da Roma. [...] Nei riguardi del Suo libro non c'è dubbio che io cercherò tutti i modi per agevolare la pubblicazione. Mi dica con precisione di quante pagine verrebbe nel formato solito delle pubblicazioni fatte da Zanichelli sotto l'auspicio del Consiglio Nazionale Ricerche. L'Istituto di Alta Matematica non ha attualmente disponibilità per pubblicazioni. Io mi sto ora occupando per avere un assegno provvisorio per quest'anno, intanto che si conduce a fondo lo schema di riforma della legge costitutiva dell'Istituto, al quale il Ministero ha già dato in massima il suo assenso. Saluti nostri in famiglia e una cordiale stretta di mano a Lei, col vivo rammarico che vada sfumando il mio desiderio immediato! Suo aff.mo F. Severi

P.S. *Hodge non ha per ora risposto alla mia lettera. Ha scritto a Lei?*

Chapter 11
An Episode of Partial Professional Retraining: Alessandro Terracini in Argentina

Born in Turin in 1889, Alessandro Terracini grew up at the School of E. D'Ovidio, G. Fano, G. Fubini, and C. Segre (the latter acting as supervisor of his degree thesis), all of whom inspired his research in projective-differential geometry and left a lasting impression on him. The quintessentially Italian style of that School to which Terracini would always claim to belong emerges both in his early research on tangent and osculating spaces to a manifold, in relation to their possible singularities, and in his later works on the geometric meaning of the projective normal, W-congruences and the concept of approximation order in the incidence of two planes or spaces.[1] It is equally apparent in the essay *Esposizione di alcuni risultati di geometria proiettiva differenziale negli iperspazi*, published as an Appendix to the treatise by Fubini and E. Čech, *Geometria proiettivo-differenziale* (Bologna, Zanichelli, II, 1927, pp. 729–69), in which, despite using differential equation systems, Terracini accompanied each analytical result with a clear geometric interpretation in order to demonstrate how analytic and synthetic procedures can be integrated and illuminate each other.

Trained during the era of international congresses and the *mobilité savante*, Terracini saw his academic career take off in the years between the two wars. The assistantship to Fano (1919) and the chair of Algebraic Analysis in Modena (1919–1920) were followed by the offer of two chairs of Analytic Geometry in Cagliari and Catania (he chose Catania), in February 1925, and finally by the call to Turin in the autumn of that same year.

This was a wonderful time for him. From a professional point of view, Terracini was appreciated locally, nationally, and internationally; he published about 40 scientific works in less than 16 years and a volume of lessons in analytic and projective geometry, in collaboration with Fano, which demonstrated his teaching talent and would be republished four times (in 1929, 1940, 1948, and 1957). Family life also provided him with great joy. On April 16, 1924, Terracini married Giulia Sacerdote,

---

[1] Bompiani (1970).

Fig. 11.1 A. Terracini,
16.4.1924, Terracini Coll.

Fig. 11.2 A. Terracini and his wife Giulia in the late twenties, Terracini Coll.

and in the following years, they had three children: Lore (1925–1995), Cesare (1927–1972), and Benedetto (1931) (Figs. 11.1 and 11.2).

Faculty colleagues were also good friends, and F. Tricomi, E. Persico, B. Segre, and E. Togliatti all describe Terracini as a somewhat reckless skier, a passionate mountaineer, and a companion of countless trips, both on foot and with skis.

Then, the events of 1938 interrupted all this. Terracini, a man brought up within the tenets of duty and dignity, suffused with a subtle temperament, a sort of Calvinian lightness, almost immediately decided to flee racist Italy. These qualities—the same that in the post-war period would enable him to resume relations with colleagues who had shamefully taken advantage of the exclusion of Jews—led him to see emigration not as an escape but as a choice. They led him to see his stay in Argentina not as a hiatus but as a new phase in his professional journey. The period in Argentina signified for Terracini the beginning of a new stage of scientific activity as a cultural entrepreneur and ambassador of Italian mathematics. If evoking a cultural bilingualism is perhaps excessive, it is not like speaking of a 'happy encounter between openings, ferments, local fervours and available European energies'.[2] Terracini would always feel he had received a lot from the Argentinian milieu: hospitality, affection, and cultural stimuli. On the contrary, it was a great fortune for this country to benefit from the scientific, teaching, and organisational talent of a scholar who greatly honoured his chair in Tucumán and helped to lead the *Universidad Nacional* in the international arena of mathematical research.[3]

11.1 The Last Months in Turin

Traumatic shock was Terracini's first reaction to a series of laws with which the State deprived itself of some of its servants: 'so it happened – wrote Terracini to his brother Benvenuto – and much more than what all of us expected. There's nothing to do but take the blow, as philosophically as possible, and think about what it will be necessary to do'.[4]

Bewilderment was particularly strong because Terracini had remained essentially detached from the political life of the country and, having grown up in a perfectly integrated family, he had witnessed very few demonstrations of anti-Semitism:

> If I try to remember episodes of 'anti-Semitism' in my earliest youth, there is practically nothing, which confirms how far the movement has gone in Italy many years later. I only remember that my classmate at primary school, the bottom of the class, sometimes and with clear intentions, turned to me and folding the edge of his jacket slightly like a pig, accompanied his gesture with the words «ebreu früst» [ragged Jew] to which I responded with «cristian quader» [dumb Christian].[5]

---

[2] Terracini L. (1989), p. 360: *encuentro feliz entre aerturas, fermentos, fervores locales y energías europeas disponibles.*

[3] Luciano (2020, 2022a).

[4] A. Terracini to B. Terracini, Lido di Venezia 3.9.1938, in Terracini L. 1990, p. 444: *ecco dunque avvenuto . . . e assai più di quello che si aspettava! Non c'è che da incassare il colpo prendendosela il meno che si può, e pensare a quanto sarà necessario.*

[5] Terracini A. (1968), p. 3: *Se cerco di ricordare episodi di "antisemitismo" negli anni della mia prima gioventù, essi si riducono a ben poca cosa, il che conferma quanto artificiosamente tale movimento sia stato creato in Italia molti anni dopo. Soltanto ricordo che un mio compagno delle scuole elementari, l'ultimo della classe, talvolta, con intenzione evidente, si rivolgeva a me, piegando il lembo della sua giacca a mo' di maiale, accompagnando quel segno con le parole «ebreu früst»; alle quali mi fu suggerito di rispondere «cristian quader».* Terracini was enrolled in

Terracini was relieved from the chair, effective from 29 September and 11 November, respectively. On 21 October, he was expelled from the National Union of Italian Officers on Leave; on 10 November, he was removed from the Academy of Sciences of Turin and the Italian Mathematical Union.

He immediately submitted a request for counter-discrimination for military merits. Called to arms in the battalion of railway engineers in Rome, Alessandro had been sent to the front, first to Gorizia, then to Gemona, and finally to Breganze, as a reserve officer of the 22nd Miners' Company. He had obtained his teaching qualification during a furlough, developed a periscope variant, and demonstrated a formula linked to the compilation of mountain-shooting tables. His brother Benvenuto had been seriously injured shortly before the battle of Caporetto. Both Alessandro and his brother Benvenuto were decorated with the silver medal of military valour in 1917. Their applications would be accepted, but they would be informed of the outcome where they had already left Italy.

In the meanwhile, Alessandro dedicated his energies to tackling the most painful aspect of persecution: the fact that the racial laws had stolen the future of young generations, preventing them from studying. To guarantee his children Lore and Cesare the education they deserved after being driven out of the lyceum, he then began to reflect on the reorganisation of the Jewish school in Turin, a task in which he was to be strongly involved. In particular, Alessandro pursued modifications to make the school more secular and aligned with the standards of national institutions; he strived to guarantee that the professional and technical classes were flanked with a complete classical curriculum; he strongly emphasised 'the necessity for our schools to be organised with programmes that were in no way inferior to those of the State institutions' and 'the consequent need to minimise the religious part';[6] he endeavoured to equip physics, chemistry and science cabinets with instruments and scientific collections. His efforts played a fundamental role in the modernisation of the school. Terracini did not confine himself to playing an advisory role but, at the friendly insistence of M. Falco and M. Tedeschi, embraced the idea of creating an advanced curriculum aimed at ensuring a preparation to those young people who were prevented from enrolling at the University in the autumn of 1938. In doing so, he exploited his professional skills as far as the mathematical curriculum was concerned, obtained the collaboration of some colleagues (Fano and Guido Ascoli), and personally collected applications from several excluded freshers of Piedmont and Lombardy; however, finally, he left the project.

Terracini spent the winter of 1939 in Turin, in a sort of voluntary isolation, mitigated only by the affectionate presence of Persico and Tricomi. To bring him out of the depression into which he was slipping, friends lent him the books and

---

the PNF from 1933 and had belonged to the Fascist School Association since its founding but had never indulged in political activity.

[6] A. Terracini to B. Terracini, Lido di Venezia 4.9.1938, in Terracini L. (1990), p. 446: *necessità che la scuola sia organizzata con programma in nessuna parte inferiore a quello governativo; conseguente necessità di limitare al minimo la parte confessionale.*

articles necessary for research activity. Indeed, Tricomi invited him to publish a text under a false name: *Algebra elementare ad uso dei licei*. The handbook, which would come out in 1940, constituted Terracini's first foray into the field of mathematics education and represented a unique work in his portfolio. In fact, showing a taste for refined logical-deductive rigour, which one hardly expected from a geometer belonging to the Italian School of Geometry, here he fully developed the modern theory of real numbers according to Dedekind's construction. The fundamental aim to which every author of school texts aspires—claimed Terracini in the preface—is clarity. Thus, in order to meet the demands for rigour that are compulsory in introducing the first elements of infinitesimal analysis, traditional ancient premises have to be reconsidered. Modern pupils cannot ignore the concept of contiguous classes, which 'threaten to reappear at all times'.[7] In order to overcome this obstacle, the most valid alternative is to introduce real numbers by passing through the notion of couples of Cauchy-convergent sequences of rational numbers, not-decreasing and not-increasing, respectively, which approximate the given number by default and excess. In its attention paid to foundational aspects, the *Algebra* by Tricomi (*alias* Terracini) was a very atypical work, both with respect to contemporary educational literature and in relation to the output of its virtual author, who would actually confess 'the slight discomfort' experienced in seeing his name on the cover of a book inspired by didactic tenets completely different from his own and in which the treatment of some parts was more developed than in his university lectures.[8]

Apart from his commitment to the renewal of the Jewish school and along with some editorial duties,[9] the last months spent by Terracini in Turin were characterised by the search for a position abroad and preparation for departure. He enjoyed adequate means of livelihood, having been dismissed with a decent pension after 10 years of seniority as a full professor,[10] and he could rely on the support of top figures in solidarity chains. Consequently, immediately after the promulgation of racial laws and on his return from holidays at the Lido of Venice, Terracini visited T. Levi-Civita in Padua, where he consulted him on this matter, evaluating possible destinations. Levi-Civita put him in contact with Veblen, in an attempt to obtain a position for him in the United States.

A large number of mathematicians—O. Veblen, O. Zariski, V. Snyder, S. Lefschetz, E.P. Lane, W.C. Graustein, W.P. Milne, J.G. Semple, and P. Sperry—took action to help Terracini and presented his case to the SPSL and ECADFS, but both the societies and solidarity networks failed in their attempts to support

---

[7] Tricomi *alias* Terracini A. (1940), p. VII: *minacciano di riapparire a ogni momento*.

[8] Tricomi (1967), p. 66: *un lieve disagio vedere il mio nome sulla copertina di un libro ispirato a criteri didattici completamente diversi dai miei. Al punto che in quel libro pei Licei, la trattazione di alcune cose è più "elevata" che nelle mie* Lezioni *per l'Università!*

[9] In addition to the textbook in *Algebra*, Terracini also worked with Gino Fano to prepare the second edition of *Lezioni di Geometria analitica e proiettiva*. This treatise was not primarily addressed to teaching, as a consequence, despite its being authored by two Jewish scholars, it escaped withdrawal. It would be published in Turin by Paravia in February 1940.

[10] Terracini Coll.: Decree of liquidation of retirement pension by Terracini and determination of retirement allowances, 11.2.1939 and 12.9.1939.

him.[11] His applications for chairs vacant in the Universities of Aberdeen and Durham were rejected.[12] A combination of various factors is behind such failures. Terracini moved quickly but with some disorganisation, first writing to Veblen, then to SPSL, and some months later to the ECADFS.[13] In the first stage, he was probably waiting for the conclusion of the counter-discrimination practice. Second, Terracini was a good mathematician, but not an excellent one. In the United Kingdom, both Milne and Semple, who brought his case to the attention of SPSL, considered him a very distinguished man and scholar, though not in the same class as B. Segre.[14] Overseas, the evaluation was basically the same. Writing to Zariski, Veblen confessed that he was 'very much concerned about Segre and also, perhaps to a lesser extent, about Terracini'.[15] After talking the problem over with Lefschetz, Graustein, and Lane, Veblen thus did not insist on the plan of bringing Terracini to Princeton, even if for a short time only. He continued trying to find an opening for him in institutions that had not yet received many of the refugees but without any real conviction of success.

The fact that none of these scholars personally knew Terracini had an impact. As a matter of fact, Terracini had spent very little time abroad: in Cambridge in 1912 for the ICM, in Germany, Switzerland, and Belgium, but little else. Furthermore, there was the crux of language skills, because he declared that he could read and write English with ease, but he did not speak it fluently, which constituted a serious handicap with regard to obtaining a teaching position in an English-speaking country. Finally, in the documents provided by Terracini (*curricula*, commented lists of his publications, etc.), he always explicitly asserted his membership of the Italian School of Algebraic Geometry and postulated his scientific affiliation with the Italian geometric tradition.[16] Such clear claims of his own cultural roots were probably not appropriate, especially among the American milieu, where advocates of the algebraic-topological approach would support Terracini, but in very mild terms.[17]

---

[11] See, for example, ECADFS, *AT*: P. Sperry to ECADFS, Berkeley 30.1.1939; B. Drury [ECADFS] to P. Sperry, New York 9.2.1939; TERRACINI.1; *OVP*, *AT*: O. Veblen to E.P. Lane, [Princeton] 30.9.1938; TERRACINI.2; SPSL, *AT*: W.P. Milne to E. Simpson [SPSL], Leeds 20.1.1939; E. Simpson [SPSL] to W.P. Milne, London 21.1.1939; D.C. Thomson to J.G. Semple, London 3.3.1939; TERRACINI.5-7. Pauline Sperry (1885–1967), was the first female tenure-track mathematics faculty member at the University of California at Berkeley (1917–1950). Sperry was an active Quaker and involved in various humanitarian and political causes.

[12] TERRACINI.4.

[13] SPSL, *AT*: A. Terracini to SPSL, Turin 11.12.1938 (D3.20.1); E. Simpson [SPSL] to A. Terracini, London 22.12.1938; A. Terracini to SPSL, Turin 4.1.1939; E. Simpson [SPSL] to A. Terracini, London 4.1.1939; ECADFS, *AT*: A. Terracini to ECADFS, Turin 1.5.1939 and B. Drury [ECADFS] to A. Terracini, New York 19.5.1939.

[14] SPSL, *AT*: W.P. Milne to E. Simpson [SPSL], Leeds 20.1.1939; SPSL, *BS*: G. Semple to SPSL, London 2.3.1939.

[15] OVP, *BS*: O. Veblen to O. Zariski, [Princeton] 8.11.1938.

[16] TERRACINI.3 and D3.20.1.

[17] TERRACINI.6.

Fig. 11.3 Family of Alessandro Terracini and Giulia Sacerdote leaving for Argentina, Terracini Coll.

Faced with the difficulties and due to the situation making his emigration and that of his family more urgent with every day that passed, Terracini extended the set of possible destinations, investigating openings in Latin America, which was among the very few possibilities he still espied.[18] Interviews with Brazil seemed to be promising when a letter from A. Guzmán, Dean of the Engineering Faculty of Tucumán, reached Terracini, inviting him to occupy the chair of Projective and Descriptive Geometry for the degree programme in Architecture and that of Higher Mathematics within the *Profesorado*. The proposal was accepted by Terracini with warm emotion: 'I can't believe I will resume the life of teaching!'[19] he wrote to Levi-Civita on the day after the call (June 9, 1939).

After taking leave of their families, the Terracinis were expected to flee Italy aboard the Augustus liner on 24 August 1939. However, due to the international situation, they were able to travel to Argentina only three weeks later, on 16 September 1939, after an obligatory stay in Quinto al Mare and in Sant'Alluccio (Fig. 11.3).

Landing in Buenos Aires on October 3, they travelled by train for 24 hours through the dusty fields to reach their new home at no. 417 Calle Salta, Tucumán[20] (Figs. 11.4a, b).

---

[18] D3.15.3.

[19] A. Terracini to T. Levi-Civita, Turin 10.6.1939, in Nastasi and Tazzioli (2000), p. 401: *non mi par vero di poter riprendere la vita dell'insegnamento!*

[20] Terracini A. (1968), p. 123–125.

Fig. 11.4 (**a**, **b**) Tucumán, Argentina, in the thirties, courtesy of Erika Luciano

On October 11, three days after settling, Terracini held his inaugural lecture. Alessandro's brother Benvenuto, nephew Eva, and grandmother Eugenia would reach Tucumán in 1940.

11.2 'Sower of Ideas in a Land That Was Virgin, but Eager to Produce'

Revealing remarkable linguistic abilities, Terracini quickly managed to integrate into his workplace. Thrilled by the new surroundings, he resumed publishing and took up research, despite the feeling he had had in the last months spent in Italy that his career was now over. Production, in quantitative terms, was impressive with 30 papers published in less than 10 years in all Latin American journals, both major and minor: *Revista de la Unión Matemática Argentina*, *Boletín Matemático*, *Revista Electrotécnica*, the proceedings of the Academies of Sciences in Rio de Janeiro and Lima, and others.

Terracini combined research with a demanding teaching charge (three courses per year, both at the undergraduate level and for the master's degree programme) and with a frenetic commitment to science popularisation activities, through the organisation of series of lectures, public conferences, and radio broadcasts.[21]

Still more influential, in terms of cultural legacy, was Terracini's commitment to the publishing field. Less than a year after Terracini's settling in Tucumán, in December 1940, the first issue of his *Revista de Matemáticas y Física Teórica* affiliated with the *Universidad Nacional* was published. Papers were submitted to the *Revista*'s editorial board from all the world and, in his role of chief editor, Terracini built himself a network of international contacts much larger than the one he had woven before emigration. Striking evidence of this fact can be found in the

---

[21] For example, the Terracini Coll. preserves the typewritten text of the radio podcast devoted to *La Geometría descriptiva y su valor formativo*, 1942, fols. 1–5.

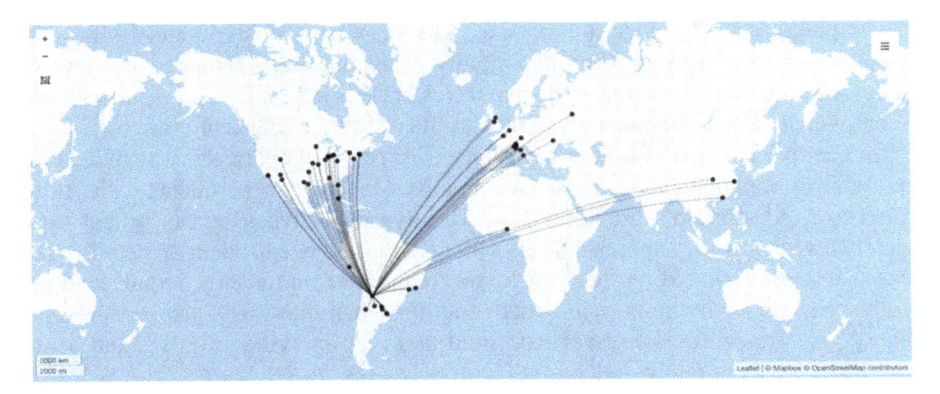

Fig. 11.5 Terracini's network of international contacts as editor of *Revista*, courtesy of Erika Luciano

geographical distribution of the 9000 offprints constituting Terracini's personal library[22] (Fig. 11.5).

For Terracini, emigration to Latin America, begun as an alien mourning the loss of his cultural roots and national identity, turned into a true professional twist. In a very short time, he achieved considerable influence and established himself as one of the 'fathers' of mathematics in Latin America. He who had been expelled from all Italian learned societies was co-opted as a member of the *Sociedad Científica Argentina* the day after his arrival in Tucumán (9 October 1939).[23] Shortly thereafter, he joined the *Asociación argentina para el progreso de las ciencias* which charged him with the task of carrying out a survey on 'what should be done for the progress of science in Argentina', with special reference to facilities for mathematical studies.[24]

Appointed fellow of the American Mathematical Society in April 1942,[25] Terracini did not miss any favourable opportunities to promote wide-ranging cultural projects and to expand partnerships with their American colleagues at a time when international relationships were boycotted. For example, by taking advantage of the meetings with G.D. Birkhoff and M. Stone, who visited Tucumán in the autumn of 1942, he highlighted the difficulties of those scholars who, living far from study centres and libraries, had to depend on the help of colleagues and friends. Hence, a proposal was made (although unfortunately aborted due to the war) to establish a central committee for bibliographical information in mathematics and physics under the aegis of the American Mathematical Society.[26]

---

[22] See Luciano and Scalambro (2020).

[23] Terracini would be elected *socio activo* on December 12, 1940.

[24] Terracini A. (1942b).

[25] Ayres (1942), p. 499.

[26] Terracini A. (1942a).

Despite the major recognition obtained, the years between 1940 and 1943 were very painful for Terracini. The impossibility of helping other refugees, pessimism about the war, and the horror of the massacres of Jews across Europe—information about which, despite fragmentary, began to filter—were permanent concerns.

In the summer of 1944, after a prolonged isolation, the Terracinis managed to re-establish communications with their relatives in Italy, in particular with Aldo Sacerdote, Alessandro's brother-in-law. The five collective letters they sent him clearly reveal how they apprehensively observed the war events, their anxiety for the fate of friends deported by the Nazi rogues (*canaglie*), and even a certain kind of guilty remorse for living in such better conditions than those who had remained in Italy.[27] As Terracini's son, Benedetto, told us in an interview in 2018, and as is confirmed by these family letters:

> After 8 September 1943, contact with relatives in northern Italy was lost. Somehow, via the Red Cross, even before the end of the war we heard of the cousins deported to Germany. But I can't say (perhaps because parents didn't want to talk about it in front of their children) when and how they became aware of the size and modalities of the murders committed in the extermination camps. In retrospect, my father's optimism expressed in a letter from Tucumán dated 27 April 1945 amazes me: «unfortunately, we have heard that many people have been deported. The only thing we can hope is that the devil was not so bad for all of them and that soon we will hear about it.»[28]

In June 1943, meanwhile, a revolutionary *coup d'etat* led to the presidency of Argentina general P.P. Ramírez, then succeeded by E.J. Farrell (February 1944) and J.D. Perón (October 1945 to February 1946). Terracini did not suffer any persecution by the Peronists but being afraid of losing his individual freedom again, he turned to the American colleagues. Once again, Veblen, Snyder, A. Einstein, J. von Neumann, and H. Weyl tried to help him find a position, even if temporary, in the United States, Uruguay, or Peru.[29] Despite the support of five institutions (IAS, Guggenheim, Rockefeller and Carnegie foundations, and the Pan-American Union), the search was not successful, but, in the meantime, the day finally came for the liberation of Italy.[30]

---

[27] See D7.2.1, TERRACINI.10, TERRACINI.14, and TERRACINI.15.

[28] Benedetto Terracini, interview released to L. Giacardi and E. Luciano in Turin, winter of 2018: *Dopo l'8 settembre 1943 si persero i contatti con i parenti in Italia settentrionale. In qualche modo, via Croce Rossa, ancora prima della fine della guerra si seppe dei cugini deportati in Germania. Ma non saprei dire (forse perché i genitori non volevano parlarne davanti ai figli) quando e come si prese coscienza delle dimensioni e delle modalità degli omicidi commessi nei campi di sterminio. A posteriori, mi stupisce l'ottimismo da mio padre espresso in una lettera da Tucumán del 27 aprile 1945: «Di tante persone purtroppo si è saputo che sono state deportate. La sola cosa che possiamo sperare è che il diavolo non sia stato per tutti loro tanto brutto e che presto se ne possa avere notizie.»*

[29] See TERRACINI.9, TERRACINI.11–13.

[30] TERRACINI.8.

11.3   The Bitter-Sweet Return to 'the Italy of the Stunning Amnesty'

With the end of the war and news of the reintegration of academics, Terracini posed himself the dilemma of return. It is likely that he had very ambivalent feelings in this connection. On the one hand, there was the utmost gratitude for the country that hosted him with affection and generosity, 'never made him feel a stranger'[31] and gave him the highest honours, including the presidency of the *Unión Matemática Argentina* (1945–1947).[32] On the other hand, the argument of a potential reconciliation played in favour of his return.

Finally, his doubts dissipated in a pattern of friendly exchanges of opinions with Persico, Tricomi, and G. Castelnuovo, at the beginning of 1946 Terracini started the procedure for rehabilitation.[33] Apparently unaware of the betrayal of intellectuals which had occurred in 1938, he re-established relations with M. Picone, who organised a petition. A total of 86 mathematicians (practically all the Italian university professors of this discipline) signed a statement wherein they requested aid to support the return of Terracini, whose presence was considered essential for the reconstruction of the country.[34] The Turin Faculty of Sciences, in turn, recommended that the government allocate funds to finance the return trip of Terracini 'a teacher of high value, an exemplary citizen, beloved by all'.[35]

The moment of coming back home finally arrived in 1948. Preceded by his wife and children, Terracini left Argentina in February, after finishing the teaching semester, bitterly disappointed about the so-called Togliatti amnesty, but at the same time comforted by the understanding 'of having performed quite well his scholar responsibilities and duties, and thus having contributed to a good evaluation of the work accomplished by Italians abroad'.[36]

At his departure, students and colleagues suffered the feeling of a great loss.[37] Terracini, however, maintained ties with his host land, and for about 10 years, he

---

[31] Terracini Coll.: typewritten document *La scuola in Argentina*, Turin 4.4.1948, p. 1: *Nessuno mi ha mai fatto sentire la mia qualità di straniero.*

[32] During his term, Terracini established the by-laws of the Society and convened the *Primeras Jornadas Matemáticas Argentinas*, which would greatly contribute to the construction of a national spirit in the Argentinian mathematical community. See Terracini Coll.: typewritten text of the keynote lecture delivered at the *Primeras Jornadas Matemáticas Argentinas*, July 1945, fols. 1–3 and Santaló 2001, pp. 6–9.

[33] TERRACINI.16–22.

[34] See D7.1.2, D7.1.3.

[35] ASUT-*Terracini*: excerpt of the meeting of the full professors of the Faculty of Sciences, 14.11.1945, in footnote 9 in Chap. 7.

[36] Terracini A. (1968), p. 152: *coscienza di avere compiuto abbastanza bene il mio dovere di professore, e di avere così contribuito a una favorevole valutazione del lavoro compiuto dagli italiani in Argentina.*

[37] *Evolución de las ciencias en la República Argentina . . . 1979, p. 203: experimentaron la penosa impresión de que algo muy de ellos les era quitado.*

continued to edit the *Revista* from a distance. Many of his protégés, some of whom became colleagues and friends over the course of time, inherited Terracini's ideas and research projects both in differential projective geometry and in mathematics education. Several of his former alumni would sit on the honours committee for the publication of his selected works and would curate the Spanish edition of the volume *Ricordi di un matematico. Un sessantennio di vita universitaria.*[38]

On being reinstated, Terracini returned to teaching and resumed the previously disrupted path of his scientific and professional career. He held the chairs of Geometry I and Higher Geometry until his retirement in 1963. From 1952, he enlarged his research interests to include studies in history of mathematics, publishing, among other things, a beautiful essay on A.L. Cauchy's stay in Turin (1957). However, he devoted most of his efforts to institutional tasks, holding prestigious positions at both national and international levels. Convinced by L. Berzolari to accept reinstatement to the Italian Mathematical Union, he was Vice-President (1952–1958) and then President for the subsequent two three-year terms. He was also elected to both the executive committee of the *Groupement des mathématiciens d'expression latine* (1955) and the board of Directors of INDAM (1958); in addition, he was the Italian delegate to the International Mathematical Union. In the 60s, he created the Italian groups of seminars and mathematical institutes, which were one of the driving forces of Italian scientific research. Finally, as its Director, he contributed to the rebuilding of the Special Mathematics Library in Turin, which had been heavily damaged by the bombings.

Documents

Terracini.1 O. Veblen to W.C. Graustein, Princeton 30.9.1938

OVP, *AT*, fol. 1r[39]

Dear Graustein: I have received appeals from and on behalf of two geometers who have been displaced from their positions by recent events. One of them is Alessandro Terracini, who has been Professor of Analytic Geometry at Turin and has written very extensively on algebraic and projective differential geometry. The other is the differential geometer A. Duschek of Vienna, who is, among other things, the co-author with Walther Mayer of a well-known book on differential geometry.[40] I haven't been able to think of anything effective to do for either of them. Most of the possibilities which would occur to me have been used up. I

---

[38] de D'Angelo and Herrera (1994).

[39] A very similar letter is sent to Ernest Preston Lane (1886–1969), professor of Mathematics at the University of Chicago. William Caspar Graustein (1888–1941) was Assistant Dean of the Faculty of Arts and Sciences of Harvard University (1939–1941).

[40] Adalbert Duschek (1895–1957), *Privatdozent* at the Technical University in Vienna, had been dismissed shortly after the *Anschluss*. He had published with W. Mayer a *Lehrbuch der*

wonder whether you could do anything, or make any suggestion? I should think there might be among the smaller colleges some which might profit by the chance to add men of genuine scientific distinction to their Faculties without great financial strain. I am sure that many of the refugees would be glad to take very modest positions. I have quite a long list before me, but I am only mentioning these two who I thought might interest you particularly. [...] Yours sincerely, Oswald Veblen

Terracini.2 E.P. Lane to O. Veblen, Chicago 3.10.1938

OVP, *AT*, fol. 1r
Dear Professor Veblen: I have received your letter of September 30 with regard to the misfortune which has befallen Professors Terracini and Duschek. I do not know Duschek personally, but I became well acquainted with Terracini while I was in Italy and know his writings very well. Mr. Bliss and I discussed this matter this morning but could only conclude that we do not know of anything that we can do at the moment for these men.[41] However, I shall keep the matter in mind and will inform you at once if I learn of anything that would seem to alleviate the situation of these men. I am especially interested, of course, in the fate of Terracini. Very sincerely yours, E.P. Lane

Terracini.3 Some Account of My Scientific Papers

ECADFS Record, *AT*, fols. 1r-9r; SPSL, *AT*, fols. 369–377
1 [*Nota su una classe di determinanti*, Giornale di Matematiche (Battaglini), XLVII, 1909]. A juvenile paper on the determinants whose elements a_{rs} and a_{sr} have a constant sum. Some of the properties of the skew-symmetric determinants are preserved, with some modifications.

2 [*Sulle V_k per cui la varietà degli S_h (h+1)-seganti ha dimensione minore dell'ordinario*, Rendiconti del Circolo Matematico di Palermo, XXXI, 1911]. Del Pezzo had proved that Veronese surfaces and cones are the only surfaces in S_r ($r \geq 5$) whose tangent planes are mutually intersecting. On the other hand Severi had proved that the same are the only surfaces whose chords do not fill the space. I have found that the two properties, a priori, coincide; more generally for a locus

---

Differentialgeometrie: Kurven und Flächen im euklidischen Raum (Berlin, B.G. Teubner, vol. 1, 1930).

[41] Gilbert Ames Bliss (1876–1951) was the chair of the Mathematics Department of the University of Chicago (1927–1941).

V_k in the space S_r $(r > 2k)$ the properties: (1) its chords fill a locus having the dimension $2k - i$ $(i \geq 0)$ and (2) two tangent S_k have always a common S_i are equivalent.

3 [*Di alcune superficie del terz'ordine, che sfuggono ad una generazione data da Steiner*, Giornale di Matematiche (Battaglini), XLIX, 1911]. I determine what classes of cubic surfaces are obtainable as loci of contact points of the tangent lines from a fixed point to the quadrics of a pencil.

4-8-15-19-20-21 [*Sulle* V_k *che rappresentano più di* $k(k-1)/2$ *equazioni di Laplace linearmente indipendenti*, Rendiconti del Circolo Matematico di Palermo, XXXIII, 1911; *Alcune questioni sugli spazi tangenti e osculatori a una varietà. Nota I*, Atti della R. Accademia delle Scienze di Torino, XLIX, 1914; *Alcune questioni sugli spazi tangenti e osculatori a una varietà. Nota II*, Atti della R. Accademia delle Scienze di Torino, LI, 1915; *Alcune questioni sugli spazi tangenti e osculatori a una varietà. Nota III*, Atti della R. Accademia delle Scienze di Torino, LV, 1919; *Sulla varietà degli spazi tangenti a una data varietà. Nota I e Nota II*, Rendiconti della R. Accademia dei Lincei, s. 5, vol. XXIX, 1920]. The moment I begun to undertake my researches on projective differential geometry happened to coincide with the years in which this branch had just left its initial period. Some of the methods were already formed and had been put to the test through the easier problems which always present themselves at the dawn of a new theory; the opportunity of contriving other methods was still kept for the future. Among the first the most important was doubtless the method based on the use of linear partial differential equations. This method was already classical for the curves, and Wilczynski had successfully employed it in the theory of surfaces. But with the consideration of loci in hyperspaces new interesting problems arise. I have endeavoured to use such a method to confront several problems which presented themselves. If the number of linearly independent equations (of second order) exceeds a certain limit, Corrado Segre had shown that the V_k belongs to certain well determined classes, and precisely that the dimension of the locus W of its tangent spaces S_k is $< 2k$. But this condition remained, so to say, merely nominal, until the V_k for which the dimension of the locus W is $< 2k$ were effectively known. I succeeded in specifying the whole class of such V_k (4). Moreover, the same condition even being necessary is not always sufficient that the number of linearly independent partial equations exceeds the mentioned limit. What are the cases when it is not sufficient? This is the main question which I have studied in 8, 15. But to this purpose it has been necessary for me to reach many fundamental results about the manner of interfering of the structure of the system of partial equations with the geometric nature of the V_k. I have also observed that the number of linearly independent equations represented both by a V_k and by its generical prime sections may present some irregularities: for what loci does it happen so? (8, 19). As to 20, 21 I have found in them some further relations between a locus V_k and the mentioned locus W_{2k}, when this has a regular dimension $2k$.

5 [*Sul carattere invariantivo di alcune espressioni vettoriali*, Rendiconti del Circolo Matematico di Palermo, XXXIII, 1911] Remarks on the invariantive character of some vectorial expressions relating to a former paper of Boggio.

6 [*Sulle varietà di spazi con carattere di sviluppabili*, Atti della R. Accademia delle Scienze di Torino, XLVIII, 1913] What are the systems of ∞^2 or more S_k, such that two "consecutive" S_k always intersect? With such questions deals 6. I prove also theorems like this: a locus V_{2k} containing ∞^{k+1} S_k (with some further conditions) is necessarily the locus (of Grassmann) which represents the straight lines of a space S_{k+1}.

7 [*Alcune considerazioni sul teorema del valor medio*, Giornale di Matematiche (Battaglini), LI, 1913] Some elementary results (relating to a former paper of L. Galvani) about the law of the mean in the differential calculus.

8 [sic! read 9] [*Su alcune superficie rigate razionali*, Rendiconti del R. Istituto Lombardo di Scienze e Lettere, XLVIII, 1915] An extension to the rational normal surfaces of the Clifford's theorem on rational normal curves. Points of hyperosculation of a ruled surface. Every rational ruled surface R^4 (in the space S_3) contains four points of hyperosculation; if they are not independent, they are collinear. This takes place only when the linear complex of the nodal cubic and the linear complex containing the generators are in involution.

10-14 [*Sulla rappresentazione delle coppie di forme ternarie mediante somme di potenze di forme lineari*, Annali di Matematica pura e applicata, s. 3, XXIV, 1915; *Sulla rappresentazione delle forme quaternarie mediante somme di potenze di forme lineari*, Atti della R. Accademia delle Scienze di Torino, LI, 1915]. Both deal with canonical forms. Except for $n = 3$, two ternary forms of degree n are always linear combinations of the nth powers of the same N linear forms, the value of N being given by computation of constants. An analogous result holds for a quaternary form, except for $n = 2, n = 4$. (I believe that proofs of this theorem given by some other Authors lack the necessary rigour).

11 [*Su una questione che si presenta nello studio delle omografie tra spazi sovrapposti*, Giornale di Matematiche (Battaglini), LIII, 1915] A relation between homographies and a covariant system of quadrics.

12 [*Su alcune particolari quartiche piane di genere 3*, Rendiconti della R. Accademia dei Lincei, s. 5, XXV, 1916] The Klein's C^4 $x_1^3 x_2 + x_2^3 x_3 + x_3^3 x_1 = 0$ has its 24 points of inflection arranged in 8 chains such that the tangent line at each of these points meets again the C^4 at a point in the same chain. I prove the existence of a new kind of plane C^4 having their points of inflection arranged, instead, in 6 chains consisting each of 4 points. There are 5 species of such C^4, projectively distinct.

13 [*Sulle varietà trasversali delle rigate algebriche di uno spazio pari*, Rendiconti della R. Accademia dei Lincei, s. 5, XXV, 1916] Algebraic geometric results on the transversal loci of an algebraic ruled surface.

17 [*Sulle congruenze W di cui una falda focale è una quadrica*, in *Scritti matematici offerti a Enrico D'Ovidio*, Torino, Bocca, 1918] A simple solution of the problem of finding the W congruences having a quadric as a focal surface.

18-26 [*Sui sistemi coniugati permanenti nelle deformazioni di una superficie*, Rendiconti della R. Accademia dei Lincei, s. 5, XXVIII, 1919; *Osservazioni sui sistemi isotermo-coniugati che sono permanenti nelle deformazioni di una superficie*, Annali di Matematica pura e applicata, s. 3, XXX, 1921] Isothermal-conjugate nets which are permanent in the deformation of a surface.

22 [*Eine Bemerkung über die Funktionalgleichungen der isomorphen Abbildung*, Mathematische Annalen, LXXXII, 1921] Relating to a former paper of Emmy Noether I prove that every discontinuous solution φ of the system $\varphi(z + z') = \varphi(z) + \varphi(z')$, $\varphi(zz') = \varphi(z)\varphi(z')$, if existing, is "extremely" discontinuous.

23 [*Di una superficie del sesto ordine e della sesta classe le cui asintotiche sono cubiche sghembe*, Rendiconti della R. Accademia dei Lincei, s. 5, XXIX, 1920] A surface of the 6th order and class whose asymptotic lines are twisted cubic.

24 [*Sulle superficie le cui asintotiche dei due sistemi sono cubiche sghembe*, Atti della Società dei naturalisti e matematici di Modena, s. 5, V, 1919-20] Particular examples of surfaces whose asymptotic lines are twisted cubic (that is the most simple twisted lines, from the algebraical point of view) were still known. I succeeded in a general method for finding the whole class of these surfaces. More generally I found (in finite terms) all the surfaces whose asymptotic lines of both systems belong to linear complexes (a particular case is that of the Tzitzeica-Wilczynski surfaces).

25 [*Sull'esistenza di polarità ordinarie che mutano l'una nell'altra due quadriche non degeneri*, Annali di Matematica pura e applicata, s. 3, XXX, 1921] Being given two quadrics in a space S_r, it is very easy to state that "generally" there is some polarity which carries them into each other. But it is less easy to prove that it happens not only "generally", but always. This is what I did in 25.

27 [*Sul modulo delle forme contenenti una varietà di Segre*, Rendiconti della R. Accademia dei Lincei, s. 5, XXX, 1921] I find the modulus of the forms containing a Segre locus (which represents groups of points belonging to two or more spaces).

28 [*Su due problemi concernenti la determinazione di alcune classi di superficie considerati da G. Scorza e da F. Palatini*, Atti della Società dei naturalisti e matematici di Modena, s. 5, VI, 1921-22] I needed, to use them in some other researches, the answers to both questions: (1) for what surfaces of the space S_r ($r > 5$) the S_5 determined by two tangent planes always contains ∞^1 tangent planes? (2) for what surfaces the locus of the S_h which intersect them at $h+1$ points has a dimension smaller than usual? These questions had been answered before by some other Authors, but only in the hypothesis of algebraic surfaces, and with some mistakes which had let them miss some solutions. I have found the complete solution.

30 [*Sulle superficie i cui spazi osculatori presentano particolari incidenze coi piani tangenti o fra loro*, Atti della Società dei naturalisti e matematici di Modena, s. 5, VI, 1921-22] As before mentioned, the tangent planes of a Veronese surface intersect each other. I have studied the surfaces for which either tangent planes intersect osculating spaces, or osculating spaces intersect each other.

31 [*Una nuova proprietà del cilindroide di Cayley e di un altro conoide retto*, Torino, Doyen, 1922] A new property of Cayley's cylindroid and another ruled surface. These are the only right conoids which have that property.

32 [*Su una proprietà caratteristica della superficie di Veronese*, Atti della Società dei naturalisti e matematici di Modena, s. 5, VII, 1922] A characteristic property of Veronese surface.

33 [*Su una presunta decomponibilità delle trasformazioni di contatto*, Atti della Società dei naturalisti e matematici di Modena, s. 5, VII, 1922] C. Rabut had believed that every contact transformation is always obtainable as a product of some elementary transformations. I prove that he was wrong.

34 [*Correlazioni geodetiche tra due superficie*, Bollettino dell'UMI, II, 1923] What are the couples of surfaces such that it is possible to let the points of each correspond to the geodesic lines of the other in such a way to preserve incidence? They are necessarily couples of surfaces of constant curvature.

35 [*Sui punti di flesso delle quartiche piane generali*, Atti della R. Accademia delle Scienze di Torino, LIX, 1924] Some attempts had been made to find out whether the configuration of the points of inflection of a plane C^3 has its analogue for a C^4. I have proved, instead, that the plane C^4 having six of their points of inflection lying on a conic constitute a totality which do not cover the totality of the plane C^4.

36 [*Sulle superficie con un Sistema di asintotiche in complessi lineari*, Atti della R. Accademia delle Scienze di Torino, LIX, 1924] A systematical geometrical research about the surfaces having a family of asymptotic lines belonging to linear complexes. A theory of the transformations of such surfaces through W congruences, and a theorem of permutability.

37 [*Esposizione di alcuni risultati di geometria proiettiva differenziale negli iperspazi*, Torino, Sten; appendice al *Trattato di Geometria proiettiva differenziale di G. Fubini ed E. Čech*, Bologna, Zanichelli, 1926] An account of many results of projective differential geometry obtained by Corrado Segre,

Bompiani and myself. It was published, in another edition, as an Appendix in the Treatise of projective differential geometry of Fubini and Čech.

38 [*Corrado Segre*, Jahresbericht der Deutschen Mathematiker Vereinigung, XXXV, 1926] A detailed account of the geometrical work of Corrado Segre (upon invitation of the Deutsche Mathematiker Vereinigung).

39 [*Sul significato geometrico della normale proiettiva*, Rendiconti della R. Accademia dei Lincei, s. 6, III, 1926] A geometrical interpretation of the projective normal.

40 [*Sull'elemento lineare proiettivo di una superficie*, Rendiconti della R. Accademia dei Lincei, s. 6, IV, 1926] In Fubini's theory of surfaces, based on the systematical use of differential forms, the projective linear element of a surface, F_3/F_2, remained the most important differential form, but only from the analytical point of view, not having till then any geometrical signification. I have found that $(F_3/F_2)^2$, except for a constant factor, is the cross ratio of the asymptotic tangents in two "consecutive" points x, x' (on the intersection of the tangent planes at x, x').

41 [*Caratterizzazione dei sistemi del Bianchi di ∞^1 superficie*, Rendiconti della R. Accademia dei Lincei, s. 6, IV, 1926] In Bianchi's theorem of permutability for the asymptotic transformations of surfaces there are two ∞^1 systems of surfaces which play the most important part. I have found the distinctive characters of those systems.

42 [*Un'osservazione sugli invarianti di un'equazione di Laplace*, Bollettino dell'UMI, IV, 1927] A simple geometrical interpretation of the well known invariants of a Laplace equation.

43-46 [*Sulla teoria delle congruenze* W, Rendiconti del R. Istituto Lombardo di Scienze e Lettere, LX, 1927; *Nuove ricerche sulle congruenze* W, Atti del R. Istituto Veneto di Scienze, Lettere ed Arti, LXXXVII, 1927-28] A new systematizing of the theory of the W congruences. Fubini's theory is powerful, but in many points its geometrical signification remained concealed. On the other hand, Tzitzeica-Ribaucour's theory is geometrically perspicuous, but limited in its capacity. I have tried constructing a new theory combining the advantages of both. Besides, some relations between a Laplace equation and its adjoint through integro-differential transformations.

44 [*Sulle superficie aventi un sistema o entrambi di asintotiche in complessi lineari*, appendice al *Trattato di Geometria proiettiva differenziale di G. Fubini ed E. Čech*, Bologna, Zanichelli, 1926] An account (printed as an Appendix in Fubini and Čech's Treatise of projective differential geometry) of my former results about surfaces having one or both systems of asymptotic lines belonging to linear complexes.

45 [*Sulla geometria proiettiva differenziale delle ipersuperficie*, Rendiconti della R. Accademia dei Lincei, s. 6, VI, 1927] Some differential forms relating to the theory of primals.

47 [*Sulle superficie coniugate a un complesso quadratico*, Atti del R. Istituto Veneto di Scienze, Lettere ed Arti, LXXXVII, 1927-28] A first contribution to the general theory of surfaces being "conjugate to a quadratic complex." The definition of such surfaces traces its origin back to Sophus Lie; but - except for a Memoir of Rouyer - only particular cases had been considered by him and others. I have found also a theory of the transformations of those surfaces (which has some relations with the transformation of the surfaces applicable to a quadric).

48 [*Un nuovo problema di geometria proiettiva differenziale*, in *Atti del Congresso Internazionale dei Matematici*, Bologna, 1928] The ordinary projective applicability of two surfaces applies only to a restricted class of surfaces. An attempt of substituting it by a new kind of applicability (pseudo-applicability).

49 [*La quasi-applicabilità proiettiva di una superficie sul piano*, Rendiconti della R. Accademia dei Lincei, s. 6, XI, 1930] The surfaces being pseudo-applicable (s. 48) on a plane are (except for trivial cases) the so called R surfaces.

51 [*Su alcuni sistemi ∞^3 di rette dello spazio a quattro dimensioni*, Rendiconti del R. Istituto Lombardo di Scienze e Lettere, LXIV, 1931] A system of ∞^3 lines in the space S_4 has three focal loci V_3. If these three loci are each generated by planes, when do these planes correspond? A general solution in finite terms.

50-52-54 [*Su una classe di curve dello spazio a quattro dimensioni*, Bollettino dell'UMI, X, 1931; *Sur la réductibilité de certaines correspondances algébriques*, Comptes rendus de l'Académie des Sciences de Paris, 193, 1931; *Sulla riducibilità di alcune particolari corrispondenze algebriche*, Rendiconti del Circolo Matematico di Palermo, LVI, 1932] The question if there is any algebraic twisted curve whose tangent lines meet again the curve has been the origin of my research (this question is not answered even now, as far as I know). In 54 I have brought a first contribution to the following problem including the former. Let us consider the correspondence which arises on an algebraic curve in the space S_r, if we let correspond every point P of the curve with the residual intersections of the curve with the osculating S_{r-1} at P: when such a correspondence happens to be reducible? 52 is a previous communication on the same subject to the Académie des Sciences of Paris. 50 deals with a class of curves in the space S_4 which occurred in 54.

53 [*Lettera del prof. Terracini al prof. Burgatti*, Bollettino dell'UMI, XI, 1932] A small contribution to a theorem stating the reality of the roots of a class of polynomials.

56 [*Sullo scarto dalla normalità delle congruenze rettilinee*, Bollettino dell'UMI, XII, 1933] I have defined a manner of measuring the "deviation" of general rectilinear congruences from normal congruences.

57 [*La lunghezza proiettiva di un arco di curva piana*, Periodico di Matematiche, XIII, 1933] An account on the projective length of a plane curve.

58 [*Su alcuni elementi lineari proiettivi*, Annali della R. Scuola Normale Superiore di Pisa, Bologna, 1933] This Memoir deals with two subjects. Fubini had defined the projective linear element of a rectilinear congruence. But, as I showed, its preservation is not a necessary and sufficient condition that two congruences are projectively applicable. I have found a different projective linear element which has no longer this inconvenience and which has a very simple geometrical interpretation. Researches on the figure constituted by a surface and a congruence having this surface as a surface of reference.

59-60 [*Sulle congruenze associate rispetto a una superficie*, Rendiconti della R. Accademia dei Lincei, s. 6, XVIII, 1933; *Sulle congruenze di rette più volte associabili rispetto ad una superficie*, Rendiconti della R. Accademia dei Lincei, s. 6, XVIII, 1933] In the former paper the notion of "associability" of two congruences had presented itself. Further researches on this notion.

61 [*Osservazioni sulla geometria proiettiva differenziale delle congruenze di rette*, Atti del R. Istituto Veneto di Scienze, Lettere ed Arti, XCIX, 1934] I distinguish between two different species of projective applicability of two congruences, and give a set of conditions for both.

62 [*Le origini dei primi concetti della geometria differenziale*, Periodico di Matematiche, XV, 1935] A lecture on the origin of the first notions of differential geometry: besides well known works, I employed some less improved sources.

63 [*Un procedimento per la risoluzione numerica dei sistemi di equazioni lineari*, Ricerche di Ingegneria, anno III, 1935] The numerical resolution of a system of algebraic linear equations is always practically complicated if the number of unknowns is a large one. I propose the use of "simultaneous elimination" which I have found very suitable, above all with the use of calculating machines.

64 [*Sul criterio di Plücker-Clebsch*, Rendiconti della R. Accademia dei Lincei, s. 6, XXI, 1935] A contribution to the foundation of the so-called Plücker-Clebsch's criterion.

65 [*Sulle linee proiettive di una superficie*, Rendiconti della R. Accademia dei Lincei, s. 6, XXII, 1935] A remarkable system of ∞^3 curves lying on a surface had just been found by Beniamino Segre: they are the "projective lines" of the surface. I have found these lines from another point of view.

66. [*Sulla deformabilità proiettiva delle congruenze rettilinee*, Rendiconti della R. Accademia dei Lincei, s. 6, XXII, 1935] It is known (Cartan) that the degree of generality of the projectively deformable congruences is the same as for a function of two variables. But very little is known about such congruences. I have succeeded in writing the conditions in order that a congruence is deformable in a relatively simple form.

68-69 [*Densità di una corrispondenza di tipo dualistico ed estensione dell'invariante di Mehmke-Segre*, Atti della R. Accademia delle Scienze di Torino, LXXI, 1935-36; *Invariante di Mehmke-Segre generalizzato e applicazione alle congruenze di rette*, Bollettino dell'UMI, XV, 1936] Tricomi had defined the "density" of a correspondence between points of a space S_3 and planes of another space, which is intended to measure, so to say, how close to each other are planes corresponding to points lying in the neighbourhood of a given point. But he had confined himself to the analytical expression of the density. I have found its geometrical significance, and also a relation with the notion of the total curvature. Moreover, as the metrical notion of curvature leads to the projective invariant of Mehmke-Segre, so two correspondences as before mentioned give rise to a projective invariant. (68). An application to rectilinear congruences (69).

67-70-74 [*Sulle varietà luoghi di ∞^1 spazi*, Rendiconti della R. Accademia dei Lincei, s. 6, XXIII, 1936; *Sull'incidenza di spazi infinitamente vicini*, in *Scritti matematici offerti a Luigi Berzolari*, Pavia, 1936; *Sui sistemi semplicemente infiniti di piani nello spazio a cinque dimensioni*, Atti della R. Accademia delle Scienze di Torino, LXXIII, 1938] Two "consecutive" generators of a developable ruled surface are intersecting. When, instead of ∞^1 lines, ∞^1 S_k ($k > 1$) are considered whose two "consecutive" always have a common point, this may be true in many degrees of approximation σ, which I defined and studied in 70. This Memoir is intended, so to say, to discover various complicated things which may take place in the infinitesimal world which play a part in the mentioned intersecting. The first case presenting an effective interest is that of ∞^1 planes lying in S_5: the number σ must be even and its possible values are 2, 4, 6, ... 14, 16, unless $\sigma = \infty$. In every case I give necessary and sufficient conditions, both analytical and geometrical (70). 67 is a previous communication. 74 deals with the geometrical interpretation of some analytical processes which I had employed in 70. In the same 74, moreover, I have found what are the surfaces of the space S_5 having a family of plane lines being multiple as principal lines of the surface (s. below).

71-72 [*Su una possibile particolarità delle linee principali di una superficie. Nota I e Nota II*, Rendiconti della R. Accademia dei Lincei, s. 6, XXVI, 1937] Blaschke has found an unexpected relation between topological differential Geometry and projective differential Geometry. His "5-Gewebe" lead to a special class of surfaces in the space S_5: Bol had indicated an example of such surfaces. The question remained open, whether this was the only example. I have found that there are many other surfaces in the same conditions.

73 [*Superficie particolari dello spazio a cinque dimensioni in relazione con le loro linee principali*, Annali di Matematica pura e applicata, s. 4, XVII, 1938] As well known, on a surface of the space S_5 the most important lines are the principal lines (constituting 5 systems). Their importance was increased of late by Blaschke's discovering their relations with Topology (s. 71, 72). But they interfere also with the theory of intersecting of "consecutive" planes in the space S_5 (s. 70). In 73 I have studied many particular circumstances which may affect the principal lines, especially from the last point of view.

75-76 [*Nuove ricerche sull'incidenza di piani infinitamente vicini*, Atti della R. Accademia delle Scienze di Torino, LXXIII, 1938; *I sistemi* Θ *di piani con congruenza sostegno parabolica*, Atti del R. Istituto Veneto di Scienze, Lettere ed Arti, XCII, 1938] Many years ago systems of ∞^2 (or more) planes whose two "consecutive" planes always intersect had been considered (s. 6). Having now at my disposal the notion of the order of approximation in which two "consecutive" planes may intersect, the problem has arisen to put in relation the old theory with the new notion. I could give a complete enumeration of the systems of ∞^2 or more planes whose consecutive planes always intersect in an order of approximation greater than usual. Projective applicabilities of a higher order play also a part: isothermal-asymptotic surfaces of S_3 appear from a new point of view.

77-79 [*Sur l'existence de surfaces ayant des lignes principals données*, Comptes rendus de l'Académie des Sciences de Paris, 1939; *On the existence of surfaces whose principal curves are given*, in the press] The interest which has again risen about principal lines and our relatively scarce knowledges about them make it opportune to sound more deeply their theory. For instance, as Blaschke points out in his new book *Geometrie der Gewebe* [W. Blaschke, G. Bol, *Zur Geometrie der Gewebe, Topologische Fragen der Differentialgeometrie*, Berlin, Springer, 1938], it is not yet known whether – the differential equation of the principal lines being arbitrarily given a priori – the existence of a surface having those principal lines may be asserted. I have occupied myself with this problem in these last months and arrived to an affirmative conclusion. I found also that – an arbitrary surface being given – it is always possible to map it on several others with preservation of the principal lines (79). 77 is a previous communication of my results to the Académie des Sciences of Paris [while I am drawing up a longer paper on the subject].

78 [*Sur l'interprétation géométrique des caractéristiques des équations aux dérivées partielles du premier ordre*, Bulletin de la Société Royale des Sciences de Liège, 1939] A partial differential equation of the first order involving the unknown function $z = z(x, y)$ may be interpreted on the Klein's quadric of the space S_r as a ∞^4 system of lines. Such a system gives rise to many interesting properties: for instance, it spontaneously offers an arrangement of its ∞^4 lines in ∞^3 ruled surfaces, which are the images of the characteristics. A simple geometrical definition of the characteristics is so obtained.

80 [Gino Fano e Alessandro Terracini, *Lezioni di geometria analitica e proiettiva*, 1 vol. di pp. VIII-630, con 211 figure, Torino, Paravia, 1930] (by Prof. Fano and myself) The purpose of this book was to assist our students in their introductory studies of analytical and projective geometry.

Terracini.4 A. Terracini to SPSL, Turin 13.3.1939

SPSL, *AT*, fol. 363
The Society for the Protection of Science and Learning – London. Referring to my previous correspondence with you, I let you know that – having been apprised of vacancies of a Lectureship of Mathematics in the University of Aberdeen and of a Chair of Mathematics in the University of Durham, I attempted sending my letters of application for both Offices. I send you (separately) copies of both, to acquaint you with all particulars. If eventually you are aware of other similar vacancies, I should be very grateful to you to let me know something. At the same time, I beg you to note that my address is no longer "corso Francia 19 bis" but "corso Gabriele D'Annunzio 19 bis, Turin (Italy)". I beg you to agree my most sincere thanks and best regards. Yours faithfully Alessandro Terracini

Application for a Lectureship in Aberdeen, 3.3.1939, SPSL, *AT*, fols. 378–379
Mr. H.J. Buthchart, Secretary to the University, Aberdeen. Dear Sir, According to my former letter of February the 17th, being desirous to be considered for the Lectureship of Mathematics in the University of Aberdeen, I direct you the present letter of application for that Office. I have been full Professor (*Professore ordinario*) of Mathematics in the University of Turin, Italy, and have been dismissed last December, being a Jew.

I was born on October the 19th 1889 in Turin.

I have studied in the University of Turin, having as Professors Fano, Peano, D'Ovidio, Fubini, and above all Corrado Segre (Higher Geometry):[42] doctor in Mathematics, July 1911; assistant of Prof. Fano (Projective and Descriptive Geometry) 1911-1919.

Officer during the war from 1916 till 1918.

Assistant Professor (*Professore incaricato*) at the University of Modena 1919-23 (Algebra and Descriptive Geometry), and at the University and Polytechnicum of Turin (Analytical Geometry) 1923-24.

Associate Professor (*Professore straordinario*) of Analytical Geometry at the University of Catania 1925; the same at the University of Turin 1925-28.

Full Professor (*Professore ordinario*) at the University of Turin (Analytical Geometry) 1928-38. Moreover in the years 1919, 1925, 1926, 1928, 1929, 1931, 1932, 1933, 1934, 1935, 1936, 1937, 1938 I taught Higher Geometry, on the following subjects:

---

[42] Gino Fano taught Descriptive and Projective Geometry; Giuseppe Peano (1858–1932) Infinitesimal Calculus; Enrico D'Ovidio (1843–1933) Algebraic Analysis and Analytic Geometry; Guido Fubini (1879–1943) taught Higher Analysis and Corrado Segre (1863–1924) Higher Geometry.

Ruled surfaces;

Metric differential Geometry;

Projective differential Geometry in the hyperspaces;

Algebraic correspondences;

Functions of a complex variable and Abelian integrals;

Geometry on an algebraic curve from the geometrical point of view;

Geometry of the minimal surfaces and connected questions;

Sophus Lie and Luigi Cremona as Geometers;

Theory of groups and the first principles of Topology;

The geometrical work of Corrado Segre;

Selected Topics in differential Geometry;

Algebraic and differential Line-Geometry.

As original production, I have hitherto published 76 Notes or Memoirs on different subjects chiefly of algebraic and especially differential Geometry; further (together with Prof. Fano) a Textbook on analytical and projective Geometry (Turin, Paravia, 1930).[43] You have already received from me a list of my printed publications; I send now a second copy (not inclosed). I send also some of them. I add to the present letter a testimonial of *Prof. Guido Castelnuovo* (Rome) and another of *Prof. Tullio Levi-Civita F.M.R.S.* (Rome). Besides I shall send, as soon as available, a document of the Italian Ministry of Public Instruction stating my having been Professor in the University of Turin; but this will take some weeks. Scientific references about me may be obtained by:

Prof. H.F. Baker, Cambridge (Storey's Way 3);

Prof. W.P. Milne, The University, Leeds.

As for personal references: *Captain Guy Hamilton*, Pine Close, Woodlands, Harrogate even not being directly acquainted with me, has a sister married in Turin (*Mrs. Helen Ruth Corti Hamilton*, via Maria Vittoria 52, Turin) through whom he will be able to give you any information. The same informations you may have also through: *William M^c. Lellan* Esq. (former Director of P.a.W.Mc Lellan Iron a. Steel Works, Glasgow), Bryanston Square London W. or *Sir Maurice Denny* of Denny Bros. Shipbuilders, Dumbarton, who are respectively uncle and cousin to Mrs. Corti Hamilton. If necessary, for general informations about me, you may also ask the Society for the Protection of Science and Learning, Gordon Square 6, London W.C.1. I hope that, though I am a foreigner,

---

[43] G. Fano, A. Terracini, *Lezioni di geometria analitica e proiettiva*, Torino, Paravia, 1930.

the present application may be taken in benevolent consideration, and beg you to agree my best regards. Yours faithfully Alessandro Terracini

Testimonial by T. Levi-Civita, Rome 28.2.1939

SPSL, *AT*, fols. 381–382

To the Secretary of the Faculty of Science - Aberdeen. I am not a specialist in algebraic geometry. I have had however many occasions of perusing and appreciating several papers of Professor Alessandro Terracini especially his interesting contributions to differential geometry and to the theory of differential equations. He has wide, thorough and well harmonized knowledges also in analysis and in those selected chapters both of geometry and analysis which have the most fruitful connexion with elementary mathematics from a higher standpoint. Furthermore, I am aware that Professor Terracini is an excellent and brilliant teacher. I cordially recommend him for a chair (or eventually lecturer-ship) of pure mathematics in the illustrious University of Aberdeen. Sincerely yours Tullio Levi-Civita Former Professor of Rational Mechanics in the University of Rome, For. Mem. Royal Society

Testimonial by G. Castelnuovo, Rome 1.3.1939

SPSL, *AT*, fol. 366

Alessandro Terracini has taught many years, and till last October, at the R. University of Turin, as ordinary Professor of Analytical and Projective Geometry, charged also of courses of Superior Mathematics. Disciple of Corrado Segre, he published many and worthy works, in which are prosecuted the researches of his Master, obtaining important results. The greatest part of these works concern the differential Geometry, and particularly the Projective Differential Geometry, of which Terracini is one of the best representatives; a resuming article of many results obtained by Terracini in this field till 1927 is published as Appendix of the second book of the Differential Projective Geometry of Fubini and Čech.[44] Some other very interesting works of Terracini belong to the Algebraic Geometry and other ones to the Theory of the Algebraic Forms. Scientist of large culture, Terracini can teach any branch of the Mathematics. He is a clear and efficacious teacher, has a very good character, and enjoys the friendship and the affection of his colleagues and of his numerous disciples. I think, so, that it is highly

---

[44] A. Terracini, *Esposizione di alcuni risultati di geometria proiettiva differenziale negli iperspazi*, Appendix to G. Fubini, E. Čech, *Geometria proiettiva differenziale*, 2 vols., Bologna, Zanichelli, 1926–1927.

recommendable the choice of Terracini as Professor or Lecturer of any branch of the Mathematics. Prof. Guido Castelnuovo of the R. University of Rome

Testimonial by T. Levi-Civita, Rome 6.3.1939

SPSL, *AT*, fol. 367

To the Registrar of the University of Durham. The mathematical production of Professor Terracini seems to me very fertile, ingenious, and remarkably manifold. He has mainly cultivated geometry in the widest meaning of this term, especially algebraic geometry, but also a great deal of questions belonging to differential geometry, both metric and projective, and to the theory of partial differential equations, which draw from algebraic-geometric thinking and methods elegance and simplicity. Professor Terracini possesses large knowledges not only in general analysis, but also in those selected chapters, which have the most fruitful connexion with elementary mathematics from a higher standpoint. Furthermore, I am aware that he is an excellent and brilliant teacher. Therefore, I very cordially recommend him for the chair of mathematics which he applies. Sincerely yours Tullio Levi-Civita Former Professor of Rational Mechanics in the University of Rome, For. Mem. of the Royal Society

Testimonial by G. Fano, Turin 8.3.1939

SPSL, *AT*, fol. 368

Prof. Alessandro Terracini was a pupil of Corrado Segre and of myself in Turin 1907-11. He was afterwards my assistant 1911-19; and was in military service, as officer in the Engineers during the whole period of the war 1915-18. Since 1925 he has been my colleague at the University of Turin, and was always much appreciated as a teacher by all colleagues and pupils. He was recently dismissed, as belonging to the Jewish race. As a mathematician he worked principally in Geometry, and has a very important place among Italian Geometers. The most of his Memoirs concern projective differential geometry, with analytical methods: he should be therefore the right man for an University chair in which Analysis and Geometry should have both their part. The University which would resolve to give a chair to Prof. Alessandro Terracini should make - no doubt - a very good choice, and be in the future very satisfied of it. Gino Fano Professore at the University of Turin 1901-38

Terracini.5 A. Terracini to ECADFS, Turin 1.5.1939

ECADFS Records, *AT*, fol. 1r

The Emergency Committee in Aid of Displaced Foreign Scholars.
Prof. P. Sperry, of the University of California (Berkeley, Calif.) informed you about my case in last February. I think you have received my curriculum through the same Prof. Sperry: it contains some indications about my Professorship here before my dismissal. I hear now by Prof. Arnold Reichenberger[45] that some further indications may be useful. Therefore, I send you (not enclosed) a list of my printed papers and also (enclosed) a testimonial by Prof. G. Castelnuovo, of the University of Rome (now in retirement). As to references about me, I think you may have them by Prof. Oswald Veblen (The Institute for Advanced Study, School of Mathematics, Fine Hall, University of Princeton, N.J.), Prof. G. Fubini (Fine Hall, Princeton, N.J.); Prof. Pauline Sperry (The University of California, Berkeley, California); Prof. E.B. Stouffer,[46] University of Kansas (Lawrence); Prof. J.L. Coolidge[47] of the Harvard University, Lowell House, 50 Holyoke St., Cambridge Mass. I beg you to note my new address "corso Gabriele D'Annunzio 19 bis Turin (Italy)." With best thanks, Sincerely yours Alessandro Terracini

Enclosed 1 testimonial. If required, I may send you some other testimonials.

Terracini.6 S. Lefschetz to ECADFS, Princeton 7.7.1939

ECADFS Records, *AT*, fol. 1r

My dear Sirs: The attached material is the curriculum vitae of Professor Alessandro Terracini, formerly Professor at the University of Turin. Professor Terracini is one of the outstanding geometers of Italy and likewise very well-known abroad. He has worked in all phases of this difficult speciality which has very considerable contact with the theory of relativity. While I am not personally acquainted with Professor Terracini, I understand from various sources that he was a very good teacher and that in general, his personality leaves nothing to be desired. I therefore strongly urge that your Committee consider doing something for this worthy scientist. Yours very sincerely, S. Lefschetz

---

[45] Arnold Gottfried Reichenberger (1903–1977), reader in German at the University of Milan (1934–1938). In 1939–1940 taught at the New School for Social Research in New York City.

[46] Ellis Bagley Stouffer (1884–1965), professor at the University of Kansas (in Lawrence) from 1914, and Head of the Graduate School (1921–1955). Stouffer had spent long study trips in Italy in 1926 and 1928.

[47] Julian Coolidge (1873–1954), chairman of the Mathematics Department at Harvard (1927–1940). Coolidge had completed his mathematical training in Turin under the guidance of C. Segre in 1904.

Terracini.7 O. Veblen to A. Terracini, [Princeton] 20.11.1939

OVP, *AT*, fol. 1r
Dear Professor Terracini: I was very pleased to receive a notice which assures me that you have found a new academic home. I had been regretting that my own efforts in this direction on your behalf had not been successful. Possibly you may now be in a position to help other victims of the disorderly state of affairs in Europe. In particular, I have in mind one of the most brilliant younger French mathematicians. If you think that there would be any use in my doing so I should be glad to write you further about him. Yours sincerely, Oswald Veblen

Terracini.8 A. Terracini to MEN, Tucumán 9.5.1944

ACS-*Terracini*, fol. 1r
I, the undersigned, Alessandro Terracini appointed full professor in the RR. Italian Universities from 16 February 1925, promoted to full professor on 16 February 1928, provided uninterrupted service in this quality (as professor of Analytic Geometry at the Royal University of Catania from 15-2-1925 to 1-12-1925, and as professor of Analytic Geometry, then of "Analytic Geometry with elements of projective and descriptive geometry" in the R. University of Turin from 1-12-1925 to the date of my exemption from the service) until 14 December 1938, the date on which, being a Jew, I was exempted from the service pursuant to the royal decrees law 17-11-1938, no. 1728 and 15-11-1938 no. 1779. I was beneficiary of pension no. 3405790 for the role, granted by ministerial decree of 29 August 1939, no. 9323. I received (with letter dated 4 May 1944, no. 97 of the royal vice consulate in Tucumán) an invitation to present myself to the aforementioned vice consulate to take note of the decree dated 20 January 1944 concerning the abrogation of the racial laws and the modalities of the applications for re-admission to roles of Public Administration. I was informed of the aforementioned decree from this royal deputy consulate (sent by circular telegram no. 2 of the Ministry of Foreign Affairs) and of the aforementioned procedures, according to the provisions issued, by the royal embassy in Buenos Aires. Under such considerations, I ask this honourable Ministry that I be re-employed in the organic roles as full professor in RR. Universities, with due seniority. Not possessing a copy of the decree of dispensation from service, as required in the provisions of the royal embassy to prove the reasons for my exemption (which, according to a note written by myself, should be a ministerial decree dated 30-11-1938), I enclose in replacement a photographic copy of the record issued to me by the Ministry of National Education on March 16, 1939, proving among other things the cause of aforesaid exemption (Attachments B, C); I also enclose a two-page photographic copy of my pension certificate (Attachments D, E). I enclose a declaration signed by me to

comply with the orders of the Royal Government, according to the text accompanying the provisions of the royal embassy (Annex A). With high esteem, Alessandro Terracini

Io sottoscritto, Alessandro Terracini, nominato professore di ruolo nelle RR. Università italiane dal 16 febbraio 1925, promosso ordinario il 16 febbraio 1928, avendo prestato in tale qualità servizio ininterrotto (come professore di Geometria Analitica nella R. Università di Catania dal 15-2-1925 al 1-12-1925, e come professore di Geometria Analitica, poi di "Geometria Analitica con elementi di geometria proiettiva e descrittiva" nella R. Università di Torino dal 1-12-1925 alla data della dispensa dal servizio) fino al 14 dicembre 1938, data in cui per la mia qualità di ebreo fui dispensato dal servizio ai sensi dei RR.DD.LL. 17-11-1938, n. 1728 e 15-11-1938 n. 1779; fruente di pensione n. 3405790 del ruolo concessa con Decreto ministeriale del 29 Agosto 1939, n. 9323; avendo ricevuto (con lettera in data 4 Maggio 1944, n. 97 del R. Vice Consolato in Tucumán) invito a presentarmi al predetto Vice Consolato per prendere conoscenza del Decreto in data 20 gennaio u.s. concernente l'abrogazione delle leggi razziali e delle modalità delle domande per la riammissione nei ruoli della Pubblica Amministrazione; avendo preso conoscenza presso questo R. Vice Consolato del predetto Decreto (trasmesso con telegramma circolare n. 2 del Ministero Affari Esteri) e delle menzionate modalità, secondo le disposizioni emanate dalla R. Ambasciata in Buenos Aires; rivolgo domanda a cotesto onorevole Ministero per essere riassunto nei ruoli organici come professore ordinario nelle RR. Università, con la dovuta anzianità. Non possedendo copia del Decreto di dispensa dal servizio, come richiesto nelle disposizioni della R. Ambasciata per comprovare le ragioni del mio esonero (il quale decreto secondo una annotazione in mio potere dovrebbe essere un decreto ministeriale 30-11-1938), accludo in sostituzione di esso copia fotografica dello stato di servizio rilasciatomi dal Ministero dell'Educazione Nazionale in data 16 marzo 1939, comprovante tra l'altro la causa di detto esonero (Allegati B, C); nonché copia fotografica di due pagine del mio libretto di pensione (Allegati D, E). Accludo dichiarazione da me firmata di ottemperare agli ordini del Regio Governo, secondo il testo che accompagna le disposizioni della R. Ambasciata (Allegato A). Con distinta stima Alessandro Terracini

Terracini.9 A. Terracini to A. Einstein, Tucumán 6.8.1944

AEA, *AT*, fol. 1r-v
Much-respected Maestro, remembering the kindness that you showed me by sending me your contribution *Demonstration of the non-existence of gravitational fields with a non-vanishing total mass free of singularities* three years ago for the *Revista de Matemáticas y Física Teórica*, I allow myself to ask for your

authoritative advice on a matter that is of fundamental importance to me.[48] In these months, some contracts of this university with foreign professors have been cancelled (e.g., that of my brother [Benvenuto], previously professor of Linguistics at the Universities of Padua and Milan). My contract (which was signed in 1939, when I came here after leaving Italy) has not been touched yet: however, I cannot exclude that my turn will come also, and I would not like to find myself at the last moment without having any prospects. For this reason, it would be of vital importance for me to know if, in the event of the aforementioned eventuality, the probability that I would find a job, or any position to continue my mathematical work in the United States, could arise. If this be absolutely not possible, I would like to know if there could be the possibility that an institution in the United States would grant me a scholarship to continue my mathematical research in a South American University, such as that of Montevideo. I suppose that later I will be able to return to my chair in Italy; but at this moment it seems difficult to predict when this may happen; so, not having own resources, I have to concern myself with finding a solution that allows me to live with my family (wife and three children aged 19, 17 and 13) in the event that I cannot continue with my teaching here.[49] Please, allow me to briefly indicate the following information: born in Turin 19.10.1938; doctor of Mathematics, Turin 1911; Assistant at the University of Turin 1911-19; professor (assistant, then extraordinary, finally full professor) at the Universities of Modena, Catania and Turin 1919-38: at the University of Turin I taught Analytical Geometry (also at the Engineering School), and Higher Geometry from 1925 to 1938, when I had to leave my professorship due to the anti-Semitic laws. My works (about 90) mainly concern differential geometry (in particular, differential projective geometry and algebraic geometry): until 1938 they were mostly published in Italian journals (*Rend. Lincei, Atti Acc. Torino, Rend Circ. Mat. di Palermo, Ann. di Matematica*, etc.) and since 1939 in Argentine journals. I have published together with G. Fano the book *Lezioni di geometria analitica e proiettiva* and two of the Appendices of the treatise *Geometria proiettiva differenziale* by Fubini and Čech.[50] I beg you, illustrious *Maestro*, please forgive me for the inconvenience that I take the liberty of giving you. I am henceforth extremely grateful to you for any suggestion and support on your part. Please accept my cordial greetings, Alejandro Terracini

---

[48] A. Einstein, *Demostración de la no existencia de campos gravitacionales sin singularidades de masa total no nula*, Demonstration of the non-existence of gravitational fields with a non-vanishing total mass free of singularities, Revista. Serie A. Matemáticas y Física Teórica, II, 1941, pp. 5–15.

[49] Giulia Sacerdote (1899–1974), Lore (1925–1995), Cesare (1927–1972), and Benedetto (1931).

[50] G. Fano, A. Terracini, *Lezioni di geometria analitica e proiettiva*, Torino, Paravia, 1930; A. Terracini, *Esposizione di alcuni risultati di geometria proiettiva differenziale negli iperspazi*, Appendix to G. Fubini, E. Čech, *Geometria proiettiva differenziale*, Bologna, Zanichelli, 2 vols., 1926–1927.

Muy apreciado Maestro, recordando la bondad que Ud. demostró conmigo al enviarme hace tres años su colaboración Demonstration of the non-existence of gravitational fields with a non-vanishing total mass free of singularities *para la* Revista de Matemáticas y Física Teórica, *por mí dirigida, me permite dirigirme a Ud. para pedirle su autorizado consejo en un asunto que tiene para mí importancia fundamental. En estos meses han sido anulados algunos contratos de esta universidad con profesores extranjeros (p. e. con mi hermano, anteriormente profesor de lingüística en las universidades de Padua y Milan). Mi contrato (que rige desde 1939, cuando yo vine aquí al dejar Italia) hasta este momento no ha sido tocado: sin embargo, no puedo excluir que llegue también mi turno, y no quisiera encontrarme al último momento sin tener ninguna perspectiva. Por tal razón tendría para mí importancia básica saber si en el caso de producirse la aludida eventualidad, podría presentarse la probabilidad que yo encuentre una colocación, o cualquier manera de continuar mi trabajo matemático en Estados Unidos. Si esto no fuera absolutamente posible quisiera saber si existiría la posibilidad de que alguna institución de Estados Unidos me otorgara una beca para continuar mis investigaciones matemáticas en alguna Universidad sudamericana, como podría ser p. e. la de Montevideo. Yo supongo que más adelante podré volver a ocupar mi cátedra en Italia; pero en este momento me parece difícil prever cuándo esto podrá ocurrir; de manera que, no teniendo recursos propios, debo preocuparme para encontrar una solución que me permita vivir, con mi familia (esposa y tres hijos de 19, 17 j 13 años) si se presentara el caso de no poder continuar con mi enseñanza aquí. Me permito indicarle brevemente los siguientes datos: nacido en Turin 19.10.1889; doctor en Matemática Turin 1911; Asistente en la Universidad de Turin 1911-19; profesor (encargado, extraordinario, y ordinario) en las Universidades de Modena, Catania y Turin 1919-38: en la Universidad de Turin he enseñado Geometría Analítica (también en la escuela de ingenieros) y Geometría Superior desde 1925 hasta 1938, cuando tuve que dejar mi cátedra por les leyes antisemíticas. Mis trabajos (unos 90) conciernen principalmente la geometría diferencial (en particular geometría proyectiva diferencial y la geometría algebraica): hasta 1938 están publicados en su mayoría en revistas italianas* (Rend. Lincei, Atti Acc. Torino, Rend. Circ. Mat. di Palermo, Ann. di Matematica, ecc.) *y desde 1939 en revistas argentinas. He publicado junto con G. Fano el libro* Lezioni di geometria analitica e proiettiva *y dos de los Apéndices de la obra* Geometria proiettiva differenziale *de Fubini y Čech. Ruégale, ilustre Maestro, tenga Ud. a bien disculparme la molestia que tomo la libertad de darle. Le estoy desde ahora sumamente agradecido por cualquier sugerencia y apoyo por su parte. Reciba mis cordiales saludos, Alejandro Terracini*

Terracini.10 Terracini Family to A. Sacerdote, Tucumán 6.8.1944

ATCET, fol. 1r, 2r

Dearest Aldo, we continue living in absolute silence from you and I am quite worried about it.[51] I can't convince myself that you didn't find a way to let us hear from you after 17 May, and especially that you didn't send us a message from Rome, where you could certainly have gone after 4 June, by Vatican Radio, as Giulia and Dario did![52] Sandro says that communications with Rome may have been very difficult, but now two months have passed, and the difficulties will have diminished. I look forward to hearing about your meeting with Alma and Micky and hearing from them.[53] I wrote to Alma's father in Rome, I wrote to Giulia, to Angelo [Treves], to Valentina and we hope to get an answer from someone. In the meantime, once again we are writing to you too, but this irregular correspondence is really unnerving. While I am writing this letter, Lore, who turns 19 today, is chatting in the park with her friends.[54] I think a lot about Cicì: where she will be on this day!! And I think of poor Giacomo, who would have turned 42 today.[55] Poor dear Giacomo!! The thought of his children, of Rita, of all our loved ones of whom we do not have any news fills me with sadness and I would like these months to pass in a flash to have news of them ... But sometimes I'm so afraid to know!! Nothing new about us: our future is somewhat uncertain. I hope you have received our previous letters and are aware of our desire to return. We do not doubt that you will do everything possible to ensure that Sandro is reinstated in his post, indeed we thank you for what you have already written to us. Unfortunately, however, this will take a long time. Our hearts are already fully set on the return. The children have a lot of plans, and we hope they won't all be messed up!! [...] The boys, as usual, are busy with their studies after the July holidays. Sandro continues his three courses, and he has enough to do. We follow the progress of the war with the interest and enthusiasm that you can well imagine; and we are very confident of a good end in the near future. Dear Aldo, write to us soon and at length. With Alma and Micky, I embrace you with all my heart. Giulia [...]

Dear Aldo, here with me, as I write to you, there is also Antonio, who has come to visit us after a long absence, and he sends you his greetings. He has a mania for rebus, as he once had; but regardless of this otherwise harmless passion, he is still a great boy. We had not seen him anymore because he's in the country, very far from here: he's very well. Giulia has already told you about us and our anxiety

---

[51] Aldo Sacerdote (1897–1979), engineer, was Alessandro's brother-in-law. See also D7.2.1.

[52] Giulia Vitale Dalmonte and Dario Dalmonte. Giulia Vitale's mother was first cousin of Giulia Sacerdote Terracini.

[53] Alma Belloni Sacerdote (1905–1986) and Micaela Sacerdote (born in 1939), wife and daughter of Aldo.

[54] Lore Terracini (1925–1995), Alessandro's first daughter.

[55] Giacomo Sacerdote (1902–1940), brother of Giulia, and Alessandro's brother-in-law.

about not having yet heard of your reunion, in June, I believe, with the rest of your family. We hope to hear from you soon. We continue as usual, I do my lessons, the boys their studies. [...] A few days ago I wrote to old Castelnuovo, who I hope is well after a long time since I have heard from him: I think you would do well to get in touch with him to hear what you can do to make my call to some Faculty effective. In the meantime, we have seen in the newspapers about the reintegration in the liberated part of Italy, as of 1 August as far as Rome is concerned, of primary teachers and other "professors": this is what the news said. We get Italian news mainly through the newspaper *Italia Libre* that comes out in Buenos Aires: it was a weekly journal until last year and now it is a daily publication. [...] That's enough, dear Aldo and all of you: I recommend that you do what you can. An affectionate kiss to Micaela and to you an equally affectionate hug from Sandro.

P.S. Finally it seems that the new volume of my mathematical journal is coming down.[56] I beg you to take the necessary steps so that our foregone pensions do not become time-barred (mine as a professor; Benvenuto's pension as a professor and the additional sum as war privilege and medal surcharge) as well as the famous arrears. Benvenuto's war pension has not been paid since June 1941; the civilian pensions from June 1943.

Carissimo Aldo, continuiamo nel più assoluto silenzio da te e ne sono alquanto preoccupata. Non riesco a convincermi che tu non abbia trovato il modo di farci avere tue notizie dopo il 17 maggio e specialmente che da Roma, dove certamente avrai potuto recarti dopo il 4 giugno, non ci abbia mandato un messaggio a mezzo Radio Vaticana come hanno fatto Giulia e Dario! Sandro dice che le comunicazioni con Roma saranno state difficilissime, ma ormai sono passati due mesi e le difficoltà saranno diminuite. Sono ansiosissima di sapere del tuo incontro con Alma e Micky e di notizie loro. Ho scritto al papà di Alma a Roma, ho scritto a Giulia, a Angelo, a Valentina e speriamo da qualcuno di avere risposta. Intanto ancora una volta scriviamo anche a te, ma questa corrispondenza così irregolare snerva veramente. Mentre scrivo Lore, che compie oggi i suoi 19 anni, chiacchiera nel parco con le sue amiche, penso tanto a Cicì e dove sarà in questo giorno!! e penso al povero Giacomo che avrebbe compiuto oggi appena 42 anni. Povero caro Giacomo!! Il pensiero dei suoi bimbi, di Rita, di tutti i nostri cari di cui non sappiamo mi riempie di tristezza e vorrei che questi mesi passassero in un baleno per avere loro notizie ... ma alle volte ho tanta paura di sapere!! Di noi nulla di nuovo: il nostro avvenire è alquanto incerto. Spero che tu avrai ricevuto le nostre precedenti lettere e sarai al corrente del nostro desiderio di ritornare. Non dubitiamo che tu farai tutto il possibile perché Sandro sia reintegrato nel suo posto, anzi ti ringraziamo di quanto ci hai scritto di aver già fatto. Purtroppo però la cosa andrà molto per le

---

[56]The *Revista. Series A: Matemáticas y Física Teórica*, created by Terracini in 1940 and published by the Universidad Nacional de Tucumán.

lunghe. Il nostro animo è già completamente preparato al ritorno. I ragazzi hanno moltissimi progetti e speriamo che non siamo poi tutti delusi!!! [...] I ragazzi, come al solito, occupati con i loro studi dopo le vacanze di luglio. Sandro continua i suoi tre corsi, ed ha abbastanza da fare. Seguiamo l'andamento della guerra con quell'interesse e quell'entusiasmo che tu ben puoi immaginare; e siamo assai fiduciosi in una buona fine prossima. Caro Aldo scrivici presto e a lungo. Con Alma e Micky ti abbraccio con tutto il cuore. Giulia [...]

Caro Aldo, qua con me, mentre ti scrivo, c'è anche Antonio, che è venuto a trovarci dopo tanto tempo che non si faceva vedere, e ti manda i suoi saluti. Ha la mania dei rebus, come aveva una volta; ma prescindendo da tale passione del resto innocua è sempre un gran bravo ragazzo. Non lo avevamo più visto perché lui sta in campagna, e assai lontano da qua: sta benissimo. Di noi e della nostra inquietudine per non avere ancora avuto notizia del tuo ricongiungimento, in giugno, credo, con il resto della tua famigliuola ti ha già detto Giulia. Speriamo tra poco ricevere qualche tua notizia in proposito. Noi continuiamo regolarmente come sempre, io a fare le mie lezioni, i ragazzi ai loro studi. [...] Ho scritto giorni fa al vecchio Castelnuovo, che spero stia bene dopo tanto tempo che non ho più avuto notizie di lui: credo faresti bene a metterti in comunicazione con lui per sentire cosa potete fare così perché si faccia effettiva la mia chiamata a una qualche Facoltà. Intanto abbiamo visto sui giornali di quella reintegrazione nella parte liberata d'Italia a partire dal 1° agosto per quanto concerne Roma, dei maestri e altri "docenti": così diceva la notizia. Le notizie italiane le abbiamo soprattutto per mezzo del giornale Italia Libre *che esce a Buenos Aires: era settimanale fino all'anno passato e ora è giornaliero. [...] Basta, caro Aldo e cari voi: mi raccomando che tu faccia quanto puoi. Un affettuoso bacio a Micaela e a voi un altrettanto affettuoso abbraccio da Sandro.*

P.S. Finalmente pare stia uscendo il nuovo volume della mia rivista matematica. Ti prego [di] fare i passi necessari perché non vadano in prescrizione le somme non esatte delle nostre pensioni (la mia di professore; quella di Benvenuto idem e inoltre la previlegiata di guerra e soprassoldo medaglia) così come i famosi arretrati. La pensione di guerra di Benvenuto non è pagata dal giugno 1941; le civili dal giugno 1943.

Terracini.11 V. Snyder to O. Veblen, Ithaca 22.8.1944

OVP, *AT*, fol. 1r, 2r, 3r
Dear Veblen: Yesterday I received a letter from Professor Alexander Terracini, of Tucumán, Argentine, a crude translation of which I am enclosing.[57] It is marked

---

[57]Terracini had probably written to Snyder in Spanish. However, only the English version of the letter is preserved in the file. Virgil Snyder, professor at Cornell University (1910–1938), had been in close contact with Italian geometers. He had studied in Turin with C. Segre in 1922.

"translation". This matter is so far outside of my ken that I cannot proceed without help. My first step was to write a tentative draft of a reply to Terracini, which I am also enclosing. My plan is to hold this reply until I hear you, and then to incorporate such revisions as you may suggest. I had thought of The Institute for Advanced Study, The Guggenheim Foundation, The Rockefeller Foundation, The Carnegie Foundation, The Pan American Union as these to send to Terracini, but that is still purely tentative. I met Terracini at the Bologna Congress in 1928.[58] He was one of four to be sent to South America by the Italian government in 1939.[59] One of his jobs at Tucumán was to found the *Revista*. Later, he invited me to send a contribution to it, which I did.[60] That is about the extent of my association with him. Any suggestion as to possible later procedure will be gratefully received. Mrs. Snyder joins me in sending cordial greetings to you and to Mrs. Veblen. Sincerely yours, Virgil Snyder

Translation [from Spanish into English]
Tucumán, August 6, 1944

Dear Professor Snyder: Permit me to address you in order that I may have authentic advice concerning a matter of prime importance to me.[61] A number of contracts made by this University with foreign professors are being annulled. My own (which dates from 1939 when I came from Italy) has thus far not been affected. But such steps may be taken at an any time. I do not wish it to happen at the last moment without any further perspective. For that reason, it is of prime importance for me to know, if anything should happen here, of some possible employment, of whatever nature, to enable me to continue my mathematical work in the United States. If this is not possible, is there any institution in the United States from which it might be possible to obtain a fellowship to enable me to continue my mathematical studies at some South American institution, for example, at Montevideo? I suppose that later I may be able to return to Italy and resume my position there, but at present that would be very difficult; I do not see how or when that could occur. I am without means and I must occupy myself to find a solution that will enable me and my family to live, in case it becomes impossible to continue my present position here. I shall be very grateful to you for any suggestions or for such help that you may be able to offer. Let me beg your

---

[58] The International Congress of Mathematicians which had been held in Bologna, from 3 to 10 September 1928.

[59] See Sect. 3.15.

[60] V. Snyder, *Cremona involutions belonging to the Bordiga surface in (4)*, Revista. Serie A. Matemáticas y Física Teórica, 2, 1941, pp. 203–210.

[61] See also TERRACINI.9.

indulgence for this intrusion. Accept my most sincere regards. Alejandro Terracini

First draft of a tentative letter
Ithaca, New York, August 1944

Dear Professor Terracini: A letter from you is always a pleasure but I was shocked and grieved by your last epistle. As no steps had been taken in regard to your position when your letter was written (Aug. 6) let us hope that your association with the Universidad of Tucumán may continue for some time. If, however, it should terminate, let me urge you to make a real effort to find further employment in South America, even of a nature different from that for which your interests and your training most properly fit you. In regard to finding fitting employment in the United States, I am sorry to say that there are no early prospects. During the last few years, the whole scheme of instruction in mathematics in the United States has been greatly confused: many new positions were created on account of the war, and many persons from other countries were provided with temporary positions. This has taxed the available financial sources to such an extent that they are now completely exhausted; further help is simply impossible, although the influx of worthy appeals does not diminish. I am sending you the addresses of some of the agencies that have been active in such cases. No one of them is a probable source of early relief, but it is possible that among them you may succeed in finding some relief. I hope that something worthwhile may come out of the situation. At any rate, permit me to send best wishes, and the assurance of my best personal regards. Sincerely yours, Virgil Snyder

Terracini.12 J. von Neumann to V. Snyder, [Princeton] 26.8.1944

OVP, *AT*, fol. 1r
Dear Professor Snyder: Professor Veblen is absent on a Government mission for about a month. In his absence Mrs. Veblen[62] showed me your letter concerning Professor Terracini and the attached correspondence. Much to my regret I am personally unacquainted with Professor Terracini's work, so I am not at all the proper person to take any positive action in this matter. I did discuss the subject with Professor Lefschetz of Princeton University who, I think, among those now accessible in Princeton is best informed on the subject. His opinion on Terracini's chances in the United States was unfavourable. However, I know from past correspondence and from previous actions connected with Terracini that Veblen, and possibly Weyl, might take positive interest in his case. Both of them are absent from Princeton now, and both are due back here sometime between

---

[62] Elizabeth Richardson Veblen (1882–1974).

September 15 and October 1. It seems to me therefore that the subject of Terracini's letter might best be discussed by us at that time. I may not be in Princeton in the second half of September, but I am sure that Veblen will at that time write to you concerning this. I am sorry that I cannot for the moment contribute anything more constructive than this. Very truly yours, John von Neumann

Terracini.13 O. Veblen to V. Snyder, [Princeton] 7.11.1944

OVP, *AT*, fol. 1r

Dear Snyder: On my return I found the correspondence with regard to Terracini. My impression is that it will be difficult, if not impossible, to find an academic job for him in the United States. I should think, however, that it will not be long before the Italian Government will be wanting to restore some of its professors to academic positions. My suggestion would be that you wait until you hear from Terracini again, and then let me know what he has to say. At that time, we might also communicate with Marshall Stone, who might know how to get some help from the State Department in arranging any diplomatic difficulties which might arise.[63] With best greetings to Mrs. Snyder and yourself, Yours sincerely, Oswald Veblen

Terracini.14 Terracini Family to A. Sacerdote, Tucumán 27.1.1945

ATCET, fol. 1r, 2r

My dearest, the letters of 10 and 16 September from Aldo, that of 10 September from Alma and that of 30 August from Giulia, as well as one from Valentina (6 September) and another from the same of 23 October, arrived all together. I can't tell you how happy I am to see our correspondence reactivated, and to see the time between your letters and our answers diminished!! I begin by replying to Alma, whose letter we have read and re-read with great interest. She so well describes, in a few words, the arrival of Aldo, your meeting after so many months of separation and after all the war passed so near to you, and the appearance of the two New Zealanders, that we were really moved to read it!! I hope you are all gathered in Naples and relatively serene!! [...] We have recently heard on the radio that the trains between Rome and Naples have been reactivated, and this will make it much easier for you to communicate. I hope that the difficulties for food in the city will not bring you discomfort and that you will not have to suffer

---

[63] Marshall Stone (1903–1989), full professor at Harvard from 1937. During World War II, Stone did classified research as part of the Office of Naval Operations and the Office of the Chief of Staff of the U.S. Department of War.

too many difficulties and too many deprivations. You can't imagine how much remorse I have for living in such abundance here!! and how I would heartily wish to send you something essential. From here, until a few days ago, it was not possible to send parcels to Italy, and I had written to Camillo to ask him if he could do that. I have not yet received an answer, but two days ago we were informed that the Red Cross is in charge of sending parcels privately to Italy; so, I immediately instructed Eva, who is in Buenos Aires, to send you something; and I hope it will come to you not too late.[64] [...] I'm writing to Giulia today. I was very sorry for her sad letter! We tried from Italy, and we charged Angelo Treves in the United States to take all steps at the various committees in order to see if it is possible to know something about Ada and the others, but I doubt whether we will achieve any results!! God grant that we will soon be at the end of the war and that we can find them!! These days we are all hopeful for the success of the Russians! Very good news about Tricomi and Castelnuovo!! Maybe they will help us in due time for the return!! [...] As for our return, we strongly consider what you write about the difficulties of life there, but I believe that it is still premature to think about it concretely, although our desire always remains that of resuming our former life as soon as possible. And now I'm going to tell you about us. [...] Again there is nothing new. Sandro works on his journal that is about to be published. Lore gets stuck in Newton's translation, which she is expected to submit to the publisher in March or April.[65] Cesare studies a little for the Physics exam that he will take in March and devotes himself more and more to tennis, swimming, etc. Bene[detto] gets scolded because he is the opposite of his brother and does not want to do gymnastics, he studies a little French and Latin, but he does not get very agitated. I am always very worried about our loved ones. And the mood isn't always very uplifted. I try to be brave and not too pessimistic; but sometimes I'm really tired of this anguished thought, and often I can't take my mind off all this tragedy!! Write to us, write to us. [...] Giulia

Dear friends, it is with great pleasure that we have read all your letters, including that of 6 August, in the last few days. All's well that ends well, and we can only hope that Rita and Olga and our other relatives in the North were also able to escape the persecutions of the fascists and the Germans. Unfortunately, there are

---

[64]Eva Terracini (1914–1986), Benvenuto's daughter.

[65]Newton's translation, for Editorial Yerba Buena of Tucumán, was not completed. Terracini recalled (1968, p. 206): 'My daughter [Lore], my brother Benvenuto, and finally I myself spent many afternoons discussing page by page the original text of Isacco Newton's *Principia Mathematica Philosophiae Naturalis*, of which we intended to carry out a Spanish translation, which did not reduce to a mere translation of the well-known translation by Florian Cajori. [...] After a long time and various fruitless attempts [...] we gave up, and the translation of the *Principia* never came out'. (*Passammo molti pomeriggi mia figlia, mio fratello Benvenuto, e infine io stesso a discutere pagina per pagina l'originale dei* Principia Mathematica Philosophiae Naturalis *di Isacco Newton, dei quali si trattava di eseguire una traduzione diretta in spagnolo, che non si riducesse dunque a una traduzione della nota opera dovuta a Florian Cajori. [...] Dopo parecchio tempo e varii tentativi infruttuosi [...] desistemmo, e la traduzione dei Principia non uscì mai.*)

often reports of acquaintances deported by those scoundrels. Your Vatican message never arrived. Fortunately, we can hope, a little, in Jewish organisations that have saved and continue to save people. I'll tell you quickly about us. Giulia, overall good. I'm fine too, and everyone else fine. I have a lot to do. Aldo [Sacerdote] asks me about our economic conditions: I have (hoping it will last) at my disposal 1060 pieces a month, equal to about 260 dollars at the official exchange rate, but maybe double as purchasing power. Let's hope it lasts. For Benvenuto [Terracini] there are good chances of renewing his contract. And so, I hope that in due time we will be able to renew mine. [...] We have not heard anything about our reinstatement. By the end of December, our colleagues in North America had already been asked about whether they wanted to return. But here, so far, everything is silent, I have no doubt that Aldo will keep up with it. I had no communication from Castelnuovo Amaldi, Tricomi. I am amazed by the silence, especially from the latter. I hope Aldo doesn't work too hard and be wares of further burnouts. I understand that life is difficult, but why not live it? A hug. Sandro

Carissimi, giunsero tutte insieme le lettere del 10 e 16 settembre di Aldo, quella del 10 settembre di Alma e quella del 30 agosto di Giulia, nonché una del 6 settembre di Valentina e altra della stessa del 23 ottobre. Non vi dico quanto mi fa piacere vedere riattivata la nostra corrispondenza, e vedere diminuito l'intervallo di tempo fra quando voi scrivete e quando noi riceviamo!!! Incomincio a rispondere a Alma, la cui lettera abbiamo letta e riletta con moltissimo interesse. In essa è così ben descritto con poche parole l'arrivo di Aldo, e il vostro incontro dopo tanti mesi di separazione, e dopo tutta la guerra passata a così breve distanza da voi e l'apparizione dei due neozelandesi, che ci siamo proprio commossi nel leggerla!! Vi spero ora tutti riuniti a Napoli e relativamente tranquilli!! [...] Abbiamo sentito per radio, in questi giorni, che sono stati riattivati i treni fra Roma e Napoli, e questo renderà molto più facile per voi il poter comunicare. Spero che le difficoltà per il vitto in città non vi portino molto disagio e che non dobbiate avere troppe difficoltà e troppe privazioni. Non potete immaginare quanto rimorso ho io di vivere così nell'abbondanza qui!! e con quanto piacere vorrei potervi fare avere qualche cosa di essenziale. Da qui, fino a qualche giorno fa non si poteva fare spedizioni in Italia di pacchi, e avevo scritto a Camillo per chiedergli se non aveva difficoltà a farvi lui una spedizione. Non ne ho ancora ricevuto risposta, ma da due giorni è venuta la notizia che la Croce Rossa di qui si incarica di spedire pacchi privatamente per l'Italia; così ho subito incaricato Eva, che è a Buenos Aires, che vi mandi subito qualche cosa; e spero che vi giunga non troppo tardi. [...] A Giulia scrivo oggi stesso. La sua lettera tanto triste mi ha fatto proprio molta pena! Abbiamo fatto di qui ed incaricato Angelo Treves negli Stati Uniti, di tutti i passi presso i vari comitati per vedere se sarà possibile sapere qualche cosa di Ada e di tutti, ma temo che si possa giungere a qualche risultato!! Dio voglia che si sia presto alla fine della guerra e che possiamo subito ritrovarli!! In questi giorni siamo tutti speranzosi per i successi dei Russi! Molto bene le notizie riguardanti Tricomi e Castelnuovo!! Chissà che non ci aiutino a suo tempo per il ritorno!! [...] In quanto al nostro ritorno, teniamo in alta considerazione quanto scrivete sulle difficoltà della vita costì, ma credo che sia ancora prematuro pensarci concretamente, sebbene il nostro desiderio rimanga sempre quello di riprendere appena sarà possibile la nostra vita di prima. Ed ora vi dirò di noi. [...] Di nuovo non c'è nulla. Sandro lavora, e si occupa della rivista che è in stampa. Lore ci dà dentro alla traduzione di Newton che dovrebbe consegnare all'editore per marzo o aprile. Cesare studia un poco per l'esame di Fisica che darà in marzo e si dedica sempre più al tennis, al nuoto ecc. Bene[detto] si fa sgridare perché è l'opposto di suo fratello e non vuole fare ginnastica, studia un poco di francese e di latino, ma non si agita molto. Io sono sempre molto in pensiero per i nostri cari. E l'umore non è sempre molto sollevato. Cerco di farmi coraggio e di non essere troppo pessimista; ma qualche volta sono proprio stanca di questo angoscioso pensiero, e spesso non riesco a distogliere il mio pensiero da tutta questa tragedia!! Scriveteci, scriveteci. [...] Giulia

Carissimi, con molto piacere abbiamo letto tutte le vostre lettere e anche quella del 6 agosto, in questi ultimi giorni. Tutto è bene quello che finisce bene, e ci resta da sperare solo che anche Rita e Olga e gli altri nostri parenti del Nord abbiano potuto sfuggire alle persecuzioni dei fascisti e dei tedeschi. Purtroppo spesso giungono notizie di conoscenti deportati da quelle canaglie. Il vostro messaggio vaticano non è mai giunto. Per fortuna si può anche sperare, un poco, in organizzazioni ebraiche che hanno salvato e continuano a salvare gente. Vi dico rapidamente di noi. Giulia complessivamente bene. Io anche benino, e tutti gli altri bene. Ho un lavoro enorme. Aldo mi domanda delle nostre condizioni economiche: ho (sperando la duri) a mia disposizione 1060 pezzi al mese, pari a circa 260 dollari al cambio ufficiale, ma forse al doppio come potenza di acquisto. Speriamo la duri. Per Benvenuto ci sono buone promesse per rifargli il contratto. E così spero che a suo tempo si potrà rifare il mio. [...] Della riammissione in ruolo non abbiamo saputo nulla. A fine dicembre i nostri colleghi in Nord America erano stati già interpellati in proposito se intendevano tornare; ma qua finora tutto tace, Non dubito che Aldo starà dietro alla cosa. Nessuna comunicazione ho avuto da Castelnuovo Amaldi, Tricomi. Mi stupisce il silenzio specie di quest'ultimo. Spero Aldo non lavori troppo e si guardi da altri esaurimenti. Capisco che la vita costì è difficile, ma perché non farla? Un abbraccio. Sandro

Terracini.15 Extract of a Letter from the Terracini Family to A. Sacerdote, Tucumán [January-February 1945]

ATCET, fol. 1r

We are following events in Europe with great interest, and we hope for a very close end to the war. Is that going to be a great day!!! We have sent you two parcels containing food, and I have also sent Camillo some boxes of powdered milk, powdered eggs, butter, etc., so that he can send it to you. I sent another parcel to Giulia and to Giorgio Falco who, poor guy, miraculously escaped capture by the Germans, while he was a refugee in the convent of San Paolo; now he is free in Rome with his sick wife, and he has written to us asking for some food help.[66] He must be in very difficult financial straits. Now it seems that there are other possibilities to send stuff to Italy through North America, through a committee that was formed in Montevideo, while it is increasingly difficult to send things through the Red Cross. I have already written to you that from Buenos Aires we have sent a telegram with prepaid answer to the World Jewish Congress in Geneva so that they can investigate and communicate news of the Vitale family.[67] Hopefully, with the liberation of so many German cities, we might be able to track them!! I think very often of these dear ones, and I hope to know soon that they have been saved. I received a very affectionate letter from Giulia, with messages also from Dario, and I will reply in the next few days. We also had Amadio and Umberto's address in Switzerland and we wrote to them.[68] Aurelia hasn't heard anything about Arrigo. Ermanno is still in Switzerland but without money, because he can't get it sent.[69] Aurelia sends you special greetings and thanks for the good wishes etc. University things are going very well here. The new authorities are very satisfactory, and we hope very well for Benvenuto and for us. The mood is much better than it was a few months ago. But our desire is always to return. A thousand kisses to all.

Seguiamo con vivissimo interesse gli avvenimenti europei e speriamo in una prossimissima fine della guerra. Sarà quello un gran giorno!!!! Vi abbiamo spedito due pacchi contenenti viveri, inoltre ho mandato a Camillo alcune scatole di latte in polvere, uova in polvere, burro ecc. perché ve lo faccia avere. Altro pacco ho spedito a Giulia e altro a Giorgio Falco che, poveretto, sfuggito miracolosamente dalla cattura per parte dei tedeschi, mentre era rifugiato nel convento di San Paolo, è ora libero a Roma con la moglie ammalata, e ci ha scritto chiedendoci qualche aiuto alimentare; deve essere in pessime condizioni finanziarie. Ora pare che ci siano altre possibilità per mandare roba in Italia per mezzo del Nord America, passando per un comitato che si è formato a Montevideo, mentre sempre più difficile riesce mandare per mezzo della Croce Rossa.

---

[66] Giorgio Falco (1888–1966), professor of Medieval History at the University of Turin from 1930 to 1938, when he was dismissed because of the racial laws. On the night between 3 and 4 February 1944, he was able to escape one of the last German raids in the monastery of S. Paolo fuori le Mura (Rome).

[67] Dario Dalmonte and Giulia Vitale Dalmonte were cousins of Giulia Sacerdote. Giulia Vitale's mother was killed in Auschwitz with her husband, her son Achille, her daughter-in-law and two children.

[68] Umberto Terracini (1895–1983), freed from confinement after the fall of fascism, had escaped in Switzerland. Amadio was his brother.

[69] Aurelia, Ermanno, and Arrigo Tedeschi were Terracinis' cousins living in Ferrara.

Vi ho già scritto che da Buenos Aires abbiamo fatto mandare un telegramma risp. pagata al congresso mondiale ebraico di Ginevra perché facciano ricerche e ci comunichino notizie dei Vitale. Chissà che con la liberazione di tante città tedesche, non si riesca a rintracciarli!!! Io penso spessissimo a questi nostri carissimi, e mi auguro di sapere presto che si sono salvati. Di Giulia ho avuto una affettuosissima lettera, con scritti anche di Dario, e risponderò in questi giorni. Abbiamo anche avuto l'indirizzo in Svizzera di Amadio e Umberto e abbiamo loro scritto. Aurelia non sa nulla di Arrigo. Ermanno è sempre in Svizzera ma senza soldi, perché non riesce a farseli mandare. Aurelia ti manda speciali saluti e ringraziamenti per gli auguri ecc. Qui le cose universitarie vanno assai bene. Le nuove autorità sono soddisfacentissime, e speriamo molto bene per Benvenuto e per noi. Si respira un'aria assai migliore che qualche mese fa. Ma il nostro desiderio è sempre quello di ritornare. Mille baci a tutti.

Terracini.16 A. Terracini to the Ministry of the Treasury, Tucumán 13.9.1945

ACS-*Terracini*, fol. 1r-v

Some time ago, I received the following communication from the regent of the consulate in Cordoba, dated 12 May 1945: "We are pleased to make it known that the Royal Ministry of the Treasury has just informed us, through the Royal Ministry of Foreign Affairs, of the present impossibility of paying pensions due to compatriots residing abroad, because of lack of liquidity and the inexistence of an official exchange rate. For the moment, the settlement of the amounts due for this reason may be carried out in favour of the entitled people in lire and in Italy. Please kindly make known your possible approval regarding this payment in Italy and in Italian lire of the pension due to you." On 8 June, I replied on the matter by directing to the *Chargé d'Affaires* in Buenos Aires. Subsequently, on 9 August 1945, the vice consulate of Italy in Tucumán informed me that the *Chargé d'Affaires* in Buenos Aires could not proceed with my answer, as the embassy was unable to give more details on previous communication about the payment of pensions; and invited me to directly address the Ministry of the Treasury. I do this now, also considering the changes in my situation of which I have been informed in the meantime:

1° On 31 July 1945, the vice consulate of Italy in Tucumán informed me that with an ongoing decree of the Minister of Public Instruction I was re-admitted to service as full professor of Analytic Geometry with elements of Projective and Descriptive Geometry with Design at the University of Turin, effective from 1 January 1944, for economic purposes. Therefore, starting from 1 January 1944 my rights will not concern the pension, but the salary, and this is obviously to be paid in lire and in Italy. To this end, I inform you that my general attorney in Italy is Eng. Aldo Sacerdote, vice-manager of the Southern Electricity Company in

Naples (via P. Emilio Imbriani 42), resident in Naples, via Egiziaca in Pizzofalcone 43.

2° The question seems to be quite different as far as the payment of the four-year allowance (difference between salary and pension from December 1938 to December 1942) is concerned, a payment recognised by the sentence of the Council of State of 24 September 1941, no. 302, on the basis of law of 23 May 1940 no. 587, and issued by the Ministry on 19 October 1942, for which as early as May 1943 the extract of the individual role was transmitted by the Tax Office of Rome and now, as far as I know, is deposited at the consulate of Buenos Aires. For this cheque, which had already matured in full since December 1942 and whose payment order from the Ministry of the Treasury had to be appropriately made in Buenos Aires before the suspension of the payment of pensions in Argentina, I ask that it be paid here, applying the official exchange rate in force at the time when the measure should have been carried out.

3° Finally for the arrears of the pension (instalments overdue and not paid from June to December 1943) I ask – also for the reasons given later – that they be paid to me here, applying to me the equity criteria and exchange conditions adopted for diplomatic and consular staff remaining in Argentina.

I am convinced that at this moment every Italian citizen must manfully bear his share of the weight of the tragic conditions into which the fascist government has thrown our country; but at the same time I am confident that the superior authorities will judge my requests legitimate and will also take into due account the fact that I did not leave my homeland for personal interests, but was obliged to take the path of exile from an inequitable law that had undermined my rights and dignity as a citizen, precluding me from any activity as a teacher and scholar to which I had dedicated my life; an unfair law which also prevented my children from completing their studies. Added to this is the fact that the sums requested are even more necessary because at the time of my departure I had to forcibly leave all my personal assets in Italy.

I take this opportunity to confirm my previous letter of 8 May 1945 to the Ministry of Public Instruction with which I intended to reaffirm my rights in this regard and eliminate even any remote doubt of a possible expiry of the statute of limitations for all sums owed to me in any capacity. Thanking you sincerely and with perfect observance, Alessandro Terracini

Ho ricevuto a suo tempo dal Sig. Reggente il Consolato di Cordoba la seguente comunicazione in data 12 maggio u.s.: "Si ha il pregio di far conoscere che il R. Ministero del Tesoro ha testé reso nota, pel tramite del R. Ministero degli Affari esteri, la presente impossibilità di eseguire il pagamento di pensioni relative a connazionali residenti all'estero, a causa della mancanza di una disponibilità spendibile in divisa e della inesistenza di una quotazione ufficiale di cambi. La liquidazione degli importi dovuti a tale titolo potrà per il momento essere effettuata a favore degli aventi diritto in lire ed in Italia. Si prega di voler cortesemente far conoscere il suo eventuale gradimento in merito a tale liquidazione in Italia ed in lire italiane della pensione a Lei spettante." Il giorno 8 giugno risposi in merito dirigendo al Signor Incaricato d'Affari in Buenos Aires. Successivamente in data 9 agosto 1945 il Vice Consolato d'Italia in Tucumán mi comunicò che l'Incaricato d'Affari in Buenos Aires non poteva dar corso alla mia risposta, non essendo l'Ambasciata in grado di dare maggiori precisazioni su quanto era già stato comunicato a suo tempo in merito al pagamento delle pensioni; e m'invitò a dirigermi direttamente al Ministero del Tesoro. Ciò faccio attualmente, tenendo anche conto dei mutamenti della mia situazione che mi sono stati resi noti nel frattempo: 1° - In data 31 luglio 1945 il Vice Consolato d'Italia in Tucumán mi comunicò che con decreto in corso del Ministro della Pubblica Istruzione sono stato riammesso in servizio quale ordinario di Geometria Analitica con elementi di Proiettiva e Geometria Descrittiva con disegno nella Università di Torino, con decorrenza ai fini economici dal 1 gennaio 1944. Pertanto, a partire dal 1 gennaio i miei diritti verteranno non sulla pensione, ma sullo stipendio, e questo è ovvio che sia liquidato in lire ed in Italia. A tale scopo comunico che mio procuratore generale in Italia è l'Ing. Aldo Sacerdote, Vicedirettore generale della Società Meridionale di Elettricità in Napoli (via P. Emilio Imbriani 42), residente in Napoli, via Egiziaca a Pizzofalcone 43. 2° La questione si presenta ben diversamente per il pagamento dell'assegno quadriennale (differenza fra stipendio e pensione dal dicembre 1938 al dicembre 1942), riconosciutomi da sentenza del Consiglio di Stato del 24 settembre 1941, n° 302, in base alla legge 23 maggio 1940 n° 587, e liquidato dal Ministero in data 19 ottobre 1942, per cui già nel maggio 1943 deve essere pervenuto al Banco di Napoli di Buenos Aires l'estratto del ruolo individuale trasmesso dall'Intendenza di Finanza di Roma ed ora, a quanto mi consta, giacente presso il Consolato Generale di Buenos Aires. Per questo assegno, già maturato per intero fin dal dicembre 1942 e il cui ordine di pagamento da parte del Ministero del Tesoro doveva trovarsi in Buenos Aires presso chi di dovere prima che avvenisse la sospensione del pagamento delle pensioni in Argentina, domando mi sia pagato qui applicando il cambio ufficiale vigente nel momento in cui il provvedimento avrebbe dovuto essere eseguito. 3° Finalmente per gli arretrati della pensione (rate scadute e non pagate dal giugno al dicembre 1943) domando – anche per le ragioni addotte in seguito – che mi siano pagati qui, applicandosi anche a me i criterii di equità e le condizioni di cambio adottate per il personale diplomatico e consolare rimasto in Argentina. Sono convinto che in questo momento ogni cittadino italiano deve sopportare virilmente la sua parte del peso delle tragiche condizioni in cui il governo fascista ha gettato il nostro Paese; ma al tempo stesso sono fiducioso che le superiori autorità giudicheranno legittime le mie richieste e terranno anche nel debito conto la circostanza che io non ho lasciato la mia Patria per un interesse personale, bensì indotto a prendere con tutta la mia famiglia la via dell'esilio da una legge iniqua che aveva minorato i miei diritti e la mia dignità di cittadino e mi aveva precluso ogni attività di docente e di studioso cui avevo dedicata la mia vita; legge iniqua che impediva inoltre ai miei figli di portare a compimento i loro studi. A ciò si aggiunge che le somme da me richieste mi sono maggiormente necessarie perché nella mia partenza ho dovuto forzatamente lasciare in Italia i miei averi personali. Colgo l'occasione per confermare la mia precedente lettera dell'8 maggio u.s. diretta al Ministero della Pubblica Istruzione, con la quale intesi con una nuova affermazione dei miei diritti in proposito eliminare anche ogni remoto dubbio di una possibile decorrenza di termini di prescrizione per tutte le somme dovutemi a qualsiasi titolo. Ringraziando sentitamente e con perfetta osservanza, Alessandro Terracini

Terracini.17 MPI to MFA, Rome 21.12.1945

ACS-*Terracini*, fol. 1r-v

The Rector of the R. University of Turin [Mario Allara] has transmitted to this Ministry an extract of the minutes of the meeting of that Faculty of Sciences with which it has been voted that Prof. Alessandro Terracini, full professor of Analytic Geometry at that University and currently resident in Tucumán (Argentina), can return to Italy as soon as possible to resume his teaching.[70] Therefore, this Dicastery is requested to take the appropriate steps with the Argentine Government to facilitate prompt return to the homeland of the aforesaid professor and his family, consisting of his wife and three young children. The Minister Molé

Il Rettore della R. Università di Torino ha trasmesso a questo Ministero un estratto di verbale di adunanza di quella Facoltà di Scienze con il quale si fanno voti affinché il Prof. Alessandro Terracini, ordinario di Geometria Analitica presso detto Ateneo, attualmente residente a Tucumán (Argentina), possa rientrare al più presto in Italia per riprendere il suo insegnamento. Si prega, pertanto, codesto Dicastero di voler svolgere gli opportuni passi presso il Governo argentino affinché sia facilitato il sollecito ritorno in Patria di detto professore e della sua famiglia, composta della moglie e di tre figli minorenni. Il Ministro Molé

Terracini.18 Memorandum for the Minister of Public Instruction, [November 1946]

ACS-*Terracini*, fol. 1r-v, 2r

Prof. Alessandro Terracini, already exempted for the so-called racial laws by the ceased fascist government, was re-admitted for service as a professor of Analytic

---

[70] ASUT-*Terracini*: excerpt of the meeting of the full professors of the Faculty of Sciences, Turin 14.11.1945, in footnote 9 in Chap. 7.

Geometry with elements of Projective Geometry and Descriptive Geometry with Design in the Royal University of Turin, in accordance with current provisions. The professor carried out his activity at the University of Tucumán (Argentina), where he emigrated to escape persecution by the fascist government. The professor has a very strong desire to return to Italy to resume his activity and his mission as a scientist; but the fulfilment of his desire is impeded by the cost of the return trip which, due to the number of family members (mother, wife, and three children), reaches a sum that is too high in light of his economic resources. With the joint request, the professors holding chairs of mathematical teachings in the Italian Universities vote that this dicastery provide Prof. Terracini with the means for returning to his homeland, considering the merits of the professor who with his works and teaching promoted the University of Turin, and (also abroad) has not ceased to spread the knowledge of Italian mathematical production.[71] In this regard, I would like to point out that, following a vote expressed by the Faculty of Sciences of the Royal University of Turin, with letter no. 15583 of 21 December 1945, the Ministry of Foreign Affairs was asked to take appropriate steps with the Argentine government to facilitate the return of Prof. Alessandro Terracini to his homeland.[72] As regards the possibility of reimbursing travel expenses to the aforementioned professor, it should be noted that this office, on 22 November 1945 with letter no. 13995, contacted the Ministry of the Treasury to find out whether the current transfer allowance for State personnel can be paid to professors previously dispensed for political or racial reasons and now readmitted to service, when they move together with their family from an Italian city, where they went when they were dispensed, to the newly assigned seat. In the letter no. 141063 of 11 February 1946, the Ministry of the Treasury expressed the opinion that professors who find themselves in the conditions set out above will be entitled to reimbursement of transfer costs to the new place of employment only, i.e., only the reimbursement of travel expenses. Under current provisions, the reimbursement of personal travel expenses must also be limited, in the case of people coming from abroad, to travel from the first border station or landing place to the assigned workplace. Therefore, in the current state of legislation, there would be no possibility of reimbursing the travel expenses that Prof. Terracini should encounter in returning to Italy with his family. Nor would it be possible to provide them in the form of a subsidy, both for the sum (about 1 million), and because if this facility was granted to the aforementioned professor, also all the others who are abroad, as well as those few who have already returned, would not fail to request it.[73] On the other hand, the professors who went abroad to escape the persecutions of the fascist government, and who in the years of exile kept and still hold high the name of Italy are worthy of every help. Moreover, also officials employed by other State Administrations could be found in the same conditions

---

[71] See D7.1.2.

[72] ASUT: Verbale del Consiglio della Facoltà di Scienze, 14.11.1945, footnote 9 in Chap. 7.

[73] See D7.1.7.

of university professors and therefore you, Mr. Minister, will decide whether it is appropriate or not to involve the Presidency of the Council of Ministers, so that a specific provision of law be issued that allows reimbursement of travel expenses to those who return from abroad to Italy to resume their roles, from which they were removed in the past. I would like to point out that there are currently about thirty professors abroad. The General Director

Il Prof. Alessandro Terracini, già dispensato per le cosiddette leggi razziali dal cessato governo fascista, è stato riammesso in servizio quale ordinario di Geometria Analitica con elementi di Proiettiva e Geometria Descrittiva con Disegno nella R. Università di Torino, ai sensi delle vigenti disposizioni. Il predetto professore ha svolto la sua attività presso la Università di Tucumán (Argentina), dove emigrò per sfuggire alle persecuzioni del governo fascista. Il professore ha vivissimo desiderio di ritornare in Italia per riprendervi la sua attività e la sua missione di scienziato; ma all'esaudimento di questo suo desiderio si oppone il costo del viaggio di ritorno che, per il numero dei famigliari (madre, moglie, e tre figli), raggiunge una cifra troppo superiore alle sue risorse economiche. Coll'unita istanza i professori titolari di cattedre di insegnamenti matematici delle Università Italiane fanno voto perché questo dicastero fornisca al Prof. Terracini i mezzi per il ritorno in Patria, tenuto conto delle particolari benemerenze del professore che con le opere e l'insegnamento illustrò l'Università di Torino ed anche all'estero non ha cessato di diffondere la conoscenza della produzione matematica italiana. Al riguardo faccio presente che a seguito di un voto espresso dalla Facoltà di Scienze della R. Università di Torino fu già provveduto, con lettera n. 15583 del 21 dicembre 1945, ad interessare il Ministero degli Affari Esteri perché svolgesse gli opportuni passi presso il Governo Argentino per facilitare il ritorno in Patria del Prof. Alessandro Terracini. Per quanto concerne la possibilità di rimborsare le spese di viaggio al professore anzidetto, si fa presente che questo ufficio, il 22 novembre 1945 con lettera n. 13995, si rivolse al Ministero del Tesoro per conoscere se ai professori già dispensati per motivi politici o razziali, ora riammessi in servizio, quando si trasferiscano insieme alla famiglia da una città Italiana, ove si erano recati allorché furono dispensati, alla sede nuovamente loro assegnata, possa essere corrisposto il trattamento di trasferimento in vigore per il personale statale. Il Ministero del Tesoro, con lettera n. 141063 dell'11 febbraio u.s., ha espresso il parere che ai professori che si trovino nelle condizioni su esposte, potrà essere corrisposto solo il trattamento previsto per il raggiungimento della sede di servizio in occasione di una nuova nomina, e cioè il solo rimborso delle spese di viaggio. In base alle vigenti disposizioni anche il rimborso delle spese personali di viaggio deve essere limitato, nel caso di persone provenienti dall'estero, al viaggio dalla prima stazione di frontiera o posto di sbarco alla sede assegnata. Pertanto, allo stato attuale della legislazione, non vi sarebbe alcuna possibilità di rimborsare le spese di viaggio che il prof. Terracini dovrebbe sostenere per rientrare in Italia con la famiglia. Né sarebbe possibile provvedervi sotto forma di sussidio, sia per

l'entità della somma (si tratta di circa 1 milione), sia perché ove si concedesse
tale agevolazione al predetto professore, anche tutti gli altri che si trovano
all'estero, nonché quei pochi che sono già tornati, non mancherebbero di
richiederla. D'altra parte i professori che si recano all'estero per sfuggire
alle persecuzioni del governo fascista, e che negli anni dell'esilio tennero e
tuttora tengono alto il nome dell'Italia sono degni di ogni aiuto. Peraltro, nelle
condizioni dei professori universitari potrebbero trovarsi funzionari dipendenti
da altre Amministrazioni dello Stato e pertanto vorrà Ella, Signor Ministro,
decidere se non sia il caso di interessare la Presidenza del Consiglio dei Ministri,
perché sia emanata un'apposita disposizione di legge che consenta il rimborso
delle spese di viaggio a coloro che dall'estero rientrino in Italia per riprendere il
loro ufficio, da cui furono a suo tempo dispensati. Le faccio presente che i
professori attualmente all'estero sono circa trenta. Il Direttore Generale

Terracini.19 A. Terracini to B. Segre, Tucumán 29.11.1946

CA, *BSP*, fol. 1r

Dear Beniamino, It was with great pleasure that I received your letter dating back
to the end of July. When it reached me, I was in poor health, due to infectious
angina, stomatitis, etc. that kept me in bed and then in the house for some time.
Then I recovered, but now I feel unwell again, for another reason: this time I don't
know if it is a myalgia or neuralgia or rheumatic pain in a shoulder. I can lead my
normal life, but with some limitations. As you know, Fano was very ill, following
circulation problems. He had to be at almost absolute rest for a long time; but now
he is returning more or less to his normal life. I am very happy with you for your
return to Italy, even though as you tell me, it is a prostrate country with serious
destruction. But I have full confidence that, with everyone's hard work, it will
recover. You told me in your letter that your family would soon join you, which I
suppose has long since happened. I can give you excellent news apropos the
publication of your work; I have seen the first and then the second and the third
drafts, and the extracts should already be ready.[74] I don't know if I can have them
sent to you before the volume comes out; but even if you had to wait until the
termination of the issue, it is now a matter of a few weeks. Generally speaking, it
seems to me that the printing has turned out very well, although the new
typography that composed the volume is poor in fonts. This, for example, forced
me to replace the gothic characters with common straight characters, so that they
can be distinguished from the current slanted ones. Mahler's job is also definitely

---

[74] B. Segre, *Equivalenza ed automorfismi delle forme binarie in un dato anello o campo numerico*,
Revista. Serie A. Matemáticas y Física Teórica, 5, 1946, pp. 7–68.

approved.[75] If you don't write to me with anything to the contrary, I will send you the offprints according to your instructions. I always have one foot in America and one in Europe, and I would like to reunite them both in Europe as soon as possible. These days I am overloaded with exams, because of the numerous *cessantie* of professors which have been arranged. In my commission, we must examine students on at least twelve different subjects! Now a student strike is in the pipeline as a protest against such *cessantie*; but selfishly I hope nothing is done about it, because I don't want my kids to miss the chance to finish their exams. From all of us, warm regards to you and your family. Yours sincerely Alessandro

Caro Beniamino, Con molto piacere ho avuta a suo tempo la tua lettera della fine di luglio. Quando essa mi giunse, stavo poco bene di salute, a causa di un'angina infettiva, stomatite ecc. che mi ha tenuto a letto e poi in casa per un certo tempo. Poi sono guarito, e ora di nuovo sto poco bene, però per altra causa: stavolta è non so se una mialgia o nevralgia o dolore reumatico a una spalla. Posso fare la mia vita normale, ma con qualche limitazione. Come saprai chi è stato assai poco bene è Fano, in seguito a disturbi di circolazione. Ha dovuto stare a riposo quasi assoluto parecchio tempo, ma ora sta tornando a fare la sua vita quasi normale. Mi rallegro molto con te per il tuo ritorno in Italia, anche se come tu mi dici si tratta di un paese prostrato e con gravi distruzioni. Ma ho piena fiducia che, con la buona volontà di tutti, si riprenderà. Nella tua lettera mi dicevi che presto ti avrebbe raggiunto la tua famiglia, cosa che suppongo ormai sarà avvenuta da tempo. Ti posso dare ottime notizie della pubblicazione del tuo lavoro; ho visto le prime e poi le seconde e anche le terze bozze, e dovrebbero essere pronti già gli estratti. Non so se potrò farteli mandare prima che esca il volume, ma anche se dovessi aspettare fino alla terminazione del medesimo, ormai è questione di poche settimane. In complesso mi pare la stampa sia venuta assai bene, per quanto la nuova tipografia che ha composto il volume in questione, sia povera di caratteri. Ciò, per esempio, mi ha obbligato a sostituire le gotiche con caratteri comuni, diritti perché si distinguano da quelli – inclinati – di tipo corrente. Anche il lavoro di Mahler è definitivamente licenziato. Se tu non mi scrivi nulla in contrario, ti faccio mandare gli estratti secondo le tue indicazioni. Io sono sempre con un piede in America e uno in Europa, e vorrei al più presto riunirli tutti [e] due in Europa. In questi giorni sono sovraccarico di esami, come conseguenza delle numerose cessantie *di professori che sono state disposte. Nella mia commissione dobbiamo esaminare in ben dodici materie diverse! Ora è alle viste uno sciopero di studenti come protesta per tali* cessantie*; ma egoisticamente spero non se ne faccia niente, perché non vorrei che i miei ragazzi perdessero la possibilità di terminare i loro esami. Da tutti i miei molti affettuosi saluti a te e famiglia. Tuo aff.mo Alessandro*

---

[75]K. Mahler, *Lattice points in n-dimensional star bodies*, Revista. Serie A. Matemáticas y Física Teórica, 5, 1946, pp. 113–24.

Terracini.20 U. Rodinò to G. Gonella, Naples 3.1.1947

ACS-*Terracini*, fol. 1r

Dear Gonella, I will be very grateful if you would request the granting by the Ministry of the Treasury of the funds necessary for the repatriation from Tucumán (Argentina) of prof. Alessandro Terracini, previously forced to emigrate for racial reasons. As you know, he is a scholar whose return to Italy, in the interest of school and science, is invoked by the mathematicians who teach at all Italian Universities and by the Accademia dei Lincei.[76] I look forward to your kind reply and please accept many thanks and greetings. Hon. Ugo Rodinò[77]

Caro Gonella, ti sarò assai grato se vorrai sollecitare la concessione da parte del Ministero del Tesoro dei fondi necessari per il rimpatrio da Tucumán (Argentina) del prof. Alessandro Terracini a suo tempo costretto ad emigrare per ragioni razziali. Si tratta, come sai, di uno studioso il cui ritorno in Italia, nell'interesse della scuola e della scienza, è invocato dai matematici docenti in tutte le Università italiane e dalla Accademia dei Lincei. Resto in attesa di cortese riscontro e ti prego gradire molti ringraziamenti e saluti. On. Ugo Rodinò

Terracini.21 Memorandum for the Ministry of the Treasury. Aid to Prof. Alessandro Terracini for repatriation, 6.2.1947

ACS-*Terracini*, fol. 1r-2r

The Ministry of Public Instruction in November 1945 proposed to the Ministry of the Treasury the opportunity to award university professors, removed from their professorships following the racial laws and reinstated in the same professorships, with the reimbursement of expenses incurred by themselves and families in moving from the place where they had resettled after being dispossessed to their newly assigned posts.

The Ministry of the Treasury in February 1946 pointed out that it could not accept the request, based on current regulations. The Ministry itself did not rule out the possibility of granting some financial aid, in the form of a subsidy, after evaluating the individual cases and costs incurred in reaching the place of employment. By letter dated 6 Nov. 1946, the Minister of Public Instruction pointed out to the Treasury the usefulness, in the interest of Italian education and science, of the return to Italy of Prof. Alessandro Terracini, reinstated in the chair of Analytic Geometry at the University of Turin, a return hoped for by the President of the *Accademia dei Lincei*, by the National Research Council, and by the University of

---

[76] See D7.1.2 and D7.1.6.

[77] Ugo Rodinò di Miglione (1904–1949) was elected deputy to the Constituent Assembly in 1946 and deputy of the first legislature (1948–1949).

Turin, as well as by 86 professors of Italian Universities who presented a complaint to the Ministry of Public Instruction to ask that this return be made possible.[78] Professor Terracini, now resident in Tucumán in Argentina, can return to Italy, as he wishes and, as mentioned above, is requested by Italian cultural bodies and scientists, only if he is given the financial means – that he does not have – to cover the travel expenses for himself and his family (his wife and three children). These expenses are estimated at 3500/4000 dollars. The State should grant Prof. Terracini, whose repatriation can be considered one of the unique cases envisaged by the Ministry of the Treasury in February 1946, a subsidy for this sum.

The subsidy should be made available to the interested scholar in dollars or in local Argentine currency.

If this should not be possible, this subsidy, in an adequate measure, could be granted in Italian lire and paid in Italy to Eng. Aldo Sacerdote, prof. Terracini's attorney general, but at the same time his transfer to Argentina should be authorised at the official exchange rate. You could also:

– grant the part of the subsidy (to a very small extent) corresponding to the expenses that prof. Terracini will incur before boarding in Argentina in dollars or in local currency, either in Italy or transferring it to Argentina as mentioned above;
– directly pay for the transport by sea of Prof. Terracini and his family on steamers of the Italian shipping company operating between Argentina and Italy;
– grant in Italian lire the part of the subsidy corresponding to the expenses that Prof. Terrracini will have to bear in order to reach Turin from the port of landing with his family.

In place of the payment by the State of the transport by sea, the shipping company could be authorised to collect in Italy, in Italian lire, the price of the above-mentioned transport. The payment could be made at the time of disembarking, or before, by Eng. Sacerdote, as attorney of Prof. Terracini. Only the part of the subsidy corresponding to the expenses that Prof. Terracini will have to bear before embarkment would remain to be paid in Argentina or Italy, with the option of transferring the sum to Argentina, which would significantly reduce the use of foreign currency. The Minister of the Treasury is asked to examine the proposal made by the Minister of Public Instruction for the granting of a subsidy for the repatriation of Prof. Terracini and regarding the aforesaid methods of its paying and to decide favourably and promptly so that Prof. Terracini may return as soon as possible, as he wishes, after 8 years of exile.

Il Ministero della Pubblica Istruzione ha prospettato nel novembre 1945 al Ministero del Tesoro l'opportunità di concedere ai professori universitari,

---

[78] See D7.1.7.

allontanati dalle loro cattedre a seguito delle leggi razziali e reintegrati nelle cattedre stesse, il rimborso delle spese sostenute per sé e per la famiglia per trasferirsi dal luogo in cui si erano stabiliti dopo la dispensa dal servizio, alla sede loro nuovamente assegnata. Il Ministero del Tesoro nel febbraio 1946 faceva presente di non potere accogliere, in base alle norme vigenti, l'accennata richiesta. Lo stesso Ministero non escludeva che si potesse concedere qualche aiuto finanziario, sotto forma di sussidio, previa valutazione dei singoli casi e degli oneri da sostenere per il raggiungimento della sede di servizio. Con lettera in data 6 nov. 1946, il Ministro della Pubblica Istruzione ha fatto presente a quello del Tesoro l'utilità, nell'interesse della scuola e della scienza italiana, del ritorno in Italia del Prof. Alessandro Terracini, reintegrato nella cattedra di Geometria Analitica dell'Università di Torino, ritorno auspicato dal Presidente dell'Accademia dei Lincei, dal Consiglio Nazionale delle Ricerche, dall'Università di Torino, nonché da 86 professori delle Università Italiane che hanno presentato un esposto al Ministero dell'Istruzione per chiedere che detto ritorno sia reso possibile. Il prof. Terracini, oggi residente a Tucumán in Argentina, può rientrare in Italia, come egli desidera e, come detto sopra, è richiesto da enti culturali e da scienziati italiani, solo se gli vengono dati i mezzi finanziari, di cui egli non dispone, per sostenere le spese per il viaggio suo e della famiglia, composta dalla moglie e tre figli. Queste spese sono preventivate in 3500/4000 dollari. Lo Stato dovrebbe concedere al Prof. Terracini, il cui rimpatrio può considerarsi uno dei singoli casi previsti dal Ministero del Tesoro nel febbraio 1946, un sussidio per detta somma. Il sussidio sarebbe opportuno fosse messo a disposizione dell'interessato in dollari o in valuta locale Argentina. Se ciò non fosse possibile, detto sussidio, in misura adeguata, potrebbe essere concesso in lire italiane e versato in Italia all'Ing. Aldo Sacerdote, del prof. Terracini procuratore generale, ma contemporaneamente ne dovrebbe essere autorizzato il trasferimento in Argentina al cambio ufficiale. Si potrebbe anche:

- *concedere in Argentina in dollari o in valuta locale, od in Italia per essere trasferito in Argentina come detto sopra, la parte del sussidio (in misura molto ridotta) corrispondente alle spese che il prof. Terracini dovrà sostenere prima dell'imbarco;*
- *effettuare da parte dello Stato il pagamento del trasporto per via mare del Prof. Terracini e famiglia su piroscafi di compagnia di navigazione italiana che fanno il servizio fra l'Argentina e l'Italia;*
- *concedere in lire italiane la parte del sussidio corrispondente alle spese che il Prof. Terracini dovrà sostenere per raggiungere, con la famiglia, Torino dal porto di sbarco.*

In luogo del pagamento da parte dello Stato del trasporto via mare, potrebbe essere autorizzata la compagnia di navigazione a riscuotere in Italia, in lire italiane, il prezzo di detto trasporto. Il pagamento potrebbe essere fatto dall'Ing. Sacerdote, come procuratore del Prof. Terracini stesso al momento dello sbarco, o prima di esso dall'Ing. Sacerdote. Rimarrebbe da corrispondere in Argentina o in Italia, con la facoltà di trasferimento in Argentina, la sola parte di sussidio corrispondente alle spese che il Prof. Terracini dovrà sostenere prima dell'imbarco, il che ridurrebbe sensibilmente l'impiego di valuta estera. Si chiede che il Ministro del Tesoro prenda in esame la proposta fatta dal Ministro della Pubblica Istruzione per la concessione di un sussidio per il rimpatrio del prof. Terracini e quanto sopra esposto circa le modalità di corrispondere detto sussidio e decida favorevolmente e sollecitamente in modo che il prof. Terracini possa al più presto rimpatriare, come egli desidera, dopo 8 anni di esilio.

Terracini.22 G. Gonella to U. Rodinò, Rome 12.5.1947

ACS-*Terracini*, fol. 1r-v

Dear Rodinò, in reference to your endorsement of Prof. Alessandro Terracini, I would like to inform you that the Presidency of the Council of Ministers (with a letter dated 24 April 1947, no. 103628 / III04.5 / I.3.I) has again confirmed that he cannot adhere to the proposal, which I have again advanced, to grant reimbursement of the expenses that the aforementioned professor should meet for his repatriation. I would also like to point out that the Ministry of Finance and the Treasury (General Accounting of the State Div. XVII) by letter dated 11 March 1946, no. 179879 communicated that he does not believe he can adhere to the proposal of this Ministry to grant reimbursement of personal and family travel expenses to university professors, now readmitted to service and who are abroad, where they went when they were exempted from service for political or racial reasons during the fascist period. However, with letter no. 5705 of 22 April 1947, I have recalled the attention of the aforementioned Dicastery to the importance of a provision invoked in the best interest of Italian science and I once again drew the special attention of the same Dicastery to the particular situation of Prof Alessandro Terracini and the interest on his return that rightly arises from the Italian mathematical School. I reserve the right to give you further news as soon as possible. Cordial regards, Gonella

Caro Rodinò, in riferimento alle tue premure in favore del prof. Alessandro Terracini ti faccio presente che la Presidenza del Consiglio dei Ministri con lettera del 24 aprile u.s. n. 103628/III04.5/I.3.I ha nuovamente confermato di non poter aderire alla proposta, da me di nuovo avanzata, di concedere il rimborso delle spese che il predetto professore dovrebbe sostenere per il suo rimpatrio. Ti faccio inoltre presente che anche il Ministero delle Finanze e del Tesoro Ragioneria Generale dello Stato Div. XVII) con lettera in data 11 marzo u.s. n. 179879 ha comunicato che non ritiene di poter aderire alla proposta di questo Ministero di concedere il rimborso delle spese di viaggio personali e della famiglia ai professori universitari, ora riammessi in servizio e che si trovano all'estero, ove si recarono o in seguito a dispensa dal servizio per motivi politici o razziali durante il periodo fascista. Ho tuttavia con lettera, n. 5705 del 22 aprile u.s., richiamato l'attenzione del predetto Dicastero sull'importanza che riveste il provvedimento invocato nel superiore interesse della Scienza italiana ed ho ancora una volta richiamato la speciale attenzione dello stesso Dicastero sulla particolare situazione del prof. Alessandro Terracini e l'interesse che al suo ritorno giustamente si pone dalla Scuola matematica italiana. Mi riservo di darti appena possibile ulteriori notizie. Cordiali saluti, Gonella

Chapter 12
Beppo Levi, a Leader in His Host Country

A high-profile mathematician, Levi is a representative of the Italian School of Algebraic Geometry, known for his relevant contributions not only in the field of geometry but also in other sectors of mathematics, including logic, foundational studies, analysis, number theory (with a paper on Minkowski's conjecture), theoretical physics, and methodological and didactic issues.[1]

Born in 1875 in Turin, the fourth of ten brothers and sisters, he was trained in Turin under the guidance of two exceptional Masters: C. Segre and G. Peano. He graduated in 1896 (supervised by Segre), discussing a thesis on the resolution of singularities of algebraic surfaces, a leading theme in the research of Italian geometers of the time. After graduation, he worked as an assistant. In 1898, his father, Giulio Giacomo, died and Beppo took over the leadership of the family. He resigned from the role of assistant and then began teaching in various Italian secondary schools (high schools, technical institutes, schools for the training of primary teachers). The experience 'favoured the formation of a personal vision of mathematics and its teaching'.[2] At the same time, Levi devoted himself intensely to research, alternating the studies of algebraic geometry with those of logic, which were clearly influenced by the mentorship of Peano and stimulated by the logical-mathematical Turin environment (Fig. 12.1).

In December 1906, he obtained the chair of Projective and Descriptive Geometry at the University of Cagliari. From this moment onward, he focused mainly on analysis, obtaining perhaps his most important results: a theorem (that still bears his name) on the integration of monotonous sequences and the introduction of functional spaces (Levi spaces) in two works on Dirichlet's principle. Some of Levi's essays on the arithmetic theory of ternary cubic forms date back to the same period

---

[1] For an analysis of the mathematical content of Levi's works, see Viola (1961), Santaló (1962), Schappacher and Schoof (1996), Coen (1999), and Lolli (1999). On Levi and the didactics of mathematics, see Giacardi and Raspitzu (2017) and Giacardi (2019).

[2] Giacardi (2019), p. 54: *favorì il formarsi di una personale visione della matematica e del suo insegnamento.*

E. Luciano, *The Jewish Mathematical Diaspora from Fascist Italy*, Science Networks. Historical Studies 64, https://doi.org/10.1007/978-3-031-64896-0_12

Fig. 12.1 B. Levi in the late nineteenth century, courtesy of Emilia Levi

and, prior to those of B. Segre, were practically the only Italian works of arithmetic geometry.

In 1909, he married Albina Bachi. The following year, Levi moved to Parma, where he would remain until 1928 and where his three children Giulio (1913), Laura (1915), and Emilia (1921) were born. The years spent in Parma were marked by the First World War and the painful loss of two of his brothers, Decio and Eugenio Elia, who was a mathematician of the highest quality. However, those years were also stimulating in terms of multifaceted academic and institutional experiences and for the positions of responsibility he assumed. In 1918, he was elected Dean of the Faculty of Science, founded the Mathematical Institute, and created a section of Mathesis, the national association for mathematics teachers. From the point of view of research, he extended the spectrum of his interests to physics (electrotechnics, relativity, statistical applications of quantum mechanics, etc.) and began to deal with the dissemination of science and didactics, self-publishing a textbook of infinitesimal and algebraic analysis, which included and expanded the course he held, and a small book entitled *Abbaco da 1 to 20* dedicated to the teaching of arithmetic for pre-school children.[3]

The Bolognese period (1928–1938), equally intense, was characterised mainly by Levi's tireless teaching commitment and impressive work for UMI as treasurer (1931–1938) and as a member of the Scientific Commission (1933–1938) and of the editorial board of the *Bollettino* (1929–1938).[4] In this journal, he published 11 articles and reviewed 53 books, both Italian and foreign. At the same time, he collaborated on the Treccani encyclopedia, for which he wrote the entries *Logica matematica* and *Giuseppe Peano*.

---

[3]Levi B. (1922, 1937). On Levi's *Abbaco* see Coen (1998).
[4]Coen (2002).

12.1 Hunted by the UMI

Levi realised almost immediately the true nature of the fascist movement, and in 1925, he signed the manifesto of anti-fascist intellectuals.[5] Some personal episodes (including his failure to be appointed co-director of the *Annali di Matematica pura e applicata*) probably warned him about the interference of the regime in the Italian mathematical society. It is certain that well before the racial laws were promulgated, Levi was systematically the subject of an anti-Jewish campaign within the Italian Mathematical Union, with accusations so crude and gratuitous by E. Bortolotti and E. Bompiani that President L. Berzolari intervened in his defence:

> it would seem to me a sin against honesty not to say a word in his favour. You alluded to Bologna, but without specifying details, to discontent that circulated on his account. Now he certainly has flaws, but he also has great qualities. He is a person of smart intelligence and of a very broad mathematical culture; in his conversations with me, I have always found him attentive, deferential, precise. He never failed to inform me of every little thing about the Union; he has always read with me all the works submitted to the *Bollettino*, and if any of these works do not contain errors, it is all thanks to Levi: I can assure you that I will never again find a person so variously agile, patient, impartial, as he is. In judging a person, one must not only emphasise the defects, but also consider the good aspects.[6]

It is not plausible to think that Levi did not detect the slightest hint of this envy towards him. Probably, in a certain way, the opportunist anti-Semitism by part of the UMI leadership prepared him for the storm to come (Fig. 12.2).

Dismissed on 30 November 1938, Levi applied for counter-discrimination but preferring expatriation to a life in Italy in a humiliating condition of marginalisation, immediately began to think about emigration. On 29 December 1938, he already wrote to his brother Giulio Augusto that he was reading *Don Quixote* and other books in Spanish to become familiar with the language he would have to use in Latin America. Levi had, and would maintain, a very strong concept of his dignity as a scientist and man, which he expressed in these terms: 'the aim of life is a worthy life and the end of science is a worthy science; also, the judgment of dignity comes only from our conscience'.[7]

At this juncture, he reacted differently from all his companions in misfortune. For idealist and political reasons (anti-imperialism and anti-Zionism), Levi did not want

---

[5] See Levi L. (2000), p. 25; Momigliano Levi (2016), pp. 273–274.

[6] AS-UMI: L. Berzolari to E. Bompiani, Pavia 7.11.1938: *mi sembrerebbe un peccato contro l'onestà non dire una parola in suo favore. Tu hai alluso a Bologna, ma senza specificare, a malcontenti che c'erano sul suo conto. Ora egli ha certamente dei difetti, ma ha pure delle grandi qualità. È persona d'ingegno e di larghissima cultura matematica; nei rapporti con me l'ho sempre trovato premuroso, deferente, preciso. Non ha mai tralasciato d'informarmi di ogni più piccola cosa relativa all'Unione; ha sempre letto con me tutti i lavori mandati per la stampa nel* Bollettino, *e se di tali lavori qualcuno non contiene errori, è tutto merito del Levi: ti posso assicurare che non troverò mai più una persona così variamente agile, paziente, disinteressata come lui. Nel giudicare una persona, non si deve soltanto porre in rilievo i difetti, ma si deve pur tener conto dei lati buoni.*

[7] Levi B. (1940), p. 109: *El fin de la vida es la vida digna y el fin de la ciencia es una ciencia digna; más el juicio de la dignidad sale solo de nuestra conciencia.*

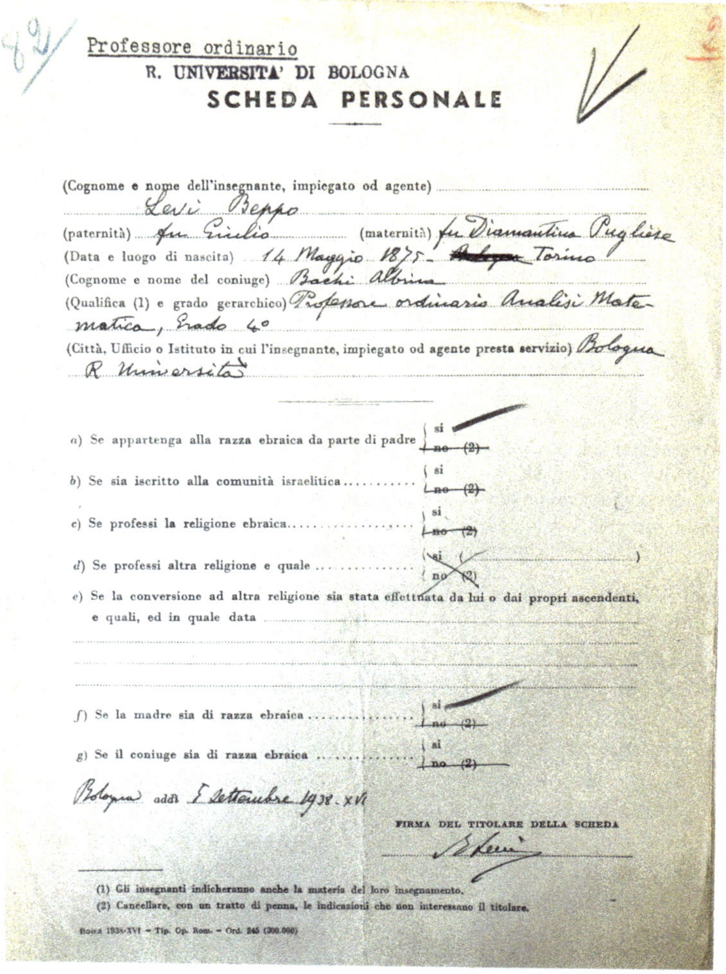

82

Professore ordinario

R. UNIVERSITA' DI BOLOGNA

SCHEDA PERSONALE

(Cognome e nome dell'insegnante, impiegato od agente) Levi Beppo

(paternità) fu Giulio (maternità) fu Diamantina Pugliese

(Data e luogo di nascita) 14 Maggio 1875 ~~Bologna~~ Torino

(Cognome e nome del coniuge) Bachi Albina

(Qualifica (1) e grado gerarchico) Professore ordinario Analisi Matematica, Grado 4°

(Città, Ufficio o Istituto in cui l'insegnante, impiegato od agente presta servizio) Bologna R Università

a) Se appartenga alla razza ebraica da parte di padre: si ~~no (2)~~

b) Se sia iscritto alla comunità israelitica: si ~~no (2)~~

c) Se professi la religione ebraica: si ~~no (2)~~

d) Se professi altra religione e quale: ~~si~~ ( ) ~~no (2)~~

e) Se la conversione ad altra religione sia stata effettuata da lui o dai propri ascendenti, e quali, ed in quale data

f) Se la madre sia di razza ebraica: si ~~no (2)~~

g) Se il coniuge sia di razza ebraica: si ~~no (2)~~

Bologna addì 5 Settembre 1938. XVI

FIRMA DEL TITOLARE DELLA SCHEDA

B. Levi

(1) Gli insegnanti indicheranno anche la materia del loro insegnamento.

(2) Cancellare, con un tratto di penna, le indicazioni che non interessano il titolare.

Roma 1938-XVI - Tip. Op. Rom. - Ord. 245 (300.000)

Fig. 12.2 Racial census form of Levi. ACS-*Ebrei*

to turn to either the SPSL or the ECADFS and was even less disposed to the idea of an *aliyah* in Palestine. His case was therefore not submitted to any rescue agency for displaced scholars, and for this reason, his name appears comparatively less than others in this volume. Levi asked for help only and exclusively from some colleagues with whom he was in contact: T. Levi-Civita and the Argentine mathematician J.C. Vignaux, with whom he had started a correspondence from 1937 at the time of the publication of a paper by Vignaux in the UMI *Bollettino*, and whom he had met at the ICM in Bologna in 1928.[8] Vignaux put him in touch with C. Pla, Dean of the Faculty of Sciences of the Universidad Nacional del Litoral. On 23 December 1938, the board of this Faculty had approved the creation of an Institute of Mathematics in the headquarters of Rosario. Pla had included the clause that an authority of world prestige and undisputed ability be engaged to direct it.[9] Levi-Civita was in contact with the Argentine mathematical environment, and the deal was soon complete: people from Rosario asked Levi-Civita to propose a name to direct the Institute, among the Italian scholars who had lost their positions due to the racial laws.[10] Levi-Civita considered recommending B. Segre or Levi. Segre declined the invitation, as he was already settled, at least temporarily, in England. Levi accepted immediately and with immense gratitude. The telegram announcing his appointment as Director of the Institute arrived from Buenos Aires on the morning of 12 April 1939. The night before, Levi had accompanied his son Giulio to the railway station, as he was leaving for Palestine.

The choice of emigrating was certainly not easy for him. At the age of 63 years, Levi was not unaware of the untold situations and difficulties that the future held for him—a new job in a totally unknown country—but he was ready to face them:

> I wrote to Vignaux that, as for the material conditions, I did not doubt they would be corresponding to the role and such as to allow me to maintain the due decorum by coming there with my wife and two daughters; therefore, in principle I certainly accepted. Among us, I understand that I am faced with other unknown issues [. . .] But what is sure, in the problem of our near future?[11]

Sensing that the international political situation would degenerate in a short time, Levi did not 'waste time' trying to obtain an entry visa to Argentina. After several weeks of attempts, the Argentine consul in Milan made it clear that the delay in issuing the passport was due to the fact that he had declared himself to be Jewish; it would have been better for him to declare himself Catholic. The suggestion was repugnant to Levi:

> You know what I think: classifying myself in the Israelite religion is just as false as classifying myself in the Catholic one. I have no scruples when men force me into a classification, by contrast I have scruples when they force me to be irreligious by making an act of faith that is not sincere and that I don't feel the need for. [. . .] As I have said other

---

[8]The reading of the published works, and those to be published, led him on many occasions to intervene with personal observations on the works submitted to the *Bollettino* and to establish relationships with authors of the articles he was in charge of reviewing.

[9]*Antecedentes de la creación del Instituto* 1940, p. 67; Pla (1962), pp. XIII–XIV.

[10]Levi B. (1942b), p. 155.

[11]B. Levi to T. Levi-Civita, [Bologna, December 1938-January 1939] in Nastasi and Tazzioli (2000), p. 312: *Ho scritto al Vignaux che, quanto alle condizioni materiali, non dubitavo sarebbero state corrispondenti al posto e tali da permettermi di mantenere il dovuto decoro recandomi colà con mia moglie e due figlie, e che quindi in massima accettavo senz'altro. Fra noi, comprendo che ho di fronte altre incognite* [. . .]: *ma che cosa è cognito nel problema del nostro avvenire prossimo?*

times, the repugnance would be identical if I had to make a statement of faith with respect to any religion, including the Jewish one.[12]

In the end, since the situation did not resolve itself, Levi bought a tourist visa with a procedure that was not totally legal. The decision was wise. Enrico Volterra, who had been called at the same time as Levi to direct the Institute of Stability, did not want to go down this route and could no longer reach Argentina. As Levi wrote to Levi-Civita: 'having understood from the course of facts and from correspondence with the Dean that the regular entry permit was postponed every 15 days and would perhaps never arrive, I decided to obtain a tourist permit. I do not need to tell you about the difficulties and how I overcame them. Now I can say that I am happy with the decision'.[13]

Taking leave of relatives, Levi, together with his wife Albina and daughters Laura and Emilia, embarked on Oceania on 21 October 1939. The daughter recalled:

> Everyone knew that normal communications between the two continents could be interrupted at any time, so the ship was overloaded with passengers of all kinds, who wanted to flee Europe before being trapped by the war. These included some Argentines who had to urgently interrupt their tourist journey to Europe, and emigrants like us, who had managed to obtain a visa in time to enter some of the countries where the ship docked, namely Brazil, Uruguay, and Argentina. As a result, there was a certain atmosphere of concern on board. What would the arrival in the New World, at that time so unknown to us, have reserved, especially for new immigrants? What would happen in this Europe that we were leaving behind? Despite this, life on board retained some traditional features of the cruise, including the hours spent on deck sunbathing, the feast of the crossing of the equator and a first tourist visit to the American cities, Rio de Janeiro, Santos, Montevideo, where you could go ashore for a few hours.[14]

---

[12]B. Levi to G.A. Levi, [Autumn of 1939], in Momigliano Levi (2016), p. 279: *Tu sai quale è il mio pensiero: il classificarmi nella religione israelitica è altrettanto falso come classificarmi nella cattolica e quindi non posso avere scrupoli una volta che gli uomini mi obbligano a una classificazione. Lo scrupolo l'ho quando mi obbligassero ad essere irreligioso con fare un atto di fede non sincero e di cui non sento la necessità. [...] Come ho detto altre volte la ripugnanza sarebbe identica se dovessi fare una dichiarazione di fede per una religione qualunque, compresa la ebraica.*

[13]B. Levi to T. Levi-Civita, Rosario 14.11.1939, in Nastasi and Tazzioli (2000), p. 313: *avendo compreso dallo svolgimento dei fatti e dalla corrispondenza col Decano che il regolare permesso d'ingresso si rimandava di 15 giorni in 15 giorni e non sarebbe forse mai arrivato, ho preso la risoluzione di ottenere un ingresso turistico. Non occorre le dica le difficoltà e come le abbia vinte. Della risoluzione presa posso al momento dirmi contento.*

[14]Levi L. (2000), p. 62: *Todos sabían, sin embargo, que en cualquier momento las comunicaciones normales entre los dos continentes podían interrumpirse, de manera que el barco estaba sobrecargado de pasajeros de todo tipo, que querían dejar Europa antes de ser atrapados por la guerra. Entre ellos se contaban algunos argentinos que habían debido interrumpir con urgencia su viaje turístico por Europa, y emigrantes como nosotros, que habían podido conseguir a tiempo una visa para ingresar a alguno de los países cuyos puertos abordaba el barco, es decir Brasil, Uruguay y la Argentina. En consecuencia, flotaba a bordo cierta atmósfera de preocupación. ¿Que nos depararía, especialmente a los nuevos inmigrantes, la llegada al Nuevo Mundo, entonces tan desconocido para nosotros? ¿Que sucedería en esa Europa que estábamos dejando atrás? Pese a ello, la vida de a bordo mantenía ciertas características tradicionales de crucero, entre ellas las horas pasadas en cubierta tomando sol, la fiesta del cruce del Ecuador y una primera mirada*

On 6 November, after 15 days of voyage and a stop in Montevideo, the Levis landed in Buenos Aires.[15] Waiting for them on the quay of the port was Pla and a delegation of future colleagues: C. Isella, J. Olguín, and F. Gaspar. After a two-day stop at the Hotel Colon, they left for Rosario, accompanied by L.A. Santaló, a refugee from Spain who had recently been hired as a researcher at the Institute of Mathematics. Apparently, the authorities of the Faculty of Rosario, who had not yet obtained the definitive residence permit for Levi and his family, wished not to make his entry into the country too well known and considered it more appropriate that their guests immediately reach the destination.[16] The welcome was splendid. This text comes from the first letter written to his brother upon his arrival in Argentina:

> Arrived in Buenos Aires the day before yesterday, it seems like a dream [...]. I hoped that someone would come to meet me in Montevideo; since nobody came, I was starting to get a little nervous. However, the welcome in Buenos Aires was splendid: our friend Parise, Dante Bachi, a friend of Giulio, Laudbanc, a Polish Jew, and three delegates from Rosario were waiting for us upon disembarkment. The Faculty of Rosario had already arranged rooms for us in the hotel where I am fully paid for and a number of colleagues of all specialties repeatedly visited me. Dean Cortés Pla came from Rosario in the afternoon and the next day we were brought here by car. I wasn't even able to deal with some chores that I wanted to deal with in Buenos Aires. The welcome continued here, where today I have already spent three hours in the faculty building, taking the lay of the land, so to speak, and talking with many people. The friendliness is so great here that I too launch myself in talking with great ease, half in Spanish and half in Italian. These colleagues repeatedly congratulate me on the decision I took to arrive on a tourist visa, which was the only possible solution. "Radicación" will come when it comes, but – they say – I have now entered and everything will go as if it had been already granted.[17]

The city of Rosario had grown considerably since the beginning of the century and, with its population of almost one million inhabitants, was three times the size of

turística a las ciudades americanas, Río de Janeiro, Santos, Montevideo, donde se podía bajar a tierra por algunas horas.

[15]LEVI.1.

[16]The practice of the residence permit, the so-called _radicación_, was completed without further difficulty shortly after arriving in Rosario, on 15 January 1940.

[17]B. Levi to G.A. Levi, [Buenos Aires 6.11.1939], in Momigliano Levi (2016), p. 279: _Giunti jer l'altro a Buenos Aires, sembra un sogno [...]. Avevo la speranza che qualcuno mi venisse incontro a Montevideo; non essendo venuto alcuno incominciavo ad essere un po' nervoso. In compenso l'accoglienza a Buenos Aires è stata splendida: erano ad attenderci allo sbarco l'amico Parise, Dante Bachi, un amico di Giulio, Laudbanc, ebreo polacco, e tre delegate di Rosario. La Facoltà di Rosario aveva già provveduto a fissarmi le camere in albergo ove sono completamente spesato e ripetutamente visitato da una quantità di colleghi di ogni varietà. Il Decano Cortés Pla è venuto nel pomeriggio da Rosario e il giorno dopo siamo stati portati in automobile qui. Non mi è stato possibile nemmeno fare qualche incombenza che avrei inteso a Buenos Aires. Le accoglienze hanno continuato qui, dove oggi ho già passato tre ore nella sede della Facoltà a prendere in certo modo possesso e a discorrere con molte persone. La cordialità è qui tanto grande che anch'io prendo slancio a discorrere con grande disinvoltura metà in spagnolo e metà in italiano. Questi colleghi si congratulano ripetutamente per la risoluzione che ho preso di arrivare col visto turistico, che era l'unica soluzione possibile. La "Radicación" verrà quando verrà, ma - dicono - ormai sono entrato e tutto andrà come se essa fosse stata concessa._

Bologna. For its development, it was considered the second town of Argentina but, in the eyes of the Levis, appeared a provincial place, with 'straight streets, lined with houses generally with only the ground floor or one floor'.[18] The Levi family initially settled in the Savoy Hotel and then moved to a small apartment in Calle San Luis, one of the chic streets of the city, where they waited for the arrival of the containers with the furniture, books, and objects that had been shipped from Italy. In 1940, they settled in a house in Via S. Lorenzo 2133 where Beppo Levi would remain until his death.

12.2   A Bi-Cultural Scholar: Levi's Second Life in Rosario

Despite the moments of nostalgia and the concern for friends and relatives far away (several of Levi's grandchildren had emigrated to America, his son Giulio in Palestine, a part of the family group had remained in Italy), life in the new environment was immediately very positive. As Levi's wife wrote to her sister-in-law, Itala: 'Beppo arrives home cheerfully as a child, many times after continuing the discussions started in the Institute having a coffee in a café and often using paper napkins to make notes'[19] (Fig. 12.3).

Various factors contributed to this rapid adjustment, including the friendliness with which Levi was received and the satisfaction of being surrounded by consideration and appreciation after the discrimination suffered at home.

The Argentine era represented a new season in Levi's professional trajectory. Levi was in fact hired as Director of the Institute of Mathematics and was assigned some specific tasks: promoting mathematical research, establishing new scientific relationships with other similar institutions abroad, and delivering courses, conferences, and specialised seminars aimed at organising in Rosario a PhD program in mathematics.[20] This was particularly timely for Levi. The interest in performing a role he had never held in Italy compensated for the difficulties he would encounter in developing individual research activity, in a context very different from Italy, and also allowed him to develop some aspects of his personality, in particular his skills as a populariser of mathematical thought. Levi himself, on the inauguration of the

---

[18] Levi L. (2000), p. 64: *las calles rectas, flanqueadas por casas en su mayor parte de planta baja o un piso.*

[19] Levi L. (2000), p. 66: *Beppo llega a casa alegre como un chico, muchas veces después de haber continuado las discusiones iniciadas en el Instituto, tomando café en una confitería y utilizando a menudo las servilletas de papel para anotaciones.*

[20] As for specialized courses and seminars, Beppo Levi held, until 1948, courses of Projective Geometry, Modern Analysis and Analytical Functions (1940), Integral equations (1942), Partial Differential Equations (1941), and Invariant Theory of Algebraic Forms (1946). After this date, he went back to teaching standard courses. In particular, he lectured on Analytic Geometry, Infinitesimal Analysis and Rational Mechanics.

Fig. 12.3 B. Levi in the 40s, courtesy of Emilia Levi

Institute (18 May 1940), affirmed that the new work came at a time in his life in which he had realised that he was driven:

> towards a social love for mathematics, or rather, for mathematical thought; because mathematics [as] constituted science, as a science *of* and *for* specialists, interests me much less.[21]

Upon Levi's arrival, the Faculty of Exact Sciences had already reached a point of stability and was flowering, with numerous students in the three specialties (Engineering, Architecture, and Surveying). The Institute of Mathematics was already equipped with a large room, and initial nucleus of a library, with some collections of books and journals (*Mathematische Zeitschrift, Mathematische Annalen, Transactions of the American Mathematical Society, Journal des Mathématiques*, etc.), gathered thanks to the intervention of some professors, including J. Olguín, S. Rubinstein and C. Dieulefait who since 1935 had been buying books and periodicals with the financial support of the Faculty and also with their own personal funds.[22] For Levi, it was a happy discovery that the Institute, although only recently established, had adequate infrastructures and equipment and that the local community was animated by goodwill and enthusiasm.

After the period of forced idleness that had followed his dismissal in Italy, Levi was looking forward to returning to work. A few days after landing, he immediately organised two postgraduate courses in Projective Geometry and Theory of Functions, which he personally held and which he proposed again in January 1940. At the

---

[21]Levi L. (2000), p. 67: *Hacia un afecto humano por la matemática, o no, diré mejor, por el pensamiento matemático; porque la matemática, ciencia constituida, ciencia de especialistas, me interesa mucho menos.*

[22]Levi B. (1940), p. 101; Levi L. (2000), p. 68.

same time, he launched the series *Publicaciones* of the *Instituto de Matemática* whose first volume, with two essays by Levi, one by Santaló and one by F. Amodeo (translated by N. and J. Babini), came out at the end of 1939. The periodical, whose publication was discontinued in 1948, after eight volumes, would reach a high scientific level and boast the collaboration of mathematicians from all over the world (M. Cotlar, J. Rey Pastor, and A. Rosenblatt, in addition to the Italians G. Fubini and A. Terracini). *Publicaciones* would be joined in 1944 by the series *Monografías*, consisting of treatises on specific scientific themes. In 1941, Levi founded another journal, *Mathematicae Notae*, a creation entirely moulded by this scholar in every aspect, from the title (which derives from his love of classic culture and languages) to the form and content.[23] Destined to a relatively wide audience, this journal was the natural expression of Levi's conception of mathematics:

> Mathematics, although it is a science in the ordinary sense of the positive sciences, that is, a system of knowledge, - and in this sense it is without further ado the first science to which a child approaches - and although it interests the engineer, physicist, chemist as a powerful instrument for applications, more than anything it is a way of thinking, a philosophy.[24]

Levi not only contributed his articles but also checked in detail all the material received for publication and personally took charge of the graphic layout of each dossier by going weekly to Santa Fe to supervise the typographic work. Since even the articles that do not bear his name were completely reviewed and examined by him, one issue had not yet even been published and he already had to think about the next. Levi not only reviewed the articles but also edited the three sections - *Ejercicios y problemas*, *Cuestiones*, and *Flores y Hojas* - which constituted the heart of the journal and which, in his intentions, were to constitute a bridge between professors and students, motivating the latter to approach mathematics and investigate its contents, foundations, and connections with other sciences. The feedback was excellent: in the period of Levi's direction (1941–1961), 47 'friends of the Institute near and far away' contributed to *Mathematicae Notae*. Levi published 42 articles, Santaló 25, and Cotlar 9.[25] As for Levi, in addition to his contributions to

---

[23] See Sect. 6.5.

[24] Levi B. (1941a), p. 7: *la matemática, si bien es una ciencia en el sentido ordinario de las ciencias positivas, es decir, un sistema de conocimientos - y en este sentido es sin más la primera ciencia a la que se acerca el niño - y si bien ella interesa al ingeniero, al físico, al químico como instrumento poderoso para las aplicaciones, más que todo es un modo de pensar, es una filosofía.*

[25] The authors who gave their contribution to the journal were (in brackets, provenance and number of articles published in the period 1941–1961): J. Babini (Argentina, 1); M. Balanzat (Spain, 1); G. Beck (Czechoslovakia, 1); C. Birindelli (Italy, 1); J. L. Brenner (Massachusetts, 1); R. P. Boas (Washington, 1); P. Hair (N.D., 1); M. Cotlar (Ukraine, 9); A. Daghetto (Argentina, 1); C. E. Dieulefait (Argentina, 1); A. Durañona y Vedia (Argentina, 1); G. Fernández (Argentina, 1); E. O. Ferrari (Argentina, 2); R. Frucht (Germany, 3); G. Fubini (Italy, 1); F. Gaeta (Spain, 1); F. Gaspar (Argentina, 1); L. Godeaux (Belgium, 2); F. González Asenjo (Argentina, 1); A. González Dominguez (Argentina, 1); H. Greppi (Italy, 1); B. Gross (Germany, 1); V. Inglada (Spain, 1); G. Knie (Germany, 1); R. Laguardia (Uruguay, 1); B. Levi (Italy, 42); Laura Levi (Italy, 1); F. Marsicano (Argentina, 1); J. L. Massera (Italy, 1); J. Mateo (Argentina, 1); T. Mihailescu (Romania, 1); L. Nachbin (Brazil, 1); J. Olguín (Argentina, 1); A. Petracca (Uruguay, 1); P. Pi

algebraic geometry and function theory, he published works devoted to modern theoretical physics; the formation of the concept of number in children and primitive societies; the development of mathematical thought in the period 1830–1930; and the future political and social effects of the applications of atomic energy. *Mathematicae Notae* also featured his tributes to Italian mathematicians who had disappeared, such as V. Volterra, T. Levi-Civita, F. Enriques, and L. Berzolari, revealing Levi's strong affection for those who had been his colleagues and Masters.[26] In addition to bringing about a fluid and open contact with the scientific community that was being formed in Argentina, the Institute's publications represented a hub of interest and, together with the *Revista* directed by Terracini, were the only periodicals in the country that appeared regularly on topics of mathematics and physics.

Because of his enthusiasm, the reliability of his work, and the attitude with which he presented himself, the scientific activity of Levi in Argentina gave great impetus to the local cultural milieu, strongly influencing mathematicians. However, his influence also extended both geographically and scientifically beyond Rosario and mathematics. Levi was invited to lecture in the Republic of Uruguay and he, who had followed with interest the developments of quantum and undulatory mechanics, joined the *Asociación Física Argentina* and was in excellent relations with various physicists including R. Laguardia (Uruguay), A. Levialdi (Cuba), E. Gaviola, and G. Beck (Argentina).[27] His students joined in praise of the deep impression left by his work:

> Prof. Levi has radiated warmth, advice, and enthusiasm to all Argentine mathematical centres. his was the work of a teacher, an adviser, an indefatigable champion of mathematics in the pure and highest sense of the word.[28]

This photograph (Fig. 12.4), taken in the 40s, restitutes in an iconic way the atmosphere that reigned in the Institute during its first years of life. Levi sits at the centre, surrounded by a group comprising some of the professors who had contributed most to the project of creating the Institute (S. Rubinstein, J. Olguín, F. Gaspar, H. Greppi, and G. Liserre) and his closest collaborators, L.A. Santaló, P.E. Zadunaisky, and R. Laguardia, the latter belonging to the Department of Mathematics of the Faculty of Engineering of the University of Montevideo (Uruguay). Behind the group, one can see the shelves that made up the library, formed then

---

Calleja (Spain, 5); C. Plá (Argentina, 1); L. D. Porta (Argentina, 1); R. A. Ricabarra (Argentina, 1); J. Ricagno (Argentina, 1); A. E. Sagastume Berra (Argentina, 2); L. A. Santaló (Spain, 25); A. Terracini (Italy, 2); S. Toni (1); J. V. Uspensky (Mongolia, 3); G Valiron (France, 1); A. Wintner (Hungary, 2); and P. E. Zadunaisky (Argentina, 3). The journal was sent to 168 institutions and established exchanges with 146 journals from 31 countries.

[26] Levi (1941b, 1942b, 1946a, 1949).

[27] Levi.2 and Levi.3.

[28] *Homenaje a Beppo Levi 1955: el Prof. Levi ha irradiado a todos los centros matemáticos argentinos calor, consejo y entusiasmo; su [es] labor de Maestro, de orientador, de infatigable paladín de la matemática en el sentido puro y superior que esta palabra adquiere.*

Fig. 12.4 Beppo Levi and colleagues from the Institute of Mathematics of the Universidad Nacional del Litoral, ca. 1940, in L. Levi 2000, p. 70

by the first collections acquired by the Faculty and, in large part, by the personal library brought from Italy by Levi.[29] Among these scholars, both senior and junior, a scientific and personal chemistry was immediately created. Santaló, who was already a recognised mathematician at the time, remembered how carefully Levi revised his works and corrected them according to his own criteria. Zadunaisky, who was a young student, appreciated the patience with which Levi helped him in interpreting E. Persico's book on atomic mechanics (1936) while revealing his typical reservations in this field. Olguín and Laguardia, as well as collaborators, were friends with whom Levi could converse freely, on political and social as well as scientific issues. Cotlar recalled similarly:

> Personally, I remember Beppo Levi with deep admiration and gratitude. The year of his arrival, I was just beginning my mathematical research and he encouraged me in a most generous way. Thanks to his mentoring, I was able to publish most of my work of the 1940–1950 period. He patiently suggested corrections and offered advice.[30]

For this reason, evaluating overall Levi's work as Director of the Institute of Mathematics, it can be said that he was, among the Italian mathematicians refugees, the only one who came to re-constitute a School.

The political events that agitated Argentina after the military coup also upset the activities of the Institute. Pla, who had been one of the signatories of the Declaration for Democracy, was fired.[31] Levi was never directly affected, except for the fact that

---

[29] Levi L. (2000), p. 70.

[30] Cotlar (1991), p. 147.

[31] Galles (2003), pp. 278–281.

he had to return to teaching basic courses (Analytic Geometry and Infinitesimal Calculus II, later Rational Mechanics), renouncing the specialist courses. However, this did not bother him too much. Just as he had held basic courses in Italian Universities, he could do so in Rosario, and 'it cannot be excluded that contact with young students did not bring him a certain satisfaction'.[32]

12.3   Life After the Second World War

In September 1945, Levi received news that he had been readmitted to service in Bologna on 9 June 1945, with a supra-numerary position.[33] He would be reinstated among the members of the UMI in early 1946.

He was immediately and strongly divided over the two options: to stay or return to Italy. At the age of 72 years, he was frightened by the prospect of returning 'as a foreigner, to disturb vested interests'[34] to cover his chair for only three or four years at most. On the contrary, the Universidad Nacional del Litoral did not place any constraint on age for the position of Director of the Institute.

As can be deduced from his article *Dopo la bomba atomica*, there was also a political reason behind his perplexities.[35] At the beginning of 1946, despite a general fear of the beginning of the Cold War and the very existence of atomic bombs, the opinion was being imposed that nuclear reactions would soon offer extraordinary sources of energy, turning into a useful achievement for humanity. Levi was deeply pessimistic about it and believed that future peaceful applications of atomic energy were exalted by political and journalistic propaganda only to distract public opinion from the consciousness of the horror that their war applications had produced. Viewing with distrust the international tension between the Western powers and the Soviet Union, and increasingly dehumanised progress, in the face of the prospect of a third world war fought with accumulated nuclear weapons, Levi hoped that Argentina, by its geographical position, was much safer than the old Europe.

At the root of his decision to stay in Rosario, however, there was above all another motivation: 'I would leave here something that I started, and that would be destined to die'.[36] Levi did not feel like abandoning the Institute, both for sentimental reasons and because he considered it his moral duty to keep alive what he had

---

[32]Levi L. (2000), p. 72: *no puede excluirse que el contacto con los jóvenes estudiantes no le produjera cierta satisfacción.*

[33]LEVI.4-5.

[34]Levi L. (2000), p. 77: *como un extranjero, a molestar intereses constituidos.* See also D7.2.10.

[35]Levi B. (1946b). At the beginning of 1946, Levi had met in Montevideo with Luce Fabbri, who directed the anarchist anti-fascist magazine *Studi Sociali* published in Uruguay but intended for Italian anarchist anti-fascist emigrants residing in different countries. Although Levi did not belong to this political movement, a strong mutual understanding was established between them, which led him to publish the article *Dopo la bomba atomica.*

[36]Levi L. (2000), p. 77: *yo dejaría aquí algo que empecé y estaría destinado a morir.*

created and of which he felt himself an integral part. It should be borne in mind that both the Institute and its publications, the *Mathematicae Notae* above all, were all due to his efforts, and Levi considered them in all respects his creations, so much so that he often used the language of parents to talk about them:

> My Institute is truly a son who grows up and perhaps in a few months he will also grow spatially.[37]

> the *Mathematicae Notae* are a daughter who is always hungry.[38]

On the occasion of a conversation in Turin, in January 2016, P. Momigliano Levi added:

> perhaps a feeling of revenge against Italy which had forced him into exile also played a role, and the feeling that, despite the great changes that had occurred after the end of the Second World War, university chairs continued to be assigned not only on the basis of teachers' merits, but also on the basis of their links with government parties. As the assignment of university chairs in Italy, Beppo Levi realistically noted that during fascism many had risen to the chair because they were loyal to the regime and now, despite the purge, they remained in the chair; others will aspire to do the same «if not today, then tomorrow, due to their relations with the Christian Democrats».[39]

For these concomitant causes, at the beginning of 1947, Levi broke the delay and asked for a paid leave to Argentina, so as to complete his teaching career in Rosario. After a long bureaucratic process, due to the inertia of the Ministry of Foreign Affairs and the partial opposition of the Ministry of Public Instruction, the request was finally approved. The Faculty of Bologna unanimously confirmed the extension of Levi's secondment until his retirement.[40]

Levi, however, promptly and cordially resumed relations with the Bolognese and Italian mathematical communities. At the end of 1948, he returned to Italy for the first time, accompanied by his wife, to visit relatives and friends. On that occasion, he also met his colleagues, first in Bologna, where he was invited to give a series of lectures, and then in Ferrara, at a meeting of the Mathesis association. He would return to Italy one last time in 1951, shortly before the death of his wife Albina, which occurred in Torre Pellice.

As evidence of the complete restoration of cultural ties with Italy, and with the mathematical environment in particular, Levi was involved in a UMI commission

---

[37] Momigliano Levi (2016), p. 280: *Il mio Istituto è veramente un figlio che cresce e forse fra qualche mese crescerà anche di [un] locale.*

[38] Levi L. (2000), p. 82: *la Revista [...] es una hija que tiene siempre hambre.*

[39] Momigliano Levi (2016), p. 282: *forse ebbero un ruolo anche un sentimento di rivalsa contro l'Italia che lo aveva costretto all'esilio e la sensazione che, nonostante i grandi cambiamenti intervenuti con la fine della seconda guerra mondiale, le cattedre universitarie continuassero ad essere assegnate non soltanto per i meriti dei docenti, ma anche in base a legami con i partiti di governo. Sull'assegnazione delle cattedre universitarie in Italia Beppo Levi notava realisticamente che durante il fascismo molti erano saliti in cattedra perché fedeli al regime e ora, nonostante l'epurazione, restavano in cattedra; altri vi aspireranno «se non oggi, domani, per le loro relazioni con la Democrazia Cristiana.»*

[40] See LEVI.6 and LEVI.7.

charged with studying the project of a Latin Mathematical Union. Left a widow, Levi devoted the last ten years of his life entirely to teaching and to the Institute of Mathematics, surrounded by the affection of collaborators, students, and colleagues.[41] In this last period, he had important recognition from both his 'two homelands': in 1951, he was declared professor emeritus of the University of Bologna, in 1955, the *Revista de la Unión Matemática Argentina* published a volume in his honour, and in 1956, the Accademia dei Lincei awarded him the Feltrinelli prize for Mathematics.[42] On 8 August 1961, feeling his energies weaken, he resigned from the direction of *Mathematicae Notae*. He died suddenly some days later, without seeing the special issue of the journal (XVIII, 1962) that colleagues had dedicated to the 'father of mathematics in Rosario'.[43]

Referring retrospectively to his expulsion from the University of Bologna, he was to declare with subtle irony: 'I didn't pass the race exam, but I managed to do much better'.[44]

Documents

Levi.1 MFA to the Ministry of the Interior, Directorate General for Foreign Italians, 19.1.1940

ACS-*Levi*, fol. 1r

For due knowledge, a report by the Ambassador in Buenos Aires is transcribed below about the landing in that port of well-known Italian intellectuals of Jewish race. On the current 6th November, with the motorboat Oceania, a group of Italian Jews arrived and obtained from these Authorities the permission to disembark in Argentina. Among these are Prof. Benedetto Morpurgo, former holder of the chair of General Pathology at the University of Turin, Prof. Beppo Levi, who will hold a course in Higher Mathematics at the University of the Litoral in La Plata, the doctor in Physics Laura Levi, Prof. Marcello Finzi, Eng. Silvio Fubini and the former Hon. Prof. Gino Olivetti, who declared that he is here in Argentina for pleasure and has no intention of settling here.[45] It does not appear that any of them released statements or granted interviews upon arrival, with the exception of

---

[41] Levi.8. See also Levi.9.

[42] *Homenaje a Beppo Levi* 1955.

[43] Pessino and Marangunic (2017), p. 62.

[44] Levi [1950 ca.] in Momigliano Levi (2016), p. 282.

[45] Benedetto Morpurgo (1861–1944), professor of General Pathology at the University of Turin from 1903 to 1935. Laura Levi (1915–2003), Beppo's daugher. Marcello Finzi (1879–1956), jurist, professor at the University of Modena until 1938. He emigrated to Argentina, where he held a professorship in Cordoba. Jacob Angelo Gino Olivetti (1880–1942), lawyer and deputy of the Kingdom of Italy (1919–1938), left Italy with his wife for Argentina.

Prof. Morpurgo. The latter, interviewed by *Crítica*,[46] refrained from making political statements but talked about his work as an oncology researcher and his love for Italy. He finally claimed to have come to Argentina 'for purely personal reasons'. The press has naturally given considerable coverage of photographs and titles to the disembarkment of the passengers from Oceania, emphasising their Semitic origin.

Per opportuna conoscenza si trascrive qui di seguito un rapporto del R. Ambasciatore in Buenos Aires, circa lo sbarco in quel porto di noti intellettuali italiani di razza ebrea. Il 6 corrente [Novembre] con la Motonave Oceania è giunto un gruppo di ebrei italiani che ha ottenuto da queste Autorità il permesso di sbarco in Argentina. Tra questi figurano il prof. Benedetto Morpurgo, già titolare della cattedra di Patologia Generale nell'Università di Torino, il Prof. Beppo Levi, che terrà un corso di Matematica Superiore all'Università del Litorale in La Plata, la dottoressa in Fisica Laura Levi, il Prof. Marcello Finzi, l'Ing. Silvio Fubini e l'ex On. Prof. Gino Olivetti, il quale ultimo ha dichiarato di venire in Argentina per diporto e di non avere intenzione di stabilirsi qui. Non risulta che nessuno dei predetti abbia fatto dichiarazioni o concesso interviste al momento dell'arrivo, ad eccezione del prof. Morpurgo. Quest'ultimo, intervistato da Crítica, *si astenne dal fare dichiarazioni di politica, parlò del suo lavoro come cancerologo e del suo amore per l'Italia. Affermò infine di essere venuto in Argentina 'per ragioni puramente personali.' Questa stampa ha naturalmente dato un notevole risalto di fotografie e di titoli allo sbarco dei suddetti passeggeri dall'Oceania, sottolineandone l'origine semita.*

Levi.2 R. Legation of Italy in Montevideo to the Ministry of Foreign Affairs and the Ministry of Popular Culture, Montevideo 28.9.1940

ACS-*Levi*, fol. 1r

Invited by the Faculty of Engineering of the local University, in accordance with the intellectual exchange plan in force with the Argentine Universities, Prof. Beppo Levi, Director of the Institute of Mathematics of the Faculty of Exact Sciences of Rosario di Santa Fe, has arrived here. The aforementioned professor will give three scientific conferences in this University. In giving news of his arrival, the press recalls that Prof. Beppo Levi trained in Italy and asserts that he is one of the greatest contemporary figures in the field of Exact Sciences. Signed Bellardini Ricci

Invitato dalla Facoltà di Ingegneria della locale Università in base al piano di intercambio intellettuale vigente con le Università Argentine è qui giunto il Prof.

---

[46] *Crítica de la Argentina* was a daily newspaper from Buenos Aires.

Beppo Levi, Direttore dell'Istituto di Matematica della Facoltà di Scienze Esatte di Rosario di Santa Fe. Il predetto professore terrà tre conferenze di carattere scientifico in questa Università. Nel dare la notizia del suo arrivo la stampa ricorda che il prof. Beppo Levi ha fatto la sua formazione in Italia, e asserisce che egli è una delle più grandi illustrazioni contemporanee nel campo delle Scienze Esatte. F.^{to} Bellardini Ricci

Levi.3 MEN to the Ministry of Popular Culture-Directorate General for Propaganda Services, Rome 22.12.1940

ACS-*Levi*, fol. 1r

With reference to your request, we inform you that Prof. Beppo Levi, former professor of Mathematical Analysis at the Royal University of Bologna, was discharged from service by a decree of 30 November 1938-XVI, no. 1779. With regard to the scientific value of the aforementioned professor, it is known that, due to the variety of courses held, Levi has shown possession of an uncommon mathematical culture, and has achieved results of considerable importance in the field of analysis and algebra. The Minister signed Giustini

In relazione a quanto è stato richiesto con la nota sopraindicata, si comunica che il Prof. Beppo Levi, già ordinario di Analisi Matematica nella Regia Università di Bologna, è stato dispensato dal servizio, con decreto del 30.11.1938 XVI, n. 1779. In merito al valore scientifico del predetto professore, si fa conoscere che il medesimo, per la varietà degli insegnamenti impartiti, ha mostrato di possedere una cultura matematica non comune, ed ha conseguito nel campo dell'analisi e dell'algebra risultati di notevole importanza. Il Ministro, F.^{to} Giustini

Levi.4 B. Levi to the Minister of National Education of the Kingdom of Italy, Rosario 19.5.1944

ACS-*Levi*, fol. 1r

The undersigned, Beppo or Beppino Levi, former full professor of Mathematical Analysis (Algebraic and Infinitesimal) at the R. University of Bologna, discharged from service on 14 December 1938, in application of the racial laws, received communication from the consul general of Italy in Rosario (Argentina) of the acts aimed at repealing the aforementioned laws and reinstating individuals who were affected. He thus asks to be readmitted to the functions corresponding to his role as a university professor. As documentation he presents, in addition to the well-known facts that are easily verifiable by testimonies in any place of Italian culture, the copy signed and authenticated by the consul of a document issued in January 1939 by the Rector of the University of Bologna, Prof. Alessandro Ghigi[47] re. the facts previously mentioned, and the preamble of the decree of issue of the pension, which also states the reason for this. In view of the practical conditions, by which some time would necessarily pass before the new provisions become effective, he considers it his duty to point out that, born on 14 May 1875, his eligibility for remaining in office under the University Law (1936) should cease at the end of the academic year 1944–45, provided that a new law does not overturn this rule. He therefore expresses his desire for information on this matter. With observance, Beppo Levi

Il sottoscritto, Beppo o Beppino Levi, già professore ordinario di Analisi Matematica (Algebrica e Infinitesimale) nella R. Università di Bologna, dispensato dal servizio il 14 dicembre 1938 in applicazione delle leggi sopra la razza, avendo ricevuto comunicazione dal Sig. Console Generale d'Italia in Rosario (Argentina) delle disposizioni dirette ad abrogare le dette leggi e rendere possibile la riammissione in servizio dei funzionari che ne erano stati colpiti, chiede di essere riammesso nelle funzioni corrispondenti alla sua qualità di professore universitario. Come documentazione presenta, in più del fatto notorio e facilmente verificabile per testimonianze in qualunque sede di cultura italiana, la copia da lui firmata e autenticata dal sig. Console di un documento rilasciato nel gennaio 1939 dal Rettore dell'Università di Bologna Prof. Alessandro Ghigi e relativo ai fatti accennati e del preambolo del decreto di liquidazione della pensione, che ugualmente ricorda la causale di essa. In vista delle condizioni di fatto, per le quali dovrebbe necessariamente intercedere un certo tempo prima che le nuove disposizioni possano farsi effettive, considera suo dovere far notare che, essendo nato il 14 maggio 1875, la sua possibilità di restare in carica a norma della legge universitaria del 1936 dovrebbe cessare colla fine dell'anno accademico 1944–45, sempre che altra disposizione di legge non venga ad interrompere la validità di detta norma. Esprime pertanto il desiderio di ottenere informazioni al riguardo. Con tutta osservanza, Beppo Levi

Levi.5 B. Levi to L. Nicolai, Consul General of Italy, Rosario 14.6.1944

ACS-*Levi*, fol. 1r-v
Mr. consul, I am pleased to acknowledge the verbal communication *re.* the remark made by the Royal embassy of Buenos Aires about the fact that, attached to my request for reinstatement, there is no declaration of fidelity and obedience to the R. Government. I would ask you to convey to the Royal embassy the following considerations in this regard. As far as I know, the declaration formula

---

[47]Alessandro Ghigi (1875–1970), Rector of the University of Bologna (1930–1943).

dates to February of the current year or earlier. Italy and its government are in a process of transformation. For a request of this nature to have value, it would need to be issued by the government currently in force; even in this case it would be of supremely questionable value due to the declared precariousness of the government itself. Furthermore, the current government has declared that, in the shortest time possible following complete liberation of the homeland, the ballots for election of the Constituent Assembly will be convened; and no citizen has obligations to a government, but only to the country, its Constitution, its laws, or possibly to the symbols that represent them. The current government was the first to give us the example of such interpretation, if the news received in the newspapers are true that the symbolic oath of loyalty to the King has been replaced by an oath of loyalty to the country. If that were the meaning to be given to the request you made to me, I would have no difficulty in adhering and I ask that this statement be considered equivalent. But I must note that the text of the formula does not permit this interpretation in any way and is so absolute and indeterminate that its purpose and origin cannot be established. About my situation, I must also point out that, in any case, the text of the proposed declaration speaks of a royal official; a university professor has never been considered as such and can even less be considered so in a regime of liberty such as that which is going to be voted. For these reasons I ask the Royal embassy to attach this statement in place of the requested declaration. I thank you for your kindness and I present my most respectful greetings, B. Levi

Ill.mo Sig. Console, con piacere le do atto della comunicazione verbale circa il rilievo fatto dalla R. Ambasciata di Buenos Aires per il fatto che, in appoggio della mia domanda di riassunzione in servizio, manca una dichiarazione di fedeltà e obbedienza agli ordini emanati dal R. Governo. Mi permetto di pregarla di voler trasmettere alla R. Ambasciata le seguenti considerazioni al riguardo. Per quanto mi consta, la formula della dichiarazione rimonta al febbraio dell'anno corrente o prima. L'Italia e il suo governo sono in via di trasformazione. Perché una richiesta di questa natura potesse aver valore dovrebbe provenire dal governo attualmente in funzione; ed anche in tal caso sarebbe tal valore sommamente discutibile per la necessaria e dichiarata precarietà del governo medesimo. D'altra parte, il governo attuale ha dichiarato che, nel più breve tempo dalla completa liberazione del suolo patrio, sarebbero convocati i comizii per la elezione della Costituente; ed è chiaro che nessun cittadino ha obblighi verso un governo, ma solo verso il Paese, la sua Costituzione, le sue leggi, od eventualmente verso i simboli che li rappresentano. Di tale interpretazione è stato primo l'attuale governo a darci l'esempio, se sono esatte le notizie pervenute coi giornali che al simbolico giuramento di fedeltà al Re sia stato proposto sostituire un giuramento di lealtà verso il Paese. Se tale fosse il significato da attribuire alla richiesta che mi viene fatta, non avrei nessuna difficoltà ad aderirvi e chiedo che la presente dichiarazione sia considerata equivalente. Però debbo notare che il testo della formula non si presta in nessun modo, così assoluto ed indeterminato da non potersene intendere lo scopo e la provenienza. Nel merito della mia particolare situazione, debbo anche far notare che, in qualunque caso, il testo della dichiarazione proposta parla di regio funzionario e tale non è mai stato considerato il professore universitario; e meno ancora potrà esserlo in un regime di libertà quale è ai voti. Per queste ragioni prego la R. Ambasciata di voler allegare la presente in sostituzione della richiesta dichiarazione. La ringrazio per la sua gentilezza e le presento i miei distinti saluti, B. Levi

Levi.6 Memorandum for the Minister of Public Instruction, April–July 1947

ACS-*Levi*, fol. 1r-v

By cable, dated 15 February 1947, the Ministry of Foreign Affairs pointed out the desire of Beppino Levi, professor of Mathematical Analysis at the University of Bologna but still resident in Rosario (Argentina), where he took refuge following the well-known measures known under the name of racial laws, to be called to that 'University of Litoral'. In answering the Ministry, taking up the suggestion already supported *re.* a similar question previously advanced by Prof. Renzo Ravà,[48] I suggested postponing (for the moment) the placement at the disposal of the Ministry of Foreign Affairs (pursuant to Article 96 of the Laws on Higher Education), of the teachers rehired in service, who were abroad at the time of the rehiring: and instead to wait for the position of the teachers in question to be regulated in a uniform way, according to the legislative plan known to the Ministry. In this regard, I recalled that Article 2 of that scheme proposed exempting detached scholars from teaching and to consider them as being on research trips abroad, without, however, having to attribute any special allowance to them, apart from their salaries due as tenured university professors. The Ministry of Foreign Affairs now reports that the reasons set out above caused some objections on the part of our embassy in Buenos Aires, mainly because of the different treatment proffered to the professors Lattes and Viterbo, whose applications for placement at the disposal of the Ministry of Foreign Affairs, pursuant to art. 96, have been approved.[49] The embassy observes that rejecting

---

[48] Renzo Ravà (1906–1994), professor of Labour Law at the University of Florence from 1937. In November of 1939, Ravà had emigrated to the United States. Reinstated on April 2, 1946, he obtained a period of leave from 1946 to 1949, under the aegis of the Ministry of Foreign Affairs, as legal advisor to the Italian representation at the ONU; this arrangement was extended until 1951 and was followed by leave of absence for personal reasons until 1952.

[49] Leone Lattes (1887–1954), professor of Forensic Medicine at the University of Pavia, had emigrated to Argentina with his family in 1939. He returned to Italy in 1946. Camillo Viterbo (1900–1948), extraordinary professor of Commercial Law at the University of Cagliari, had emigrated first to Brazil, then to Argentina, where he had obtained a position in Cordoba. Viterbo came back to Italy in 1946, for a very short period, before returning, seriously ill, to Argentina.

Prof. Levi's request could give rise to interpretations that should be avoided, 'all the more so since Prof. Levi, both for his position in the Argentine scientific arena, and for the seniority of his services, is just as deserving of the requested treatment as his aforementioned colleagues'. In this regard, it should be noted that Prof. Lattes was placed at the disposal of the Ministry of Foreign Affairs after he had already returned to Italy and had regularly resumed his lectures. As for Prof. Viterbo, for whom this Ministry had raised the same exception as for Prof. Levi, it should be noted that he was made the subject of the provision only because of the repeated insistence of the Ministry of Foreign Affairs. However, given these two precedents and considering the care shown by the Dicastery in favour of Prof. Levi, we are pleased to submit the case to the examination of your authority for those decisions that you, Minister, will deem appropriate to adopt in this regard.

Con telespresso in data 15 febbraio c.a. [1947], il Ministero degli Affari Esteri faceva presente il desiderio del Prof. Beppino Levi, ordinario di Analisi Matematica nell'Università di Bologna ma tuttora residente in Rosario (Argentina), dove si era rifugiato in seguito ai noti provvedimenti noti sotto il nome di leggi razziali, di essere comandato presso quella Università del Litorale. Nel rispondere al suddetto Ministero, lo scrivente, riprendendo la tesi già sostenuta a proposito di analoga domanda precedentemente avanzata dal Prof. Renzo Ravà, sosteneva l'opportunità di soprassedere per il momento al collocamento a disposizione del Ministero degli Affari Esteri (ai sensi dell'art. 96 del T.U. delle leggi sull'istruzione superiore), dei docenti riassunti in servizio, che si trovavano all'estero al momento della riassunzione stessa: ciò in attesa che la posizione dei docenti in parola venisse regolata in modo uniforme, secondo lo schema di provvedimento legislativo noto anche al suddetto Ministero. A tal proposito si ricordava che l'art. 2 del citato schema di provvedimento proponeva di esonerare quei docenti dall'obbligo dell'insegnamento e considerarli in missione all'estero, senza che peraltro si dovesse attribuire loro nessuna speciale indennità, oltre gli emolumenti loro spettanti quali professori universitari di ruolo. Il Ministero degli Affari Esteri riferisce ora che le ragioni su esposte hanno sollevato qualche obiezione da parte della nostra Ambasciata a Buenos Aires, e ciò principalmente a motivo del diverso trattamento usato ai professori Lattes e Viterbo, le cui domande di collocamento a disposizione del Ministero degli Affari Esteri, ai sensi del citato art. 96, hanno trovato favorevole accoglimento. La suddetta Ambasciata osserva che il non accogliere ora la richiesta del Prof. Levi potrebbe dare adito ad interpretazioni che conviene evitare, 'tanto più che il Prof. Levi, sia per la sua posizione nel campo scientifico argentino, sia per l'anzianità dei suoi servizi, è altrettanto meritevole dei predetti suoi colleghi del trattamento richiesto'. Al riguardo, giova osservare che il Prof. Lattes venne collocato a disposizione del Ministero degli Affari Esteri dopo che era già rientrato in Italia e aveva regolarmente ripreso le sue lezioni. In quanto al Prof. Viterbo, per il quale questo Ministero aveva sollevato la stessa eccezione che per il Prof. Levi, va notato che non fu fatto oggetto dell'accennato provvedimento che in seguito alle reiterate insistenze del Ministero degli Affari Esteri. Comunque, dati questi due precedenti e considerata la premura dimostrata dal suddetto Dicastero a favore del Prof. Levi, si ha il pregio di sottoporre il caso all'esame della S.V. per quelle decisioni che Ella, Signor Ministro, riterrà opportuno adottare al riguardo.

Levi.7 Extract from the Minutes of the Session of the Faculty of Physical, Mathematical and Natural Sciences of the University of Bologna, 9.10.1948

ACS-*Levi*, fol. 1r

The Dean conveys a letter from the Minister which asks for the Faculty's consent for prof. Beppo Levi to extend his teaching activity in South America for this year. The Faculty, although sorry to be unable to enjoy the teaching and advice of their illustrious colleague, is proud that Mr. Levi keeps up the name and prestige of the Bolognese Faculty and Italian Science in the distant lands of the Argentine Republic and therefore agreed that Prof. Levi still be available for this year to the Ministry of Foreign Affairs for educational and scientific assignments in South America. On this occasion, the Faculty wishes to express to Prof. Levi the warmest congratulations for the successes he has met, and the honours received from the major centres of South American higher education. Prof. [Beniamino] Segre takes the floor and informs colleagues that Prof. Levi will most likely return to Italy during the Argentine university holidays and therefore in a period of school activity in Italy. In this period, Prof. Levi wishes to hold a course of lectures on advanced topics in our university.[50] The Faculty is very pleased to hear this news provided by Prof. Segre and authorises him to arrange with Prof. Levi the concrete details of this project and to report to the Faculty regarding the case.

Il Preside comunica quindi una lettera del Ministro nella quale si chiede alla Facoltà il consenso perché il Prof. Beppo Levi possa prolungare ancora per quest'anno la Sua attività didattica nell'America del Sud. La Facoltà pure essendo dolente di non poter usufruire dell'insegnamento e del consiglio dell'illustre collega Levi è d'altra parte orgogliosa che il collega Levi possa tenere alto il nome ed il lustro della Facoltà bolognese e della Scienza italiana nelle lontane terre della Repubblica argentina e concede perciò il suo consenso a che il Prof. Levi possa ancora per quest'anno restare a disposizione del Ministero degli Esteri per incarichi didattici e scientifici nell'America del Sud. La Facoltà tiene anzi ad esprimere in questa occasione al Prof. Levi le più vive

---

[50]Levi held a series of eight lectures at the Mathematical Seminar of Bologna, starting from 20 January 1949.

felicitazioni per i successi avuti e gli onori da Lui ricevuti nei maggiori centri dell'alta cultura sud-americana. Il Prof. Segre prende la parola per informare i colleghi che molto probabilmente il Prof. Levi ritornerà in Italia nel periodo delle ferie universitarie argentine e quindi in un periodo di attività scolastica in Italia. Il prof. Levi desidera in detto periodo di tenere un corso di conferenze su argomenti superiori nella nostra Università. La Facoltà è ben lieta nell'udire queste notizie fornite dal Prof. Segre e dà mandato allo stesso di voler trattare con il Prof. Levi circa la realizzazione concreta di questo progetto e riferire poi alla Facoltà per le delibere del caso.

Levi.8 B. Levi to MFA, Rosario 28.11.1948

ACS-*Levi*, fols. 1r-2v
Final report by Dr. Beppo Levi, Director of the Mathematical Institute of the Facultad de Ciencias Matemáticas, Físico-Químicas y Naturales aplicadas a la Industria—Universidad Nacional del Litoral.

Having already given, in my first report, extensive information on the tasks and activities of the Institute, I limit myself to answering, as closely as possible, the individual questions listed in your cable no. 5883

1° In the academic year that closes next December, I have had the opportunity to add to the aforementioned main functions of my activity, other teaching assignments of considerable importance:

a) I was invited by the Faculty of Physical-Mathematical Sciences of the University of La Plata, at the end of March, to held a course of Foundations of Mathematics which was addressed, for strategic reasons, both to the students of the *Profesorado* and to the 5th year students of the degree programme in Mathematics; I accepted the charge, despite the considerable distance, under the condition (immediately accepted) of unifying the weekly teaching into 3 hours to be delivered all on the same day, namely Saturday. The first two hours of each day were allocated to a course of a more general and elementary character intended simultaneously for the two groups of students, while the third hour was devoted to a monographic course for students of Pure Mathematics; I chose as topic, in this case, transfinite arithmetic.

b) In June, the Faculty of Rosario (on which the Institute depends) invited me to give a course of Analytic Geometry and Infinitesimal Calculus II, for student engineers, who until that date, for administrative and other temporary reasons, had been suspended due to the lack of professor. To recover the lost months, I intensified the lessons, teaching 6 hours a week (three groups of two hours); the exams are currently taking place, but I have the impression that the outcome was excellent and fully appreciated by the Faculty [. . .]

In the course in Foundations of Mathematics held in La Plata, starting from basic notions of Mathematical Logic and the axiomatics of natural numbers, in which the name of Giuseppe Peano necessarily features, the various characterisations and transformations of the notion of number were treated; a final connection with the foundations of Geometry was developed, necessarily following the personal thought of the teacher, who by training and feeling, aspires to represent a little Italian culture. The monographic course was also founded mostly on personal research.

The course for Engineering students is held in accordance with a local statutory distribution of the contents of courses; the labelled title of Analytic Geometry and Infinitesimal Calculus II refer to the fact that it should develop analytic geometry in 3 dimensions and infinitesimal analysis in several variables, together with the theory of differential equations; the course was structured by making the first part depend on vector notions, constituting a fairly essential variant to the address followed previously, and bringing the analytical part closer to the features developed during my teaching in Italian Universities. [. . .]

The academic authorities of the Universidad Nacional del Litoral are particularly favourable to Italian culture, with the Dean Eng. Ángel Guido of the Faculty of Sciences and the Rector Eng. Luciano Micheletti both being of Italian origin.[51] At the same time, Italian culture is also highly appreciated by the population of Spanish origin, if only for linguistic affinity. The Institute, as I said in my previous report, is attended by many colleagues who, also for personal relationships, do not fail to demonstrate their sympathy. In relation to the 7th and following questions, I believe that the essentially scientific direction of my work dispenses me from more detailed answers. Nevertheless, considering the problem of diffusion of the Italian scientific book, and declaring a priori my complete ignorance on commercial facets that can exert a decisive influence on the subject, it seems useful to note the lack of presentation of these publications in the large bookshops, which are not lacking in the country, and which are sufficiently equipped with foreign scientific and technical literature. Undoubtedly, these stores are mainly concentrated in Buenos Aires, where I have limited opportunities to stay for enough time to know many details; I know that in Buenos Aires there are some Italian bookshops such as *Il libro italiano*, another recently established (named S. Marco, if I am right), the old Vallardi publishing house, etc., but I can point out that, for the scientific books to be sold, in a country where the offer of the English and particularly North American books is not huge, either in original version or in translation, and where the cities are enormously vast, it is necessary for the client to find these books among others, without looking for a specialised national bookstore (the issue may be different for literary and artistic works). Several Italian books have been kindly sent to me by Italian publishers and colleagues, to make them known through reviews in my journal:

---

[51] Ángel Francisco Guido (1896–1960) architect, engineer, and writer.

Mathematicae Notae. I must apologise because, due to administrative and typo-graphical difficulties, the journal has interrupted publications for more than 6 months, and thus I have been unable to fulfil this welcome duty. A volume is currently in the pipeline, with two issues, in the guise of compensation for the forced silence. However, I must emphasise that I have not yet seen these volumes made available to the public. Now, to better explain the observation, just bear in mind that a large bookstore in Buenos Aires, which presents itself with the mere name of *Palacio del Libro* is notoriously a showcase for the Parisian Hachette publishing house and, despite generally stocking a broad selection of French publications, is also a place to find scientific texts from many other countries. It is my intention, immediately after completing the exams that will keep me busy next December, to go to Italy, taking advantage of the rapid current means of transport, and to resume for a few months my post at the University of Bologna, satisfying the kind requests that come to me from colleagues.

Relazione finale del Dr. Beppo Levi, Direttore dell'Istituto Matematico della Facultad de Ciencias Matemáticas, Físico-Químicas y Naturales aplicadas a la Industria—Universidad Nacional del Litoral.

Avendo dato già nella mia prima relazione informazioni ampie sopra i compiti e le attività dell'Istituto, mi limito a rispondere, nel modo più aderente possibile alle singole domande contenute nel Telespresso n. 5883

1° Alle suddette funzioni principali della mia attività ho avuto occasione nell'anno accademico che si chiude col prossimo dicembre di aggiungere funzioni docenti di notevole importanza:

a) *Invitato dalla Facoltà di Scienze Fisico-Matematiche dell'Università de La Plata, alla fine di marzo, a farmi carico di un corso di Fondamenti di Matematica da dettarsi ad uso, contemporaneamente, degli studenti dell'Istituto di Professorato e degli studenti del 5° anno per la laurea in Matematica; ho accettato l'incarico, nonostante la notevole distanza, colla condizione, che fu subito accettata, di raccogliere l'insegnamento settimanale in 3 ore da dettarsi tutte nello stesso giorno, il sabato. Le prime due ore di ciascun giorno furono destinate a un corso di natura più generale e elementare destinato simultaneamente ai due gruppi di studenti, mentre la terza ora fu destinata a un corso monografico per gli studenti di matematica pura, scegliendo in questo caso come argomento l'aritmetica transfinita.*

b) *Nel giugno, la Facoltà di Rosario, nella quale ha sede l'Istituto, mi invitò a farmi carico del corso di Geometria Analitica e Calcolo Infinitesimale II, per gli studenti ingegneri, che fino a quella data, per ragioni amministrative e altre momentanee aveva taciuto per mancanza del professore. Per rimediare ai mesi perduti ho intensificate le lezioni, in 6 ore settimanali (tre gruppi di due ore); si stanno attualmente svolgendo gli esami, ma ho tutta l'impressione che l'esito fu ottimo e di pieno gradimento per la Facoltà. [. . .]*

5° Nel programma di Fondamenti di Matematica svolto a La Plata, prendendo le mosse da nozioni di Logica Matematica e dalla assiomatica dei numeri naturali, in cui domina necessariamente il nome di Giuseppe Peano, il programma è andato svolgendosi sulle varie caratterizzazioni e trasformazioni della nozione di numero, collegandosi finalmente coi fondamenti della Geometria seguendo necessariamente il pensiero personale dello scrivente, il quale, per formazione e per sentimento, aspira a rappresentare un poco di cultura italiana. Il corso monografico fu pure fondato in massima parte sopra ricerche personali. Il Programma del corso agli studenti di ingegneria è collegato a una distribuzione statutaria locale dei corsi; il titolo indicato di Geometria analitica e Calcolo Infinitesimale II si richiama al fatto che, al corso che mi fu assegnato, sarebbe attribuita la geometria analitica in 3 dimensioni e l'analisi infinitesimale in più variabili, unitamente alla teoria delle equazioni differenziali; il corso si è svolto facendo dipendere la prima parte da nozioni vettoriali, costituenti una variante abbastanza essenziale all'indirizzo seguito anteriormente, e avvicinando la parte analitica ai lineamenti elaborati durante il mio insegnamento nelle Università italiane. [. . .]

7° Le autorità accademiche dell'Università del Litoral sono particolarmente favorevoli alla cultura italiana, essendo Rettore l'Ing. Angel Guido della Facoltà di Scienze e Decano l'Ing. Luciano Micheletti, entrambi di origine italiana. D'altronde la cultura italiana è molto apprezzata anche dalla popolazione di origine spagnola, altro non fosse che per l'affinità della lingua. L'Istituto, come già dissi nella mia relazione precedente, è frequentato da molti colleghi che, anche per i rapporti personali, non mancano di dimostrare tale simpatia. In rapporto alle domande 7° e seguenti credo che la direzione essenzialmente scientifica del mio lavoro mi dispensa da risposte più particolareggiate. Nondimeno, considerando il problema della diffusione del libro scientifico italiano, e dichiarando a priori la mia completa ignoranza sopra questioni di ordine commerciale che nell'argomento possono avere influenza decisiva, mi pare utile rilevare la mancanza di presentazione di queste pubblicazioni nelle grandi librerie del genere, che non mancano nel paese, e che sono abbastanza fornite di letteratura scientifica e tecnica estera. Indubbiamente, queste grandi librerie sono concentrate prevalentemente a Buenos Aires, dove io ho limitate occasioni di trovarmi con tempo sufficiente per conoscere molti particolari; so che a Buenos Aires esistono alcune librerie italiane come Il libro italiano, *un'altra di recente costituzione (credo di nome S. Marco), la vecchia casa Vallardi, ecc., ma credo di poter segnalare che, perché il libro scientifico si venda, in un paese dove non è enorme l'offerta del libro inglese e particolarmente nordamericano, in originale e in traduzione, e dove le città sono enormemente estese, è necessario che il cliente lo trovi fra gli altri senza cercare la casa nazionale specializzata (la cosa può essere diversa per le opere letterarie e artistiche). Vari libri italiani mi sono stati inviati gentilmente da editori e colleghi italiani perché li facessi conoscere mediante recensioni nella mia rivista:* Mathematicae Notae. *Debbo scusarmi che, per difficoltà di ordine amministrativo e tipografico, la rivista non appare da più di 6 mesi, e quindi nemmeno ho potuto compiere il grato dovere. Attualmente è in corso di stampa un fascicolo, per lo meno doppio, che cerca di compensare il forzato silenzio. Però debbo osservare che i suddetti volumi non ho visto finora a disposizione del pubblico. Ora, per spiegare meglio l'osservazione, basta tener presente che una grande libreria di Buenos Aires, che si presenta col semplice nome di* Palacio del Libro *è notoriamente della Casa Hachette di Parigi e, pur avendo un'ampia offerta di pubblicazioni francesi, è un recapito sicuro per trovare bibliografia scientifica delle più svariate provenienze. È mia intenzione, subito dopo terminati gli esami che mi occuperanno nel prossimo mese di dicembre, di recarmi in Italia, approfittando delle rapide comunicazioni attuali, e di riprendere per qualche mese il mio posto nell'Università di Bologna, obbedendo alle gentili sollecitazioni che mi vengono dai colleghi.*

Levi.9 B. Levi to A. Terracini, Rosario 14.12.1953

BSM-*Terracini*, fol. 1r

Dear Prof. Alessandro Terracini, A few hours ago, at the Institute, I received your letter of the 2nd and I hasten to answer you, as you see, in the shortest time. First of all, I warmly thank you and your colleagues for having thought of republishing, in honour of Peano, my two articles on his scientific work, which is appreciated by me in two very different ways: firstly, because you considered me for participation in the initiative and also because you share my interpretation of Peano's thought.[52] I quickly re-read the two articles to confirm my current agreement, and I am glad to give it. I noticed a few sections that refer more precisely to the time when the articles were written, but I leave to you the decision, in agreement with other similar resolutions that you can take, of whether it is better to put some brief note of justification or to delete a few words or lines. My precise opinion is that if I had to rewrite the two articles specifically for the occasion, I would extend a little some points, but at the same time I would also be less prepared for an equally complete presentation of Peano's thought. So, I answer in part to your question '... unless 20 years later you wanted to write something different'. I answer in part, I say, because if I did something different, it would no longer be a comment on Peano's thought, but a separate work, maybe aimed at summarising some of the ideas I presented in the second half of my *Correría en la Lógica.*[53] At this moment in time, I have not bothered to review the text, but I think it would coincide quite well with the last half page of my second Note on Peano (*Intorno alle vedute* ...). The final decision may depend largely on editorial information: mainly space and time, which only you can

---

[52]Terracini was editing the book *In memoria di Giuseppe Peano* (Cuneo, Liceo Scientifico Statale, 1955) which would include Levi's paper *L'opera matematica di Giuseppe Peano*, pp. 9–21. Levi's contribution derived from two previous articles: B. Levi, *L'Opera matematica di Giuseppe Peano*, Bollettino dell'UMI, 11, 1932, pp. 253–262 and *Intorno alle vedute di G. Peano circa la Logica matematica*, Bollettino dell'UMI, 12, 1933, pp. 65–68.

[53]B. Levi, *Correría en la Lógica*, Revista. Serie A. Matemáticas y Física Teórica, 3, 1942, pp. 13–78.

know. About this point, I await further clarification, always in the event that you consider a partial rewrite useful. And now I take the opportunity to give you and my colleagues some news about me. First, I would like to let you know that, if my journal is slow to arrive, it is not because it is dead. At most, I can say that it is in gestation. We are encountering two difficulties: one is a growing lack of collaboration, the other is a trend that seems to be taking root here, i.e., the worker, perhaps because he is too flattered by the government, no longer works. I just tell you that more than a month ago I travelled to Santa Fe on purpose to see what had happened in the printing firm; I was no longer receiving drafts, but I was given a promise to have them in a week. Conversely, a month has passed, and nothing has arrived. What is currently in print is a single issue that would constitute the entire work of the year; 8 or 10 pages are by Santaló; all the rest is mine and of this stuff, the essential part is research on a topic completely new for me, of which perhaps your brother [Benvenuto] told you (therefore I do not repeat myself).[54] We'll see. Your brother will have given you other news of mine. It is a fact that I live here very alone, but fortunately I continue working and for now the years do not tire me too much, even in the teaching. My brother [Augusto] pushed me, in various letters, to take a trip to Italy during the Argentine summer. The disadvantage is that the distance is great, and I would no longer want to travel by plane, although personally I have had great travel experiences. I wrote to my brother that I will think about it for next year. Please greet [Guido] Ascoli and Tricomi for me; other affectionate greetings to Prof. Benvenuto, to whom you can give, as far as I know, excellent news of his entire family in Buenos Aires; other greetings to you, your wife, and children. Yours affectionately, B. Levi

Caro Prof. Alessandro Terracini, ho ricevuto poche ore fa, all'Istituto, la sua lettera del 2 corr. e mi affretto a risponderle, com'Ella vede, nel tempo minimo possibile. In primo luogo, ringrazio di gran cuore Lei e i colleghi per aver pensato di ripubblicare, in onore del Peano, i due articoletti miei sopra la sua opera scientifica, il che mi è grato in due sensi molto diversi: quello di aver pensato alla mia partecipazione all'iniziativa e quello di averli partecipi della mia interpretazione del pensiero di Peano. Ho voluto pure rileggere rapidamente i due articoli per confermarle il mio accordo attuale per cui le do volentieri la mia completa adesione. Vi è qualche parola che si riferisce più precisamente al momento in cui gli articoli furono scritti, ma lascio a loro decidere, d'accordo con altre risoluzioni analoghe che possano prendere, se convenga porre qualche breve nota di giustificazione o sopprimere qualche parola o qualche riga. La mia precisa opinione è che se io dovessi riscrivere apposta per l'occasione i due commenti mi diffonderei un po' più sopra qualche punto, ma viceversa pure sarei meno pronto a una presentazione ugualmente completa di quel pensiero. Così

---

[54]Levi refers to volume XII–XIII, issue 2–4 of the journal *Mathematicae Notae*, 1954, which included 12 papers by Levi and one by L. Santaló: *Unas generalizaciones del teorema de los cuatro vértices* (pp. 69–78).

rispondo anche in parte alla sua domanda '... a meno che a 20 anni di distanza volessi fare qualche cosa di diverso'. In parte, dico, perché ciò che di diverso forse potrei fare sarebbe, non più come commento al pensiero di Peano, ma come collaborazione a parte, riassumere qualcuna delle idee esposte nella seconda metà della mia Correría en la Lógica *le quali in questo momento non mi sono curato di rivedere, ma che mi pare collimerebbero abbastanza bene coll'ultima mezza pagina della mia seconda Nota* (Intorno alle vedute *ecc.*). *Ciò può dipendere in massima parte dai particolari editoriali: principalmente spazio e tempo che solo loro possono conoscere. Sopra questo punto aspetto un ulteriore suo schiarimento, e sempre sotto la condizione che Ella consideri la cosa utile. Ed ora prendo l'occasione per dare a Lei e ai colleghi qualche notizia di me. In primo luogo, desidero far sapere che, se la mia rivistina tarda ad arrivare, non è perché sia morta. Tutt'al più posso dire che è in gestazione. Sono davanti a due difficoltà, una è la crescente mancanza di collaborazione, l'altra è un andazzo che pare vada crescendo qui che l'operaio, forse perché troppo adulato dal governo, non lavora più. Basta che le dica che più di un mese fa feci viaggio apposta a Santa Fe per vedere che cosa avvenisse nella tipografia per cui non ricevevo più bozze ed ottenni la promessa di averle nella settimana seguente. Viceversa, è passato, come le dissi, un mese e nulla è arrivato. Ciò che è attualmente in stampa è un fascicolo unico che rappresenterebbe tutto il lavoro dell'anno, 8 o 10 pagine sono di Santaló e tutto il resto mio e di questo resto la parte essenziale è una ricerca sopra un argomento per me del tutto nuovo, di cui forse qualcosa può averle detto suo fratello e perciò non mi ripeto. Si vedrà. Di altre notizie mie pure le avrà parlato suo fratello. È un fatto che vivo qui molto solo, ma fortunatamente continuo lavorando e per ora gli anni non mi affaticano troppo nemmeno nelle lezioni. Mio fratello mi spingeva, in varie lettere, a fare un viaggetto in Italia per l'estate argentina. Il male è che la distanza è grande e che per il momento non vorrei più rivolgermi all'aereo, sebbene personalmente l'abbia provato ripetutamente con viaggio ottimo. Così ho scritto a mio fratello che ci penserò per l'anno prossimo. La prego di salutare per me Ascoli e Tricomi ed altri saluti affettuosi fare al prof. Benvenuto dandogli, per quanto io so, ottime notizie di tutta la famiglia a Buenos Aires, ed altri saluti invio a lei e signora e figli. Aff.mo B. Levi*

Chapter 13
Bonaparte Colombo: The Inability to Return to Normal Life

Formerly a student and later a colleague of G. Fano, G. Fubini, and A. Terracini, Bonaparte Colombo was born in Turin on 29 June 1902, to Pacifico and Irma Levi. He spent his childhood and youth in Turin, graduating with honours on 14 July 1924, with a thesis in mathematical physics supervised by Somigliana. Like his fellow freshman Beniamino Segre, Colombo embarked on a university career, publishing some important works linked to the research undertaken in his thesis. In the period 1924–1938, he was an assistant at the University of Turin: first, in the Institute of Projective and Descriptive Geometry with Drawing and then in that of Rational Mechanics.[1] Those were the years of greatest creativity for Colombo, who—thanks also to his relationships of friendship and scientific collaboration with Fubini— published some articles of good value and originality, qualifying to teach Infinitesimal Analysis in 1930. In 1926, he entered the teaching staff of the Royal Military Academy of Artillery and Genius in Turin where he taught Descriptive Geometry from 1926 until the introduction of the racial laws. Contemporary, from 1932 to 1938, he was charged of the course of Complementary Mathematics at the University.

As a Jew from both maternal and paternal lineages, Colombo was removed from service in autumn of 1938 and he started working as a teacher of Mathematics and Physics at the Jewish school in Turin. In 1941, the family left Turin and moved to a small village (Moncalieri), renting a house from a Catholic cousin. In her diary, Bonaparte's daughter Marina noted:

> Every day, I take the train to go to school with my dad who goes to work. I live in fear, I have the impression that a misfortune can happen to me every time I cross the city. It is strange to go to a school where there are only Jews. We are really excluded from civilisation; it is as if we were sick and have to stay among ourselves so as not to infect anyone.[2]

---

[1] See Allasia (1999).

[2] Interview with Colombo's grand-daugher and personal documents given to E. Luciano in 2015: *La journée je prends le train pour aller à l'école avec mon papa qui va travailler. Je vis dans la peur, j'ai l'impression qu'un malheur peut m'arriver chaque fois que je traverse la ville. C'est étrange de*

E. Luciano, *The Jewish Mathematical Diaspora from Fascist Italy*, Science Networks. Historical Studies 64, https://doi.org/10.1007/978-3-031-64896-0_13

In 1942, a new resettlement arrived: Colombo's house was destroyed by bombs, and they moved to Mondovì where they lived together with grandparents Pacifico and Irma, uncle Riccardo, and their children Dino and Ezio.

In October 1943, with the beginning of the roundups in Turin, Bonaparte and his wife Adriana Treves began to think about emigration. The grandparents did not want to follow them to Switzerland, abandoning everything and starting life anew abroad. The Colombos themselves were uncertain as to what to do. Fano and Fubini were fundamental at this point in convincing Bonaparte to make such a step, and, finally, they fled at the beginning of December 1943:

> We left, my parents and I, by train from Turin to Milan. My parents hid valuable jewellery in their clothes, we only took the bare minimum, which for me meant a small suitcase with some clothes. On the train, fear invaded me, and shivers ran through my whole body when I heard passengers say: «these are Jews who are running away!» Is being Jewish written on our faces? We are no longer people like the others, we no longer have an identity, at this point we are just Jews and nothing but Jews. The fact that I can now write a diary is perhaps due to those people who did not report us. In Milan, friends lent us a room. At two o'clock in the morning, we set off on foot to Lake Como. We crossed a mountain for an hour accompanied by a smuggler who guided us. We trusted the smuggler because he had already smuggled my uncle and my cousins to Switzerland. The journey was exhausting and disquieting. We left at night, so no one noticed us. The situation reminds me of my school friends when we run after each other to catch each other. I have the impression that the war is chasing me, but if I walk as fast as possible it will not reach me. When we arrived near the border, around four o'clock in the morning, the smuggler told us: «Now you will run as fast as possible, 200 metres away there is a net; you pass under it and nothing can happen to you anymore because you will be in Switzerland.» The net represented the Swiss border; there were bells that rang when you touched it. We then ran at full speed to reach the net as quickly as possible when suddenly gunshots rang out. Surely, they were Italians or Germans trying to play the same game. I ran even faster; my body was shot through with adrenaline. Once we arrived in Ticino, we announced ourselves to a Swiss guard and he took us to the police station. We had to show our papers and we went through an interrogation in which we explained the reason for our arrival. My father explained that he was a teacher. We were accepted quite easily, I was only 10 years old, and I think the police saw that we were doing nothing wrong. I know that often young men of twenty are pushed away because they must go to fight.[3]

---

fréquenter une école où se trouvent seulement des juifs. Nous sommes vraiment exclus de la civilisation, c'est comme si nous étions malades et que nous devions rester entre nous pour ne contaminer personne.

[3] Interview with Colombo's grand-daughter and personal documents given to E. Luciano in 2015: *Nous sommes partis, mes parents et moi, en train de Turin en direction de Milan. Mes parents ont caché des bijoux de valeurs dans leurs habits, nous n'avons pris que le strict minimum, ce qui signifie pour moi une petite valise avec quelques habits. Dans le train, la peur m'a envahi et des frissons ont parcouru tout mon corps lorsque j'ai entendu des passagers dire : « ce sont des juifs qui se sauvent ! » Le fait d'être juifs se voit sur nos visages ? Nous ne sommes plus des personnes comme les autres, nous n'avons plus d'identité, nous ne sommes plus que des juifs et rien que des juifs. Le fait que je puisse t'écrire maintenant est peut-être dû à ces personnes qui ne nous ont pas dénoncés. A Milan, des amis nous ont prêté une chambre. A deux heures du matin, nous sommes partis à pied jusqu'au lac de Côme. On a traversé une montagne pendant une heure accompagnés d'un contrebandier qui nous guidait. Nous avions confiance en le passeur parce qu'il avait déjà fait passer mon oncle et mes cousins en Suisse. Le voyage était éprouvant et inquiétant. Nous sommes*

On 23 December 1943, Bonaparte Colombo was interned in Gudo (one of the worst camp in Switzerland, a camp which was described as awful by many refugees); his wife Adriana Treves and daughter Marina were inmates in Rovio au Tessin. They found themselves again quite quickly in the camp of Balerna, from whence they passed to Lauterbach and finally to Aarau where Bonaparte was hospitalised, having contracted scarlet fever (Fig. 13.1a–f).

In March 1944, Adriana urged her husband to submit his application to teach mathematics in the Italian school in Lausanne, but in the meanwhile, thanks to the intermediation of Fano, he had been invited by G. Colonnetti to IUC:

> I was pleased to hear that you would be willing to move to Lausanne to take part in our school activity. And I inform you that we are in the process of organising mathematical courses for students who aspire to admission to the *École d'Ingénieurs*, and that I would be happy to entrust you with one of these courses. I must point out to you that, to that end, however, it would be necessary for you to speed up as much as possible the procedures for the transfer because the courses in question should begin at the beginning of the summer semester, no later than the 17th.[4]

At the Italian University Camp, Colombo held a course in Special Mathematics attended by students of the Faculty of Architecture. The contents of this course, whose handouts have unfortunately been lost, included basic notions of algebra (algebraic calculus, progressions, algebraic and graphic resolution of first- and second-degree equations), geometry (fundamental properties of geometric figures, geometric transformations, and Euclidean geometry), trigonometry (trigonometric equations and resolution of problems), analytic geometry (theory of conics), and descriptive geometry (rotations, projections, representations of elementary solid

---

partis la nuit pour que personne ne nous remarque. Cela me fait penser à mes camarades d'école et moi qui nous courions après pour nous attraper. J'avais l'impression que la guerre me courait après, je marchais le plus vite possible pour qu'elle ne m'atteigne pas. Lorsque nous arrivâmes près de la frontière, vers quatre heures du matin, le contrebandier nous a dit : « Maintenant vous courrez le plus vite possible, à 200 mètres il y a un filet, vous passez dessous et il ne pourra plus rien vous arriver car vous serez en Suisse. » La barrière faite de filet, déterminait la frontière suisse, on y trouvait des cloches qui sonnaient lorsqu'on la touchait. Nous avons alors couru à toute vitesse afin d'atteindre le plus vite possible le filet quand soudain des coups de feu retentirent. C'étaient sûrement des Italiens ou des Allemands qui tentaient de gagner le jeu. J'ai couru encore plus vite, mon corps fut traversé d'adrénaline. Une fois arrivés au Tessin, nous nous sommes annoncés à un garde suisse et il nous a emmenés au poste de police. Nous avons dû montrer nos papiers et nous avons passé un interrogatoire dans lequel nous avons expliqué la raison de notre venue. Mon père a expliqué qu'il était professeur. Nous avons été acceptés assez facilement, j'ai seulement 10 ans et je pense que les policiers voyaient bien que nous ne faisions rien de mal. Je sais que souvent, les jeunes hommes de vingt ans se font refouler car ils doivent aller combattre.

[4] ACT: G. Colonnetti to B. Colombo, Lausanne 7.4.1944: *Ho appresa con piacere la notizia che Ella sarebbe disposta a trasferirsi a Losanna per prendere parte alla nostra attività scolastica. E Le comunico che noi siamo sul procinto di organizzare dei corsi di matematiche per gli allievi di questo campo che aspirano alla ammissione all'*École d'Ingénieurs, *e che io sarei ben lieto di affidare a Lei uno di questi corsi. Le debbo far presente che a tal fine occorrerebbe però che Ella accelerasse il più possibile le pratiche per il suddetto trasferimento perché i corsi di cui si tratta dovrebbero iniziarsi, coll'inizio del semestre estivo, non più tardi del 17 corrente.*

Fig. 13.1 (a–f) Refugee certificates of Colombo's Family, Colombo Coll.

figures, and shadows theory). Some knowledge in physics (statics, kinematics, dynamics, optics, acoustics, and electricity), chemistry (both organic and inorganic), and technical drawing was also included.[5]

Exactly one year after their resettlement in Switzerland, the Colombos were able to communicate with the relatives who had remained in Italy and were informed that Bonaparte's father had been arrested by the Nazis in Turin, where he had travelled to save the family's furniture. Bonaparte feared that he would never see him again. That was a trauma: he felt himself responsible for not having insisted on making his parents safe in Switzerland and guilty for having taken that chance himself. However, Bonaparte did not give up hope of re-embracing his father, and with the liberation of Auschwitz, he was trustful that he would. In March 1945, he was still confident and eager to participate in Italy's reconstruction:

> My dad went to a meeting today. Italian refugees meet every Saturday afternoon to talk about Italy. They send their news; they talk about the evolution of Italy, and they try to decide how they will rebuild it. In these days, my father returns home in a good mood, because the news about Italy is good. The Allies are winning the war and liberating Northern Italy. However, it seems that Turin is really destroyed[6] (Fig. 13.2).

This hopeful attitude was to be shattered shortly afterwards. Returning to Italy on 12 June 1945, Colombo was informed that his father had been killed in Flossenbürg. Colombo was reinstated but from that moment onwards, he completely stopped his research activity and never wrote a single line of mathematics.

In the post-war period, he was appointed adjunct professor of Analysis at the University of Turin, a position he held until 1972–1973 when he retired. From 1951 to 1968, he was Vice-Director of the Institute of Analysis and, from 1952 to 1966, he taught Graphic Statics at the School of Application in Turin. He passed away in Turin on 12 March 1989, surrounded by the affection of his students, and by the esteem of his colleagues, who had never succeeded in fully understanding why a scholar of such promising talent had not been able to really return to normal life after the tragedy of the war.

---

[5]Library of the Ecole Polytechnique, Univérsité de Lausanne, programme of the course in *Mathematiques Speciales*, Lausanne, 1944.

[6]Interview with Colombo's grand-daugher and personal documents given to E. Luciano in 2015: *Mon père est allé à une réunion aujourd'hui. Les réfugiés italiens se rejoignent chaque samedi après-midi pour discuter de l'Italie. Ils font parvenir leurs nouvelles, ils parlent de l'évolution de l'Italie et ils essayent de décider comment ils la reconstruiront. Ces temps, mon père revient de bonne humeur, car les nouvelles concernant l'Italie sont bonnes. Les Alliés sont en train de gagner la guerre et de libérer l'Italie du Nord. Cependant, il semblerait que Turin serait bien détruite.*

Fig. 13.2 Colombo and his daughter Marina, 1945, Colombo Coll.

Appendix: Database of the Italian Academic Refugees

Name	Role	Discipline	Dismissal	Destination
Algranati Mondolfo Augusta	Assistente volontaria	Clinica Medica Generale e Terapia Medica	Bologna	Argentina
Ancona Giacomo	Libero docente	Patologia Speciale Medica	Firenze	U.S.
Arias Gino	Professore ordinario	Economia Politica Corporativa	Roma	Argentina
Artom Elia Samuele	Libero docente	Lingua e Letteratura Ebraica	Firenze	Palestine
Artom Camillo	Professore ordinario	Fisiologia Umana	Palermo	U.S.
Ascarelli Tullio	Professore ordinario	Diritto Commerciale	Bologna	Brazil
Ascoli Alberto Abram	Professore ordinario	Patologia Generale e Anatomia Patologica	Milano	U.S.
Bachi Riccardo	Professore ordinario (retired)	Economia Politica	Roma	Palestine
Bachi Roberto	Professore straordinario	Statistica Metodologica ed Economica	Genova	Palestine
Bolaffi Ada	Libero docente	Chimica Biologica	Milano	Palestine
Bolaffi Aldo	Libero docente	Patologia Chirurgica	Pisa	Switzerland
Bonaventura Enzo	Assistente ordinario, Libero Docente, incaricato	Psicologia Sperimentale	Firenze	Palestine
Calef Carlo	Assistente ordinario, Libero Docente	Patologia Speciale Chirurgica	Napoli	Brazil
Camis Mario	Professore ordinario	Fisiologia Generale	Bologna	Philippines
Campos Elsa	Assistente, incaricato	Seminario Giuridico	Venezia	Palestine
Cassuto Umberto	Professore ordinario	Ebraico e Lingue Semitiche Comparate	Roma	Palestine

(continued)

E. Luciano, *The Jewish Mathematical Diaspora from Fascist Italy*, Science Networks. Historical Studies 64, https://doi.org/10.1007/978-3-031-64896-0

Name	Role	Discipline	Dismissal	Destination
Castelbolognesi Enrico	Assistente [volontario]	Diritto Civile	Modena	France, Switzerland
Castelnuovo Gina	Assistente Laboratorio Idrobiologico, Borsista	Biologia	Roma	U.S.
Colombo Bonaparte	Assistente, incaricato	Meccanica Razionale, Matematiche Complementari	Torino	Switzerland
Curiel Eugenio	Assistente ordinario, incaricato	Meccanica Razionale, Matematiche Complementari	Padova	Switzerland, France, then Italy
D'Ancona Paolo	Professore ordinario	Storia dell'Arte Medievale e Moderna	Milano	Switzerland
De Benedetti Sergio	Assistente ordinario	Istituto di Fisica	Padova	U.S.
Del Vecchio Gustavo	Professore ordinario	Economia Politica Corporativa	Bologna	Switzerland
Donati Antigono	Professore ordinario	Diritto Commerciale	Roma	Argentina
Donati Benvenuto	Professore ordinario	Filosofia del Diritto	Modena	Switzerland
Donati Donato	Professore ordinario	Diritto Costituzionale	Padova	Switzerland
Donati Mario	Professore ordinario	Clinica Chirurgica Generale e Terapia Chirurgica	Milano	Switzerland
Fano Gino	Professore ordinario	Geometria Analitica	Torino	Svizzera
Fano Ugo	Assistente volontario	Fisica	Roma	U.S.
Faldini Giulio	Libero docente	Clinica ortopedica	Parma	Perù
Fiano Alessandro	Assistente ordinario, Libero Docente	Clinica pediatrica	Firenze	Palestine
Finzi Marcello	Professore ordinario	Diritto e Procedura Penale	Modena	Argentina
Foà Bruno	Professore ordinario	Economia Politica Corporativa	Bari	U.K., then U.S.
Foà Carlo	Professore ordinario	Fisiologia Umana	Milano	Brazil
Foà Piero	Assistente ordinario	Fisiologia	Pavia	U.S.
Formiggini Benedetto	Libero Docente	Clinica chirurgica e medicina operatoria	Milano	Switzerland
Franco Salomon Enrico	Professore ordinario	Anatomia e Istologia Patologica	Pisa	Palestine
Friberti Nedda	Assistente	Geometria Superiore	Trieste	Switzerland
Fubini Ghiron Eugenio	Assistente, incaricato	Radio-ricezione, Radiotecnica, Scuola di perfezionamento in elettrotecnica G. Ferraris	Torino	U.S.
Fubini Ghiron Gino	Assistente ordinario	Costruzioni in legno, ferro e cemento armato	Torino	U.S.

(continued)

Name	Role	Discipline	Dismissal	Destination
Fubini Ghiron Guido	Professore ordinario	Analisi Algebrica e Infinitesimale	Torino	U.S.
Fubini Mario	Professore straordinario	Letteratura Italiana	Palermo	Switzerland
Gentilli Giuseppe	Assistente	Istituto di Geografia	Firenze	Australia
Ghiron Delfina	Assistente ordinario	Chimica Generale e Inorganica	Pavia	Brazil
Hajon Mondolfo Isacco Emanuele	Libero Docente	Patologia Speciale Medica	Pisa	Switzerland
Herlitzka Amedeo	Professore ordinario	Fisiologia Umana	Torino	Argentina
Jacchia Luigi Giuseppe	Assistente volontario	Istituto di Astronomia	Bologna	U.K., then U.S.
Jarach Dino	Assistente ordinario	Diritto Finanziario e Scienza delle Finanze	Pavia	Argentina
Jolles Enrico	Aiuto ordinario, Libero Docente	Chimica Farmaceutica e Tossicologica, Chimica Organica	Firenze	U.K.
Lattes Leone	Professore ordinario	Medicina Legale e delle Assicurazioni	Pavia	Argentina
Levi Alessandro	Professore ordinario	Filosofia del Diritto	Parma	Switzerland
Levi Beppo	Professore ordinario	Analisi Matematica (Algebrica e Infinitesimale)	Bologna	Argentina
Levi D'Ancona Ezio	Professore ordinario	Filologia Romanza	Napoli	Switzerland
Levi Della Vida Giorgio	Professore ordinario (dismissed in 1931 because he refused the loyalty oath)	Ebraico e Lingue Semitiche Comparate	Roma	U.S.
Deveali Levi Mario	Libero Docente	Legislazione del Lavoro	Roma	Argentina
Levi Franco	Assistente straordinario	Laboratorio di Resistenza dei Materiali	Torino	Switzerland
Levi Giorgio Renato	Professore ordinario	Chimica Generale e Inorganica	Pavia	Brazil
Levi Giulio	Assistente ordinario	Istologia ed Embriologia Generale	Bologna	Palestine
Levi Giuseppe	Professore ordinario	Anatomia Umana e Normale	Torino	Belgium
Levi Mario Giacomo	Professore ordinario	Chimica Industriale	Milano	Switzerland
Levi Montalcini Rita	Assistente volontario	Clinica Malattie Nervose e Mentali	Torino	Belgium, then Italy

(continued)

Name	Role	Discipline	Dismissal	Destination
Levi Nino	Professore ordinario	Diritto e Procedura Penale	Genova	U.S.
Levi Teodoro (Doro)	Professore ordinario	Archeologia e Storia dell'Arte Antica	Cagliari	U.S.
Levialdi Andrea	Borsista	Ottica	Arcetri	France, then Argentina, then Cuba
Liebman Enrico Tullio	Professore ordinario	Diritto Processuale Civile	Parma	Uruguay, then Brazil
Lombroso Ugo	Professore ordinario	Fisiologia Umana	Genova	France, then Italy
Lopez Roberto Sabatino	Libero docente	Storia Medioevale	Genova	U.K., then U.S.
Luria Salvatore	Borsista	Radiobiologia	Roma	U.S.
Luzzatto Fabio	Professore straordinario, libero docente (dismissed in 1931 because he refused the loyalty oath)	Diritto agrario, Diritto Civile	Milano	Switzerland
Luzzatto Gina	Aiuto ordinario	Botanica	Milano	Switzerland
Melli Guido	Professore ordinario	Patologia speciale medica	Parma	Unknown destination
Momigliano Arnaldo	Professore straordinario	Storia Romana, con Esercitazioni di Epigrafia Romana	Torino	U.K.
Mondolfo Rodolfo	Professore ordinario	Storia della Filosofia	Bologna	Argentina
Mondolfo Silvano	Assistente volontario	Clinica ortopedica	Bologna	Argentina
Mortara Franco	Assistente volontario	Clinica Ostetrica e Ginecologica	Bologna	U.S.
Mortara Giorgio	Professore ordinario	Statistica	Milano	Brazil
Mortara Nella	Assistente ordinario, Libera Docente	Istituto di Fisica, Fisica Sperimentale	Roma	Brazil, then Italy
Oppenheim Marco	Assistente ordinario	Clinica Medica Generale e Terapia Medica	Bologna	France, then Italy
Ottolenghi Michelangelo	Professore straordinario	Anatomia degli Animali Domestici, con Istologia ed Embriologia	Sassari	Ecuador
Paggi Bruno	Aiuto ordinario, Libero Docente	Clinica Chirurgica Generale, Patologia Speciale Chirurgica	Pisa	U.K., then Venezuela
Pincherle Alberto	Professore straordinario	Storia delle religioni; Storia del Cristianesimo	Cagliari; Roma	Perù

(continued)

Name	Role	Discipline	Dismissal	Destination
Pincherle Leo	Professore incaricato	Fisica teorica	Padova	U.K.
Pitigliani Fausto	Libero Docente	Economia Politica	Roma	U.S.
Pontecorvo Bruno	Borsista	Fisica	Parigi	U.S.
Pugliese Mario	Professore straordinario	Diritto Finanziario e Scienza delle Finanze	Trieste	Argentina
Racah Giulio	Professore straordinario	Fisica Teorica	Pisa	Palestine
Ravà Renzo	Professore straordinario	Legislazione del Lavoro	Firenze	U.S.
Ravà Tito	Borsa d'internato, assistente incaricato, Libero Docente	Istituto di Diritto Civile, Diritto Commerciale	Padova	Argentina
Ravenna Ferruccio	Libero Docente	Patologia Speciale Medica	Padova	Switzerland
Reichenbach Giulio	Libero Docente	Letteratura Italiana	Padova	Switzerland
Rossi Bruno	Professore ordinario	Fisica Sperimentale	Padova	U.S.
Salvadori Mario	Assistente ordinario, Libero Docente	Scienza delle Costruzioni	Roma	U.S.
Samaja Tullio	Assistente, incaricato	Industrie Agrarie	Bologna	U.S.
Schreiber Bruno	Aiuto ordinario, Libero Docente, incaricato	Istituto di Zoologia, Zoologia, Anatomia e Fisiologia Comparate, Genetica	Milano	Switzerland
Segré Angelo	Professore ordinario	Storia Economica	Trieste	U.S.
Segre Beniamino	Professore ordinario	Geometria Analitica con Elementi di Proiettiva e Descrittiva con Disegno	Bologna	U.K.
Segré Emilio	Professore straordinario	Fisica Sperimentale	Palermo	U.S.
Segre Renato	Assistente straordinario, Libero Docente	Clinica otorinolaringoiatrica	Torino	Argentina
Sereni Angelo	Professore straordinario	Diritto Internazionale	Ferrara	U.S.
Szegö Luigi	Assistente e Aiuto, incaricato, Libero Docente	Chimica Industriale, Chimica Analitica, Chimica Generale	Milano	Switzerland
Tabet Duccio	Assistente straordinario, incaricato	Agraria	Sassari	France, then U.S.
Tagliacozzo Carlo	Libero Docente	Meccanica Applicata alle Costruzioni	Roma	U.S.
Tedeschi Guido	Professore straordinario	Diritto Civile	Siena	Palestine

(continued)

Name	Role	Discipline	Dismissal	Destination
Terracini Alessandro	Professore ordinario	Geometria Analitica con Elementi di Proiettiva e Descrittiva con Disegno	Torino	Argentina
Terracini Benvenuto Aron	Professore ordinario	Glottologia	Milano	Argentina
Toaff Renzo	Assistente provvisorio	Clinica Ostetrica e Ginecologica	Pisa	Palestine
Treves Renato	Professore incaricato	Filosofia del Diritto	Urbino	Argentina
Usiglio Gino	Assistente, incaricato	Fisica	Bologna	Brazil
Viterbo Camillo	Professore straordinario	Diritto Commerciale	Cagliari	Argentina
Volterra Edoardo	Professore ordinario	Istituzioni di Diritto Romano	Bologna	France, then Italy
Volterra Enrico	Assistente ordinario, Libero Docente, incaricato	Meccanica Razionale, Meccanica Applicata alle Costruzioni, Scienza delle Costruzioni	Roma	U.K.
Volterra Mario	Aiuto ordinario, Libero Docente	Clinica Medica Generale e Terapia Medica, Chimica e Microscopia Clinica, Patologia Speciale Medica Dimostrativa	Firenze	U.S.
Zamorani Vittore (Vittorio)	Professore ordinario	Clinica Pediatrica	Pavia	Venezuela

Archives

ACS: Archivio Centrale dello Stato, Roma; Central State Archive, Rome

ACS-*Contro-discriminazione*: ACS, MPI, Direzione Generale Istruzione Pubblica, *Miscellanea di divisioni diverse*, I-II-III

ACS-*Ebrei*: ACS, MPI, Pol. Pol.: *Materia*, b. 219, *Ebrei Italiani*; *Materia*, b. 189, *Espatrio di ebrei Italiani*

ACS-*Fano*: ACS, MPI, Fascicoli personali. Professori ordinari (1940-70) 3. versamento. Busta 187, Personal dossier of Gino Fano

ACS-*Fubini*: ACS, MPI, Fascicoli personali. Professori ordinari (1940-70) 3. versamento. Busta 214, Personal dossier of Guido Fubini

ACS-*Levi*: ACS, MPI, Fascicoli personali. Professori ordinari (1940-70) 3. versamento. Busta 265, Personal dossier of Beppo Levi

ACS-*Razzismo*: ACS, MI, DGPS-DAGR: *Razzismo 1939, 1940 e 1941*

ACS-*Terracini*: ACS, MPI, Fascicoli personali. Professori ordinari (1940-70) 3. versamento. Busta 452, Personal dossier of Alessandro Terracini

ACS, MI-Demorazza: ACS, *Affari Generali*

ACS, Minculpop: ACS, *Ministero della Cultura Popolare, Gabinetto, b. 159, f. varie anno 1938*

ACT: Archivio di Stato di Torino, Archivi privati, Gustavo Colonnetti; State Archives of Turin, private collections, Gustavo Colonnetti papers

AEA: Albert Einstein Archives, Hebrew University of Jerusalem

AEA, *AT*: Albert Einstein Archives, Hebrew University of Jerusalem, *Alessandro Terracini*, mss. nos. 11128, 56279, 56280.

AFB: Swiss Federal Archives, Bern

ANL-AI: Accademia dei Lincei, Archivio della Reale Accademia d'Italia; Academy of Lincei, Archive of the Royal Academy of Italy, Rome

ANL-*Castelnuovo*: Accademia dei Lincei, Archivi privati, Archivio Guido Castelnuovo; Academy of Lincei, private Archives, Guido Castelnuovo Archive. It can be accessed at the website [Gario 2010] http://operedigitali.lincei.it/Castelnuovo/Lettere_E_Quaderni/menu.htm

E. Luciano, *The Jewish Mathematical Diaspora from Fascist Italy*, Science Networks. Historical Studies 64, https://doi.org/10.1007/978-3-031-64896-0

ANL-*Levi-Civita*: Accademia dei Lincei, Archivi privati, Archivio *Tullio Levi-Civita*; Academy of Lincei, private Archives, Tullio Levi-Civita Archive.

ANL-*Volterra*: Accademia dei Lincei, Archivi privati, Archivio *Vito Volterra*; Academy of Lincei, private Archives, Vito Volterra Archive.

APICE-*Ascoli*: As, Ap, serie 7, Cast b. 234, f. *Razza*, sf. *Personale di razza ebraica. Disposizioni generali*, Guido Ascoli

APICE-*Finzi*: As, Ap, serie 7, Cast b. 234, f. *Razza*, sf. *Dati statistici del personale. Censimento del personale di razza ebraica - Nazionalità*, Bruno Finzi

APICE-*Mortara*: As, Ap, serie 7, Cast b. 234, f. *Razza*, sf. *Personale di razza ebraica. Disposizioni generali*, Giorgio Mortara

APICE: Archivi della Parola, dell'Immagine e della Comunicazione Editoriale, Università di Milano; Archives of Words, Images and Publishing Communication, University of Milan

APR: Archivio Enrico Persico, Dipartimento di Fisica, Università La Sapienza, Roma; Enrico Persico Archive, Department of Physics, University of Rome La Sapienza.

Archivio Accademia Ligure di Scienze e Lettere: personal file of Gino Loria

Archivio Accademia Ligure di Scienze e Lettere: personal file of Alberto Mario Bedarida

AS-UMI: Archivio Storico dell'Unione Matematica Italiana, Bologna; Historical Archive of the Italian Mathematical Union, Bologna: correspondence 1938-1945 and E. Bompiani correspondence

ASPoliTo-*Fano*: Fascicolo personale di Gino Fano, Archivio Storico del Politecnico di Torino; personal file of Gino Fano, Historical Archive of the Polytechnic University of Turin

ASPoliTo-*Fubini*: Fascicolo personale di Guido Fubini, Archivio Storico del Politecnico di Torino; personal file of Guido Fubini, Historical Archive of the Polytechnic University of Turin

ASTO: Archivio di Stato di Torino; State Archives of Turin

ASUG: Archivio Storico dell'Università di Genova; Historical Archive of the University of Genoa

ASUG: personal files of Bachi, Gino Loria and Alberto Mario Bedarida

ASUT: Archivio Storico dell'Università di Torino; Historical Archive of the University of Turin

ASUT-*Fano*: Fascicolo personale di Gino Fano, Archivio Storico dell'Università di Torino, personal file of Gino Fano, Historical Archive of the University of Turin

ASUT-*Terracini*: Fascicolo personale di Alessandro Terracini, Archivio Storico dell'Università di Torino, personal file of Alessandro Terracini, Historical Archive of the University of Turin

ATCET: Archivio delle Tradizioni e del Costume Ebraici Benvenuto e Alessandro Terracini, Torino; Archive of the Jewish traditions and customs Benvenuto and Alessandro Terracini, Turin

BSM-*Fano*: Fondo *Gino Fano*, Biblioteca Speciale di Matematica, Università di Torino; Gino Fano Archive, Special Mathematics Library, University of Turin. Correspondence, 26 letters from different mathematicians, digitized in https://www.corradosegre.unito.it/fondo_fano_l.php (ed. by L. Giacardi). Manuscripts:

manuscript draft of the Lectures on Italian algebraic geometry held by Fano at the University College of Wales, Aberystwyth, Appunti vari, Scritti.4, fols. 3r-8v, 12r-v, 10r-12r, 12 bis r-v, 13r-17v, 17bis r-v, 18r-30v, 63r-70v, 82r-v, 72r-v, 71r-v, 73r-77v, 77 bisA, 77 bisB, 78r-81v, 83r-110v, 110bisA-B-C-D-E-F, 111r-117v, 119r-124v; *Aperçus général sur les surfaces du 3ème ordre*, Appunti vari, Scritti.4, fols. 36-41; *Géométrie sur les surfaces algébriques*, Appunti vari, Scritti.4, fols. 47f, 43, 47r, 48, 42, 44, 49, 50; *Les surfaces du 4ème ordre*, Appunti vari, Scritti.1, fols. 1-6; *Quelques aperçus sur le développement de la géométrie algébrique en Italie pendant le dernier siècle*, Appunti vari, Scritti.4, fols. 53f-r, 61f-r, 55r, 55, 55bisA, 55bisB, 56f-r, 57f-r, 58f-r, 59f-r, 60f-r, 54f-r, 62f-r 53; *Transformations de contact birationnelles dans le plan*, Scritti.3, fols. 1-6.

BSM-*Segre*: Fondo *Corrado Segre*, Biblioteca Speciale di Matematica, Università di Torino; Corrado Segre Archive, Special Mathematics Library, University of Turin. Documents are digitized in https://www.corradosegre.unito.it/fondo_segre.php (ed. by L. Giacardi).

BSM-*Terracini*: Fondo Alessandro Terracini, Biblioteca Speciale di Matematica, Università di Torino; Alessandro Terracini Archive, Special Mathematics Library, University of Turin. Collected papers 1947-48, 1949, 1950-51, 1952, 1953. Notebooks no. 14, *Geometria superiore 1935-36 - Argomenti vari di geometria (topologia)*, https://www.corradosegre.unito.it/doc/terraciniq14.pdf; nos. 17 e 18, *Matemáticas superiores 1940*, https://www.corradosegre.unito.it/doc/terraciniq17.pdf; nos. 19 e 20, *Metodologia (1940)*, https://www.corradosegre.unito.it/doc/quaderno19.pdf; no. 21, *Geometria analitica (1940)*; nos. 22 e 23, *Metodologia (1941)*; nos. 24 e 25, *Matemáticas superiores 1941*; nos. 26 e 27, *Matemáticas superiores 1942*, https://www.corradosegre.unito.it/doc/terraciniq26.pdf; nos. 28 e 29, *M. S. 1943 Matemáticas superiores*; nos. 30 e 31, *Matemáticas Superiores 1946 - Introduzione alla geometria superiore*

BSM-*Tricomi*: Archivio Francesco Giacomo Tricomi, Biblioteca Speciale di Matematica, Università di Torino; Francesco Giacomo Tricomi Archive, Special Mathematics Library, University of Turin

CA, *BSP*: The Caltech Archives, California Institute of Technology, Beniamino Segre Papers, box 1, folders 1 (fols. 52), 2 (fols. 25), 3 (fols. 11), 4 (fols. 1), 5 (fols. 21), 6 (fols. 49), 7 (fols. 93), 8 (fols. 84), box 7, folder 7 (Oxford University Press), fols. 33

Ceccherini Silberstein Family Archive: Tullio Levi-Civita and Libera Trevisani Levi-Civita papers, private archive of the family Ceccherini Silberstein, courtesy of Pier Vittorio and Tullio Ceccherini, Rome.

Colombo Coll.: Colombo Family Collection, Marina Colombo Fubini's private archive, Morges, Switzerland

ECADFS Records: Emergency Committee in Aid of Displaced Foreign Scholars Records 1927-1949, Manuscripts and Archives Division, The New York Public Library

ECADFS Records, *ABT*: MssCol 922, Series I, Grant files 1927-1949, I.B. Non Grantees 1927-1945, b. 123 f. 25 *Terracini, Aron Benvenuto* 1939-1941

ECADFS Records, *AT*: MssCol 922, Series I, Grant files 1927-1949, I.B. Non Grantees 1927-1945, b. 123 f. 25 *Terracini, Alessandro* 1939

ECADFS Records, *BS*: MssCol 922, Series I, Grant files 1927-1949, I.B. Non Grantees 1927-1945, b. 115 f. 37 *Segre, Beniamino* 1935, 1939-1941

ECADFS Records, *BT*: MssCol 922, Series I, Grant files 1927-1949, I.B. Non Grantees, b. 123, f. 6 *Tedeschi, Bruno* 1939

ECADFS Records, *EdV*: MssCol 922, Series I, Grant files 1927-1949, I.B. Non Grantees, b. 126, f. 20 *Volterra, Edoardo* 1939-1940

ECADFS Records, *EnV*: MssCol 922, Series I, Grant files 1927-1949, I.B. Non Grantees, b. 126, f. 21 *Volterra, Enrico* 1939-1940

ECADFS Records, *FE*: MssCol 922, Series I, Grant files 1927-1949, I.B. Non Grantees 1927-1945, b. 55 f. 25 *Enriques, Federigo* 1938

ECADFS Records, *FermiE*: MssCol 922, Series I, Grant files 1927-1949, I.B. Non Grantees 1927-1945, b. 57, f. 10, *Fermi, Enrico* 1939-1940.

ECADFS Records, *GA*: MssCol 922, Series I, Grant files 1927-1949, I.B. Non Grantees, b. 38, f. 46 *Ascoli, G.* 1941

ECADFS Records, *GC*: MssCol 922, Series I, Grant files 1927-1949, I.B. Non Grantees, b. 49, f. 1 *Castelnuovo, Gina* 1938-1942

ECADFS Records, *GDS*: MssCol 922, Series I, Grant files 1927-1949, I.B. Non Grantees, b. 51, f. 10 *De Santillana, Giorgio* 1939-1942

ECADFS Records, *GM*: MssCol 922, Series I, Grant files 1927-1949, I.B. Non Grantees 1927-1945, b. 97 f. 18 *Mortara, Giorgio* 1938-1939

ECADFS Records, *Italy*: MssCol 922, Series VIII, Internal office records 1933-1945, VIII.C. Subject files 1933-1944, b. 203 f. 3 *Italy, Situation in 1934,1938-1939*; b. 203 f. 7 *Mathematics* 1938.

ECADFS Records, *LJ*: MssCol 922, Series I, Grant files 1927-1949, I.A. Grantees 1933-1946, b. 16 f. 4 *Jacchia, Luigi Giuseppe* 1938-1944.

ECADFS Records, *MA*: MssCol 922, Series I, Grant files 1927-1949, I.B. Non Grantees, b. 38, f. 46 *Ascoli, Max* 1933-1935, 1939-1943

ECADFS Records, *NF*: MssCol 922, Series I, Grant files 1927-1949, I.B. Non Grantees, b. 60, f. 5 *Friberti, Nedda* 1939

ECADFS Records, *RF*: MssCol 922, Series I, Grant files 1927-1949, I.B. Non Grantees, b. 60, f. 55 *Frucht Roberto* 1939

ECADFS Records, *TLC*: MssCol 922, Series I, Grant files 1927-1949, I.B. Non Grantees 1927-1945, b. 88 f. 26 *Levi-Civita, Tullio* 1938-1939

ECADFS Records, *UF*: MssCol 922, Series I, Grant files 1927-1949, I.B. Non Grantees, b. 56, f. 35 *Fano, Ugo* 1939-1940

Fubini Coll.: Fubini Family Collection, Laurie and David G. Fubini's private archive, Boston

IAS, *GF*: Institute for Advanced Study, Princeton, NJ USA, Shelby White and Leon Levy Archives Center, School of Mathematics Records: Box 9, Members, Visitors, Assistant series, Alphabetical I subseries, 1933-1977, Froelicher-Fujii folder; Box 44, Director's Office Records, Member files, *Fubini, Guido*.

OVP: Oswald Veblen Papers, Library of Congress Manuscript Division, Washington, D.C.

OVP, *AD*: Subject File, 1918-1960, *Duschek, Adalbert*, 1938-39, box 31, 17 fols. n. n.

OVP, *AF*: General Correspondence, 1902-1960, box 5, *Flexner, Abraham*, 1924-1940, 4 folders, fols. nn.

OVP, *AT*: Subject File, 1918-1960, *Terracini, Alessandro*, 1938-1943, box 34, 33 fols. n.n.

OVP, *BS*: Subject File, 1918-1960, *Segre, Beniamino*, 1938-1941, box 33, 56 fols. n. n.

OVP, *ECADFS*: Subject File, 1918-1960, *Emergency Committee in Aid of Displaced German Scholars*, 1933-35, box 23-5, 52 fols. n.n.; 23-6, 62 fols. n.n.; 23-7, 52 fols. n.n.

OVP, *EdV*: Subject File, 1918-1960, *Volterra, Edoardo*, 1940, box 34, 17 fols. n.n.

OVP, *EnV*: Subject File, 1918-1960, *Volterra, Enrico*, 1939, box 34, 5 fols. n.n.

OVP, *GA*: Subject File, 1918-1960, *Ascoli, Guido*, 1939, box 31, 13 fols. n.n.

OVP, *GB*: Subject File, 1918-1960, *Bemporad, Giulio*, 1939, box 31, 20 fols. n.n.

OVP, *GC*: Subject File, 1918-1960, *Castelnuovo, Gina*, 1930-1945, box 31, 98 fols. n.n.

OVP, *JAS*: General Correspondence, 1902-1960, box 12, *Schouten, Jan Arnoldus*, 1929-1950, fols. n.n.

OVP, *JLS*: General Correspondence, 1902-1960, box 13, *Synge, John L.*, 1922-1949, fols. n.n.

OVP, *JN*: General Correspondence, 1902-1960, box 15, *von Neumann, John*, 1929-1956, fols. n.n.

OVP, *LB*: Subject File, 1918-1960, box 31, *Berwald, Ludwig*, 1939-1942, fols. n.n.

OVP, *LJ*: Subject File, 1918-1960, *Jacchia, Luigi*, 1939, box 32, 3 fols. n.n.

OVP, *RefGen*: Subject File, 1918-1960, *Refugees General 1935-41*, box 30-23, 111 fols. n.n.; 30-24, 75 fols. n.n.

OVP, *TYT*: General Correspondence, 1902-1960, box 14, *Thomas, Tracy Y.*, 1923-1950, 2 folders, fols. n.n.

SPSL: Archive of the Society for the Protection of Science and Learning, Bodleian Archives & Manuscripts, Oxford

SPSL, *AT*: Correspondence relating to individual scholars, Mathematics, Terracini, Professor Alessandro (1884-), File 1938-46, 285/5, fols. 340-87.[1]

SPSL, *BS*: Correspondence relating to individual scholars, Mathematics, Segre, Professor Beniamino (1903-1977), File 1938-60, 285/1, fols. 1-233; 441/2, fols. 214-31 (Home Office file).

SPSL, *GB*: Correspondence relating to individual scholars, Physics, Bemporad, Professor Giulio (1888-), File 1938-39, 323/6.

SPSL, *GC*: Correspondence relating to individual scholars, Biology, Castelnuovo, Dr. Gina (1908), File 1938-44, 197/2.

SPSL, *GF*: Correspondence relating to individual scholars, Mathematics, Fano, Professor Gino (1871-), File 1939-47, 278/6, Publications, fols. 296-365; 448/3, fols. 494-600.

---

[1] All material from SPSL archive has been reproduced by kind permission of Cara (the Council for At-Risk Academics).

SPSL, *GM*: Correspondence relating to individual scholars, Economics, Mortara, Professor Giorgio (1885-), File 1938-48, 235/10, fols. 344-366.

SPSL, *Italy*: Correspondence with other organisations, E.2 Correspondence with overseas organisations, E.2.2 Organisations arranged by country: Italy Shelfmark, 150/4, fols. 421-428; 150/5, fols. 429-525.

SPSL, *Mathematics*: Mathematics 277 fols. 78–79 Guido Ascoli, Mathematics 277 fols. 456–462 Guido Castelnuovo, Mathematics 281 fols. 321–322 Beppo Levi, Mathematics 282 fols. 312–313 Arturo Maroni, Mathematics 285 fols. 388–389 Scipione Treves

Terracini Coll.: Terracini Family Collection, Benedetto Terracini's private archive.

UTo-ACS: Corrado Segre Archives, University of Turin. Most of the documents can be accessed at the website (ed. by L. Giacardi) http://users.mat.unimi.it/users/gario/Elenco-Segre.html

References[2]

'Aryans' in Italy; *Nature* 142 No. 3586, 23.7.1938, 167.

Alessandro Terracini nel centenario della nascita 1889-1989; Torino: Zamorani 1990.

Allasia, G.: Bonaparte Colombo; in: C.S. Roero (ed.): *La Facoltà di Scienze Matematiche Fisiche Naturali di Torino*, volume 2: I docenti; Torino: DSSP 1999, pp. 603–606.

Almansi, D.: La progettata espulsione. Contributo alla storia delle persecuzioni razziali in Italia; *Israel* 31 No. 8 (1945), 3.

Antecedentes de la creación de las Mathematicæ Notæ; *Mathematicae Notae* 1 No. 1 (1941), 3–5.

Antecedentes de la creación del Instituto. II. Acto de inauguración oficial del Instituto; *Publicaciones del Instituto de Matemática* 2 No. 5 (1940), 1–100.

Arian Levi, G. and G. Disegni: *Fuori dal ghetto. Il 1848 degli ebrei*; Rome: Editori Riuniti 1998.

Arian Levi, G. and E. Viterbo: *Simeone Levi. La storia sconosciuta di un noto egittologo*; Torino: Ananke 1999.

Arrighi, G. (ed.): *Lettere a Mario Pieri (1884-1913)*; Milan: Quaderni Pristem (No. 6) 1997.

Arru, A., D.L. Caglioti and F. Ramella (eds.): *Donne e uomini migranti. Storie e geografie tra breve e lunga distanza*; Rome: Donzelli Editore 2008.

Artom, E.S.: *La scuola ebraica in Italia. Relazione letta al 2° convegno giovanile ebraico (Torino, 24 dicembre 1912)*; Florence: Giuntina 1913.

Ascarelli, T.: *Pensieri e lettere familiari*; Napoli: Edizioni Scientifiche Italiane 2017.

Ash, M.: *Forced Migration and Scientific Change: Emigré German-Speaking Scientists and scholars after 1933*; Cambridge and New York: University Press 1996.

Ash, M.: Forced Migration and Scientific Change After 1933. Steps Toward a New Approach; in: R. Scazzieri and R. Simili (eds.): *The Migration of Ideas*; USA: Science History Publications 2008, pp. 161–178.

Aubin, D., Hollings, C., Kennedy, S., Kent D., Luciano, E.: "An individualist of the old-fashioned American type": The informal scientific diplomacy of Oswald Veblen; *Oberwolfach Rep.* 56 (2022), 3198–3201.

Avagliano, M. and M. Palmieri: *Di pura razza italiana. L'Italia 'ariana' di fronte alle leggi razziali*; Milan: Baldini e Castoldi 2013.

Ayres, W.L.: The April Meeting in Chicago. *Bulletin of the AMS* 48 No. 7 (1942), 496–503.

Babbit, D. and J. Goodstein: Guido Castelnuovo and Francesco Severi: two personalities, two letters; *Notices of AMS* 56 No. 7 (2009), 800–808.

---

[2]References given in the critical apparatus of primary source material are not listed.

E. Luciano, *The Jewish Mathematical Diaspora from Fascist Italy*, Science Networks. Historical Studies 64, https://doi.org/10.1007/978-3-031-64896-0

Babbit, D. and J. Goodstein: A Fresh Look at Francesco Severi; *Notices of AMS* 59 No. 8 (2012), 1064-1075.

Badini Confalonieri, L. (ed.): *Testimonianze in memoria di Gustavo Colonnetti*; Torino: Stamperia Artistica Nazionale 1973.

Badini Confalonieri, L.: *Colonnetti inedito*; Biella: Sandro Maria Rosso 1978.

Badini Confalonieri, L. and G. Colonnetti: *Carissimi figlioli belli ... Lettere da Roma 1944-45*; Torino: Fondazione Alberto Colonnetti 2006.

Baker, H.F.: Non-singular cubic surfaces; *Nature* 151 No. 3819, 9.1.1943, 39–40.

Baricco, P.: *L'istruzione popolare in Torino*; Torino: Eredi Botta 1865.

Barone, V. and M. Ciardi: *Il Novecento in tasca. Scienza e cultura nel taccuino di Adriana Enriques*; Torino: Hapax 2021.

Barrow-Green, J., Fenster, D., J. Schwermer and R. Siegmund-Schultze (eds.): Emigration of Mathematicians and Transmission of Mathematics: Historical Lessons and Consequences of the Third Reich; *Oberwolfach Rep.* 8 (2011), 2891–2961.

Barrow-Green, J. and J. Gray: Geometry at Cambridge, 1863-1940; *Historia Mathematica* 33 (2006), 315–356.

Bassani, G.: L'assalto fascista alla Sinagoga di Ferrara; in: P. Pieri (eds.): *Racconti, diari, cronache (1935-1956)*; Milan: Feltrinelli 2014, pp. 453–457.

Bassi, A.: L'Università e la Scuola Matematica di Princeton; *Periodico di Matematiche* (4) 19 (1936), 57–79.

Bassi, A.: L'Università e la Scuola Matematica di Princeton. Conferenza tenuta il 21 febbraio 1938; *Conferenze di Fisica e Matematica tenute presso la R. Università e R. Scuola di Ingegneria di Torino 1938-1939* (1938), 1–24.

Batterson, S.: The Vision, Insight, and Influence of Oswald Veblen; *Notices of the AMS* 54 No. 5 2007, 606–618.

Battimelli, G. and M. De Maria (eds.): *Da via Panisperna all'America*; Rome: Editori Riuniti 1997.

Battimelli, G. and G. Paoloni (ed.): *20th Century Physics: Essays and Recollections. A Selection of Historical Writings by Edoardo Amaldi*; Singapore: World Scientific 1998.

Beer, M. and A. Foà (eds.): *Ebrei, minoranze e Risorgimento. Storia, cultura, letteratura*; Rome: Viella 2013.

Bernardini, P.: The Jews in nineteenth-century Italy: towards a reappraisal; *Journal of Modern Italian Studies* 1-2 (1996), 292–310.

Berti, D.: [Speech without title]; in: Atti della Società. Primo congresso generale tenutosi nella Regia Università di Torino i giorni 26, 27, 28, 29 e 30 ottobre 1849, XI. Sesta adunanza generale del Congresso; *Giornale della Società d'Istruzione e d'Educazione* 1 (1849–1850), 724.

Beyerchen, A.D.: *Scientists Under Hitler: Politics and the Physics Community in the Third Reich*; New Haven: Yale University Press 1977.

Boatti, G.: *Preferirei di no. Le storie dei dodici professori che si opposero a Mussolini*; Torino: Einaudi 2017.

Bompiani, E.: Alessandro Terracini; *Rend. Accad. Naz. Lincei* (1970), 1–22.

Bònoli, F. and A. Mandrino (eds.): Sotto lo stesso cielo? Le leggi razziali e gli astronomi in Italia. Numero monografico del Giornale di Astronomia, con interventi di A.A. Sermoneta, M. Procaccia, M. Zuccoli, A. Mangano, L. Schiavone, V. Zanini, A. Mandrino; *Giornale di Astronomia* 41 No. 2 (2015), 2–60.

Borgato, M.T. and C. Phili (eds.): *In Foreign Lands: the Migration of Scientists for Political or Economic Reasons*; Basel: Birkhäuser 2022.

Bottazzini, U., A. Conte and P. Gario: La Relazione di Castelnuovo ed Enriques. Documenti inediti per il Premio Reale di Matematica del 1901; *Rendiconti del Circolo matematico di Palermo*, supplemento 55 (1998), 75–156.

Bottazzini, U., A. De Benedetti and P.E. Foraciari (eds.): *Le città di mare e lo spirito scientifico. Per Federigo Enriques*; La Spezia: Agorà 2001.

Bravo, A. and D. Jalla (eds.): *La vita offesa*; Milan: Franco Angeli 1988.

Brechenmacher, F., G. Jouve, L. Mazliak and R. Tazzioli: *Images of Italian Mathematics in France. The Latin Sisters, from Risorgimento to Fascism*; Cham: Springer 2016.

Brigaglia, A.: The creation and persistence of national schools: the case of Italian algebraic geometry; in: U. Bottazzini and A. Dahan Dalmedico (eds.): *Changing Images in Mathematics. From the French Revolution to the New Millennium*; London: Routledge 2001, pp. 187–206.

Brigaglia, A. and C. Ciliberto: *Geometria Algebrica*; in: S. Di Sieno, A. Guerraggio and P. Nastasi (eds.): *La matematica italiana dopo l'unità. Gli anni tra le due guerre mondiali*; Milan: Marcos y Marcos 1998, pp. 185–320.

Brigaglia, A. and C. Ciliberto: Remarks on the relations between the Italian and American schools of algebraic geometry in the first decades of the 20th century, *Historia Mathematica* 31 (2004), 310–319.

Brigaglia, A., C. Ciliberto and C. Pedrini: The Italian School of Algebraic Geometry and Abel's Legacy; in: O.A. Laudal and R. Piene (eds.): *The Legacy of Niels Henrik Abel, The Abel Bicentennial, Oslo 2002*; Berlin: Springer 2004, pp. 295–347.

Broggini, R.: *Terra d'asilo. I rifugiati italiani in Svizzera 1943-1945*; Bologna: Il Mulino 1993.

Broggini, R.: *La frontiera della speranza. Gli ebrei dall'Italia verso la Svizzera, 1943-1945*; Milan: Mondadori 1999.

Bussotti, P. (ed.): *Federigo Enriques e la cultura europea*; Lugano: Lumières Internationales 2008.

Caffaz, U.: Mai nessuno, capro espiatorio; *Il Ponte* 11-12(1978), 1301–1532.

Campanile, B.: Robert Fano e il coraggio di vivere il 'non luogo'; *Viaggiatori. Circolazioni, scambi ed esilio* 1 No. 2 (2018), 353–386.

Camurri, R.: Idee in movimento: l'esilio degli intellettuali italiani negli Stati Uniti (1930-1945); *Memoria e Ricerca* 31 (2009), 43–62.

Capon Fermi, L.: *Atomi in famiglia*; Milan: Mondadori 1954.

Capon Fermi, L.: *Illustrious immigrants: the intellectual migration from Europe, 1930-41*; Chicago: University Press 1968.

Capristo, A.: L'esclusione degli ebrei dall'Accademia d'Italia; *La Rassegna Mensile di Israel* 67 No. 3 (2001), 1–36.

Capristo, A.: *L'espulsione degli ebrei dalle accademie italiane*; Torino: Zamorani 2002.

Capristo, A.: Tullio Levi-Civita e l'Accademia d'Italia; *La Rassegna Mensile di Israel* 69 No. 1. (2003), 237–256.

Capristo, A.: L'alta cultura e l'antisemitismo fascista. Il Convegno Volta del 1939; *Quaderni di Storia* 64 (2006), 165–226.

Capristo, A.: Il decreto legge del 5 settembre 1938 e le altre norme antiebraiche nelle scuole, nelle università e nelle accademie; *La Rassegna Mensile di Israel* 73 No. 2 (2007), 131–167.

Capristo, A.: 'Fare fagotto': l'emigrazione intellettuale ebraica dall'Italia fascista dopo il 1938; *La Rassegna Mensile di Israel* 76 No. 3 (2010), 177–200.

Capristo, A.: *Gather What You Can and Flee. Jewish Intellectual Emigration From Fascist Italy*; New York: CPL Editions 2014.

Capristo, A. and G. Fabre: *Il registro. La cacciata degli ebrei dallo Stato italiano nei protocolli della Corte dei Conti 1938-1943*; Bologna: Il Mulino 2018.

Casnati, G., Conte, A., Gatto, L., Giacardi, L., M. Marchisio and A. Verra (eds.): *From Classical to Modern Algebraic Geometry. Corrado Segre's Mastership and Legacy*; Basel: Birkhäuser 2016.

Cassata, F.: *"La Difesa della razza". Politica, ideologia e immagine del razzismo fascista*; Torino: Einaudi 2008.

Castagnola, R., F. Panzera and M. Spiga (eds.): *Spiriti liberi in Svizzera: la presenza di fuorusciti italiani nella Confederazione negli anni del fascismo e del nazismo, 1922-1945: atti del Convegno internazionale di studi, Ascona, Centro Monte Verità; Milano, Università degli studi, 8-9 novembre 2004*; Florence: Cesati 2006.

Castelnuovo, E.: Federigo Enriques e Guido Castelnuovo nel ricordo di Emma Castelnuovo; *Bollettino dell'UMI* (7) 11-A (1997), 227–235.

Castelnuovo, E.: L'Università clandestina a Roma: anni 1941-42 e 1942-43; *La Matematica nella Società e nella Cultura. Rivista dell'UMI* (8) 4-A (2001), 63–77.

Castelnuovo, G.: La Geometria algebrica e la scuola italiana; in: G. Castelnuovo (ed.): *Atti del Congresso Internazionale dei Matematici, Bologna 3-10 Settembre 1928*, volume 1: Rendiconto del Congresso. Conferenze; Bologna: Zanichelli 1929, pp. 191–201.

Castelnuovo, G.: Luigi Cremona nel centenario della nascita; *Rend. R. Accad. Lincei, Classe di Scienze FMN* (6) 12 (1930), 613–618.

Castelnuovo, G.: Vito Volterra e la sua opera scientifica; *Atti R. Accad. Lincei, Rendiconti delle Adunanze solenni* 5 (1946), 5–9.

Catalan, T., A. Di Fant and M. Perissinotto (eds.): *'Basta, qui siamo finiti!' 1938: le leggi razziste a Trieste*; Trieste: Edizioni Università di Trieste 2019.

Cattaneo, P.: Senigaglia Ermanno; *Il Bollettino di Matematica* 15 (1917-18), 51.

Cavaglion, A. and G.P. Romagnani (eds.): *Le interdizioni del duce: le leggi razziali in Italia*; Torino: Claudiana 2002.

Celli, A. and M. Mattaliano (eds.): *Eugenio Elia Levi. Le speranze perdute della matematica italiana*. Milan: Egea 2015.

Celli, A., M. Mattaliano and P. Nastasi (eds.): *Mario G. Salvadori e Mauro Picone. Un sodalizio che attraversa scienza, cultura e società del Novecento*; Rome: Consiglio Nazionale delle Ricerche 2013.

Chatterji, S. and M. Ojanguren: A glimpse of the de Rham era; *Notices of the ICCM* (2013), 117–137.

Chisini, O.: Accanto a Federigo Enriques; *Periodico di matematiche* (4) 25 (1947), 117–123.

Ciesielska, D. and L. Maligranda: Alfred Rosenblatt (1880-1947). Polish–Peruvian mathematician; *Function Spaces XII, Banach Center (Warszawa) Publications* 119 (2019), 57–108.

Ciliberto, C. and E. Sallent Del Colombo: *Francesco Severi: il suo pensiero matematico e politico prima e dopo la Grande Guerra*, arXiv (2018), 1–30. https://arxiv.org/abs/1807.05769

Clery, M.: *La théorie des probabilités et l'Institut Henri Poincaré (1918-1939): Construction d'un champ probabiliste parisien et pratique d'un transfert culturel*; unpublished Ph.D. thesis, XVIII +537 pp.; Université Paris-Saclay 2020.

Cocchi, D. and G. Favero: Gli statistici italiani e la 'questione della razza'; in: *Le leggi antiebraiche del 1938, le società scientifiche e la scuola in Italia*; Rome: Accademia Nazionale delle Scienze detta dei XL 2009, pp. 207–235.

Coen, L.W.: *The Princeton Mathematics Community in the 1930s Transcript Number 6 (PMC6). Leon W. Coen (with Albert Tucker)*; Princeton, NJ: Collections of the Seeley G. Mudd Manuscript Library 1985: https://web.math.princeton.edu/oral-history/c6.pdf

Coen, S.: Un abbaco d'antan; *La Matematica nella Società e nella Cultura. Rivista dell'UMI* (8) A (1998), 79–96.

Coen, S.: Beppo Levi: una biografia; in: B. Levi: *Opere 1897/1926*; Florence: Edizioni Cremonese 1999, pp. XIII–LIV.

Coen, S.: Beppo Levi e Beniamino Segre; in: Mirri and Arieti (eds., 2002), pp. 135–150.

Coen, S.: La vita di Vito Volterra vista anche nella varia prospettiva di biografie più o meno recenti; *La Matematica nella Società e nella Cultura. Rivista dell'UMI* (1) 1 No. 3 (2008), 443–476.

Cohen Enriques, L.: Soggiorno a Gressoney e nuovi dispiaceri; in: Bottazzini, De Benedetti and Foraciari (eds., 2001), pp. 81–85.

Collino, A., A. Conte and A. Verra: On the Life and Scientific Work of Gino Fano; *Notices of the International Consortium of Chinese Mathematicians* 2 No. 1 (2014), 43–57.

Colombo, Y.: *Il problema della scuola ebraica in Italia: relazione letta al 4° Convegno giovanile e Congresso culturale ebraico di Livorno il 3 novembre 1924*; Florence: La Poligrafica 1925.

Colonnetti, G.: *Le premesse spirituali della ricostruzione*; Campo Universitario Italiano, Università di Losanna, dispensa No. 107; republished in: Colonnetti, G.: *Pensieri e fatti dall'esilio (18 settembre 1943-7 dicembre 1944)*; Rome: Accademia dei Lincei 1973, pp. 11–20. (1944a)

Colonnetti, G.: *Due grandi problemi di vita universitaria*; Campo Universitario Italiano, Università di Losanna, dispensa No. 116; republished in: Colonnetti, G.: *Pensieri e fatti dall'esilio (18 settembre 1943-7 dicembre 1944)*; Rome: Accademia dei Lincei 1973, pp. 29–44. (1944b)

Colonnetti, G.: L'esperienza svizzera e la nostra ricostruzione universitaria; *Nuova Antologia* 184 (1945), 217–223.

Colonnetti, G.: Democrazia svizzera; *Idea* 2 No. 2 (1946), 85–89.

Colonnetti, G.: *Pensieri e fatti dall'esilio*; Rome: Accademia dei Lincei 1973.

Commissione Alleata in Italia, Sottocommissione dell'Educazione: *La politica e la legislazione scolastica in Italia dal 1922 al 1943 con cenni sui periodi precedenti e una parte conclusiva sul periodo postfascista*; Milan: Garzanti 1947.

Conforto, F.: *Il contributo italiano al progresso della geometria algebrica negli ultimi cento anni*; in: *Un secolo di progresso scientifico italiano: 1839-1939*, volume 1; Rome: SIPS 1939, pp. 125–153.

Conte, A. and L. Giacardi: Segre's University Courses and the Blossoming of the Italian School of Algebraic Geometry; in: Casnati et al. (eds., 2016), pp. 3–91.

Conte, A. and L. Giacardi (eds.): *Alessandro Terracini (1889-1968). Da Torino a Torino. A 50 anni dalla morte*; Torino: Accademia delle Scienze (Quaderni No. 36) 2020.

Conte, A., L. Giacardi and M.A. Raspanti (eds.): *Corrado Segre Lezioni inedite di due corsi universitari*; Torino: Centro Studi di Storia dell'Università di Torino, Lezioni e Inediti di 'Maestri' dell'Ateneo Torinese 2020.

Conti, A. [*alias* Severi, F.]: Il Convegno Volta 1939 dedicato alla matematica contemporanea e sue applicazioni. L'inaugurazione in Campidoglio il 22 ottobre, *Il Bollettino di Matematica* 35 (1938), 129–130.

Convegno di Scienze Fisiche Matematiche e Naturali, Matematica contemporanea e sue applicazioni; Rome: Reale Accademia d'Italia 1943.

Corinaldi, L.: La scuola nella comunità ebraica di Torino; *Ha Keillah* (2) 4 No. 66 (1988), I–VIII.

Cornell, J.: Luigi G. Jacchia (1910-1996); *Bulletin of the AAS* 28 No. 4 (1996), 2–3.

Corti, M.: Arnaldo Momigliano. Lezioni di ironia; *La Repubblica*, 3.9.1987.

Cotlar, M.: Reminiscences of Beppo Levi; in: S. Coen (ed.): *Geometry and Complex Variables*, Lecture Notes in pure and applied mathematics No. 132; New York: Dekker 1991, p. 147.

D'Agostino, G. (ed.): *Per una biografia di Carlo Somigliana*; Milan: Mimesis 2007.

D'Azeglio, M.: *Gli ebrei sono uomini!*; Florence: Le Monnier 1848.

D'Ovidio, E.: Luigi Cremona. Cenno necrologico; *Atti della R. Acc. delle Scienze di Torino* 38 (1903), 817–820.

Davi, M. and G. Simone: *Giacomo Levi Civita e l'ebraismo veneto tra Otto e Novecento*; Padova: University Press 2015.

De Benedetti, A.: Esplorazioni e viaggi nei racconti di nonno Ghigo; in: Bottazzini, De Benedetti and Foraciari (eds., 2001), pp. 55–70.

De Benedetti, S. [De Benedetti, Lydia, Vera and Gilbert (transl.)]: *Memoirs of an Anti-Fascist*; Chapman University: Sergio De Benedetti Manuscript (Book 2) 1965. https://digitalcommons.chapman.edu/debenedetti/2/

de D'Angelo, I.G. and F. Herrera (eds.): *Alessandro Terracini, Recuerdos de un matemático. 60 años de Vida Universitaria*; Tucumán: Asociación Cooperadora FACET 1994.

De Felice, R.: *Storia degli ebrei italiani sotto il fascismo*; Torino: Einaudi 1961.

Dell'Era, T. and D. Meghnagi (eds.): *"Perché di razza ebraica. Il 1938 e l'Università Italiana*; Bologna: Il Mulino 2023.

Della Pergola, S. and A. Tagliacozzo: *Gli italiani in Israele: risultati di un'indagine socio-demografica*; Milan: Federazione sionistica italiana 1978.

Della Pergola, S.: Precursori, convergenti, emarginati. Trasformazioni demografiche degli ebrei in Italia (1870-1945); in: *Italia Judaica...* (1993), pp. 48–81.

Della Peruta, F.: Le 'interdizioni' israelitiche e l'emancipazione degli ebrei nel Risorgimento; *Società e Storia* 19 (1983), 77–107.

Della Seta, F.: *L'incendio del Tevere*; Udine: Paolo Gaspari 1996.

Despeaux, S., D. Dumbaugh and J. Lorenat: History of mathematics through collaboration: Toward a composite portrait of Oswald Veblen; *Oberwolfach Rep.* 56 (2022), 1-22.

Dolza, D.: Per un contributo allo studio delle classi medie in Piemonte nei primi decenni del secolo: il caso delle insegnanti; in: U. Levra and N. Tranfaglia (eds.): *Torino fra liberalismo e fascismo*; Milan: Franco Angeli 1987, pp. 15–117.

Dresden, A.: The migration of Mathematicians; *The American Mathematical Monthly* 49 No. 7 (1942), 415–429.

Duggan, S. and B. Drury: *The Rescue of Science and Learning: The Story of the Emergency Committee in Aid of Displaced Foreign Scholars*; New York: The Macmillan Company 1948.

Eckes, C.: Organiser le recrutement de recenseurs français pour le Zentralblatt à l'automne 1940: une étude sur les premiers liens entre Harald Geppert, Helmut Hasse et Gaston Julia sous l'Occupation; *Revue d'histoire des mathématiques* 24 No. 2 (2018), 259–329.

Edallo, E.: Cattedre perseguitate. L'applicazione delle leggi antiebraiche nei confronti del corpo docente della R. Università di Milano; *Memoria e Ricerca* 26 No. 59/3 (2018), 453–472

Edallo, E.: L'applicazione delle leggi antiebraiche alla Regia Università di Milano; in: M. D'Amico, A. De Francesco and C. Siccardi (eds.): *L'Italia ai tempi del ventennio fascista. A ottant'anni dalle leggi antiebraiche: tra storia e diritto*; Milan: Franco Angeli 2020, pp. 249–261.

Einaudi, L.: *Diario dell'esilio 1943-44*; Torino: Einaudi 1997.

Enea, M.R. (ed.): *Francesco Gerbaldi e i matematici dell'Università di Palermo*; Pristem/Storia, Note di Matematica, Storia, Cultura No. 34-35; Milan: Bocconi 2013.

Enriques, G.: *Via D'Azeglio 57*; Bologna: Zanichelli 1983.

Errera Foà, L.: Quelle interrogazioni senza respiro. Scuola di Torino; *Shalom* 18/10 (1984), 30.

Errera Foà, L.: [Ricordo]; in: *Alessandro Terracini nel centenario...* (1990), pp. 19–22.

Evolución de las ciencias en la República Argentina: 1923-1972, volume 1: Matematica; Buenos Aires: Sociedad Científica Argentina 1979.

Fabre, G.: *L'elenco. Censura fascista, editoria e autori ebrei*; Torino: Zamorani 1998.

Fabre, G.: L'«Informazione Diplomatica» N. 14 del Febbraio 1938; *La Rassegna Mensile di Israel* 73 No. 2 (2007), 45–101.

Fanesi, P.R.: Gli ebrei italiani rifugiati in America latina e l'antifascismo (1938-1945); *Storia e problemi contemporanei* 7 (1994), 23–36.

Fanfani, A. [Capperucci, V., Giovagnoli, A., R. Moro and P. Roggi (eds.)]: *Diari: Volume I, 1943-1845 Quaderni svizzeri*; Rubbettino: Catanzaro 2012.

Fano, A.: L'Alijàh dall'Italia dal 1928 al 1955; *La Rassegna Mensile di Israel* 21 No. 7 (1955), 263–276.

Fano, G. [Andreotti, A. (ed.)]: Les surfaces du quatrième ordre; *Rendiconti del Seminario Matematico (Università e Politecnico di Torino)* 12 (1953–1954), 301–313.

Fano, G.: Uno sguardo alla storia della matematica. Discorso letto dal Socio Dott. Gino Fano nell'adunanza pubblica del 28 dicembre 1894; *Atti e Memorie R. Acc. Virgiliana* (1895), 1–33.

Fano, G.: *Il confine del Trentino e le trattative dello scorso aprile con la monarchia austro-ungarica. Conferenza tenuta alla Società di Cultura di Torino il giorno 11 giugno 1915*; Rome: Tipografia Armani e Stein 1915. (1915)

Fano, G.: *L'opera del Comitato regionale di mobilitazione industriale per il Piemonte: settembre 1915 - marzo 1919*; Torino: Tip. Giani 1919.

Fano, G.: *A Preface to a Series of Special Lectures on 'Italian Geometry' and Two General Lectures delivered by Professor Gino Fano of the University of Turin during the Lent Term of Session, 1922-23*; Shrewsbury: W.B. Walker 1923.

Fano, G.: Intenti, carattere, valore formativo della matematica. Conferenza tenuta alla Scuola di Guerra il 15 marzo 1924; *Alere flammam. Bollettino del Gabinetto di cultura della scuola di guerra* 2 No. 7 (1924), 9–32.

Fano, G.: Corrado Segre (1863-1924); *Annuario della R. Università di Torino 1924-25* (1925), 219–228.

Fano, G.: Trasformazioni di contatto birazionali del piano; in: G. Castelnuovo (ed.): *Atti del Congresso Internazionale dei Matematici, Bologna 3-10 Settembre 1928,* volume 4; Bologna: Zanichelli 1931, pp. 35–42. (1931a)

Fano, G.: Sulle varietà algebriche a tre dimensioni aventi tutti i generi nulli; in: G. Castelnuovo (ed.): *Atti del Congresso Internazionale dei Matematici, Bologna 3-10 Settembre 1928,* volume 4; Bologna: Zanichelli 1931, pp. 115–121. (1931b)

Fano, G.: Trasformazioni birazionali sulle varietà algebriche a tre dimensioni di generi nulli; *Rend. R. Accad. Lincei, Classe di Scienze FMN* (6) 15 (1932), 3–5.

Fano, G.: Scorrendo il volume di F. Klein: 'Vorlesungen über die Entwicklung der Mathematik im XIX Jahrhundert'. Conferenza tenuta il 27 febbraio 1934; *Conferenze di Fisica e di Matematica della Reale Università e della Reale Scuola di Ingegneria di Torino* 4 (1934), 151–171.

Fano, G.: *Complementi di Geometria, R. Università di Torino, anno accademico 1934-35*; Torino: Litografia Felice Gili 1935.

Fano G.: Sulle varietà algebriche a tre dimensioni a curve–sezioni canoniche; *Memorie della R. Acc. d'Italia* 8 (1938), 23–64. (1938a)

Fano, G.: Sulle varietà algebriche a tre dimensioni le cui sezioni iperpiane sono superficie di genere zero e bigenere uno; *Memorie della Società Italiana delle Scienze (detta dei XL)* 24 (1938), 41–66. (1938b)

Fano, G.: Sulle curve ovunque tangenti a una quintica piana generale; *Commentari Mathematici Helvetici* 12 (1940), 172–190.

Fano, G.: *Lezioni di geometria descrittiva tenute dal prof. Gino Fano, raccolte dagli studenti Roberto Ballarati e Franco Brindisi*. IUC Lezioni 37; Losanna: FESE 1944. (1944a)

Fano, G.: *Lezioni di geometria analitica*. IUC Lezioni 6; Losanna: FESE 1944. (1944b)

Fano, G.: Osservazioni varie sulle superficie regolari di genere zero e bigenere uno; *Revista de Matemáticas y Física Teórica* 4 (1944), 69–79. (1944c)

Fano, G.: Nuove ricerche sulle varietà algebriche a tre dimensioni a curve-sezioni canoniche; *Acta. Pontificia Academia Scientarum* 9 (1945), 163–167.

Fano, G.: Nuove ricerche sulle varietà algebriche a tre dimensioni a curve-sezioni canoniche; *Commentationes. Pontificia Academia Scientarum* 11 (1947), 635–720.

Fano, G.: Irrazionalità della forma cubica generale dello spazio a quattro dimensioni; *Rendiconti del Seminario Matematico (Università e Politecnico di Torino)* 9 (1950), 21–45.

Fano, R.: In Loving Memory of my Father Gino Fano; in: A. Collino, A. Conte and M. Marchisio (eds.): *The Fano Conference. Proceedings*; Torino: Dipartimento di Matematica dell'Università 2004, pp. 1–4.

Fano, U.: The memories of an Atomic Physicist for my children and grandchildren; *Physics Essays* 13 No. 2–3 (2000), 176–197.

Ferrara degli Uberti, C.: *Fare gli ebrei italiani Autorappresentazioni di una minoranza (1861-1918)*; Bologna: Il Mulino 2011.

Ferrarotto, M.: *L'Accademia d'Italia. Intellettuali e potere durante il fascismo*; Napoli: Liguori 1977.

Finzi, R.: Leggi razziali e politica accademica: il caso di Bologna; in: A. Di Meo (ed.): *Cultura ebraica e cultura scientifica in Italia*; Rome: Editori Riuniti 1994, pp. 157–172.

Finzi, R.: *L'Università italiana e le leggi antiebraiche*; Rome: Editori Riuniti 1997.

Fiorentino, G.: I ricordi di un ex-allievo dell'Università clandestine; in: *Emmatematica. Insegnamento di Emma Castelnuovo 'Vedere oltre le figure e i numeri'*; Florence: Edifir 2003, pp. 107–110.

Foà, S.: *Gli Ebrei nel Risorgimento Italiano*; Assisi-Rome: Carucci 1978.

Forman, P.: Scientific Internationalism and the Weimar Physicists: The Ideology and Its Manipulation in Germany after World War I; *Isis* 64 No. 2 (1973), 151–180.

Formiggini, G.: *Stella d'Italia stella di David. Gli ebrei dal Risorgimento alla Resistenza*; Milan: Mursia 1998.

Frucht, R.W.: How I became Interested in Graphs and Groups; *Journal of Graph Theory* 6 (1982), 101–104.

Fubini, D.G. and H. Brown: *Let me explain. Eugene G. Fubini's life in Defence of America*; Santa Fe: Sunstone Press 2015.

Fubini, G.: On a property of W-congruences; *Annals of Mathematics* (2) 41 (1940), 356–364. (1940a)

Fubini, G.: On Bianchi's permutability theorem and the theory of W-congruences; *Annals of Mathematics* (2) 41 (1940), 620–638. (1940b)

Galimi, V. and G. Procacci (eds.): *Per la difesa della razza. L'applicazione delle leggi antiebraiche nelle università italiane*; Milan: Edizioni Unicopli 2009.

Galles, C.D.: Cortés Pla: una vida universitaria; *Montalbán* 36 (2003), 271–283.

Galoppini, A.: *Le studentesse dell'Università di Pisa (1875-1940)*; Pisa: ETS 2011.

Gario, P. (ed.): *Lettere e Quaderni dell'Archivio di Guido Castelnuovo*; 1999. Web site http://operedigitali.lincei.it/Castelnuovo/Lettere_E_Quaderni/menu.htm

Gario, P.: Guido Castelnuovo: l'uomo e lo scienziato; *Rendiconti di Matematica* (7) 37 (2016), 147–183.

Geison, G.L.: Scientific Change, Emerging Specialties, and Research Schools; *History of Science* 19 No. 1 (1981), 20–40.

Geison, G.L. and F.L. Holmes: Research Schools. Historical Reappraisals; *Osiris* (2), 8 (1993), 12–48.

Gemelli, G. (ed.): *The 'unacceptables'. American Foundations and Refugee Scholars between the Two Wars and after*; Bruxelles: p.I.E. – Peter Lang 2000.

Gerbi, S.: *Giovanni Enriques dalla Olivetti alla Zanichelli*; Milan: Hoepli 2013.

Ghizzetti, A.: Aspetti dell'opera di Guido Fubini nel campo dell'analisi matematica; in: *Atti del Convegno Matematico in celebrazione del centenario della nascita di G. Fubini e F. Severi, Torino 8-10.10.1979*; supplemento al volume 115 degli *Atti della Acc. delle Scienze di Torino –* Classe di Scienze FMN; Torino: Accademia delle Scienze 1982, pp. 9–21.

Giacardi, L.: The Italian School of Algebraic Geometry and Mathematics Teaching in Secondary Schools. Methodological Approaches, Institutional and Publishing Initiatives; *International Journal for the History of Mathematics Education* 5 (2010), 1–19.

Giacardi, L. (ed.): *Corrado Segre e la Scuola Italiana di Geometria Algebrica*; 2013. Web site http://www.corradosegre.unito.it/docente.php

Giacardi, L.: The Italian School of Algebraic Geometry and the Teaching of Mathematics in Secondary Schools: Motivations, Assumptions and Strategies; in: M. Marchisio and A. Verra (eds.): *Geometry of Algebraic Varieties in Honor of Alberto Conte*; *Rendiconti del Seminario Matematico (Università e Politecnico di Torino)* 71 No. 3-4 (2015), pp. 421–461.

Giacardi, L.: Beppo Levi in Argentina (1939-1961); *Matematica, Cultura e Società. Rivista dell'UMI* (1) 4 No.1 (2019), 53–65.

Giacardi, L.: Le 'battaglie' di Federigo Enriques in difesa della 'humanitas' scientifica e le ricadute sull'insegnamento della matematica; *Scientia* 1 No. 2 (2023), 73–126.

Giacardi, L. and M. Raspitzu: *Teaching and dissemination of mathematics in Beppo Levi's work. From Italy to Argentina*: in: K. Bjarnadóttir, F. Furinghetti, M. Menghini, J. Prytz and G. Schubring, (eds.): *'Dig where you stand' 4. Proceedings of the Fourth International Conference on the History of Mathematics Education*; Rome: Nuova Cultura 2017, pp. 117–132.

Giacardi, L. and R. Tazzioli: The Unione Matematica Italiana and Its Bollettino, 1922-1928; in: Mazliak and Tazzioli (eds., 2021), pp. 31–61.

Giaconi, D.: The diaspora of Italian Economists: Intellectual Migration Between Politics and Racial Laws; in: M.M. Augello, M.E.L. Guidi and F. Bientinesi (eds.): *An Institutional History of Italian Economics in the Interwar Period*, volume 2; Basingstoke: Palgrave Macmillan 2020, pp. 211–242.

Gillette, A.: *Racial Theories in Fascist Italy*; London and New York: Rootledge 2002.

Gioberti, V.: *Delle condizioni presenti e future d'Italia*; Londra: A spese degli editori 1848.

Giribaldi Sardi, M.L.: *Scuola e vita nella comunità ebraica di Asti (1800-1930). Come ingenui agnelletti*; Torino: Rosenberg e Sellier 1993.

Gissi, A.: L'emigrazione dei 'Maestri'. Gli scienziati italiani negli Stati Uniti tra le due guerre; in: A. Arru, D.L. Caglioti and F. Ramella (eds.): *Donne e uomini migranti. Storie e geografie tra breve e lunga distanza*; Rome: Donzelli 2008, pp. 145–161.

Gissi, A.: Italian scientific migration to the United States of America after 1938 racial laws; *Österreichische Zeitschrift für Geschichtswissenschaften* 21 (2010), 100–118.

Gissi, A.: Migranti, esiliate o rifugiate? Le italiane nell''intellectual wave' (Italia-Stati Uniti, 1938-1943); in: S. Luconi and M. Varricchio (eds.): *Lontane da Casa. Donne italiane e diaspora globale dall'inizio del Novecento a oggi*; Torino: Accademia University Press 2015, pp. 97–113.

Gissi, A.: 'I should like very much to settle down in the U.S. and I will come alone'. Italian women in the 'intellectual wave' (1938-1943); in: C. de la Guardia Herrero and E. Postigo Castellanos (eds.): *Moving women and the United States: crossing the Atlantic*; Alcalá: Servicio de publicaciones de la Universidad de Alcalá 2016, pp. 63–78.

Giuliotti, D. and G. Papini: *Dizionario dell'Omo Salvatico*; Florence: Vallecchi 1923.

Goetz, H.: *Il giuramento rifiutato. I docenti universitari e il regime fascista*; Florence: La Nuova Italia 2000.

Goodstein, J.R.: A Conversation with Franco Rasetti; *Phys. Perspect.* 3 (2001), 217–313.

Goodstein, J.R.: *The Volterra Chronicles: The Life and Times of an Extraordinary Mathematician 1860-1940*; Providence: AMS 2007.

Goodstein, J.R.: To sign or not to sign: Tullio Levi-Civita, Giuseppe Levi and the Fascist Loyalty oath of 1931; *Medicina nei Secoli* 30 No. 1 (2018), 211–240.

Graffone, V.: *Espulsioni immediate. L'Università di Torino e le leggi razziali, 1938*; Torino: Zamorani 2018.

Guarnieri, P.: *Italian Psychology and Jewish Emigration under Fascism: From Florence to Jerusalem and New York*; Houndmills, Basingstoke, Hampshire, and London: Palgrave Macmillan 2016.

Guarnieri, P. (ed.): *Intellettuali in fuga dall'Italia fascista. Migranti, esuli e rifugiati per motivi politici e razziali*; 2019. Web site https://intellettualinfuga.fupress.com/ (2019a)

Guarnieri, P. (ed.): *L'emigrazione intellettuale dall'Italia fascista. Studenti e studiosi ebrei dell'Università di Firenze in fuga all'estero*; Florence: University Press 2019. (2019b)

Guerraggio, A., M. Mattaliano and P. Nastasi (eds.): *Mauro Picone e in matematici polacchi*; Rome: Accademia polacca delle Scienze. Biblioteca e Centro di Studi a Roma (Conferenze No. 121) 2007.

Guerraggio, A. and P. Nastasi (eds.): *Gentile e i matematici italiani. Lettere 1907-1943*; Torino: Bollati Boringhieri 1993.

Guerraggio, A. and P. Nastasi (eds.): *Italian Mathematics Between the Two World Wars*; Basel: Birkhäuser 2005. (2005a)

Guerraggio, A. and P. Nastasi: *Matematica in camicia nera. Il regime e gli scienziati*; Milan: Mondadori 2005. (2005b)

Guerraggio, A. and P. Nastasi (eds.): *Nazismo e fascismo. Leggi razziali. Fisici e matematici*; Pristem/Storia, Note di Matematica, Storia, Cultura No. 19-20; Milan: Eleusi 2007.

Guerraggio, A. and P. Nastasi (eds.): *Matematici da epurare*; Milan: Egea 2018. (2018a)

Guerraggio, A. and P. Nastasi: Dossier 1938-2018: a 80 anni dalle Leggi razziali; *Lettera Matematica Pristem* 104 (2018), 30–46. (2018b)

Guerraggio, A. and G. Paoloni: *Vito Volterra*; Padova: Franco Muzio 2008.

Guerraggio, A., F. Pressacco and G.I. Bischi: *Amases XL: Quarant'anni di storia dell'associazione per la matematica applicata alle scienze economiche e sociali;* Milan: Egea 2016.

Guetta Sadun, S.: La scuola ebraica dall'emancipazione alla riforma Gentile; in: Piussi (ed., 1997), pp. 169–179.

Gustavo Colonnetti per chi lo conobbe; Pollone: Fondazione Alberto Colonnetti 1973.

Herrera, F.E.: Breve Síntesis sobre la Personalidad del Profesor Doctor Alessandro Terracini; *Azzurra: revista del Instituto italiano di cultura de Cordoba* (2000), 104–108.

Homenaje a Beppo Levi. Beppo Levi en su 80 aniversario. Antecedentes personales y docentes. Publicaciones; *Revista de la Unión Matemática Argentina* 17 (1955), I–XVI.

Homenaje a la memoria de V. Volterra y J.J. Thomson; *Publicaciones del Instituto de Matemática. Facultad de Ciencias Matemáticas de la Universidad Nacional del Litoral* 3 No. 1 (1941).

Hoyle, J.: *Home is where the wind blows*; Oxford: University Press 1997.

Israel, G.: *La scienza italiana e le politiche razziali del regime*; Bologna: Il Mulino 2010.

Israel, G. and A. Millán Gasca (eds.): *The Biology of Numbers. The Correspondence of Vito Volterra on Mathematical Biology*; Basel: Birkhäuser 2002.

Israel, G. and P. Nastasi: *Scienza e razza nell'Italia fascista*; Bologna: Il Mulino 1998.

Italia Judaica. Gli Ebrei nell'Italia unita, 1870-1945, Atti del IV convegno internazionale. Siena 12-16 giugno 1989, Pubblicazione degli Archivi di Stato, Saggi 26; Rome: Ministero per i beni culturali e ambientali 1993.

Janovitz, A. and F. Mercanti: S*ull'apporto evolutivo dei matematici ebrei mantovani nella nascente nazione italiana*; Milan: Monografie di EIRIS 2008. https://diazilla.com/doc/811027/sull-apporto-evolutivo-dei-matematici-ebrei-mantovani-nella.%20Accessed%205%20January% 202022

Jewish Central Information Office (1938). Italien und die *Rassenfrage*. 12.8.1938; Die neuen antijüdischen Gesetze in Italien, 9.9.1938, 1–7.

Korn, A.: Contributi scientifici degli italiani in Argentina nel ventesimo secolo; in: *La popolazione di origine italiana in Argentina*; Torino: Fondazione Giovanni Agnelli 1988, pp. 171–196.

Krall, G.: Vito Volterra. La matematica e la scienza del suo tempo. Volterra e le istituzioni scientifiche italiane; *Civiltà delle macchine* 3 (1955), 65-77, 23–25.

Lamberti, M.: The Reception of Refugee Scholars from Nazi Germany in America: Philanthropy and Social Change in Higher Education; *Jewish Social Studies: History, Culture, Society* 12 No. 3 (2006), 157–192.

Landra, G.: La manomissione ebraica della nazione italiana. I settori più delicati dell'insegnamento monopolizzati dagli ebrei; *La difesa della razza* 2 No. 17 (1939), 20–23.

Lasserre, A.: *Frontières et camps: le refuge en Suisse de 1933 à 1945*; Lausanne: Payot 1995.

Laura e Gustavo Colonnetti: scritti di persone che li ricordano con nostalgia e affetto; Occhieppo Superiore (BI): Ecomuseo Valle Elvo-Serra 2000.

Le Ferrand, H.: *Six lettres de Giuseppe Peano à Robert Montessus de Ballore*; Publications Hal n. 00709080 (2012), http://hal.archives-ouvertes.fr/hal-00709080

Lefschetz, S.: A page of mathematical autobiography; *Bulletin of the AMS* 74 (1968), 854–879.

Lehto, O.: *Mathematics Without Borders: a History of the International Mathematical Union*; New York: Springer 1998.

Leitch, A.: *A Princeton Companion*; Princeton: Legacy Library 2015.

Levi, A.: I campi universitari italiani in Svizzera (1944–1945); *Svizzera italiana* 7 No. 62 (1947), 93–101.

Levi, A.: Ricordi di giorni penosi; in: *Scritti minori storici e politici*; Padova: Cedam 1957, pp. 391–418.

Levi, B.: Contro il "nazionalismo territoriale"; *Israel*, 25.7.1918, 2.

Levi, B.: *Abbaco da 1 a 20. Il primo libro di aritmetica*; Parma: B. Levi editore 1922.

Levi, B.: *Analisi matematica algebrica ed infinitesimale*; Bologna: Zanichelli 1937.

Levi, B.: Evolución del pensamiento matemático; *Publicaciones del Instituto de Matemática* 2 No. 5 (1940), 101–111.

Levi, B.: Prólogo; *Mathematicae Notae* 1 No. 1 (1941), 7–8. (1941a)

Levi, B.: La personalidad de Vito Volterra; *Publicaciones del Instituto de Matemática* 3 No. 3, 25–36. (1941b)

Levi, B.: El número entero y el intelectualismo (conferencia pronunciada el 3 de mayo de 1941 en el salón de actos de la Escuela Normal de profesores Dr. Nicolas Avallaneda de Rosario); in: *Ciclo de carácter general*, 1; Rosario: Asociación cultural de Conferencias de Rosario 1941, pp. 3–11. (1941c)

Levi, B.: Correría en la Lógica matemática; *Revista de Matemáticas y Física Teórica* 3 (1942), 13–78. (1942a)

Levi, B.: Tullio Levi-Civita; *Mathematicae Notae* 2 (1942), 155–159. (1942b)

Levi, B.: Cien años en la Historia de la Matemática; in: *Ciclo de Conferencias Científicas y de Carácter General de la Sociedad Científica Argentina II*; Buenos Aires: Tallères Gráficos Tomás Palumbo 1943, pp. 57–68.

Levi, B.: Federigo Enriques; *Mathematicae Notae* 6 (1946), 112–115. (1946a)

Levi, B.: Dopo la bomba atomica; *Studi Sociali – Rivista di libero esame. Montevideo III* 17 (1946), 8–11. (1946b)

Levi, B.: Pensamientos y recuerdos. En ocasión de una reunión de profesores de enseñanza media; *Ínsula* 3 (1946), 160–172. (1946c)

Levi, B.: Luigi Berzolari; *Mathematicae Notae* 9 (1949), 112.

Levi, F.: *L'ebreo in oggetto. L'applicazione della normativa antiebraica a Torino 1938-1943*; Torino: Zamorani 1991.

Levi, F.: *Le case e le cose: la persecuzione degli ebrei torinesi nelle carte dell'EGELI, 1938-1945*; Torino: Compagnia di San Paolo 1998.

Levi, F.: *Cinquant'anni prima: dalle rovine belliche alle costruzioni funzionali, Cinquant'anni dopo: il cemento armato, dai primordi alla maturità*; Torino: Testo & immagine 2002-2003.

Levi, L.: *Beppo Levi: Italia y Argentina en la vida de un matemático*; Buenos Aires: Libros del Zorzal 2000.

Levi, S.: Diario Pisano di un laureando in Matematica (1864-65); *Bollettino Storico Pisano* 52 (1864-1865) [2001], 287–295.

Levi, S.: *Complementi di aritmetica ed algebra ad uso degli aspiranti agli esami di licenza liceale, di licenza dell'Istituto Tecnico, di ammissione al corso universitario di matematiche e di ammissione alla R. Accademia Militare*; Torino: Paravia 1871.

Levi, S.: Sulle coordinate trigonali; *Giornale di Matematiche ad uso degli studenti delle università italiane* 14 (1876), 352–353.

Liceo Gioberti, Liceo D'Azeglio, Liceo Galileo Ferraris, Istituto Gobetti Marchesini Casale (eds.): *Scuola di Italiani*; Torino: Arti grafiche San Rocco 2012.

Linguerri, S.: *La grande festa della scienza. Eugenio Rignano e Federico Enriques. Lettere*; Milan: Franco Angeli 2005.

Linguerri, S. and R. Simili (eds.); *Einstein parla italiano*; Bologna: Pendragon 2008.

Lolli, G.: L'Opera logica di Beppo Levi; in: B. Levi: *Opere 1897/ 1926*; Florence: Edizioni Cremonese 1999, pp. LXVII–LXXVI.

Loré, M.: *Antisemitismo e razzismo ne «La Difesa della razza» (1938-1943)*; Soveria Mannelli (CZ): Rubettino 2008.

Luciano, E.: 'Illustrare la Nazione col senno e colla mano'. Ebraismo e istruzione nel Piemonte risorgimentale; in: C.S. Roero (ed.): *Contributi dei docenti dell'Ateneo di Torino al Risorgimento e all'Unità*; Torino: DSSP 2014, pp. 315–354. (2013a)

Luciano, E.: L'esperienza didattica di Emma Castelnuovo nelle scuole ebraiche; *La Matematica nella Società e nella Cultura. Rivista dell'UMI* (1) 6 (2013), 35–43. (2013b)

Luciano, E.: 'Ebree la cui religione si confonde con il culto dell'Italia': il caso delle insegnanti di Matematica (1848-1938); in: F. Ferrara, L. Giacardi and M. Mosca (eds.): *Conferenze e Seminari Associazione Subalpina Mathesis*; Torino: KWB 2014, pp. 323–333. (2013–2014)

Luciano, E.: «Ambasciatori di scienza e d'italianità»: l'Accademia d'Italia e la diffusione della cultura matematica all'estero; *Physis, Rivista Internazionale di Storia della Scienza* 51 (2016), 61–73.

Luciano, E.: Mathematics and Race in Turin: the Jewish community and the local context of education (1848-1945); in: K. Bjarnadóttir, F. Furinghetti, M. Menghini, J. Prytz and G. Schubring, (eds.): *'Dig where you stand' 4. Proceedings of the Fourth International Conference on the History of Mathematics Education*; Rome: Nuova Cultura 2017, pp. 189–201. (2017a)

Luciano, E. (ed.): *Scienza in esilio. Gustavo Colonnetti e i campi universitari in Svizzera (1943-1945)*; Pristem/Storia, Note di Matematica, Storia, Cultura No. 41-42; Milan: Egea 2017. (2017b)

Luciano, E.: From Emancipation to Persecution: Aspects and Moments of the Jewish Mathematical Milieu in Turin (1848-1938); *Bollettino di Storia delle Scienze Matematiche* 38 No. 1 (2018), 127–166.

Luciano, E.: 'Alla ricerca di uno spazio di sopravvivenza intellettuale': A. Terracini, le leggi razziali e il soggiorno a Tucumán (1938-1948); in: Conte and Giacardi (eds., 2020), pp. 41–64.

Luciano, E.: On Francesco G. Tricomi's heritage: Archive and miscellany; *Historia Mathematica* 56 (2021), 73–84.

Luciano, E.: Jewish intellectual diaspora and the circulation of mathematics; in: Borgato and Phili (eds., 2022), pp. 1–19. (2022a)

Luciano, E.: Su un episodio di censura internazionale: Francesco Tricomi e l'annuncio delle Mathematical Reviews; *Rivista di Storia dell'Università di Torino* 11 No. 2 (2022b), 57–76.

Luciano, E.: 'Sotto un altro cielo': l'emigrazione dei matematici ebrei dall'Italia fascista; *Matematica, Cultura e Società. Rivista dell'UMI* (1) 8 No. 1 (2023), 89–103.

Luciano, E. and C.S. Roero: From Turin to Göttingen: Dialogues and Correspondence (1879-1923); *Bollettino di Storia delle Scienze Matematiche* 32 No. 1 (2012), 7–232.

Luciano, E. and C.S. Roero: Corrado Segre and his Disciples. The Construction of an International Identity for the Italian School of Algebraic Geometry; in: Casnati et al. (eds., 2016), pp. 93–241.

Luciano, E. and E. Scalambro: La miscellanea Terracini (1908-1968); in Giacardi (ed., 2020), *Corrado Segre e la Scuola italiana di Geometria Algebrica*, http://www.corradosegre.unito.it

Luciano, E. and E. Scalambro: Gino Fano's Late Investigations on Fano-Enriques Threefolds; *Rendiconti del Seminario Matematico (Università e Politecnico di Torino)* 80 No. 2 (2022), 23–48.

Luzzato Voghera, G.: *Il prezzo dell'eguaglianza. Il dibattito sull'emancipazione degli ebrei in Italia (1781-1848)*; Milan: Franco Angeli 2000.

Luzzatti, L.: *Memorie autobiografiche e carteggi*; Bologna: Zanichelli 1933.

Mac Lane, S.: *Oswald Veblen (1880–1960). A Biographical Memoir*; Washington D.C.: National Academy of Sciences 1964.

Mac Lane, S.: Jobs in the 1930s and the views of George D. Birkhoff; *The Mathematical Intelligencer* 16 No. 3 (1994), 9–10.

Maida, B.: *1938. I bambini e le leggi razziali in Italia*; Florence: Giuntina 1999.

Maida, B.: *Dal ghetto alla città: gli ebrei torinesi nel secondo Ottocento*; Torino: Zamorani 2001.

Marbach, G. and A. Rizzi (eds.): *La Facoltà di Scienze Statistiche di Roma. Parabola di una eccellenza*; Rome: La Sapienza 2011.

Marpicati, A.: *L'Accademia d'Italia*; Milan: Mondadori 1934.

Martinelli, E.: Beniamino Segre: his life, his work; *Rendiconti Acc. Naz. delle Scienze detta dei XL. Memorie di Matematica* 6 No. 1(1978-79), 1–12.

Mazliak, L. and R. Tazzioli: *Mathematicians at war. Volterra and his French colleagues in World War I*; Dordrecht, Heidelberg, London, New York: Springer 2009.

Mazliak, L. and R. Tazzioli (eds.): *Mathematical Communities in the Reconstruction After the Great War 1918–1928. Trajectories and Institutions*; Cham: Springer 2021.

Mendel Haimovici (1906-1973), 100 Years From His Birth; *Analele ştiinţifice ale Universităţii "Alexandru Ioan Cuza" din Iaşi. Matematică (Serie nouă)* 52 No. 2 (2006), 233–240.

Menghini, M.: Guido Castelnuovo e l'insegnamento della matematica; *Rendiconti di Matematica* (7) 37 (2016), 185–197.

Meyer, F. and H. Mohrmann: Vorrede zum dritten Bande; in: F. Meyer and H. Mohrmann (eds.): *Encyklopädie der mathematischen Wissenschaften mit Einschluss ihrer Anwendungen*, vol. 3.1: Geometrie; Leipzig: Teubner 1923, pp. V–IX.

Minerbi, A.: Tra nazionalizzazione e persecuzione. La scuola ebraica in Italia, 1930-1943; *Contemporanea* 1, No. 4 (1998), 703–730.

Ministero dell'Educazione Nazionale: *Dalla Riforma Gentile alla Carta della Scuola*; Florence: Vallecchi 1941.

Mirri, D. and S. Arieti (eds.): *La cattedra negata: dal giuramento di fedeltà al fascismo alle leggi razziali nell'Università di Bologna*; Bologna: Clueb 2002.

Momigliano Levi, P.: Beppo Levi: vita e pensiero di un matematico; in: F. Ferrara, L. Giacardi and M. Mosca (eds.): *Associazione Subalpina Mathesis. Conferenze e Seminari 2015-2016*; Torino: KWB 2016, pp. 259–282.

Momigliano, A.: Gli Ebrei d'Italia; in: S. Berti (ed.): *Pagine ebraiche*; Torino: Einaudi 1987, pp. 129–142.

Montagnana, M.: I rifugiati ebrei italiani in Australia e il movimento antifascista 'Italia libera' (1942-1946); *Notiziario dell'Istituto storico della Resistenza in Cuneo e provincia* 31 (1987), 5–114.

Montel, A.: Il campo universitario italiano di Huttwil; *Bollettino della Scuola italiana e Bollettino del FESE*, No. 2 (1945), 42.

Montgomery, D.: Oswald Veblen; *Bulletin of the AMS* 69 No. 1 (1963), 26–36.

Morpurgo, P.: Le scuole e gli ebrei tra Medioevo e Risorgimento; preprint in *Educazione&Scuola* (2016) http://www.edscuola.com/archivio/didattica/scuolebrei.html

Mortara, G.: Ricordi della mia vita; in: *Omaggio a Giorgio Mortara 1885-1967: vita e opere*; Rome: Università degli Studi di Roma "La Sapienza" 1985, pp. 13–50.

Murre, J.: On the work of Gino Fano on tree–dimensional algebraic varieties; in: A. Brigaglia, C. Ciliberto and E. Sernesi (eds.): *Algebra e geometria (1860–1940): il contributo italiano*; Supplemento ai *Rendiconti del Circolo matematico di Palermo* (2) 36 (1994), pp. 219–229.

Nastasi, P.: Guido Fubini; *Lettera Matematica Pristem* 10 (1993), I–XII.

Nastasi, P.: La matematica italiana dal manifesto degli intellettuali fascisti alle leggi razziali; *La Matematica nella Società e nella Cultura. Rivista dell'UMI* (8) 1-A (1998), 317–345.

Nastasi, P. and E. Rogora: From internationalization to autarky: Mathematics in Rome between the two world wars; *Rendiconti di Matematica* (7) 41 (2020), 1–50.

Nastasi, P. and A. Scimone (eds.): *Lettere a Giovanni Vacca*; Quaderni Pristem No. 5; Palermo: Bocconi 1995.

Nastasi, P. and R. Tazzioli (eds.): *Aspetti scientifici e umani nella corrispondenza di Tullio Levi-Civita*; Quaderni Pristem. No. 12; Palermo: Bocconi 2000.

Nastasi, P. and R. Tazzioli (eds.): *Aspetti di Meccanica e di Meccanica Applicata nella corrispondenza di Tullio Levi-Civita (1873-1941)*; Quaderni Pristem No. 14; Palermo: Bocconi 2003.

Natalini, R. and M. Mattaliano: La fantasia e la memoria. Conversazione con Emma Castelnuovo; *Lettera Matematica Pristem* 52 (2004), 4–7.

Necrologio Ermanno Senigaglia, caduto in guerra; *Bollettino della Mathesis* 8 No. 2 (1916), 83–85.

Nossum, R.: Emigration of mathematicians from outside German-speaking academia 1933-1963, supported by the society for the protection of science and learning; *Historia Mathematica* 39 No. 1 (2012), 84–104.

Nossum, R. and J. Kotulek: The Society for the Protection of Science and Learning as a patron of refugee mathematicians, *BSHM Bulletin* 30 No. 2 (2015), 153–167.

O'Malley, R.E. (jr.): Singular Perturbation Theory: A Viscous Flow out of Göttingen; *Annual Review of Fluid Mechanics* 42 (2010), 1–17.

Onoranze per il giubileo scientifico del prof. Guido Castelnuovo; Città di Castello: Tipografia dell'Unione arti grafiche 1937.

Ostenc, M.: Cosa fu l'Accademia d'Italia; *Nuova Antologia* 2191 (1994), 105–138.

Palumbo, E.: Tra orgoglio e disperazione. Lettere di docenti ebrei alle comunità dopo l'espulsione del 1938; *Rivista di storia dell'educazione* 2 (2019), 173–192.

Parikh, C.: *The unreal life of Oskar Zariski*; New York: Springer 1991.

Parshall, K.: *The New Era in American Mathematics, 1920-1950*; Princeton: University Press 2022.

Parshall, K. and A.C. Rice (eds.): *Mathematics Unbound: the Evolution of an International Mathematical Research Community, 1800-1945*; Providence: AMS 2002.

Pessino, S. and P. Marangunic: *Nuestro Beppo. La historia de un gran matemático Italiano que eligió Rosario para siempre*; Santa Fe: Fundación A. Ross 2017.

Phillips, R.: Reminiscences about the 1930s; *The Mathematical Intelligencer* 16 No. 3 (1994), 6–8.

Piazza, A. (ed.): *Le leggi razziali del 1938*; Bologna: Il Mulino 2021.

Picone, M.: Guido Fubini (1879-1943); *Bollettino dell'UMI* (3) 1 No. 1 (1946), 56–58.

Pincherle, M.: *Cronaca di un esilio. Un pediatra ebreo tra persecuzione e sofferto rientro (1938-1946)*; Ancona: Affinità elettive edizioni 2011.

Piperno Beer, G.: Le scuole dei giovani ebrei di Roma durante il periodo delle leggi razziali (1938-1944); *La Rassegna Mensile di Israel* 77 No. 1-2 (2011), 227–249.

Piussi, A.M. (ed.): *E li insegnerai ai tuoi figli*; Florence: Giuntina 1997.

Pla, C.: Beppo Levi en la Argentina; *Mathematicae Notae* 18 (1962), XIII–XXII.

Polverini, L.: Albert Einstein e il giuramento fascista del 1931; *Rivista Storica Italiana* 103 (1991), 268–280.

Pontecorboli, G.: *America nuova terra promessa. Storie di italiani in fuga dal fascismo*; Milan: Brioschi 2013.

Prezzolini G.: *America in pantofole*; Florence: Vallecchi 1950.

Prof. Guido Castelnuovo, *Nature* No. 3935, 31.3.1945, 385.

Ragusa Gilli, L.: Tullio Viola e la scuola. Ricordi di un'insegnante di matematica; in: L. Giacardi and C.S. Roero (eds.): *Matematica, arte e tecnica nella storia in memoria di Tullio Viola*; Torino: KMB 2006, pp. 119–123.

Ravà, V.: Le laureate in Italia: notizie statistiche; *Bollettino Ufficiale del Ministero dell'Istruzione Pubblica*, volume 1 No. 14, 3.4.1902, 634–654.

Reid, C.: *Courant in Göttingen and New York. The Story of an Improbable Mathematician*; New York: Springer 1976.

Reingold, N.: Refugee Mathematicians in the United States of America, 1933-1941: Reception and Reaction; *Annals of Science* 38 (1981), 313–338.

Remmert, V.: Mathematicians at war. Power struggles in Nazi Germany's Mathematical Community: Gustav Doetsch and Wilhelm Süss; *Revue d'histoire des mathématiques* 5 (1999), 7–59.

Rider, R.E.: Alarm and Opportunity: Emigration of Mathematicians and Physicists to Britain and the United States, 1933-1945; *Historical Studies in the Physical Sciences* 15 No. 1 (1984), 107–176.

Rinaldelli, L.: In nome della razza. L'effetto delle leggi del 1938 sull'ambiente matematico torinese; *Quaderni di Storia dell'Università di Torino* 2 (1997-98), 149–208.

Risorgimento e minoranze religiose. Roma 14 febbraio 1997. Atti della giornata di studio; *La Rassegna Mensile di Israel* 64 No. 1 (1998), 1–94.

Roero, C.S. (ed.): Regime e dissenso. 1931. I professori che rifiutarono il giuramento fascista; numero monografico di *Rivista di Storia dell'Università di Torino* 10 No. 2 (2021).

Rogora, E.: Guido Castelnuovo e la matematica a Roma tra Risorgimento e Belle Époque; *Rendiconti di Matematica* (7) 37 (2016), 219–233.

Rollandi, M.S.: Le leggi razziali e l'Università di Genova: prime ricerche sui docenti; *Atti della Società Ligure di Storia Patria* 42 No. 2 (2002), 477–493.

Rosenblatt, A.: Vito Volterra; *Revista de Ciencias* 44 (1942), 423–442.

Rosenthal Fuà, E.: *Fuga a due*; Bologna: Il Mulino 2004.

Rota, G.C.: Fine Hall in its golden age: Remembrances of Princeton in the early fifties; in: P.L. Duren, R. Askey, U.C. Merzbach and H.M. Edwards (eds.): *A Century of Mathematics in America. Part III: History of Mathematics*, volume 3; Providence, Rhode Island: AMS 1989, pp. 223–236.

Roth, C.: Reminiscenze sugli ebrei italiani durante le loro traversie; *La Rassegna Mensile di Israel* 31 No. 5 (1965), 204–208.

Roth, L.: Obituaries. Prof. Guido Castelnuovo; *Nature* 169, 14.7.1952, 992.

Rowe, D.: 'Jewish Mathematics' at Gottingen in the Era of Felix Klein; *Isis* 77 No. 3 (1986), 422–449.

Rowe, D.: Mathematical Schools, Communities, and Networks; in: M.J. Nye (ed.): *Cambridge History of Science, volume 5: Modern Physical and Mathematical Sciences*; Cambridge: University Press 2002, pp. 113–132.

Rowe, D.: Making Mathematics in an Oral Culture: Göttingen in the Era of Klein and Hilbert; *Science in Context* 17 (2004), 85–129.

Rowe, D.: *A Richer Picture of Mathematics. The Göttingen Tradition and Beyond*; Cham: Springer 2018.

Sacchi, G.: Relazione sugli Asili d'infanzia ed altri istituti elementari, visitati nell'autunno dell'anno 1843 dall'abate Ferrante Aporti, con note di Giuseppe Sacchi; *Annali universali di statistica economia pubblica, geografia, storia, viaggi e commercio* 6 No. 16 (1845), 15–21.

Salmon, P.: Un sodalizio torinese degli anni '50; in: E. Gallo, L. Giacardi and C.S. Roero (eds.): *Conferenze e Seminari 1994–1995*; Torino: Associazione Subalpina Mathesis 1995, pp. 224–243.

Salustri, S.: Perugia and its University. Persecutions of Jews in 1938; *Trauma and Memory* 8 No. 2 (2020), 124–135.

Salvadori, M.: [A tangential life]; in: S. Peacock and T.M. Rooney (eds.): *Contemporary Authors: A Bio-Bibliographical Guide to Current Writers in Fiction, General Non-Fiction, Poetry, Journalism, Drama, Motion Pictures, Television, and Other Fields*, volume 159; Farmington Hills: Gale 1997, pp. 389–422.

Santaló, L.A.: La obra científica de Beppo Levi; *Mathematicae Notae* 18 (1962), XXIII–XXVIII.

Santaló, L.A.: Alessandro Terracini (1889-1968); *Revista de la Unión Matemática Argentina* 23 (1968), 149–151.

Santaló, L.A.: Los primeros 60 años de la U.M.A.; *Revista de la Unión Matemática Argentina* 43 (2001), 1–187.

Sarfatti, M.: Il Comitato di soccorso per i deportati italiani politici e razziali di Losanna (1944-1945); *Ricerche Storiche* 9 No. 2-3 (1979), 463–483.

Sarfatti, M.: Dopo l'8 settembre: gli ebrei e la rete confinaria italo-svizzera; *La Rassegna Mensile di Israel* 47 (1981), 150–173.

Sarfatti, M.: 1938 Le leggi contro gli ebrei; *La Rassegna Mensile di Israel* 54 No. 2 (1988), 13–18.

Sarfatti, M.: *Gli ebrei nell'Italia fascista. Vicende, identità, persecuzione*; Torino: Einaudi 2007. (2007a)

Sarfatti, M. (ed.): Numero Speciale in occasione del 70° anniversario dell'emanazione della legislazione antiebraica fascista; *La Rassegna Mensile di Israel* 54 No. 1-2 (2007). (2007b)

Scalambro, E.: *Gino Fano (1871-1952): patrimonio, ricerca e insegnamento*; unpublished Ph.D. thesis, 373 + 1698 pp.; Università degli Studi di Torino 2023.

Scarantino, L.M. (ed.): *Intorno a Enriques. Cinque conferenze*; Sarzana: Agorà 2004.

Schappacher, N.: Remarks about Intuition in Italian Algebraic Geometry; in: *History of Mathematics: Models and Visualization in the Mathematical and Physical Sciences*; Oberwolfach Rep. 47 (2015), pp. 2805–2807.

Schappacher, N. and R. Schoof: Beppo Levi and the arithmetic of elliptic curves; *The Mathematical Intelligencer* 18 No. 1 (1996), 57–69.

Schiavone, L.: Oltre l'astronomia, la vita: Giulio Bemporad e l'assistenza ai profughi ebrei; *Giornale di Astronomia* 41 No. 2 (2015), 25–41.

Schwarz, G. (ed.): *Diari di un partigiano ebreo. Gennaio 1940-febbraio 1944*; Torino: Boringhieri 2008.

Scuole Israelitiche; *La difesa della razza* 1 No. 6 (1938), 12–13.

Segal, S.L.: *Mathematicians Under the Nazis*; Princeton: University Press 2003.

Segre Fuà, E.: Un grande geometra ebreo: Corrado Segre; *Rassegna Mensile di Israel* 18 (1952), 125–127.

Segre, B.: L'analizzatore differenziale di Cambridge; *Coelum* 15 (1947), 78–79.

Segre, B.: La scuola in Inghilterra; *Il Filomate* 1 (1948), 53–58.

Segre, B.: Gino Fano; *Archimede* 4 (1952), 262–263.

Segre, B.: Commemorazione di Guido Fubini; *Rend. Accad. Naz. Lincei, Classe di Scienze FMN* (8) 17 (1954), 276–294.

Segre, B.: Nel primo centenario della nascita di Corrado Segre. Discorso tenuto presso l'Università di Torino il 20 dicembre 1963; *Rendiconti del Seminario Matematico (Università e Politecnico di Torino)* 23 (1963), 7–21.

Segre, B.: Leonard Roth; *Bulletin of the London Mathematical Society* 8 (1976), 194–202.

Segré, E.: *A Mind Always in Motion: The Autobiography of Emilio Segre*; Berkeley: University of California Press 1993.

Segre, R.: *The Jews in Piedmont*; Jerusalem: The Israel academy of sciences and humanities & Tel Aviv University 1986-90.

Servi, F.: *Gli israeliti d'Europa nella civiltà: memorie storiche, biografiche e statistiche dal 1789 al 1870*; Torino: Foà 1871.

Severi, F.: *La géométrie algébrique*; in: J.C. Fields (ed.): *Proceedings of the International Mathematical Congress held in Toronto, August 11-16*, volume 1; Toronto: University Press 1924, pp. 149–154.

Severi, F.: Fascismo e scienza; *L'Illustrazione Italiana* 60 No. 44 (1933), 643–644.

Severi, F.: Peut-on parler d'un esprit latin même dans les mathématiques? *Revue scientifique* 73 (1935), 581–589.

Severi, F.: In occasione dell'inizio dell'anno accademico 1940-41 del Reale INDAM; *Bollettino dell'UMI* (2) 3 No. 2 (1941), 130–140.

Severi, F.: Confidenze; *La scienza per i giovani* 2 (1953), 5–6 and 65–69.

Siegmund-Schultze, R.: *La légitimation des mathématiques dans l'Allemagne fasciste: trois étapes*; in: J. Olff-Nathan (ed.): *La science sous le Troisième Reich, Victime ou alliée du nazisme?*; Paris: Seuil 1993, pp. 91–102. (1993a).

Siegmund-Schultze, R.: *Mathematische Berichterstattung in Hitlerdeutschland. Der Niedergang des Jahrbuchs über die Fortschritte der Mathematik*; Göttingen: Vandenhoeck & Ruprecht 1993. (1993b)

Siegmund-Schultze, R.: *Rockefeller and the Internationalization of Mathematics between the Two World Wars*; Basel: Birkhäuser 2001.

Siegmund-Schultze, R.: *Mathematicians Fleeing from Nazi Germany. Individual Fates and Global Impact*; Princeton: University Press 2009.

Signori, E.: *La Svizzera e i fuoriusciti italiani. Aspetti e problemi dell'emigrazione politica 1943-1945*; Milan: Franco Angeli 1983.

Signori, E.: La «Conquista Fascista» dell'università: Libertà d'insegnamento e autonomia nell'Ateneo pavese dalla riforma Gentile alle leggi razziali; *Il Politico* 62 No. 3 (1997), 433–472.

Signori, E.: Una «peregrinatio academica» in età contemporanea. Gli studenti ebrei stranieri nelle università italiane tra le due guerre; *Annali di storia delle università italiane* 4 (2000), 139–162.

Signori, E.: *L'ateneo e la città tra guerre e fascismo*; Pavia, Milano and Bologna: Cisalpino 2002.

Signori, E.: Contro gli studenti. La persecuzione antiebraica negli atenei italiani e le comunità studentesche; in: Galimi and Procacci (eds., 2009), pp. 173-210.

Simili, R.: *Sotto falso nome. Scienziate italiane ebree (1938-1945)*; Bologna: Pendragon 2010.

Sittignani, M.G.: Levi Eugenio Elia; *Il Bollettino di Matematica* 15 (1917-18), 146–147.

Solow, H.: Refugee Scholars in the United States; *The American Scholar* 11 No. 3 (1942), 374–378.

Supino, G.: Gustavo Colonnetti. Discorso commemorativo pronunciato nella seduta ordinaria dell'11 gennaio 1969; *Accad. Naz. Lincei, Celebrazioni Lincee* 20 (1969), 1–10.

Sutton, G.: The Centenary of the Birth of W.H. Young (October 20, 1863); *The Mathematical Gazette* 49 No. 367 (1965), 16–21.

Synnott, M.G.: *The Half-Opened Door: Discrimination and Admissions at Harvard, Yale, and Princeton, 1900-1970*; London: Routledge 2010.

Tallini, G.: Beniamino Segre; *Annals of Discrete Mathematics* 18 (1983), 5–12.

Terracini, A.: Densità di una corrispondenza di tipo dualistico, ed estensione dell'invariante di Mehkme-Segre; *Atti della R. Acc. delle Scienze di Torino* 71 (1936a), 310–328. (1936a)

Terracini, A.: Invariante di Mehmke-Segre generalizzato e applicazione alle congruenze di rette; *Bollettino dell'UMI* (3) 15 (1936), 109-113. (1936b)

Terracini, A.: Su una possibile particolarità delle linee principali di una superficie. Nota 1 e 2; *Rend. R. Accad. Lincei, Classe di Scienze FMN* (6) 26 (1937), 84–91 and 153–158.

Terracini, A.: Sur l'existence de surfaces ayant des lignes principales données; *Comptes rendus hebdomadaires des séances de l'Académie des sciences* 208 (1939), 616–618.

Terracini, A.: El invariante de Mehmke-Segre y los sistemas lineales; *Anales de la Sociedad Científica Argentina* 3 No. 129 (1940), 97–111. (1940a)

Terracini, A.: Sobre la existencia de superficies cuyas líneas principales son dadas; *Publicación n. 16 de la Unión Matemática Argentina* (1940), 1–20. (1940b)

Terracini, A.: Sobre la ecuación diferencial $y''' = G(x, y, y')y'' + H(x, y, y')y''$; *Revista de Matemáticas y Física Teórica* 2 (1941), 245–329. (1941a)

Terracini, A.: Origines de algunos conceptos geométricos; *Publicaciones del Instituto de Matematica. Universidad Nacional del Litoral* 3 No. 6 (1941), 158–199. (1941b)

Terracini, A.: Propuesta de institución de un comité central para informaciones bibliográficas matemáticas; *Revista de Matemáticas y Física Teórica* 3 (1942), 369–379. (1942a)

Terracini, A.: Qué debe hacerse para el adelanto de la Matemática en la Argentina; *Asociación argentina para el progreso de las ciencias* (1942), 61–66. (1942b)

Terracini, A.: Guido Fubini (1879-1943); *Revista de la Unión Matemática Argentina* 10 (1944), 27–30.

Terracini, A.: Gino Fano; *Bollettino dell'UMI* (3) 8 (1952a), 485–490.

Terracini, A.: Gino Fano; *Annuario dell'Università di Torino* (1952-53), 325–328.

Terracini, A: Commemorazione del socio Gino Fano; *Rend. Accad. Naz. Lincei, Classe di Scienze FMN* (8) 14 (1953), 702–715. (1953a)

Terracini, A.: Gino Fano (1871-1952), Cenni commemorativi; *Atti dell'Acc. delle Scienze di Torino* 87 (1953), 350–360. (1953b)

Terracini, A.: Commemorazione del corrispondente Beppo Levi; *Atti Accad. Naz. Lincei, Rendiconti* 34 (1963), 590–606.

Terracini, A.: *Ricordi di un matematico. Un sessantennio di vita universitaria*; Rome: Cremonese 1968.

Terracini, Benedetto: Alessandro Terracini visto in famiglia, in: Conte and Giacardi (eds., 2020), pp. 95–103.

Terracini, Benvenuto: Il centenario della Pia Società femminile Israelitica di Torino 1832-1932; *Rassegna Mensile di Israel* 8 (1932), 93–109.

Terracini, L.: Una inmigración muy particular: 1938, los universitarios italianos en la Argentina; *Anuario del Instituto de Estudios histórico sociales* 4 (1989), 335–369.

Terracini, L. (ed.): Cacciati dalla scuola. Carteggio ebraico '38; *Belfagor* 4 (1990), 444–450.

Testimonianze in memoria di Gustavo Colonnetti; Torino: Stamperia Artistica Nazionale 1973.

Testimonianze: 75° compleanno di Franco Levi; Torino: Politecnico, Dipartimento di Ingegneria strutturale 1989.

The Position of the Jews; *Nature* No. 3739, 28.6.1941, 801.

Thomson, I.: *Primo Levi: A Life*; New York: Picador, Metropolitan Book, Henry Holt and Company 2014.

Togliatti, E.: Gino Loria; *Rassegna Mensile di Israel* 18 (1952), 499–505.

Togliatti, E.: Alessandro Terracini. Commemorazione letta dal Socio corrispondente Eugenio Togliatti nell'adunanza del 20 Novembre 1968; *Atti dell'Acc. delle Scienze di Torino* 103 (1969), 397–407.

Torchiani, F.: *Uno storico Rettore Magnifico Plinio Fraccaro e l'Università di Pavia*; Milan: Cisalpino 2010.

Toscano, M.: L'emigrazione ebraica italiana dopo il 1938; *Storia contemporanea* 6 (1988), 1287–1314.

Treves, B. (ed.): *Tre vite dall'ultimo '800 alla metà del '900: studi e memorie di Emilio, Emanuele, Ennio Artom*; Florence: Israel 1954.

Treves, R.: Incontri di culture nell'America Latina alla fine degli anni Trenta; *Nuova Antologia* 120 (1985), 90–100.

Treves, R.: [Ricordo]; in: *Alessandro Terracini nel centenario...* (1990), pp. 23–28.

Tricomi, F.G. [*alias* Terracini, A.]: *Algebra elementare ad uso dei licei*. Messina: Principato 1940.

Tricomi, F.G.: Essenza e didattica delle Matematiche in un manoscritto inedito di Corrado Segre; *Rendiconti del Seminario Matematico (Università e Politecnico di Torino)* 7 (1938-40), 103–117.

Tricomi, F.G.: Cos'è l'Analisi funzionale?; *Il Saggiatore* 1 (1940), 18–26. (1940a)

Tricomi, F.G.: Osservazioni statistiche sulle matematiche contemporanee; *Il Saggiatore* 1 (1940), 47–51. (1940b)

Tricomi, F.G.: Una nuova rivista di bibliografia matematica; *Il Saggiatore* 1 (1940), 26. (1940c)

Tricomi, F.G.: Matematici scandinavi; *Il Saggiatore* 1 (1940), 155–160. (1940d)

Tricomi, F.G.: Riflessioni sul Centenario Galileiano; *La Luce* 35 (1942), 51.

Tricomi, F.G.: Necessità pratica del perdono; *La Luce* 36 (1943), 17. (1943a)

Tricomi, F.G.: Quale dovrà essere domani il nostro compito?; *La Luce* 36 (1943), 31. (1943b)

Tricomi, F.G.: Lo scienziato e i doveri di domani; *La Festa* 22 (1943), 22. (1943c)

Tricomi, F.G.: *La mia vita di matematico attraverso la cronistoria dei miei lavori (bibliografia commentata 1916-1967)*; Padova: Cedam 1967.

Turi, G.: *Lo stato educatore. Politica e intellettuali nell'Italia fascista*; Bari: Laterza 2002.

Turi, G.: *Sorvegliare e premiare. L'Accademia d'Italia (1926-1944)*; Rome: Viella 2016.

Twardzik, S.: Le carte dei campi di internamento universitari per i militari italiani in Svizzera conservate dall'Università degli Studi di Milano; in: Castagnola, Panzera and Spiga (eds., 2006), pp. 239–252.

Varnier, G.B.: L'Accademia Ligure di Scienze e Lettere e le leggi razziali tra silenziose espulsioni e tarde reintegrazioni; *Atti della Società Ligure di Storia Patria* 42 No. 2 (2002), 495–510.

Velásquez López, R.: Alfred Rosenblatt en el Perú; in: Sociedad Peruana de Historia de la Ciencia y la Tecnología (ed.): *Hacer ciencia en el Perú. Biografía de ocho científicos*; Lima: Sophicyt 1990, pp. 109–134.

Ventura, A.: La persecuzione fascista contro gli ebrei nell'università italiana; *Rivista Storica Italiana* 109 No. 1 (1997), 121–192.

Ventura, L.: *Ebrei con il duce. 'La Nostra Bandiera', 1934-1938*; Torino: Zamorani 2002.

Vercelli, C.: *1938 Francamente razzisti. Le leggi razziali in Italia*; Torino: Edizioni del Capricorno 2018.

Veronese, G.: Commemorazione del Socio Luigi Cremona; *Rend. R. Accad. Lincei, Classe di Scienze FMN* (5) 12(1903), 664–678.

Vesentini, E.: Il caso della Matematica; in: *Conseguenze culturali delle leggi razziali in Italia. Atti del Convegno Roma, 11 maggio 1989*; Rome: Accademia dei Lincei 1990, pp. 97–105.

Vesentini, E.: Beniamino Segre and Italian Geometry; *Rendiconti Acc. Naz. delle Scienze detta dei XL. Memorie di Matematica e Applicazioni* 31 No. 1 (2009), 165–174.

Veziano, P.: *Sanremo. Una nuova comunità ebraica nell'Italia Fascista 1937-1945*; Reggio Emilia: Diabasis 2007.

Vieira Souza da Silva, L. and B. Bontempi: Gleb Vassilievich Wataghin: Physics, University and Politics in Brazil (1934-39); *RUDN Journal of Russian History* 19 No. 4 (2020), 965–978.

Vieira Souza da Silva, L. and R. Monteiro de Siqueira: An Italian mission at the University of São Paulo Science and education issues in the diplomatic relationships between Italy and Brazil in the 1930s; *Mélanges de l'École française de Rome* 130 No. 2 (2018), 407–419. https://journals.openedition.org/mefrim/4430

Vigna, L. and V. Aliberti: *Della condizione attuale degli ebrei in Piemonte, estratto dal dizionario di diritto amministrativo*; Torino: Favale 1848.

Viola, T.: Necrologio di Beppo Levi; *Bollettino dell'UMI* (3) 16 No. 4 (1961), 513–516.

Vita-Finzi, P.: *Giorni lontani*; Rastignano: Il Mulino 1989.

Voigt, K.: *Il rifugio precario. Gli esuli in Italia dal 1933 al 1945*; Florence: La Nuova Italia 1999.

Wasow, W.: *Memories of Seventy Years, 1909-1979*; Madison: Private printing 1986.

Wataghin, L.: Fundação da faculdade de filosofia, ciências e letras da universidade de São Paulo: a contribuição dos professores italianos; *Revista Do Instituto De Estudos Brasileiro* 34 (1992), 151–174.

Williams, B.: *Jews and Other Foreigners. Manchester and the Rescue of the Victims of European fascism: 1933-40*; Oxford: University Press 2013.

Wisard, F.: *L'université vaudoise d'une guerre à l'autre*; Lausanne: Payot 1998.

Zanni, A.: Mortara e Del Vecchio nel 1938; *Note economiche. Monte dei Paschi di Siena* 5-6 (1977), 70–97.

Zappa, G.: Arturo Maroni Matematico Fiorentino; *Atti e memorie dell'Accademia Toscana di Scienze e Lettere La Colombaria* 69 (2004), 1000–1018.

Zevi, T.: L'emigrazione razziale; in: A. Varsori (ed.): *L'antifascismo italiano negli Stati Uniti durante la Seconda guerra mondiale*; Rome: Archivio Trimestrale 1984, pp. 75–82.

Zund, J.D.: Oswald Veblen; in: *American National Biography* 22; Oxford: University Press 1999, p. 307–308.

Name Index